Lecture Notes on Data Engineering and Communications Technologies

Volume 171

The aim of the book series is to present cutting edge engineering approaches to data technologies and communications. It will publish latest advances on the engineering task of building and deploying distributed, scalable and reliable data infrastructures and communication systems.

The series will have a prominent applied focus on data technologies and communications with aim to promote the bridging from fundamental research on data science and networking to data engineering and communications that lead to industry products, business knowledge and standardisation.

Indexed by SCOPUS, INSPEC, EI Compendex.

All books published in the series are submitted for consideration in Web of Science.

G. Rajakumar · Ke-Lin Du · Álvaro Rocha
Editors

Intelligent Communication Technologies and Virtual Mobile Networks

Proceedings of ICICV 2023

Springer

Editors
G. Rajakumar
Department of Electronics
and Communication Engineering
Francis Xavier Engineering College
Tirunelveli, Tamil Nadu, India

Ke-Lin Du
Department of Electrical and Computer
Engineering
Concordia University
Montreal, QC, Canada

Álvaro Rocha
ISEG—University of Lisbon
Lisbon, Portugal

ISSN 2367-4512 ISSN 2367-4520 (electronic)
Lecture Notes on Data Engineering and Communications Technologies
ISBN 978-981-99-1766-2 ISBN 978-981-99-1767-9 (eBook)
https://doi.org/10.1007/978-981-99-1767-9

This Springer imprint is published by the registered company Springer Nature Singapore Pte Ltd.
The registered company address is: 152 Beach Road, #21-01/04 Gateway East, Singapore 189721, Singapore

We are grateful to dedicate the ICICV 2023 proceedings to all the participants and editors of the Intelligent Communication Technologies and Virtual Mobile Networks (ICICV 2023).

Preface

We are delighted to host the proceedings of the ICICV 2023, which was held at Francis Xavier Engineering College in Tirunelveli, Tamil Nadu, from 16 to 17 February 2023. This conference proceedings bring you the collection of articles selected and presented from the 5th International Conference on Intelligent Communication Technologies and Virtual Mobile Networks (ICICV 2023). The prime focus of this conference is to bring the academic scientists, researchers and research scholars for sharing their knowledges in future of all aspects of Engineering Technology and Innovation.

We have celebrated our 2023 version of gathering, to strive to advance the largest international professional forum on virtual mobile cloud computing and communication technologies. We have received 316 papers from various colleges and universities and selected 67 papers based on strictly peer reviewed by the reviewers and experts. All the selected papers are in high quality in technical, and it exactly matches the scope of the conference.

ICICV 2023 would like to express our gratitude towards all contributors for this conference proceedings. We would like to extend our sincere thanks to all guest editors, reviewers and experts for their valuable review comments on all papers. Also, we would like to thank our organizing committee members for the success of ICICV 2023.

Lastly, we are most indebted for the generous support given by Springer for publishing this volume.

Dr. G. Rajakumar
Professor, Department of Electronics
and Communication Engineering
Francis Xavier Engineering College
Tirunelveli, Tamil Nadu, India

Dr. Ke-Lin Du
Affiliate Associate Professor
Department of Electrical and Computer
Engineering
Concordia University
Montreal, QC, Canada

Dr. Álvaro Rocha
ISEG—University of Lisbon
Lisbon, Portugal

Contents

A DGS-Loaded Compact Super Wideband Antenna for Microwave Imaging Application 1
M. Sekhar and Suman Nelaturi

A Study on Repeated Game Theory in Wireless Sensor Networks Security 13
S. Sridevi, V. Veeramani, and S. M. Chithra

An Overview of Information-Centric Network: Concepts, Network Architecture, Comparison, and Difficulties 27
Antonio Cortés Castillo

Automated Mishap Detection and Prevention System for Vehicles 43
G. Sudha, K. Balaji, K. M. Rakesh, R. S. Roopesh, S. Saranya, and Sankari Subbiah

Autonomous Drone Using Time-of-Flight Sensor for Collision Avoidance 57
G. Naveenkumar, M. V. Suriyaprakash, and T. P. Prem Anand

Collaborative Communication Models in Non-cash Food Assistance (Bantuan Pangan Non-Tunai, BPNT) Program: Toward Community Resilience 75
Abdillah Abdillah, Ida Widianingsih, Rd Ahmad Buhari, and Rusliadi Rusliadi

Communication Technology for Information Exchange Using Short-Range Unmanned Aerial Vehicles 93
V. Kashtanov, V. Nemtinov, Yu Protasova, V. Morozov, and K. Nemtinov

Evaluative Aspects of Wireless Sensor Networks Reliability Model: A Case Study 105
D. Lingamaiah, D. Krishna Reddy, and Perumalla Naveen Kumar

Implementation of 3D Object Models for Mobile Applications in UI/UX Design Using SceneView API 119
Daniel Ryan Sunjaya, Adbul Samad Bin Shibghatullah, and Shaik Shabana Anjum

Latest Trends in Wireless Network Optimization Using Distributed Learning 129
A. Vasuki and Vijayakumar Ponnusamy

Malware Classification in Local System Executable Files Using Deep Learning 139
Pagadala Ganesh Krishna, S. Kranthi, and Ande Vijaya Krishna

New Generation 3D Optical Switch for Free Space Optical Networks ... 153
Mehman Hasanov, Khagani Abdullayev, Ali Tagiyev, Gulnar Gurbanova, and Nadir Atayev

Terrain Dimensions and Node Density Analysis of MANET Using NS2 and BonnMotion 163
Satveer Kour, Manjit Singh, Himali Sarangal, and Butta Singh

Toward Secure Fault-Tolerant Wireless Sensor Communication: Challenges and Applications 177
Radhakrishna Bhat and K. N. Pavithra

Utilizing Data from Quick Access Recorder to Predict Faulty Processing on Aircrafts 189
V. Jalajakshi and N. Myna

Vehicle Collision Warning and Accident Detection Using Raspberry Pi and SSD MobileNetV1 203
A. Vijaya Lakshmi, P. Ajay Kumar Goud, G. Mohit Raj, and K. Rahul

A Comprehensive Review on the Development of Pipe Robot 219
Md. Rawshan Habib, K. M. Monzur Rahaman, Abhishek Vadher, Tahsina Tashrif Shawmee, Md Apu Ahmed, Sibaji Roy, Md Shahnewaz Tanvir, Aditi Ghosh, and Shuva Dasgupta Avi

Design and Analysis of Efficient IoT-Based Pollution Monitoring System in Urban Area 233
M. Dhivya, N. J. Avinash, P. Bhoomika, K. S. Shanika, Trisha Chatterjee, and V. Preethisree

Internet of Things in Entrepreneurship: The Intellectual Structure Perspective 247
Agung Purnomo, Nur Asitah, Elsa Rosyidah, Gusti Pangestu, Meiryani, and Fairuz Iqbal Maulana

IoT-Based App for Commuter Travel Tracking System: A Review 259
Md. Sohaib, V. P. M. B. Aarthi, M. Naga Laxmi, G. Govind Shivaji, T. Arun Kumar, and M. Ramesh

Analysis of Block-Level Prediction-Error Expansion Approach in Data Encryption 269
Jangam Deepthi and T. Venu Gopal

Blockchain-Based Implementation on Electronic Know Your Customer (e-KYC) 281
K. S. Chandraprabha

Cryptography Using GPGPU 299
Swati Jadhav, Uttkarsh Patel, Atharv Natu, Bhavin Patil, and Sneha Palwe

Identifying and Predicting Sinkhole Attacks for Low-Power and Lossy IoT Networks 315
Animesh Giri, Abhishek Goyal, Arpit Kogta, Priyansh Jain, and Pihoo Verma

Innovative Online Ticketing Model on an Intelligent Public Blockchains 327
S. C. Prabanand, M. S. Thanabal, A. Durai Murugan, P. Mahalakshmi, S. Tarunbalaji, and T. Sriharish

Machine Learning for a Payment Security Evaluation System for Mobile Networks 347
Solleti Ramana, N. Bhaskar, M. V. Ramana Murthy, and M. Raghavender Sharma

Secure and Efficient Routing Mechanism for Healthcare Networks 357
Animesh Giri, B. V. Balaji, Bhoomika P. Bhavimath, V. Durgalakshmi, and B. Rahul

Secured Ticketless Booking System to Monuments and Museums Using Cryptography 373
Kuldeep Vayadande, Ankur Raut, Roshita Bhonsle, Vithika Pungliya, Atharva Purohit, and Varad Ingale

Security Enhanced RFID-Based Digital Locking System in Home Automation Using IoT Framework 387
M. Dhivya, A. B. Gurulakshmi, G. Rajesh, and Sanjeev Sharma

Towards a More Secure and Transparent Crowdfunding Ecosystem Using Blockchain 401
M. R. Sumalatha, Rozen Berg, B. S. Sandeep, and M. Tharunraj

A Brief Analysis on Video Compression and Deduplication Models Using Video Steganography with Optimised Feature Extraction Techniques for Integrity Preservation 419
S. K. Sameerunnisa and J. Jabez

A Study on Various Techniques of Two-Dimensional Bin Packing Problem 433
U. Prabu

Advances in Computer-Aided Diagnosis of Developmental Delay in Children Using Bioengineering Systems: A New Math Model and Algorithm 443
Igor Val Danilov

An Analysis of Real-Time Number Plate-Based Verification System with Insurance Processing Using OCR Techniques 459
P. Pandiaraja, P. Biranav Kumar, N. Jaisaran, and V. Karthick Ram

Cannotation MeasureUp to Detect Deepfake by Face Recognition via Long Short-Term Memory Networks Algorithm 475
L. Rahunathan, D. Sivabalaselvamani, A. PriyaDharshini, M. Vignesh, and G. VinithKumar

Control Strategy for Modified Quasi-Admittance Source Inverter 489
A. Angeline Esther, P. S. Manoharan, J. Nivedita, and P. Deepamangai

The Information and Communication Technology's Impact to Enhance the Tourism Industries in Indonesia 501
Kendrew Huang, William Gunawan, Rivaldi, Ford Lumban Gaol, and Tokuro Matsuo

Effect of PV Penetration on the Steady-State Voltage on Grid-Integrated PV System 509
Gaurav B. Patil, Santosh S. Raghuwanshi, and L. D. Arya

Efficient Sentiment Classification Model of Tweets Using an Adaptive Megaptera Whale Optimization LSTM Classifier 523
Priya Vinod and S. Sheeja

Electric Vehicle Charge Scheduling Based on Circle-Inspired Optimization Algorithm 539
Durga Mahato, Vikas Kumar Aharwal, and Apurba Sinha

Fuzzy Logic and ANN in an Artificial Intelligent Cloud: A Comparative Study 559
Pooja Chopra and Munish Gupta

Harnessing the Power of AI to Create Intelligent Tutoring Systems for Enhanced Classroom Experience and Improved Learning Outcomes 571
Ashraf Alam

Human Activity Recognition Using CNN-Attention-Based LSTM Neural Network 593
P. Pravanya, K. Lakshmi Priya, S. K. Khamarjaha, K. Buela Likhitha, P. M. Ashok Kumar, and R. Shankar

Impact of Online Meeting Applications in Indonesia During the Pandemic 607
Fannisa Hanin Nabilah, Nabilah Putri Intaka, Ford Lumban Gaol, and Tokuro Matsuo

Implementation Artificial Neural Network on Identification System of Neurological Disorder 619
Rismayani, Suci Rahma Dani Rachman, Sri Wahyuni, Asmanurhidayani, Joe Y. Mambu, and Martina Pineng

Implementation of HBEA for Tumor Cell Prediction Using Gene Expression and Dose Response 631
P. Selvi Rajendran and K. R. Kartheeswari

Implementing an Integrated Network Load Balancer for Minimizing Weighted Response 651
Apoorv Kumar Sinha, Sanskriti Sanjay Kumar Singh, Shreyas Sai, and M. Sivagami

Insights of Deep Learning-Based Video Anomaly Detection Approaches 663
Dipak Ramoliya and Amit Ganatra

Intelligent Indoor Positioning Systems: The Case of Imbalanced Data 677
Firuz Kamalov, Sherif Moussa, and Jorge Avante Reyes

Mechanism for Digital Transformation of Intelligent Transport Systems 687
Nozima Akhmedova

Metaverse 3C: Concept, Components, and Challenges in Travel and Tourism Sector 699
S. Poonkuzhali, J. Sangeetha Archana, and T. P. Prem Anand

The Impact of Online Technology on Increase of Learning Motivation of Information System Student During Covid-19 Pandemic 715
Jerald Glad, Ardi Satyo Ramadhan, Ikrar Gama Raditya, Ford Lumban Gaol, and Tokuro Matsuo

Modelling Sentiment Analysis on Indonesian Product Reviews Using Machine Learning ... 729
Andry Chowanda and Lasmy

Moving Pictures Feature Extraction Using Scheme Pixel Values Method ... 739
D. Saravanan and Dennis Joseph

Multi-model Essay Evaluation with Optical Character Recognition and Plagiarism Detection ... 749
Aditya Vijay Kavatage, Shishir Menon, Shreesh Ravindra Devi, Siddhi Patil, and S. Shilpa

OPSUM: An Opinion Summary Generator Model for Customer Feedback on Restaurants ... 761
C. Sindhu, Kalluri Shanmukha Sai, Akula Triyan Subramanyan, Lingamaneni Sri Nidesh, and C. Kavitha

Performance Comparison of Object Detection-Based Deep Learning Techniques: A Review ... 781
B. Sudha and Kathiravan Srinivasan

Regularized Information Loss for Improved Model Selection ... 801
Firuz Kamalov, Sherif Moussa, and Jorge Avante Reyes

Review on Use of Agile Techniques in Software Development over Traditional Customs ... 813
Himanshu Srivastava and Nitish Ojha

Smart Traffic Signal Control System Using Artificial Intelligence ... 829
G. R. P. Kumari, M. Jahnavi, M. Harika, A. Pavani, and C. Venkata Lakshmi

Strategy for Charging of Battery and Supercapacitor Combined Storage System ... 839
R. Shalini, P. S. Manoharan, N. KumaraSabapathy, and G. Kannayeram

Swarm Flight of UAV in Virtual Rigid Formation Using Olfati-Saber Algorithm ... 849
Y. Zhu and V. P. Shkodyrev

Military Aircraft Detection Using YOLOv5 ... 865
P. Ajay Kumar Goud, G. Mohit Raj, K. Rahul, and A. Vijaya Lakshmi

Developing a Curriculum for Ethical and Responsible AI: A University Course on Safety, Fairness, Privacy, and Ethics to Prepare Next Generation of AI Professionals ... 879
Ashraf Alam

The Video Conferencing Technology Implications as an Educational Method on Undergraduate Students Learning Outcomes 895
Jason Fonseca Ng, Andrew, Stanley Nathanael Irawan, Ford Lumban Gaol, and Tokuro Matsuo

Transformer-Based Medical Abbreviation Disambiguation—A Comparative Study 911
Krzysztof Pałczyński, Magda Czyżewska, Marta Gackowska, Damian Ledziński, and Tomasz Andrysiak

Vehicle Number Recognition Using General Surveillance Camera 925
Nivarthi Tushara, Voraganti Sahithi, Bushgari Haripriya, and U. Chandrasekhar

Author Index 939

About the Editors

Dr. G. Rajakumar is Controller of Examination and Professor in the Department of Electronics and Communication Engineering at Francis Xavier Engineering College, Tirunelveli. He has 14.5 years of experience and received his B.E. (Electronics and Instrumentation Engineering) from National Engineering College, Kovilpatti, under MS University, Tirunelveli, 2004, M.E. (VLSI Design) in Arulmigu Kalasalingam College of Engineering, Srivilliputhur, under Anna University, Chennai, 2006, M.B.A (Human Resource Management) under Annamalai University, 2009, and Ph.D. in Information and Communication Engineering with the specialization in VLSI Design under MS University 2014; he is Post-doctoral Fellow in VLSI Design under Delhi Technological University, Delhi, 2016 and Honorary and Doctor of Letters (D.Litt.) in Information and Communication Engineering under University of South America, USA, 2017. He has published four patents and five international engineering books. He has published more than 30 international/national journals. He actively participated in more than 17 international conferences and 31 national conferences.

Ke-Lin Du received the Ph.D. in electrical engineering from Huazhong University of Science and Technology in 1998. He was on Research Staff at Concordia University from 2001 to 2010. Prior to 2001, he was with Huawei Technologies, China Academy of Telecommunication Technology, and Chinese University of Hong Kong. He visited Hong Kong University of Science and Technology in 2008. He founded Xonlink Inc. in 2014. Dr. Du has authored four books (*Neural Networks in a Softcomputing Framework*, Springer, 2006; *Wireless Communication Systems*, Cambridge University Press, 2010; *Neural Networks and Statistical Learning*, Springer, 2014 and 2019; and *Search and Optimization by Metaheuristics*, Springer, 2016). He has published 50+ papers and has four issued US patents. Currently, his research interests include signal processing and machine learning. Dr. Du has become Senior Member of the IEEE since 2009.

Álvaro Rocha holds the title of Honorary Professor and holds a D.Sc. in Information Science, Ph.D. in Information Systems and Technologies, M.Sc. in Information Management, and B.C.S. in Computer Science. He is Professor of Information Systems at the University of Lisbon—ISEG, Researcher at the ISEG Centre for Advanced Research in Management (ADVANCE), and Collaborator Researcher at both Laboratory of Artificial Intelligence and Computer Science (LIACC) and Center for Research in Health Technologies and Information Systems (CINTESIS). His main research interests are maturity models, information systems quality, online service quality, requirements engineering, intelligent information systems, e-Government, e-Health, and information technology in education. He is also Vice Chair of the IEEE Portugal Section Systems, Man, and Cybernetics Society Chapter and Editor-in-Chief of both *Journal of Information Systems Engineering and Management (JISEM)* and *Iberian Journal of Information Systems and Technologies (RISTI)*. Moreover, he has served as Vice Chair of Experts for the European Commission's Horizon 2020 Program and as Expert at the COST—intergovernmental framework for European Cooperation in Science and Technology, at the European Commission's Horizon Europe Program, at the Government of Italy's Ministry of Universities and Research, at the Government of Latvia's Ministry of Finance, at the Government of Mexico's National Council of Science and Technology, at the Government of Polish's National Science Centre, and at the Government of Cyprus's Research and Innovation Foundations.

A DGS-Loaded Compact Super Wideband Antenna for Microwave Imaging Application

M. Sekhar and Suman Nelaturi

Abstract A super wideband compact antenna for Microwave Imaging (MI) is proposed. The antenna radiator is a truncated patch with slots and has a defected ground structure (DGS). The slots in the patch and truncation allowed the radiator to couple well with the partial ground plane loaded with DGS. This led to a super wide bandwidth of 25 GHz ranging from 5 to 30 GHz with a compact size of $0.27\lambda_0 \times 0.33\lambda_0$. Initially, with a truncated circular patch, a triple wideband antenna is achieved. To increase the impedance matching of the antenna and realise the super wideband, a circular ring slot has been etched in patch and ground is also truncated. To further enhance the bandwidth, slots in triangular shape are etched in the patch and a rectangular slot is etched in ground. The proposed antenna is a compact space-saving one. The results reveal a super wideband performance of 167% (5–30 GHz) with a consistent radiation pattern and peak gain of 10.2 dB in a compact area.

Keywords Super wideband · Truncated patch · Defected ground structure · Slotted patch · Ring slot · Partial ground

1 Introduction

Around the world, cancer is now one of the leading causes of mortality. The best strategy to properly treat patients with a high likelihood of survival is by early diagnosis of malignant tumours. Current tumour detection methods use ionic diagnosis techniques, including as magnetic resonance imaging (MRI), computer tomography (CT), and X-ray mammography (MRI). Low image resolution from these procedures necessitates interpretation, which could result in a tumour diagnosis that is incorrect. Ionising radiation is not safe for patients, though; over time, it can harm genetic material, which can lead to DNA mutation and cell damage. Additionally,

M. Sekhar (✉) · S. Nelaturi
Department of Electronics and Communication Engineering, Vignan's Foundation for Science, Technology, and Research, Guntur, India
e-mail: sekhar.snha@gmail.com

G. Rajakumar et al. (eds.), *Intelligent Communication Technologies and Virtual Mobile Networks*, Lecture Notes on Data Engineering and Communications Technologies 171,
https://doi.org/10.1007/978-981-99-1767-9_1

these detection methods are ineffective for tumours that are buried deep within the body. As a microwave imaging (MI) system is created to replace the existing technologies, a non-ionisation detection technique using microwaves is proposed and developed to address the problems with ionisation techniques for tumour identification. Despite the complexity of the device, MI technology can give a high-resolution image regardless of where the tumour is located. The MI system causes the patient very little harm and can be utilised for diagnostics over the long term [1]. Given that the input power is less than one W, attenuation is one of the main problems with the MI system. The MI system works well with microstrip antennas because they are easy to manufacture into small, portable devices with low volume and power consumption. High-resolution imaging of tumour cells is made possible by the vast spectrum of frequencies.

Patch antennas for SWB/UWB have recently been the subject of numerous publications [2–23]. The use of substrates with extremely low permittivity, increasing substrate thickness, and inserting slots or DGS into the patch or ground are only a few of the ways for developing and widening the antenna bandwidth that has been described. By utilising a U-shaped feedline, a UWB antenna with a symmetric open slot [5] was able to obtain a resonance of 3.1–13.2 GHz. The antenna in [5] is more compact than a meandering ground antenna in [7] and measures 580 mm^2. An antenna with partial ground that has several slots and a diamond-shaped patch [9], an antenna with a slotted ground plane and a fowl-shaped patch [10], a hybrid U slot antenna [11], and many others are presented in [12–23].

Microstrip line-fed antennas, on the other hand, do not completely utilise the area that is available on both sides of the substrate, which drives up the cost of production. Contrary to microstrip line-fed antennas, CPW-fed antennas occupy the most of the patch [17]. In the CPW-fed antennas, patch and ground are on the same plane.

In [4], a UWB monopole antenna with CPW feed and exposed ground stubs was shown. Due to L-shaped stubs, the antenna's working band was 3.1–12.2 GHz. [6] described an antenna with a CPW feed and a polygon-shaped patch. The patch in this design uses the fractal approach to improve bandwidth, giving it an operational range of 7.86 GHz. A very wide working band from 2.5 to 10 GHz has been created by the patch in [9] by employing fractal geometry. A meandering semi-circular antenna with CPW feed was described in [15]. The created antenna has a bandwidth of 123% and a total area of 30 by 35 mm^2. On a FR4 substrate, a wideband antenna in the shape of a crown was built [17]. This antenna's operational band ranges from 4.5 to 13.5 GHz as a result of an extended U-shaped patch. A triangular patch antenna with CPW feed was described in [19] for use in UWB applications. The two ground planes are connected by a top cross loop to widen the operating band. However, its overall dimension is 1650 mm^2. In [21], a spiral patch and coplanar strips were proposed for an antenna. But it is challenging to implement in portable communication systems because of its larger size. Telsang and Kakade [22] suggested an antenna patch that was semi-circular and had slots for CPW feed. However, due to its 40×53.3 mm^2 surface area, it is not appropriate for microwave imaging applications. Several of the above-mentioned antennas have enormous sizes, rendering them unsuitable for microwave imaging systems even if most of them have wide/ultra/super operational bandwidths.

For use in microwave imaging (MI) applications, a super wideband-slotted patch antenna is presented in this work. The ultra-wide impedance bandwidth of 25 GHz, spanning from 5 to 25 GHz with a small dimension of 16.5 × 20 × 1.59 mm^3, may satisfy the requirements of MI systems.

2 Antenna Design

Figure 1 shows the geometry of the proposed antenna. After being designed using HFSS software, an antenna with the following dimensions is constructed on a FR4 substrate: 16.5 mm × 20 mm × 1.59 mm. The total dimensions are presented in Table 1. The circular patch serves as the antenna's primary radiating element. The patch's top has been truncated. The partial ground plane has been truncated at both edges, and a rectangular slot is etched from the top side of the partial ground. The radiating element has a circular ring slot and periodically spaced triangular slots etched in it. The defective ground and slotted patch will result in the development of extremely super wide bandwidth. Additionally to enhancing impedance matching, they provide direction to the currents produced in the ground and patch.

Figure 2 depicts the antenna's development stages. Figure 3 illustrates the analysis of the impact of each antenna element on the S parameter of the antenna. Antenna 1 is a partially grounded, truncated circular patch with microstrip feed. A triple band of resonance encompassing the bands of 6.6–11.0, 12.9–21.7, and 27.4–30.6 GHz is known for this antenna's poor impedance matching. Although the antenna has a total bandwidth of 16.4 GHz, there are two intermediate notch bands and very poor impedance matching at many frequencies.

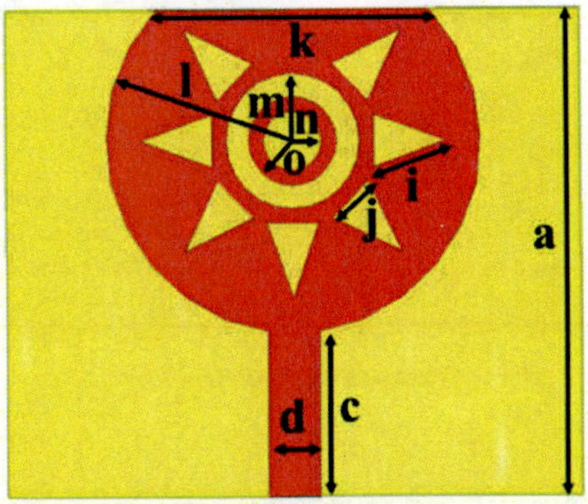

(a) Radiating Element

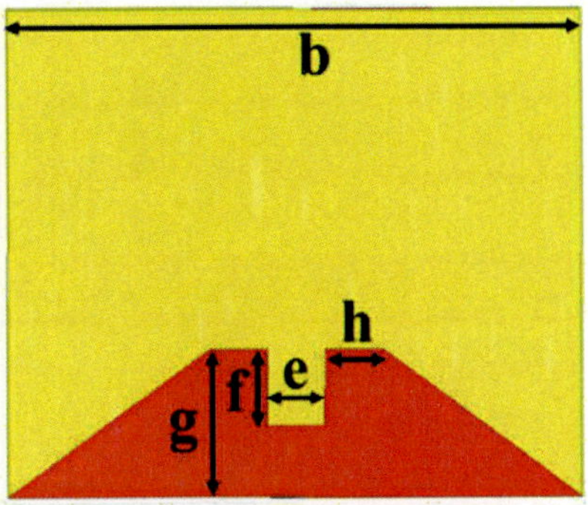

(b) Ground plane

Fig. 1 Schematic of proposed antenna

Table 1 Optimized parameter of the proposed antenna

Parameter	Value (mm)	Parameter	Value (mm)	Parameter	Value (mm)
a	16.5	f	2.6	k	9.5
b	20	g	5	l	6.5
c	5.6	h	2	m	2.3
d	1.76	i	2.7	n	1
e	2	j	1.8	o	1.5

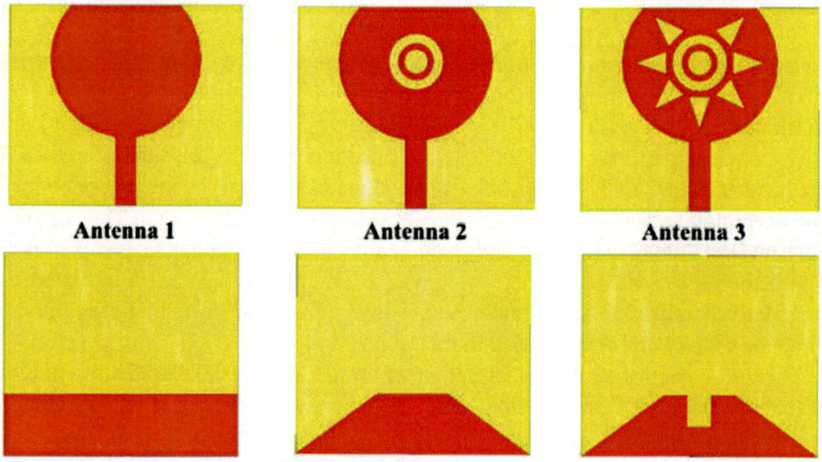

Fig. 2 Different stages of proposed antenna evolution

Fig. 3 Simulations of S_{11} at different stages

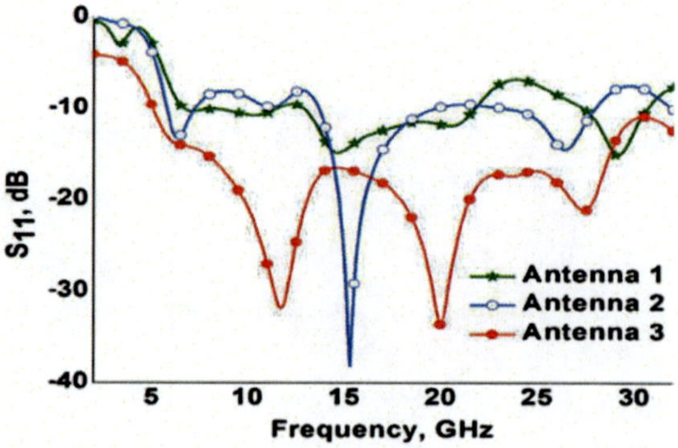

In Antenna 2, a circular ring slot has been etched in the patch and the partial ground plane has been etched at both the ends. With this, the impedance matching of the antenna has improved resulting in operational bandwidth of 24.2 GHz. Both the notch bands observed in the antenna 1 are eliminated in antenna 2. But still the impedance matching at many frequencies is very low. In antenna 3 to improve the antenna impedance matching, triangular-shaped slots are periodically etched in the patch. This improved the impedance matching which resulted in a bandwidth of 25 GHz ranging from 5 to 30 GHz. Simulations of S_{11} at different stages are shown in Fig. 3.

3 Results and Discussion

HFSS Software is used to optimise and analyse the proposed antenna. Figure 4 shows the antenna's gain. It has a peak gain of 10.2 dB, a consistent omnidirectional radiation pattern, and a small footprint of 16.5 mm × 20 mm. Due to the antenna's small size, the gain at middle frequencies is lower than compared to other working frequencies.

The notch bands between the working frequency bands of 6.6–11.0, 12.9–21.7, and 27.4–30.6 GHz were abolished with the insertion of circular ring slots in the patch and the truncation of the partial ground. The ground's truncation and the patch's slot both produced new radiating edges, which raised the associated currents in the patch and altered the antenna's impedance. By adjusting the slots' dimensions and truncation widths while generating new operating frequencies, the notch bands were reduced. The slot carved inside the patch hence boosted bandwidth by lowering the notch bands. Figures 5 and 6 demonstrate, respectively, how the truncation width and slot length in the ground plane affect antenna S_{11}, respectively. By adjusting the slot length and truncation width, the antenna's S_{11} is examined. Figure 7 illustrates

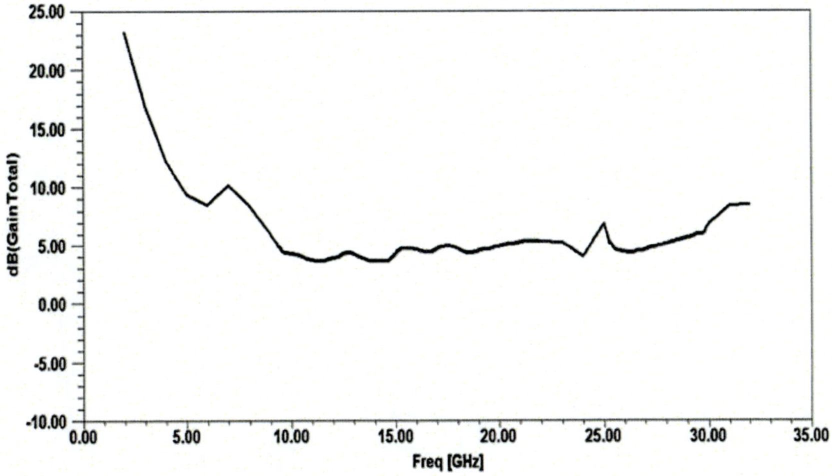

Fig. 4 Simulated and measured gain

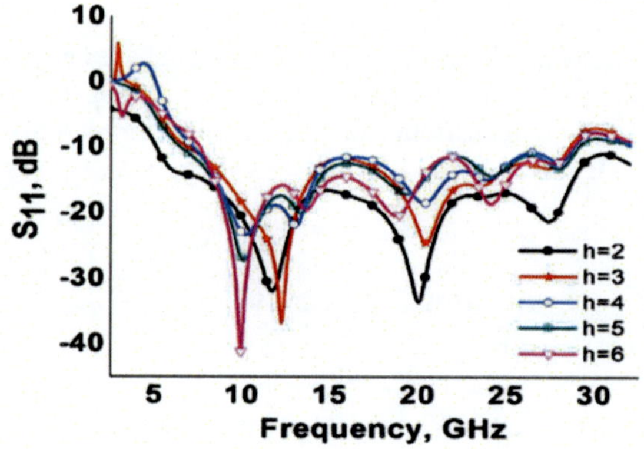

Fig. 5 Effect of ground truncation width on S_{11}

how the circular slot radius in the patch affects antenna S_{11}. With an increase in slot radius, the antenna's bandwidth and impedance matching change.

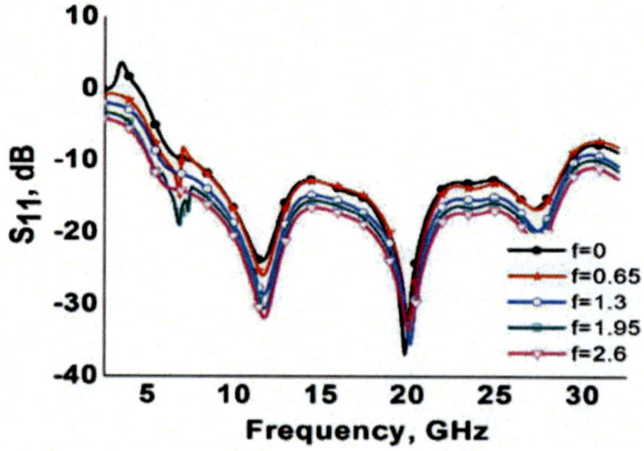

Fig. 6 Effect of ground slot length on S_{11}

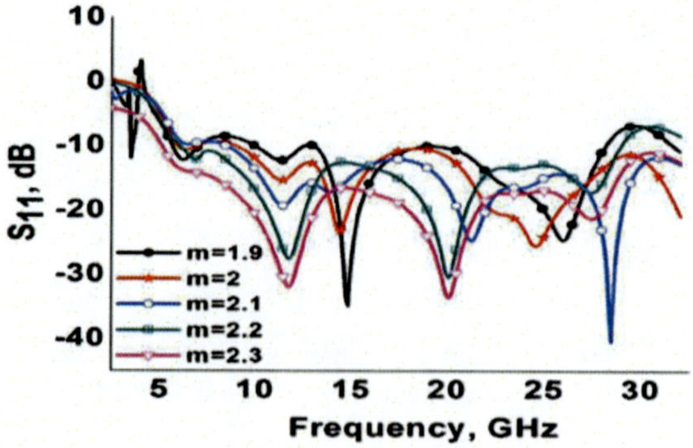

Fig. 7 Effect of circular slot radius on S_{11}

Figure 8 shows the 2D radiation patterns of the antenna at different intermediate frequencies. The E and H planes both exhibit omnidirectional patterns at all frequencies. The omnidirectional characteristics of the antenna, which are suitable for MI applications, are supported by the results.

The current distributions at 7, 11.7, 16, 20, and 27 GHz are shown in Fig. 9. The charts allow one to see how the patch and ground plane affect radiation at all frequencies.

The suggested antenna is contrasted with published ultra/super wideband antennas in Table 2. The gain, impedance bandwidth, and antenna profile of the proposed antenna are all improved.

4 Conclusion

For microwave imaging (MI) applications, a DGS-loaded small super wideband antenna is described. The first step is to create a super wideband antenna with two notch bands using a truncated circular patch. To get rid of the notch bands and create the super wideband, a circular ring slot is etched into the patch, and ground plane is truncated. Etching a slit into the ground and triangular slots into the patch improves impedance matching. The outcome shows a respectable super wideband performance of 167% (5–30 GHz). It is compact (16.5×20 mm^2), has a max gain of 10.2 dB, and radiates in all directions consistently. The antenna comprised of a slotted patch and a defective ground structure (DGS) has been designed on a 1.6 mm thick FR4 material. The antenna is having a size of $0.27\lambda_0 \times 0.33\lambda_0$ at the lowest operating frequency of 5 GHz. The slotted patch coupled well with the DGS in the ground plane leading the proposed antenna to obtain a range of operational bandwidth from 5 to 30 GHz.

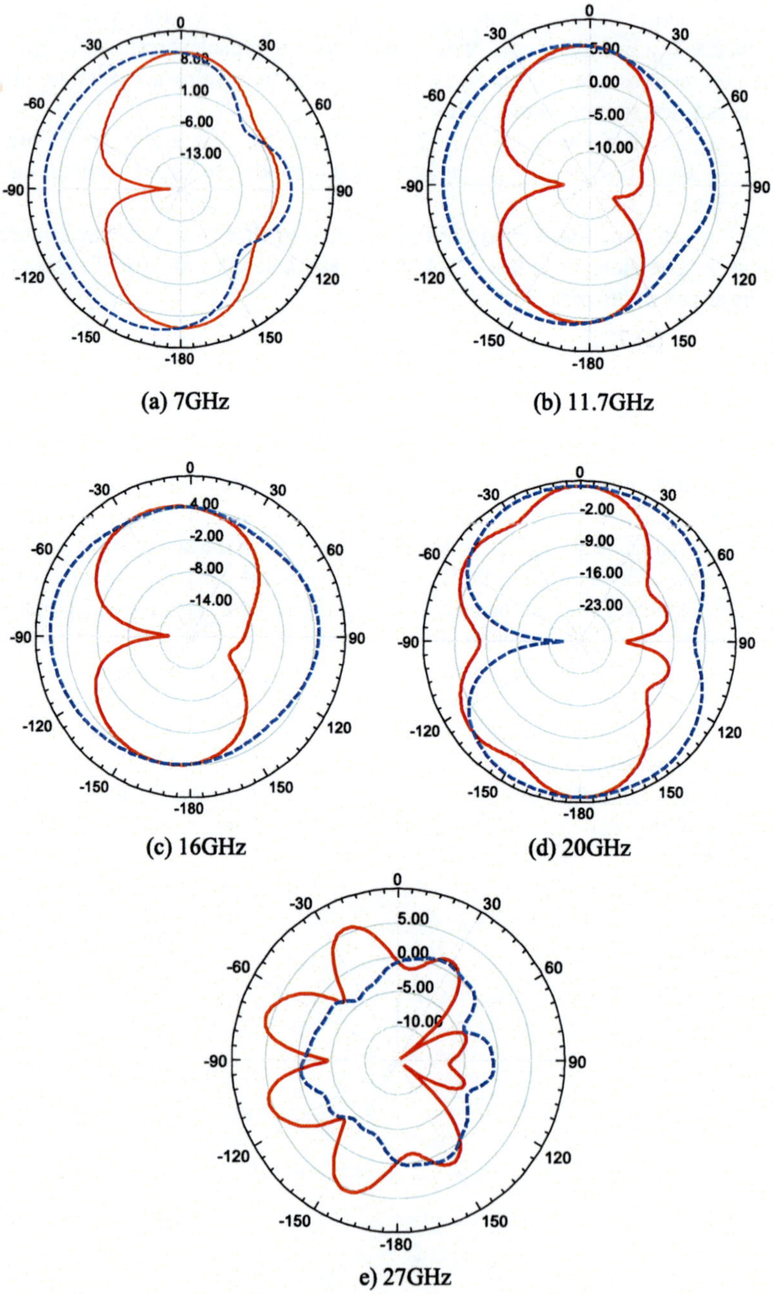

Fig. 8 Radiation patterns (E-plane and H-plane)

(a) 7GHz (b) 11.7GHz

(c) 16GHz (d) 20GHz

(e) 27GHz

Fig. 9 Current distributions at different frequencies

Table. 2. Comparison of the presented antenna with other ultra/super wideband antennas reported

References	Size (λ_0^2)	Operating band (GHz)	Bandwidth (GHz)	Peak gain (dB)
[2]	0.25 × 0.30	1.94–15.4	13.46	6
[3]	0.23 × 0.19	3.55–12.16	8.61	4.52
[4]	0.21 × 0.30	3.15–10.55	7.4	5.7
[5]	0.25 × 0.25	3.0–10.85	7.85	3
[6]	0.31 × 0.23	3.1–17.1	14	4.2
[7]	0.22 × 0.27	2.8–11.5	8.7	5.8
[8]	0.23 × 0.23	2.5–10	7.5	3.35
[9]	0.25 × 0.36	3.1–12.3	9.2	6.8
[10]	0.3 × 0.3	3.0–20	17	6
[11]	0.31 × 0.31	3.04–11	7.96	5.1
[12]	0.34 × 0.30	2.96–12	9.04	3.82
[13]	0.34 × 0.40	3.42–11.7	8.28	6
[14]	0.31 × 0.36	3.1–13.1	10	5
[15]	0.23 × 0.33	2.18–20	17.82	4.38
[16]	0.63 × 0.54	4.5–13.5	9	6.08
This work	0.27 × 0.33	5–30	25	10.2

References

1. Jalilvand M, Li X, Zwirello L, Zwick T (2015) Ultra wideband compact near-field imaging system for breast cancer detection. IET Microw Antennas Propag 9(10):1009–1014. https://doi.org/10.1049/iet-map.2014.0735
2. Bala S, Soni Reddy P, Sarkar S, Sarkar PP (2022) Small size wideband monopole antenna with five notch bands for different wireless applications. Radioengineering 31(1):23–31
3. Rizvi SNR, Awan WA, Hussain N (2021) Design and characterization of miniaturized printed antenna for UWB communication systems. J Electr Eng Technol 16(2):1003–1010. https://doi.org/10.1007/s42835-020-00630-3
4. Tan MT, Li JQ, Jiang ZY (2019) A miniaturized ultra-wideband planar monopole antenna with L-shaped ground plane stubs. Int J RF Microw Comput Aided Eng 29(11):1–7
5. Zhang S, Zhong Y, Zhou Y, Guo Y, Ji C (2019) Design of a symmetric open slot antenna for UWB applications. IEICE Electron Expr 16(20)
6. Fallahi H, Atlasbaf Z (2015) Bandwidth enhancement of a CPW-fed monopole antenna with small fractal elements. AEU-Int J Electron Commun 69(2):590–595
7. Sahoo S, Mishra LP, Mohanty MN, Mishra RK (2018) Design of compact UWB monopole planar antenna with modified partial ground plane. Microw Opt Technol Lett 60(3):578–583
8. Hossain A, Islam MT, Almutairi AF, Singh MSF, Mat K, Samsuzzaman M (2020) An octagonal ring-shaped parasitic resonator based compact ultrawideband antenna for microwave imaging applications. Sensors 20(5):1354
9. Dwivedi RP, Kommuri UK (2019) Compact high gain UWB antenna using fractal geometry and UWB-AMC. Microw Opt Technol Lett 61(3):787–793
10. Hossain MJ, Faruque MRI, Islam MM, Islam MT, Rahman MA (2016) Bird face microstrip printed monopole antenna design for ultra wide band applications. Frequenz 70(11–12):473–478

11. Elhabchi M, Srifi MN, Touahni R (2020) A novel modified U-shaped microstrip antenna for super wide band (SWB) applications. Analog Integr Circuits Signal Process 102(3):571–578
12. Mahmud MZ, Islam MT, Samsuzzaman M (2016) A high performance UWB antenna design for microwave imaging system. Microw Opt Technol Lett 58(8):1824–1831
13. Midya M, Bhattacharjee S, Mitra M (2019) Circularly polarized planar monopole antenna for ultrawideband applications. Int J RF Microw Comput-Aided Eng 29(11):1–9
14. Awad NM, Abdelazeez MK (2018) Multi-slot microstrip antenna for ultra-wide band applications. J King Saud Univ Eng Sci 30(1):38–45
15. Kundu S, Jana SK (2018) A compact umbrella shaped UWB antenna for ground-coupling GPR applications. Microw Opt Technol Lett 60(1):51–146
16. Dorostkar MA, Islam MT, Azim R (2013) Design of a novel super wideband circular hexagonal fractal antenna. Prog Electromagn Res 139:229–245. https://doi.org/10.2528/PIER13030505
17. Sharma A, Khanna P, Shinghal K, Kumar A (2016) Design of CPW fed antenna with defected substrate for wideband applications. J Electr Comput Eng 2016:1–10
18. Azim R, Islam MT, Misran N (2013) Microstrip line-fed printed planar monopole antenna for UWB applications. Arab J Sci Eng 38(9):2415–2422
19. Zhang D, Rahmat-Samii Y (2017) Top-cross-loop improving the performance of the UWB planar monopole antennas. Microw Opt Technol Lett 59(10):2432–2440
20. Kumar R, Sinha R, Choubey A, Mahto SK (2020) An ultrawide band monopole antenna using hexagonal-square shaped fractal geometry. J Electromagn Waves Appl:1–12. https://doi.org/10.1080/09205071.2020.1829094
21. Omar AA, Safia OA, Nedil M (2017) UWB coplanar waveguide-fed coplanar strips rectangular spiral antenna. Int J RF Microw Comput Aided Eng 27(7):1–6
22. Telsang TM, Kakade AB (2014) Ultra-wideband slotted semi-circular patch antenna. Microw Opt Technol Lett 56(2):362–369
23. Samsuzzaman M, Islam MT (2015) A semi circular shaped super wideband patch antenna with high bandwidth dimension ratio. Microw Opt Technol Lett 57(2):445–452

A Study on Repeated Game Theory in Wireless Sensor Networks Security

S. Sridevi, V. Veeramani, and S. M. Chithra

Abstract The performance of mathematical analyses for the learning of ad hoc wireless setups yielded limited results for the transformation of movement and traffic patterns, the theory of dynamic games, and the connection irregularity that separates these networks. Inspired by various studies, this work would theoretically influence all other researchers, addressing game theory, and generally, it would be preferred to calculate the probability of networks beforehand. This article describes several connections in advanced wireless networks that are often designed to be readily available. It allows the investigation of code of conduct and management plans for the existing resources, as well as the creation of mechanisms that promote balance and encourage individual users to work generally in a fruitful manner. The modern literature is reviewed on situational setting research game theory, proving its compatibility with power control and waveform change, mediocre input control and beat, and knot engrossment.

Keywords Ad hoc networks · Repeated game theory · Nash equilibrium · Power control · Waveform adaptation

1 Introduction

The theory of repeated games has shown that as long as households do not discount future costs and benefits, at an expensive rate and as long as behavior is mutually

S. Sridevi (✉)
Department of Mathematics, SRM TRP Engineering College, Trichy, Tamil Nadu, India
e-mail: srihasini.anbu@gmail.com

V. Veeramani
University of Technology and Applied Sciences, Salalah, Sultanate of Oman
e-mail: veeramani.sal@cas.edu.om

S. M. Chithra
RMK College of Engineering and Technology, Chennai, Tamil Nadu, India
e-mail: srilakshmivj@gmail.com

G. Rajakumar et al. (eds.), *Intelligent Communication Technologies and Virtual Mobile Networks*, Lecture Notes on Data Engineering and Communications Technologies 171,
https://doi.org/10.1007/978-981-99-1767-9_2

observable, social norms involving the use of sanctions on non-cooperative behavior, enable cooperation to be maintained. A theoretical assessment of security considerations within the Wi-Fi framework of reference networks are studied. As stated previously, mobile sophisticated channels are one type of network. Most security procedures require a password for lag channels in mobile ad networks, automobile systems, Wi-Fi systems, and verification of delay-sensitive networks. So, every mobile phone major retailer has an asymmetry key pair, and each message is agreed to sign with much of the same unique key. Although the keys may not always appear to be protected, it will be identified by third parties in addition to learning the places and patterns of network nodes. Consequently, when nodes recognize each other, it could perhaps avoid publishing private information texts [1].

Renaming the nodes was performed to achieve a higher position in cyber security on Wi-Fi advertising networks. However, if a base station keeps changing its name, its private information also isn't shielded, and the tracking system can still work properly in the situation. As a consequence, in this case, position data security necessarily is required to the help and support of neighboring network nodes. Because of these reasons mentioned, the integration of name adjustments has piqued the interest of many researchers, and various systems have been suggested. A station is used to synchronize name settings that were variables previously [2].

Furthermore, the compounding zone method is proposed to increase the security level. A combination zone strategic approach, in the lack of infrastructure, initially defines a specific area for analytical purposes. As a direct result, all connected devices within the area will use the renaming procedure to ensure data security. It lacks flexibility, however, because zoning location must be learned before trying to enter the node zone. Swing protocol is suggested to solve this type of problem [3]. It characterizes the base station entry point ahead of time in order to create a username. If a node's security feature is not at risk, then this will alter its name to enhance position safety. However, it lacks the flexibility to integrate this same show's name change with different nodes. The system of game theory was applied to the dilemma of renaming this entire framework.

As pointed out before, repeated gamification can be divided into two parts. Players in repeated cooperative games do not seem to identify. They are ready to work with neighbors and friends to reach high positions safely. Players act selfishly in non-cooperation games [4]. They are those who simply care about their privacy and don't seem too willing to cooperate when their loved ones are trying to improve their privacy. Numerous records investigate this issue in the context of sporting events cooperation. On the other hand, each node on the globe acts for selfish reasons. As a result, the repeated game is pretty realistic. Justin Freudiger was the first one to scrutinize the game-theoretical facets of privacy protection on cellular operators. A repetitive game theory technique was used to assess in multiple cases [5].

Nevertheless, because of the obvious small group of players, the technique has some problems understanding the privacy protocol. Based on the scientific hypothesis, an extreme protocol was proposed. Each node makes an attempt to fix its classification based on one's own pay and the survival rate of its environment. As a consequence, in order to solve this issue for a restricted number of players, a bluffing strategic plan that interrupts its very own path is hired, seeking to stop the phone from having to follow the monitor [6]. As a direct consequence, when trying to design good process jobs across the world, the non-expertise are considered so that video game players do not need family and friends to comprehend their different kinds.

1.1 Objective of This Paper

- Node movement in sleep mode sensor networks increases the net lifetime.
- Existing game theory-based protocols for lifetime control are not supplying sufficient.
- This research uses Expectation Extension for lifetime security effortless sensor node gathering.
- Sensor nodes are remunerated for data transmission through a community sale device.
- This research produces good outcomes in simulations-based connections with the existing rules.

1.2 Motivation

In the latest days, discussions on topics like health, naval strategy, and environmental monitoring have predominantly employed Game Theory (GT). The overwhelming acceptance of the GT with wireless sensor network (WSN) served as fuel for its growth. There are restrictions though because the rewards are greater. In GT with WSNs, choosing the best cryptographic method is encryption provides all security forces. The code size, data size, processing time, and energy consumption of cryptography techniques used in WSNs all need to be taken into consideration. When the necessity for environment sensing arises, WSN will undoubtedly find more and more possibilities given the rapid growth of both electronics and wireless technology. On the other hand, both the business sector and the scientific community have been attentive to the advances of WSN theory and systems. The usage of GT for WSNs has grown among many other alternative strategies; as a result, the focus of this research is solely on using GT for WSNs.

2 Literature Review

This section looks at the recent approach to the mobile user as a short-term and long-term benefit system of the twentieth century, which aims to provide a cost-effective plan to encourage deliberate negotiation. Sasmita Sahoo [7] has developed a set of repeated game cooperative finite databases for distributed wireless sensors.

The most widely used optimization profile in RGT is for examining distribution and mobile delivery Wi-Fi rates. Using RGT as a foundation, the proposed target line for mobile users aims to reduce costs while maximizing lifetime gains. Wu et al. [4] analyzed from an economic and RGT perspective. The study applied the most comprehensive set of game theory rules possible, using it to the furthest extent possible.

Wu et al. accepted Shapley values as the combination's payoff allocation device based on game theory. Wu et al. [8] made the WSN combinations on the foundation of WSNs and proposed the concept of attentiveness constant to size the coalitional outlines and then used NE to regulate the estimated data transmission plans of the molded coalitions [9–12].

Vanbien et al. [13] designed an integrated energetic range sharing organization to enhance the spectrum utilization and exploit the profit of operatives for cooperative wireless networks. Gehrmann and Gunnarsson [6] discussed how to use the WSN model and the resultant security structural design to gather data distribution and controller of safety–critical procedures. Khan and Ahmad [14] suggested that many IoT works are liable for driving tasks in opposite directions and making artificial corporeal objects difficult to use without social mediation.

Satapathy and Sarma [15] proposed ways to protect WSNs by certifying port security, shutting down port advancing and non-original harbors as soon as not necessary, using anti-malware software, firewalls, and intrusion detection systems, checking unsanctioned Internet rules reports and ensuring the structure is patched up and saved up to date. Lee et al. [15] proposed an occurrence tree liability quantification way based on game theory. Repeated game model theory was used to describe the communication between non-similar nodes and aid the trust level to exactly detect opponents, which effectively decreases the resource depletion [16, 17].

3 Mixing Zone and Level of Safety

Numerous techniques are being used to protect the mobile data type private information; as a consequence, the data type location data is secure, and it can be easily monitored by the opponent. This section explains the technology behind the name. The standard technique does have limitations, as backed by the professional qualifications segment. So, in order to overcome these constraints, several strategies are implemented in a highly complicated area. The pattern of complicated areas is numerous hubs, which is recognized as blended zones that connect a hard-fast region. In this zone, participating machines exchange the same names. As a consequence, this same tracker could indeed equate the new name with the old one. However, there is still the problem of monitoring one's time and site connections [18–22].

As a direct consequence, two additional approaches were developed to solve the initial made by mixing area strategy. First, the communication systems shut down their transceivers and transform the combination area into something perplexing. The participating devices could use a bluff strategy to reduce road deterioration and ensure a certain degree of safety. Consider a cellular connection largely composed of n mobile nodes. The node mix gets the drop-down box 't,' comparable to the delete procedure, and a base station can introduce the name change process by transmitting a trigger message. If the smartphone base station needs to fill in the knowledge of its neighborhood, it will choose its move based on the type and distance with its neighbors and friends [23]. Because all opponents are in the immediate area of cooperative work, all terminal points included will begin to switch name directly. If the access points around something are cracked, then it will compute the difference between both itself and its cotenant. An access point will use a fake strategic plan if the evaluated distance is less than the selected gateway. If indeed the evaluated distance surpasses, it will close its transmitter and consider the next move as a history. At the same time, the endpoints will be unable to converse with one another [6].

If any node somehow doesn't know exactly what sort of opponent it is against, it will change the name by forecasting exactly what sort of opponent it is up against. If the overall number of players in the mixing zone is much less than four, there is a way of measuring the gap as well as clearing the trail. The performance name switch tactics do not collaborate well for a comparatively tiny number of nodes since it has the relatively simple possibility of announcing the deficiency type to its opponents, and irrespective of whether or not its opponent collaborates, the level of security remains abnormally low. This motivates to narrow the extent of confidentiality to choose a bunch of players. If the name changes inside the opponent's view, the opponent will start noticing that now the participating devices change the codename in their confined environment. It will make a comparison of the codename before and after the transformation, enabling it to predict connectivity with one of the most increasing competitiveness [18]. This same task security level of the consumer with whom i used to work was at the same time the most effective name change.

$$H_i(T) = -\sum_{d=1}^{n(T)} P(d/b) \log_2 P(d/b) \tag{1}$$

where H is the consumer, i is the security flat, T is the length of time in which the user's binomial nomenclature has altered, and $P(d/b)$ is a new name for message d. Chances of this happening has something to do with the old, shortened form b. The divisor of itinerant nodes at T is $n(T)$. The possibility to achieve image privacy has been defined as the total number of endpoints and the mutability of their places in the posse area [5]. The tail will quickly find the owner's level of privacy when there is just one node as well as stops the codename, so $H_i(T) = 0$. The enthalpy change will profit from continual odds propagation for $P(d/b)$, and $\log_2 n(T)$ can be the level of security of something like the mobile cost.

3.1 Loss of Security Level

The level of protection of portable knots varies based on the scene and duration. As a consequence, the institution can be evaluated using the masterpiece affiliated with scene safety radical chin. The safe custody level deviance over time is declared as the hidden loss task $\beta_i(t, T_i^l)$ in this work, in which t is the matter of seconds and T_i^l is the moment for the most impact fake identity transition knot i. For a provided T_i^l, humans can hold

$$\beta_i(\mathrm{t}, T_i^l) = \begin{cases} \lambda * (t - T_i^l) & \text{for } T_i^l \le t \le T_i^l \\ H_i(T_i^l) & \text{for } T_i^l \le t \end{cases} \tag{2}$$

In the team's observation-based rule, anywhere and everywhere l is the tip, and i is the ego. l sets a limit on how quickly this same pain of loneliness propagates. The highly focused benefit of $\beta_i(t, T_i^l)$ just at phase from the last name change seems to be similar to the scene security plan, which can be seen [8]. So, with this cost route, the consumer node i at t's organization is

$$H_i(T) = H_i(T_i^l) - \beta_i(t, T_i^l)t \ge T_i^l. \tag{3}$$

4 Mathematical Modeling

Definition 1 The strategic interaction between competing or cooperating goals when the limitations and compensation for efforts is being taken into account is referred to as a game.

Definition 2 In a game, a player is the fundamental entity that participates in the game with a finite number of other players, represented by n, and is incharge of acting rationally for each player i represented by N_i. In a game, a player may take the place of a human, a machine, or a team of players.

Definition 3 $u_i\colon N \to H$, which evaluates the consequence for player i influenced by the activities of all players, indicates the utility/payoff, which is the favorable or negative payoff to a player for a specific action within the game. $\beta = n_{i\epsilon N} H_i$, where the Algebraic combination is represented by the notation.

Definition 4 A technique is a plan of action that a player can take while playing the game, as indicated by a strategic game $\langle N, \beta, u_i \rangle$.

Theorem 1 *An ideal sleep–wake-up algorithm for the i-barrier scope repeated game with the wireless sensor networks technique.*

Proof A sensor network N is considered. Let '*m*' be the extreme number of node-disjoint paths in the coverage graph $H(N)$ connecting the simulated nodes i and l: Additionally, a sensory node's lifespan is maintained at one. Any S-MAC sleep–wake-up algorithm for the *i*-barrier cover game can attain a lifespan of at most *l*/i.

It should be established that the repeated game with the wireless sensor networks algorithm can achieve a lifetime of the network of *l*/i in order to establish the theorem. The Stint algorithm offers *i*-barrier coverage for *i* units of time *t*. The proof is complete when n mod i equals 0. Therefore, n mod i_0 is considered. The implication is that $i < T < 2i$. For *l*/i units of time, they give *i*-barrier coverage. $T = n - li$ therefore *l*/i. So, for *l*/i units of time, the Stint algorithm offers an *i*-barrier covering game.

Theorem 2 *It is NP-complete to solve the Cluster Protective Covering Lifetime with Non-Zero Path Switching issue.*

Proof By comparing the Node-Disjoint Barrier Cover Duration with Non-Zero Path Switching issue to the Barriers Cover Life with Non-Zero Sensor Switching issue in NP, it can be proven that the barrier coverage lifetime with non-zero sensor changes problem is NP-complete.

The Node-Disjoint Barrier Coverage Duration with Non-Zero Path Switching problem is how the partitioning problem is reduced. A sensor system is built using an application of the split issue as a starting point. It is assumed that the deployment region is rectangular, with the bottom left corner at the origin, or with coordinates (0; 0). The coordinates are set for the right-bottom corner (2*i*; 0). Let $\in\, > 0$. A sensor is installed at coordinates $[H, (n - i)(2n + \lambda)]$ for each integer $H_i \in N. l = \sum_i^n \frac{n_i}{2}$ and $i = 2$ are set. All n paths connecting the two virtual nodes i and l in the covering graph of this sensor network are node disjoints.

If the split problem has a 'sure' response, then $T_i, i = 1, 2, 3 \ldots\ldots n$ such that $\sum_{i\in n} T_i = \sum_{l\notin n} T_i$. Now that the set of sensors can be divided into two sets according to T_i and $1, 2, 3 \ldots n - l$, where each set provides i-barrier coverage game for l units of time, the sensor network may accomplish k barrier coverage for l units of time. As an approach, the sensors can be divided into two disjoint groups so that each set gives i-barrier covered game for l period of time if the sensor network provides i-barrier cover game for l period of time. This occurs as a sensor is turned on, and it remains that way till its lifetime (T_l, for some l) is over. Additionally, if the network delivers i-barrier coverage for l time units, each sensor has completely used up its lifetime.

4.1 A Secure System

Security is typically made an essential part of an organization's strategy, with the protection offered by a separate segment. After dividing these shared keys, the modified version of the key trying to set up a path, on the other side, precludes the cumulative total of the latest nodes just after lean management. Aside from the utilization of modern sensing devices, cryptographic protocols (such as Daffier Hellman's flagship) are an extra option. One of most major improvements is that one base station can see the retreat crucial with opposing base station inside the network [8]. Some other point of view is to reorganize the infrastructure using a single geometrical force held in common by each node pair, regardless of the fact that this force is not correctly evaluated. Each node inside n-node sensor nodes must recite the $(n - 1)$ key, and $\left(\frac{n(n-1)}{2}\right)$ key should be installed inside the system. Some other alternative is to use shoe keys that link up to a trusted ground station. Each base station just wants to give one key with both the lower terminals, and the keys are installed with opposite stations via the lesser station.

The lesser station continues to suffer some failure as a result of the agreement, and because there is just one station, the network connects harm packaging to the lower depot and eradicates the risk of force attack [18]. Random key redistributing wealth methods [2] are developed in recent years, in which an outer group of symmetric key encryption is selected and dispersed to a randomly selected subset of the collective at each sensor network. The network entities that want to interact, look for their organizations to see if they have public keys; if so, they utilize seeing the session key. Every pair of nodes does not really have public keys, but if the opportunity of uncovering the mystery is high sufficiently, the hubs can find buttons with a huge number of nodes to shape a densely integrated network.

Correlation to a reliable base station is precluded by the key distribution automated system. The disadvantage of this approach would be that aggressors who interact on various fronts can sidetrack the strategic plan by reconstituting the whole core team. It is hoped to start investigating simplified arbitrary key redistributing wealth strategies in the long term, as well as hardware support for general core cryptography and much more efficient overall encryption polynomial kernel (such as elliptic curve cryptography).

Finally, a secure and efficient key distribution mechanism for large-scale sensor networks that allows for simple key setup is preferred. Many detector application programs, like traditional networks, require ear bud, touchpads, and glove compartment modification protection. Cryptography is a common security measure [24]. Because, once crypto is interconnected to sensor networks, interesting commercial transactions emerge for instant interaction, with end-to-end data encryption, a high degree of protection. The keyboards, on the other hand, have seemed to be conflicting with the silent ability to participate and must be assembled in smaller classes. The key design has been simplified and is supported by correlated encrypted communications, with such a network level able to share key.

4.2 Network Security Services

The most basic security sources to safeguard smart sensors are studied. Secure wallet or purse management, intercept identification, and safe data collection are just a few of the security tools represented. The networking and telecommunication abilities of each terminal in most wireless sensor nodes are restricted. However, organizations of nodes are utilized to collect and evaluate interesting network information. One node, for instance, is in control of the show's merged tracking of a means of transportation. Specific knots in a bunch can change rapidly and unpredictably. Groups perform many of the key functions in sensor networks. The group calculation's effects are typically sent to a ground station [25]. To make sure that it happened to come from such a capable gathering, the weekly magazine must be lawful. Wireless sensor networks seem to be prone to a range of threats. Traffic as well as prolonged activity are generally monitored and analyzed for discrepancies in wired broadband at fence posts apart from similar marks. It is often expensive in terms of network memory and power consumption, by its inherently restricted bandwidth. In terms of declaration, power, and memory usage, wireless sensors necessitate a highly decentralized and expensive reaction. As a consequence, it is necessary to comprehend inconsistencies, apps, and standard types of threats [11]. Recognizing how combined opponents can strike a scheme is critical for researchers and practitioners. The use of secure groups could be a hopeful way of detecting distributed intrusion.

For example, to prevent wrong alarm systems in true action recognition, the system could indeed estimate the temp or humidity of the segment, or merge sensors data to quantify the verity and speed of an object moving, or accumulated data. Build-up can occur at numerous places inside the wireless sensor nodes in expectation of the wireless network's setup. All points of the gathering must be shielded. For the users with reliable acceptance, powerful approaches are useful. Underneath the right set of assumptions, sampling a tiny fraction of the endpoints and deciding whether they must function nicely can help detect various types of attacks [5].

4.3 Simulation Setup

The NS2 simulation model is being used to assess the hypothesized Power Effective Speed Transmission scheme. The simulation parameters are provided in Tables 1 and 2 shows the radio power consumption.

4.4 Simulation Results

The NS2 is used to accomplish the timetabled Strength Effective Speed MAC protocol. The network size is fixed for the demonstration and independent inquiry of

Table 1 Simulation parameters

Parameters	Value
The quantity of loops	50
Location of the system	100 × 100
Scope of signaling	16 m
The preliminary electricity	2 kJ
Hooks for transmission and reception	60 ms
Spectrum of transmitting	20 m
Message measurements	64 Bytes
The restrict of power generation E_{th}	0.001 mJ
Recurrence of the channel	2.4 GHz
Prototype of propagation loss	Two ray model

Table 2 Radio power consumption

Mode	Transmit	Receive	Sleep	Idle
Electricity (mW)	42	29	100 μW	12.36

PELLMAC practitioners with S-MAC practitioner virtual environment boundaries. A modeling setup with five levels, ten endpoints in each stack, and four starts to sink can be used for fast S-MAC and PELLMAC consumption. This comprehensive chart illustrates that PELLMAC requires less power than S-MAC. As a direct consequence, the PELLMAC procedure uses up to 65–73% less energy than the existing S-MAC with regular intervals of sleep procedure. The corresponding results are shown in Figs. 1, 2, 3 and 4.

In Fig. 2, S-MAC's consumption with periodic sleep is significantly greater than PELLMAC's and lesser than S-MAC's snooze method of operation. As a direct consequence of the S-MAC method of operation, the sensor networks have devoured more power and the ability to remember waking up regularly, even when there is no traffic congestion inside the sensor net. The PELLMAC requires less power than the S-MAC rules due to the spaced scheduling. Whenever the number of knots is 10, the suggested PELLMAC mode of operation consumes 32% less electricity and saves 88.89% of its energy than the established S-MAC with 10% of overall responsive snooze.

Due to the obvious hybrid technique used in the PELLMAC procedure, the capacity factor of the scheduled procedure is 71% significantly greater than that of the S-MAC procedure.

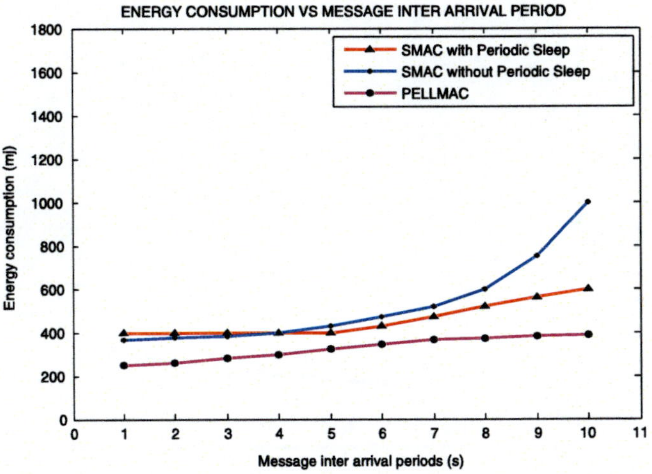

Fig. 1 Comparison of energy consumption based on message interval arrival period

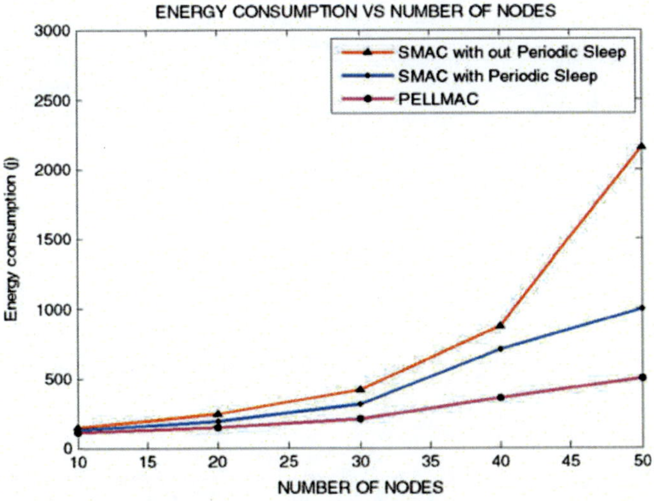

Fig. 2 Comparison of energy consumption based on nodes

4.5 Advantages and Disadvantages

- Taking into account the multi-dimensional and full-scenario worldwide connection that will continue to grow in the future networks will become a new direction for future network planning, operation, management, and running as well as a crucial way to achieve network intelligence and efficiency.

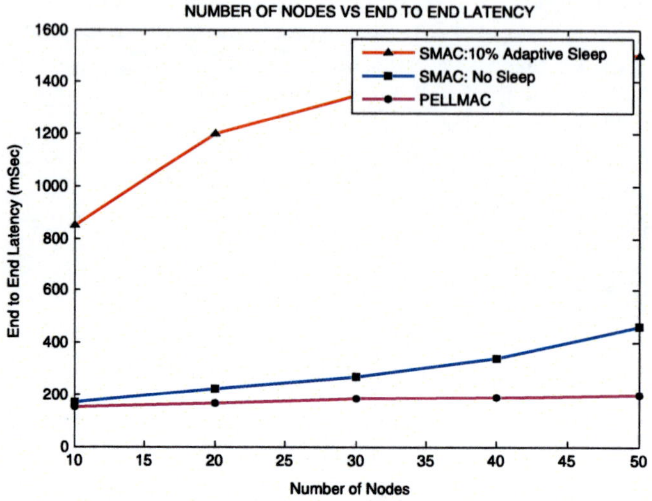

Fig. 3 Latency comparison of PELLMAC with S-MAC

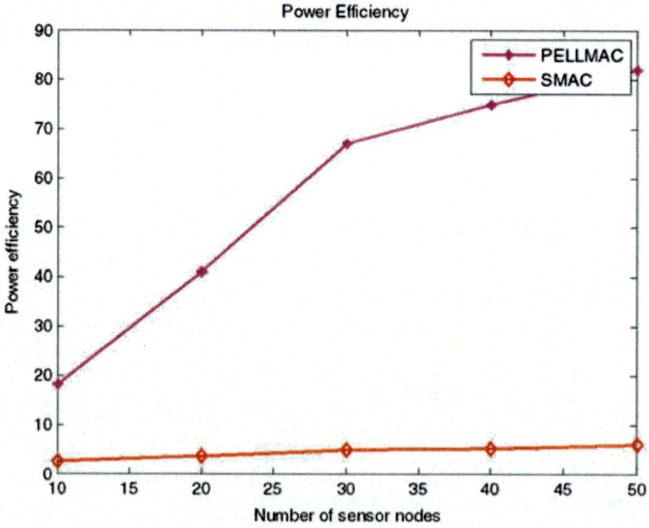

Fig. 4 Power efficiency comparisons with S-MAC

- Users will be better able to understand the network state, easily mine important network data, and explore cutting-edge network applications with a more amiable immersive interactive interface, thanks to the projected display of the real world and virtual interface map of the network.

- Users may also be shown the dynamic change process, real-time status, and future development direction of the whole service life of the network together with the wireless network model. Different network elements and structure information in the network can be dynamically shown.
- In order to promote the efficient repair of weaknesses in the wireless network, this study investigates the mining and repair of vulnerabilities based on the evolutionary game theory from the perspective of the lifespan network security vulnerabilities.
- This work has made some advances in choosing a data security approach for commercial networks based on repeated game theory. In contrast, there are numerous random interference variables in the attack and defensive process in the actual network environment, such as a change in the system operating environment that will cause a change in the game model. To further improve the game style, stochastic aspects should be taken into account in the future study.

5 Conclusion

On the basis of energy lifespan, link quality energy usage on a per-packet and per-individual basis, and network throughput, PELLMAC, and S-MAC performance are assessed. The focus is on discussing PELLMAC and S-MAC's performance in comparison with other network architectures. When compared to PELLMAC, S-MAC performs better since it increases network throughput and longevity while using less energy. The energy consumption per packet is decreased to less than 30%, the throughput is raised by two times, and the network lifetime is improved by more than 70%. The comparison of PELLMAC with S-MAC reveals that PELLMAC boosted network throughput by 1.82 times, increased network lifetime by more than 70%, and decreased energy usage by more than 30%.

References

1. Cagalj M, Ganeriwal S, Aad I, Hubaux J (2005) On selfish behavior in CSMA/CA networks. In: Proceedings of INFOCOM 2005 24th annual joint conference of the IEEE computer and communications societies, Miami, FL, USA, 13–17 Mar 2005
2. Qiu Y, Chen Z, Xu L (2010) Active defense model of wireless sensor networks based on evolutionary game theory. In: Proceedings of 2010 6th international conference on wireless communications, networking and mobile computing, Chengdu, China, 23–25 Sept 2010
3. Saad W, Zhu H, Debbah M, Hjorungnes A, Basar T (2009) Coalitional game theory for communication networks: a tutorial. IEEE Sign Process Mag 26:77–97
4. Wu T, Yue K, Liu W (2011) An energy-efficient coalition game model for wireless sensor networks. In: Proceedings of 2011 30th Chinese control conference (CCC 2011), Yantai, China, 22–24 July 2011
5. Apt K, Witzel A (2006) A generic approach to coalition formation. In: Proceedings of the international workshop computational social choice (COMSOC), Amsterdam, The Netherlands, 6–8 Dec 2006

6. Gehrmann C, Gunnarsson M (2019) A digital twin based industrial automation and control system security architecture. IEEE Trans Industr Inf 16(1):669–680
7. Sahoo S, Acharya BM (2011) Energy efficient routing protocol in wireless sensor network. Int J Eng Sci Technol 3(7):6084–6089
8. Wu T, Yue K, Liu W, Xu J (2011) An energy-efficient data transfer model of wireless sensor networks based on the coalitional game theory. In: Proceedings of 2011 8th international conference on fuzzy systems and knowledge discovery (FSKD 2011), Shanghai, China, 26–28 July 2011
9. Satapathy S, Sarma N (2006) TREEPSI: tree based energy efficient protocol for sensor information. In: Wireless and optical communications networks, IFIP international conference. https://doi.org/10.1109/WOCN.2006.1666530
10. Mehrotra SS, Rajan M (2011) Comparative analysis of encryption algorithm for data communication. Int J Comput Sci Technol 2(2):292–294
11. Gebremichael T, Ledwaba LPI, Eldefrawy MH (2020) Security and privacy in the industrial internet of things: current standards and future challenges. IEEE Access 8:152351–152366
12. Kiran V, Rani S, Singh P (2020) Towards a light weight routing security in IoT using non-cooperative game models and Dempster-Shaffer theory. Wireless Pers Commun 110(4):1729–2174
13. Vanbien L, Feng Z, Zhang P, Huang Y, Wang X (2008) A dynamic spectrum allocation scheme with interference mitigation in cooperative networks. In: Proceedings of 2008 IEEE wireless communications and networking conference, Las Vegas, NV, USA, 31 March–3 Apr 2008
14. Khan SU, Ahmad I (2006) Non-cooperative, semi-cooperative, and cooperative games based grid resource allocation. In: Parallel and distributed processing symposium, IPDPS
15. Lee S, Kim S, Choi K (2018) Game theory-based security vulnerability quantification for social internet of things. Futur Gener Comput Syst 82:752–760
16. Bernhard P (1979) Linear-quadratic, two–person, zero-sum differential games: necessary and sufficient conditions. J Opto Theory Appl 27:51–69
17. Perkins C, Belding-Royer E (1999) Ad-Hoc on-demand distance vector routing. In: Proceedings of 1999 mobile computing systems and applications, New Orleans, LA, USA, 25–26 Feb 1999
18. Charles DE, Panagopoulos AD (2010) A survey on game theory applications in wireless networks. Comput Netw 54(18):3421–3430
19. Fudenberg D, Tirole J (1983) Game theory. MIT Press
20. Kong F, Zhou Y, Xia B (2019) A security reputation model for IoT health data using S-Alex net and dynamic game theory in cloud computing environment. IEEE Access 7:161822–161830
21. Srivastava G, Parizi RM, Dehghantanha A (2020) The future of blockchain technology in healthcare internet of things security. Blockchain Cybersecur Trust Privacy:161–184
22. Nash J (1951) Non-cooperative games. Second Ser 54(2):286–295
23. Khayatian H, Saadat R, Mirjalily G (2010) Distributed power allocation based on coalitional and noncooperative games for wireless networks. In: Proceedings of 2010 5th international symposium on telecommunications, Tehran, Iran, 4–6 Dec 2010
24. Triggiani R (2012) Min-max game theory and non-standard differential Riccati equations for abstract hyperbolic-like equations. Nonlin Anal Ser A 75:1572–1591
25. Romp G (2001) Game theory: introduction and applications. OUP Catalogue

An Overview of Information-Centric Network: Concepts, Network Architecture, Comparison, and Difficulties

Antonio Cortés Castillo

Abstract In the past decades, there have been traditional network architectures that were responsible for routing data between various nodes using different types of IP addresses through a router using different types of routable protocols such as TCP/IP. But these networks presented a problem with respect to the management of large volumes of information. Faced with this situation, the Telco industry proposes a network structure called Information Centered Networks (ICN), which is a part of the Content Centered Network paradigm. The ICN help distribute and retrieve content on the network once they are requested by the user, which allows independence in the location and storage of this cached content in its various ICN nodes. In turn, the ICN architectures use protocols that allow the unique assignment and sending of data objects when making use of the respective network. Hence, the goal of the ICN topology is to deliver these data objects to users located in the network, in contrast to networks that use TCP/IP protocols and in which communications is carried out between the various nodes of the network. The main objective of this article is to familiarize users by comparing different proposed ICN architectures, as well as network caching and ICN transformation.

Keywords Information-centric networks · Information-centric networks topologies · Named data networks · Network caching · Information-centric networks transformation · Service identifier · Data-driven

1 Introduction

Today, people use the latest generation of networks to view content through various devices (e.g., mobile devices, tablets, etc.) rather than having a direct connection to some person, device, or server [1]. This situation meant that the network operator providing the service had to have a new optimized network architecture to allow

A. C. Castillo (✉)
Department of Computer Science, University of Panama, Panama City, Panama
e-mail: antonio.cortes@up.ac.pa

G. Rajakumar et al. (eds.), *Intelligent Communication Technologies and Virtual Mobile Networks*, Lecture Notes on Data Engineering and Communications Technologies 171,
https://doi.org/10.1007/978-981-99-1767-9_3

content retrieval, thus allowing the optimization and increase of the network topology performance [2, 3]. At the same time, the idea behind the ICN structure is that more users they can diversify their needs through the request of diverse massive content that can be stored on a server or on a content router. As such, these are considered within the context of the ICN as a unit, the named data object (NDO), which consists of a unique identifier, data, and metadata. This NDO uses name resolution, and its main function is to be able to send content over the ICN network, using routing, forwarding, and clients as participants. For example, Fig. 1, you can have two clients, in the first stage you run with the origin location, and then use the step to request the forwarding of the message, which is the name resolution system. In the second stage, routing through the ICN network carrying out by using the name of the content, which requires connecting to a content router (CR), thus allowing the storage of requests made by users in the form of object contents. Therefore, there are two ways to route the information for this content through the ICN. First, a Distributed Batch Table (DHT) using for routing, and the second way is to use unstructured routing, which is IP routing. Thus, the performance of NDO is achieving through the storage of the content from the closest router in the ICN network; thus, there is a variety of copies that distributing in the network and can share between the nodes of the net. The storage is done in the content router (CR) of the target node, which allows the ICN architecture to use its own storage method.

On the other hand, in an information-centric network, the client manages the recovery of segmented NDO by retransmitting them as NDO are replicated in networks nodes. Also, in terms of network efficiency and optimization, content-centric networks make peer-to-peer communication an inefficient process in an IP network environment, where user management of applications is proposed.

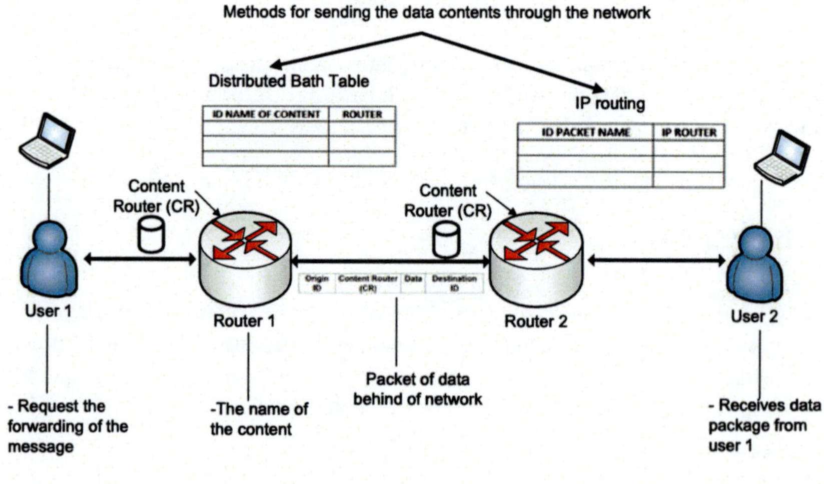

Fig. 1 Named data object in the context of ICN

2 Related Work

However, to efficiently utilize network resources and be able to deliver content to users on time, information-centric networks use network layer-level content names [4] and caching techniques in the network [5, 6]. In turn, these requests are made by the various clients and are processed and stored by a content router on the network-centric on information, instead of using proprietary servers of companies in charge of storing this content. Correspondingly, content names have no link to any location on the network, so their management carrying out separately in the communication session, which allows [7] to simplify the ubiquity of the user without the need to establish new communication sessions. Therefore, instead of protecting the communication medium between two nodes [11–13], the information-centric network protects these content names through the user's digital signature [8, 10].

So, information-centric networks allow routing based on content names, the storage of this content carrying out through routers that locating closer to the network and data security, allowing an analysis of some problems encountered in those networks based on IP protocols, which allows predicting a new future with respect to Internet Architectures [14, 15].

However, ICN allows for better control over the performance of network content flow control, as well as establishing content-oriented access methods and simplifying security procedures through autonomous data management and processing. Like any new network architecture that arises, as is the case with ICN, [16] it is still in an early stage and there is still aspect to consider. For example, although the storage of content through a router that is closer to the network helps to improve network performance considerably, and the techniques used for content storage present problems with respect to the veracity and authentication of these contents. In this same direction, the content that storing in the ICN in a more generalized way in the network allows for achieving an optimal and efficient delivery for the users. The privacy of users on the network where the content storing is important to considerer the integrity of data objects, as these data objects can be stored in content routers located anywhere in the network by allowing an NDO to authenticate. Likewise, it is important to have methods that allow validating the authenticity of named data object. Another aspect to consider is the name of the data object since these are identified by a unique name that using as an identifier of the data objects.

Needless to say, projects related to the ICN propose some solutions to the problems that may occur at the time of construction of a network structure centered on information [17, 18].

For this article, comprehensive information is provided on the different information-centric network topologies, taking Named Data Networks (NDN) as a reference, which allow converging with various network structures using single data-centric network information. Information-centric networks are more secure and flexible since they allow eliminating redundancy now in which the client makes a request for the content, which is represented by means of content objects that storing in the closest router from the node destiny. In the context of data-driven, this type of model is more oriented to applications and organizational culture, where companies focused on applications present problems related to data replication, costs, non-real time, vision not integrated, departments divided by silos and partial intelligence. While data-centric is more focused on the robustness of the network architecture itself, the optimization of bandwidth and the improvements in the performance of the balance of data load in the network. In this way, the ICN presents a set of aspects that will help improve the problems of current and future networks in the context of Internet network architecture.

Some of the limitations in this work are specifically presented in the use of empirical data for the simulation of the proposed network architectures in the ICN field, which would allow us to better observe the behavior of these network models. Some of the free software tools used to simulate these environments involve using very specific versions of the Ubuntu operating system with updates to the PHP programming language, which sometimes results in incompatibilities with the packages used to run the applications.

Followed by the introduction, we have Sect. 2, in which essential concepts of information-centric networks are presented. In this same direction, in Sect. 3, some of the network structures most used in information-centric another issue that must be considered and evaluated. In addition, its networks are described. Similarly, in Sect. 4, an essential discussion of information-centric networks is performed. However, in Sect. 5, the article conclusions are outlined.

3 Proposed Work

3.1 Essential Concepts of ICN

The management of large volumes of information is one of the obstacles of current traditional networks that do not allow offering better services to their users, so new alternatives to this problem are presented, such as information-centric networks that allow managing these contents of data in a more efficient and timely manner, making the network architecture more cost-effective to have more optimized bandwidth with better information flow as we can see in Fig. 2, aspects that current Internet IP network slack. In view of this situation, access to data content is proposed, through forwarding and storage in content routers, which allows better distribution of data and an improvement in the bandwidth of networks focused on information. This is achieved through the unique identification of data objects (a technique known as naming), which locates named data content in the network structure based on information referenced by content routers bandwidth are subsequently sent to their respective recipient nodes (procedure known as routing), which allows establishing a level of security for those data contents that travel through the information-centric network (which we will call security), thus guaranteeing the independent mobility of users through the ICN (known as ubiquity).

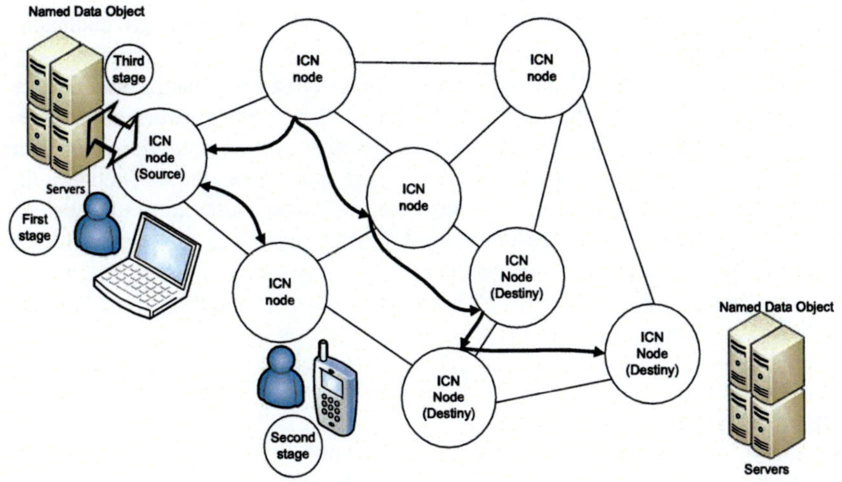

Fig. 2 Essential elements of an information-centric network

Figure 1 shows the first stage which consists of storing the data contained in the content router (CR). Subsequently, in the second stage, the various user requests are served from any node of the information-centric network, and finally, in the third stage, before responding to client requests, we proceed with the optimization of the data objects data in the information-centric network node.

Therefore, network architectures such as information-centric networks are composed of a series of elements, as we will show later.

3.1.1 Naming Data Object (NDO)

The identification of data objects in information-centric networks, regardless of which content router they are in, is carried out through a fundamental component such as the naming data object. To identify these data objects, the route firstly uses a hierarchical name structure and a similarly flat name structure. The hierarchical naming structure is related to the information that is used to route content the network, which helps improve performance in the various processes that are used to route content. To improve routing performance, hash tables are used that point to a data object such as a data name or data content. According to the value that these hash tables may have on the network, the flat name structure produces several names with a fixed length for each data object, thus allowing to speed up the sending of the data content through the network.

3.1.2 Routing Data Object (RDO)

To carry out the routing of the data objects, the information-centric network must have to its credit the resolution of names, followed by a route between the nodes and finally the delivery of this data object to the client. Name resolution allows the data object to be sent in a predetermined coverage area, while the route between the nodes allows the investigation of the route that the router must carry out to deliver the data object according to the user request. On the other hand, delivering this data object to the user allows the requested data object to be routed to its respective user. In turn, routing in information-centric networks dominate routing-by-name (RBNR) and location-by-name routing (LBNR).

3.1.3 Security

Information-centric networks to be able to store data objects in the internal memories of content routers see the need to validate these data objects in such a way that they maintain integrity and reliability in what they contain. Due to this, security mechanisms must be incorporated to guarantee that these data objects have not been previously modified by another entity. Therefore, making the data objects robust and reliable is a process that is considered in the ICN. If any of these security techniques fail, the compass is opened for DoS-level attacks to emerge in which malicious code is released on the network.

3.1.4 Ubiquity

The idea that one has with respect to the ubiquity of objects in a pre-established space and time is intrinsic to the mobility of users since it makes the uninterrupted reception of content by clients more flexible, regardless of where they are located, and initially without a perceptible interruption in the applications that are in information-centric networks. In turn, the retrieval of content proposed by the receiver of the information-centric network allows *n* clients, located in different parts, not to see the continuous reception of content interrupted. However, if we take the idea of mobility to the context of servers, the scenario is a bit more complicated, because mobility in information-centric networks requires working collaboratively with the various routing process, which allows identifying data objects that bear a level of similarity to the data object being transmitted.

3.2 Types of Architectures for Information-Centric Networks

Some of the most common network structures in the context of ICNs are detailed below.

3.2.1 MobilityFirst Architecture

The MobilityFirst architecture arises as an alternative to Internet IP addressing networks, where communication is carried out between the source and destination node through a pre-established route. In this same direction, this type of network architecture divides the names, identified by conglomerates of data content of the network addresses, in which there are various access points that allow users to connect when they have various locations connected and reconnected to the network structures more than once (multi-homing). In turn, the MobilityFirst network uses a unique identifier called Globally Unique Identifiers (GUID), which allows any object in the MobilityFirst network to be identified. This network identifier helps attach objects, which come from routers, nodes, servers, content, etc. In MobilityFirst networks, the globally unique identifier manages network services and the service identifier (SID), which is used in multicast, unicast and multi-homing services. The use of the SID in the previously mentioned IP transmission modes is intended for the data packet to travel through the network in such a way that it reaches the addressee. These SIDs help conserve bandwidth and reduce network workload by getting data packets to their destination in such a way that they don't get lost in the network and create unwanted bottlenecks.

MobilityFirst network architectures route information across the network in a hybrid fashion, using GUIDs and Named Addresses (NA), allowing for network growth. It is important to mention that GUIDs manage network services through the network mac address, which uses an IP address assigned to the network port, which allows packets to travel on the network and reach their destination. In this way, the data that is routed in the network is stored in a Generalized Storage Aware Routing (GSTAR), since if the network presents problems, the data can be temporarily stored in the router using this GSTAR protocol, since these have the capacity to store data and allow caching in the router for the transmission of content, which helps to resolve the instability that may occur in mobile networks and avoids retransmission of lost data from the origin node thus avoiding network congestion. In turn, the routers that allow temporary storage of data have the capacity to send large data files on the network. In addition, routers allow the temporary storage of data, with routing tables that are stored in a database of the same router, in which there is the name of the data packet, the source and destination addresses for this packet of data and a unique identifier that allows the packet to reach its destination.

In turn, the data transmission in a network structure, MobilityFirst, is carried out between the nodes of the network, using the hop-by-hop technique in packet routing, as seen in Fig. 3, and in which the distribution units of (PDU), receive and store at each network hop and transmit the data at the next hop, which provides better performance in supporting network mobility.

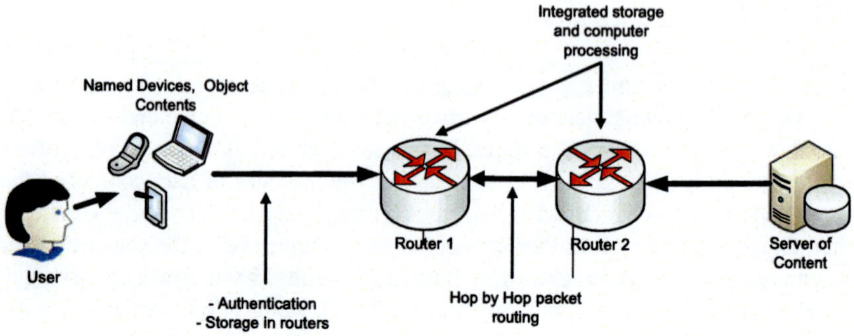

Fig. 3 MobilityFirst basic network structure

3.2.2 Named Data Networking (NDN)

The Named Data Networking network architecture is based on data, as opposed to IP networks, which base their data routing on the host, are responsible for managing name data, and in which communication carrying out using data packets and interest packs. Each of these packets, in turn, has a unique data identifier (UID), which allows the data to be identified from its name. In contrast, the packets of interest have a selector that allows a concise description of the packets of data that transmitting on the network; therefore, a random sequential number is generated that we are going to call Nonce, which guarantees that the data contents can be encrypted between users who communicate through the network. On the other hand, packets also have a very specific way to ensure their transmission in the network, namely by assigning a signature in the same packet. In network topologies (NDN), data delivery basing on names, while in traditional IP networks it is making up of IP addresses. These packets consist of addresses and names that allow the user to be identified. In the following figure, Fig. 4, the internal structure of a data packet and that of a packet of interest are shown.

In this same direction, the forwarding processes of the data packets of interest, stages 1–3, that carry through the routers, as shown in Fig. 4. To locate these packets, the FIB uses a name prefix, and packets that cannot be located are stored in the PIT. So, temporary data is stored in the CS, which allows users to send a packet of interest, after verification in the CS. If the data finding in the (CS), they are sent to the user, otherwise, it is verified in the table (PIT).

At the same time, the content of the data that rests in the routers is not lost in the path of the network structure, so, Fig. 5, the data is returned through the corresponding router, steps 4–6, to the user.

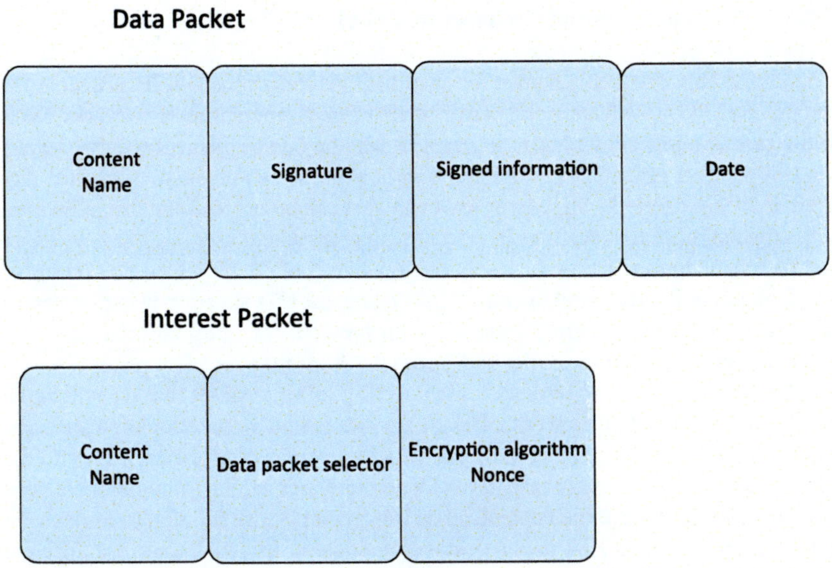

Fig.4 Elements of a data package and an interest package

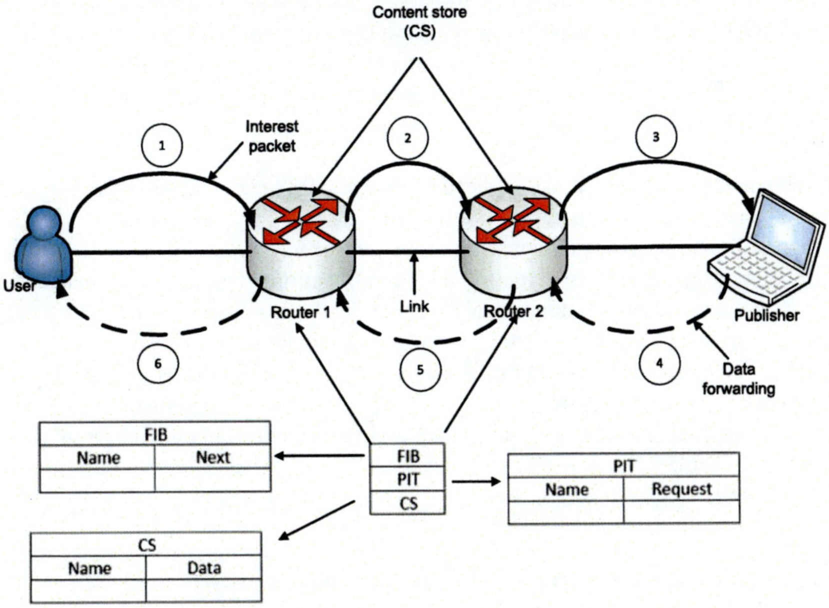

Fig. 5 Elements of a named data networking architecture

3.2.3 Expressive Internet Architecture (XIA)

Traditional network architectures, for example, networks based on IP protocols, do not promote the growth and good performance of new generations of networks. Such is the case of networks that want to adopt IP version 6 to implement security levels, since it presents efficiency and reliability problems, which give rise to the fact that there is no standardizing and normalized way that can be adopted and that allows the implementation of new features in current networks, where users will have to adopt the IP header format, using the same version 4 or 6 of IP. Presented this problem, the Expressive Internet Architecture (XIA) addresses this situation using addresses of both IP versions that are compatible with previous versions and that make the information that is sent in the network reach the respective recipient node.

Thus, the nodes use to retrieve content (CID), using a packet that we will call a packet in the destination network (NASTY), which is sent to the destination node and in which the host refers to a process that provides service, to forward it back to the origin node (sender). In this case, the XIA network architecture provides a pleasant and safe way to retrieve the content, since it is not a priority for XIA to retrieve the NASTY packet, since XIA uses a destination address, in which there are different routes or paths to travel and allows to recover the package in a satisfactory way. However, this type of network topology XIA collects information and thus obtains the identity of the service and the permission to bind it to the source of the service ID, on which the NASTY packet is routed, using the Service ID to reach more nodes in the network, making the sender use the same address to obtain the respective support for the service. Below are some of the essential components of the XIA network structure, as we can see in Fig. 6.

In the ecosystem of XIA networks, there are three essential stages. The first of these aspects is related to those applications that require a certain type of functionality to be able to process certain types of data packets since each application uses new types of function, which allows the introduction of new types and even before the structure of the network undergoes further modifications. The second aspect is that the XIA networks can introduce new functionality for applications. In the last stage, the encryption of the data contents between the various nodes that make up the network is considered, in which the security level must be the most reliable and optimal.

A comparison between the various network architectures in the context of ICNs is given in Table 1.

3.3 A Critical Perspective of Information-Centric Networks

In the ICN ecosystem, data content is an essential aspect that helps better network growth and performance. This situation allows for the distribution and storage of the data objects in the data contents of the routers. In this same address, the data

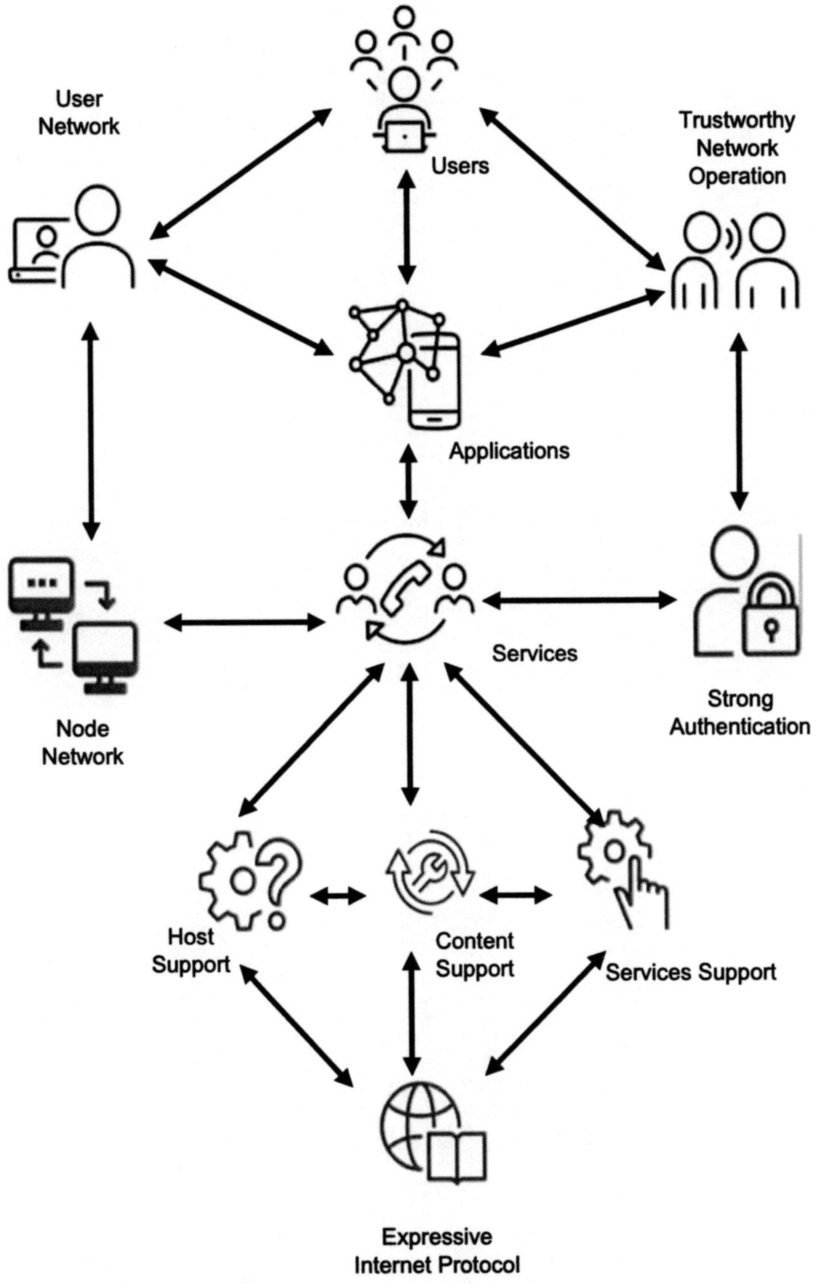

Fig. 6 Expressive internet architecture

Table 1 Comparison between of information-centric network topologies

Attributes of information-centric networks versus network topologies	Naming and address	Security	Network caching strategies	Transport of data object	Privacy between users	Application program interface (API)
MobilityFirst architecture	Use a global name resolution service (GNRS)	Use an encrypted public key	Allows it	Hop-by-hop transport between nodes	Used for data content and users	Used API to access services
Named data networking (NDN)	Unique data identifier (UID) for these packets data	The data contents are linked to the name	Network caching occurs automatically	In the transport layer of the network, the contents of the data to be transported in the network can be held	The signature is used for the data contents in the same container as the router contents	Forwarding information base (FIB)
Expressive internet architecture (XIA)	The contents are identified by means of a content retriever (CID)	It is implemented through the certification of certificates	Storage of data content in routers	Over hybrid network	It is implemented using the identity secret	An API for XIA

contents have their own identifier and attributable to the ICN. In turn, ICN outperforms TCP/IP network structures, in that they allow mobility (ubiquity) at the user level, optimization of data transfer, and the establishment of security measures that guarantee interaction between users from a private point of view.

In this same direction, some authors in take some parameters related to performance, transfer of data content, caching of data content in router containers, and bandwidth. Regarding the transfer of data content, these are more secure and reliable in the ICN, with respect to networks based on TCP/IP protocols. In the case of ICN, the contents of the data packets are stored permanently in the router container so that they can be accessed later by users, thus guaranteeing responses to user requests and network distribution of data contents. However, the user, through the transport protocol, the size for caching, the sending and receiving of data packets in the network and the level of popularity that these data packets can reach, allow the ICN to establish better user mobility, through its mobility, thus improving the performance of the network and optimizing the use of network resources, those nodes in the network that have the link down are omitted, as well as those redundant data contents that can create bottlenecks which can lead to network malfunctions.

Some authors analyze the problem of request forwarding and data congestion in the network in the context of information-centric networks. An optimization model is proposed, based on the NDN network structure, which helps to reduce network costs and maximize performance at the user level, which generates a subdivision at the request forwarding level and management of congestion using alternate routes in the network. Thus, the ICN manages network congestion based on the coordination, prior to its establishment, of congestion control, where there are mechanisms for forwarding data content and handling popular content and vice versa, allowing the performance of the network improve. The model based on an NDN network topology allows evaluating the behavior of the network in the face of a presumed demand for network data traffic, as well as different scenarios and storage of data contained in the router containers. In the ecosystem of the IoT, several aspects are considered for and against that the IoT network architectures may present and where users retrieve data contents from the network and not from the Internet node location. Therefore, the naming of data, regardless of its location, is essential for information-centric networks to help simplify the process of storing data content in router containers. On the Internet of Things, first is the service to the user and the sending of data, leaving in second place the interconnection between devices, so that the data contents are distributed by the information-centric network from the data packets that are generated in IoT devices, which provide data content and services, where labeling these content and service names is much more relevant than labeling the naming of the devices. The procedures used for caching can be more effective for IoT networks, since they have fewer resources available, such as power supplies and wireless bandwidth.

About bandwidth and optimization of battery power use in the context of IoT networks, this content circulating on the network can be stored through data packets in those routers closest to the network. The MobilityFirst network allows you to manage data content for ICNs when considering the containers of the routers (CR).

Simulation experiments show that caching may not be the most beneficial over edge caching. Therefore, if you have a basic caching system, and the routing of the data is done from the nearest node in the perimeter network and it is enabled, the performance of the network improves.

4 Results Analysis

4.1 Difficulties of Information-Centric Networks

In the architectures of the Internet of Things (IoT), to identify the various artifacts that are interconnected, labels are used to identify each of these devices by assigning a name. Another aspect to consider is the size of the files, since they are larger in information-centric networks, unlike IoT network structures, where the size of data files is smaller. However, this name assigned to the file must be unique to proceed with the identification of the data contents. So, it is not recommended to use a very long file name with reference to the actual data, since the latter tends to be smaller. In addition, the fact that routers can cache the contents of the data that will later be sent to various users is useful and necessary, since it avoids congestion and loss of data packets in the network, unlike IoT networks that require implementation plus details of network storage processes. On the other hand, the security implemented through security keys, digital signatures, etc., requires the separation of the user from the provider, which becomes quite a challenge, since, in the context of the ICN, this is a very important factor.

In turn, in the context of NDN networks, the security aspect is carried out through encryption processes that are applied to the data content and data packets that are sent through the network; therefore, the exchange of encrypted data using codes, for example, a public key, between the hosts that make up the network is not permitted. In any case, the issuer requests a public key using packets of interest, which are sent through data packets that carry a public key encapsulated within them. Similarly, users can use the exchange of packets of interest and data content to safeguard their data, using digital signatures, which allows the data content to be retrieved by name. Therefore, the link between the data content and the file name must be guaranteed to ensure secure and reliable delivery over the network.

In ICN networks, the storage capacity of the data content in the containers of the routers is limited, which affects the various services that are offered through the providers. When something like this happens, it is more relevant to have heterogeneous devices where the data flow is different, according to the user's needs and requirements. Another alternative solution is to be able to avoid the storage of data content in a certain time, which allows the router container to be released. Another aspect that must be considered is that caching is not cheap, so it requires extensive computational processing.

Table 2 Difficulties of information-centric networks

Parameters	Difficulties
Integration of ICNs in the IoT ecosystem	File sizes in IoT architectures are smaller than ICN network structures where data contents are large, so it is recommended to use small file names with reference to actual data
Security	It is important to ensure security at level of data content rather than considering the communication medium through which data packets travel
Storage mechanisms	The storage capacity at the level of user requests is reduced in the scope of the CR
PIT table	The PIT tables have a limitation that is the storage capacity of the data contents that come from the CR

However, in ICN architectures, there are tables called PIT that are responsible for storing various user requests sent to the server. Similarly, caching that occurs through router containers is limited, as the PIT tables can only store a limited number of requests because the router containers become saturated with requests from the router's different users. To provide a solution to this problem to use packages of interest that allow to attend multiple requests from users, but with a higher bandwidth consumption that allows sending the data packages to the users and that these packages of data remain in the PIT table until its execution time is exhausted.

Below in Table 2, some of the difficulties in the context of the ICN are presented, taking as reference some of its parameters.

5 Conclusion

Information-centric network architectures provide relevant aspects to be considered for future research, thus allowing networks with better performance and optimization. One of the virtues that ICNs have is that they can guarantee the various network services that providers offer to their users, and where the data contents are distributed in the network in a satisfactory manner and the recovery of data packets is reliable and optimal. In the context of information-centric networks, the network structures related to MobilityFirst, Named Data Networking (NDN), and Expressive Internet Architecture (XIA) are outlined. The use of information-centric networks in the context of the Internet of Things is also raised, and the advantages that these can represent. At the same time, some of the difficulties that arise in information-centric networks are addressed.

As future work, some of the previously mentioned network architectures can be taken and tested in a discrete event simulator such as OMNet++, NS-3, or others to observe the behavior of this type of network and thus obtain numerical values that allow validating this type of network grid structures.

References

1. Cisco VNI (2018) Cisco visual networking index: Forecast and trends, 2017–2022. Cisco, San Jose, CA, USA, white paper, 2018
2. Ahlgren B, Dannewitz C, Imbrenda C, Kutscher D, Ohlman B (2012) A survey of information-centric networking. IEEE Commun Mag 50(7):26–36
3. Xylomenos G, Ververidis CN, Siris VA, Fotiou N, Tsilopoulos C, Vasilakos X, Katsaros KV, Polyzos GC (2014) A survey of information-centric networking research. IEEE Commun Surveys Tuts 16(2):1024–1049
4. Bari MF, Chowdhury SR, Ahmed R, Boutaba R, Mathieu B (2012) A survey of naming and routing in information-centric networks. IEEE Commun Mag 50(12):44–53
5. Zhang M, Luo H, Zhang H (2015) A survey of caching mechanisms in information-centric networking. IEEE Commun Surveys Tuts 17(3):1473–1499
6. Din IU, Hassan S, Khan MK, Guizani M, Ghazali O, Habbal A (2018) Caching in information-centric networking: strategies, challenges, and future research directions. IEEE Commun Surveys Tuts 20(2):1443–1474
7. Fang C, Yao H, Wang Z, Wu W, Jin X, Yu FR (2018) A survey of mobile information-centric networking: research issues and challenges. IEEE Commun Surveys Tuts 20(3):2353–2371
8. AbdAllah EG, Hassanein HS, Zulkernine M (2015) A survey of security attacks in information-centric networking. IEEE Commun Surveys Tuts 17(3):1441–1454
9. Tourani R, Misra S, Mick T, Panwar G (2018) Security, privacy, and access control in information-centric networking: a survey. IEEE Commun Surveys Tuts 20(1):566–600
10. Wang L, Zhang Z, Dong M, Wang L, Cao Z, Yang Y (2018) Securing named data networking: attribute-based encryption and beyond. IEEE Commun Mag 56(11):76–81
11. Wu J, Dong M, Ota K, Li J, Guan Z. FCSS: fog computing-based content-aware filtering for security services in information centric social networks. IEEE Trans Emerg Topics Comput. https://doi.org/10.1109/TETC.2017.2747158
12. Li G, Wu J, Li J, Wang K, Ye T (2018) Service popularity-based smart resources partitioning for fog computing-enabled industrial Internet of things. IEEE Trans Ind Inform 14(10):4702–4711
13. Zhou Z, Guo Y, He Y, Zhao X, Bazzi WM (2019) Access control and resource allocation for M2M communications in industrial automation. IEEE Trans Ind Inform 15(5):3093–3103
14. Zhou Z, Feng J, Gu B, Ai B, Mumtaz S, Rodriguez J, Guizani M (2018) When mobile crowd sensing meets UAV: energy-efficient task assignment and route planning. IEEE Trans Commun 66(11):5526–5538
15. Zhou Z, Liao H, Gu B, Huq KMS, Mumtaz S, Rodriguez J (2018) Robust mobile crowd sensing: when deep learning meets edge computing. IEEE Netw 32(4):54–60
16. Jacobson V, Smetters DK, Thornton JD, Plass MF, Briggs NH, Braynard RL (2009) Networking named content. In: Proceedings of 5th international conference on emerging network experiments and technologies (ACM CoNEXT), Rome, Italy, Dec 2009, pp 1–12
17. Qi X, Wen Z, Tsuda T, Kameyama W, Shibata K, Katto J, Sato T (2016) Content oriented surveillance system based on information-centric network. In: Proceedings of IEEE Globecom workshops (GC Wkshps), Washington, DC, USA, Dec 2016, pp 1–6
18. Qi X, Wen Z, Yu K, Murata K, Shibata K, Sato T (2019) Content-oriented disaster network utilizing named node routing and field experiment evaluation. IEICE Trans Inf Syst E102.D(5):988–997

Automated Mishap Detection and Prevention System for Vehicles

G. Sudha, K. Balaji, K. M. Rakesh, R. S. Roopesh, S. Saranya, and Sankari Subbiah

Abstract The proposed accident prevention system decreases the likelihood of accidents on roads and also send its location and notify people who can take immediate action. The system combines detection of alcohol, driver's fatigue, fuel level, overheating of fuel tank and if the appropriate conditions are not satisfied, the vehicle stops. Additionally, the vehicle's speed is tracked and in case of over speeding, the vehicle stops automatically. In case of an accident, the Global Positioning System locates the geographic coordinates and using Global System for Mobile (GSM), it will send a message to the registered mobile number. A microprocessor and an ultrasonic sensor are deployed to determine the distance between two driving vehicles in the same lane and notify the driver coming behind the vehicle using RF technology.

Keywords Vibration sensor · Fatigue · Accident prevention · Danger zone · Geographic coordinates

---

G. Sudha (✉) · K. Balaji · K. M. Rakesh · R. S. Roopesh · S. Saranya
Department of ECE, Sri Sairam Engineering College, Chennai, India
e-mail: sudhaganesh74@gmail.com

K. Balaji
e-mail: balaji28krishnan@gmail.com

K. M. Rakesh
e-mail: kmrakesh23@gmail.com

R. S. Roopesh
e-mail: rsroopesh123@gmail.com

S. Saranya
e-mail: saranya.ece@sairam.edu.in

S. Subbiah
Department of Information Technology, Sri Sairam Engineering College, Chennai, India
e-mail: sankari2705@gmail.com

G. Rajakumar et al. (eds.), *Intelligent Communication Technologies and Virtual Mobile Networks*, Lecture Notes on Data Engineering and Communications Technologies 171,
https://doi.org/10.1007/978-981-99-1767-9_4

1 Introduction

A road accident is any injury caused by collisions that start, end, or involve a vehicle that is partially or entirely on a public road. By 2024, it is expected that road traffic accidents will rise to the third spot among the main causes of diseases that burden worldwide. Consequently, the road safety statistics for the year 2018 states 1.24 million people die in motor vehicle mishaps every year, an unacceptable number worldwide [1]. Young persons aged 15–29 dies most frequently from traffic related injuries. The majority of persons who suffer and die on roads around the world are older people, children, pedestrians, cyclists, and children, making up the most vulnerable road users.

Rapid urbanization, motorization, inadequate road engineering, low levels of knowledge, and a lack of injury prevention programmes have made the situation worse. The expensive, flashy, sporty, and massive muscle automobiles that once dominated the automotive industry have long since lost their lustre. Instead, the majority of the automobiles on the road today are distinguished by characteristics like safety systems and very well they can endure collisions without harming the driver and occupants. With some of these safety features now considered standard, automobile manufacturers can easily create exceptionally safe cars for anyone willing to pay top dollar. However, they are still very expensive. Safety has been and will remain a major priority for car manufacturers for as long as cars are driven on public roads. It has appeared to be both a gift and a blast with the advancement of technology, according to all accounts. Innovation has made daily living more enjoyable, but it has also come to represent a danger to human life.

The proposed work being presented is based on Internet of Things (IoT) to locate vehicles and guard against accidents by sounding an alarm when something goes wrong. For instance, if the driver is sluggish while driving and the car is about to be hit, the alarm will ring, alerting the driver to his situation. Global Positioning System (GPS) is used by this programme to determine the location of the car. Through this, it is also possible to use the vehicle's coordinates to calculate the distance it has travelled in 'X' seconds. The user must first register using the credentials provided upon registration in order to start sending location data to the server from his phone. By specifying the emergency contact numbers to which the alert message will be transmitted, the person's geographical coordinates will be sent to those registered numbers. The suggested method's primary goals are the detection and prevention of accidents.

2 Literature Survey

In [2], the authors claim that the primary reason for fatalities in car accidents is a delay in receiving medical attention. By promptly notifying the relevant authorities and emergency contacts, this can be avoided. The system is made up of an Arduino,

GPS, and Global System for Mobile (GSM). Unless a crash is observed, the gadget maintains the sensitive value for measuring the parameters throughout the accident and in case of a crash, GSM is activated to send a pre-stored Short Message Service (SMS) to the emergency contact of the victim.

System provided in [3] indicates that traffic accidents are one of the major causes of fatalities. Accident survival rates are significantly impacted by the time between the occurrence of an accident and the deployment of emergency services to the site. A vibration sensor detects a collision and GSM module sends a message to the emergency contact and the GPS module locates the geographic coordinates where the accident occurred. The Arduino, GPS, GSM, and vibration sensor are all part of this system.

From [4], inattention, sleepiness, and intoxicated driving are the main factors contributing to traffic accidents. This article suggests developing a method to prevent these situations. The proposed solution here uses a night vision camera to avoid and manage accidents. When the automobile starts, this system watches the driver's face, which mostly aids in continual observation. It has two uses: the first is to read blinking and the second is to detect eye blinking. Additionally coupled with a controlling system are automatic braking and driving systems. The suggested method warns the driver based on his condition and checks to see if he is sleepy.

According to the assertion made in [5], speed is one of the most important and essential risk variables when driving. It increases the likelihood of crashes as well as their severity. Through a microcontroller chip, the GPS monitors the vehicle's speed, and if an accident happens, the gadget will then transmit the GPS-assisted accident position through the GSM network.

Vibration sensors, a GPS module, and a GSM module are the three primary parts of this system in [6]. When a car crashes, the sensor detects the impact, and Arduino compares it to the threshold value specified in the software. If the reading is higher than the threshold value, Arduino will use GPS to determine the current location and GSM alerts the appropriate authorities. Accident detection involves the use of a vibration sensor.

In order to reduce the likelihood of an accident, authors in [7] have developed a system that would not only notify users of an accident and also assess whether the driver is feeling tired and whether he is fit to drive. The suggested device makes use of a smartphone with a front camera to monitor the driver's posture.

A system suggested in [8] uses a smartphone to determine whether a mishap has occurred. The suggested solution makes advantage of the phone's GPS receiver to identify an abrupt, quick change in deceleration that occurs during an accident. Additionally, a smartphone's accelerometer and pressure sensor are used to read the angle of tilt and change in pressure, respectively. The Android app would notify the emergency contact after identifying these three circumstances. The gadget has a switch that, if touched, will stop notifying the emergency contact, preventing any erroneous alarms. This could be helpful if the sensors give up incorrect readings or if the collision is small and not fatal.

In [9], a technology is devised to avoid accidents by identifying the driver's condition, such as if the motorist is feeling tired or is not fit to drive. The position of the eyes is determined using infrared (IR) reflective obstacle sensors. When the eyes are opened, the sensor sends to the controller a low signal, and when eyes are closed for 4 s or more, the sensor sends the controller a high signal. The controller receives the high or low signal. The sensor sends input to the Peripheral Interface Controller (PIC), which processes it to turn on or off the buzzer and display "Driver Slept" on the Liquid Control Display (LCD) if the buzzer is ON. As a result, the driver is informed, thereby preventing a serious accident. This technology offers a very effective means of reducing such incidents because drowsiness is a significant contributor to accidents.

A GPS/GSM-based approach was proposed in [10, 11], that is used for vehicle tracking along with a display for the indications. A system for drowsiness detection was proposed in [12], with an alert technology to the driver and a similar system was proposed in [13] for night driving.

A system on chip (SoC) was employed in a model in [14] to precisely monitor the driver's state while driving, and upon detecting incapacity to drive, the system switched to auto drive mode and safely parked in the left lane.

For quick rescue operations, real-time accident detection and location tracking are essential. As a result, in [15], a low-cost automatic accident prevention, post accidental rescue, and a black-box system have been constructed using Arduino UNO and nano. With the aid of this system, drivers will be able to prevent collisions. In the worst-case situation, if the collision is unavoidable, it will benefit the rescue crew by giving them a precise position of the accident site.

An IoT-based system was presented in [16], to enable the ceaseless monitoring of a running vehicle's parameters, such as vehicle speed, number of hard braking, and rolling, which can be a measure of the driving quality. In order to immediately monitor the driver's performance from a centralized server, the parameters are sent to a server at every moment. This increases the driver's responsibility to refrain from hazardous driving.

3 Existing System

There are existing methods for preventing both drunk driving and accident detection separately. A GPS and GSM focused accident detection system continuously records latitude, longitude, and the speed of moving objects in relation to time. A signal would be sent as an SMS using Arduino and a GSM module with the aid of the accelerometer when the vehicle's velocity crosses the defined value for the specific geographic locations determined from a real-time Google Map. In order to warn the drivers of the vehicle of the hazard, an LCD display and a micro-SD card module are also used. If the vehicle has been in an accident, an alarm will be sent to the salvage crew if they don't hear back from them in the allotted period.

When the MQ3 sensor detects alcohol, the driver is disallowed from operating the car. An eye blink sensor has already been used by driver drowsiness systems. The driver is required to wear the eye blink sensor frame while driving, and the blink must last for a few seconds in order to identify tiredness. Wheel speed decreases whenever the steering movement varies randomly. It is possible to adjust the vibration sensor's threshold and behave accordingly. As a result, if the driver nods asleep, the LCD shows the warning messages and the vibrator linked to the eye blink sensor's frame vibrates. Depending on the situation, the wheel is either slowed down or stopped. In the proposed method, both the approaches are combined along with the detection of vehicles on same lane and whether they are about to collide by measuring the distance between the vehicles.

4 Proposed Architecture

4.1 Architectural Block Diagram of the Proposed Design

The proposed block diagram is shown in Fig. 1 which comprises

- Arduino UNO
- Power supply
- GPS module
- GSM module
- Control switch (DC motor)
- Buzzer
- LED
- LCD display
- Accelerometer
- Vibration sensor
- Temperature sensor
- MQ3 sensor
- Camera
- IR sensor.

A microcontroller is a reasonably priced microcomputer that is designed specifically to carry out the functions of embedded systems. The Central Processing Unit (CPU), memory, serial ports, peripherals, and other components make up a generic microcontroller.

GPS is an in-space satellite navigation system that delivers position and time information on the Earth. The most common uses of GPS receivers are for time dissemination, positioning, and other types of research. Satellites that orbit the Earth make up the GPS. With an orbital period, equal to that of the Earth's rotation, these satellites are geosynchronous. The GPS calculates the distance between a GPS satellite and a GPS receiver by measuring the time it takes for a radio signal (the GPS signal) to travel from a GPS satellite to a GPS receiver. As a result, they remain in exactly the same location with respect to the Earth below. The receiver and the

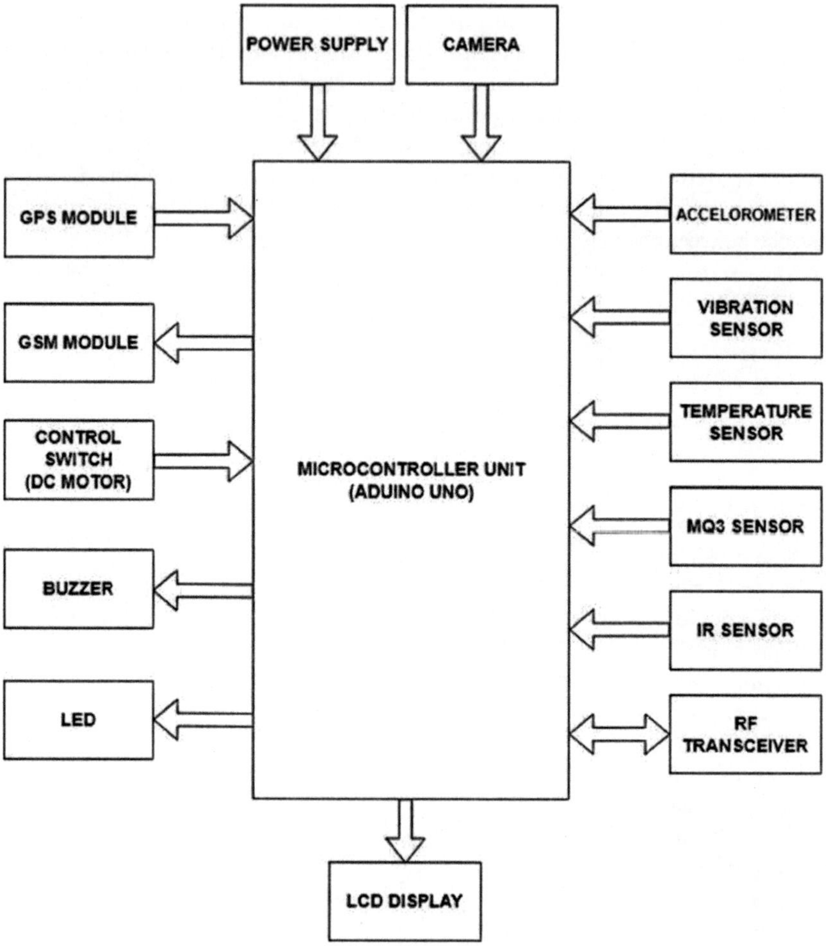

Fig. 1 Proposed block diagram

satellites utilize extremely precise clocks that are synchronized so that they produce the same code at the exact same time in order to collect reliable information.

A GSM modem of modem that works with a SIM card and requires a mobile operator subscription in order to function, much like a mobile phone. Since the sender is responsible for the cost of message transmission, GSM modems are an affordable option for receiving SMS texts. A GSM modem must support an enhanced ATtention (AT) command set for transmitting and receiving SMS messages, in order to carry out these functions.

In order to pinpoint an object's location in space and track its movement, an accelerometer monitors the acceleration forces that are acting on the object. The rate at which an object's velocity changes, or acceleration, is a vector quantity. Static forces and dynamic forces are the two different categories of acceleration forces. Constantly acting forces on an object are called static forces. Dynamic forces are forces that are moving towards the object at varying rates. This is the reason why accelerometers are employed in automotive crash safety systems. The quantity and frequency of vibration in a system, machine, or piece of equipment are measured by a vibration sensor. These metrics can be used to identify imbalances or other problems with the asset and detect upcoming failures. A vibration sensor can either be wirelessly monitored or directly connected to an asset. Depending on the type of sensor used, it will use different methods to detect vibrations coming from the asset after it is in place. A temperature sensor is a device used to check whether the vehicle's engine get heated or not.

The Alcohol Gas Sensor MQ3 is a low-cost semiconductor sensor that can detect alcohol gases at concentrations between 0.05 and 10 mg/L. The sensitive element in this sensor is SnO_2, a material whose conductivity is lower in pure air. Alcohol gas conductivity increases along with its concentration. It can tolerate smoke, mist, and gasoline well, but is highly sensitive to alcohol. Analogue and digital outputs are provided by this module. The MQ3 alcohol sensor module is simply programmable and may be connected to microcontrollers, Arduino boards, Raspberry Pi, and other devices. This alcohol sensor can recognize the presence of alcohol in your breath, much like a typical breathalyser. An analogue signal is produced by the sensor based on alcohol content.

An electrical device that monitors and detects infrared radiation in its environment is called an IR sensor. A Light Emitting Diode (LED) and a receiver are the two components of an active IR sensor. The receiver detects the infrared light from the LED that reflects off an object as it gets close to the sensor. Active IR sensors serve as proximity sensors, and obstacle detection systems frequently employ them (such as in robots). Passive Infrared (PIR) sensors do not produce infrared radiation themselves; they only detect it when it is present. The most typical application for PIR sensors is motion-based detection, such as in home security systems.

This method suggests a relatively effective method of control. Accident detection is the following action, though, if a collision does take place. A GSM-based accident detection and notification system will follow the collision with the help of deployed impact sensors, process the data through a microcontroller unit, and then notify the victim's relatives. The microcontroller then sends a signal to the Radio Frequency (RF) transmitter whenever an accident is discovered. The RF transmitter then transmits an accident alert message to an RF receiver in another vehicle at a distance of 300 m and a wavelength of 0.69 m, and the RF receiver screen then indicates that an accident has been discovered in the specific location. A RF is any electromagnetic wave frequency that falls within the range of 3 kHz–300 GHz, including those utilized for radar signals and communications. A RF transmitter module makes up the RF transmitting circuit. A 12-V/1 A battery powers the voltage regulator circuit, giving the motor an unregulated 12-V supply, while the Arduino and the receiver module each receive a 5-V regulated supply. The user-set data is encoded and delivered to the receiver module when the RF transmitter is turned on. In order to compare the embedded data, the receiver module decodes the data and provides it to the Arduino.

The Arduino uses RF wireless technology to transmit the alert message to the following car if the separation between the vehicles is less than the limit zone. The GPS uses data from both base stations on Earth and satellites in orbit to establish its precise location. The MQ3 sensor is used to detect alcohol. Seat belt detection uses an infrared sensor. By employing the camera and face recognition, drowsiness in the driver can be identified.

Employing a temperature sensor to identify engine system failures. Liquid-level sensors calibrate the accurate amount of fuel in the vehicle's tank and warn the system if the tank is empty while also showing the exact amount of fuel that is available. A microcontroller and an ultrasonic sensor can be used to construct a collision avoidance system by measuring the distance between two moving vehicles in the same lane and heading in the same direction, and notifying the driver whenever they are approaching a dangerous distance.

Ultrasonic sensors use their sensor head to transmit ultrasonic waves while simultaneously receiving the reflected waves from an object. It determines the object's position by measuring how long it takes for the sonic wave to travel from transmission to reception. The latitude, longitude, and speed of moving vehicles with regard to time are all continuously measured by the vehicle speed limit detection system. The speedometer enable us to determine the vehicle's speed and when it exceeds the predetermined value for the specific geographic locations is determined from a real-time Google Map. The danger was warned to the people within the automobile via an LCD display. The haversine of the central angle between any two points on a sphere is defined as follows:

$$\mathrm{hvs}\left(\frac{d}{r}\right) = \mathrm{hvs}(\varphi_2 - \varphi_1) + \cos(\varphi_2) * \cos(\varphi_1) * \mathrm{hvs}(\lambda_2 - \lambda_1) \tag{1}$$

The haversine function's representation, 'hvs', is given as

$$\mathrm{hvs}(\theta) = \sin^2\left(\frac{\theta}{2}\right) = \frac{1 - \cos(\theta)}{2} \tag{2}$$

- d is the distance between the two points
- r is the sphere's radius
- φ_1, φ_2 are latitudes of point 1 and point 2
- λ_1, λ_2 are the longitudes of point 1 and point 2
- d/r is the central angle, assuming angles are measured in radians

Solving for 'd' by applying the inverse haversine,

$$D = \mathrm{hvs}^{-1}(h) = 2r \arcsin\left(\sqrt{h}\right) \tag{3}$$

where $h = \mathrm{hvs}(d/r)$, should not exceed 1 and

$$D = 2r \arcsin\left(\sqrt{\mathrm{hvs}(\varphi_2 - \varphi_1) + \cos(\varphi_2) * \cos(\varphi_1) * \mathrm{hvs}(\lambda_2 - \lambda_1)}\right) \quad (4)$$

$$D = 2r \arcsin\left(\sqrt{\frac{\sin^2(\varphi_2 - \varphi_1)}{2} + \cos(\varphi_2) * \cos(\varphi_1) * \frac{\sin^2(\lambda_2 - \lambda_1)}{2}}\right) \quad (5)$$

Once the distance has been determined, calculating speed only requires dividing the distance by the time. Speed is determined for the proposed system every second.

4.2 Process Flow of the Proposed Architecture

The process flow used is depicted in Fig. 2. An alcohol sensor is activated and searches for the drunken state of the driver, when the key is inserted into the car. If he is drunk, the car's engine won't start. If not, the engine will start. Seat belt detection is performed using IR sensor. If the seat belt is not detected, the car's engine won't start. Face recognition algorithms are used to identify driver drowsiness while driving. The device will issue an alarm message if the driver has closed his eyes for more than one minute owing to fatigue, otherwise the vehicle continues its run.

A temperature sensor is be used to identify any engine failure that may have taken place. The mechanism will warn the driver if the fuel level in the tank drops below the required level. Vibrational sensors can identify accidents using vibrations. A vibrational sensor recognizes the vibration in the case of a mishap.

If the value is higher than the threshold limit, the system concludes that an accident has happened. If the driver presses the panic button within the authorized time, the accident alert message won't be sent to the registered mobile number. If not, the system will communicate the accident alarm message and the position coordinates to the registered mobile phone using a GSM and GPS module.

Liquid-level sensor is used to detect the amount of fuel available in the tank. If it is less, then the system will issue the warning to the driver, otherwise the vehicle will continue its run. The accident detection message will be delivered to the registered mobile number along with the latitudinal and longitudinal coordinates with the help of GPS and GSM. Using the RF Transceiver and Zigbee module, this accident detection message will also be broadcast to the neighbouring vehicles.

Accident prevention mechanism is implemented in this system by using the collision avoidance mechanism. With the help of this mechanism the accidents occurring due to collision between the vehicles will be avoided. In this mechanism, the RF transceiver is mounted on all the vehicles. When the following vehicle reaches the danger range, the first vehicle acts as a transmitter and the following vehicle acts as a receiver. Thus, the first vehicle will send the collision avoidance message to the following vehicle through an RF transmitter. The following car receives the collision avoidance message through RF receiver and intimates the driver to slow down the vehicle.

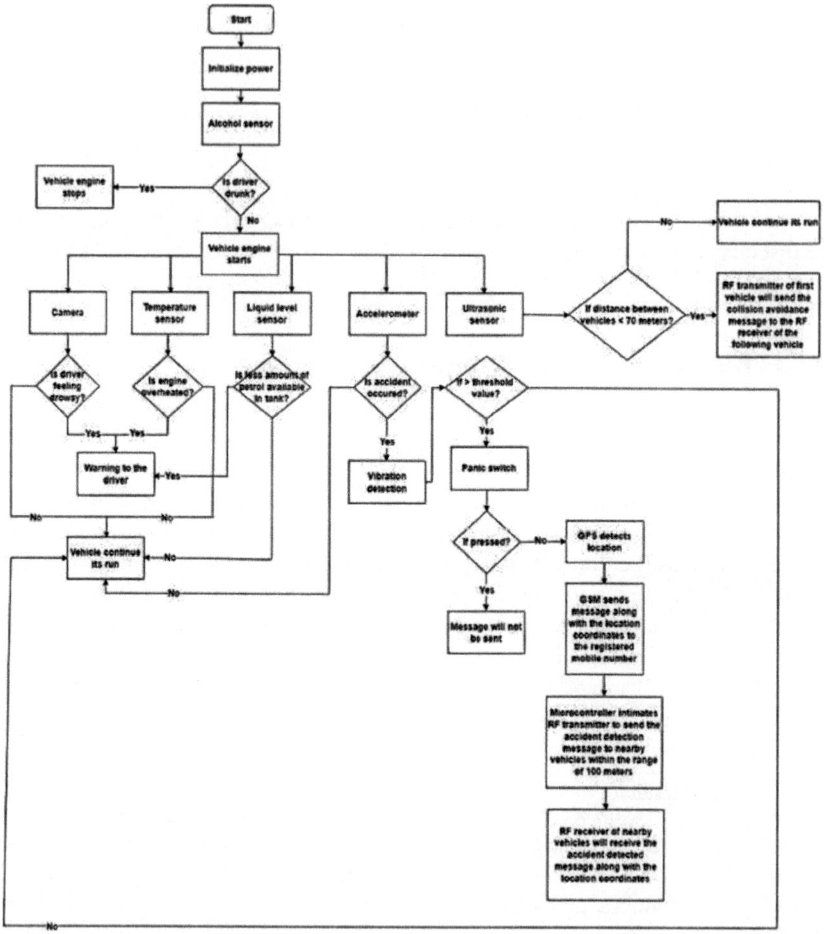

Fig. 2 Process flow diagram

5 Results and Discussion

It is our responsibility to control it since as the number of vehicles rises, so is the number of traffic accidents. Most often, drunken drivers, sleepy drivers, and overheating engines that catch fire seem to be the reason for accidents. The proposed system's implementation will contribute to a reduction in accidents caused by the key reason. The technology is automatic, inexpensive, and energy efficient, making it simple to install in cars. Unfortunately, if an accident does occur, the system will detect it and use GPS to identify its exact location so that the GSM module may alert the emergency unit. By alerting rescuers in time, this contributes to the saving of numerous lives.

Overall, this system targets normal people, is relatively economical, and is simple to install in all kinds of automobiles. These systems aid in obtaining medical assistance as soon as an injury is discovered, but the majority of them waste a great deal of time and resources when the damage is minor and no medical assistance is necessary. The danger of an accident will be lower if the system alerts the driver to impending inclement weather and hazardous road conditions. The simulated results are shown in Figs. 3, 4 and 5.

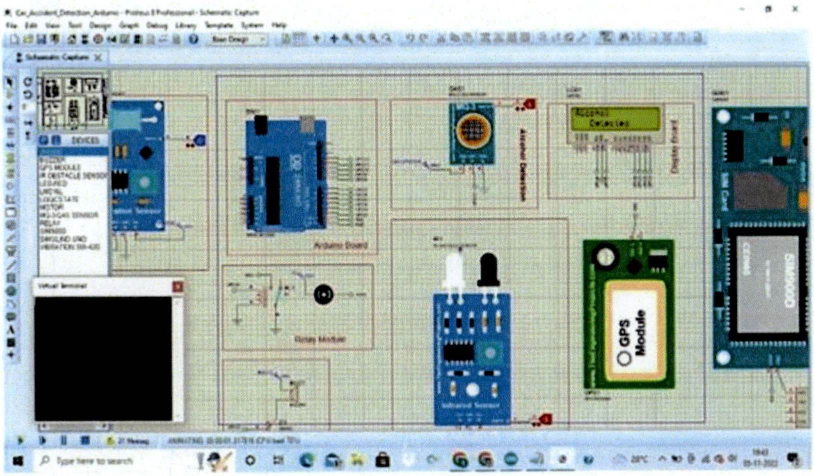

Fig. 3 Simulated result for alcohol detection

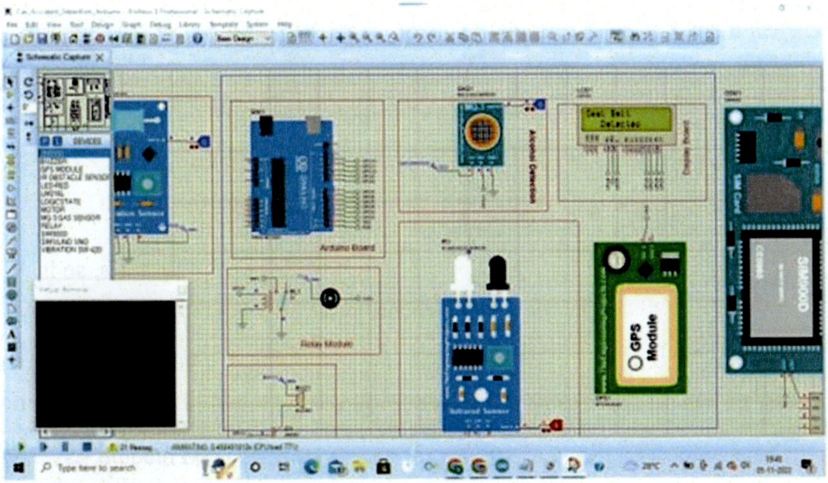

Fig. 4 Simulated result for seat belt detection

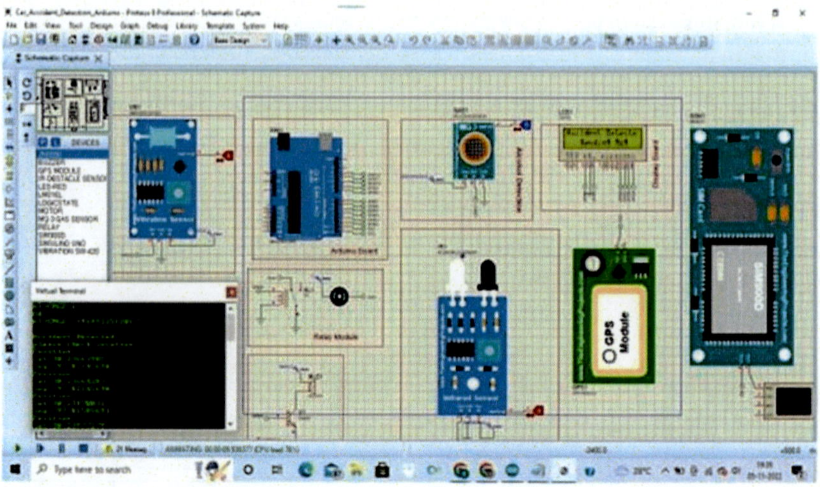

Fig. 5 Simulated result for sending messages

This is the working prototype of the proposed model. In this model, an Arduino is used as a microcontroller. Alcohol sensor will be activated first when the vehicle engine starts. When the alcohol is detected, the input state of the alcohol sensor becomes high which in turn stops the vehicle engine and prevents drunken driving. If the alcohol is not detected, the seat belt detection system will be activated. The seat belt detection system using infrared sensor is activated, only if alcohol is not detected. When the seat belt is inserted in its buckle, then the input state of the IR sensor becomes high. This implies that the driver is wearing a seat belt.

Figure 6 shows the simulated results of how the alert messages are transmitted between the vehicles. The accident may be located with the aid of the vibrational sensor by the vibrations. When an accident occurs, the GPS will find the location's geographic coordinates and, using the GSM, it will send an SMS to the registered mobile number.

6 Conclusion and Future Scope

This system is utilized to successfully develop a vehicle accident detection and tracking system using GSM and GPS. With their benefits, such as preventing accidents on curvy, mountainous, and hill roads, fully automated, cheap cost, and simple integration into the current roads, the various technologies used in this study are integrated. The GPS-derived accident location coordinates are transmitted through the GSM network to a user specified mobile number. Although the cost of vehicles will go up with the system's installation, it is still preferable to have some level of safety

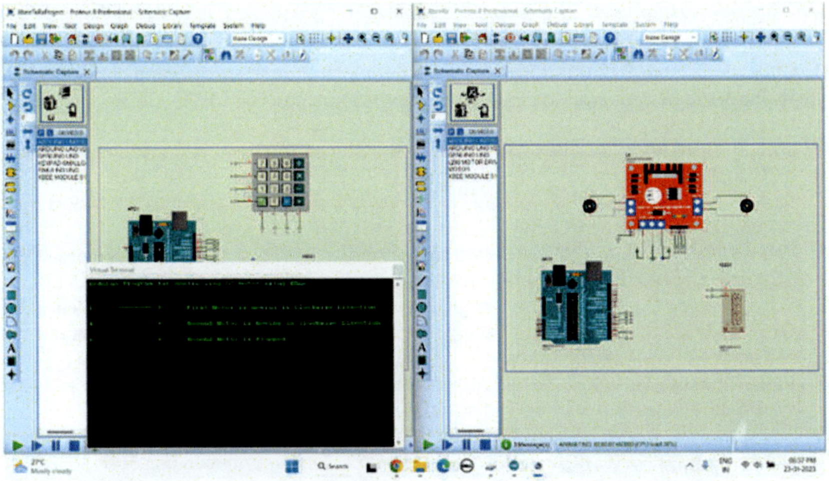

Fig. 6 Simulated result for transmission of alert messages between the vehicles

than none at all. It has been demonstrated that the suggested strategy is quite advantageous to the automotive sector. This proposed model can be modified to be used for traffic estimation and accident surveys throughout the nation. The primary goals of the proposed system are to prevent accidents and provide warnings when they do occur. The development of an intelligent vehicle that will drive at the safest speed in high-risk areas and notify the family and emergency services of any accidents is addressed in an efficient way. Thus, it can be inferred from the aforementioned study that using automatic vehicle speed control and accident alarm systems will significantly reduce accidental collisions. As a future work, analysis is being done to protect the sensors from being damaged in critical places and to decrease the size of the unit so that it occupies a small place and is easily kept in narrow roads.

References

1. https://www.who.int/publications/i/item/9789241565684
2. Rishi R, Yede S, Kunal K, Bansode NV (2020) Automatic messaging system for vehicle tracking and accident detection. In: Proceedings of the international conference on electronics and sustainable communication systems, ICESC
3. Aarya DS, Athulya CK, Anas P, Kuriakose B, Joy JS, Thomas L (2018) Accident alert and tracking using Arduino. Int J Adv Res Electr Electron Instrument Eng 7(4)
4. Kailasam S, Karthiga M, Kartheeban K, Priyadarshani RM, Anithadevi K (2019) Accident alert system using face recognition. In: IEEE international conference on intelligent techniques in control, optimization and signal processing (INCOS)
5. Syedul Amin M, Jalil J, Reaz MBI (2012) Accident detection and reporting system using GPS, GPRS and GSM technology. In: International conference on informatics, electronics and vision (ICIEV), Dhaka, Bangladesh

6. Kalyani T, Monika S, Naresh B, Vucha M (2019) Accident detection and alert system. Int J Innov Technol Exploring Eng (IJITEE)
7. Guksa B, Erkmen B (2017) Smartphone application for drowsiness detection during driving. In: International conference on frontiers of sensors technologies (ICFST), China
8. Faiz AB, Imteaj A, Chowdhury M (2015) Smart vehicle accident detection and alarming system using a smartphone. In: International conference on computer and information engineering (ICCIE), Bangladesh
9. Sankpal PM, More PP (2017) Accident avoidance system using IR transmitter. Int J Res Appl Sci Eng Technol (IJRASET) 5(IV)
10. Wukkadada B, Fernandes A. Vehicle tracking system using GPS and GSM technologies. IOSR J Comput Eng (IOSR-JCE):05–08
11. Lee S, Tewolde G, Kwon J (2014) Design and implementation of vehicle tracking system using GPS/GSM/GPRS technology and smartphone application. In: IEEE World Forum on Internet of Things (WF-IoT), South Korea
12. Ruthu R, Irshad Hussain B (2022) Arduino based automated drowsiness detection and alert system. Int J Innov Res Sci Eng Technol (IJIRSET) 11(6)
13. Aishwarya SR, Rai A, Charitha PMA, Savitha SC (2015) An IoT based accident prevention and tracking system for night drivers. Int J Innov Res Comput Commun Eng 3(4)
14. Hyder G, Chowdhry BS, Memon K, Ahmed A (2020) The smart automobile (SAM): an application based on drowsiness detection, alcohol detection, vital sign monitoring and lane based auto drive to avoid accidents. In: 2020 global conference on wireless and optical technologies (GCWOT)
15. Khan MHU, Howlader MM (2019) Design of an intelligent autonomous accident prevention, detection and vehicle monitoring system. In: 2019 IEEE international conference on robotics, automation, artificial-intelligence and internet-of-things (RAAICON), Bangladesh
16. Kader MA, Eftekhar Alam M, Momtaj S, Necha S, Alam MS, Kadar Muhammad Masum A (2019) IoT based vehicle monitoring with accident detection and rescue system. In: 2019 22nd international conference on computer and information technology (ICCIT), Dhaka, Bangladesh

Autonomous Drone Using Time-of-Flight Sensor for Collision Avoidance

G. Naveenkumar, M. V. Suriyaprakash, and T. P. Prem Anand

Abstract The most complicated problems in the field of self-driving aircraft are the design of a strong real-time obstacle discovery and evasion framework. The issue is becoming a little complex due to the size and shape of the vehicle, and it becomes a little complex to accomplish the mission. As a result, we will be using a Flight Time Sensor (TOF sensor—time of flight) which is a lightweight sensor, and it is known as a MicroLiDAR sensor. The man's behavior of detecting the collision status of the obstacles to the approach using the time-of-flight sensor is proposed here. The control board has related to a time-of-flight sensor and Artificial Intelligence (AI) in order to recognize the obstruction and prevent a collision. During the motion of the aerial vehicle (UAV), the detection set of rules estimates modifications inside the length of the upcoming obstruction zone. The strategy primarily recognizes the characteristic focuses of the deterrents and after that the impediments which are likely to approach the UAV. Another, by comparing the obstacle surface ratio, the UAV's position helps determine whether an obstacle can cause a collision. The algorithm was tested by doing actual flights, and the results show that it is accurate. With a focus on unmanned aerial vehicles, the study offers a thorough investigation of collision evasion methods for unmanned vehicles (UAVs); it could be a careful examination of a few collision avoidance strategies that are categorically characterized, besides a comparison of the ways taken into thought in connection to different circumstances and specialized contemplations. Also covered are how various sensor types are used in the context of UAVs to avoid collisions. It is highly accurate and collision avoidance ultimately happens.

G. Naveenkumar (✉) · M. V. Suriyaprakash · T. P. Prem Anand
Rajalakshmi Engineering College, Chennai, India
e-mail: naveenkumar.g.2019.aero@rajalakshmi.edu.in

M. V. Suriyaprakash
e-mail: suriyaprakash.mv.2019.aero@rajalakshmi.edu.in

T. P. Prem Anand
e-mail: premanand.tp@rajalakshmi.edu.in

G. Rajakumar et al. (eds.), *Intelligent Communication Technologies and Virtual Mobile Networks*, Lecture Notes on Data Engineering and Communications Technologies 171,
https://doi.org/10.1007/978-981-99-1767-9_5

Keywords Unmanned aerial vehicle · Self-driving · Obstruction detection · Artificial intelligence · Avoidance system · Time-of-flight sensor · Surface ratio

1 Introduction

The drone is defined as any aerial craft that uses software to fly autonomously or that can be controlled remotely by a pilot is referred to as a drone. In addition to having propellers to stabilize flying patterns, many drones incorporate cameras for obtaining visual data.

Drone technology has been incorporated into industries like transportation, agriculture, search and rescue, and cinematography.

1.1 Types of Drones

(1) Single-Rotor Helicopter Drones

Single-rotor helicopters can be propelled by gas or electricity and resemble miniature helicopters identically. Its steadiness and ability to fly farther are aided by its single-blade and gas-powered propulsion. The LiDAR systems, which can be used to survey land, study storms, and map degradation brought on by global warming, are typically transported by these UAVs along with heavier items.

(2) Fixed-Wing Drones

When the wings, rather than the rotors, provide lift, fixed-wing drones, which resemble regular airplanes, are particularly effective. To hover for more than 16 h, these drones frequently run on gasoline rather than electricity. These drones need runways to take off and land because of their size, which is normally much larger, and their design. Fixed-wing UAVs are used by the military to carry out strikes, scientists to transport massive amounts of equipment, and even by non-profit organizations to deliver food and other supplies to inaccessible locations.

(3) Fixed-Wing Hybrid VTOL Drones

Fixed-wing hybrid VTOL drones incorporate rotors that are affixed to the wings, combining the advantages of rotor-based and fixed-wing drones.

This technology combines a fixed-wing design's endurance with a rotor-focused strategy to provide consumers with the best of both worlds.

(4) Multi-Rotor Drones

Among the tiniest and lightest drones available are frequently multi-rotor models. They are not the best flying vehicles for amateurs and aerial photographers because of their restricted range, speed, and height.

These drones can typically fly for 20–30 min with a small payload, such as a camera.

Utilizing software-controlled flight plans in its implanted frameworks, a ramble could be a flying robot which will be remotely commanded or fly independently while utilizing onboard sensors and a worldwide situating framework (GPS). The ramble could be a sort of unmanned ethereal vehicle (UAV), which is an airplane without a human pilot, group, or travelers.

The unmanned ethereal vehicles (UAVs) are utilized more as often as possible in farther detecting applications such accuracy farming, biosecurity, catastrophe checking, and reconnaissance as a result of their upgraded independence.

An onboard UAV cognitive capability for comprehending and responding in situations with incorrect or deficient perceptions, for things of intrigued interior complex scenes.

Are, all things considered, limited, and have not however been completely inspected. The suggested system architecture has a modular structure and devotes most of its processing resources to achieving high levels of realism in virtual worlds. The landing will be more exact and the destination will be reached with the aid of the time-of-flight sensor.

An advanced driver assistance system called a collision avoidance system (CAS), regularly alluded to as a pre-crash framework, forward collision caution framework, or collision relief framework, is aiming to anticipate or reduce the seriousness of a collision. In its least difficult frame.

A forward collision caution framework keeps track of a car's speed, the speed of the car before it, and the space between the two so that it can alarm the driver on the off chance that the cars are getting to be as well near and conceivably anticipate a mishap.

To identify a looming mischance, an assortment of innovations and sensors are used, such as radar (all-weather), sometimes laser (LiDAR), and cameras (utilizing picture acknowledgment). Through a position database, GPS sensors can distinguish settled dangers.

1. The drone can approach the destination using a different path if the impediment is stationary since it can see a different way around it.
2. The drone will report back to the Command Center for manual guiding if the obstacle's position is dynamically changing and the route around it is very difficult to calculate.

Advanced drones, including several DJI drones, have an obstacle avoidance safety feature. Drones with this skill may scan their surroundings and immediately identify any obstacles in their path. When your drone aircraft becomes aware of the obstruction, it quickly takes action to avoid. Overall working process of obstacle detection is shown in Fig. 1.

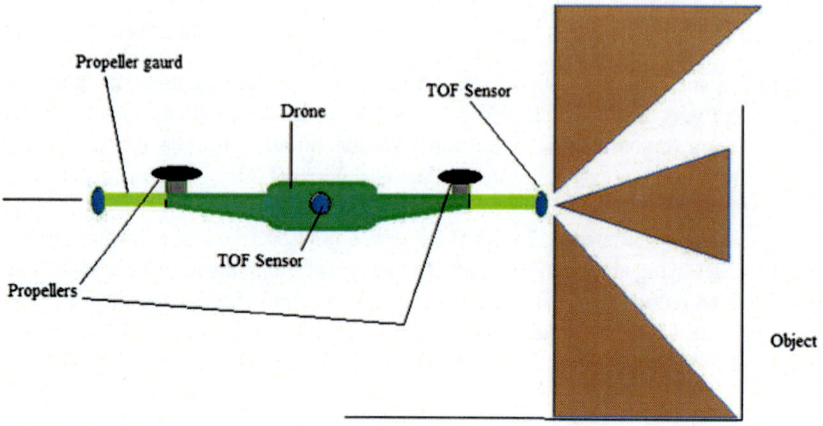

Fig. 1 Overall working process of obstacle detection

1.2 Estimation of Obstacle Position

TOF employs travel time in two different approaches to calculate distance and depth, including

(1) Using Timed Pulses

For event, it works by to start with edifying the target with laser light and after that measuring the reflected light with a scanner, where the object's partitioned is initiated utilizing the speed of light to completely calculate the evacuate voyage. View of TOF sensor working is shown in Fig. 2.

To calculate the distance of the object: (speed of light × time of flight)/2.

(2) TOF Time Pulses

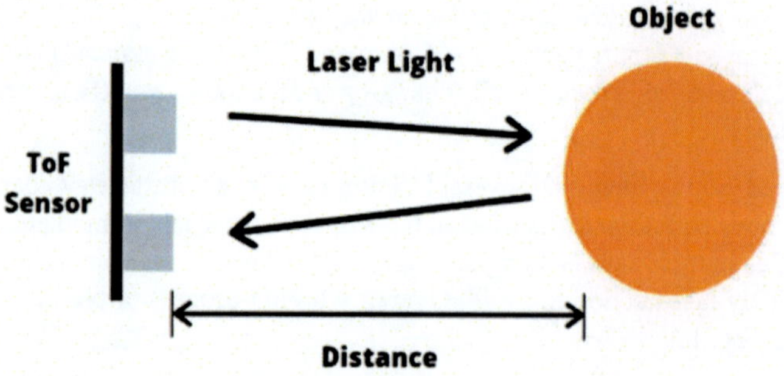

Fig. 2 View of TOF sensor working. *Source* SEEED Studio(1)

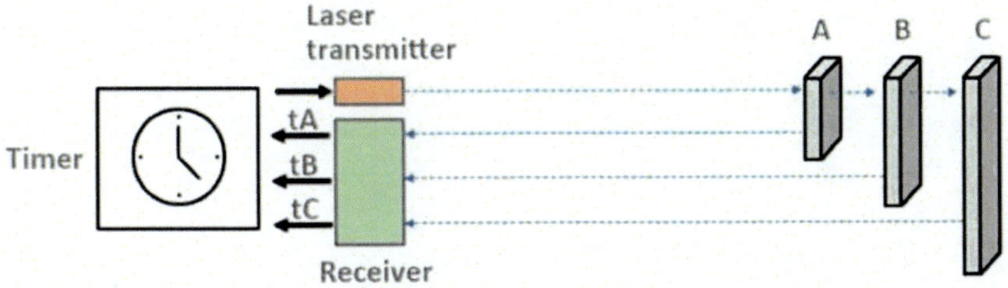

Fig. 3 Measurements of distance and depth of the obstacle. *Source* SEEED Studio(2)

When the receiver detects the return light, a timer will start as the light is departing and return the time when it has finished. The distance may be easily calculated using the method above because light travels at a constant speed, and the "time of flight" of the light is obtained by dividing the two times. This allows for the determination of every point on the surface of the item. Measurements of distance and depth of the obstacle are shown in Fig. 3.

2 Methodology

A decision tree is used to describe the implementation scheme. When the UAV reaches a waypoint on the trajectory, the same process is used. The drone first collects samples from the space to check for obstacles; if any are present, it notes its position for a set amount of time; otherwise, the planned course is followed. If a new obstacle is discovered, a polynomial regression is performed to fit the sensor readings to predict its future position. In addition, a security distance is established based on the size and speed of the object.

The optimization algorithm is given the location of this new obstacle, which causes it to recalculate the new trajectory profile considering its location and utilizing the scheduled trajectory as a starting point for the iterative process. The operation is finished when the UAV reaches the last waypoint on the trajectory. Flowchart for obstacle avoidance process is shown in Fig. 4.

Flight controller, BLDC motor, electronic speed controller, frame, propeller, battery, sensor, transmitter, and receiver are the hardware requirements.

This project utilizes the following methodology:

Simultaneous localization and mapping (SLAM),
Relative/differential GPS measurements,
Path changing algorithm,
Collision avoidance (time-of-flight sensor—TOF).

The following capabilities have been included in the UAV:

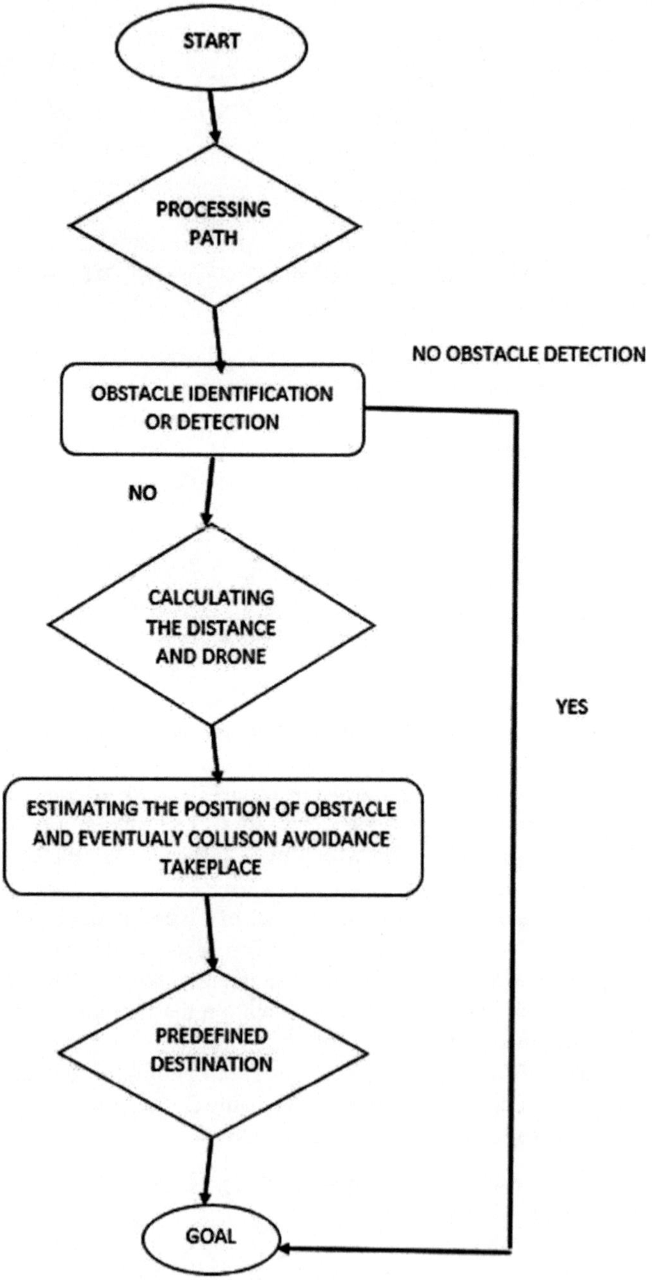

Fig. 4 Flowchart for obstacle avoidance process

(1) Real-time access to TOF sensor data that provides knowledge about the UAV's immediate environment.
(2) Calculating the location of an obstacle using a UAV board's sensor
(3) While the UAV is flying toward the target, detecting obstacles, and creating a path to avoid them.

When localizing the location of the UAV on the map, SLAM builds a map of the immediate area. In key frame-based techniques, the tracking and mapping procedures were divided into two threads.

The bundle adjustment process causes the mapping thread to compute slowly. Since this mapping thread is started each time, a new key frame is added to the map, applying empirical to key frame selection which might speed up this laborious process. In filtering-based SLAM techniques, the point attributes are triangulated and the entire map is inspected for each frame. Point features are only added to the map for freshly formed key frames when utilizing key frame-based approaches and only after a bundle adjustment operation has been finished. Finding techniques for obstacle identification that are nearly real-time has advantages.

3 Literature Survey

3.1 A Survey of Autonomous Vision-Based See and Avoid for Unmanned Aircraft Systems

McFadyen and Mejias [1] presented data fusion using simple and inexpensive ultrasonic sensors, an autonomous flying quad copter's method to obstacle detection and collision avoidance. A self-developed quad copter was used to implement and test the strategy, and evaluation revealed that it was generally realizable.

3.2 A Survey of Motion Planning Algorithm Rhythms from the Perspective of Autonomous UAV Guidance

Goertzen et al. [2] presented that utilizing a monocular camera in a bio-inspired manner, UAVs may identify obstacles and avoid them in a manner like humans. Using an obstacle detection algorithm and input images from the front camera, vision-based navigation and guidance are the method used in this method.

3.3 Search and Tracking Algorithms for Swarms of Robots

Senanayake et al. [3] demonstrated combining image sensors and a colorable tracking system to achieve collision avoidance on a quad rotor unmanned aerial vehicle (QUAV). To gather depth information from flat object surfaces, two HD stereo cameras were selected as the stereo vision sensor. To project a high-contrast tracking spot for depth estimation using standard triangulation, a laser transmitter was used. The control algorithm was created to alter the response of the QUAV based on the depth estimated, and the stereo vision algorithm was created to acquire the distance from the tracked point to the QUAV. With the aid of the optic track motion tracking system, a nonlinear model was used to design an attitude and position controller.

3.4 Impact of Drone Swarm Formations in 3D Scene Reconstruction

Milani and Memo [4] presented an autonomous navigation system for creating settings with moving and immovable objects that avoids collisions in real time. The current implementation was made to allow an unmanned aerial vehicle (UAV) to autonomously navigate between waypoints on a predetermined flight trajectory while carrying out duties like inspections or keeping track of the progress of construction. It used data from 3D sensors to locate impediments like people or other UAVs and detect them, which are not logged in the initial cloud. Ifan obstruction is found, and the algorithm calculates an avoidance path while taking the geometry of the surrounding area into account and estimates its mobility. The approach was computationally possible to execute onboard a UAV because execution times were measured.

3.5 Summary of Literature

The literature reviews on sensors and autonomous drones. Many autonomous drones have poor response times and limited range accuracy. Putting a time-of-flight sensor [TOF] in its place, it belongs to the same family as IR sensors, but the key distinction is that they have a range of up to 15 m, which is sufficient to do the previous action in response to the reaction [5–10]. This sensor is utilized in robots for object identification and navigation, and it will be employed in a quadcopter for path tracking and completely autonomous flight [11–15].

Fig. 5 TOF sensor. *Source* Flyrobo website [(3)]

4 Components and Validation

4.1 Sensor

The infrared light produced by a tiny laser used by TOF sensors (Fig. 5) is fired at any object, bouncing off it, and then returning to the sensor. The distance between an object and the sensor can be calculated using the time difference between the light's emission and its return to the sensor after being reflected by the object.

4.2 Flight Controller

The PX4 open-hardware project created the sophisticated autopilot system known as Pixhawk (Fig. 6). It offers outstanding performance, adaptability, and dependability for managing any autonomous vehicle thanks to cutting-edge processor and sensor technologies from ST Microelectronics and a NuttX real-time working framework. The Pixhawk framework offers focal points like as built-in multithreading, a programming environment taking after Unix/Linux, completely modern autopilot highlights counting Lau scripting of missions and flight conduct, and a bespoke PX4 driver layer giving exact synchronization over all forms.

4.3 Telemetry

Data collected by a drone and delivered back to the operator or ground control station are known as telemetry (GCS). The drone's autopilot, sensors like accelerometers, gyroscopes, and GPS, as well as other components like the aircraft's power source may all provide these data. When an unmanned aerial vehicle is in operation, the telemetry system is used to transmit data such as position (coordinate point), altitude, direction, and some information that reveals the state of the aircraft in real time; both hardware and software make up the telemetry systems of UAVs.

The ability of light to combine higher speed, greater range, less weight, and eye safety makes it the ideal carrier for TeraRanger sensors. It can ensure less signal

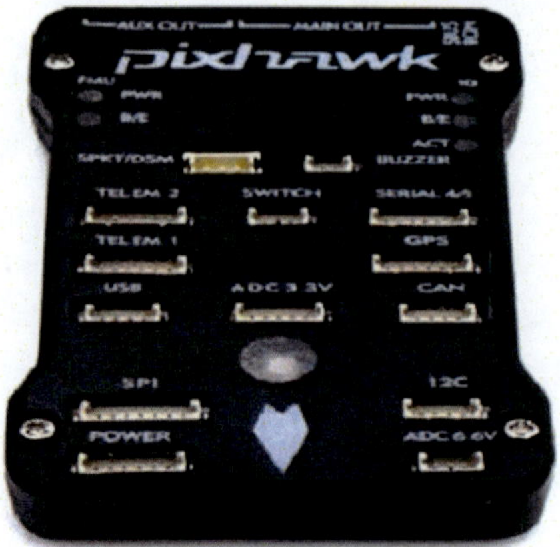

Fig. 6 Pixhawk (flight control board). *Source* Flyrobo website(4)

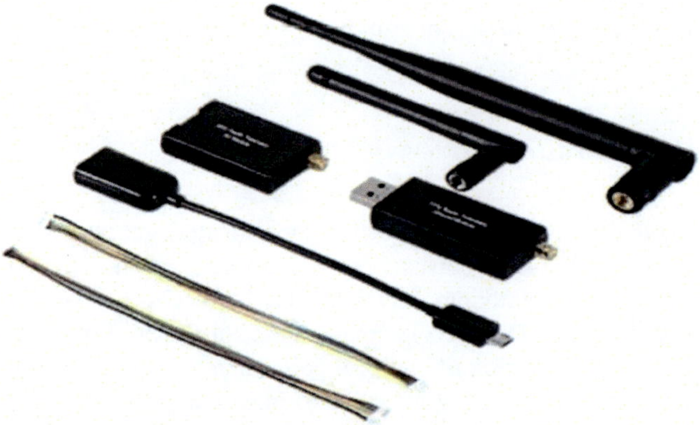

Fig. 7 3DR radio telemetry 433 MHz 500mW for APM and PIXHAWK. *Source* Flyrobo website(5)

interference and simpler separation from ambient light by using infrared light, which produces the highest-performing distance sensors for their given size and weight. A 3DR Radio Telemetry 433 MHz 500 mW for APM and PIXHAWK is shown in Fig. 7.

4.4 *Transmitter and Reciever*

An electronic gadget called a drone radio transmitter sends commands to the radio receiver wirelessly over a predetermined radio frequency; this is related to the remotely operated drone.

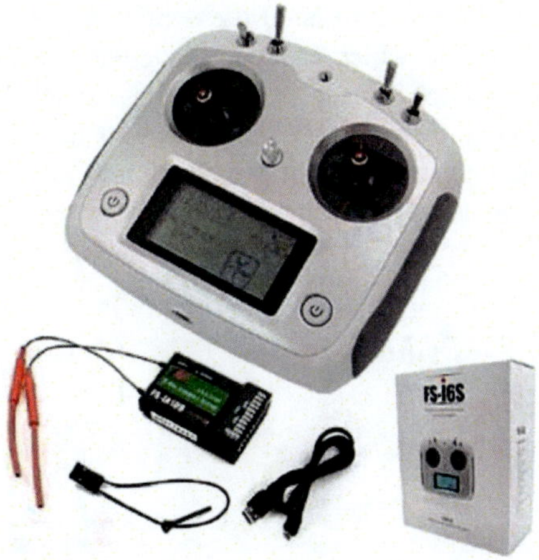

Fig. 8 Flysky 10-channel. *Source* Flyrobo website (6)

Thus, the mechanism oversees translating the pilot's commands into multi-rotor movement. Channels are used by an FPV drone radio transmitter to transmit orders. Each channel conveys to the aircraft a certain action.

(1) Pitch—forward or backward tilt of your drone.
(2) Roll—literally "rolls" your drone by moving it left or right in the air.
(3) Throttle—controls how much control is conveyed to your ramble, permitting you to alter how rapidly or gradually it moves.
(4) Yaw—turn your drone in either a clockwise or counterclockwise direction to create patterns or circles in the air.

A radio receiver is a device that can receive instructions from a radio transmitter and then use that information to control the flight of an aircraft. Subsequently, it is significant that your ramble transmitter and recipient are congruous with one another in terms of recurrence and other components when choosing them.

Moreover, it is essential that both components meet the measurements and highlights required for the given application. Flysky 10-channel is shown in Fig. 8.

5 Execution and Simulation

5.1 Dropbox for Python

Using the Dropbox Python SDK, we have moved the object to a certain spot where it could recognize and avoid obstacles before making its way to the desired destination. We used Python Arena, the repulsive navigator, and sprites, all of which are tools for programming in Python, to achieve this approach. Using various pieces of Python code, which each plays a specific role in the simulation, such as the arena function, which is used to build the simulation's background and outer layer, the repulsive navigator, which functions to guide the drone along a predetermined path to its destination, and lastly the sprites, which carry out the majority of the command functions such as obstacle detection and avoidance, is all represented. As a result of our integration with the flight control board, the flight will still make it to its intended destination despite the presence of several obstacles.

5.2 *Python*

```
def PlotBearing(pos1, pos2):
        '''Returns the bearing in degrees of pos2 [x2, y2]
        from pos1 [x1, y1] on an equal scale plot'''
        dx = float(pos2[0] - pos1[0])
        dy = float(pos2[1] - pos1[1])
        b = math.atan2(dy,dx)

        if b <=0:
            final = math.pi/2 + abs(b)
        elif b > 0:
            final = (math.pi/2 - b)%(math.pi * 2)

        return final

def FindCentreBearing(hdg, rec):
    '''Given two bearing lines, find the central bearing'''

    h = (math.sin(hdg), math.cos(hdg))
    r = (math.sin(rec), math.cos(rec))

    f = (h[0] + r[0], h[1] + r[1])

    theta = math.atan2(f[1], f[0])

    if theta <=0:
        final = (math.pi/2 + abs(theta))
    elif theta > 0:
        final = (math.pi/2 - theta)%(math.pi * 2)
    return final


def Distance2D(p1, p2):
    '''Returns the 2D distance between two points'''
    x1, y1 = p1[0], p1[1]
    x2, y2 = p2[0], p2[1]
    d = math.sqrt(math.pow(x1 - x2,2) + math.pow(y1 - y2,2))
    return d

class Sprite:
    '''Parent class for all plotting objects'''
    def __init__(self, ID, pos, radius = 20, lineThick = 3, color = "Black"):
        self.ID = ID
        self.pos = pos
        self.color = color
        self.radius = radius
        self.lineThickness = lineThick
```

Obstacle

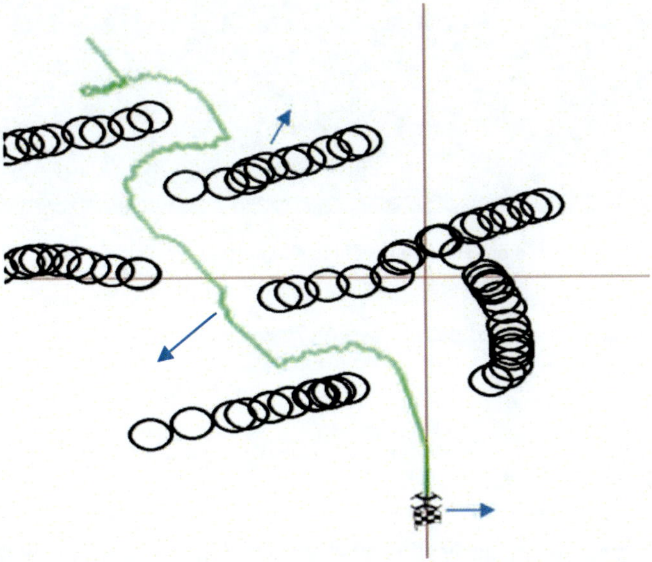

Flow path of drone

Drone is reaching goal, after avoiding the entire obstacle.

Coordination with a time-of-flight sensor and AI has been integrated into the control board to detect the collision.

5.3 Arena Python

You can define the appearance and behavior of objects in a scene or a background using Python programs, which take advantage of the fact that all objects in a scene as with the help of MQTT bus algorithm. This Python program can be used anywhere with the access to the MQTT bus algorithm, and it can be executed in the program which can be programmed by the Arena itself using ARTs.

5.4 Repulsive Navigator Python

If given sufficient time, graph searching methods such as a breadth-first search might find a path, but other methods that "explore" the graph would often get at the goal sooner. An illustration would be a person crossing a room in the direction of their

Fig. 9 DRL simulator of drone and testing of obstacle collision avoidance in DRL simulator

target, only veering off the path to avoid obstacles and keeping veers as brief as feasible.

5.5 Sprites Python

Sprites are things having many different characteristics, such as height, width, color, and others, in addition to movements to the right, left, up, and down jumped. In this post, we will discuss how to make a controllable object that readers may use to navigate up, down, forward, and backward by using the arrow keys.

5.6 Simulation and Visualization

We ran the simulation on the DRL simulator to test path planning algorithms and fine-tune flight control models. In a virtual setting that is more like the real world, high-fidelity drone simulators test UAV applications.

To determine performance and stability, the aforementioned simulation was run the DRL simulation (Fig. 9).

6 Result and Discussion

Modern computer technology is used by drones, including analogue controls, microcontrollers, on-chip systems, and single-board computers.

A digital spatial model generated utilizing aerial photographs obtained by a drone is referred to as a drone 3D model. Aerial photogrammetry is the process of using aerial pictures to create models and maps. A UAV-based independent movement

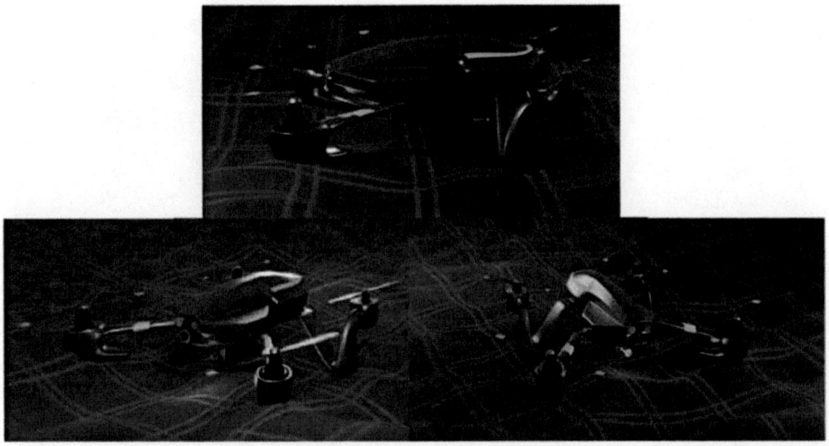

Fig. 10 Side, top, and isometric views of drone blender model

arranging and protest finding framework beneath instability and halfway discernibleness in exterior settings is displayed through this ramble DRL test system. Side, top, and isometric views of drone blender model are shown in Fig. 10.

Unmanned aircraft was shown, which used TOF sensors to emit infrared light that eventually reflected from surfaces and objects. A Sense-and-Avoid (SAA) architecture for small- to medium-sized unmanned aerial vehicles (UAVs) is proposed, with the system serving as one of the key non-cooperative sensors. A typical simulation case study was presented together with the algorithms for calculating the avoidance volumes related to obstacles and for creating the best avoidance trajectories.

All potentially conflicting impediments are avoided thanks to the proven detection, warning, and avoidance performances, total uncertainty volumes estimation, and avoidance trajectory generation algorithms.

The analysis demonstrates the demonstrated solution's ability to fly autonomously and avoid collisions in a variety of situations. This work's primary contribution is to debunk the misconception that only expensive sensors, such laser scanners, are appropriate for such a task and inexpensive range finders are not. The test revealed that the system can adjust its distance from barriers like walls and people in order to avoid colliding with them. Experimental setup is shown in Fig. 11.

7 Conclusion and Future Works

The drone has avoided collisions with the aid of a TOF sensor, and we have restricted the distance to 100 cm, or 1 m, even though the particular TOF sensor we have chosen can detect objects up to 15 m away. As a result, the TOF sensor will shift to the left or right depending on the object or obstacle it detects inside its 1 m field of view. In

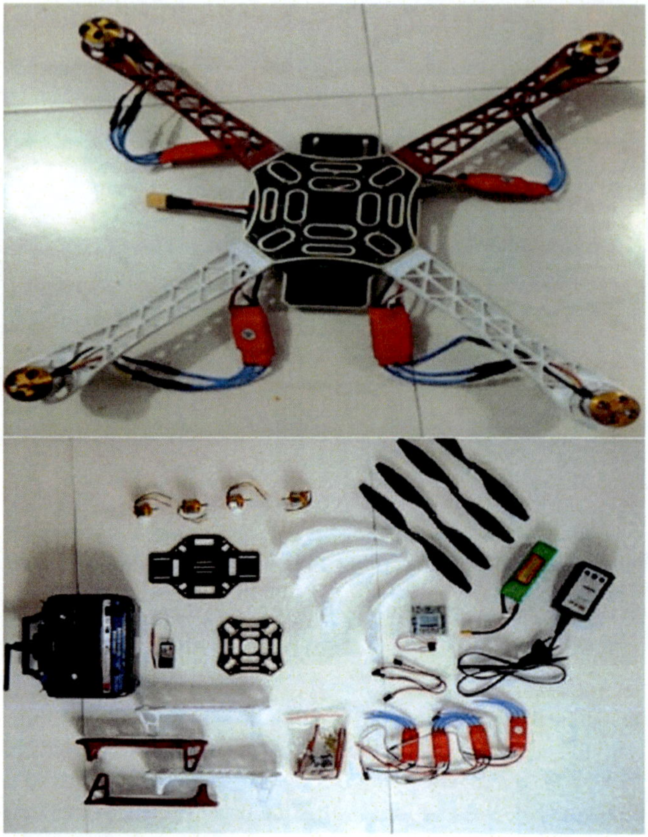

Fig. 11 Experimental setup

the future, we will endeavor to prevent collisions in all 360° directions. If the drone is stuck in a space, it must be stable in one place and maintain its altitude until the way is clear before it may begin to go ahead.

Acknowledgements The authors would like to thank Centre of Excellence in Assistive Technology of Rajalakshmi Engineering College for providing the infrastructure to develop UAV systems.

References

1. Mcfadyen A, Mejias L (2016) A survey of autonomous vision-based see and avoid for unmanned aircraft systems. Prog Aerospace Sci 80:1–17
2. Goerzen C, Kong Z, Mettler B (2009) A survey of motion planning algorithms from the perspective of autonomous UAV guidance. Nov 2009

3. Senanayake M, Senthooran I, Murshed M (2016) Search and tracking algorithms for swarms of robots: a survey. Robot Auton Syst 75:422–434
4. Milani S, Memo A (2016) Impact of drone swarm formations in 3D scene re-construction. In: Proceedings of IEEE international conference on image process. (ICIP), Sept 2016
5. Gageik N, Benz P, Montenegro S (2015) Obstacle detection and collision avoidance for a UAV with complementary low-cost sensors. IEEE Access 3:599–609
6. Mohamed N, Al-Jaroodi J, Jawhar I, Idries A, Mohammed F (2020) Un-manned aerial vehicles applications in future smart cities. Technol Forecasting Social Change 153
7. Huang H-M, Messina ER (2007) Autonomy levels for unmanned systems (ALFUS) frame work volume II: framework models initial version. National Institute of Standards Technology, Gaithersburg, MD, USA, Tech. Rep. 1011-I-2.0
8. Chen H, Wang X-M, Li Y (2009) A survey of autonomous control for UAV. In: Proceedings of international conference on artificial intelligence computational intelligence, pp 267–271
9. Zhang W, Zelinsky G, Samaras D (2007) Real-time accurate object detection using multiple resolutions. In: Proceedings of IEEE 11th international conference on computer vision, Oct 2007
10. Shinde SN, Chorage SS (2016) Unmanned ground vehicle. Int J Adv Eng Manage Sci 2(10):1–4
11. Iqbal J, Pasha SM, Baizid K, Khan AA, Iqbal J (2013) Computer vision inspired real-time autonomous moving target detection, tracking and locking. Life Sci J 10(4):3338–3345
12. Chao H, Cao Y, Chen Y (2010) Autopilots for small unmanned aerial vehicles: a survey. Int J Control Autom Syst 8(1):36–44
13. Zhuge C, Cai Y, Tang Z (2017) A novel dynamic obstacle avoidance algorithm based on collision time histogram. Chin J Electron 26(3):522–529
14. Mizumachi M, Kaminuma A, Ono N, Ando S (2014) Robust sensing of approaching vehicles relying on acoustic cue. In: Proceedings of international symposium on computational consumption control, Jun 2014, pp 533–536
15. van Leeuwen MB, Groen FCA (2005) Vehicle detection with a mobile camera: spotting midrange, distant, and passing cars. IEEE Robot Autom Mag 12(1):37–43
16. Recchia G, Fasano G, Accardo D, Moccia A, Paparone L (2007) An optical flow based electro-optical see-and-avoid system for UAVs. In: Proceedings of IEEE aerospace conference, Mar 2007, pp 1–9
17. Kóta F, Zsedrovits T, Nagy Z (2019) Sense-and-avoid system development on an FPGA. In: Proceedings of international conference on unmanned aircraft system (ICUAS), June 2019, pp 575–579
18. Mcfadyen A, Durand-Petiteville A, Mejias L (2014) Decision strategies for automated visual collision avoidance. In: Proceedings of international conference on unmanned aircraft system (ICUAS), May 2014, pp 715–725
19. Beard R, Saunders J (2008) Reactive vision-based obstacle avoidance with camera field of view constraints. In: Proceedings of AIAA guidance navigation control conference exhibition, Aug 2008, p 7250
20. Saha S, Natraj A, Waharte S (2014) A real-time monocular vision-based frontal obstacle detection and avoidance for low-cost UAVs in GPS denied environment. In: Proceedings of IEEE international conference on aerospace electronics remote sensing technology, Nov 2014, pp 189–195
21. https://www.seeedstudio.com/blog/2020/01/08/what-is-a-time-of-flight-sensor-and-how-does-a-tof-sensor-work/
22. https://www.mouser.in/new/dfrobot/dfrobot-sen0413-tof-ir-distance
23. https://www.flyrobo.in/pixhawk_px4_32bits_flight_controller
24. https://www.flyrobo.in/3dr-radio-telemetry-433mhz-500mw-for-apm-and-pixhawk
25. https://www.flyrobo.in/flysky-fs-i6x-2.4ghz-6ch-afhds-2a-rc-transmitter-with-fs-ia10b-2.4ghz-10ch-receiver

Collaborative Communication Models in Non-cash Food Assistance (Bantuan Pangan Non-Tunai, BPNT) Program: Toward Community Resilience

Abdillah Abdillah, Ida Widianingsih, Rd Ahmad Buhari, and Rusliadi Rusliadi

Abstract The goal of this study is to analyze the dynamics of the collaborative communication model for the implementation of the Non-Cash Food Assistance (*Bantuan Pangan Non-Tunai*, BPNT) Program in Takalar Regency, Indonesia, to fight poverty toward community resilience. In order to go more into the subject issue, this study employs a qualitative exploratory methodology. Based on the results of the study, it shows that the implementation of the Non-Cash Food Assistance Program is intended to increase community resilience in the midst of a global crisis in Indonesia. Because outreach to the community is lacking, the implementation of the Non-Cash Food Assistance (*Bantuan Pangan Non-Tunai,* BPNT) aid program in Takalar Regency, Indonesia, has not been sufficiently successful. This is evident from a number of program targets that were missed. The distribution of the non-cash food Assistance (*Bantuan Pangan Non-Tunai,* BPNT) aid program must promote collaborative communication with all Takalar Regency stakeholders in order to resolve this issue. This will have an impact on the extent to which distribution policies and management will be implemented in supporting the success of the program.

Keywords Collaborative communication · BPNT · Community resilience · Poverty · Indonesia

A. Abdillah (✉)
Graduate Program in Public Administrasion, Universitas Padjadjaran, Bandung, Indonesia
e-mail: abdillah18001@mail.unpad.ac.id

A. Abdillah · I. Widianingsih · R. A. Buhari · R. Rusliadi
Center for Decentralization and Participatory Development Research, Faculty of Social and Political Sciences, Universitas Padjadjaran, Bandung, Indonesia
e-mail: ida.widianingsih@unpad.ac.id

R. A. Buhari
e-mail: ahmad.buchari@unpad.ac.id

R. Rusliadi
e-mail: rusliadi21001@mail.unpad.ac.id

G. Rajakumar et al. (eds.), *Intelligent Communication Technologies and Virtual Mobile Networks*, Lecture Notes on Data Engineering and Communications Technologies 171,
https://doi.org/10.1007/978-981-99-1767-9_6

1 Introduction

One of the objectives of the SDGs reflects the moral principle if there is no country that is in a lagging position, while other countries face the welfare of each person and the state is responsible for playing their position in delivering the global vision of the SDGs [1]. The structural differentiation of countries and international systems is presented as a way to explain the limitations and possibilities in the eradication of poverty [2, 3] and global justice which later became the first goal of the SDGs (at Point 1), namely eliminating all forms of poverty [4, 5]. One of the points that is considered the most important and still entrenched in various countries is related to poverty. The state must be present to reduce even by presenting regulations to overcome the poverty that exists in its country.

In following up on sustainable development, it is necessary to have strict regulations and the paradigm of sustainable development must be the way of life of the Indonesian people [6–8]. In response to the Global agenda, the Government of Indonesia has produced The Sustainable Development Goals Implementation Presidential Regulation No. 59 of 2017 [9]. To date, a mix of economic, social, structural, capability, and learning perspectives has been applied to understand poverty and planning strategies to accelerate its reduction [8, 10]. Poverty in Indonesia is still the main problem and a heavy burden for each government, most notably when it comes to the widening significant wealth gap between rich and poor. As part of the United Nations members, Indonesia is certainly committed to alleviating ongoing poverty [11, 12].

The impact of poverty in Indonesia is inseparable from Covid-19. The Covid-19 pandemic has also had an impact on the poverty rate through employment. Covid-19 has affected almost all countries on all continents around the world. Covid-19 has caused exceptional economic and public health responses, both in Indonesia and many other places, as a result of its proliferation and difficulties [13, 14]. The estimated economic growth rate for 2020 is expected to drop from around 5% to between 4.2 and − 3.5% as a result of the severe economic effect [15–18]. Village and urban poverty rate 2018–2022 is shown in Table 1.

Despite having a well-known track record of reducing poverty, only in September 2014 and March 2015 did the percentage of the people living in poverty rise [18, 19]. In reality, despite state revenues, all countries have struggled to escape the impact of Covid-19. However, Indonesia is still struggling to maintain its positive economic

Table 1 Village and urban poverty rate 2018–2022

Semester (year)	September (2018)	September (2019)	September (2020)	September (2021)	March (2022)
Urban	10,13	9,86	12,04	11,86	11,82
Rural	15,54	14,93	15,51	14,64	14,34
Total	25,67	24,79	27,55	26,50	26,16

Source Processed from various sources, 2023

growth [20]. The economic impact of Covid-19 is assumed to have started as a negative supply shock [21]. During Covid-19, the poverty rate in Indonesia touched 10.19% in September 2020. This makes the number of poor people in Indonesia increase by 2.76 million people when compared to the previous same period [16, 17, 19]. Due to the prevalence of poverty in Indonesia, the government created the BPNT/basic food program, a non-cash social food assistance that is distributed to benefit families (*Keluarga Penerima Manfaat*, KPM) on a monthly basis using banking methods. The Channeling Bank will provide non-cash support kits to beneficiary households (*Keluarga Penerima Manfaat*, KPM) in the form of electronic coupons (e-vouchers) (like *Bank Negara Indonesia* [BNI]) [22]. Non-Cash Food Assistance for BPNT is worth IDR 110.000/KPM per month. Meanwhile, the amount of the basic food program for the January–February period is IDR 150,000, but since the period of March–August 2020, it has been increased to IDR 200,000. The assistance cannot be taken in cash, and if the assistance is not spent within the month, the value of the assistance will still be stored and accumulated. KPM can use the e-voucher to buy rice and other foodstuffs such as eggs, according to the desired quantity and quality in e-warong [12, 18].

Takalar Regency, Indonesia, which is one of the regencies in South Sulawesi Province, has a poverty rate of 24,600 people in 2021, while in 2020 it was 25,380 people. Meanwhile, in 2019 the number of poor people was 25.93 people. This is something that is still considered positive because Takalar Regency is able to reduce the number of its population who fall into the poor category. Whereas Takalar Regency had a population of 300,447 people in 2022 [17]. In Indonesia, one of the districts that has a fairly high poverty rate is Takalar Regency. Takalar Regency is located in South Sulawesi Province which consists of nine sub-districts and has an area of $\pm$ 566.51 km^2 with a population of 250,000 people who currently have a poverty rate of up to 42 thousand households or approximately 124 thousand people. With this data, it means that the number of poor people in Takalar Regency is around 65% of the total population [12, 16–18]. Because of this, the support of all stakeholders is needed in communicative collaborative action in overcoming the problem of poverty and social vulnerability in Takalar Regency.

Based on Fig. 1, Non-Cash Food Assistance Programs (Non-Cash Food Assistance, BPNT) are identified and analyzed which are distributed to the poor in Indonesia. The purpose of providing this assistance is to increase community resilience in facing social and economic vulnerabilities due to the pandemic and food crisis in Indonesia. This phenomenon is divided into several issue clusters. In the first cluster, it is explained how the implementation of programs aimed at overcoming the problem of poverty is related, both from the perspective of policies, factors that cause poverty and poverty programs. The second cluster describes the role of government in relation to policy, assistance, poverty, and effectiveness. The third cluster describes the non-cash food social assistance program (*Bantuan Pangan Non-Tunai*, BPNT), which is one of the programs in poverty alleviation, related to issues of implementation, effectiveness, the role of e-warong and in what form the assistance is provided.

Fig. 1 Research topics related to Indonesia's non-cash food assistance program (Bantuan Pangan Non-Tunai, BPNT) are identified and analyzed. *Source* Processed via Nvivo 12 Pro, 2023

Based on the explanation above, this study goals to examine the dynamics of the implementation of the Non-Cash Food Assistance Program (*Bantuan Pangan Non-Tunai*, BPNT) in a model of collaborative communication between the central government and local governments in overcoming poverty for community resilience in Takalar Regency, Indonesia. The implementation approach in the Non-Cash Food Assistance Program is strongly influenced by the collaborative communication model carried out in the field in stimulating the reduction of community poverty levels. This can be achieved through support and cooperation from the central government, local governments, and other stakeholders to improve the community's economy in solving poverty. Other approaches include distributing cashless food assistance to residents and providing knowledge and objectives of the program to help overcome poverty. The Non-Cash Food Assistance (*Bantuan Pangan Non-Tunai,* BPNT) Program's implementation seeks to combat poverty and aid those who are less fortunate in a challenging economic this time. The welfare and freedom of people can both be enhanced by the Non-Cash Food Assistance (*Bantuan Pangan Non-Tunai,* BPNT) Program. This study helps to improve the coordination of communication among stakeholders in the implementation of non-cash basic food assistance program policies in Takalar Regency, Indonesia. It also seeks to increase both public and private involvements in the fight against poverty.

2 Methodology/Approach

The goal of this study is to identify how the Non-Cash Food Assistance Program (*Bantuan Pangan Non-Tunai*, or BPNT) would be implemented in Takalar Regency, Indonesia, as part of a cooperative communication strategy to combat poverty and promote community resilience. This research was conducted in Sanrobone Sub-district, Takalar Regency in South Sulawesi, Indonesia, as an area with a fairly severe poverty rate in the South Sulawesi region, Indonesia, and at the same time implemented the Non-Cash Food Assistance Program (*Bantuan Pangan Non-Tunai,* BPNT) which is not optimal in Indonesia. This qualitative approach is intended to make a deep understanding, a systematic, factual, and accurate picture of an object, a set of present conditions, and the relationships between the phenomena investigated. This research complements the approach of policy implementation theory which focuses on policy standards and objectives/measures and policy objectives, existing resources, communication between organizations, attitudes of implementers, characteristics of implementing agents, and what are the conditions of the social environment, political economy in the area in a collaborative communication model for the implementation of Non-Cash Food Assistance (*Bantuan Pangan Non-Tunai*, or BPNT) Programs in the Takalar Regency, Indonesia. Data are gathered by observation, literary studies, and other methods, including reading and studying books, journals, government papers, and other pertinent sources. Data analysis through collection, reduction, presentation, and verification is the processes proposed by Miles et al. [23] that are used to evaluate and understand the acquired data in order to provide conclusions in the form of fresh findings that are helpful to the reader. Additionally, NVivo 12 Pro [24] can be used to delve deeper into the issue to reach the best conclusion. This study's limitation is that it depends on the researcher's interpretation of the meaning that was conveyed during the interview; therefore, bias is still a possibility. By cross-checking data with information from several study observations, the triangulation technique is used to eliminate bias. The steps of this research can be explained in a simple way as in Fig. 2.

3 Results and Discussion

*3.1 National Program, Local Response: Dynamics of Non-cash Food Assistance Program (*Bantuan Pangan Non-Tunai, *BPNT) in Takalar Regency, Indonesia*

Reducing the rate of poverty is one of the objectives for national development. Because poverty is one of the economic diseases, it needs to be cured or at the very least alleviated. Indonesia's persistent poverty remains a major issue and a significant burden for all governments, particularly in light of the growing gap between

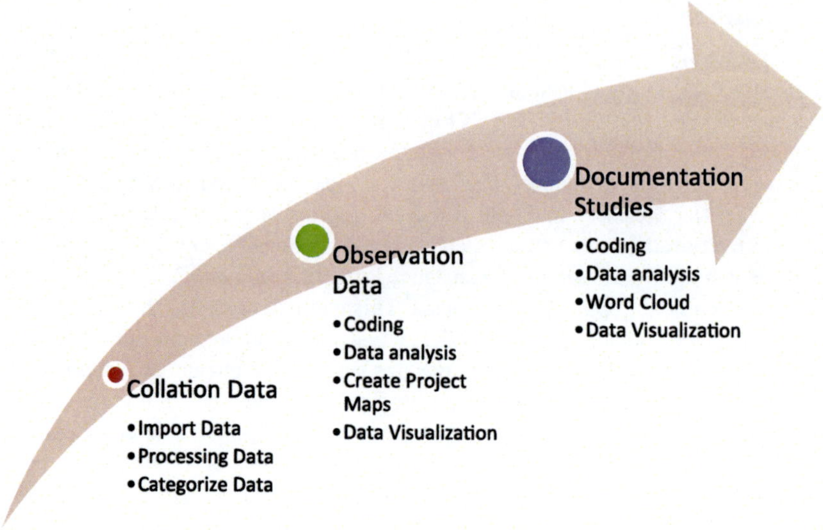

Fig. 2 Data collection and analysis techniques with NVivo 12 Pro. *Source* processed from Woolf and Silver [24], 2023

the rich and the poor. As part of the United Nations members, Indonesia is certainly committed to alleviating ongoing poverty [11]. There are also research results related to the SDGs in the world showing Points 1, 6, and 7 which strongly show that poverty and the lack of cleanliness of water and energy are still the main problems. The analysis of the ranking of 17 SDGs sourced from the state group said that SDGs 1 and 4 are very meaningful [4, 9, 11]. In line with this goal, a Non-Cash Food Assistance Program was created which was intended to increase food security at the Beneficiary Family level, as well as a mechanism for social protection and poverty alleviation in Indonesia [16–19]. In the Non-Cash Food Assistance Program, Hope Recipient Families (KPM) receive non-cash food social support from the government every month through banking procedures under the Non-Cash Food Assistance Program (BPNT). The program targets is underprivileged communities intended for BPNT Participants/beneficiaries are Families, hereinafter referred to as Beneficiary Families (*Keluarga Penerima Manfaat*, KPM) of Non-Cash Food Assistance [25]. Beneficiary Families (*Keluarga Penerima Manfaat,* KPM) represented the population in the implementation area with the lowest 25% of socioeconomic situations in 2017. The Integrated Data of the Poor Handling Program, also known as DT-PFM, is the outcome of the Integrated Database Update in 2015 and serves as the source of (*Keluarga Penerima Manfaat,* KPM) data for non-cash food assistance [18]. The Integrated Data Management Working Group of the Poor Handling Program, also known as the Data Working Group, was established by the Decree of the Minister of Social Affairs No. 284/HUK/2016 dated September 21, 2016. It is responsible for managing Integrated Data Program for Handling the Poor and Disadvantaged

Table 2 Beneficiary family (*Keluarga Penerima Manfaat,* KPM), non-cash food assistance (*Bantuan Pangan Non-Tunai,* BPNT) Program at the Village Level, Takalar Regency, Indonesia, from 2021 to 2022

No.	Village names in Takalar Regency	Number of beneficiaries (*Keluarga Penerima Manfaat,* KPM)	
		2021	2022
1	Sanrobone Village	334	430
2	Tonasa Village	286	449
3	Paddinging Village	239	314
4	Lagaruda Village	222	304
5	Ujung Baji Village	315	413
6	Banyuanyara Village	238	331
Total		1.634	2.241

Source Processed from Sub-District Social Welfare Workers (*Tenaga Kesejahteraan Sosial Kecamatan,* TKSK) Sanrobone Sub-District, Kabuapten Takalar, Indonesia, 2022

People (DT-PFM) [12]. If the (*Keluarga Penerima Manfaat,* KPM) whose name is registered in the list of beneficiaries (*Daftar Penerima Manfaat*, DPM) already has an account for the distribution of other Social Assistance programs, then the account can be used to receive the Non-Cash Food Assistance Program. For each Beneficiary Family (*Keluarga Penerima Manfaat,* KPM), list of beneficiaries (*Daftar Penerima Manfaat*, DPM) contains information on the name of the head of the family, the name of the spouse of the head of the family, the name of other family members, the family's residential address, the national identity number (NIK) (if any), and the unique family code. The following is Table 2 of the number of Beneficiary Families (*Keluarga Penerima Manfaat,* KPM) in Sanrobone Sub-District, Takalar Regency.

Table 2 shows that the Non-Cash Food Assistance Program (*Bantuan Pangan Non-Tunai,* BPNT) in Takalar Regency, Indonesia, has been distributed through Combo Cards since October 2018 by Bank Negara Indonesia (BNI) as a Channeling Bank appointed by the Minister of Social Affairs in Takalar Regency. There are also additional data on recipients of Non-Cash Food Assistance (*Bantuan Pangan Non-Tunai,* BPNT) during the Covid-19 pandemic [14, 22, 25]. Implementing the Non-Cash Food Assistance (*Bantuan Pangan Non-Tunai,* BPNT) Program which is based on the large number of poor people in Takalar Regency is a form of government concern from both the regional government, local government, and the Indonesian national government, to the large number of people who are unable to meet food needs in their daily lives. The following can be seen about the information network regarding the Non-Cash Food Assistance (*Bantuan Pangan Non-Tunai,* BPNT) Program scheme in Takalar Regency, Indonesia, to increasing food security at the Beneficiary Family level as well as social protection and poverty alleviation (see Fig. 3).

Based on the Non-Cash Food Assistance Program distribution scheme in Takalar Regency, Indonesia, the requirements are set by the Channeling Bank (*Bank Negara*

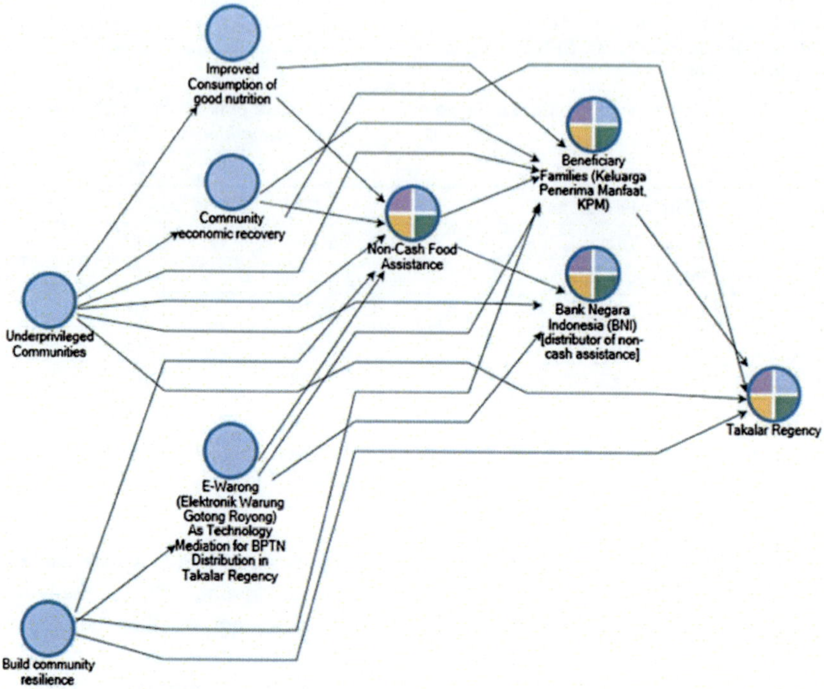

Fig. 3 Non-cash food assistance (*Bantuan Pangan Non-Tunai,* BPNT) Program scheme in Takalar Regency, Indonesia. *Source* Processed through Nvivo 12 Pro, 2023

Indonesia, BNI) to establish an e-warong, and this is intended so that the implementation of the Non-Cash Food Assistance Program (BPNT) will run smoothly and be easily monitored by the Channeling Bank (*Bank Negara Indonesia*, BNI) so that nothing can hinder the distribution of the program (Non-Cash Food Assistance, BPNT) to the community. Distribution of the program (Non-Cash Food Assistance, BPNT) to the community through the Prosperous Family Card (KKS) to predetermined e-warongs. Beneficiary Families (KPM) are not allowed to buy goods such as cigarettes; if it is known that a KPM can buy these goods, then sanctions will be given to the e-warong that sells them [16, 18, 22, 25]. E-warong is a shop/trader recommended by the Village Government and determined by Bank Negara Indonesia (BNI) and as a place for transactions for the distribution of the BPNT Program. However, in 2022 the term e-warong agent has changed to a free market and it has been handled directly by PT. POS means that KPM no longer has to do the shopping process at e-warong. A person who owns and manages an e-warong is referred to as a Bank Agent, in this case a BNI Agent In addition, for the duties and functions of Sub-district Social Welfare Workers (TKSK) and BPNT Assistants in each village/kelurahan is to provide socialization to KPM regarding the BPNT Program and prepare KPM data, as well as coordinate the place and time of BPNT disbursement to KPM every

Table 3 Data on the number of e-warong villages in Takalar Regency, Indonesia, 2019–2021

Village name	Number of e-warong
Sanrobone Village	1
Tonasa Village	2
Paddinging Village	2
Lagaruda Village	2
Ujung Baji Village	2
Banyuanyara Village	1
Total	**11**

Source Processed from sub-district social welfare workers (*Tenaga Kesejahteraan Sosial Kecamatan,* TKSK) Sanrobone Sub-District, Takalar Regency, Indonesia, 2023

month so that the implementation of distribution can be runs in accordance with the objectives of the BPNT Program [16, 18, 22, 25].

Table 3 shows that there are 11 e-warongs in Takalar Regency, Indonesia, to support the cashless food assistance program. The e-warong was formed by the distributing bank with the mechanism of the stall to register its stalls to become partners in distributing the program to support the success of the poverty reduction program while increasing MSMEs in the community [16, 18, 22, 25]. In 2022, beneficiaries of the Non-Cash Food Assistance Program (*Bantuan Pangan Non-Tunai,* BPNT) no longer have to make buying and selling transactions on e-warong, and in the end, the beneficiary communities of the assistance are free to use the money they receive to spend according to community needs and in accordance with technical instructions from the Ministry of Social Affairs, Indonesia [16, 18, 22, 25]. The Non-Cash Food Assistance Program (*Bantuan Pangan Non-Tunai,* BPNT) helps improve the community's economy and realizing community resilience in the time of Covid-19. This makes it possible to reduce the poverty rate in support of the SDGs agenda in 2030. The Non-Cash Food Assistance Program (*Bantuan Pangan Non-Tunai,* BPNT) in Indonesia is poverty alleviation as a step for implementing social welfare policies in society that is sustainable, to increase the consumption of both nutrition and the community's economy which is still classified as poor.

3.2 *Non-cash Food Assistance (*Bantuan Pangan Non-Tunai, *BPNT) Program in Overcoming Poverty: A Collaborative Communication Model Toward Community Resilience*

To administer the Non-Cash Food Assistance Program (*Bantuan Pangan Non-Tunai,* BPNT) in Takalar Regency, Indonesia, effective collaboration between governmental and private sector players is required so that the effectiveness of program implementation is realized [7, 26, 27] through good cooperation. Each stakeholder has a duty to implement the Non-Cash Food Assistance Program (BPNT) policy to overcome poverty in the community. Therefore, good communication among stakeholders is very important. This can be done through collaborative communication between the government and the community which provides fast, interesting, easy to understand, and accountable information [25, 28, 29] regarding the Non-Cash Food Assistance Program (BPNT) and its distribution.

The development target of Indonesia's 2020–2024 National Medium-Term Development Plan (RPJMN) is to create an independent, developed, just, and prosperous Indonesian society through accelerating development in various fields, including in overcoming poverty in Indonesia. The government targets the extreme poverty rate in the country to reach zero percent by 2024 [15, 20]. This is the government's target because of the many programs that have been given by the government to the community in order to stabilize their economy to avoid poverty. Meanwhile, in Takalar Regency, the phenomenon of poverty in general can be seen through several indicators of poverty. First, The number and percentage of the population living below the poverty line. In 2020, there are still around 25.38 thousand poor people in Takalar Regency or about 8.44% of the total population of Takalar Regency. They are classified as poor because the average monthly expenditure for both food and non-food consumptions is still below the poverty line which at that time reached IDR 364,378/person/month [16, 22, 25]. According to the 2020 South Sulawesi Provincial Poverty Data and Information, trends of a high number of poor people in South Sulawesi, Indonesia, intersect with the percentage of poverty in Takalar Regency.

Figure 4 shows the development of the number and percentage of poor people in Takalar Regency in 2016–2020. It can be seen that the number and percentage of poor people in Takalar Regency have a tendency to decline. When viewed from the 2016–2020 period, the number of poor people has decreased from 27.05 thousand people to 25.38 thousand people. In the same period, the percentage of poor people also decreased from 9.35 to 8.44%. This has to do with the non-cash food aid program's implementation in Takalar Regency to combat poverty. The decrease in the number of poor people is inseparable from the government's efforts in providing assistance to the community through the Non-Cash Food Assistance Program (*Bantuan Pangan Non-Tunai*, BPNT) [16, 22, 25].

It should be underlined that a public policy will be effective and directed in its implementation, if all implementing parties can carry out the program according to the guidelines and deadlines set to strengthen the local economy and reduce poverty. The phenomenon related to the corruption case of the Non-Cash Food Assistance Program (BPNT) in Takalar Regency was caused by the implementation of policies that did not comply with the guidelines and communicative of each stakeholder to align goals in encouraging effective and efficient program distribution that had not been carried out properly. Support from the government, both central and regional governments, as well as other stakeholders for the implementation of this program can increase citizen participation. This can overcome the problem of corruption in the distribution of the Non-Cash Food Assistance Program (BPNT) in Takalar Regency, because it in communicative, collaborative, and transparent actions will create an

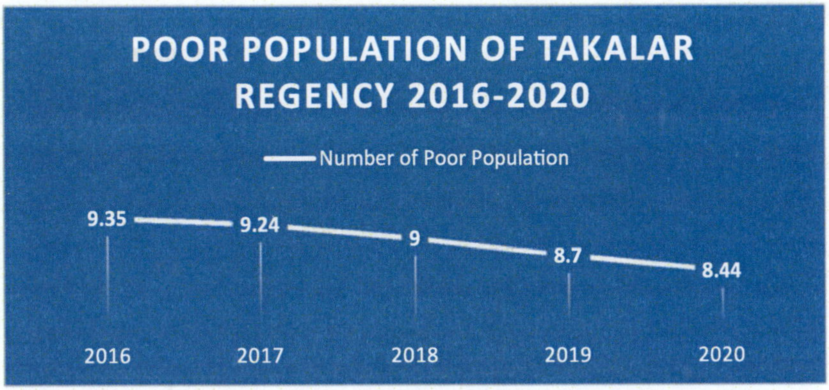

Fig. 4 Trends in the level of poor people in Takalar Regency, Indonesia. *Source* Regency/City Poverty Data and Information 2016–2020, 2023

attitude of responsibility and adherence to policies and implementation guidelines [10, 13, 16, 22, 25, 27–29].

Based on the research findings in Table 4, the results of the analysis of the results of this study with reference to the policy implementation approach of Van Metter and Van Horn [30] show that through the aspect of "standards and policy objectives/targets," the Non-Cash Food Assistance Program (BPNT) in Takalar Indonesia has not yet been implemented quite effectively because the beneficiaries of BPNT are still constrained in terms of uncertain distribution schedules, including communication between stakeholders (National Government, Local Government, and program implementing agencies, and the community) which is still not optimal [28, 29, 31]. Based on the "resource" aspect, implementing the Non-Cash Food Assistance (BPNT) Program in Takalar Regency, Indonesia, from a resource perspective (human, infrastructure, and financial) has not been effective. In the aspect of "character of the implementing agent" as the capacity of all implementing agents has been given education and understanding related to the BPNT program both in terms of regulations and technical guidelines in its implementation. In the aspect of the "attitude of the implementers" of the implementation of the BPNT program, the implementing agent has carried out their duties well. From the Takalar Regency Social Service, Sub-District Social Welfare Workers (*Tenaga Kesejahteraan Sosial Kecamatan,* TKSK), Village Government, e-warong owners and beneficiaries of the BPNT program. In the aspect of "social environmental conditions, political economy," the presence of the Non-Cash Food Assistance Program was very well responded by the local government in Takalar Regency, Indonesia, and the aspect of "communication between organizations" communication between organizations in the Non-Cash Food Assistance Program (BPNT) in Takalar Regency, Indonesia, was seen communication between these organizations is already going well. This is evidenced when there are obstacles in the field, the Sub-District Social Welfare Workers (*Tenaga Kesejahteraan Sosial Kecamatan,* TKSK) will directly review and will coordinate this matter to the Social

Service, Empowerment of the Poor, Takalar Regency, Indonesia. Based on the policy implementation approach, it can be seen that there is a gap in the implementation of the Non-Cash Food Assistance Program (BPNT) in Takalar Regency, Indonesia.

Table 4 Goals and successes of the non-cash food assistance program in Takalar Regency, Indonesia (*Bantuan Pangan Non-Tunai*, BPNT)

No.	Indicators	Achievements
1	Policy standards and objectives/objectives	Based on the aspects of standards and policy goals/objectives in the non-cash food assistance program (*Bantuan Pangan Non-Tunai,* BPNT) in Takalar Regency, Indonesia has not been running effectively enough because the obstacles experienced by (*Bantuan Pangan Non-Tunai,* BPNT) beneficiaries are still constrained in terms of erratic distribution schedules. In fact, this makes KPM-BPNT even wait 2–3 months to be able to disburse the aid funds that will be spent on e-warong. When viewed from the accuracy of the target, it has been classified as effective, and it is because the village government is in direct contact to record its underprivileged communities to be proposed for assistance
2	Resources	In running the non-cash food assistance program in Takalar Regency, Indonesia, in terms of resources (human, program infrastructure, and finance), it has not been running effectively. The support of good resources can make it easier for the community to disburse and spend the assistance obtained, so the channeling bank will appoint e-warong which will be a partner of 11 e-warongs and will be controlled by the sub-district social welfare worker (*Tenaga Kesejahteraan Sosial Kecamatan,* TKSK) of the existing sub-district of Takalar Regency. The provisions for recipients of the (*Bantuan Pangan Non-Tunai,* BPNT) program in spending the assistance obtained are only valid until 2021. Regulations changed in 2022 that beneficiaries of the BPNT program do not have to spend the assistance they get on e-warong. It is just that it is not socialized to the public, so people still make transactions in e-warong even if the recipient wants to buy something that not all e-warong has

(continued)

Table 4 (continued)

No.	Indicators	Achievements
3	Characteristics of the executing agent	In this study, it was found that in capacity all implementing agents have been given education and understanding related to the BPNT program both in terms of regulations and technical guidelines in its implementation. The field of empowerment of the poor and assistant for sub-district social welfare workers (*Tenaga Kesejahteraan Sosial Kecamatan,* TKSK) of Takalar Regency is very pro-active in providing direction to both e-warong and beneficiaries. This is evidenced by the regular holding of meetings every month. Monitoring of the distribution carried out by Dinsos and sub-district social welfare workers (*Tenaga Kesejahteraan Sosial Kecamatan,* TKSK) has been maximized
4	Attitude of the policy implementing agent	In field observations, it was found that the implementation of the BPNT program was that the implementing agent had carried out their duties properly. Starting from the Takalar Regency social service, sub-district social welfare workers (*Tenaga Kesejahteraan Sosial Kecamatan,* TKSK), village government, e-warong owners, and beneficiaries of the BPNT program. It is just that it is still considered not optimal because the price of staples in e-warong is considered not transparent. This is because BPNT beneficiaries are only allowed to take the packages that have been provided. The estimated price of staples in e-warong is also when compared and calculated by the beneficiaries, the price is higher than the price in the ordinary and market markets
5	Social, economic, political environment	The presence of the non-cash food assistance program was very well responded by the local government in Takalar Regency. The majority of people's jobs in Takalar Regency, Indonesia, are farmers and fishermen, with the number of unemployed people 1858 people in 2020. With this assistance, the economic condition of the community begins to be helped and that is what will improve the community's economy and will reduce existing poverty

(continued)

Table 4 (continued)

No.	Indicators	Achievements
6	Inter-organization communication	Based on the data in this study, the communication aspect between organizations in the non-cash food assistance program (*Bantuan Pangan Non-Tunai,* BPNT) in Takalar Regency, Indonesia, it can be seen that communication between these organizations has been running well. This is proven when there are obstacles in the field, the sub-district social welfare workers (*Tenaga Kesejahteraan Sosial Kecamatan,* TKSK) will review directly and will coordinate this matter with the social service, empowerment of the poor of Takalar Regency. The same thing when the village government proposes that its people who are classified as underprivileged will be followed up responsively. Socialization is also still running to BPNT beneficiaries so that they can use the assistance wisely and not spend the assistance on things that are not in accordance with the provisions. But, this is different from the communication of the e-warong owner to the (*Bantuan Pangan Non-Tunai,* BPNT) recipient which is categorized as not optimal. Because in the implementation of the (*Bantuan Pangan Non-Tunai,* BPNT) program, beneficiaries are still required to spend the money they get in e-warong. The food products provided have also been packaged by the owner of the e-warong for (*Bantuan Pangan Non-Tunai,* BPNT) beneficiaries

Source Processed by authors from various sources, 2023

This is due to various problems such as non-synergistic working relationships between the central government and local governments in Takalar Regency. There are misaligned policy gaps at each level between the central and regional levels. The tasks of the program implementing agents at each central and regional level sometimes overlap, until communication between the central and regional governments does not go well so that this research finds that a collaborative communication model is needed between the Indonesian national government that synergizes with the local government of Takalar Regency in implementing an effective and targeted Non-Cash Food Assistance (BPNT) Program. This is in line with research from Schultz and Evans [29] and Mohr et al. [28] which mentions that collaborative communication in organizations is key to effectivization, control, and strategy in achieving organizational goals. Based on the results of this study, shows that the implementation of the Non-Cash Food Assistance Program (BPNT) is intended to increase community resilience in the midst of a global crisis in Indonesia. Because outreach to the community is lacking, the implementation of the Non-Cash Food Assistance (*Bantuan Pangan Non-Tunai,* BPNT) aid program in Takalar Regency, Indonesia, has not been sufficiently successful. This is evident from a number of program targets that were missed. The distribution of the non-cash food Assistance (*Bantuan Pangan Non-Tunai,* BPNT) aid program must promote collaborative communication with all Takalar Regency stakeholders in order to resolve this issue. This also has an impact on the extent to which policies and programs will be implemented in supporting the success of the program.

4 Conclusion

The results of this Non-Cash Food Assistance Program's deployment have been positive. But because community socialization is still regarded as being subpar, the implementation of the non-cash food aid program based on standards and policy goals/objectives, resources, inter-organizational communication, and the implementers' attitudes has not been sufficiently effective. This is a result of a lack of effective communication processes that encompass the dimensions of information transformation, transparency, and consistency, as well as a lack of monitoring and supervision of the actions and duties of each policy implementer. Policy objectives will also be impacted by the impact of policies and programs whose degree of implementation is subpar due to late delivery. The non-cash food assistance aid program is being implemented in Takalar Regency with the help of the local administration as a way of showing care for the less fortunate. During the social, economic, and health challenges brought on by the Covid-19 pandemic in Indonesia, this is meant to strengthen community resilience. Therefore, the Ministry of Social Affairs of the Indonesian National Government and the Takalar Regency Government must develop and improve collaborative communication to carry out programs in achieving community welfare. Local governments are also capable of carrying out the government's socialization program to ensure its long-term viability. In order to combat poverty in the regions through effective implementer communication, this research advises local governments to implement a non-cash food assistance (*Bantuan Pangan Non-Tunai,* BPNT) aid program. Problems with the implementation of the (*Bantuan Pangan Non-Tunai,* BPNT) program in overcoming poverty in Takalar Regency, Indonesia, are caused by actor communication and supervision that are not optimal. Therefore, it is necessary to increase supervision and monitoring from both the government and other institutions in the implementation process to be the key to the successful implementation of this program throughout Indonesia.

References

1. Osborn D, Cutter A, Ullah F (2015) Universal sustainable development goals understanding the transformational challenge for developed countries report of a study by stakeholder forum May 2015. https://sustainabledevelopment.un.org/content/documents/733FutureWeWant.pdf
2. Plagerson S (2021) Mainstreaming poverty, inequality and social exclusion: a systematic assessment of public policy in south Africa. Dev South Afr:1–17. https://doi.org/10.1080/0376835X.2021.1993793
3. Bekenova A (2022) Cross-country analysis of the impact of corruption and other factors on youth presence in parliament. J Contemp Govern Publ Policy 3(2):71–87. https://doi.org/10.46507/jcgpp.v3i2.73
4. Cimadamore AD (2016) Global justice, international relations and the sustainable development goals' quest for poverty eradication. J Int Comparative Social Policy 32(2):131–148. https://doi.org/10.1080/21699763.2016.1198267
5. Liu QQ, Yu M, Wang XL (2015) Poverty reduction within the framework of SDGs and post-2015 development agenda. In: Advances in climate change research. Elsevier. https://www.sciencedirect.com/science/article/pii/S1674927815000489
6. Abdoellah OS (2016) Pembangunan berkelanjutan di Indonesia: Di persimpangan jalan. Gramedia Pustaka Utama
7. Abdillah A, Ahmadi Z (2022) Tata Kelola Berkelanjutan: Studi Tentang Idealitas Model Tata Kelola Daerah Sulawesi Selatan, Indonesia. Kybernology J Gov Stud 2(1):44–69. https://doi.org/10.26618/kjgs.v2i1.8433
8. Abdillah A, Widianingsih I, Buchari RA, Mustari N, Saleh S (2022) Governance and Quintuple Helix innovation model: Insights from the local government of East Luwu Regency, Indonesia. Front Climate 4(1012108):1–15. https://doi.org/10.3389/fclim.2022.1012108
9. Irhamsyah F (2019) Sustainable development goals (SDGs) dan Dampaknya Bagi Ketahanan Nasional. Jurnal Kajian Lemhannas RI 38:45–54. https://jurnal.lemhannas.go.id/index.php/jkl/article/view/71
10. Sofo F, Wicks A (2017) An occupational perspective of poverty and npoverty reduction. J Occup Sci 24(2):244–249. https://doi.org/10.1080/14427591.2017.1314223
11. Ishartono O, Raharjo ST (2016) Sustainable development goals (SDGS) Dan Pengentasan Kemiskinan. Soc Work J 6(2):154–272. https://doi.org/10.24198/share.v6i2.13198
12. Permensos (2018) Peraturan Menteri Sosial tentang Penyaluran Bantuan Pangan Non-tunai [online]. Available at https://peraturan.bpk.go.id/Home/Details/129449/permensos-no-20-tahun-2019. Accessed on Jan 2023
13. Gibson J, Olivia S (2020) Direct and indirect effects of covid-19 on life expectancy and poverty in Indonesia. Bull Indones Econ Stud 56(3):325–344. https://doi.org/10.1080/00074918.2020.1847244
14. Putera PB, Widianingsih I, Ningrum S, Suryanto S, Rianto Y (2022) Overcoming the COVID-19 pandemic in Indonesia: a science, technology, and innovation (STI) policy perspective. Health Policy Technol 11(3):100650. https://doi.org/10.1016/j.hlpt.2022.100650
15. Suryahadi A, Al Izzati R, Suryadarma D (2020) Estimating the impact of covid-19 on poverty in Indonesia. Bull Indones Econ Stud 56(2):175–192. https://doi.org/10.1080/00074918.2020.1779390
16. Dithen Kependudukan dan Catatan Sipil (2021) Daftar Desa & Keluarahan di Kecamatan Sanrobone Kabupaten Takalar [online]. Available at https://www.nomor.net/_kodepos.php?_i=desa-kodepos&sby=000000&daerah=Kecamatan-Kab.-Takalar&jobs=Sanrobone. Accessed on Jan 2023
17. Badan Pusat Statistik (2022) Jumlah penduduk (Jiwa) Sulawesi Selatan, 2020–2022 [online]. Available at https://sulsel.bps.go.id/indicator/12/83/2/jumlah-penduduk.html. Accessed on Jan 2023
18. Badan Pusat Statistik (2023) Penerima Bantuan Pangan Non Tunai (BPNT)/Program Sembako [online]. Available at https://sirusa.bps.go.id/sirusa/index.php/variabel/8625. Accessed on Jan 2023

19. Fauzia M (2021) Penduduk Miskin Indonesia Naik Jadi 27,55 Juta akibat Covid-19, Tren Penurunan Kemiskinan. kompas.com. https://money.kompas.com/read/2021/02/16/073400926/penduduk-miskin-indonesia-naik-jadi-27-55-juta-akibat-covid-19-tren-penurunan?page=all
20. Bahri M (2020) The impacts of covid-19 on Indonesian poverty and unemployment (no. mz849). Center for Open Science
21. Hausmann R (2020) Flattening the COVID-19 curve in developing countries, project syndicate, 24 Mar
22. Anggela NL (2023) 6 Bansos yang lanjut pada 2023: PKH Hingga Kartu Pekerja [online]. Available at https://ekonomi.bisnis.com/read/20230101/9/1613867/6-bansos-yang-lanjut-pada-2023-pkh-hingga-kartu-prakerja. Accessed on Jan
23. Miles MB, Huberman AM, Saldaña J (2018) Qualitative data analysis: a methods sourcebook. Sage Publications
24. Woolf NH, Silver C (2017) Qualitative analysis using NVivo: the five-level QDA® method. Routledge
25. Komenko PMK (2019) Konsisten Penyaluran BPNT [online]. Available at https://www.kemenkopmk.go.id/konsistensi-penyaluran-bpnt. Accessed on Jan 2023
26. Harini S, Paskarina C, Rachman JB, Widianingsih I (2022) Jogo Tonggo and Pager Mangkok: synergy of government and public participation in the face of COVID-19. J Int Women's Stud 24(8):5. https://vc.bridgew.edu/jiws/vol24/iss8/5
27. Hupe P (2021) Implementing public policy: an introduction to the study of operational governance. http://digital.casalini.it/9781529766523
28. Mohr JJ, Fisher RJ, Nevin JR (1996) Collaborative communication in interfirm relationships: moderating effects of integration and control. J Mark 60(3):103–115. https://doi.org/10.1177/002224299606000307
29. Schultz RJ, Evans KR (2002) Strategic collaborative communication by key account representatives. J Personal Selling Sales Manage 22(1):23–31. https://doi.org/10.1080/08853134.2002.10754290
30. Van Meter DS, Van Horn CE (1975) The policy implementation process: a conceptual framework. Administr Soc 6(4):445–488. https://doi.org/10.1177/009539977500600404
31. Kinyondo A, Pelizzo R (2022) Evaluating the performance of parliamentary committees. Otoritas: Jurnal Ilmu Pemerintahan 12(1):1–15. https://doi.org/10.26618/ojip.v12i1.7319

Communication Technology for Information Exchange Using Short-Range Unmanned Aerial Vehicles

V. Kashtanov, V. Nemtinov, Yu Protasova, V. Morozov, and K. Nemtinov

Abstract In this article, a review on the use of unmanned aerial vehicles (UAVs) for retransmission has been conducted. The description of the basic network architecture and the main characteristics of communication channels are given. Various types of data transmission using UAVs, an approach to the organization of communication using unmanned aerial vehicles—short-range repeaters, which allows for information exchange in emergency situations, are considered. The general principles of UAV operation when used for weather monitoring, forest fire detection, traffic jam tracking, cargo transportation, use in rescue operations, etc., are analyzed. The advantages and disadvantages of such use of UAVs, the general architecture of communication systems and the place of the UAV in it, as well as the general characteristics of communication channels and types of data transmission are described. The technology was implemented using the example of establishing communication between separate units during training in a mountainous area using a low-flying UAV of the DJI Mavic Pro 2 type with suspended equipment for retransmission of radio signals SURECOM SR-112.

Keywords Architecture of signal transmission · Unmanned aerial vehicles of short range · Organization of communication in emergency situations · Retransmission of signals · Characteristics of communication channels

V. Kashtanov · V. Nemtinov (✉) · V. Morozov · K. Nemtinov
Department of Computer-Integrated Systems in Mechanical Engineering, Tambov State Technical University, 106 Sovetskaya Street, Tambov 392000, Russian Federation
e-mail: nemtinov.va@yandex.ru

Y. Protasova
Department of Management, Service and Tourism, Tambov State University Named After G. R. Derzhavin, 33 Internatsionalnaya Street, Tambov 392000, Russian Federation

G. Rajakumar et al. (eds.), *Intelligent Communication Technologies and Virtual Mobile Networks*, Lecture Notes on Data Engineering and Communications Technologies 171,
https://doi.org/10.1007/978-981-99-1767-9_7

1 Introduction

Currently, various services that ensure the national security of the Russian Federation are showing great interest in complexes of aerial surveillance and monitoring of natural environment and technical facilities using small-sized unmanned aerial vehicles (UAVs).

Due to the high degree of mobility and low cost, UAVs have been widely used in various fields over the past few decades [1–3]. First of all, UAVs have always been used in the military industry, performing various tasks in hostile territory and reducing losses among pilots. With a decrease in the cost and miniaturization of their components, small-sized UAVs (with a mass usually not exceeding 25 kg) have become available for wider use for civil and commercial purposes: weather monitoring, forest fire detection, traffic jam tracking, cargo transportation, rescue operations, retransmission of communication signals, etc. [4]. Unmanned aerial vehicles can be divided into two categories: aircraft and helicopter type. Each of them has its advantages and disadvantages. For example, aircraft-type UAVs have high speed and are capable of carrying a lot of weight, but they must maintain constant forward movement in order to retain altitude. As a result, they cannot be used for a number of tasks, such as a thorough inspection of the terrain. Helicopter-type unmanned aerial vehicles, on the contrary, have a limited degree of mobility and portable weight, but are able to move in any direction, as well as to stay at one point for a long time. Thus, the choice of an UAV depends on the purpose of its application.

Among the various areas of UAV application, their use as repeaters in wireless communication systems is one of the most important (Fig. 1) [5–7]. Communication systems using UAVs have the most important advantage—unlike ground-based signal relay systems, their operation cannot be hindered by shading from urban or mountainous terrain or damage to the communication infrastructure caused by natural disasters [8, 9].

It is worth noting that in addition to the use of UAVs, an alternative solution for providing wireless communication is the use of other objects, for example, balloons. They usually function in the stratosphere, which is about tens of kilometers above the earth's surface. However, UAVs (usually operating at an altitude not exceeding several kilometers) have a number of advantages over similar systems and their

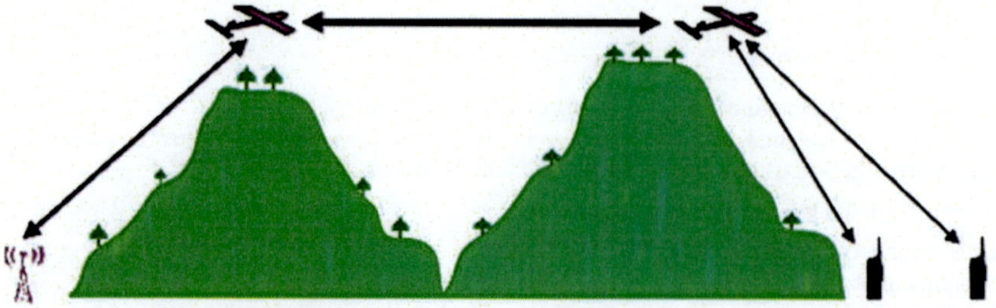

Fig. 1 Use of UAVs for signal retransmission

terrestrial and satellite counterparts. Firstly, their use is associated with lower costs and faster preparation, which makes them a suitable option for short-term use. In addition, the operation of UAVs for such purposes often provides direct visibility of communication facilities for UAVs, which leads to an increase in the quality of communication.

In addition, the high degree of maneuverability of such devices is also reflected in the improvement of communication due to the ability to adapt to the peculiarities of the terrain in which they operate. Moreover, such an option of UAV functioning is possible, when the quality of communication signal can influence the process of their flight. For example, if there is a stable connection with ground objects, the flight speed of the device will decrease to ensure the transmission of more data. Thanks to these advantages, UAVs can become an essential component of wireless communication systems in the future.

2 Areas of Application

Recently, UAVs have been used in many areas that involve the use of the device as a repeater in rescue operations, extinguishing forest fires, aerial photography, protection of territories, logistics, and many others. In all such applications, UAVs perform tasks in places remote from the ground communication infrastructure, which requires the presence of one or more devices performing relay functions.

The use of UAVs in rescue and search operations can significantly reduce the search time and reduce the number of victims in disasters and accidents. Unmanned aerial vehicles, due to their maneuverability, speed, and autonomy of actions, are able to perform tasks that cannot be performed by manned vehicles (Fig. 2). In the most typical scenario, UAVs are sent to a given territory, and data on the number of detected victims are collected using photo and other types of shooting and transmitted to rescue teams or signal relay systems, which can be presented by another group of UAVs.

Unmanned aerial vehicles have already proved their effectiveness in this field by helping rescue teams to narrow down the search area while avoiding possible dangers. In 2006, when searching for victims after the Hurricane Katrina, two UAVs were used to detect people trapped under the rubble.

Another promising option for using UAVs is to perform various tasks when extinguishing fires. Fire services highly appreciate the potential of using UAVs. Upon arrival of the fire brigade to the place of ignition, UAVs can be used to thoroughly examine the fire area, while reducing the risk for fire service workers. Unmanned aerial vehicles can be equipped with heat-resistant cameras which allow them to detect fires. In addition, such factors as smoke and poor visibility cannot interfere with their work. Also, these devices can use flashlights and spotlights to help the fire brigade when extinguishing a fire in low visibility conditions. After extinguishing the fire, UAVs are able to facilitate the assessment of fire damage. In addition,

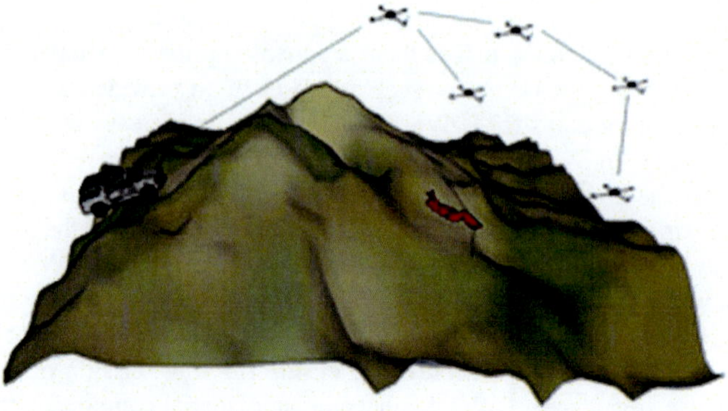

Fig. 2 Use of UAVs during the search operation

UAVs are able to detect forest fires at early stages, which facilitate the task of extinguishing them. So, in 2017, the UAV was used by the Los Angeles Fire Department to extinguish huge forest fires raging near the city [10, 11].

When extinguishing fires in remote places, it is also necessary to use relay systems. In the absence of appropriate infrastructure on the ground, the use of UAVs for these purposes is the most rational solution.

3 Basic Architecture of Signal Transmission

Figure 3 shows the general architecture of wireless communication with the use of UAVs, consisting of two types of information exchange channels—control and data transmission channels [5, 6].

Control channel. This communication channel is necessary for the successful and safe operation of the UAV. Such channels should provide reliable and secure two-way transmission of information with low latency, since the information transmitted with their help is critically important for the operation of the UAV. At the same time, the exchange of such information can occur both between different UAVs and between the UAV and the ground control point. The information transmitted via this channel can be divided into three main categories: commands transmitted by UAVs from ground control points, reports on the status of the device sent from the UAV to the ground, and information about obstacles and dangers transmitted between UAVs. Control channels are necessary even in the case of using autonomous unmanned aerial vehicles capable of performing work using on-board computers without the need for constant monitoring by the operator, as in emergency cases, there is still a need for manual control.

Since the information transmitted through the control channel is critically important, it must be transmitted using protected frequency spectra. To date, two frequency bands are used—the range L (960–977 MHz) and the range C (5030–5091 MHz) [12,

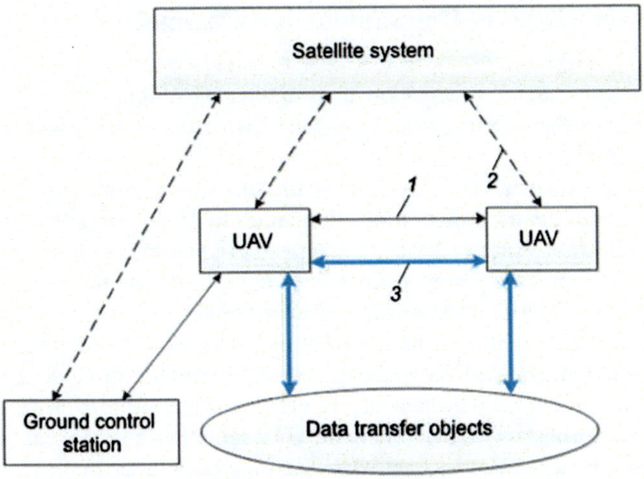

Fig. 3 General architecture of communication using UAVs: *1*, *2*, *3*—control channels: main, additional, and data transmissions, respectively

13]. It is important to note that despite the fact that UAVs are controlled primarily from ground stations, satellite systems are also used for this purpose, which increases reliability and quality of communication. Another key issue is security of control channels. The lack of effective security mechanisms can lead to the so-called "ghost control" of the UAV—a situation in which the control of the UAV is carried out by an unauthorized operator. As a consequence, reliable authentication methods are needed to gain access to UAV control systems.

Data transmission channel. These channels, on the contrary, are designed to transmit information related to the tasks performed by the UAV. Information is transmitted to ground terminals, which, depending on the task, can be ground stations, mobile terminals, data storage nodes, wireless sensors, etc. Data transmission channels must support the following communication modes: direct communication between UAVs and mobile terminals in cases of overload or failure of ground stations; communication between UAVs and ground stations and between UAVs and data storage nodes, as well as communication between two UAVs. Data transfer rate requirements depend on the target and can range from several kB/s for transmission between UAVs and sensors to tens of GB/s for communication between UAVs and data storage nodes. In comparison with control channels, data transmission channels do not have such strict requirements for possible delays and transmission security. As for the frequency range, it usually depends on the specific tasks of the UAV. For example, when providing cellular coverage, UAVs use LTE band frequencies. However, it is also possible to use other frequency ranges, for example, the millimeter wave range to ensure stable communication between two UAVs [14, 15].

4 Characteristics of Communication Channels

There are two types of information transmission: UAV–ground and UAV–UAV. Both types have characteristics that distinguish them from the transmission between ground systems.

UAV–earth transmission. This type of transmission is somewhat different, in comparison with the transmission of information to ground objects from manned aircraft [6, 8]. Unlike similar devices that transmit signals to ground stations located in an open area and have powerful antenna systems, UAVs are forced to transmit in areas where such infrastructure is usually absent. Despite the advantage of being able to establish a line of sight from the transmission object, various obstacles in the form of terrain, buildings, or the UAV frame can make transmission difficult. In particular, studies show that if a UAV maneuvers, the delay in data transmission caused by the appearance of an obstacle in the form of the UAV frame can be up to tens of seconds, which should be taken into account when planning UAV tasks. In the case of using low-flying UAVs, communication can also be difficult due to refraction, scattering, and reflection of the signal from mountains, the earth's surface, tree foliage, etc. When the UAV performs its tasks over the desert or the sea, only reflections from the surface interfere with signal transmission. At the same time, the resulting attenuations can be described by a Ricean attenuation model that characterizes the effect of the scattered component of the signal on transmission in the presence of a line of sight between transmission objects. Depending on the conditions of the area where the communication facilities are located and the frequency used, the degree of impact of the Ricean factors may vary. Thus, the ratio between the transmitted signals and their scattered components can take a value of about 15 dB for the L range and 28 dB for the C range when the UAV is operating in hilly terrain [15].

UAV–UAV-type transmission. When transmitting data between UAVs, the transmission objects are most often in line of sight. Despite the fact that the reflections of the signal from the surface can cause attenuation to some extent, their effect is much weaker than when transmitting from a UAV to the ground or when transmitting between objects on the ground. However, in this case, there is a much stronger influence of the Doppler effect than when transmitting a signal between a UAV and ground objects, due to the high values of the relative speed of the UAV. These factors must be taken into account when choosing the frequency range for communication between UAVs. On the one hand, due to the almost constant presence of a line of sight between transmission objects, it is possible to use the millimeter range in order to ensure stable communication between UAVs. On the other hand, when using this range, a high degree of Doppler frequency shift can be observed due to the high relative speed of the UAVs. Due to its peculiarities, this type of transmission is currently the subject of many studies aimed at finding effective methods of its application.

5 Organization of Communication When Using Short-Range Unmanned Aerial Vehicles as Repeaters

Joint actions of search teams and short-range (SR) UAVs make it possible to reduce the risks of losses when searching for missing people in hard-to-reach areas. At the same time, the key aspect is to ensure the required quality of information exchange between the members of the search expedition through the use of airmobile communication networks (ACNs) of the SR UAV. In this regard, the field of research related to aeronautical radio communications using UAVs is relevant. The basis of the information exchange between UAVs is the communication technologies which are being developed and constantly improved in Russia and abroad. The paper [9] presents the theoretical and practical results of research in the field of UAV application as repeaters of non-tethered and tethered types (aerostats and helicopter-type UAVs). The use of single UAV repeaters has become a new step in the development of radio communications and allows to quickly provide communication services in conditions where it is impossible or inappropriate to use other means of communication. The effect of using such repeaters on SR UAVs is an increase in the communication range.

Single relay platforms are further developed to form regional communication networks using UAVs, which are considered as an addition to the ground part of the communication system, as well as independent systems for relaying information flows. It is possible to single out a separate particular case when there is a need to ensure the information exchange between UAVs and interacting objects over long distances. In this case, the resources of single UAVs may not be enough, and the deployment of zonal communication networks is redundant. In this regard, it is required to develop methods of organizing communication using UAV groups in selected information directions, which make it possible to increase the possibilities of relaying. The main correspondents are participants in search operations with the use of SR unmanned aircrafts. The disadvantages of this method of organizing communication are the high vulnerability of the data transmission system to the failure of a single UAV repeater and insufficient communication range.

A typical situation that may arise while using of a group of UAVs may be the loss of radio communication with a separate group of UAVs or the ground control point (GCP) of the UAV. To solve the problem of restoring radio communications, it is proposed to use a method in accordance with which a radio link is built using UAVs as elements of the ACN. The radio link is a set of technical devices and a medium for the propagation of radio waves that ensure the transmission of messages from a source to a recipient.

The way to organize a radio link using UAV repeaters is to perform the following sequence of actions to solve a number of typical tasks:

(1) determination of the Azimuth to correspondent No. 2, relative to correspondent No. 1;
(2) launch of the UAV 1 in the direction of correspondent No. 2 under the control the operator;

(3) flight toward correspondent No. 2 carried out within the zone of reliable reception of radio signals via two channels: a command-telemetry radio link (KTR) from the UAV's GCP and an information exchange channel from correspondent No. 2 (receiving equipment is located together with the UAV's GCP);
(4) during the flight, the on-board communications of UAV 1 constantly search for correspondent No. 2 on the air to establish radio contact. If it is possible to establish communication with correspondent No. 2 within the zone of reliable radio reception between UAV 1 and GCP, the loitering zone is optimized taking into account the provision of the required communication quality in the sections of the radio link of GCP (correspondent No. 1)—UAV 1 and UAV 1—correspondent No.2;
(5) in the case when it is not possible to establish radio contact with the second correspondent within the zone of reliable radio reception of the GCP-UAV 1, UAV 1 proceeds to loitering at the maximum distance from the GCP in the direction of correspondent No. 2 and the launch of UAV 2 in the direction of UAV 1 is done;
(6) during the flight of UAV 2, its on-board means constantly search for correspondent No. 2 on the radio to establish radio communication and maintains contact with UAV 1, as UAV 2 approaches UAV 1, the KTR and information exchange channels of UAV 2 connect with the GCP and correspondent No. 1 via UAV 1;
(7) after the establishment of a radio exchange of UAV 2 with GCP and correspondent No. 1 through UAV 1, UAV 2 is flown toward correspondent No. 2 within the zone of reliable reception of radio signals from UAV 1.

The inclusion of new UAVs in the process of establishing radio communications continues until radio communications are established between correspondents No. 1 and No. 2. To determine the trajectory parameters of UAV repeaters, it is proposed to use geometric relationships taking into account the effect of wind disturbances, as well as radio engineering parameters of the communication equipment of control points, remote UAV repeaters, ground correspondents, and on-board equipment for relaying signals from UAV repeaters, such as directional diagram and antenna gain power, power of the receivers and transmitters, signal-to-noise ratio.

The developed method provides for the organization of communication between the correspondents by deploying a radio link in a limited period of time (the period of time required to search for people in hard-to-reach areas) in areas that are not equipped with respect to communications, areas in which it is difficult to deploy and use ground-based communications (difficult terrain, swampiness, fires, impassable water and mountain areas, forest blockages, etc.).

The transfer of information between the correspondents is carried out via a composite channel organized with the help of standard devices located on the UAV. The advantage of the proposed method is the possibility of information exchange between the correspondents who are outside the line-of-sight zone in the required period of time.

6 Adaptation and Practical Testing of Methods of Using UAVs for Relaying Communication Signals

The methods of using UAVs presented above are adapted for relaying communication signals and were applied during military exercises, the main purpose of which was to check the degree of readiness of the leadership and military management bodies of the joint strategic command of a military district, the command staff of military contingents, tactical (operational) groups to manage groups of troops (forces), formations and units in the course of joint operations to localize and resolve armed conflicts related to countering terrorism.

The main part of the exercises took place in a mountainous area. The presence of mountainous terrain often makes it difficult to use traditional communication schemes, since mountain structures are usually an obstacle in the way of radio waves. In such conditions, the use of UAVs for these purposes is the most rational step (Fig. 4).

During the tests of UAVs, the task was to establish communication with the units on the other side of the mountain range. Attempts to establish a connection by traditional methods did not bring any significant results. Obviously, the reason was in the mountain structures reflecting and scattering radio waves. For this reason, a low-flying UAV of the DJI Mavic Pro 2 type with attachments for relaying SURECOM SR-112 radio signals was used.

Such a step ensured the successful completion of the task—stable and clear (speech intelligibility 100%) radio communication was established between the units separated by the mountain range.

An experimental evaluation of the proposed method of communication using standard radio communication channels of the SR UAV was also carried out for the case of restoring information exchange with a remote search group in hard-to-reach areas. The experiment was carried out under the following meteorological conditions: wind speed at different height levels from 3 to 25 m/s with gusts up to 28 m/s; the minimum air temperature is – 9 °C. Heavy icing was observed on one of the UAV relays.

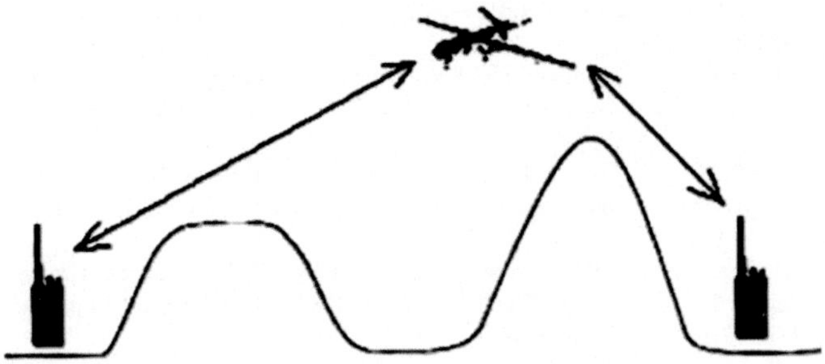

Fig. 4 Using UAVs to relay signals in mountainous terrain

The implementation of the developed method provided reliable communication using the SR UAV at a distance of several hundred kilometers in adverse weather conditions. The experimentally measured communication range is at least 2.5 times higher than the range of a standard SR UAV communication channel.

Thus, the result of applying the method of organizing communication based on composite communication channels built using UAV repeaters can be the implementation of an additional possibility of carrying out information exchange of subscribers in the required period of time (the period of time required to perform a search operation in hard-to-reach areas) in areas not equipped for communications, as well as in areas where it is difficult to deploy and use ground-based communications, as well as in case of emergency restoration of communications with UAV groups.

7 Conclusion

The general principles of the operation of UAVs when they are used for weather monitoring, detection of fires in forests, tracking traffic jams, cargo transportation, rescue operations support, etc., are analyzed. The advantages and disadvantages of such use of UAVs, the general architecture of communication systems and the place of UAVs in it, as well as the general characteristics of communication channels and types of data transmission are described.

One of the promising areas of UAV use is the retransmission of signals during data transmission. The methods of using UAVs described above are adapted for relaying communication signals and have been tested when establishing communication between separate units during training in a mountainous area. Further research in this subject can ensure the improvement of wireless communication systems and more efficient use of UAVs in such systems.

References

1. Valavanis KP, Vachtsevanos GJ (2015) Handbook of unmanned aerial vehicles. Springer, Netherlands, 3022 p
2. Shafique A, Mehmood A, Elhadef M (2021) Survey of security protocols and vulnerabilities in unmanned aerial vehicles. IEEE Access 9:46927–46948. https://doi.org/10.1109/ACCESS.2021.3066778
3. Zheng Y, Li S, Xing K, Zhang X (2021) Unmanned aerial vehicles for magnetic surveys: a review on platform selection and interference suppression. Drones 5(3). https://doi.org/10.3390/drones5030093
4. Unmanned aircraft system (UAS) service demand 2015–2035: literature review & projections of future usage. US Department of Transportation, 2013, 151 p. https://info.publicintelligence.net/USAF-UAS-ServiceDemand.pdf
5. Zeng Y, Zhang R, Lim TJ (2016) Wireless communications with unmanned aerial vehicles: opportunities and challenges. IEEE Commun Mag 54(5):36–42. https://doi.org/10.1109/MCOM.2016.7470933

6. Filatenkov PA, Sergun TA, Vinokurov AA, Morozova EV (2019) The main technical problems of unmanned aerial vehicles on the example of multicopters. In: Proceedings of the 2019 IEEE conference of Russian young researchers in electrical and electronic engineering, ElConRus 2019. Institute of Electrical and Electronics Engineers Inc., Saint Petersburg, Moscow, pp 500–504. https://doi.org/10.1109/EIConRus.2019.8656812
7. Kovbasiuk S, Kanevskyy L, Chernyshuk S, Romanchuk M (2020) Detection of vehicles on images obtained from unmanned aerial vehicles using instance segmentation. In: Proceedings—15th international conference on advanced trends in radio electronics, telecommunications and computer engineering, TCSET 2020: 15, Lviv-Slavske, pp 267–271. https://doi.org/10.1109/TCSET49122.2020.235437
8. Merwaday A, Guvenc I (2015) UAV assisted heterogeneous networks for public safety communications. In: Proceedings IEEE wireless communications and networking conference workshops, 09–12 Mar 2015, New Orleans, LA, USA, pp 329–334. https://doi.org/10.1109/WCNCW.2015.7122576
9. Parfiryev AV, Dushkin AV, Dubrovin AS, Stepanov LV (2019) Control of unmanned aerial vehicles based on the detection algorithm. J Phys Conf Ser 1202:12014. https://doi.org/10.1088/1742-6596/1202/1/012014
10. Vimalkular R, Shaw KK (2020) Design and development of heavy drone for fire fighting operation. Int J Eng Techn Res 9(6):572–576. https://doi.org/10.17577/IJERTV9IS060433
11. Ahmed H, Talib MA, Abbas S et al (2022) Unmanned aerial vehicles (UAVs) and artificial intelligence (AI) in fire related disaster recovery: analytical survey study. In: International conference on business analytics for technology and security, ICBATS 2022, Dubai. https://doi.org/10.1109/ICBATS54253.2022.9759056
12. Matolak DW, Sun R (2015) Unmanned aircraft systems: air-ground channel characterization for future applications. IEEE Vehic Technol Mag 10(2):79–85. https://doi.org/10.1109/MVT.2015.2411191
13. Petrov ID, Shkodyrev VP (2021) Processing echo signals reflected from unmanned aerial vehicles and received by radar. T-Comm 15(9):48–54. https://doi.org/10.36724/2072-8735-2021-15-9-48-54
14. Rappaport TS, Heath RC, Daniels J, Murdock JN (2014) Millimeter wave wireless communications. Prentice Hall, 704 p
15. Denisov A, Levonevskiy D (2021) Algorithm for radio survey of the cyber-physical systems operating areas using unmanned aerial vehicles. Lecture notes in computer science. 12998 LNAI, pp 40–49. https://doi.org/10.1007/978-3-030-87725-5_4

Evaluative Aspects of Wireless Sensor Networks Reliability Model: A Case Study

D. Lingamaiah, D. Krishna Reddy, and Perumalla Naveen Kumar

Abstract Reliability (information management) and lifetime (energy management) are the two most important factors for the real-time application of WSNs. The reliability and lifetime must be managed properly as there are on-demand individual networks, subnets, extended networks, and other real-time applications of WSNs. The overall reliability of a wireless sensor network (WSN) depends on its lifetime, which in turn depends on battery-powered nodes, node-sink distances, and traffic near the sink. The referred literature proved that energy efficiency and energy balancing depend upon optimized node-sink distances that are obtained by a proactive heuristic ACO (ant colony optimization) routing algorithm. This algorithm also finds the minimum value of average nodal energy consumption. The efficient and balanced energy paths lead to good packet delivery. The majority of the referred literature uses a programmed random node deployment strategy and a clustered WSN model. This work uses a programmed manual node deployment strategy with uniform and random procedures and a simple sectored network model. The performance comparison results are placed with respect to different strategies such as node deployment planning, different transmission planning, different routing methods, different network sizes, different energy levels, and iterations (rounds). This comparison paves the way for futuristic applications of WSN. The applications include subnets, extended networks, real-time patient monitoring systems, and driverless cars.

Keywords Ant colony algorithm · Balanced energy · Cloud computing · Energy efficiency · Heuristic algorithm · Optimization · Pheromone intensity · Quality of service · Routing protocols · Network lifetime · Wireless sensor networks

D. Lingamaiah (✉) · P. N. Kumar
Department of ECE, University College of Engineering, Osmania University, Hyderabad, Telangana, India
e-mail: lingam22daggu@gmail.com

D. Krishna Reddy
Department of ECE, Chaitanya Bharathi Institute of Technology, Hyderabad, Telangana, India
e-mail: dkrishnareddy_ece@cbit.ac.in

G. Rajakumar et al. (eds.), *Intelligent Communication Technologies and Virtual Mobile Networks*, Lecture Notes on Data Engineering and Communications Technologies 171,
https://doi.org/10.1007/978-981-99-1767-9_8

1 Introduction

Reviewing the initial article [1] that claimed proactive methodology was not appropriate for WSN due to the numerous nodes that necessitate a corresponding routing table with memory, and the most recent articles [2, 3] that created an algorithm to enhance WSN's performance based on prior observations. Moving further and reviewing articles [4–6], this paper establishes the fact that, using efficient and balanced energy concepts, the proactive method performs better by employing programmed manual node deployment with uniform and random procedures. The network has a number of nodes and a single sink with unlimited energy resources. It collects the aggregated data and then transmits it to the external network. The idea is to apply the proposed protocol to a single individual network, and then the idea can be extended to other networks to get greater scalability, stability, reliability, and network lifetime for the overall network configuration. Going by the referred literature [2–6], a disk-shaped network model (Fig. 1) having a radius R, uniform nodal density, nodes with limited energy, and a single stationary sink with enough power acting as a base station. Each model can be further divided into a sectored model having N disconnected concentric coronas with width $w = R/N$. The angle $\theta = 30°$ can be seen dividing the area into sectors S_i, S_2, S_N [2–6], resulting into a simple sectored model.

Nodes are assumed to have initial energy ε_0 and self-generating data of φ bits. Multiple stationary nodes are located on the network with finite battery power. Nodes are supposed to have a maximum transmission range which is divided into K levels with a unit range corresponding to corona thickness ω. Self-generated data volume of the nodes is computed, which is assumed to be the same for every sector. An ant

Fig. 1 Network model

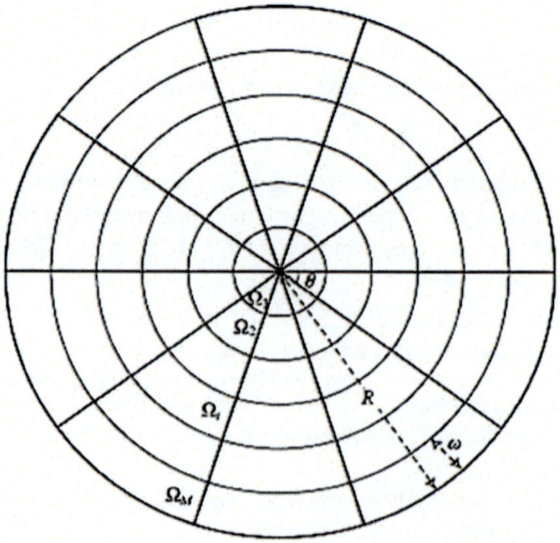

is assumed to be placed in every sector and the movement of the ant is in accordance with a specific probability. The farthest ant moves first and the movement stops with the last ant's last hop. Every hop is assessed for less energy use, thereby setting an optimal transmission distance d of x hops. The energy-efficient and energy-balanced distances (EEAEB), for every sector, are estimated to minimize the total energy usage along the transmission path. If all the nodes use the same transmission distance, then the overall energy usage will be reduced. Total data volume is estimated which is comprised of self-generated data volume [4, 5] and received data volume [4, 5]. The average value of energy consumption per node (ECPN) [4–6] is evaluated for every sector. Pheromone intensity is estimated on every path and it is updated after each iteration. And the final result will be a path having minimum value of average maximum value of ECPN which governs the optimal path leading to maximization of network lifetime.

Efficient communication represents the maximum information delivery that is possible if energy is handled properly for the intended distances. The problem of having an optimum path from source to sink is solved by using the ant colony algorithm. The spread of pheromones is an ant's natural intelligence and depending upon pheromone intensity and visibility, the ant lays a path that will be optimized in its own way. The basic ant colony algorithm (7) is suitable for smaller networks but suffers over the larger network area, and thus a proposed heuristic approach [7, 8] is used which is suitable for extended network areas.

For efficient running of WSN and to have a better network lifetime [9–11], energy conservation of sensors is a must, such as to have reliable delivery of detected information to the destination. The transmission and traffic handling mechanisms are to be handled favorably in order to achieve energy efficiency, load-traffic balancing, and network lifetime. Distances and areas of the network do matter the most as traffic is seen more near to the sink as compared to the farthest part of the network. Many passive methods are used to address these issues. The distance-based approaches such as single-hop (SH), a direct transmission scheme, and a multi-hop (MH), with fixed transmission distances, are presented in [12, 13], and a clustering-based low-energy adaptive hierarchy (LEACH) is presented in [14]. These algorithms are designed to address the network load, energy consumption, residual energy, and packet collisions. Some of the methodologies take into account secondary cluster head selection using residual energy and minimum average differences, thereby distributing the load for better efficiency and the lifetime of the network. Addressing the clustering problem, a hybrid energy-efficient distributed clustering approach (HEED) [15] is presented which uses a hybrid methodology comprising the node's residual energy and its closeness with the neighbors and sink. Lifetime is improved by data aggregation. Ant colony optimization (ACO) addresses the multi-objective optimization problem of WSN [15]. An energy-aware ACO protocol (EAACA) is presented in [16], in which the transmission is in accordance with the nodal average energy and residual energy. The energy-efficient ant-based routing (EEABR) algorithm [17] establishes forward and backward trails of ants using pheromone probability and sets up an optimized route tree. All these are the passive methods and lag the traffic handling mechanisms [18]. The distances between the source node to the sink, transmission range, and traffic matter the most in addressing the network lifetime. The proposed methodology addresses nodal energy consumption, energy efficiency, and traffic balancing through efficient and balanced energy distances and improves the overall network lifetime.

Main contributions and innovations of this paper include the following aspects: (1) The works lays a foundation for futuristic aspects such as WSN-Cloud integration (driverless cars, real-time patient monitoring), subnets, and extended networks. (2) Information transmission and reception are based on a simple sectored-based approach instead of a clustered approach. (3) Node and base station distances are optimized using ACO, resulting in topology optimization and energy reliability. (4) Programmed manual nodal deployment strategy is used and compared with the default programmed random procedure. (5) Networks lifetime is maximized by preserving the data flow (self + incoming) and optimized energy routes. (6) Each node is assumed to be delivering the same amount of information at any given time.

The rest of the paper layout is as follows: Sect. 2 presents the mathematical modeling with relevant equations, Sect. 3 validates the results, and Sect. 4 concludes the work with a note on future aspects.

2 Network Model

Four mathematical aspects, viz., transmission model, distance model, energy model, and routing model lead to network lifetime maximization. Finding the optimized distances through ACO and thereby finding the minimum value of maximum average nodal energy consumption (ECPN) are the algorithmic needs. The circuit and network parameters used in the modeling and simulations are assumed, viz., network radius range 25–35 m, no. of nodes 100, nodal density 5, initial nodal energy 10 J, unit transmission range 5 m, evaporation rate 0.05, and control parameters $\alpha = 5, \beta = 5$. The nodal energy required (consumed) to transmit and receive m bits of information at a distance d with path loss γ is given by [4, 5, 19, 20]:

$$E_{Tx}(m, d) = m\varepsilon_{\text{elec}} + m\varepsilon_{\text{amp}}d^{\gamma} \tag{1}$$

$$E_{Rx}(m) = m\varepsilon_{\text{elec}} \tag{2}$$

where $\varepsilon_{\text{elec}}$ (50 nJ/bit) is the energy consumed in information transmission involving coding, modulation, filtering, etc. ε_{amp} (13 pJ/bit/m^2) denotes energy consumed by transmitting amplifier [4, 5]. The energy model can be had from Eqs. (1) and (2) that gives, the average value of the energy consumed per node $(\overline{E_i})$ for a communication of sector S_i:

$$\overline{E_i} = \frac{E_{Rx}(\Phi_i) + E_{Tx}\left(\Xi_i, d_{ij}\right)}{C_i} \tag{3}$$

$$\overline{E_i} = \frac{\Phi_i \varepsilon_{\text{elec}} + (\Phi_i + \Xi_i)\varepsilon_{\text{elec}} + (\Phi_i + \Xi_i)\varepsilon_{\text{amp}}[(i - j)\omega]^{\gamma}}{C_i} \tag{4}$$

where C_i is the sector node number, Ξ_i is the initial data volume, and Φ_i is the incoming data volume and d_{ij} is the distance of sectoral movements of ants from S_i to S_j that sets up a trail $T(i, j)$ and is expressed as:

$$d_{ij} = (i - j)\omega \tag{5}$$

where ω is the corona width.

The distance model can be arrived at by assuming the values of $\varepsilon_{\text{elec}}$, ε_{amp}, γ, ω, σ, θ, and article [3–5], and the equation for the energy efficient (Eq. 6) and energy-balanced distances (Eq. 7) for every sector are given by:

$$d = \sqrt[\gamma]{\frac{2\varepsilon_{\text{elec}}}{(\gamma - 1)\varepsilon_{\text{amp}}}} \tag{6}$$

$$d_{ij} = \sqrt[\gamma]{\frac{C_i\vartheta - (2\Phi_i + \Xi_i)\varepsilon_{\text{elec}}}{(\Phi_i + \Xi_i)\varepsilon_{\text{amp}}}} \tag{7}$$

where ϑ is the nodal energy consumption in every sector and is given by:

$$\vartheta = \frac{\Phi_i \varepsilon_{\text{elec}} + (\Phi_i + \Xi_i)\varepsilon_{\text{elec}} + (\Phi_i + \Xi_i)\varepsilon_{\text{amp}} d_{ij}^{\gamma}}{C_i} \tag{8}$$

In the routing model [4, 5, 21], the movement of the ant starts with the farthest ant first and then terminates with the last ant's last hop. The data is transmitted from the source node toward the sink and the algorithm utilizes heuristic information [8, 21], pheromone intensity, and probability of transition in order to find a best routing abstraction. In the referred methodology, the pheromone and the heuristics are related to data volume and transmission range. The transition probability of the ant for its movement from sector S_i to sector S_j after ith iteration is given by:

$$\mathrm{P}_{ij}(t) = \frac{[\mathrm{I}_{ij}(t)]^{\alpha}[\mathrm{H}_{ij}(t)]^{\beta}}{\sum_{S_r \in DIS_j}[\mathrm{I}_{ij}(t)]^{\alpha}[\mathrm{H}_{ij}(t)]^{\beta}} \tag{9}$$

where $I_{ij}(t)$ is the pheromone concentration, $H_{ij}(t)$ is the heuristics (visibility) of the trail $T(i, j)$. α is the pheromone control parameter, β is the heuristics control parameter, ψ_1 and ψ_2 are the energy-efficient, and the energy-balanced distance controllers and constants $\lambda_1 > 0$ and $\lambda_2 > 0$ ensure each denominator is not zero.

The network evaluation is performed by assessing the average energy consumption per node, which is the ratio of initial energy to the maximum average energy consumption per node. The network lifetime is usually expressed in terms of the ratio of the sum of the total initial energy and average nodal energy consumption. Mathematically given as: (after ith iteration) [4, 5, 7]:

$$f(t) = \frac{\varepsilon_0}{\max\{\overline{E_i}(t), i = 1, 2, ...M\}} \tag{10}$$

where ε_0 is the initial energy, $\overline{E_i}(t)$ is the average energy consumption per node [4] of a sector S_i, and max $(\overline{E_i}(t))$ is the maximum average energy consumption per node [4] for different sectors.

2.1 Algorithmic Steps

The algorithmic steps are given below and the respective results are got by running the m-script.

Set the physical and network parameters Initialize the parameters Set the sector model and the network model Deploy the nodes
Get simulate values for the energy- efficient and energy-balance distances
Move the ants
Get simulated transmit data volume Get simulated received data volume Get simulated total data volume
Get simulated transition probability Get simulated nodal energy consumption Get simulated overall network lifetime Get the simulated graphs
End

3 Simulation and Validations

Having the algorithmic mathematical aspects and the circuital network parameters in hand, the simulations are done with respect to node deployment and transmission strategies. The simulation results are with respect to energy consumption, network lifetime, throughput and number of live nodes. MATLAB-2014b is used for simulations running the m-script.

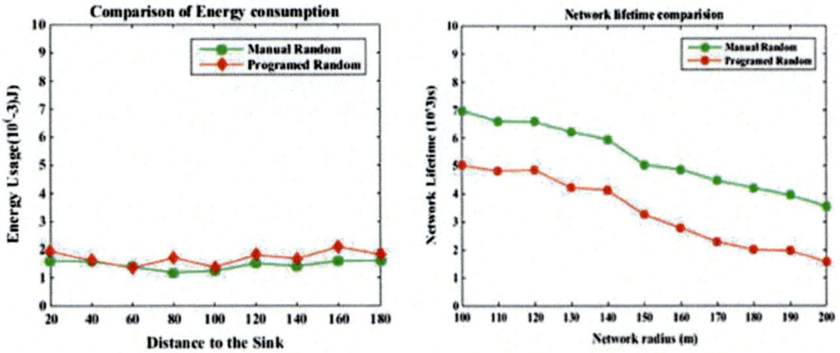

(a) Energy consumption (b) Network Lifetime

Fig. 2 Simulation results to show impact of node deployment strategy. **a** Energy consumption. **b** Network lifetime

3.1 Case 1—Impact of Node Deployment Strategy

Most of the literature [2–17, 19, 20, 22–24] is based on conventionally programmed random nodal deployment but, this work opts for programmed manual deployment in which nodes are placed manually at the stroke of a mouse. The comparative results consider energy consumption and network lifetime. The comparison plots for energy consumption and network lifetime are shown in Fig. 2a, b. As can be seen from the plots, the programmed manual node deployment performs better than the programmed random node deployment because it follows the optimum energy-efficient and stable paths that are arrived at during the considered iterations (rounds), whereas energy consumption is comparatively higher in programmed random nodal deployment due to repetitive iterations. These iterations are for optimum path findings. This impacts the overall network lifetime, as shown in Fig. 2b. The programmed random strategy has a shorter lifetime as compared with the manual random node deployment strategy.

3.2 Case 2—Impact of Routıng Methods on Network Lıfetıme and Energy Consumptıon.

The literature finds routing protocols classified as direct, flat, and cluster based. The protocols considered for the validation, in this case, are single-hop (direct) [12], EAACA (flat) [16], and LEACH (clustered) [14], and the respective comparative plots are shown in Fig. 3. Figure 3a shows an initial five-sectored model [4, 5, 24], and as per the algorithmic insights brought out in introduction part, Fig. 3b shows manual placement of nodes. Figure 3c, d show movements of the ant with the farthest ant's first move and last ant's last hop which terminates the movement, respectively.

The distance estimation parameters such as θ, γ, σ, ω, φ, μ, $\varepsilon_{\text{elec}}$, and ε_{amp} (Eqs. 9 and 10) applied are same for all the protocols but, the difference is with respect to estimation of the distance from source to sink. The proposed algorithm establishes the path with the application of ACO, which gives out the optimal path with the help of the ant's intelligence (pheromones and heuristics), whereas the path in the other protocols is established by using nodal residual energy, which differs along the path. This difference in energy along the path creates an energy-hole problem, and thus the network cuts off. The simulated distance estimations are shown in Table 1, which shows that the value of the balanced distance is the lowest for the proposed protocol, resulting in less energy consumption and a longer lifetime.

It is a known fact that the overall network lifetime is the ratio of sum of the initial energy to the total energy dissipated in one round. The initial energy is kept uniform and the average energy consumption per node is evaluated. Table 2 presents the overall network lifetime.

The derived fact is that the transmission power is proportional to the square of the distance [4, 5] and it can be seen from Fig. 3e, SH (from 1.2×10^3 to 8.5×10^3 J) [12] has more energy consumption due to long distance, basic LEACH (from 3.2×10^3 to 1.5×10^3 J) [14] and EAACA (from 2×10^3 to 2.5×10^3) [16] also perform below the proposed protocol due to unbalanced energy resulting into unbalanced traffic handling capacity near the sink. The proposed methodology gives balanced energy consumption (from 1.5×10^3 to 1.5×10^3 J) and better lifetime (Table 2) because of balanced energy all along the transmission path. If all the nodes follow these balanced distances to send information, then energy consumption will be minimum resulting into a greater network lifetime (Fig. 3e, f).

3.3 Case 3—Impact of Different Energy Levels (K)

Figures 4a–d illustrates the impact of change in energy levels from minimum ($K = 5$ J) to maximum ($K = 12$ J) levels on the network with a manual nodal layout. Figure 4d illustrate that the proposed algorithm consumes less energy and shows balanced energy consumption as compared to other algorithms. This balanced energy consumption of the proposed algorithm predicts a load balancing in the network which eventually gives greater network lifetime as shown in Fig. 4d. The simulated graph in Fig. 4d depicts lifetime for different transmission methodologies at different network radius. The comparison is with respect to transmission strategies such as single-hop (SH) [12], LEACH [14], EAACA [16], and the proposed protocol. The simulated graph shows that, as the network radius increases, lifetime decreases because of increase in traffic load. At a network radius of 100 m, the proposed protocol reaches a network lifetime of 8.92×10^3 s and others reaching 1.5×10^3 s (SH), 3.2×10^3 (LEACH), 3.8×10^3 s (EAACA), thus proving that the proposed strategy performs far better than the rest. The fall out of other transmission

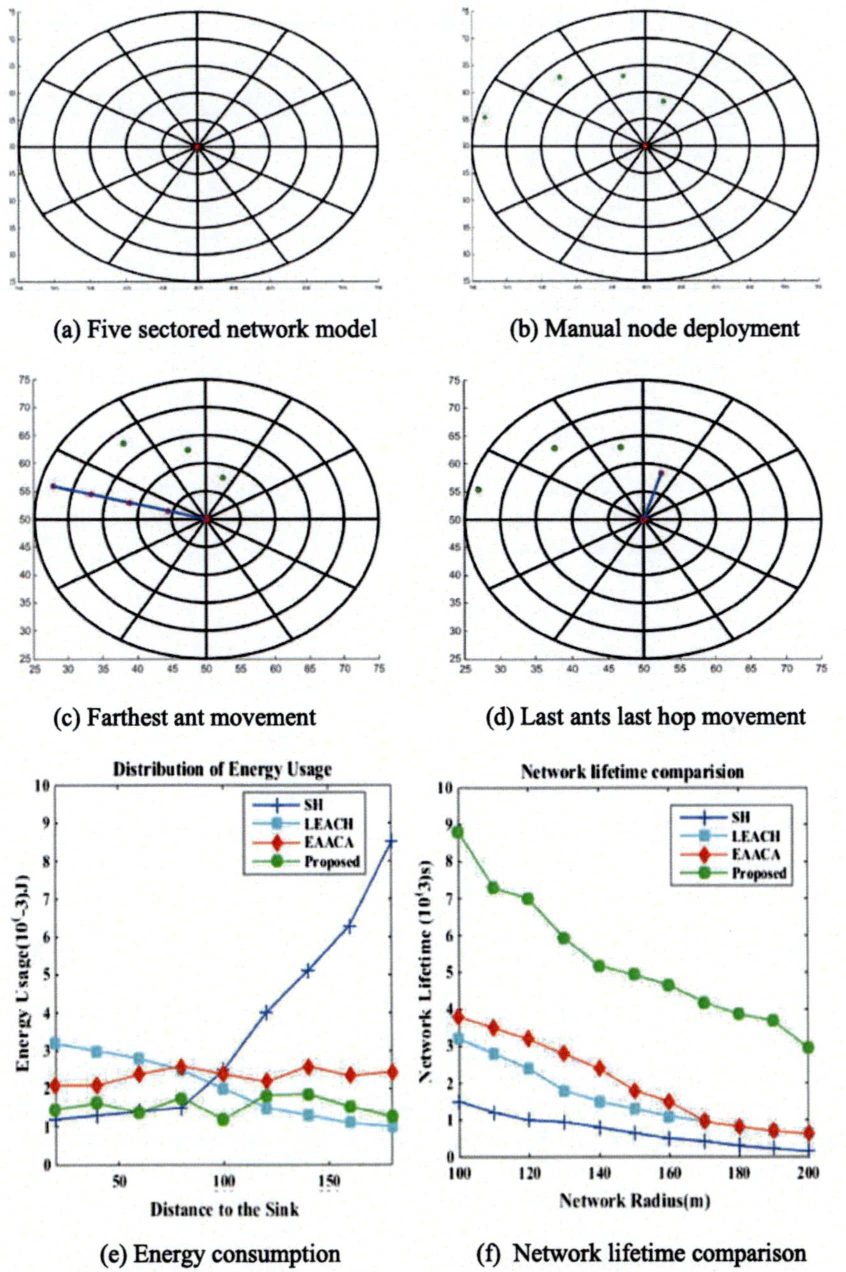

Fig. 3 Comparative plots for different routing strategies. **a** Five sectored network model. **b** Manual node deployment. **c** Farthest ant movement. **d** Last ants last hop movement. **e** Energy consumption. **f** Network lifetime comparison

Table 1 Average of best efficient and balanced distances for 5 sectors

Energy efficient Distance	Average of balanced energy distance			
	SH	LEACH	EAACA	Proposed
2.7735	3.58	3.13	2.68	1.43

Table 2 Overall network lifetime

Protocols	Network lifetime										
SH	1.5	1.2	1	0.9	0.8	0.6	0.5	0.4	0.3	0.2	0.1
LEACH	3.2	2.8	2.4	1.8	1.5	1.3	1.1	0.9	0.8	0.7	0.6
EAACA	3.8	3.5	3.2	2.8	2.4	1.8	1.5	0.9	0.8	0.7	0.6
Proposed	8.9	7.5	7.0	6.3	5.5	4.7	4.2	4.0	3.8	3.3	3.1

strategies is because of long distanced communication (SH), less load balancing in LEACH, EAACA. The proposed algorithm works on optimum distances both for energy efficiency and load balancing.

3.4 Case 4—Impact of Number of Rounds

To analyze the effect of rounds on live nodes and throughput, an initial simple energy-efficient routing protocol (SEER) [1] is added in the comparison domain. SEER is a simple, scalable, and source-initiated routing protocol. Routes are selected as per the distances to the base station and on the residual energy of nodes. SEER is an event-based protocol because of which transmissions are reduced and, so, energy consumption. Comparisons are made with SEER, LEACH [14], EEACA [16], and the proposed one.

Figure 5 demonstrates why energy consumption and its management are necessary for a small and extended WSN. As can be seen from Fig. 5a, for the proposed protocol, the first node dies at the 650th round and sees a progressive fall in the number of nodes till all nodes die at the 1600th round. The rest protocols SEER [1], EEACA [16], and LEACH [14] consume more energy because of which nodes die quickly. As seen from the plot, the nodes die completely at 750 for SEER, at 900 for LEACH, and at 1200 for EAACA. These protocols are residual energy based and have traffic handling problems. Thus, the proposed method gives a prolonged lifetime because of an optimum energy-efficient and energy-balanced path and uniform load-energy distribution.

The network throughput is shown in Fig. 5b. Initially, the throughput is less for all the protocols because the initial stage is for finding a probabilistic route. The throughput and hence the message delivery ratio (no. of received packets/ no. of transmitted packets) tends to stabilize after certain iterations (rounds). The proposed

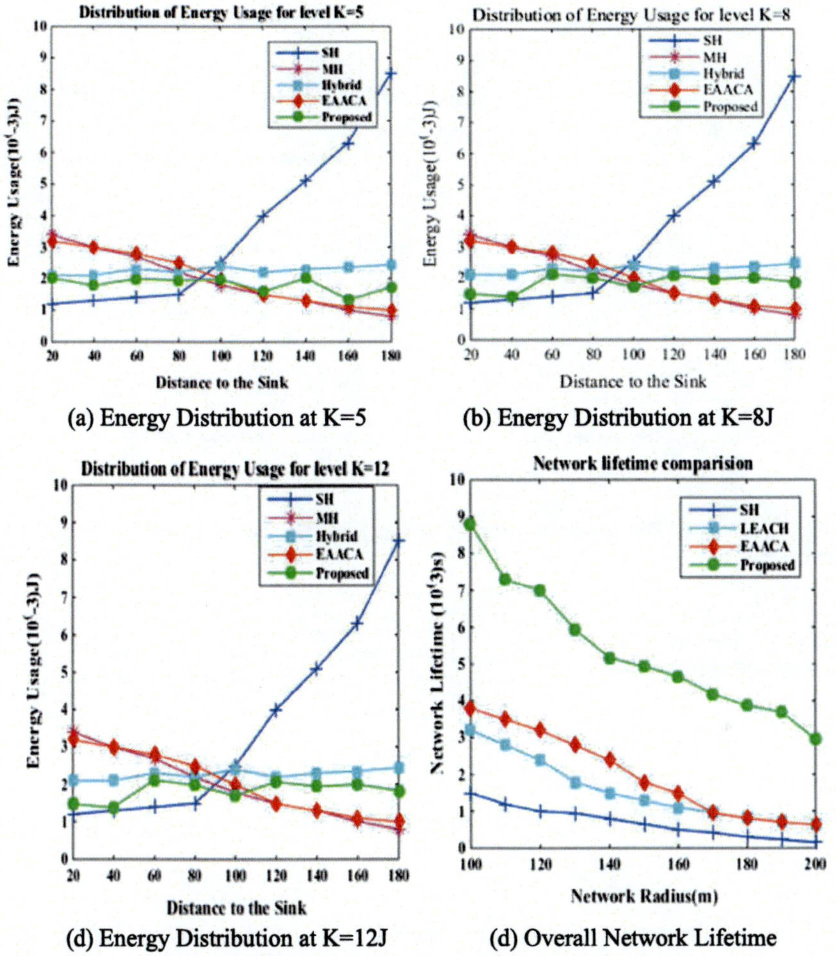

Fig. 4 Impact of energy levels on network lifetime. **a** Energy distribution at $K = 5$. **b** Energy distribution at $K = 8$ J. **c** Energy distribution at $K = 12$ J. **d** Overall network lifetime

protocol follows the optimum path because of which there is effective data transmission. The proposed method achieves transmission stability early as compared to other protocols. The network has total number of nodes: 100, initial energy of nodes: 10 J, nodal density: 5/m^2, no. of iterations: 10, deployment area of 100 m × 100 m (width × height), and radius considered is 25 m. Table 3 compares the various protocols with respect to different network aspects.

The case studies established the design requirements for the proposed programmed manual node deployment using a simple, sectored network model. The case studies are for an individual network, but the same can be applied to extended networks, subnets, and time-bound networks as well. Also, the established facts of

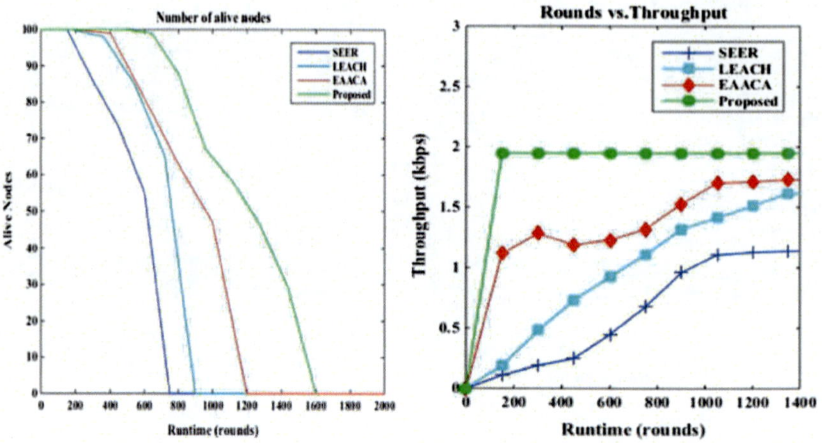

(a) Number of Live nodes (b) Network Throughput

Fig. 5 Number of live nodes and network throughput. **a** Number of live nodes. **b** Network throughput

Table 3 Parametric comparison of protocols

S. No.	Parameters	Proposed Algorithm	Other Algorithms
1	Stability	High	Low
2	Lifetime	High	Low
3	Energy Consumption	Minimum	Maximum
4	Throughput	High	Low

the case studies can be applied to the WSN-Cloud integration system applications (driverless cars and real-time patient monitoring). A detailed real-time patient monitoring system is under consideration, and the challenges like, bandwidth, traffic load, hop count, cost and worst case delay, coverage, etc., are considered for the work. The results of the work will be published in due course of time.

4 Conclusion

This work lays the foundation for futuristic application aspects of WSN, such as the WSN-Cloud integration, subnets, and extended networks. Energy and packet delivery are the factors that define the overall reliability of WSNs. This work finds efficient and balanced energy routes using a proactive ACO algorithm. The work applies to an individual single network and investigates the proposed and the basic but best-in-field routing protocols with respect to various performance metrics and validates the effectiveness of the proposed algorithm. The literature is full of direct, flat, and clustered protocols, but the work proposes a simple sectored network topology. The algorithm considers a programmed manual nodal deployment with uniform and random distribution and a fixed nodal density. The optimized sector network model enables the nodes to communicate with their counterparts. The ACO algorithm is designed to construct an optimal and efficient path between the nodes to minimize the total number of transmissions. The proposed protocol aspects can be applied to any network that needs improvement. The applied aspects not only extend the life cycle of WSN but also effectively improve network congestion and minimize the average delay. Finally, the effectiveness of the suggested approach has been validated through numerous performance comparisons. Additional futuristic aspects may include a non-uniform nodal density model, applicability for IoT platforms, and energy harvesting techniques.

References

1. Hancke GP, Leuschner CJ (2007) SEER: a simple energy efficient routing protocol for wireless sensor networks, vol 2007, no 39, pp 17–24. https://doi.org/10.10520/EJC28039
2. Mukase S, Xia K, Umar A (2021) Optimal base station location for network lifetime maximization in wireless sensor network. Electronics 10(22):2760. https://doi.org/10.3390/electronics10222760
3. Chakraborty S, Goyal NK, Mahapatra S, Soh S (2020) Minimal path-based reliability model for wireless sensor networks with multistate nodes. IEEE Trans Reliab 69(1):382–400. https://doi.org/10.1109/tr.2019.2954894
4. Liu X (2015) An optimal-distance-based transmission strategy for lifetime maximization of wireless sensor networks. IEEE Sens J 15(6):3484–3491. https://doi.org/10.1109/jsen.2014.2372340
5. Liu X (2014) A transmission scheme for wireless sensor networks using ant colony optimization with unconventional characteristics. IEEE Commun Lett 18(7):1214–1217. https://doi.org/10.1109/lcomm.2014.2317789
6. Nguyen HT, Van Nguyen L, Le HX (2017) Efficient approach for maximizing lifespan in wireless sensor networks by using mobile sinks. ETRI J 39(3):353–363. https://doi.org/10.4218/etrij.17.0116.0629
7. Kavitha V, Ganapathy K (2018) Efficient and optimal routing using ant colony optimization mechanism for wireless sensor networks. Period Eng Nat Sci (PEN) 6(1):171. https://doi.org/10.21533/pen.v6i1.274
8. Abd-Alsabour N, Hefny H, Moneim A (2013) Heuristic information for ant colony optimization for the feature selection problem. IEEE Conf Anthol. https://doi.org/10.1109/anthology.2013.6784795
9. Jones CE, Sivalingam KM, Agrawal P, Chen JC (2001) A survey of energy efficient network protocols for wireless networks. Wireless Netw 7(4):343–358. https://doi.org/10.1023/a:1016627727877
10. Engmann F, Katsriku FA, Abdulai JD, Adu-Manu KS, Banaseka FK (2018) Prolonging the lifetime of wireless sensor networks: a review of current techniques. Wirel Commun Mob Comput 2018:1–23. https://doi.org/10.1155/2018/8035065
11. Bhardwaj M, Garnett T, Chandrakasan AP (2001) Upper bounds on the lifetime of sensor networks. In: ICC 2001. IEEE international conference on communications, vol 3, no 3, pp 785–790. Conference Record (Cat. No. 01CH37240). https://doi.org/10.1109/ICC.2001.937346

12. Mark Perillo ZC (2004) On the problem of unbalanced load distribution in wireless sensor networks. In: IEEE global telecommunications conference workshops, 2004. GlobeCom workshops 2004. https://doi.org/10.1109/glocomw.2004.1417552
13. Mhatre V, Rosenberg C (2004) Design guidelines for wireless sensor networks: communication, clustering and aggregation. Ad Hoc Netw 2(1):45–63. https://doi.org/10.1016/s1570-8705(03)00047-7
14. Tayal D, Varshney D, Gupta RK (2014) A lifetime improvement approach using load distribution in LEACH protocol. In: National conference on synergetic trends in engineering and technology (STET-2014)
15. Younis O, Fahmy S (2004) Heed: a hybrid, energy-efficient, distributed clustering approach for ad hoc sensor networks. IEEE Trans Mob Comput 3(4):366–379. https://doi.org/10.1109/tmc.2004.41
16. Cheng D, Xun Y, Zhou T, Li W (2011) An energy aware ant colony algorithm for the routing of wireless sensor networks. In: Chen R (eds) Intelligent computing and information science. ICICIS 2011. Communications in computer and information science, vol 134. Springer, Berlin, Heidelberg. https://doi.org/10.1007/978-3-642-18129-0-62
17. Camilo T, Carreto C, Silva JS, Boavida F (2006) An energy-efficient ant base routing algorithm for wireless sensor networks. In: ANTS 2006—fifth international workshop on ant colony optimization and swarm intelligence, vol 4150, pp 49–59. https://doi.org/10.1007/11839088-5
18. Getahun M, Azath M, Sharma DP, Tuni A, Adane A (2022) Efficient energy utilization algorithm through energy harvesting for heterogeneous clustered wireless sensor network. Wirel Commun Mob Comput 2022:1–17. https://doi.org/10.1155/2022/4154742
19. Słabicki M, Wojciechowski B, Surmacz T (2012) Realistic model of radio communication in wireless sensor networks. In: Kwiecień A, Gaj P, Stera P (eds) Computer networks. CN 2012. Communications in computer and information science, vol 291. Springer, Berlin, Heidelberg. https://doi.org/10.1007/978-3-642-31217-5-35
20. Giacomin JC, Correia LHA, Heimfarth T, Pereira GM, Silva VF, de Santana JLP (2010) Radio channel model of wireless sensor networks operating in 2.4 GHz ISM band. INFOCOMP J Comput Sci 9(1):98–106
21. Abidoye AP, Ochola E, Ibidun C, Desmond W (2020) An improved ant colony optimization algorithm: a technique for extending wireless sensor networks lifetime utilization. Int J Adv Comput Sci Appl 11(8). https://doi.org/10.14569/ijacsa.2020.0110855
22. Okdem S, Karaboga D (2006) Routing in wireless sensor networks using ant colony optimization. In: First NASA/ESA conference on adaptive hardware and systems (AHS'06). https://doi.org/10.1109/ahs.2006.63
23. Li X, Keegan B, Mtenzi F (2018) Energy efficient hybrid routing protocol based on the artificial fish swarm algorithm and ant colony optimisation for WSNs. Sensors 18:3351. https://doi.org/10.3390/s18103351
24. Rahman AU, Alharby A, Hasbullah H, Almuzaini K (2016) Corona based deployment strategies in wireless sensor network: a survey. J Netw Comput Appl 64:176–193. https://doi.org/10.1016/j.jnca.2016.02.003

Implementation of 3D Object Models for Mobile Applications in UI/UX Design Using SceneView API

Daniel Ryan Sunjaya, Adbul Samad Bin Shibghatullah, and Shaik Shabana Anjum

Abstract The improvement of mobile application has improved over the years with the number of libraries and APIs that are shared for the developers. Resources such as APIs and libraries will brings more items that developers can use and utilized in their software projects mainly mobile applications. Right up to this moment, finding a complete documentation of integrating a 3D object model to a mobile application is very difficult. This brings up the question on to "why developers are not using 3D models to be used in their mobile applications?" It is a valid question because developers will always try and use the newest and latest thing on the market to build their mobile applications, for a 3D object model not included in that category is surprising. For this paper, a deep dive into 3D visualizations API called SceneView will be conducted to see if the 3D model technologies for mobile application is relevant for the current style of mobile application design and understand on what a 3D model can do for a mobile android application when used correctly.

Keywords 3D object · 3D visualization · User interactions · SceneView · Android

1 Introduction

In this phase of technology evolving in mobile application design, there are not much mobile application that incorporates a 3D model. Many of the known software brands are not adopting the use of 3D model in their design. Understanding on the base of

D. R. Sunjaya (✉) · A. S. B. Shibghatullah · S. S. Anjum
Institute of Computer Science and Digital Innovation (ICSDI), UCSI University, Kuala Lumpur, Malaysia
e-mail: ryansunjaya.sunjaya@gmail.com

A. S. B. Shibghatullah
e-mail: abdulsamad@ucsiuniversity.edu.my

S. S. Anjum
e-mail: shabana@ucsiuniversity.edu.my

G. Rajakumar et al. (eds.), *Intelligent Communication Technologies and Virtual Mobile Networks*, Lecture Notes on Data Engineering and Communications Technologies 171,
https://doi.org/10.1007/978-981-99-1767-9_9

what makes a great UI is the fact that user interactivity between the software and the user is required, thus why a UX is part of the design on a mobile application.

A SceneView API is an API that focuses on the usage of 3D object model rendering on android mobile application. This includes the use of 3D model in Augment Reality (AR) and 3D visualization. SceneView API provided the developer with many tools that they can use to produce a unique set of android mobile applications. One of the most used functions of SceneView is the integration of using ".glb" files for their 3D object assets. This opens up a wide variety of designing tools that are available today, such as Blender.

Using Blender to design a 3D object is confusing at first as for most people who are not familiar with designing tools will finds it hard to use. Other popular software such as Unity can also be used to render a 3D object model. Understanding this, developers that are able to utilize the use of 3D object model can take advantage of SceneView API, that gives the developer all of the tools required to integrate their 3D model from various 3D designing tools to their android mobile application.

SceneView also offers the flexibility on 3D objects that we have rendered in our mobile application. The addition of ARCore can further boost the usability of SceneView in the field of UI/UX design of mobile application. With all of it is benefits, there seem to be a lack of usage from developers in using SceneView API for their projects. One of the problems is the lack of documentation that SceneView have. In order to kick start a project based on SceneView, developers have to go through a big list of raw codes in a GitHub page to properly understand how they can use the API is for.

This paper will discuss on the reason that developers nowadays should adopt a 3D object model to their mobile application. We will also look at the things that can make SceneView such an exceptional API to have by all developers who are interested in incorporating 3D object model in their android mobile application. In this paper, we will also see how to add SceneView to a personal android mobile application project with few simple lines of codes. At the end of this paper, we can also see how a mobile application, such as Event Hotspot Management System, can utilize SceneView and 3D object model to meet the user's needs and boost the UI/UX design of the mobile application.

2 Related Works

With the help of ArcGIS, many people can utilize them to fit the needs of the user. Some are able to use ArcGIS to model the buildings of their campus and make a 3D map, while others use the technology to replace other technologies such as LIDAR or using air-borne vehicles, such as drone [1]. ArcGIS gives the developer an easier time when building 3D models of an area, compares to other technologies. This gives the developer the flexibility for them to use the ArcGIS or GIS according to their needs [1, 2].

The Yunnan Normal University has been using an outdated mapping system used for camping and the university navigation. It lacks crucial information and not accurate. The paper takes on the project to build an android application where it will give a real-time update and a more accurate map [3]. They incorporated ArcGIS as the tools to make the map layout as it is accurate and gives more detail information; this includes the ideal route between university and camps, bus lines, nearby buildings, and crossroads. The aim is to build a mobile application that will help students, teachers, and visitors. Compared to the previous system in the university, their functions are better navigation, nearby search, vehicle navigation, and location search. The use of ArcGIS is to build the overall map layout of their university area and ArcGIS will acts as the storage server for all of the map assets [4, 5].

ArcGIS plays a role in the making of the 3D map as well as the hosting service of the assets. This allows the student and lecturers to access their campus map using their browser. This means that there is not much processing done through the smartphone as it is all done through the web and the ArcGIS hosting service [6–11]. ArcGIS is very useful in small applications, such as small area like university. But it is also mostly used to cover a much larger space, where the terrain of the area is taken into account for certain applications. This is one of the reasons that ArcGIS is widely used for any application that involves the use of a map [12–14]. The city map is a large map that can span up to thousands of kilometer squares [15]. One of the uses of a city map is to accurately mark territories, and this requires a high accuracy map where ArcGIS, with it is large database, can come in handy for this kind of project. The project was undergone in the aim of precisely mark the area of the city. This will consider the land elevation, terrains, and other aspect of the land. ArcGIS provides a powerful tool to 3D render a high precise map of the terrain, which allows the user to clearly see the changes in elevation or area of the city [16].

Improvement in technology is inevitable, making it mandatory for a system to adapt and adopt their system with the new technology. This is true in the case of map. Nowadays, 80–90% of system have not yet adopt and change their map from 2 to 3D [17]. One of the reasons of the slow adoption is due to the lack of support since back then, smartphone is still unable to process complex 3D models and it is ineffective for companies to redirect their goals and forcing a change [1]. Compared to this current day, smartphone is becoming more and more powerful as the hardware improves and the software is supporting wider technologies [3, 15, 18, 19].

The 3D Campus Map is a steppingstone for the current system to adopt new technologies. In their paper, their goal was to recreate a 3D map model of the UPNM campus in the hope that it can replace the current traditional 2D map [20]. Comparing a 2D map with 3D map can be categorize to 3 aspects: information load, functionalities, and performance. In terms of information load, 3D map is able to capture and give a more accurate information as there is a vertical height given to each building rendered in the map. This allows them to add specific point of interest on the building, while it is impossible for them to do when using a 2D map. A 3D map also opens the door for more functionalities, where additional functions such as locating a specific room in the campus is possible with the help of 3D map. Lastly, in terms of performance, based on the paper, it is possible for people with current smart devices to run the application as their application is running using a web through ArcGIS online [20, 21].

2.1 Limitations

To summarize, there are 3 main factors that should be taken into consideration when creating the user application: Utilizing 3D map models, user interaction with the model, and all of it can be access through a mobile application. One of the main limitations of some of the existing work is that the integration of mobile application and the 3D model is not fully utilized. The other limitation of the current systems implementation is that they are not interactive and uses a 3D model for the user to interact with as shown in Table 1. Based on the system mentioned above, there are some advantage and disadvantage from each of the system, where Event Hotspot Mobile Application System will come in is to incorporate all of the technologies and create a more improved version of the current mobile map application. The goal of making this project to develop a mobile application that incorporated a 3D map, where user can interact with it and view the information displayed.

3 Proposed Work

SceneView API has brought many possibilities for the developer too use. Incorporating a 3D object model is a very neat way of giving the user on a 3D visualization of the object. Developers should also take some important things that SceneView and 3D object model can bring into the table when it comes into UI/UX design of an android mobile application. There are 3 main reasons on to why SceneView is such a powerful API that developers should take notes: Functionalities, quality of life (QoS), and UI/UX design [18].

Table 1 System comparisons

System name	Fully 3D map	Interactive	Type of application
Camps navigation system	No	Yes	Mobile
A-mobile 3D city map	Yes	No	Web
3D campus map	Yes	Yes	Web

3.1 Functionalities

SceneView 3D object model can give developer some functionalities that the user can interact with. One notably function is when an object it tapped. This can present the user with additional information. Since SceneView will listen to each individual 3D object, the developers can add multiple 3D objects into a mobile application and have the user tap on different 3D object to display or run a certain function based on the clicked/tapped object.

Being able to use 3D object model in a mobile application will bring some quality of life for the user where depending on the 3D object that is rendered. In the case of using a 3D map as the object, the user is able to get a clear visualization on the map that the developer wants them to see. A complete 3D view of the building or a landscape can be rendered with SceneView for user to interact with. The ease of use when a map is shown in 2D or 3D visualization makes a big difference in how user will use and interact with the map itself. 3D object provides the additional quality of life where user can interact directly with the map rather than interacting with a static image/2D.

Incorporating 3D object model to a mobile application will also add into it is appeal and stand out more than other mobile application that uses a conventional 2D assets. Since the 3D object can be interact with by the user, it adds to the user experience and can give a much better user interactivity. As we know that user experience also consists of the UX, which is where user can interact with the mobile application more personally.

3.2 Using Blender to Make a 3D Model

In this step, we can start designing the 3D objects that we wish to integrate into an android mobile application. The view of Blender and it is toolbox is shown in Fig. 1.

By using Blender as our 3D designing tool of choice, we are able to make any 3D object that we wish for. For this case, we will be rendering UCSI University, complete with each individual building. We want to make sure that the 3D model can be individually clicked for our case. To do this, we will have to manually save each single building to their separate asset file in ".glb".

3.3 Integrating 3D Object Models into Mobile Applications

After the 3D objects are ready, we can then move onto Android Studio, where we will start making out mobile application and integrating SceneView API. To add a 3D object into the view of a mobile application, we can simply use the Builder function

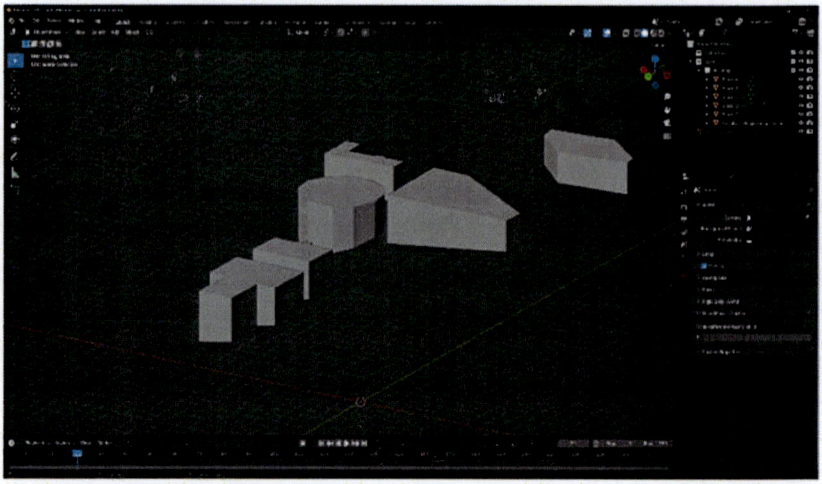

Fig. 1 Blender application on windows

given to ModelRenderable under SceneView API and refer the path of our 3D object assets as shown in Fig. 2.

Figure 2 is basically to prepare the 3D object for it to be rendered. In Fig. 3, we will then set any attributes that the 3D object will be holding. In here, we can set the size of the 3D object, the name of the 3D object, and it is location on the screen of the user's device. Setting everything up correctly will result in a well-placed 3D object as shown in Fig. 4. In the 3D object configuration, we will also need to set a listener to listen for any tap from a user. A set of function within it will be run when the user tapped on one of the 3D objects.

Clicking one of the 3D objects of UCSI University building will show a dropdown menu with the list of events that is held in that place. One of the neat features of SceneView in android is the fact that user is able to rotate and zoom-in using their 2 fingers. We can see the result of this in Fig. 5. As we can see that the list of events is shown here when the user taps on one of the 3D objects.

With this, we have completed a simple mobile application that fully utilize the SceneView API. The Event Hotspot Management System (E.H.M.S) runs properly

```
blockA_stage = ModelRenderable.builder()
        .setSource( context: this,Uri.parse("glbFiles/a.glb"))
        .setIsFilamentGltf(true)
        .setAsyncLoadEnabled(true)
        .build();
```

Fig. 2 ModelRenderable for the given ".glb" assets

```
private void addOtherNodetoUniversity(String name, ModelRenderable modelRenderable){
    Node otherBuilding = new Node();
    otherBuilding.setParent(university);
    otherBuilding.setRenderable(modelRenderable);
    otherBuilding.setName(name);
    otherBuilding.setLocalRotation(Quaternion.multiply(
            Quaternion.axisAngle(new Vector3( x: 0.5f,  y: 0f,  z: 0f),  degrees: 50),
            Quaternion.axisAngle(new Vector3( x: 0f,  y: 1f,  z: 0f),  degrees: 170)));
    otherBuilding.setOnTapListener(new Node.OnTapListener() {
        @Override
        public void onTap(HitTestResult hitTestResult, MotionEvent motionEvent) {
            if(isLogin && isOnMap){
                nodeTap(name);
            }
        }
    });

    Log.d( tag: TAG+" [addOtherNodetoUniversity]",  msg: "Building "+name+" Rendered");
}
```

Fig. 3 Creating a node for each individual 3D object

and follows all of the functions that we have placed in it. The result can be seen in Figs. 4 and 5.

4 Result Analysis

By understanding the benefits of using a 3D object model in a mobile application, we can now start implementing the 3D object to an android mobile application. Android Studio will be used as the primary IDE for this case. The other reason is that SceneView now supported android mobile application. The other tool that will be used any 3D designing tools in this case is Blender. Blender is chosen because of it is wide popularity by many developers and it is 3D object asset compatibility with SceneView.

For this example, we will be building Event Hotspot Management System (E.H.M.S) mobile application. This is the application that will be used for the UCSI University students to check on the list of events in the specific area of the UCSI University. This will make it easier to locate and pinpoint the exact place of the event, hence the name. The 3D object will be used to represent all of the UCSI University buildings. We want to make sure that whenever the student clicked on one of the 3D objects on their mobile device, the correct information of the event list in that particular building is displayed.

The ability to see the 3D model in real time through the simple use for our mobile device and improve many tasks that involved the need of 3D model. There are many application that can utilize the use of 3D object models such as maps, architectural

Fig. 4 Main page layout of E.H.M.S

rendering, and model showcase. With many use cases, the development of this kind of technology and innovation is necessary and could improve many quality of life things in the use of our mobile device and helps many professionals in their respective field of work.

5 Conclusion

With everything wrapped up, we have come to the conclusion for this paper. From the introduction, the existing of SceneView as an API for android mobile development is very useful and sought after. This allows the technology to be used in a more diverse way and developers are able to utilize it is functions and features. In this current day and age, improving and experimenting on ways to attract more user to use a mobile application is a skill. Utilizing SceneView to it is maximum potential can greatly boost the usability of a mobile application and allows it to stand out more from the rest. Throughout the paper, we also discussed on the benefits of integrating a 3D

Fig. 5 Dropdown menu of the list of events

object model to a mobile application. The main thing to understand that more of then than not, the usability of a mobile application is the most important.

We have also seen that using the SceneView API for android mobile development is as simple as adding a few lines of codes and a 3D object model can be rendered into a mobile application. SceneView being able to use the ".glb" format as the 3D object is huge as developers can now use a popular designing tool like Blender do design their object and displaying them into a mobile application.

To conclude this paper, 3D object model visualization for android mobile development is a nice feature that developers should look into it further and experiment with the usability of it. An improved documentation on SceneView as an API is also important to attract more developer in using the API. In the future, we hope that improvements in updates for SceneView or other much better API or library for 3D object rendering in android mobile application can make 3D visualization in android mobile application much more accessible.

References

1. Kulawiak M (2022) A cost-effective method for reconstructing city-building 3D models from sparse LIDAR POINT CLOUDS. Remote Sensing 14(5):1278
2. Ellul C, Altenbuchner J (2014) Investigating approaches to improving rendering performance of 3D City models on mobile devices. Geo-Spat Inf Sci 17(2):73–84
3. Schuckert N (2021) 3D visualization of geological layer—a case study of New South Wales, Australia and the Wadi Feiran Basin, Sinai. In: International conference of young professionals «GeoTerrace-2021»
4. Wang C, Wei S, Du S, Zhuang D, Li Y, Shi X, Jin X, Zhou X (2021) A systematic method to develop three dimensional geometry models of buildings for urban building energy modeling. Sustain Cities Soc 71:102998
5. Zhang E, Peng S, Zhai Y (2019) Design and application development of the camps navigation system based on ArcGIS runtime SDK for android: taking the Yunnan Normal University as an example. In: 2019 IEEE 4th advanced information technology, electronic and automation control conference (IAEAC)
6. Hamid HM, Al-Azawi FW (2021) Production of a three-dimensional triangulated irregular network (3d Tin) for Al-Karkh University of Science, Baghdad-Iraq. Iraqi J Sci:4529–4536
7. Fu J, Fan L, Roimela K, You Y, Mattila VV (2014) A 3D map augmented photo gallery application on mobile device. In: 2014 IEEE international conference on image processing (ICIP)
8. He J (2021) Analysis and application of digital campus 3D landscape modeling method. In: 2021 2nd international conference on computer science and management technology (ICCSMT)
9. Noguera JM, Barranco MJ, Segura RJ, Martínez L (2012) A mobile 3D-GIS hybrid recommender system for tourism. Inf Sci 215:37–52
10. Dedhia K, Dagli S, Konkar M, Katre N (2021) A proposed approach to indoor navigation and store recommendation. In: 2021 smart technologies, communication and robotics (STCR)
11. Hamza MH, Chmit M (2022) GIS-based planning and web/3D web GIS applications for the analysis and management of MV/LV electrical networks (a case study in Tunisia). Appl Sci 12(5):2554
12. Apollonio FI, Fantini F, Garagnani S, Gaiani M (2021) A photogrammetry-based workflow for the accurate 3D construction and visualization of museums assets. Remote Sensing 13(3):486
13. Moioli G (2022) Building a 3D environment. Intro Blender 3:167–228
14. Herath HMCK, Wanniarachchi A, Fernando R (1970) Research direction for Android based indoor navigation solution for shopping malls through augmented reality-EasyMap. IR@KDU Home, 01 Jan 1970 [online]. Available http://ir.kdu.ac.lk/handle/345/5249. Accessed 01 Dec 2022
15. Pepe M, Costantino D, Alfio VS, Vozza G, Cartellino E (2021) A novel method based on deep learning, GIS and geomatics software for building a 3D city model from VHR satellite stereo imagery. ISPRS Int J Geo Inf 10(10):697
16. Nurminen A (2006) M-loma—a mobile 3D city map. In: Proceedings of the eleventh international conference on 3D web technology—Web3D'06
17. Sze Mei AW, Ling Hong P, Keikhosrokiani P, Xin CH, Xue Ying T, Samat N (2021) A GIS-based mobile application to improve tourism experience: a case study of Terengganu, Malaysia. In: 2021 international congress of advanced technology and engineering (ICOTEN)
18. Henry P, Krainin M, Herbst E, Ren X, Fox D (2014) RGB-D mapping: using depth cameras for dense 3D modeling of indoor environments. Exper Robot:477–491
19. Song P, Sun G, Liu X (2021) Research on 3D campus integrated management based on ArcGIS pro and CityEngine. In: Proceedings of the 2021 5th international conference on electronic information technology and computer engineering
20. Syafuan WM, Husin RM, Azizi MI (2022) 3D campus map towards sustainable development and infrastructure management in UPNM. IOP Conf Ser Earth Environ Sci 1019(1):012035
21. Liu C, Yu X (2022) Construction of 3D design model of urban public space based on ArcGIS water system terrain visualization data. Math Probl Eng 2022:1–11

Latest Trends in Wireless Network Optimization Using Distributed Learning

A. Vasuki and Vijayakumar Ponnusamy

Abstract The demand for machine learning (ML) algorithms in wireless communication have increased over the last decade. Anyhow to enhance the prediction quality in complex problems, a significant amount of training is needed. ML requires centralized training which consumes more processing power. In the case of massive networks, consequently, it is necessary to train and test the algorithms in a distributed manner, which results in distributed learning. Recent development and methodologies used for distributed learning in network optimization of wireless networks are presented in a comprehensive manner in this paper.

Keywords Machine learning · Distributed learning · Wireless networks · Optimization

1 Introduction

Real-time interactive services and advanced applications like the Internet of Things (IoT) have increased network traffic significantly in next-generation wireless communication [1]. Therefore, advancements in next-generation wireless communication (6G) have become inevitable [2]. At the same time, machine learning (ML)-based algorithms have been applied to solve/simplify the processes like channel estimation, interference cancelation, resource management, and signal detection thereby maintaining the quality of service in the wireless networks. Most ML algorithms require training in a centralized manner. That is, the gathered information has to be transferred to the central data processing unit and examined. Due to the restricted available resources, it is hard to examine all gathered information from various devices at a

A. Vasuki
Department of ECE, SRMIST, Chennai, India
e-mail: vasukia@srmist.edu.in

V. Ponnusamy (✉)
Department of ECE, SRMIST, Kattankulathur, India
e-mail: vijayakp@srmist.edu.in

G. Rajakumar et al. (eds.), *Intelligent Communication Technologies and Virtual Mobile Networks*, Lecture Notes on Data Engineering and Communications Technologies 171,
https://doi.org/10.1007/978-981-99-1767-9_10

single point and analyze it for large networks. Therefore, the deployment of conventional ML algorithms becomes impossible in such situations. Even so, some emerging wireless applications such as device-to-device communication, cognitive radio, and self-automated services are working in a distributed manner. Moreover, the beyond 5G (B5G) or 6G applications require a minimal delay, and fast and appropriate decisions, which could not be achieved in a centralized manner.

Thus, a distributed learning mechanism [3] is more suitable to address these enormous challenges like reliable communication, delay issues, and protecting the data. It uses proper optimization methods based on the communication network requirements in an intelligent manner. Devices in the network have the capability to train the gathered data by creating a model in a distributed structure. In the recent communication era, distributed learning mechanisms are recognized as promising tools in industry and business.

This paper provides a thorough overview of the most recent research on performance optimization based on distributed learning in wireless communication.

2 Conventional Machine Learning and Distributed Learning

2.1 Types of Distributed Learning Models

The ML model could be divided into training and prediction stages [4]. In general, the training stage requires a huge amount of training data to generate appropriate training data. During the prediction stage, the generated training model is deployed in practice. The implemented model generates an output in response to new input called prediction. In the case of distributed learning, as illustrated in Fig. 1, the problem could be solved by dividing the data or machine, termed data-divide and machine-divide methods.

Data is split and applied simultaneously to the nodes in the machine. The nodes then process different data by applying the same algorithm. The machine will be the same for all nodes. In the machine-divide method, entire data is handled by the nodes in different parts of the machine. Then the machine can sum up all parts of the model. This method could not be applied to all ML algorithms, since the machine parameters cannot be divided. The distributed ML system's topology is the final architectural choice. The distributed system's various nodes must be connected using a particular architectural design to successfully complete a task. The pattern chosen affects the role a node can play, the level of communication between nodes, and the failure resilience of the deployment as a whole.

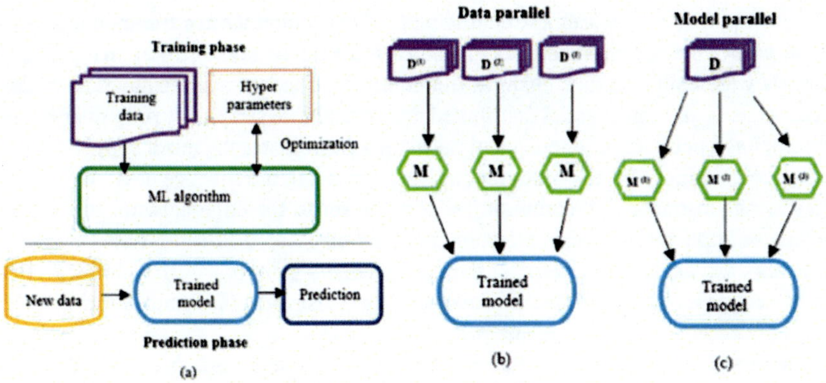

Fig. 1 **a** ML model, **b** data parallel model, **c** model parallel

2.2 *Structure of Distributed Learning Model*

The structure shows how the model nodes are organized and interconnected. The different possible structures are illustrated in Fig. 2.

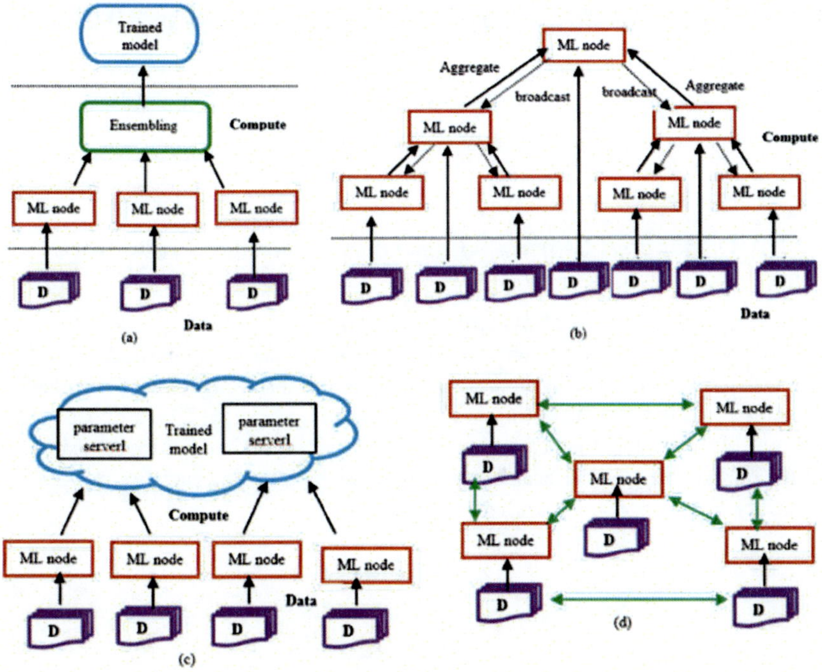

Fig. 2 **a** Ensemble, **b** tree, **c** parameter server, **d** distributed servers

A centralized model uses a structural method to accumulate at a common location. In the case of the decentralized model, neutral retrieval methods are used in replicated models were collected data changes frequently as it becomes available at all nodes in the tree, or in a split model distributed across multiple servers. A fully distributed system has a network of independent nodes that coordinates the result cooperatively.

A combination of edge computing and artificial intelligence (AI) technologies brings resource-rich computing servers closer to the edge network to support advanced applications. Recent developments in distributed learning-based wireless networks are analyzed comprehensively [5]. Merits of hybrid distributed learning architectures are proposed. In the proposed hybrid distributed learning algorithm, the users are classified into active and passive users based on their abilities to use either conventional learning or federated learning. Passive users send their local datasets to the parameter server to make use of them for training the ML model. At the same time, the active users upload the locally calculated gradient information based on their datasets. The proposed hybrid algorithm has a limitation in which the active users need to wait for the passive users to finish their data transmission. To tackle this, a sequential dataset transmission method is proposed, where the passive users are split into smaller blocks thereby maintaining both users to perform gradients and data transmission during the same communication round.

3 Recent Applications of Distributed Learning in Wireless Networks

This section discusses different applications in various stages of wireless networks.

3.1 A Distributed Learning-Based Virtual Reality Framework

An appealing application of 5G network is virtual reality. This has a congestion problem while rendering service. Using a distributed learning approach, an adaptive virtual reality framework is used to maintain the scalability and transformation capability in millimeter Wave (mmWave) wireless networks [6]. The proposed work combines deep reinforcement learning for offline training and a game theory approach for the online phase. Better scalability is achieved by allowing each user to forecast the utility from every action with different states.

3.2 A Distributed Learning-Based Self-Organized Wireless Networks

This section discusses different applications in various stages of wireless networks. Independent controls of mobile devices are used in modern applications like the internet of things and mobile edge computing. The next-generation wireless network has a primary limitation in managing independent devices. These networks undergo a critical deployment issue that requires different levels of cooperation among devices. To tackle this issue, the work [7] recognizes distributed procedures among self-organized devices, such as dispersed estimations and communication principles to manage random intercommunication configurations. The neural network design grasps wide intercommunication policies for self-organized models. Energy-efficient connectivity is an essential parameter in massive IoT networks. Distributed learning is applied with self-coordination.

3.3 Distributed Training Over Wireless Networks

The amount of information gathered and transferred in wireless networks is increasing exponentially with the increase in mobile devices. The transferring of the video content influences data traffic, which is also deliberated for machine analysis. The application of intelligence in IoT devices is to offload all related information to a cloud server with proper training with all available datasets and processing power. This centralized solution is not suitable because of the introduced latency. This problem could be solved by collaborative training [8]. This article applies distributed ML algorithms, and the local data is available on every device.

The article [9] provides efficient distributed learning communication frameworks in different applications. This work illustrates how communication efficiency is enhanced by optimizing types of payloads, transfer techniques, scheduling, and ML architecture. The role of distributed learning in different scenarios like communication energy reduction, enhancing convergence speed in a dynamic network, reduction of communication overhead, predicting future mmWave channels using received signal strength, etc.

3.4 Distributed Learning-Based Computational Offloading

Technical challenges in integrating wireless communication and multi-access edge computing are addressed in [10] using distributed learning. It utilizes a multi-agent Markov decision process to solve the computational offloading problem. The potential benefits of distributed learning are presented in the proposed framework as shown in Fig. 3.

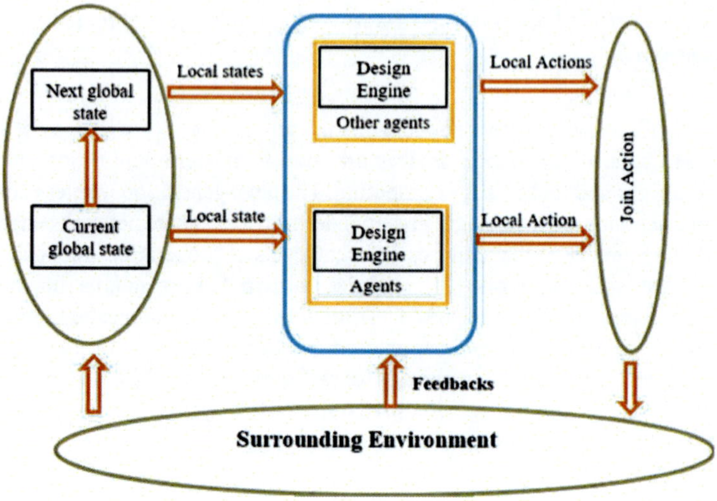

Fig. 3 Proposed multi-agent Markov decision process

3.5 Distributed Learning-Based Secure Network

The requirements to protect the information in multi-agent using distributed learning are studied in [11]. A distributed algorithm is proposed to achieve differential privacy by concerning both states and directions. Each agent should preserve its objective from other agents and eavesdroppers in a multi-agent network. The work achieves differential security by perturbing states in the proposed algorithm. With the step size of 19, the maximum-security level of 14 is achieved.

A privacy-preserving distributed architecture is presented in [12] for telecom networks in the 6th generation to secure the user's data. The quality of transmission is estimated in optical networks, wherein no user information is shared with the network. The secure framework is illustrated as shown in Fig. 4.

3.6 Distributed Learning Based on IRS-Aided Wireless Networks

Distributed learning is introduced and analyzed in IRS-aided wireless networks [13]. The parallel processing in wireless networks is obtained with the alternating direction method of multipliers. In the proposed method, users in the system update their learning model in the global model. The network base station takes care of user scheduling and broadcasting the updated information to all users.

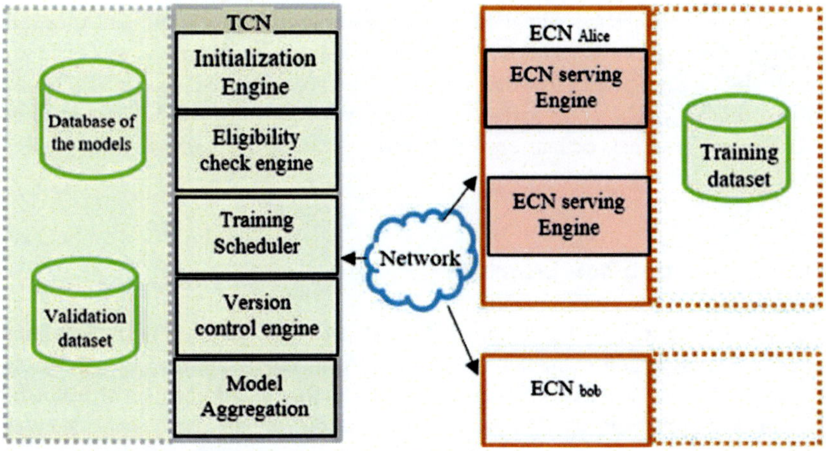

Fig. 4 Distributed learning-based secure network

3.7 *Distributed Learning Based on Dynamic Spectrum Access*

The paper [14] examines the dynamic access to the spectrum in the practical environment. The interlinkage among users in the dynamic system is considered a non-cooperative game, thereby achieving effective capacity. The problem is considered a multi-agent problem to reach the Nash equilibrium condition.

3.8 *Convergence of Distributed Learning*

In the distributed network, multiple nodes need to interact and compress the information into bits for transferring over a digital channel. The design of quantizers for these types of communication for these networks is illustrated in [15] thereby maintaining linear convergence. The communication time of the distributed learning algorithms is characterized as a function of the available rate.

3.9 *Energy-Efficient Low-Complex IoT Network*

Energy efficiency is an essential parameter to enable large-scale IoT networks. The wide network mainly depends on coordination, energy consumption, and exponential growth of IoT devices, making centralized learning unsuitable for next-generation networks. In [16], a low-complex distributed learning algorithm is analyzed for IoT networks to enable self-coordination through learning from past communications. A learning solution that adapts the environment parameters to ensure energy efficiency and reliability. Low-power, low-data rate faraway transmission is made possible with LoRa technology. The chirp spread spectrum modulation is used, which transmits and receives the signals concurrently with different spreading gains without causing any discrimination.

3.9.1 Hybrid Distribute Learning Approach

The combined effect of distributed, federated, and split learning is implemented to reduce transmission overhead while transferring large datasets [17]. A hybrid architecture is introduced by gaining parallel model training mechanism of federated learning and split learning. This approach trains part of the users with federated learning, and other parts of users with the base station are trained using split training. The performance summary of the discussed article in this section is listed in Table 1.

4 Future Scope

Since the industry is particularly concerned about the accuracy of the distributed learning model, distributed learning must create privacy-enhancing strategies that do not give up accuracy to provide tight privacy guarantees. Another important concern is to examine the distributed learning convergence to assess the optimality of distributed learning solutions as well as the time and effort expended in training distributed learning algorithms.

5 Conclusion

In this study, we have provided a comprehensive study on distributed learning-enhanced wireless communication systems in recent research works. We have discussed various algorithms, architectures, and frameworks utilized in wireless network applications to optimize performance compared to ML techniques. We have also highlighted the performance comparison of distributed learning with the conventional machine learning algorithms in the different categories of wireless applications. As a promising technique, distributed learning gets more attention and progress in research for optimizing the performance parameters such as energy efficiency, security, system capacity, and transmission quality in large-scale wireless networks.

Table 1 Performance summary on the recent application of distributed learning

Ref. No.	Application	Parameter analyzed	Obtained results
[6]	Wireless virtual reality framework	Latency	Achieves 14.4 and 29.2% more average latency than Q-learning and greedy algorithms with 100 users
[7]	Self-organized wireless networks	Sum-throughput	Gains 3.4 nats/s/Hz at 20 dB SNR
[8]	IoT networks	Latency	The proposed method obtains 5–7 times fast processing than conventional federated learning
[9]	Wireless networks	Test accuracy	95% test accuracy is achieved with 0.8 GB data transmission in distributed learning, wherein stochastic gradient has 55% accuracy for 2 GB data transmission
[10]	B5G networks	Average transmit and CPU energy	Requires average energy of 12 J at a 3.9 Mbps data arrival rate, whereas the queue-aware method needs 13 J
[11]	Multi-agent systems	Security	Achieved secure level: 14 Step size:19
[12]	Optical networks	Quality of transmission	The proposed method achieves 89.31% transmission quality
[13]	IRS-aided wireless networks	Learning efficiency	Achieved optimal weighted sum 4.5 at 6 number of users with 60 number of reflective elements
[14]	Time-varying network	System capacity	Achieves 90% performance in about 500 iterations Achieves 5 more capacity of packets/slot with 20 channels than the random approach
[15]	Wireless networks	Convergence behavior	With $b = 12$ bits/dimension, got almost the same convergence rate as without quantization
[16]	IoT networks	Energy consumption	The proposed algorithm consumes 0.03 J/packet less energy than the conventional method

(continued)

Table 1 (continued)

Ref. No.	Application	Parameter analyzed	Obtained results
[17]	UAV networks	Accuracy	The proposed method obtains 90% test accuracy within 50 communication rounds

References

1. Tong W, Li GY (2022) Nine challenges in artificial intelligence and wireless communications for 6G. IEEE Wireless Communications
2. Qian L, Yang P, Xiao M, Dobre OA, Renzo MD, Li J, Han Z, Yi Q, Zhao J (2022) Distributed learning for wireless communications: methods, applications and challenges. IEEE J Sel Top Signal Process 16(3):326–342
3. Muscinelli E, Shinde SS, Tarchi D (2022) Overview of distributed machine learning techniques for 6G networks. Algorithms 15(6):210
4. Verbraeken J, Wolting M, Katzy J, Kloppenburg J, Verbelen T, Rellermeyer JS (2020) A survey on distributed machine learning. ACM Comput Surv (CSUR) 53(2):1–33
5. Liu X, Yu J, Liu Y, Gao Y, Mahmoodi T, Lambotharan S, Tsang DHK (2022) Distributed intelligence in wireless networks. arXiv:2208.00545
6. Guo F, Richard Yu F, Zhang H, Ji H, Leung VCM, Li X (2020) An adaptive wireless virtual reality framework in future wireless networks: a distributed learning approach. IEEE Trans Vehic Technol 69(8):8514–8528
7. Lee H, Lee SH, Quek TQS (2022) Artificial intelligence meets autonomy in wireless networks: a distributed learning approach. IEEE Netw 36(6):100–107
8. Ozfatura E, Gündüz D, Vincent Poor H (2022) 12 collaborative learning over wireless networks: an introductory overview. Mach Learn Wirel Commun:353
9. Park J, Samarakoon S, Elgabli A, Kim J, Bennis M, Kim S-L, Debbah M (2021) Communication-efficient and distributed learning over wireless networks: principles and applications. Proc IEEE 109(5):796–819
10. Chen X, Wu C, Liu Z, Zhang N, Ji Y (2021) Computation offloading in beyond 5G networks: a distributed learning framework and applications. IEEE Wirel Commun 28(2):56–62
11. Ding T, Zhu S, He J, Chen C, Guan X (2021) Differentially private distributed optimization via state and direction perturbation in multiagent systems. IEEE Trans Autom Control 67(2):722–737
12. Safari P, Shariati B, Fischer JK (2020) Privacy-preserving distributed learning framework for 6G telecom ecosystems. arXiv:2008.07225
13. Ma X, Zhao J, Gong Y, Li Q (2021) Distributed learning over IRS-assisted intelligent wireless networks. In: ICC 2021-IEEE international conference on communications. IEEE, pp 1–6
14. Xu Y, Wang J, Wu Q, Zheng J, Shen L, Anpalagan A (2017) Dynamic spectrum access in time-varying environment: distributed learning beyond expectation optimization. IEEE Trans Commun 65(12):5305–5318
15. Magnússon S, Shokri-Ghadikolaei H, Li N (2020) On maintaining linear convergence of distributed learning and optimization under limited communication. IEEE Trans Signal Process 68:6101–6116
16. Azari A, Cavdar C (2018) Self-organized low-power IoT networks: a distributed learning approach. In: 2018 IEEE global communications conference (GLOBECOM). IEEE, pp 1–7
17. Liu X, Deng Y, Mahmoodi T (2022) Wireless distributed learning: a new hybrid split and federated learning approach. IEEE Trans Wirel Commun

Malware Classification in Local System Executable Files Using Deep Learning

Pagadala Ganesh Krishna, S. Kranthi, and Ande Vijaya Krishna

Abstract One of the biggest and most severe risks on the Internet today is malicious software, generally known as malware. Attackers are producing malware that has the ability to change its source code as it spreads and is polymorphic and metamorphic. Furthermore, the variety and quantity of their variants seriously compromise the effectiveness of current defences, which frequently rely on signature-based techniques and are unable to identify malicious executables that have not yet been detected. Variants from different malware families have behavioural traits that are indicative of their function and place in society. Utilizing the behavioural patterns obtained either statically or dynamically, deep learning techniques can be utilized to discover and classify novel viruses into their recognized families. In this digital age, security failures brought on by malware attacks are on the rise and pose a serious security concern. Malware detection is still a strongly contested academic topic because of the significant implications that malware attacks have on businesses, governments, and computer users. For the real-time identification of unknown malware, the efficacy of current malware detection techniques, which entail the static and dynamic analysis of malware signatures and behaviour patterns, has not been shown. For classifying malware, we mostly utilize CNN and ELM deep learning algorithms.

Keywords Classification · Convolutional neural network · Extreme learning machine · Malware · Image processing

P. G. Krishna (✉) · S. Kranthi · A. V. Krishna
Velagapudi Ramakrishna Siddhartha Engineering College, Vijayawada, Andhra Pradesh, India
e-mail: ganeshkrishnapagadala@gmail.com

S. Kranthi
e-mail: kranthisri41@gmail.com

A. V. Krishna
e-mail: andevijaykrishna26@gmail.com

G. Rajakumar et al. (eds.), *Intelligent Communication Technologies and Virtual Mobile Networks*, Lecture Notes on Data Engineering and Communications Technologies 171,
https://doi.org/10.1007/978-981-99-1767-9_11

1 Introduction

Malware binary is a malicious application that could damage computer operating systems. It typically has a file name extension of ".exe" or ".bin." It occasionally may have numerous variations with frequently recycled fundamental designs. This suggests that malware binaries may be divided into a number of families (classes), and that each variation would inherit the traits of its own family. Therefore, it is critical to accurately identify malware binaries and understand potential variations. This is difficult but not trivial. A digital grey image can be used to represent a malware binary file. After visualization, the detection of malware binary transforms into a deep learning issue of multi-class image classification [1].

The objective of this study is primarily focuses on:

- Developing CNN and ELM models that can categorize malware images.
- Comparing and evaluating the performance of both models.
- Constructing a CNN-ELM model by hybridizing the CNN and ELM algorithms.
- Reviewing the effectiveness and performance of three models.

2 Related Work

This section primarily focuses on the resources that helped us understand the philosophy behind the field of research on malware detection and classification approaches. The research publications went into great detail about the many algorithms and frameworks that may be used to categorize malware data. In this study, we discovered that more modern approaches, such as the ELM-CNN ensemble, outperform CNN and ResNet-50 in terms of performance and that the ELM can be trained in less than 30% of the time required by the fastest CNN architecture. We must therefore develop a model that performs better while requiring less training time every epoch [1–15].

Jain et al. [1] primarily concentrates on convolutional neural networks (CNNs) and extreme learning machines as two machine learning models (ELM). Surprisingly, we discover that ELMs can attain accuracy levels comparable to CNNs while using less than 2% of the time to train an equivalent CNN. Although ELMs have numerous critics, the experimental findings strongly suggest that they can, at least in certain circumstances, compete in the malware realm.

By applying deep CNN and ResNet-50 to their data set, they were able to classify malware samples with 96% and 97% accuracy, respectively, using their models [2]. Recent methods like the ELM-CNN ensemble outperform CNN and ResNet-50 in terms of performance. Both data sets are trained and tested using deep CNN and ResNet-50 models. Additionally, packed and previously undiscovered malware samples are used in the studies, laying the groundwork for further research.

Marastoni et al. [3] demonstrated how two deep learning models (a CNN and an LSTM) perform in a classification test on two data sets of actual malware as well as a produced data set. The models were developed with the goal of classifying images, and they were then refined using a specially created data set of programmes. The results unmistakably demonstrate that the LSTM model outperforms other studies in the state of the art, with an average accuracy of 98.5% on a hold-out set of the Malimg data set. They were able to confirm that the characteristics learnt from each data set may be utilized to classify malware from the other using the accuracy of both of these classifiers created using transfer learning.

Abolaji et al. [4] explain malware's origins, classifications, methods of obfuscation, methods of detection, and suggestions for protecting all computer users from malware events. The proverbial saying goes, "Prevention is better than cure." By taking deliberate corrective actions, malware infestation can be consistently reduced or entirely avoided. It is obvious that the virus problem will not go away anytime soon as new malware joins the ones already in existence every day. This document provides guidelines for reducing malware events for all computer users while also describing the history of malwares and their types, obfuscation tactics, detection, and history.

Pinhero et al. [5] have demonstrated the capability of CNN-based MDS to quickly and accurately classify malware files while being resistant to repeated API injections. Malware files can be effectively represented as images by using a straightforward byte to-pixel method, and tests have shown that greyscale images are more resistant to redundant API injection. Our network can accept images of any size as input thanks to Spatial Pyramid Pooling, but because of physical memory limitations, large files must be handled carefully and the "divide and conquer" strategy fails horribly. As a resizing technique, however, straightforward bi-linear interpolation yields an encouraging outcome. It would be fascinating and instructive to investigate this topic in more detail.

Falana et al. [6] created a system called Mal-Detect for analysing and categorizing malware because malware authors continue to use a variety of evasion strategies. First, RGB pictures were used to represent both malicious and good files. In order to stop the adversarial attack, new malware was then created using DGAN using train malware data sets. The model was trained using both the generated malware samples and the original malware data set from train samples. For training and classifying malicious and benign features into several malware families, DCNN is used. Finally, Mal-precision Detects were evaluated against other current state-of-the-art malware visualization tools. The outcomes demonstrated that Mal-Detect increases the precision and effectiveness of every trial.

Qiang et al. [7] presented a reliable and efficient deep learning-based malware detection method. Without the requirement for laborious feature extraction, they used the instruction-level control flow traces as the feature set and transformed the traces into behavioural byte sequences as the input of the neural network. High-level information can be captured in local settings using the suggested CNN-LSTM and CNN-BiLSTM architectures while maintaining their semantic relationships. The evaluation findings demonstrate that our method has good flexibility to the evolution of malware and can accurately detect hidden samples with a 94.6% detection rate. Additionally, we discovered that our method based on control flow traces is more reliable than the one based on API call sequences, which is not as reliable as popularly believed.

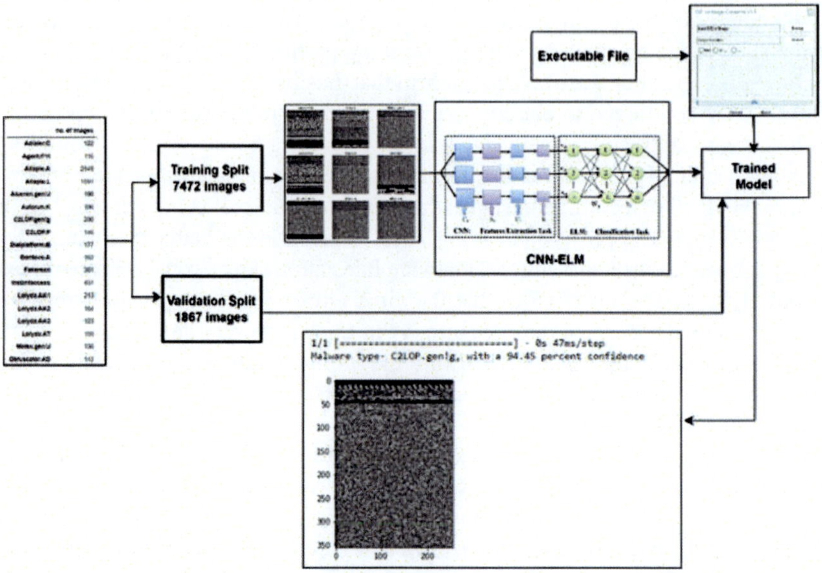

Fig. 1 System architecture

3 Proposed Work

We found that the ELM can be trained in less than 30% of the time needed by the quickest CNN architecture. Therefore, we must create a model that produces better results while yet requiring less training time every epoch. So we are going to propose a hybrid model of CNN and ELM algorithms which can be trained quickly and can perform classification task efficiently. CNN-ELM model can be able to perform both feature extraction task as well as classification task.

4 System Architecture

System architecture of the proposed work is shown in Fig. 1.

5 Design Methodology

In this research, we propose a hybrid model combining CNN and ELM algorithms that, when compared to separate CNN and ELM models, produced optimum outcomes. The ELM classifier receives the features taken from CNN's convolution layer as input, and ELM classifies the output based on the characteristics retrieved from the images using CNN. Later, we can categorize an executable file as malware or not using that model.

Two algorithms, CNN and ELM, are used in the construction of our hybrid model as shown Fig. 2. Convolution and pooling layers in CNN are responsible for feature extraction. And the retrieved features are supplied to the ELM algorithm's input layer as input. The ELM algorithm has three layers: input, hidden, and output. It uses these layers to classify the features that are sent to it as input.. Exe to Image converter is shown in Fig. 3.

With this technology, an executable (.exe) file is transformed into binary image format. When we train our model with transformed photos later, identifying whether a file is malicious or benign will be a simple image classification task for us. This converter is a java-based application.

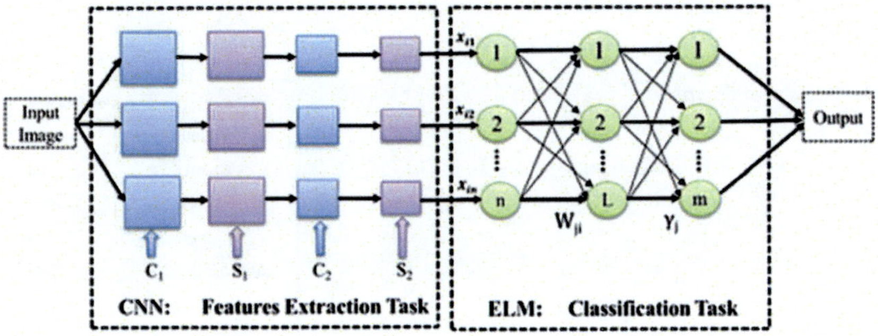

Fig. 2 Design methodology

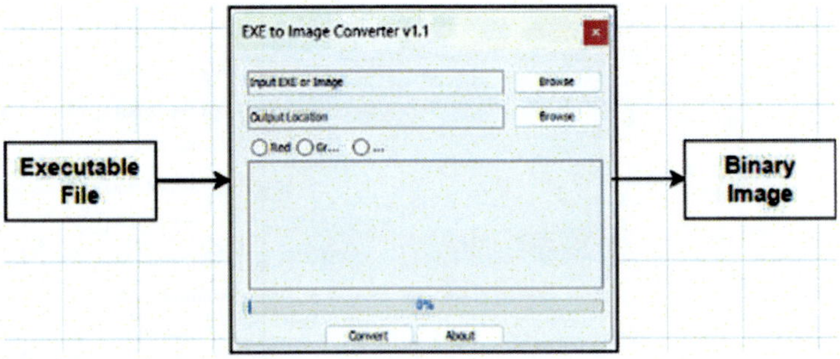

Fig. 3 Exe to Image converter

6 Data Set Description

The data set used for training and validation of our proposed model is described in detail here:

- Source: Malimg data set
- Total no. of classes: **25**
- Total no. of images: **9339**
- No. of images used for training: **7472**
- No. of images used for validation: **1867**.

We take into account malware classification using the Malimg data set [1]. In this data set, there are more than 9339 greyscale images from 25 different malware families. Several earlier research studies based their findings on this data set as shown in Fig. 4.

Fig. 4 Data set description

	no. of images
Adialer.C	122
Agent.FYI	116
Allaple.A	2949
Allaple.L	1591
Alueron.gen!J	198
Autorun.K	106
C2LOP.gen!g	200
C2LOP.P	146
Dialplatform.B	177
Dontovo.A	162
Fakerean	381
Instantaccess	431
Lolyda.AA1	213
Lolyda.AA2	184
Lolyda.AA3	123
Lolyda.AT	159
Malex.gen!J	136
Obfuscator.AD	142
Rbot!gen	158
Skintrim.N	80
Swizzor.gen!E	128
Swizzor.gen!I	132
VB.AT	408
Wintrim.BX	97
Yuner.A	800

7 Implementation and Results

In the phases that follow, we'll show how we implemented our proposed system. The hyperparameters for the models we implemented are as follows:

- Train-test split ratio—4:1
- Optimizer—Adam optimizer
- Activation Function—ReLu
- Loss function—Categorical cross entropy
- Number of epochs—30
- Batch size—32
- Image dimensions—(180, 180)

7.1 Phase-1

- We trained our CNN model with the entire data set during this phase.
- After that, we calculated accuracy to assess our model's performance.
- We also got a 97.38% accuracy from evaluating our model on validation data.
- The input data set was successfully used to train the CNN model, and each epoch lasts up to 700 ms.

7.2 Phase-2

- We trained our ELM model with the entire data set during this phase,
- After that, we calculated accuracy to assess our model's performance.
- We also got a 96.04% accuracy from evaluating our model on validation data.
- The input data set was successfully used to train the ELM model, and each epoch lasts up to 125 ms and it is faster in every epoch and requires less training time.
- The plots after plotting training and validation accuracy are as follows:

7.3 Phase-3

- We trained our CNN-ELM model with the entire data set during this phase.
- After that, we calculated accuracy to assess our model's performance.
- We also got a 98.18% accuracy from evaluating our model on validation data.
- The input data set was successfully used to train the ELM model, and each epoch lasts up to 280 ms.
- The CNN-ELM had given optimized performance when compared to individual CNN and ELM. The plots after plotting training and validation accuracy are as follows.

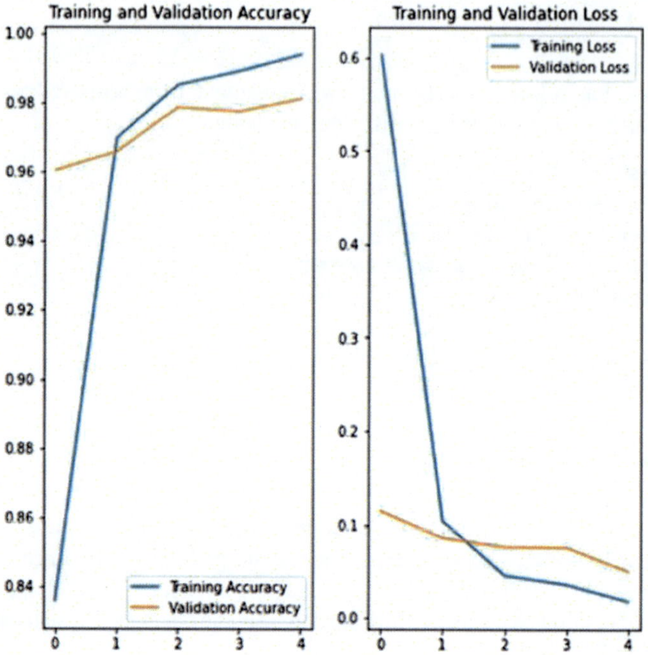

Fig. 5 Plotted results of CNN

In three phases from Figs. 5, 6 and 7, we can observe how accuracy was getting improved and loss getting decreased in every epoch. Also CNN-ELM (Fig. 7) model was getting trained in less time when compared to individual CNN (Fig. 5) and ELM (Fig. 6) models. We also found that whereas the accuracy of our ELM model was 96.04% that of our CNN model was 97.38%. Additionally, we found that although ELM takes 125 ms for each epoch, CNN takes roughly 700 ms. In order to achieve better results, we integrated the layers of both models to develop the CNN-ELM model, which delivered 280 ms for each epoch and 98.18% accuracy (Table 1).

From above table, we can easily observe that the CNN-ELM model was providing optimized performance in accuracy as well as training speed. Hence, we can say that our proposed method of CNN-ELM model is obviously a better performer than many existing methods.

7.4 Test Case Results

Now we must determine how effective our model is at predicting new data. So, we took a few test cases and created a model to predict the disease's class.

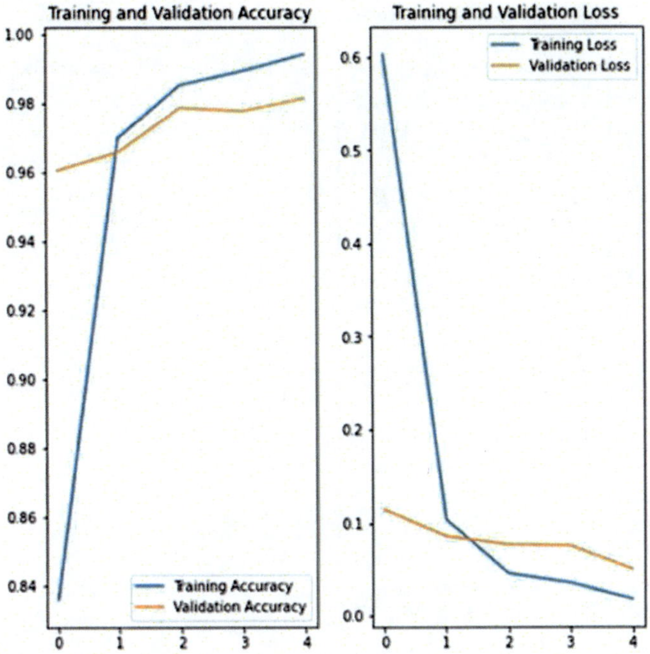

Fig. 6 Plotted results of ELM

Test case-1

Expected output: Fakerean.
Actual output:

Test case-2

Expected output: Wintrim.BX.
Actual output:

Test case-3

Expected output: Instantaccess.
Actual output:

Test case-4

Expected output: VB.AT.
Actual output:

Figures 8, 9, 10 and 11 show that the binary images extracted from exe files are classified in to their respective malware family and also it shows the confidence per cent. If the confidence per cent is high, then the exe file is can be considered as a malware file and belongs to one of the malware family of our data set or else it may be benign.

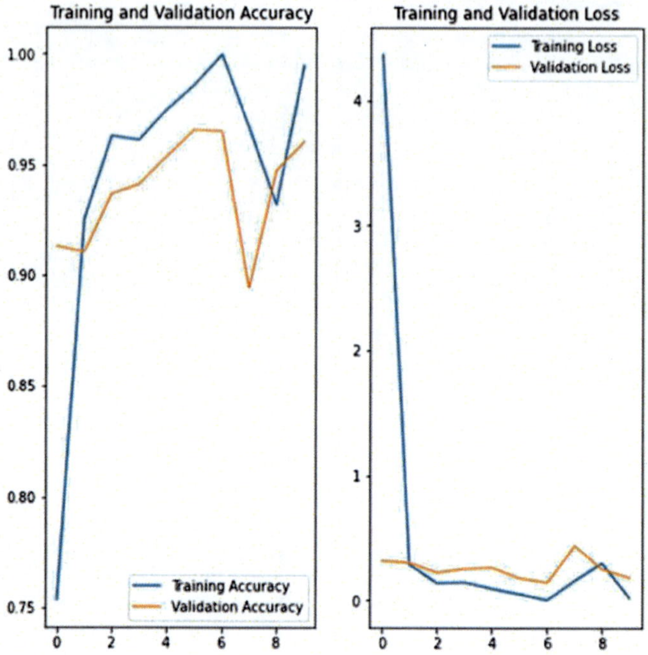

Fig. 7 Plotted results of CNN-ELM

Table 1 Performance comparison of CNN, ELM, and CNN-ELM models

Model	Accuracy (%)	Speed/epoch (ms)
CNN-ELM	98.18	280
ELM	96.04	125
CNN	97.38	700

8 Conclusion and Future Work

Images extracted from malware samples were classed by our model. The finished item is highly competitive. The ability to view viruses as images has been a significant achievement. Many researchers have used this technique in order to recognize and categories malware. However, additional research has shown that this method is easily vulnerable to adversarial attack and can produce false positives. By incorporating new malware classes into the data set and enhancing the framework and procedures employed in this study, we can expand it in future and produce more accurate and effective findings. Future work will focus on developing effective new deep learning techniques. As is common knowledge, 560,000 new instances of malware are found daily. There are currently more than one billion malware programmes available, thus this area of security has a lot of potential. Modern techniques can also be used

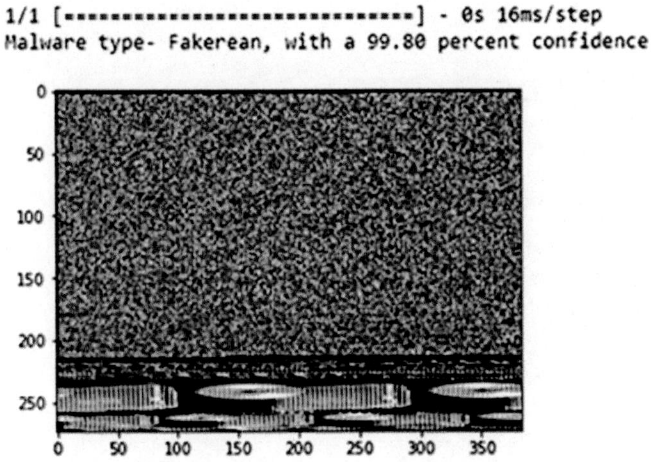

Fig. 8 Fakerean

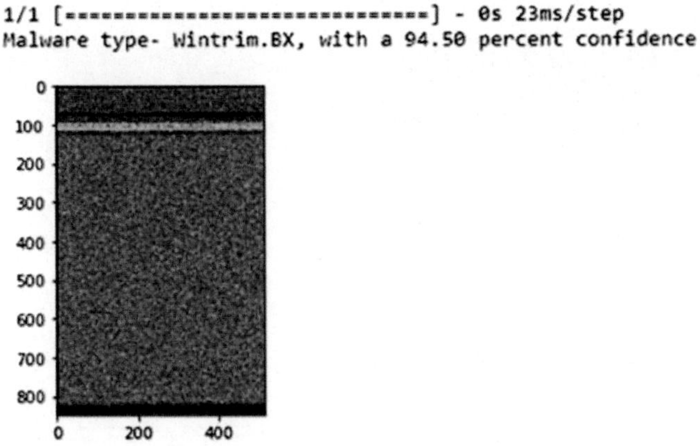

Fig. 9 Wintrim.BX

to improve procedures like the conversion of executable files to images. Additionally, hybridization is important in creating a model with numerous computational advantages. Therefore, we continue to develop effective algorithms that will provide optimal results in performing image classification task.

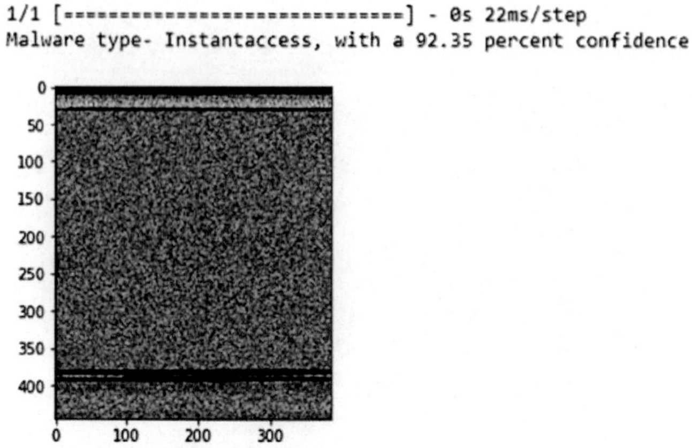

Fig. 10 Instantaccess

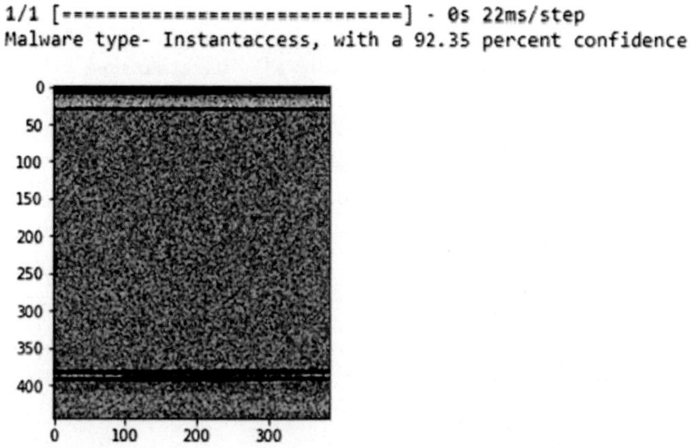

Fig. 11 VB.AT

References

1. Jain M, Andrpoulos W, Stamp M (2021) CNN vs ELM for image-based malware classification, Kolkata, India
2. Singh A, Handa A, Kumar N, Shukla SK (2020) Malware classification using image representation. Kanpur, India
3. Marastoni N, Giacobazzi R, Preda MD (2021) Data augmentation and transfer learning to classify malware images in a deep learning context. France
4. Aslan O, Samet R (2019) A comprehensive review on malware detection approaches. Computer Engineering Department, Ankara University, Ankara, Turkey
5. Akinde OK, Ilori AO, Afolayan AO, Adewuyi OB (2021) Review of computer malware: detection and preventive strategies. Comput Electron Agricult
6. Chatterjee JM, Jhanjhi NZ, Brohi SN (2020) Convolutional neural networks and extreme learning machines for malware classification. Vellore, India

7. Pinhero A, Anupama M, Vinod PB, Visaggio CA (2020) Malware detection employed by visualization and deep neural network. Appl Artif Intell
8. Qianga W, Yanga L, Jin H (2022) Efficient and robust malware detection based on control flow traces using deep neural networks. China, Apr 2022
9. Kannojia SP, Jaiswal G (2020) Ensemble of hybrid CNN-ELM model for image classification. In: 2020 IEEE conference Lucknow, India
10. Baek S, Jeon J, Jeong B (2020) Two-stage hybrid malware detection using deep learning, vol 170, 105220
11. He K, Kim DS (2021) Malware detection with malware images using deep learning techniques. IEEE
12. Kranthi S, Kanchana M, Suneetha M (2022) A study of IDS-based software-defined networking by using machine learning concept. Lecture notes in networks and systems, 318, pp 65–79
13. Kranthi S, Kanchana M, Suneetha M (2022) An intelligent intrusion prediction and prevention system for software defined internet of things cloud networks. Peer-to-Peer Netw Appl. https://doi.org/10.1007/s12083-022-01374
14. Tekerek A, Yapici MM (2022) A novel malware classification and augmentation model based on convolutional neural network. Comput Secur 112. ISSN 0167-4048
15. Chaganti R, Ravi V, Pham TD (2022) Image-based malware representation approach with EfficientNet convolutional neural networks for effective malware classification. J Inf Secur Appl 69:103306. ISSN 2214-2126

New Generation 3D Optical Switch for Free Space Optical Networks

Mehman Hasanov, Khagani Abdullayev, Ali Tagiyev, Gulnar Gurbanova, and Nadir Atayev

Abstract The technical problems during the application of the optical satellite communication network (FSON) in free space were analyzed, with the aim of creating new optical photon network switches for the construction of the optical communication network between nanosatellites and terminals in ground stations (GS) of the FSON nanosatellite-nano-satellite and nanosatellite-ground terminals, whereas a new generation 3D optical photon switch (OS) is proposed for switching optical channels between transmitters, and the principle of its operation at different wavelengths is explained appropriately. Based on the working principle of the new generation 3D optical switch, which is intended to be installed on nanosatellites and optical terminals, the reliability indicators of its elements were calculated and the sustained operation period was investigated.

Keywords Optical satellite communication network · Optical switch · Nanosatellite · Sustained duration · Optical terminal

1 Introduction

The wide range of optical satellite communication network (FSON) application areas in free space, including broadcasting of satellite-to-satellite, terrestrial Internet and other multimedia services, conducting remote sensing, application in radio astronomy, restoration of natural and other disaster areas, mobile and stationary military application in technologies and fields, research of oceans and seas, application

M. Hasanov · A. Tagiyev · G. Gurbanova · N. Atayev (✉)
Azerbaijan Technical University, Baku, Azerbaijan
e-mail: nadiratayev@gmail.com

M. Hasanov
e-mail: mhasanovnew@gmail.com

K. Abdullayev
National Aviation Academy, Baku, Azerbaijan
e-mail: khabdullayev@naa.edu.az

G. Rajakumar et al. (eds.), *Intelligent Communication Technologies and Virtual Mobile Networks*, Lecture Notes on Data Engineering and Communications Technologies 171,
https://doi.org/10.1007/978-981-99-1767-9_12

of 5-6G technologies due to fast broadcasting and broadband spectrum capabilities and high-speed transmission, organization of reception, reliability, high economic and technical indicators in alternative to radio frequency satellites and radio wave frequencies, having a small mass, energy efficiency, having broadband and high transmission capacity, resistance to external interference and obstacles, ensuring information security, at the same time organizing communication between several nanosatellites and terminals of stations on the surface of the earth, etc., reaching inaccessible also difficult communication places it creates valuable scientific and applied interest points for the organization of multimedia services [1].

At the same time, the use of the radio frequency (RF) spectrum is limited, but by using new wavelength division multiplexing techniques, optical operators have an attractive role for high throughout and capacity [2–7] applications without requiring a spectrum licensing process. The application of the FSO network in the next generation of 6G mobile GSM networks has great prospects [8]. The free space optical communication network (FSON) is designed in the wavelength range (500–2000 nm) and uses the propagation and reception of a laser beam as a carrier signal that provides wireless communication between the transmitting optical terminals of the satellites and the ground station terminals using the spectral compression method [2, 3]. Other advantages of the FSON communication network include easily scalable and reduced segment network sizes, lightweight, and compactness [9–15]. Based on the analysis, the development of space technologies and the creation of new complex space service areas open new perspectives for optical space communication. High-speed data processing between spacecraft, the creation of broadband optical transmission medium architecture (inter-satellite, satellite-ground-satellite) in FSO is of great interest to researchers [16].

Challenges ahead. Ahead problems: Currently, in order to expand and accelerate the application of FSO, the study and application of the composition of the materials used in the preparation of nanosatellites, the problems of monitoring optical beams, the study and creation of devices that switch optical beams in nanosatellites and ground stations, located at different distances from each other theoretical study of atmospheric parameter changes and turbulence affecting the effective indicators of optical signals during the application of FSON for the wavelength range (500–2000 nm) between nanosatellites, nanosatellite-ground-nano-satellite, between terminals of ground stations, studying the optical properties of new materials creation of optical devices with parameters, training of specialists in this field, etc. The lack of standards for nanosatellite parameters and indicators, the investigation of technological shortcomings to increase the service life of nanosatellites, the application of alternative energy sources in nanosatellites, the investigation of management problems in full FSO optical networks, the application of FSO technologies in 6G mobile network projects [15, 16], monitoring of FSO networks, intermediate receiver and creation of transmitting optical transponder beacon systems, implementation, and organization of PON system of passive optical networks over FSO, etc., problems have not yet found their solutions.

Objective. Until now, only two missions have successfully flown optical communication systems: NICT's SOTA (Small Optical TrAnsponder) and DLR's OSIRIS

(Optical Space Infrared Downlink System). These systems are still not suitable for CubeSat platforms, but they have been the closest successful attempts in this direction, with only satellite to ground communication application.

With that in the near future, there is a great need for the establishment of FSON in low earth orbit using long-lived, stable nanosatellites, the creation of new optical network switches for the establishment of an optical communication network between nanosatellites and terminals in ground stations that meet the requirements of space conditions [17, 18]. In this regard, this paper proposes the application of FSON whereas a new generation 3D optical switch (OS) for nanosatellites, terrestrial receiver, and transmitter terminals in LEO.

2 Arrangement of Optical Communication Between Nanosatellites and Nanosatellite-Ground Station Terminals

Figure 1 shows the arrangement scheme of optical communication between nanosatellite and nanosatellite-ground station terminals through optical switches.

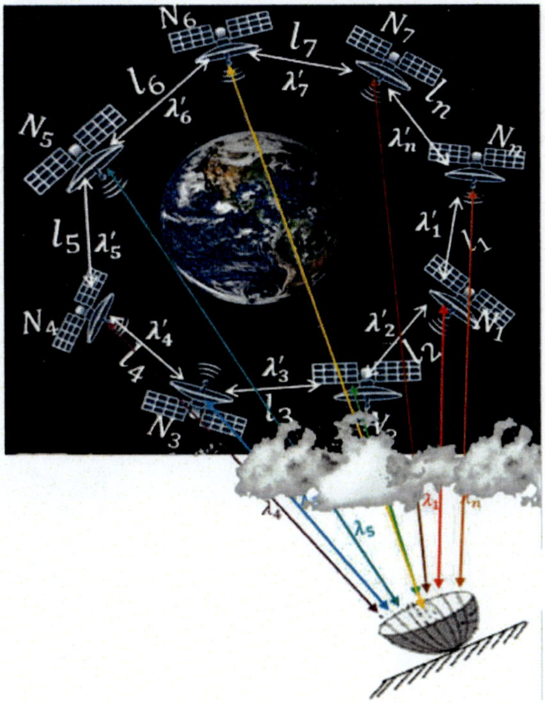

Fig. 1 Constellation scheme of optical communication between nanosatellite and nanosatellite-ground station terminals

As can be seen from Fig. 1, for the organization of the FSON, the application of the topology of the communication between the terminals of the ground station and the number of nanosatellites at an altitude of H (H = 500 – 2000 km), N_n ($N = N_1 - N_n$)n the optical switch LEO is envisaged. The receiver and transmitter terminals of each nanosatellite, located at different distances $L_1 - L_n$ from each other, are provided with an optical switch, and the incoming optical information stream for the reception and transmission of an optical laser beam is transferred to the wavelengths $\lambda_{xx}(\lambda_n^{;})$ of optical channels with a capacity of $n \times m$ according to the switching carried out by the pre-program. With each nanosatellite, the ground terminal interacts with the corresponding FSO channel of λ_1 to λ_n wavelengths.

2.1 The Working Principle of the Optical Switch (OS)

In this regard, we developed a new switch for use between nanosatellites located in LEO and the terminals of the ground station. As shown in Fig. 2, the optical photon switch has input/output λ_1 to λ_{nm}. input and output optical ports. Because wavelength multiplexing is used, each optical port has the ability to perform input and output functions.

The optical part of the optical switch that can interact in space receives signals from n number of optical channels with a wavelength of λ_{nm}, which are compressed according to the spectrum. $\sum_{n=1}^{m} \lambda_{nm} = \lambda_{11} + \lambda_{12} + \lambda_{13} + \cdots + \lambda_{1n} + \lambda_{21} + \cdots + \lambda_{2n} + \cdots + \lambda_{31} + \cdots + \lambda_{3n} + \cdots + \lambda_{nm}$ from the optical channels entering from the 8-optical waveguide, choosing the other 9-required λ_{nn}, , 3-freely interactive microelectromechanical part to direct to switched optical output channels and 5-micromotors (attached to electromechanical parts of the joint cross) based on the

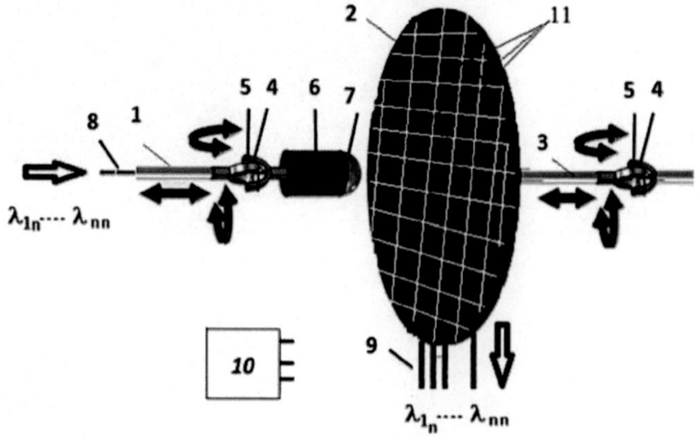

Fig. 2 Optical-photon switch

appropriate command given by the software control system attached to microelectromechanical with 6-scanning laser 3D axis movement in space inclined to the intersection of the necessary channels and is directed to the 11-semitransparent mirrors (SM) of the 2-hemispherical switching surface through the 7-lens. With the help of the 11-semitransparent mirror located at the intersection, channels are switched and the optical information stream is transmitted to the switched 9-output optical channels. The structural block diagram showing the working principle of the new generation 3D optical switch is shown in Fig. 1, and the working principle is the same as in the devices described in [17, 18].

Each 11-semitransparent mirror that switches optical channels can be in two states. One of the states is the initial or passive state of the 11-semitransparent mirrors. In this case, the semitransparent mirror-11 does not block the optical light stream and the optical beam enters the channel. In the second case, the 11-semitransparent mirror is raised in the upward direction. In this case, the 11-semitransparent mirror cuts the front of the optical stream and directs the optical beam into two perpendicular directions to separate optical fibers. 11-the active working parts of the semitransparent mirrors are covered with a layer capable of reflecting and transmitting a part of the optical light stream. The state of the semitransparent mirror-11 is controlled based on the command given by the control system with the program-10 corresponding to the state of the optical signal entering the network from different directions.

In order to increase the number of switching channels of the presented optical switch, the surface of 2 half-spherical switching is equipped with k moving 11-semitransparent mirrors located at the intersection of n vertical rows and m horizontal columns in the form of a matrix. In this case, it is possible to switch the number of channels corresponding to the combination $2 \cdot \left(2^{n-1} - 1\right) + 2$ at the $n \times m$ output of the switch, and the direction of the incoming optical input stream can be changed in different directions, including the opposite direction (wavelength selection and control according to the program) will be able to direct. The light diode of the switch moves in microns by choosing the closest and optimal distance up and down and right and left or at different angles and provides $n \times m$ optical channel switching. Semitransparent mirrors of the optical switch channels are all monitored and controlled by a 10-program control system from the ground station.

The structural block diagram showing the working principle of the new generation 3D optical switch is shown in Fig. 3, and the working principle is the same as in the devices described in [17, 18].

In the new model of the presented OS, 1-micromotor provides rotation of the 2-micromotor for 360° around the central axis of the system. Micromotors 1 and 2 direct the optical signals from the optical input/output channel onto the SM through the directional laser and the head of the light-emitting diodes. SM is made in the form of a matrix, and each matrix is calculated for one of the wavelengths $\lambda_{1n} \div \lambda_{nn}$, and optical signals separated by wavelengths in separate vertical and horizontal directions are directed to the output/input optical channels. In the reverse process, it is carried out according to the same rule. To reduce channel switching time, SM is placed on 7-hinges, and SM is moved by 3-micromotor to direct the optical beam to the required channel. Control of all micromotors and regulation of the state of the SM is

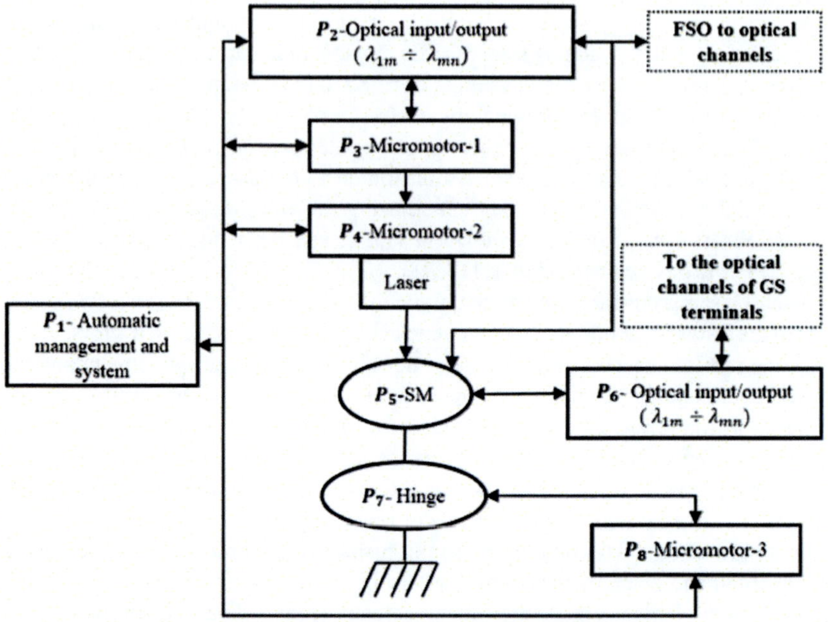

Fig. 3 Structural block diagram of a new generation 3D optical switch

carried out with the help of a 10-automatic control system that provides OS software and operation through a feedback loop.

The new generation 3D optical switch prototype has the following advantages compared to other optical switches:

- Depending on the number of light sources, laser light sources and light diodes can be placed together or separately in the 360° rotating head, compact assembly of lasers and light diodes, thereby reducing their size to a minimum, reducing the amount of material used and the final OS dimensions and financial cost;
- Controlling the heads of final OS lasers and light-emitting diodes by means of a high-precision micromotor, the switching time of optical signals is increased many times, which includes controlling the movement of lasers and light-emitting diodes in the programmed direction;
- By means of the control and measurement system integrated into the switch, the possibility of permanent remote monitoring of the optical switch and the effective diagnosis of the problems that have arisen.

Assume that all the active elements of the 3D optical switch shown in Fig. 2 belong to non-recoverable systems [9]. Let us denote the probability of unrejected operation of the elements by generalizing similar elements (micromotors) from P_1 to P_8. Here let us take P_1 as—automatic control, P_2 as—optical input/output, P_3.

as—micromotor 1, P_4 as—micromotor 2, P_5 as—MP, P_6 as—optical output/input, P_7 as—joint, P_8 as the probability of unrejected work of 3rd micromotor.

If we denote the probability of continuous operation of each element in time t by $P_i(t), i = \overline{1,8}$ and the intensity of rejection by $\lambda_i(t), i = \overline{1,8}$

$$P_i(t) = e^{-\int_0^t \lambda_i(\tau)d\tau}, \quad i = \overline{1,8} \tag{1}$$

then this can be written for each element [9, 10]. This will be the case if the rejection intensity of the elements is constant

$$\lambda_i(t) = \lambda_i = \text{const} \tag{2}$$

and in this case,

$$P_i(t) = e^{-\lambda_i t}, \quad i = \overline{1,8} \tag{3}$$

it will be written as formula (1).

According to Scheme 2, we can write for the probability of unrejection operation of the device as a whole—$P(t)$ according to the method of connection of each element in the device.

$$\begin{aligned} P(t) = {} & [1 - P_1(t) \cdot P_7(t) \cdot P_8(t)] \cdot [1 - P_1(t) \cdot P_5(t)] \\ & \cdot [1 - (1 - P_2(t)) \cdot (1 - P_3(t)) \cdot ((1 - P_4(t))) \cdot (1 - P_2(t) \cdot P_3(t)) \cdot (1 - P_2(t) \cdot P_4(t))] \\ & \cdot [1 - (1 - P_6(t)) \cdot (1 - P_7(t)) \cdot (1 - P_7(t) \cdot P_8(t))] \end{aligned} \tag{4}$$

In this case, it will be calculated as the average time of unrejection operation of the proposed device.

$$T_c = \int_0^\infty P(\tau)d\tau \tag{5}$$

The work case of unrejection dispersion will be calculated as follows:

$$D_c = 2\int_0^\infty \tau P(\tau)d\tau - T_c^2 \tag{6}$$

It will be calculated as the probability of unrejection operation of the device in a limited time interval (t_1, t_2).

$$P(t_1, t_2) = \frac{P(t_1)}{P(t_2)} \tag{7}$$

There will be a probability of unrejection operation of the device within the (3) condition.

$$P(t) = \left[1 - e^{-(\lambda_1+\lambda_7+\lambda_8)t}\right] \cdot \left[1 - e^{-(\lambda_1+\lambda_5)t}\right]$$
$$\cdot \left[1 - \left(1 - e^{-\lambda_2 t}\right) \cdot \left(1 - e^{-\lambda_3 t}\right) \cdot \left(1 - e^{-\lambda_4 t}\right) \cdot \left(1 - e^{-(\lambda_2+\lambda_3)t}\right) \cdot \left(1 - e^{-(\lambda_2+\lambda_4)t}\right)\right]$$
$$\cdot \left[1 - \left(1 - e^{-\lambda_6 t}\right) \cdot \left(1 - e^{-\lambda_7 t}\right) \cdot \left(1 - e^{-(\lambda_7+\lambda_8)t}\right)\right] \quad (8)$$

If we assume that all elements have the same rejection intensity,

$$\lambda_i = \lambda, \quad i = 1, 8$$

then formula (8) will be simplified to show the following:

$$P(t) = 6 \cdot e^{-2\lambda t} - 2 \cdot e^{-3\lambda t} - 22 \cdot e^{-4\lambda t} + 11 \cdot e^{-5\lambda t} + 29 \cdot e^{-6\lambda t}$$
$$- 14 \cdot e^{-7\lambda t} - 23 \cdot e^{-8\lambda t} + e^{-9\lambda t} + 26 \cdot e^{-10\lambda t} - 23 \cdot e^{-12\lambda t}$$
$$+ 9 \cdot e^{-13\lambda t} + 6 \cdot e^{-14\lambda t} - 5 \cdot e^{-15\lambda t} + e^{-16\lambda t} \quad (9)$$

If we take into account the formula (9) in (5), the average duration of the unrejected operation of the proposed device of the variance of the unreject operation is:

$$T_c = \frac{0.636}{\lambda} \quad (10)$$

The work case of unrejection dispersion will be:

$$D_c = \frac{1047}{\lambda^2} \quad (11)$$

Assume that all elements of the optical switch have the same non-rejection duty cycle. The period of operation of the experimental elements is assumed to be 10 years.

As illustrated in Fig. 4, the probability of system failure increases until the average working time of the elements reaches its maximum value of 0.5. Increasing the unrejected operation time of the elements does not affect the maximum probability of unrejected operation. Since the unrejected operation process for each element occurs with the same constant rejection intensity, the reliability indicators of the system will vary only depending on the way how these elements are combined. The obtained results prove that the proposed new switch is suitable for FSON (with a lifetime of nanosatellites is about 3 years).

The results of the calculations show that the maximum value of the probability of fail-safe operation of the presented switch takes a relatively different value than if the reliability indicators of the elements are the same. This confirms the above-mentioned idea once again, as the fail-safe operation time of the system determines

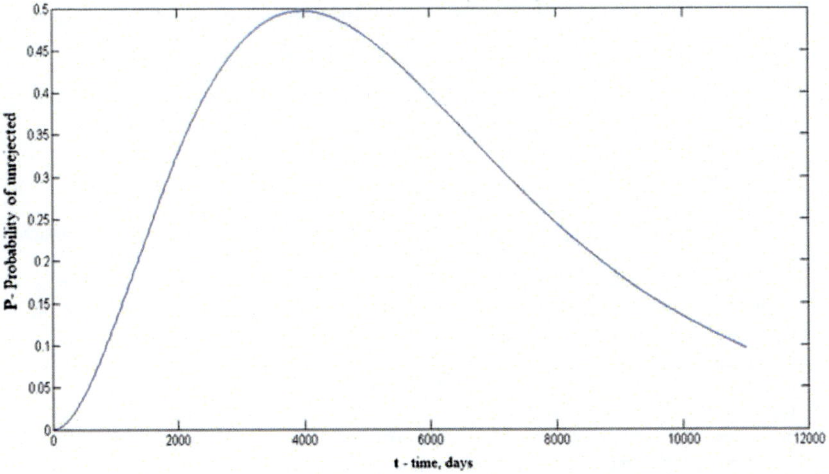

Fig. 4 Graph of the probability of failure of OS if the warranty period of the elements of the 3D optical switch is 10 years

the dependence of the method of their combination in the switch, rather than the fail-safe operation times of the elements separately.

3 The Result

Technical problems during the implementation of the free space optical satellite communication network (FSON) were analyzed, and new optical network switches were created for the design of the free space optical communication network (FSO) between nanosatellites and between terminals in satellite-ground stations (GS), FSON receiver and transmitter were created. A new generation of a 3D optical switch for terminals, and its principle of operation at different wavelengths is justified. Based on the working principle of the new generation 3D optical switch, which is intended to be installed on nanosatellites and optical terminals, the reliability indicators of its elements were calculated and the uninterrupted operation period was investigated. The maximum value of the probability of non-failure operation of the proposed commutator is relatively different from the case where the reliability indicators of the elements are the same, and it has been confirmed that the non-failure operation time of the system is more dependent on the method of their connection in the commutator than the time periods of the individual non-failure operation of the elements. But currently, analyzed research paper relies on only computational mathematic and modeling aspects with early determined parameter values, which limits some determination characteristics. After all detailed computational analyses initial prototype model of CubeSat platform will be developed and its appropriate real-time implementations are to be detailly determined.

References

1. Kaushal H, Kaddoum G (2017) Optical communication in space: Challenges and mitigation techniques. IEEE Commun Surv Tutor 19(1):57–96
2. Tarasenkov MV, Belov VV, Poznakharev ES (2022) Estimation of optimal wavelengths for atmospheric non-line-of-sight optical communication in the UV range of the spectrum in daytime and at night for baseline distances from 50 m to 50 km. J Opt Soc Am A 39:177–188
3. Maharjan N, Devkota N, Kim BW (2022) Atmospheric effects on satellite-ground free space uplink and downlink optical transmissions. Appl Sci 12:10944
4. Nikbakht-Sardari N, Ghiamy M, Akbari ME (2022) A charmin. Novel adaptive hard-switching based hybrid RF-FSO system architecture using two threshold values in the presence of atmospheric turbulence and pointing error. In: Results in engineering
5. Kumari M, Arya V (2023) Design of ring-based 1 Tbps hybrid PON-FSO fault protection system using add/drop multiplexer. Opt Quant Electron
6. Hasanov MH, Atayev NA (2022) Early conceptual model of nanosatellite with laser beam control and active transponder system. In: Systems of signal synchronization, generating and processing in telecommunications (SYNCHROINFO). IEEE, pp 1–4
7. Hasanov MH, Atayev NA (2022) Algorithm design nanosatellite based on radio frequency and optical communication. In: Problems of information technology, vol 13, no 2, pp 61–68. https://jpit.az
8. Carrasco-Casado A, Mata-Calvo R (2020) Free-space optical links for space communication networks. Japan, 66 p
9. NASA (2017) CubeSat 101 basic concepts and processes for first-time CubeSat developers. California, 66 p
10. Balaji KA, Prabu K (2018) Performance evaluation of FSO system using wavelength and time diversity over malaga turbulence channel with pointing errors. Optics Commun 410:643–651
11. Carrasco-Casado A, Mata-Calvo R (2012) Free-space optical links for space communication networks. In: Springer handbook of optical networks, pp 1057–1103
12. Zhou Y, Gu Z, Zhang J, Ji Y (2023) Efficient deployment of aerial relays in FSO-based backhaul networks. J Opt Commun Netw 15(1):29–42
13. Kim SJ, Han SK. (2022) Power efficiency analysis of spatial diversity based vertical FSO links with pointing error in multiple beam transmissions
14. El-Nahal F, Xu T, AlQahtani D, Leeson M (2022) A bidirectional wavelength division multiplexed (WDM) free space optical communication (FSO) system for deployment in data center networks (DCNs). Sensors 22(24):9703
15. Nath S, Singh SP, Sengar S (202) Interference and noise analysis for hybrid FSO/RF-based 6G mobile backhaul. ETRI J 44(6):966–976
16. Hosseini N, Jamal H, Haque J, Magesacher T, Matolak DW (2019) UAV command and control, navigation and surveillance: a review of potential 5G and satellite systems. In: IEEE aerospace conference, Big Sky, MT, USA, pp 1–10
17. Hasanov MH (2019) Photon switch of full optical networks, vol 13, № 8. t-Comm., Moscow, pp 47–51
18. Hasanov MH, Hacıyeva KR, Qodjaeva SF (2018) Multifunctional adaptive piezoelectric switch of optical channels. T-Comm Telecommun Transp 13(29):64–69

Terrain Dimensions and Node Density Analysis of MANET Using NS2 and BonnMotion

Satveer Kour, Manjit Singh, Himali Sarangal, and Butta Singh

Abstract Different reasons make data packet to be lost in Mobile Ad-hoc Networks (MANETs) over wired networks, in which packet loss is typically brought on by network congestion. Packet losses in MANETs can either be attributed to the dynamic character of such networks or to the wireless communication situations (fading, interference, multi-path routing, link failures, network partitioning). This latter issue might be brought on by the node's mobility or by its battery running out. Terrain dimensions, speed, and density of nodes are the important features that set apart MANETs from wired networks. The major objective is to reduce packet loss and delay and improve packet delivery ratio and throughput. In this paper, we have implemented MANET by using AODV, DSR, and DSDV and simulated on network simulator. The significance of terrain areas and node density is analyzed over reactive and proactive routing protocols. The expansion of terrain dimensions and node density is analyzed through simulation experiments by using NS2 and BonnMotion tools. The comparative outcomes are analyzed with a pictorial evaluation of five QoS performance metrics through node density under various terrain dimensions.

Keywords Mobile ad-hoc networks · Routing protocols · Network simulator 2 · BonnMotion

S. Kour
Department of CET, Guru Nanak Dev University, Amritsar, India
e-mail: satveer.dcet@gndu.ac.in

M. Singh · H. Sarangal · B. Singh (✉)
Department of Engineering and Technology, Guru Nanak Dev University, Regional Campus, Jalandhar, India
e-mail: bsl.khanna@gmail.com

M. Singh
e-mail: manjit.ecejal@gndu.ac.in

G. Rajakumar et al. (eds.), *Intelligent Communication Technologies and Virtual Mobile Networks*, Lecture Notes on Data Engineering and Communications Technologies 171,
https://doi.org/10.1007/978-981-99-1767-9_13

1 Introduction

Since the 1970s, wireless network technologies have been able to give mobile users access to pervasive computing without worrying about their location [1]. A computer network employing wireless data links among network nodes is stated as a wireless network. The cell phone networks, wireless local area networks, wireless sensor networks, satellite communication networks, and terrestrial microwave networks are some examples of wireless networks [2]. A wireless network covers with two types including infrastructure and no infrastructure. The concept of guided media is existed in wireless infrastructure networks to make communication between the fixed connected base stations with other fixed base stations. A cell defines as the communication range of a base station. The moving nodes existed in the cell communicate with the nearest base station. If a moving node leaves the previous base station's range, it connects to the subsequent base station. This circumstance qualifies as a handoff. There are no fixed base stations in the network with no infrastructure. The nodes do communicate with one another and are mobile in nature. The system's organization and management are entirely under the control of a terminal. The entire system is portable, and each terminal is free to travel however it pleases [3]. For making communication with long distance users, the multi-hop property is used. A network device will serve as a router during the construction of a route. Any moving objects, including boats, cars, trucks, and planes, can serve as a node [4]. Besides the homogenous and heterogeneous types, some of the advantages of MANETs are cost, convenience, mobility, productivity, easy setup, expansion, security. Some performance issues like distributed operation, loop freedom, demand-based operation, security, topology change, terminal problems, and traffic patterns also affect the performance of MANETs. The process of routing [5] is performed to discover a path between mobile nodes. The node-by-node utilizes the MANETs current routing protocols. The path establishment and path maintenance are two main functions of a routing protocol. MANET routing protocols are classified into three categories: reactive (AODV, DSR), proactive (DSDV, WRP, CGSR), and hybrid (TORA, ZRP) [6]. Ad-hoc On Distance Vector (AODV) creates a link on demand only due to its reactive property. It creates the concept of sequence number which is accommodated with packet's header. It assures the cycle-free links through these sequence numbers. The path with the lower sequence number is considered as the fresh path. An entry for the terminal node's single path can be found in a node's table. The address of a target, its sequence number, its neighbors, all of its neighbors, its lifetime, and any flags are all contained in the specified row of the table [7]. Transmission is done using the origin routing idea by Dynamic Source Routing (DSR). The whole path must be discovered first before initiating the transmission. The route making procedure is similar as to AODV (means path discovers on need only) [8]. The mobile node has its own cache memory to save the links. Transmission will start if the path is already found in the routing table, else, route discovery will start. The Destination Sequenced Distance Vector (DSDV) algorithm is built on the Bellman–Ford algorithm. The technique states that each node holds a next neighbor entry in its routing table. The table also holds the entry for all reachable nodes. It has the concept of duplex link instead of simplex links. Just to make the network cycle-free, it also applies the concept of sequence numbers. A terminal node creates the sequence number. At the periodical intervals, DSDV floods advertisement message of the advertising node. This flooding procedure helps to keep routing information accurate and stable [9].

The research work is based on the comparative study of proactive and reactive MANET routing protocols. On the basis of previous research work, we have chosen two reactive protocols (DSR and AODV) and one proactive routing protocol (DSDV). The most of the research works related to MANET routing protocols are restricted to 100 nodes only. We try to explore the congested scenario (up to 300 nodes) and further analyze the performance of the network.

2 Algorithm and Work Flow of Proposed Work

As per the node's movements concern, Algorithm 1 is showing the actual node movements. The designed scenario is executed for n seconds (i.e., 100 s). Initially, a source node and a destination node are selected. The iterative process starts in step 2 with a counter variable, *i*. The selected mobility model is random waypoint model which selects speed for a specific node 0 and max. The moving node travels with a selected speed. After every interval (i.e., specified by *i*), the new location is find out and new location is compared with the destined location. If the new location is the destined location, then the moving source node pauses for a while. For the next iteration, it will again select a new destination. The selection of a new destination depends upon the value of flag parameter.

Algorithm 1: Node Movements

```
Select a source and destination node.
   Initially, set flag = 0.
   For i = 0 to n-1 do
   {
   If flag = 0 then
   {
   Choose speed from 0 to max (as per random waypoint model)
   Set flag = 1.
   }
   Travel to destination node with chosen speed.
   Update the current location with new location for time interval i.
   If new location is the destination location then
   {
   Take a pause.
   Select a new destination node.
   Set flag = 0.
       }
   }
```

Pseudo Code for Sender:
S

1. Network Establishment Process.
2. Is Network Establishment successful? Then send the packet.
   Else continuously checking for a network establishment.
3. Is the network still alive? Then again send the packet.
   Else stop.

Pseudo Code for Forwarder Node:
F:

1. If a packet available, then receive the packet.
   a) Is this the nearest node to the sender/receiver?
   i) If yes, the pause transmission; forward the packet.
   ii) If no, then find out the nearest forwarder node to the sender/receiver; the pause transmission; forward the packet to the forwarder.
   Else waiting for the packets from the receiver side.

Pseudo Code for the Receiver Side:
R:

1. Routing Control: Establish Network.
2. Is network establishment successful? Then receive the packet and stop.
   Else checking for establishing a network.

Fig. 1 Pseudo code for sender, intermediate, and receiver nodes

A pseudo code for data transmission at sender side, at intermediate node, and at receiver side is given here in Fig. 1:

3 Simulative Tools and Setup

3.1 NS2

All types of networks (having different topologies) can be simulated on NS2 to examine the routing protocols. C++ is used as a programming language in NS2, and for making simulation interface, this scripting language is used. NS2 primarily uses UNIX as a platform, but a virtual machine named as VMWare Workstation is created. For the experiment, VMWare Workstation is run on Fedora20 operating system. The actual movements of nodes are shown on Network animator (NAM). Some buttons like next, run, exit/stop, and pause are existed in NAM as shown in Fig. 2. NS2 gives facilities to use different protocols like CBR, FTP, HTTP, TCP, and UDP. It embeds Object-oriented Tool command language (OTcl) and C++ languages. The components of NS2 are depicted in Fig. 3 [10].

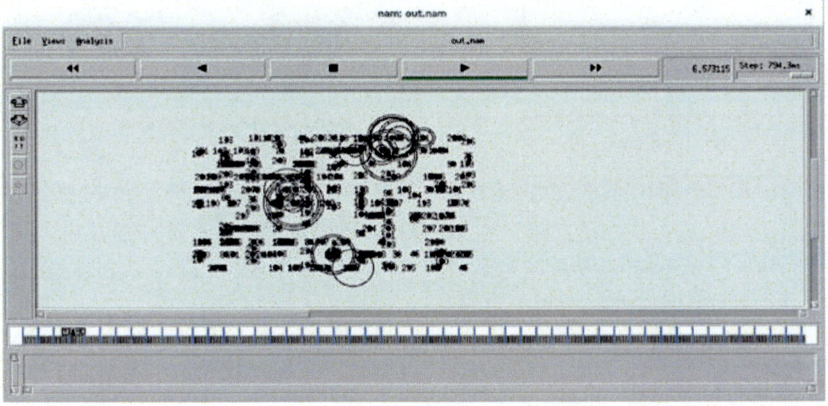

Fig. 2 NAM file

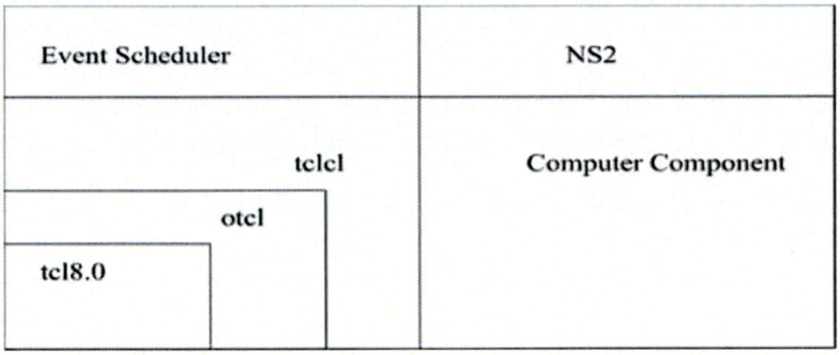

Fig. 3 Components used by NS2

The tcl files are created in Tcl and C++ languages in NS2 because C++ language carries simulative facilities like algorithm implementation, manipulation of byte, packets processing as shown in Fig. 4.

3.2 BonnMotion

The mobility scenarios are created and analyzed through BonnMotion tool which is Java software. The movements of a mobile node can be recorded in.bm file of a specific chosen model. Further, this.bm file is exported (embedded) with the tcl file. The generated scenarios can also be compatible with other simulators like ONE, QualNet, COOJA, NS3, etc. A VMware Workstation machine is initiated on Windows 10 operating system. Fedora20 is acted as a dual-mode operating system on virtual

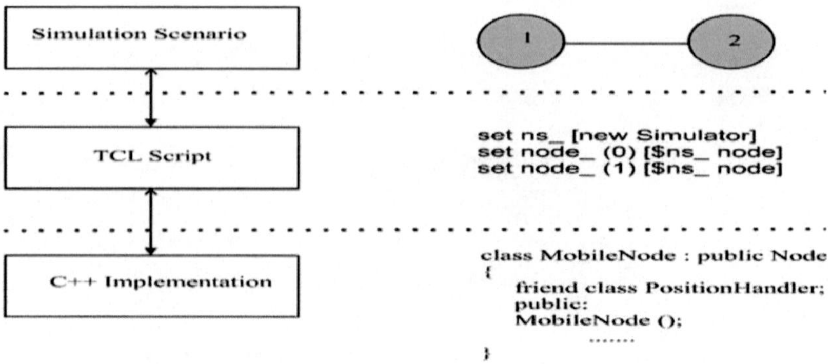

Fig. 4 tcl file creation in NS2

machine. NS2 and BonnMotion are operated on Fedora20 for this research. BonnMotion tool is started and compiled through bm package [11]. The BonnMotion-3.0.1 file structure is shown in Fig. 4. There are a total of 24 mobility models in BonnMotion-3.0.1. The file structure of BonnMotion is shown in Fig. 5. A command line input or a file input are both acceptable methods of input. Because settings are also preserved when using a file approach, parameter values can be changed without causing any changes.

The syntax of movement generation in BonnMotion tool is:

$simulator_interface at time "$node_numbersetdest X Y speed".

Here, ns_ is simulator interface, node_(0) is the node number, setdest is the location of destination node (in X, Y coordinates) to which it is traveling, and speed is the generated speed by mobility model. The below statement is showing as an example of speed generated by the tool:

$node_(0) set X_ 924.0
$node_(0) set Y_ 319.85461187918474
$ns_ at 0.0 "$node_(0) setdest 924.0 324.2429657630937 1.7805484663717528"
$ns_ at 2.464607937828937 "$node_(0) setdest 924.0 324.2429657630937 0.0"
$ns_ at 3.4384419477476342 "$node_(0) setdest 924.0 329.2429657630937

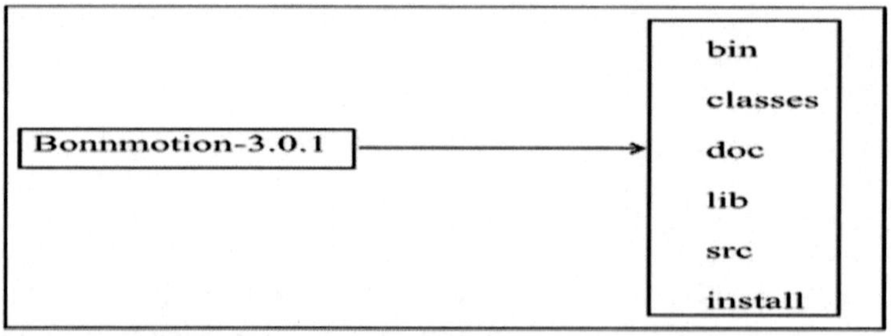

Fig. 5 File organization of BonnMotion-3.0.1

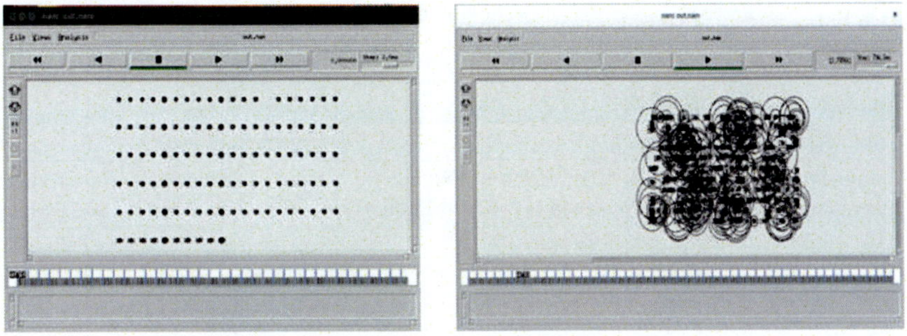

Fig. 6 Initial node distribution and node's movements during simulation

2.4542655358129593"

The movements (before and after) of nodes are shown in Fig. 6.

4 Results and Discussions

The different QoS performance metrics (packet delivery ratio, packets delay, dropped packets, throughput, packet overhead) are studied on four different networks: (1) a network with 25 nodes in 900 m × 50 m, (2) a network with 50 nodes in 1966 m × 100 m, (3) a network with110 nodes in 5300 m × 800 m, and (4) a network with 300 nodes in 15229 m × 1300 m. Table 1 is covered the list of parameters used.

The mentioned simulator NS2.35 is used for simulative experiments. The experiment is performed for 100 s for the node density for [900 × 50]m (one node per 1800 m), for [1966 × 100]m (one node per 3932 m), for [5300 × 800]m (one node per 38,545 m), and for [15229 × 1300]m (one node per 65,992 m). The number of used mobile nodes is 25, 50, 100, and 300 acted over random waypoint mobility model. The queue is Drop Tail queue for AODV and DSDV protocols and CMU Priority queue for DSR routing protocol. The buffer capacity of mentioned queues is 5.

4.1 Packet Delivery Ratio (PDR)

PDR is the value of received packets over the total sent packets (represented in percentage) [12]. DSR has an excessive PDR followed by AODV and DSDV in lower node density and terrain area as shown in Fig. 5, because DSR permits many paths to any target and permit each sender to select and control the paths used in sending its packets. But, its output is diminished when it works in a huge territory (with high density) because every node must spend more time to obtain the path information even if it is not the respective receiver. Hence, AODV and DSDV are consistently

Table 1 Description of variables

Variable	Value	Detail
Simulator	NS2.35	Network simulator
Experiment time	100 s	Experiment time
Terrain dimensions	[900 × 50]m, [1966 × 100] m, [5300 × 800]m, [15229 × 1300] m	X and Y directions for movements
Node density	25; 50; 100; 300	Total number of nodes in defined terrain area
Node arrangement	Grid	Grid topology
Mobility model	Random waypoint	Movement method
Connections	15; 31; 61; 156	Number of links
Routing protocols	DSDV, DSR, AODV	Path finding
MAC protocol	802_11 MAC	Protocol for wireless communication
Queue	CMU Priority, Drop Tail	Types of used queues
Length of a queue	5	Queue capacity
Antenna	Omni	Omni-directional antenna
Propagation method	Two-ray ground	Propagation model

Table 2 Packet delivery ratio, average end-to-end delay, and throughput analysis

	Packet delivery ratio			Average end-to-end delay			Throughput		
Nodes	AODV	DSDV	DSR	AODV	DSDV	DSR	AODV	DSDV	DSR
25	89.823	92.182	97.718	35.227	24.476	79.306	242.18	291.59	243.63
50	91.286	93.083	98.373	24.378	29.482	9.1336	495.84	546.68	472.05
100	93.374	94.964	98.276	10.226	10.133	27.854	969.09	1268.37	769.31
300	89.612	93.853	10.904	12.943	14.924	202.77	907.07	1591.15	2.1

playing better in all cases, because AODV does not add any extra workloads on data packets. DSDV is quite suitable for creating ad-hoc networks with lesser nodes. The performance of AODV and DSDV is slightly degraded when number of nodes is 300 due to congestion on the network and this thing takes more delay for transmitting packets. The quantitative and graphical results of PDR are depicted in Table 2 and Fig. 7, respectively.

4.2 Packets Delay

Packets' delay is the time difference of two times of packets, i.e., received and sent times of a data packet over the total packets [13]. DSR has a longer delay in comparison with DSDV and AODV. AODV is acting better in all scenarios due to

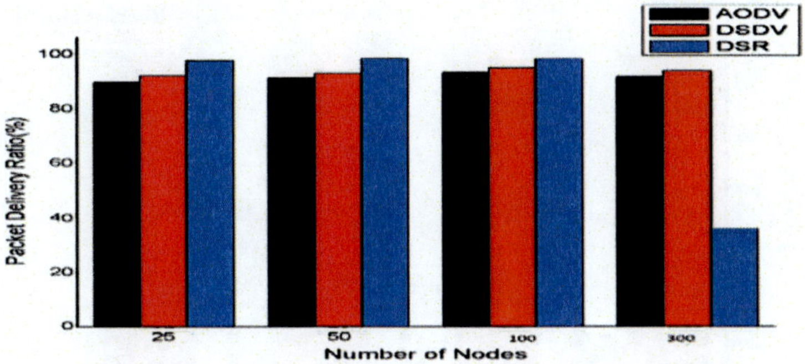

Fig. 7 Packet delivery ratio

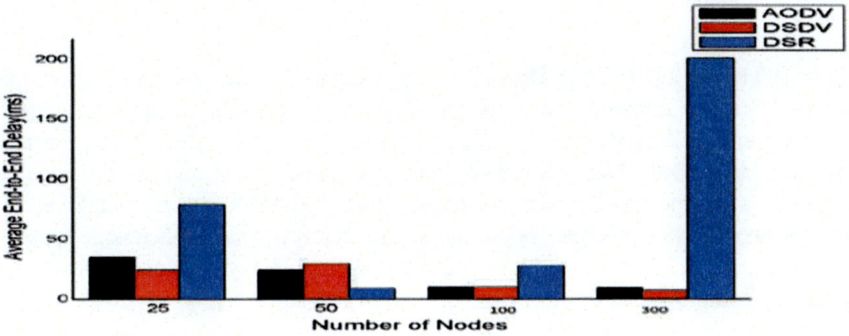

Fig. 8 Packets' delay

its suitability in dynamic nature of MANET. It has continuously decreasing delay values up to 300 nodes (from 35 to 9 ms), whereas DSR has the extensive delays (in low-density networks) because the route maintenance cannot repair a damaged path and the delay is more than in table-driven protocols. The delay value is increased for DSR when the number of nodes are more than or equal to 100 nodes. DSDV is consistently performing better in all networks because it is quite suitable for creating ad-hoc networks with lesser nodes and needs a periodical update of its routing tables (consumes power and a little amount of bandwidth). The delay value for DSDV is continuously decreased up to 300 nodes (from 24 to 10 ms) as shown in Fig. 8.

The quantitative results of average end-to-end delay is depicted in Table 2.

4.3 Throughput

Throughput is the successful transmission of packets from origin to target node in per second [14]. For DSDV, throughput is raised when the node density is increased

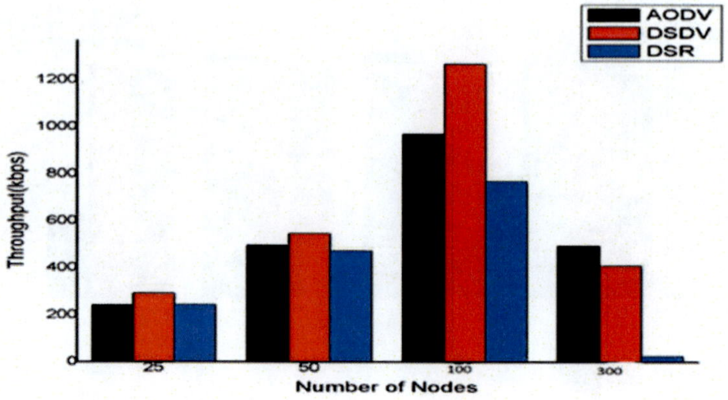

Fig. 9 Throughput

up to 100 (from 291 to 1268 kbps). But, its throughput value is decreased for 300 nodes due to its suitability for implementing ad-hoc networks with lesser nodes. The outcome of AODV is also upgraded with respect to node density, and terrain area increased (from 242 to 969 kbps) up to 100 nodes. But, the performance is degraded when the number of nodes is more than 100. DSR is performing well in short network (243–769 kbps), but its output drastically degrades when node density and area are increased due to higher connection setup delay. The quantitative and graphical comparison is shown in Table 2 and Fig. 9, respectively.

4.4 Packet Overhead

Packet overhead is the summation of packets generated for rerouting and data packet [15]. DSDV is generating more packets because it requires a periodical checking for its routing tables and it consumes a battery power, whereas DSR has less generation of packets because a node proceeds a HELLO message only if the destination address is not existed in the header part of the message. It reduces the count of route requests' spreader in the environment. The packet overhead rate is increased in case of AODV up to 100 nodes (from 12,440 to 48,737 packets). But, packet overhead rate is decreased when it is more than 100. The same conclusion has been made for the DSDV routing protocol. The value is increased up to 100 nodes (14,724–61,521 packets), but decreased for 300 nodes. The packet overhead rate is increased for DSR up to 100 nodes, but decreased for 300 nodes due to connection setup delay as shown in Fig. 10. The quantitative comparison of packet overhead is shown in Table 3.

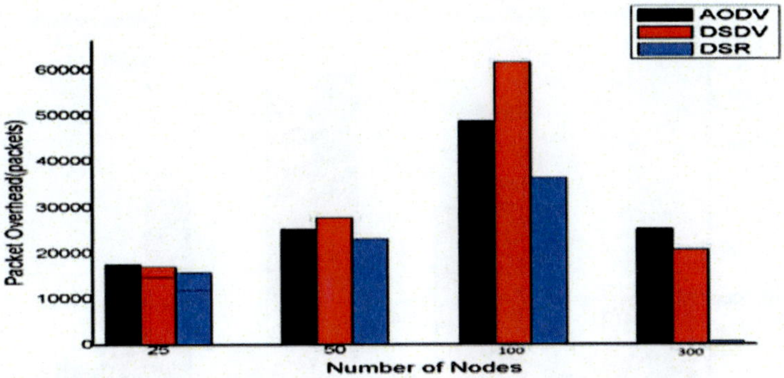

Fig. 10 Packet overhead

Table 3 Packet overhead and packet loss analysis

	Packet overhead			Packet loss		
Nodes	AODV	DSDV	DSR	AODV	DSDV	DSR
25	12,440	14,724	12,009	1266	1151	274
50	25,258	27,747	23,114	2202	1919	376
100	48,737	61,521	36,270	3229	3098	625
300	46,759	79,373	1073	4857	4879	956

4.5 Total Dropped Packets

Total dropped packets depict the loss rate of packets because of congestion or unavailability of routes [16]. The packet drop rate has been raised by protocols if the network size is fall off or substantial with 25, 50, 100, and 300 nodes except DSR. Due to on-demand route finding ability, AODV performs worst in all scenarios. It did not investigate path in periodic times, so transmission of data is taking more delay. Same as, performance of DSDV is also degraded because it requires a regular checking of its routing tables (consumes bandwidth). The performance of all routing protocols for drop packets is degraded up to 100 nodes, but improved when the node density is equal to 300 as shown in Fig. 11. The quantitative comparison of packet loss is depicted in Table 3.

5 Conclusion

As the results shown, a packet delivery ratio is raised for DSR, when there is reduction in nodes and areas. Overall, DSDV and AODV are playing superior in all scenarios. DSR's delay rate is longer as compared to other two protocols. In metric

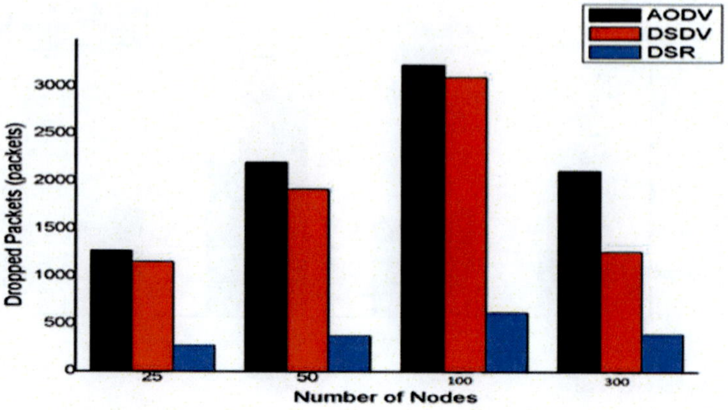

Fig. 11 Dropped packets

throughput, DSDV played better and DSR performed worst when the density and area are increased. In dropped packets' ratio, DSR played well and poorly performed by AODV. We observed that DSDV is supercilious in the packet delivery ratio than other two protocols.

The influence of variation in terrain and node density in MANET is clearly investigated and compared over routing protocols. The results vary due to congestion when variation comes in node density. A stable path may have broken; the finding of an expiry time is laborious. We can try to analyze the energy-efficient and mobility-efficient factors in future. We can try to decrease the ratio of dropped packets and packet overhead. Only one-side linking routing scheme is applicable in this scenario for making links between nodes. In future, we can apply multicasting scheme also. In future, we can try to analyze the load balance and power consumption factors. We can attempt to minimize the delay and packet overhead. As mentioned, unicasting routing scheme is followed by all the connected nodes.

References

1. Kaur J, Singh A (2019) A review study on the use of MANET for wireless devices. In: Proceedings of the international conference on advances in electronics, electrical & computational intelligence (ICAEEC)
2. Rashidah FO, Burhan UK, Farhat A, Roohie NM (2020) MANET security appraisal: challenges, essentials, attacks, countermeasures & future directions. Int J Recent Technol Eng 8(6):3013–3024
3. Alvarez Aldana JA, Maag S, Zaïdi F (2018) MANETs interoperability: current trends and open research. In: 32nd International conference on advanced information networking and applications workshops (WAINA). https://doi.org/10.1109/WAINA.2018.00132
4. Sobti R (2015) A study on challenges and issues on MANET. Int J Adv Res Electr Electron Instrum Eng 4(9):7854–7861

5. Panday N (2014) A comparative study of MANET routing protocols. STM J Mob Comput Commun Mob Netw 1(9)
6. Kaur G, Thakur P (2019) Routing protocols in MANET: an overview. In: 2nd International conference on intelligent computing, instrumentation and control technologies (ICICICT). https://doi.org/10.1109/ICICICT46008.2019.8993294
7. Fahad TA, Naseer S, Salim MS, Fouad S, Aljunid SA (2018) MANET routing protocols evaluation: AODV, DSR and DSDV perspective. In: Malaysia technical universities conference on engineering and technology (MUCET)
8. Yassine M (2014) Comparative analysis of routing protocols AODV DSDV and DSR in MANET. Int J Comput Sci Mob Comput 3:379–385
9. Raza N, Aftab MU, Akbar MQ, Ashraf O, Irfan M (2016) Mobile Ad-hoc networks applications and its challenges. Commun Netw 8(3):131–136
10. Fan-Shuo K, Bei-Bei C (2017) Performance evaluation of AODV, DSR and DSDV in mobile ad-hoc network using NS-2. In: The 4th annual international conference on information technology and applications (ITA)
11. Aschenbruck N, Ernst R, Schwamborn M (2010) BonnMotion—a mobility scenario generation and analysis tool. In: SIMUTools 2010—3rd international ICST conference on simulation tools and techniques, vol 51
12. Ching TW, Aman AHM, Azamuddin WHM, Attarbashi Z (2021) Performance evaluation of AODV routing protocol in MANET using NS-3 simulator. In: 3rd International cyber resilience conference (CRC), pp 1–4
13. Malany A, Sarma VR (2009) Throughput and delay comparison of MANET routing protocols. Int J Open Prob Comput Sci Math 2:461–468
14. Priyambodo T, Wijayanto D, Gitakarma M (2020) Performance optimization of MANET networks through routing protocol analysis. Computers 10(2)
15. Tamilselvi P, Ravi TN (2018) A survey on various techniques to minimize routing overhead in MANET. Int J Comput Sci Eng 6(9):374–379
16. Khan MS, Waris S, Ahmed I, Khan MI (2018) A comprehensive analysis of packet loss in MANETs. In: 17th IEEE international conference on trust, security and privacy in computing and communications/12th IEEE international conference on big data science and engineering, pp 118–123

Toward Secure Fault-Tolerant Wireless Sensor Communication: Challenges and Applications

Radhakrishna Bhat and K. N. Pavithra

Abstract The wireless sensor network (WSN) is gaining huge importance these days. It forms the backbone of IoT systems, a cutting-edge technology. As the applications are rapidly increasing, it is important to address the security concerns around this network. The fault-tolerant algorithms form an integral part of the wireless sensor distributed network. These algorithms make the wireless sensor network act as fault-tolerant, thus making the sensor readings more reliable. Even today, many sensor nodes are transmitting the data within the network in a wireless mode, with inappropriate security mechanisms in place. To overcome this gap, elliptic curve cryptography (ECC) and its variants are integrated with fault-tolerant sensor communications recently. In this study, we focus on presenting the challenges involved in implementing ECC-based secure sensor communication in the existing systems and the suitability of ECC-based secure communication for many real-time applications.

Keywords Wireless sensor network · Elliptic curve cryptography · Internet of things · Distributed computing

1 Introduction

The advancements in wireless technology, digital electronics, and micro-electro-mechanical systems have made wireless sensor networks to be of great interest. The formation of sensor-based wireless network is highly useful in many recent applications including healthcare, military, astrophysics, etc. There are be hundreds to thousands of sensor nodes in a wireless sensor network, forming a distributed system [1]. The sensor-based information plays a vital role in many Internet of things (IoT) applications [2].

R. Bhat (✉) · K. N. Pavithra
Department of Computer Science and Engineering, Manipal Institute of Technology, Manipal Academy of Higher Education, Manipal, Karnataka 576104, India
e-mail: rsb567@gmail.com

G. Rajakumar et al. (eds.), *Intelligent Communication Technologies and Virtual Mobile Networks*, Lecture Notes on Data Engineering and Communications Technologies 171,
https://doi.org/10.1007/978-981-99-1767-9_14

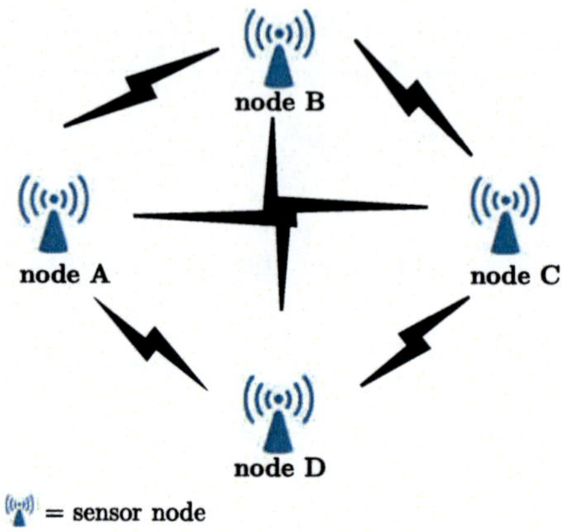

Fig. 1 Plain wireless communication

The widely known issue with these sensors is that the value obtained from sensors is not always accurate. Also, the expected precise values are not obtained [1]. Once the sensor senses the data, it needs to send this information to its neighbors or to the fusion center for advanced processing as shown in Fig. 1. Sometimes, a sensor node might also fail to respond due to unexpected exceptions like failure of link, energy depletion, radio interference, environmental calamities, and synchronization mismatch. When a faulty sensor node starts giving out erroneous data and the underlying network is not aware of it, this results in failure of estimation of sensor data or produces inaccurate data [3]. This issue has gained a lot of interest among the researchers, and finally, sensor fusion technique was defined as the best possible solution for getting acceptable results from the sensor networks. Among the set of algorithms defined for the fault-tolerance, the widely used algorithm is the Brooks-Iyengar algorithm. This is a hybrid distributed algorithm which combines the approach of Byzantine agreement protocol along with sensor fusion. The algorithm works even in the presence of noisy data making it useful for real-time applications. Every processing element of the distributed network runs this algorithm. The results from each such processing element are put together, and the weighted average over the midpoints of the region is calculated. The algorithm efficiency is defined as $\mathcal{O}(N \log N)$ where N corresponds to the number of processing elements.

The Brooks-Iyengar algorithm gained a wide popularity because of its seamless approach in real-time applications [2]. The practicality of this algorithm was such that the prominent defense agency, Defense Advanced Research Projects Agency (DARPA), has used it for collecting sensor data in real-time. Also, another prominent UK defense manufacturer, The Thales Group, has also used this algorithm. Both the precision and accuracy have been improved in this distributed algorithm, since each sensor communicates with other sensors using energy efficient methods. The Brooks-Iyengar algorithm has made a great impact in wireless sensor technology and is playing a vital role in automation process. The WSNs are deployed in a variety of areas varying from personal applications to commercial applications and goes on

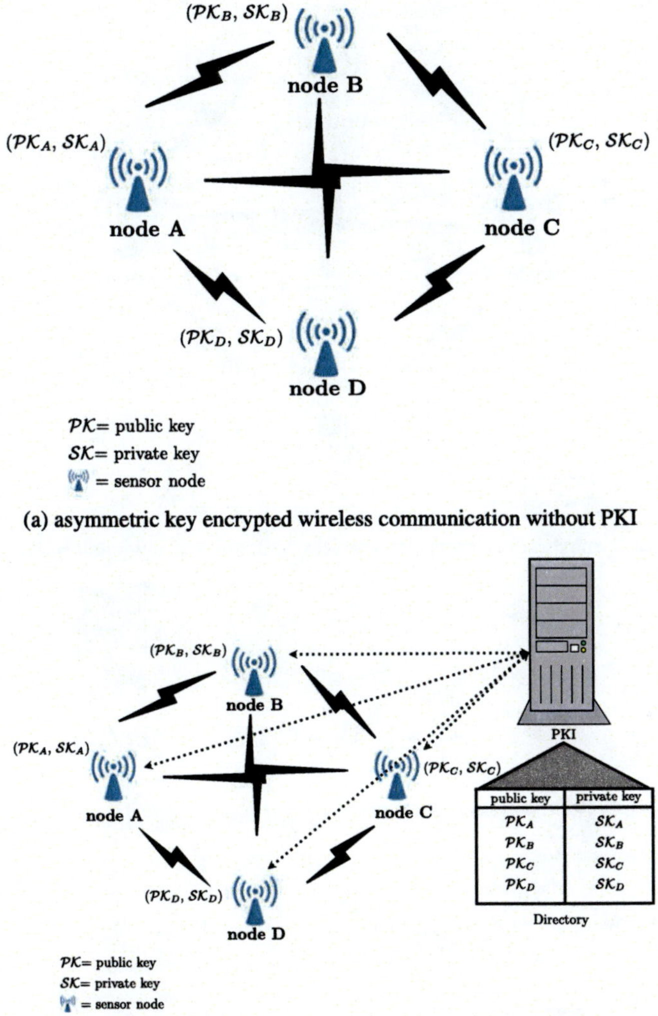

(a) asymmetric key encrypted wireless communication without PKI

(b) asymmetric key encrypted wireless communication with PKI

Fig. 2 Asymmetric key-based wireless communication

up to the deployment in critical environments like battle areas. The WSN consists of numerous amounts of sensor nodes. The deployment area can happen to be very harsh or hostile environment and remote [4]. This attracts the attackers to exploit the vulnerabilities in the system if they are left unattended. There could be a lot of attacks involved in the WSNs like eavesdropping the traffic, injecting the malicious packets into the network, compromising a sensor node, etc. The major security issues that are to be addressed at the earliest in any WSN are to preserve the privacy of the data

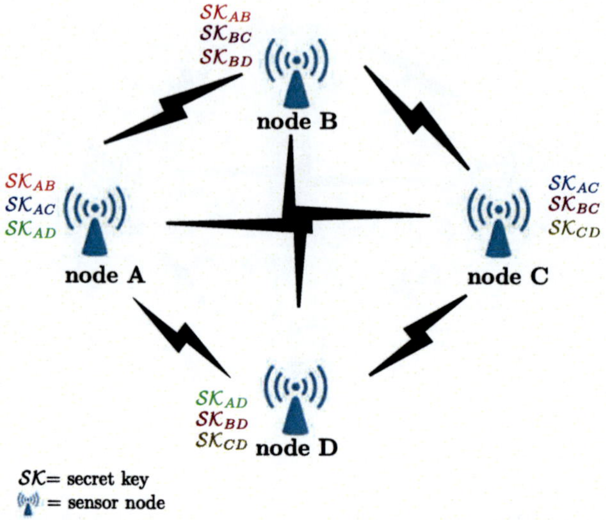

(a) symmetric key encrypted wireless communication without KDC

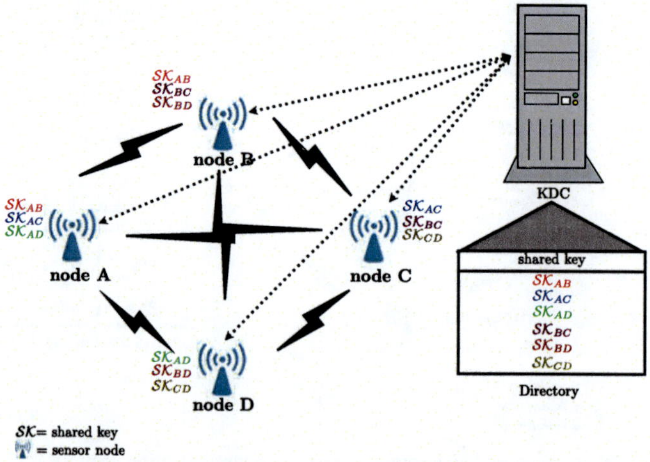

(b) symmetric key encrypted wireless communication with KDC

Fig. 3 Symmetric key-based wireless communication

and authenticate the nodes. The privacy enables a secured communication among sensor nodes or between sensor node and control site as shown in Figs. 2 and 3. The authentication ensures no illegal sensor node or any other device can fraudulently involve in the WSN communication [5]. Due to the resource constraint nature of the sensor nodes, it is impossible to employ the traditional cryptographic techniques as it is to the WSN. This might eventually cause an overhead in the system [4].

The general requirement for the WSN can be thought as (i) data confidentiality which provides assurance that only the authorized receiver is reading the data, (ii) data integrity which is resistant toward unauthorized alterations, (iii) availability which makes sure that the data is available at any given time, (iv) data freshness to make sure that the data sent by the sensors is the recent ones and does not contain any old values (v) self-organization to enable the sensors to have self-healing, (vi) secure localization helping to quickly locate the sensor, and (vii) time synchronization for obtaining a good performance and authentication to prove the identity of sender [4].

The attacks can happen in order to breach the confidentiality or network integrity. With reference to open systems interconnection (OSI) model, there can be various possibilities of attacks [4] which can happen at different layers. The sensor system needs to equip with appropriate security mechanisms to mitigate the security risks. The main reason which greatly constitutes the security risk is because of the advantage which the WSN provides to its users and is the in-network processing of the collected data [6]. This creates an aggregated information which constitutes a major point of security threat. The attacker can straight away steal or modify the useful intelligible information itself. Building a secure model for the WSN is restricted because of the constraints which are part and parcel of WSN such as limited resource availability and unreliable communication. The sensors generally equipped with less memory, less CPU, and less energy. Also, the wireless channel creates an open platform for the attackers to get involved with the malicious intents. Designing the security model for such system would require the designer to choose a relevant technique [6].

2 Related Work

The large-scale wireless sensor networks are resource constrained systems and are mostly battery powered. Therefore, energy conservation is also an important factor to be considered in these networks along with the fault-tolerant mechanism. A distributed fault-tolerant routing algorithm (DFTR) is proposed in [7] to achieve the goal of enhancing the energy efficiency and provide fault-tolerance in case of cluster head failures. The cluster formulation helps in reducing the energy consumption. However, the single point of contact in each such cluster is the cluster head, mostly a relay node; a battery powered resource. Failure occurring in cluster heads results in fault in the network. The proposed algorithm was tested in three different scenarios comprising 350–500 sensor nodes, 35–50 relay nodes in 300×300 or 500×500 square meter area. Many rounds are performed with different counts of dead cluster heads and were tested against energy consumption and amount of packets received to the base station in each round. Each scenario consists of a different position of the base station mimicking the real-time scenario but with the stationary setup of sensor nodes. The proposed algorithm is applicable for the large-scale WSN scenarios where all the sensors in the network are required to communicate with the base station through cluster heads. Relay nodes acting as cluster heads are different from the sensor nodes and are required to forward the aggregate results of the sensor nodes present in its cluster to the next cluster head or the base station. Only the permanent faults occurring in the cluster heads are taken into account and not any other type of faults like transient fault or intermittent fault.

A fault-tolerant mechanism for heterogeneous WSN, *k*-fault-tolerant FTMRT scheme, has been proposed in [8]. Each sensor node in the network contains a father node and k-1 uncle nodes. The initial phase of construction is involved in determining the father and uncle nodes. The reconstruction phase activates based on the occurrence of failure. The tolerance scheme starts reconstructing the path in case of failure occurring only if the failed node is its father node or among its uncle nodes. The route path is updated based on this new algorithm. The failed father node or criteria set by this algorithm. The proposed algorithm is simulated using 2 fault nodes in the network consisting of 100–300 sensor nodes. The proposed algorithm is focused toward reconstructing the routing path as a solution to fault-tolerance, and there is further scope for optimization of the protocol.

A comprehensive survey on the fault-tolerant techniques is described by Shyama and Pillai [9]. The increased use of WSN has found its implementation in many critical applications. This network can even be implemented in remote areas or in any hazardous environments too. The fault-tolerance has become a desirable part of WSN. In this work, different criteria are defined for classifying the fault-tolerance techniques based on time, objective, size, implementation, and fault-region. The tolerance technique can be applied at the initial stage in the network as an alternative implementation to act as preventive measure, or the measures can be applied only after occurrence of an error. The objective of a tolerance mechanism is to minimize the energy consumption, increase the data flow, or to improve the connectivity. The large-scale network can tolerate some amount of node failure since the nodes are huge whereas each node is given equal importance in case of a small-scale network. The techniques also differ if the implementation is redundancy-based or clustering-based or deployment-based. Also, the fault can occur at the node itself or at the link connecting the nodes, or it can occur due to some malfunctioning of the node in itself. The detection of these faults happening at the WSN can be managed centrally or it can be distributed. The fault recovery can be either active or passive depending on the functionality of the nodes. Also, to prevent the fault occurrence, the network needs to be monitored throughout for status of critical components and for the occurrence of any error.

To overcome the shortcomings of the directed diffusion technique, a new routing protocol called as fault-tolerant and storage efficient directed diffusion (FTSDD) scheme has been proposed by Liu and Ping [10]. The proposed protocol uses a packet header HC which helps in the routing. In this, node need not store information of all its neighbors, instead stores the information from any one of nodes based on the HC value. During the time of fault, instead of rediscovering the entire path, the decision is made using the detection of health check of nodes based on the exchange of exploratory and reference data. Only if the node receives the reinforcement, the neighbor node is considered as healthy. The proposed algorithm focuses on link failures and finds a new path in a cost-effective manner.

As IoT devices are finding great advancement due to a wide variety of applications being deployed. The WSNs form the core part of IoT devices. The wireless communication nature of these devices makes it prone to security vulnerabilities and causing huge implications. This makes it important to have appropriate security mechanisms to be embedded into it. Since these networks are dynamic in nature and moreover the wireless nature makes it more vulnerable, secure data aggregation has become a primary requirement in these networks in order to prevent the security attacks. The basic security requirements include data availability, data confidentiality, data integrity, access control, authentication, data freshness, non-repudiation, and privacy.

Keerthika and Shanmugapriya [11] highlight the security vulnerabilities that can occur in wireless sensor nodes when left unattended. They explain briefly the basic architecture of the wireless sensor network. The wireless nature of the network makes it prone to a variety of security threats. These are broadly classified as active and passive and further into sub-categories under them. A generalized approach on countermeasures is aslo discussed while intending toward providing a brief insight on different attacks which can occur in WSNs.

Yang et al. [12] focus on sensor networks which are deployed in underwater applications. The threats and attacks which can occur in these networks are highlighted. Even underwater sensor networks are prone to security threats. The attacks are broadly classified as active and passive. The additional security requirements specific to these systems are described in [12]. It describes the existing security measures and its applicability to underwater sensor networks. The encryption and other key management schemes are considered costly in terms of bandwidth and energy consumption. The proposed scheme stresses upon the system to follow a hierarchical scheme instead of pure centralized or pure distributed, since the applications are more resource constrained.

The synchronization of time in WSN also plays a vital role. Analysis of different time synchronization techniques in fault-tolerance has been studied in [13]. All the sensor nodes in the network need to be in synchronization with respect to a common time-frame. The loss of synchronization may result in the error in time synchronization algorithms running in the network. The change in frequency of clock in the sensor nodes results in the variance in time-frame. In this work, three kinds of popular strategies are analyzed. In master and servant type time synchronization, the sensor nodes directly get the synchronization from the server making it a simple direct approach. But, this technique is considered to have a low ability of fault-tolerance and also might lead to congestion in the network because of many requests trying to contact the server. The Byzantine agreement type of synchronization strategy using majority decision strategy is another method of synchronization but results in exchange of multiple messages within the network by adding additional load. One of the techniques is the convergence function which is further evolved with the use of sliding window algorithm.

Alami [14] throws light on the crucial role of security in smart home healthcare systems. The WSN plays an important role in these systems enabling it to transmit or to monitor health of a patient remotely. The data received over WSN is used by hospitals to monitor patient's health in real-time. These networks capture minute details which are crucial for saving patient lives during emergency situation. The general critical attacks which can occur in WSNs are interception, modification, spoofing, etc., which are applicable for healthcare system too. The ECC is a preferred asymmetric cryptosystem suggested in [14]. Along with the encryption, it also highlights the implementation of access control mechanisms in preventing unauthorized access to the network. Specifically, it puts forward the access control for a specific set of nodes chosen from the election algorithm. Similarly, the user authentication is also considered as another important factor in these systems.

The smart grid applications also make use of WSN to enable communications like monitoring and controlling between the user and the utility provider. Even this communication is vulnerable to attacks. Alohali and Vassilakis [15] provide an analysis on significant security threats around this network. The major attacks listed are jamming, tampering, collision, spoofing, DoS, sybil, and flooding which occur at different layers. Yousefpoor et al. [16] highlight a secure data aggregation methods in WSNs, which are the integral part of IoT devices. The kind of attacks which can happen in WSNs are broadly classified as active, passive, internal, and external. In this work, the secure data aggregation schemes are classified in detail into seven types. Each of the proposed schemes are analyzed, and a detailed observation on its strengths and weaknesses are highlighted along with evaluating them in terms of security requirements. The countermeasures are also provided by each method against the various security attacks.

The system security must be ensured when used in critical environments. A hybrid cryptography model is proposed in [17] to ensure the security of WSNs. The hybrid model is constructed using the symmetric encryption AES and cryptographic hash function HMAC. The proposed method is implemented and tested against man-in-the-middle and brute force attack, and the model has proven results against these. Teguig and Touati [18] proposed an ECC-based security mechanism to protect WSN against man-in-the-middle attack by providing a secure way for key authentication and key management. The drawback of symmetric key encryption is it utilizes more power and storage space. The discrete logarithm calculation feature of ECC ensures security in the system and helps in establishing a secured key among sensor nodes. The proposed model comprises three phases, namely network deployment, neighborhood recognition, and ciphering-deciphering process along with signature verification.

A new mechanism for key management of secret keys within WSN has been proposed in [19]. The method makes use of a trusted-server scheme and key pre-distribution scheme within a new centralized system architecture wherein there is a central control unit, several small base stations per cluster, and a main base station. A single trusted server implements the task of key distribution and management. The system uses the SHA-2 hash function to generate encryption keys. Also a wired connection requirement is put in place between the trusted server and the main base station. The resource limitations on sensor devices may cause the implementation of ECC directly on the sensor nodes a challenging task since it consumes considerable amount of ROM as well as RAM on the devices. The additional optimizers will further increase the memory consumption in the network causing more bottlenecks on the resources. Liu and Ning [20] present a flexible and easily integrable open-source software called TinyECC for the implementation of secure WSN. This library can be implemented using the nesC language, runs on TinyOS, and also includes an optimizer which is software configurable and provides multiple options to suit the requirements. The library includes the support for elliptic curve Diffie-Hellman (ECDH) key agreement scheme, elliptic curve digital signature algorithm (ECDSA), and elliptic curve integrated encryption scheme (ECIES). The library was tested in different sensor platforms such as MICAz, TelosB, Tmote Sky, and Imote2.

The secret key establishment between the communicating sensors is an important factor before initiating any communication. The authenticated key establishment in WSNs [21] is achieved using transport methods implemented using TinyAKE. This key establishment runs on TinyOS platform and uses TinyECC library APIs. The proposed method has two stages of communication known as pre-distribution stage which occurs prior to the key exchange and key establishment stage which occurs afterward. The TinyKGC part of the proposed protocol handles the generation of private and public keys as well as the certificate for authentication. The pre-distribution stage takes care of filling all the information pertaining to individual sensor nodes to be embedded onto it. In key establishment phase, the challenge-response mechanism is used with nounce. Along with this, queue buffering, re-transmission, and invalid clean-up mechanisms are also implemented. The proposed protocol is tested for correctness, confidentiality, and authentication. The feasibility of the implementation is proven by its practicality and trustability over other existing protocols.

The detection of attacks in WSN also plays an important role. The detection of certain attacks itself is challenging. Al Ahmadi [22] proposed a machine learning technique based on a random forest classifier to detect DoS attacks. It compares the results for techniques implemented using different models such as Naive Bayes, neural networks, support vector machines, and decision tree. The proposed method is focused toward detecting DoS attacks in WSN which uses the LEACH protocol. The analysis is conducted on the Waikato environment for various performance metrics. For the random forest classifier, the tool has shown 99.72% detection accuracy with time complexity of approximately five minutes. Even though the time complexity of Naive Bayes was 1.25 seconds, the classifier has detection accuracy of 95.35%. Other classifier accuracy lies within these ranges. The WSNs are prone to attacks due to their deployment in hostile areas. Various machine learning techniques have found its way in overcoming the attacks on WSNs. Kaur and Sandhu [23] gave a brief insight on layer-wise attacks on WSNs based on the OSI model and throw light on machine learning techniques to determine these attacks. The major security requirements for a WSN are confidentiality, availability, integrity, authenticity, and time synchronization. The authenticated data from sensor nodes needs to reach the destination without any tampering in between. Some of the layer-wise attacks are DoS, spoofing, monitoring, hello flood, and clone attack. Few machine learning techniques have been successful in fighting against these attacks such as random forest approach, KNN, ANN, SVM, and Q-Learning. In this work, the detailed study on each layer attacks and the corresponding machine learning technique used is highlighted.

A new algorithm for authenticated key exchange in WSN has been proposed in [24]. The protocol is based on ECDH. The model works on the basis of using asymmetric encryption for secret key establishment followed by encryption–decryption by securely established symmetric key. The proposed scheme starts with separate registration for each user and the sensor with the gateway. The user registration happens using a user id, biometric parameter along with a smart-card, whereas the sensor registration happens using sensor id. The password received by the user can also be changed as per user's request by the gateway. The key authentication and exchange are done with a minimum propagation delay. Habeeb and Muhajjar [25] proposed a new key management system using a symmetric mechanism. Also, the randomization technique is implemented in the algorithm by making use of pseudorandom number generator (PRNG). Before deployment of WSN, each node is pre-loaded with a key. Each node is assumed to have loaded with different initial keys. The PRNG algorithm is responsible for establishing a common shared key between the node and the base station. This algorithm makes use of simple operations such as XOR to achieve low computation cost. It uses 4 registers as basic building blocks to construct the key. To provide improved security, the proposed scheme uses CBC mode with RC5 encryption technique. Also, periodic key updates are done using the same PRNG algorithm. The drawback of the proposed scheme is that the algorithm assumes a direct communication between all sensor nodes with the base station.

Mallick and Bhatia [26] provide a comprehensive analysis of impact of the cryptography algorithms on WSN and help the designer while choosing the best algorithm for the development. The passive attacks are considered as most challenging things to detect since there are no visible or obvious effects seen in the system behavior. This leads to pouring out of confidential information to the unauthorized users. Not only these, there are other various kinds of possible attacks in the WSNs. In a resource constrained WSN, it is always challenging to integrate energy efficient security algorithms with effective computational cost. An energy efficient secure pattern-based data aggregation (ESPDA) scheme has been proposed in [26] as an efficient way for energy management. When key exchange and management are considered, ECC is a preferred choice compared to other asymmetric or symmetric algorithm since ECC would not require digital signature to provide authenticity. It also provides a considerable amount of security as compared to other algorithms. Even though TinyECC is more useful in resource constrained systems, due to its poor scalability, ECC outperforms TinyECC.

In light of resource constraint systems, lightweight cryptosystems would fit well into the system. Further, this work takes into consideration the popular lightweight block cipher techniques like AES, PRINT, PICCOLO, and KATAN and highlights their special features. The WSN network prefers mostly block ciphers as compared to stream ciphers. In order to optimize the security, a blend of both symmetric and asymmetric cryptography can be used. A three mega security concept is proposed using ECC, AES, MD5, and XOR-DUAL RSA. Making use of these multiple algorithms helps in making the system work with less computational overhead with improved security. There are few other models developed as well using hybrid cryptographic technique. The major factors which can help in shielding an information system from security threats are to retain confidentiality, integrity, availability, authentication, and non-repudiation in the system. Karakaya and Akleylek [27] focus on analyzing the security attacks, key management issues, and authentication methods. In a general WSN architecture, a sink node can be considered as the mediator node between base station and the sensor nodes in the network. The base station then takes care of transmitting the information further to the data center. There could be a variety of attacks that can occur in WSN using which attackers can compromise WSN. Therefore, ECC is considered as the best key generation protocol. This algorithm provides protection against attacks of impersonation. Also, this work suggests the use of protocols like SPINS, LEAP, RKP, and TinySec toward prevention against different attacks.

3 Conclusion and Future Scope

The wireless sensor networks have become part and parcel of increasingly growing IoT and other automation systems around. Adding appropriate security to the sensor network communication protects them from malicious attacks. To overcome security vulnerabilities, ECC-based cryptography algorithm and its variants are integrated with fault-tolerant sensor communications. This work highlights the challenges involved in implementing ECC-based secure sensor communication in the existing systems and the suitability of ECC-based secure communication for many real-time sensor network applications.

References

1. Sniatala P, Amini MH, Boroojeni KG (2020) Fundamentals of Brooks-Iyengar distributed sensing algorithm
2. Kumar V (2013) Impact of Brooks-Iyengar distributed sensing algorithm on real time systems. IEEE Trans Parall Distrib Syst 25(5):1370
3. Panigrahi T, Panda M, Panda G (2016) Fault tolerant distributed estimation in wireless sensor networks. J Netw Comp Appl 69:27–39
4. Sen J (2012) Security in wireless sensor networks. Wire Sens Netw Curr Stat Fut Trends 407:408
5. Li CT (2010) Security of wireless sensor networks: current status and key issues. Smart Wire Sens Netw 299–313
6. Manavendra S, Rohan V, Mittal S, Secur Wire Sens Netw
7. Azharuddin M, Jana PK (2015) A distributed algorithm for energy efficient and fault tolerant routing in wireless sensor networks. Wire Netw 21(1):251–267
8. Ma G, Yang Y, Qiu X, Gao Z, Li H (2017) Fault-tolerant topology control for heterogeneous wireless sensor networks using multi-routing tree. IFIP/IEEE Symp Integr Netw Serv Manage 620–623
9. Shyama M, Pillai AS (2018) Fault tolerance strategies for wireless sensor networks—a comprehensive survey. IEEE Int Conf Invent Comput Technol (ICICT) 707–711

10. Liu J, Ping Z (2010) Fault tolerant and storage efficient directed-diffusion for wireless sensor networks. IEEE Int Conf Inform Theo Inf Secur 884–887
11. Keerthika M, Shanmugapriya D (2021) Wireless sensor networks: active and passive attacks-vulnerabilities and countermeasures. Glob Trans Proc 2(2):362–367
12. Yang G, Dai L, Si G, Wang S, Wang S (2019) Challenges and security issues in underwater wireless sensor networks. Proc Comp Sci 147:210–216
13. Dong J, Gu L, Zheng C (2011) Research on fault-tolerant strategy of time synchronization for industrial wireless sensor network. IEEE Int Conf Measur Technol Mechatron Autom 2:1146–1149
14. Alami A, Benhlima L, Bah S (2018) A study of security requirements in wireless sensor networks for smart home healthcare systems. In: Proceedings of the 3rd international conference on smart city applications, pp 1–8
15. Alohali BA, Vassilakis VG (2017) Security of wireless sensor network (WSN) in smart grid. In: Proceedings of the second international conference on internet of things, data and cloud computing, pp 1–6
16. Yousefpoor MS, Yousefpoor E, Barati H, Barati A, Movaghar A, Hosseinzadeh M (2021) Secure data aggregation methods and countermeasures against various attacks in wireless sensor networks: a comprehensive review. J Netw Comp Appl 190:103–118
17. Ardiansyah F, Setia Budi A, Primananda R (2021) Hybrid cryptography for data security in wireless sensor network. Int Conf Sustain Inf Eng Technol 221–225
18. Teguig EH, Touati Y (2018) Security in wireless sensor network and IoT: an elliptic curves cryptosystem based approach. IEEE Ann Ubiquit Comput Electron Mob Commun Conf (UEMCON) 526–530
19. Yang S, Liu J, Fan C, Zhang X, Zou J (2010) A new design of security wireless sensor network using efficient key management scheme. IEEE Int Conf Netw Infrast Digi Cont 504–508
20. Liu A, Ning P (2008) TinyECC: a configurable library for elliptic curve cryptography in wireless sensor networks. In: 2008 IEEE international conference on information processing in sensor networks, pp 245–256
21. Sun F, He S, Zhang X, Shen F, Li Q, He Y (2021) TinyAKE: a more practicable and trustable scheme for authenticated key establishment in WSNs. arXiv preprint arXiv:2104.01907
22. Al-Ahmadi S (2021) Performance evaluation of machine learning techniques for DOS detection in wireless sensor network. Int J Netw Secur Its Appl (IJNSA) 13
23. Kaur R, Sandhu JK (2021) A study on security attacks in wireless sensor network. In: 2021 IEEE international conference on advance computing and innovative technologies in engineering (ICACITE), pp 850–855
24. Baghbanijam S, Sanaei H (2021) An improved authentication and key exchange protocol based on ECDH for WSNs. arXiv preprint arXiv:2109.11450
25. Habeeb IJ, Muhajjar RAA (2016) Secured wireless sensor network using improved key management. In: Proceedings of the fifth international conference on network, communication and computing, pp 302–305
26. Mallick BB, Bhatia A (2021) Comparative analysis of impact of cryptography algorithms on wireless sensor networks. arXiv preprint arXiv:2107.01810
27. Karakaya A, Akleylek S (2018) A survey on security threats and authentication approaches in wireless sensor networks. In: 2018 6th IEEE international symposium on digital forensic and security (ISDFS), pp 1–4

Utilizing Data from Quick Access Recorder to Predict Faulty Processing on Aircrafts

V. Jalajakshi and N. Myna

Abstract A quick access recorder (QAR) that is carried on the aircraft captures hundreds of parameter data generated during flight. However, these information are not increasingly being applied well. The use of QAR data for fault prediction is becoming more and more essential. First, an illustration of the technological path for civil aviation defect prediction is shown. 2, 4 commercial airplane defect prediction procedure rely on QAR information is presented. These methodologies include advanced methodology based on Enhanced Grey Prototype, performance analysis method based on Template Matching, Time-series data Extrapolation, and Non-parametric Correlation and assessment on predicated efficiency regarding the trend prediction Technique. Finally, a thorough description of the forecasting system's architecture is provided. As a result, it can keep an eye on how the aircraft's systems and components are functioning in order to promptly identify any defect symptoms, devise an appropriate routine maintenance, and guarantee flying security. Finally, using the approach outlined in this study, the temperatures' characteristics of the air conditioning systems of Boeing are anticipated. The outcome predictions validate the approach's efficacy.

Keywords Query processing · QAR · Sensors

1 Introduction

Currently, the three principal types of defect prediction systems used by the airline companies for civil aircraft are the classic model, looking over the model of the QAR and the ACARS model. The traditional civil device failure prognostication innovation [1, 2] mainly relies on the data from the aircraft maintenance record and documents to be resolved and evaluated, along with engineer understanding and the airplane failed hypothesis and characteristics, to predict the likelihood of

V. Jalajakshi (✉) · N. Myna
Department of CSE, Navkis College of Engineering, Hassan, Karnataka, India
e-mail: jalajakshi.prakash@gmail.com

G. Rajakumar et al. (eds.), *Intelligent Communication Technologies and Virtual Mobile Networks*, Lecture Notes on Data Engineering and Communications Technologies 171,
https://doi.org/10.1007/978-981-99-1767-9_15

system failure. Since the skills and knowledge and limited knowledge necessary for this traditional deficiency estimation technique, its prediction performance is also constrained. There are numerous different types of ACARS packets, but the primary form that is helpful for failure prediction is aircraft operation control messages.

There seems to be a development and functioning working prototype model for intercepting and evaluating Aircraft Communication Addressing and Reporting System (ACARS) data link information, for both flying aircraft monitoring and for predicting the beginning of certain flight malfunctions [3, 4].

The implementation of ACARS header information in defect prediction technologies is somewhat constrained by the disadvantages of the ACARS mode, which include data discontinuity (interval transmission) and fewer record variables. The airplane defect prediction system employs methods such as flight data query, cockpit instruments panel replication, and analysis of the rich QAR data gained via decoding to make its forecasts, flight parameter curves, overflow reports, and other metrics are used to track how the aircraft's systems and components are functioning so that defect symptoms may be identified quickly. This can reduce the amount of time needed for upkeep and increase operational effectiveness.

With both the aim of cutting down on bug fixing and boosting operational capabilities, the following specific path of air transport defect prediction is demonstrated. Then, four different categories of fault prediction systems are presented. After that, an information in connection with the prognostication program's execution follows.

Numerous parametric information acquired while the airplane is in flight are recorded on a quick access recording device (QAR) positioned on the flight. The timing of breakdown of important system components is anticipated in this study using QAR data to examine variations in sensors' collection settings. To rationalize the creation of maintenance plans that will assure the security of aircraft manually add the databases with the processed information. Due to the defect prediction program's necessity to compile a high number of data sample and the fact that QAR data includes a variety of parameter kinds, a custom databases must be set up to handle information. Develop parameters' template and list all QAR variables that have an impact on aviation system failures as shown in Fig. 1 flowchart depiction.

Data processing Because of the effects of the workplace environment, QAR information is frequently unstable and absent in practice. The data is preprocessed before being examined, which improves the accuracy of the processing and ensures the precision of the failure detection.

More emphasis is being placed on the civil aircraft health management system in an effort to increase maintenance effectiveness and lower the likelihood of catastrophic events. An important component of the study is the airborne quick access recorder (QAR), which records data on aircraft position, status, and performance parameters. The quick access recorder data function is presented in this paper, along with the applications before suggesting that it can be used to diagnose faults and evaluate the state of a typical airborne system. In one instance, the failure mode and corresponding QAR parameter are investigated to diagnose the environment control system (ECS) heat exchanger.

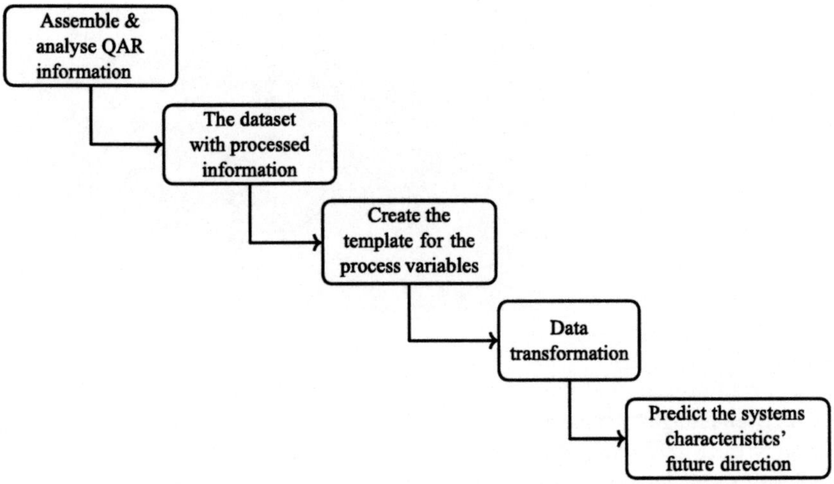

Fig. 1 Technical route flowchart diagram

The airplane health service regarding its primary management's program and design, algorithms, and rapid prototyping have lately been receiving more and more focus. The vast majority of current studies in health managerial technology is founded on floor test, modeling, and aircraft information because it is challenging to obtain raw flight information. QAR is gaining popularity between many healthcare study technicians because of its widespread use in Airport Operations Quality Management and Efficient Maintenance Quality standards. The airplane's security and well-being depend heavily on the situation, status, and performance criteria that are documented in quick access recorder.

2 QAR Data Feature and Application

For something like the planning of the well-being of civilian aviation, QAR data is a crucial source of information. The civilian aircraft well-being management system can obtain a large amount of data regarding aircraft operation, situation, and efficiency through the quick access recorder for detecting defects, key equipment prognosis, and ground preservation. The airborne quick access recorder is a piece of airborne equipment used to record data from aircraft data acquisition systems and transmit that data manually or using the Global System for Mobile Communications (GSM) network to the air carrier data processing center. The QAR system varies between various aircraft models.

Nevertheless, the idea is largely identical and comprised a system for collecting data and a rapid recollection recording device. The data captures common rules so that financial and sends airborne system information to a swift recollection recorder's disk

Fig. 2 Quick access recorder

drive in conformance with customer request. The pretty standard aspect is shown in Fig. 2. Even though their main quick access recorder storage is moderate, air carriers currently use 3-inch rewritable electric magnetic optical (MO) memory cards from Rockwell and Teledyne. These magnetic disks can store large numbers of airborne process variables for a maximum of 100 h and have a large capacity for storage which is enough to make sure that the data can be quickly analyzed and used for further detection purposes.

There are notable aspects of QAR data. First off, the amount of data is incredible. QAR data is a stream dataset that is collected from a variety of sensors during flight, one sample at a time. The knowledge must be used immediately. One reason the information is genuine is so that airplane earlier than usual processes can respond to oddities as those who arise. However, the information is serial because of the close connection among both time and the quick access recorder data. Thirdly, the information includes first-hand knowledge. Maintenance experts can perform data analysis to identify fault isolation and create maintenance plans. Fourth, there are a lot of noise and disturbance in the data, making accurate analysis difficult [2].

3 Method of Civilian Aircraft Predicting Centered upon QAR Information

The very first observed results to the heritage of a specific model is typically the foundation for the failure predictive model based on quick access recorder data, followed by the prediction of design variables, get the correlating forecast. It is essential to create a sound prototype in order to improve prediction accuracy. For various types of data and performance criteria, a variety of techniques can be used to anticipate the advantages and disadvantages of conventional forecasting models.

3.1 Based on Curve Fitting the Performance Forecasting Analysis

By matching the variable data and examining the dependence relationship between both the variable sequences, the variable sequences are expressed by an approximative way. For instance based on the curve fitting method. You may anticipate the trend of recent data with the this indicates how efficiently with ease.

This approach works well for predicting data that changes slowly over time, making it more appropriate for predicting parameters with periodic variation.

3.2 Methodology for Predicting Job Performance Using Time Series

The time-series prediction technique [5, 6] is a way of predicting the extrapolation's future direction in addition to the evolution and regularity of failure occurrences that really can represent time-series data. Through the creation and assessment of time series, in conformance with the time-series data mirroring the development process and patterns, similarities, or extensions, it is possible to anticipate the potential condition of the hardware in the forthcoming interval of time.

Less historical data and labor are needed for time-series analysis, but it also demands that the variables affecting the projected object do not change. It is ideal for short-term estimating but just not lengthy prognostication due to the fairly uniform genetic variations.

3.3 Using Non-parametric Statistical Regression for Achievement Prediction

Non-parametric statistical regression models are excellent for forecasting nonlinear information with clear periodicity since they do not require the creation of a prediction model and can forecast provided that there is a substantial amount of historical information. The approach is appropriate for short-term forecasting.

3.4 Innovation for Defect Prediction Based on a New Grey Prototype

Grey system theory is a novel approach to researching the issue. With fewer data points and pieces of information [7], it is predicated on "poor information" and a "tiny proportion" that the grey systems' approach is valid. The phrase "a few specifics are knowledge are recognized and portion of it is unidentified" is another term for the unsure scheme. The conception and growth of already current knowledge, and the information extraction of valuable information to extract beneficial facts and to arrive at an accurate modeling approach behavior, the law of evolution, and efficient oversight. The grey framework that is now utilized for forecasting is mostly the GM (1,1) approach, but this prototype is best suited for series forecasts with steadily over time increasing or declining trends because non-monotonic sequences have low predictive performance. We enhance the prediction model and provide fluid forecasting using the dynamic equal dimensional GM (1,1) framework. The four forecasting techniques mentioned in Table 1 each have advantages and drawbacks that are specific to the type of variable being forecasted, and it helps to get a detailed and enhanced view of each method and its various detailed information to take into overview.

4 Execution of Prediction Systems

4.1 Initiating a System Repository

Monitoring and fault prediction both depend on gathering flight data. For the mathematical formalism of civil failure detection to be realized, a essential contribution is a reliable, reliable data origin.

User account modules, flight parametric library, and system database and defect prediction method library primarily store system integration data, decrypted flight information, and the data needed by the fault prediction subsystem. This information is required to quantify and monitor the complete procedure of the way sensors operate and immediately diagnose the sensor which is failed in its operations. We collect a sample dataset of famous airlines sensor information across the world to gain a view of the sensor working timeline and understand the failure rate.

4.2 Determination of Quality and Performance Process Variables

Certain variables can be directly acquired by aerial devices, and others may represent modifications in the operation of the aircraft, component, and appendices. The data are captured independently at each phase of the aircraft.

The data are captured independently at each phase of the aircraft. With this, we get a detailed analysis of time-to-time information of the sensor data with time stamps which is very much required for analyzing it precisely. If required, these variables are changed to their typical state settings and then contrasted with their default values

Table 1 Comparison metrics of various parameters

S. No.	Method	Benefits	Drawback	Real application
1	Performance analysis using the method of interpolation	Can estimate sudden changes in the sensors' data. Closely packed spaced evenly scores are well interpolated and can change cell values by increasing or decreasing the number of data points	Because of their rigidity (due to smoothness), they tend to over fit the data	Mathematical Trigonometric and polynomial value
2	Performance forecasting method based on time series	Prognostication of static variants or primary tendencies for information and criterion keeps a constant value	The prediction effect is poor for nonstationary trends	Sales' forecasting. Budgetary analysis. Stock Market Analysis
3	Performance forecasting method based on non-parametric regression analysis	(1) There is no requirement to create a precise statistical equation for forecasting complex and time-varying networks (2) The method is highly efficient and executes quickly	(1) It is impossible to foresee lengthy trends (2) Manual input is required for design variables to assure the individual's completion	Forecasting the weather and future price predictions of the currency
4	Fault prediction technology based on Fuzzy method	It is a reliable system that does not require precise input. Such algorithms can accept a variety of ingredients, include hazy, skewed, or inaccurate data	Occasionally, the fuzzy rationale is mistaken for likelihood hypothesis	Image segmentation to help doctors diagnose

or specially created normal evaluation metrics. So, it is possible to determine the increase in variance.

It is possible to evaluate the sensor's overall health and carry out the monitoring of the attachment through the assessment of the divergence and the changing pattern of the deviance. With this deviance, we can analyze the range of deviation, and if it crosses a certain threshold, it gets alerted and now we shall analyze on finding the basis for evaluation of the failure and health rate of the sensor.

These sensors can provide information on key readings such as temperature, pressure, vibration, and other measurements of the plane's physical performance. This data is not only available to the pilots, but through being connected, the sensors also connect with support facilities on the ground and are fed through analytical programs to flag any information which may represent an issue that needs to be addressed.

The foundation for prevention and troubleshooting can be given by quickly identifying the range of deviation abnormalities between a parameter's accepted value or its trend and analyzing the causes for anomalous behavior. In the parametric selection procedure (Fig. 2), the variables tied closely to the efficiency of the network, subsystem, and divisional connections are selected in accordance with both the choosing principles of each, and then, the filter is sent to make a sensible decision, resulting in the performance characteristics that the scheme needs to maintain.

As we can see in Fig. 3, there will be an overall system overview from the top design to monitor the performance evaluation to understand if the system if working as per the accepted rate or not. Then, there is another methodology where each subsystem's performance will be evaluated and its metrics will be considered from time to time.

Flight operation are sensitive and time critical detection, so we there is also micro molecular detection and monitoring of the sensors, and hence, every single sensor performance is evaluated with the standard readings and report is generated, and later, all the performance analyses of three different levels, those are whole system view, subsystem, and every sensor, are combined to form a singular performance metrics of the machine [10]. This report generated will have the complete information of the sensors data and its timestamps of various detection procedures.

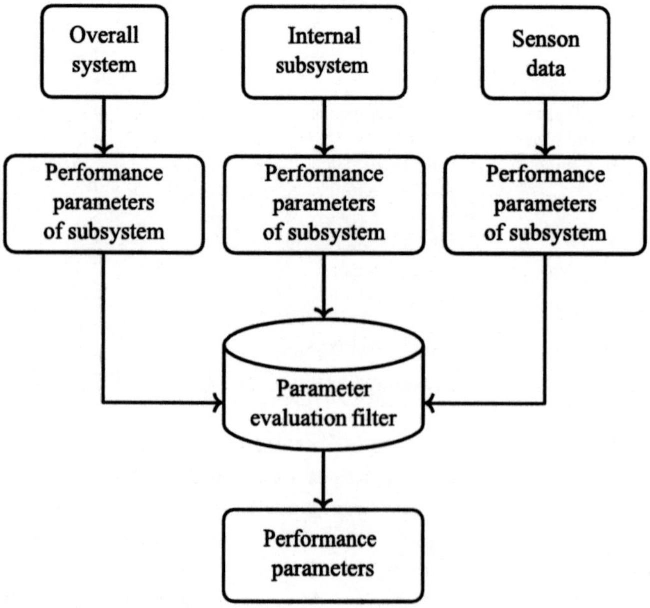

Fig. 3 Parameter evaluation flowchart

4.3 Threshold Validation

There is a prefixed selection of the threshold where the parameter's efficiency and performance will be analyzed, also there is a certain functioning of the overshoot alarm based on various scenarios depending on the performance of the parameters observed, and there will be certain determination of the threshold as observed in Fig. 3.

If parts of the QAR information are valid, the system takes them. Data should be available to shift significantly in magnitude, and its control parameters are not altered simultaneously [9]. Or when necessary, it is decided that the information available at the moment is the wild point is eliminated. Data following the elimination of smoothing of the wild spot and typical characteristics are chosen.

The areas that are important in this context could either be the parameter values in its original state or the analytical performance-related values, such as average value, RMS value, excessive valuation. Initial data format for the variable or possibly the statistical values for the performance characteristics, like mean value, root mean square value, extreme value as depicted in Fig. 4.

In Fig. 4, we can observe a complete analysis of data flow and its threshold taken into consideration; here, we use a comparator where we try to compare the value and also the set threshold which in turn helps to decide whether the alarm have to be triggered or not.

Real-time comparisons can be made between the determined characteristic parameter data and the limit. The limit must indicate the distinctive feature parameter settings' standard operating variety and sound an alert if it is exceeded.

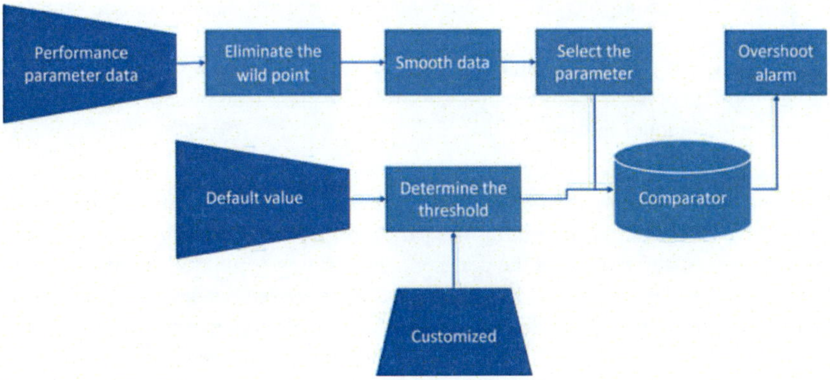

Fig. 4 Overshoot alarm

4.4 Choosing and Examining a Mathematical Model

The implementation of four different statistical equations prognostication methodology based on various advanced methods like curve fitting, ML algorithms, time-series data forecasting according to the time sequence, achievement analysis of responses methodology based on non-parametric linear regression, and forecast technique utilizing improved grayish prototype GM—in civil device, fault prediction is the main focus of this paper (1,1). The main purpose of the mathematical formula utilizing time series and linear regression is to forecast the trend of critical aircraft components and variables.

Whenever the change rate abruptly changes when there are no additional conditions or events, it is appropriate for forecasting [8]. The main components of the realization function are parametric time-series modeling, singular value recognition, modeling of real-time forecasts, parametric forecasting, and so forth. In practical applications, specific steps include:

1. Gather data about the status and plane incidents, such as the aircraft's identification, subsystems, components, parameters, anticipated standard errors, range linear discriminate functions, and sampling points for the prediction model.
2. A likely to fit graph or auto-regressive (AR) model is developed to account for parameter variation.
3. Singularity identification: Establish the range discriminant function to check for the presence of a exception. Here, we straight are using the Euclidean distance feature [9]. The duration when the irregularity happens and the valuation as per the moment in time are noted in ahead of time if the standard error for the worth of the neighboring points of the series is constantly judged to be exceeded.
4. A new forecasting model is created, new findings are decided to add, and the prototype is improved using measured data close to the point of failings.
5. Prediction of parameter value: Using supposition forecasting from polynomial appropriate or other methodologies, provide the parameter prediction value after a specific amount of time.

On the basis of time series and curve fitting, a dynamic trend prediction, and alarm flowchart as depicted in Fig. 5, the primary procedure is: If the time-series method is not chosen after reading the parameter data to be analyzed, preprocess the information and figure out the pertinent curve fitting technique variables. When using the series data approach, data preprocessing is carried out, and the smoothness of the processed data is assessed. The chosen AR model's parameters are determined using the stable data after which the model is tested and improved [11].

The model is anticipated based on threshold value that has set in order to see whether or not a solitary valuation outside of the threshold level is constantly show-cased. If so, it means that the inflection point has already been reached. Depending on the method chosen, after the inflection point manifests, gather data close to it, create a model, and use it to make predictions.

In order to increase the prediction's accuracy and ensure that the cycle is completed to its conclusion, model parameters are optimized during the prediction process. The

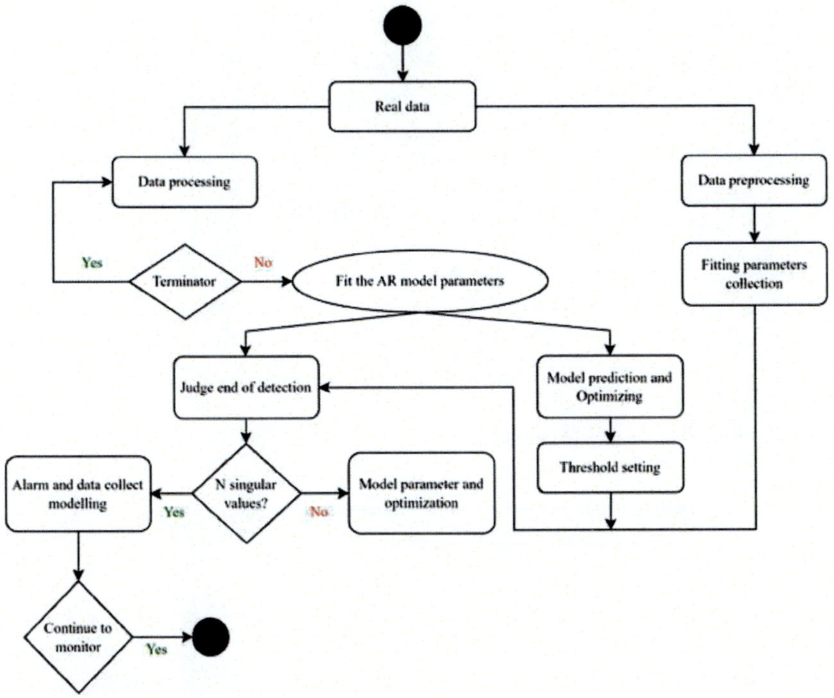

Fig. 5 Diagram of the prediction process logic

quick access recorder data of civil flights is used to validate the fault prediction methods [12].

The heat of a Boeing air conditioning system is depicted in Fig. 6 as the original quick access recorder data changes trend graph, and Fig. 7 represents the correlating forecasting. The low-temperature failure can be anticipated and alerted to by forecasting the trend of the temperature parameter and comparing it with the parametric limit.

The augmented GM (1,1) technique for nonlinear multivariate regression regulation of complicated system parameters or the non-parametric regression analysis can be utilized to start creating the factors for the suggested method.

5 Conclusion

The technique used in this article to anticipate civilian aircraft faults is predicated on quick access recorder data. First, a diagram of the design methodology for defect prediction in civilian aviation is given. On the basis of quick access recorder data, stages for fault prediction are then presented. After that, an information in connection

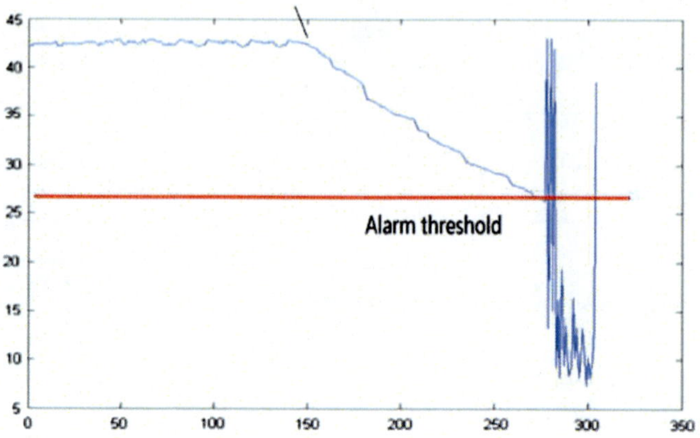

Fig. 6 Data inflection point and threshold

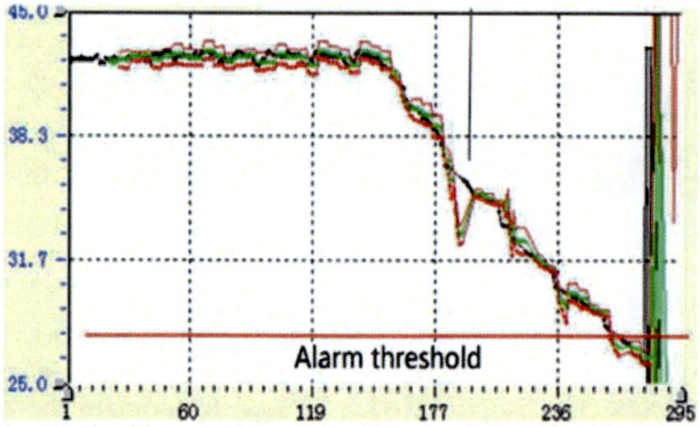

Fig. 7 Parameter trend prediction and alarm

of the prognostication system's execution follows. This is very crucial to collect the error data to diagnose and repair the faulty component to avoid potential threat to the functioning of the aircraft.

References

1. Wang H (2002) A survey of maintenance policies of deteriorating systems. Eur J Oper Res 139(3):469–489
2. Cho DI, Parlar M (1991) A survey of maintenance models for multi-unit systems. Eur J Oper Res 51(1):1–23
3. Zhang X (2001) Aircraft maintenance support system based on ACARS and MEDMS. Aviat Maintenance Eng
4. Wen H (2005) Application of ACARS system in aircraft maintenance practice. Jiangsu Aviat I
5. Hu Z, Xiao M (2013) Research on fault prediction based on time series model. Comput Meas Control VI

6. Zuo Z, Xue Z (2004) Research on fault prediction of aircraft equipment based on time series. In: Proceedings of the 23rd China control conference, vol II
7. Zhu H, Cui Y, Gao F (2014) Application of grey forecasting model in fault prediction of military flight control system. J Sichuan Ordnance VI
8. Wang Y, Zhao G, Liu Z, Qing X (2012) The review of the integrated vehicle health management. Civil Aircr Technol Highlights 3:4–6
9. Yang G (2011) Introduction to QAR and flight safety. Jiangsu Airlines 3:15–16
10. Tang Y, Sun Q, Lv Z, Qiu J (2014) Signal processing and health assessment techniques in structural health monitoring. Vibroengineering Procedia 4:11–16
11. Kumar KP, Rao KV, Krishna KR, Theja B (2012) Neural network based vibration analysis with novelty in data detection for a large steam turbine. Shock Vib 19(1):25–35
12. Wang Z, Qin X, Liu Z, Wang Y (2017) Verification and validation of integrated vehicle health management system. Vibroengineering Procedia 4:324–328

Vehicle Collision Warning and Accident Detection Using Raspberry Pi and SSD MobileNetV1

A. Vijaya Lakshmi, P. Ajay Kumar Goud, G. Mohit Raj, and K. Rahul

Abstract The rise in fatal and permanently disabling road accidents is a severe public health issue. Human lives are frequently lost in road accidents when medical assistance arrives too slowly. Therefore, deaths from road accidents are more common. Several accident prevention systems can, to some extent, avoid accidents. Still, they lack in giving collision warning while driving and the ability to contact emergency services in the event of an accident. A collision warning is one of the great difficulties regarding active safety for automobiles on the road. In this paper, the proposed system is to design and implement collision warning and accident detection for automobiles. In the proposed work, the collision warning is done using object detection, and accident detection is implemented by Raspberry Pi, a vibrator sensor, GPS, and a Telegram bot. The detection of the vehicles is based on the Single-Shot Detection (SSD) MobileNetV1 algorithm. Suppose the vehicles did not maintain a minimum distance between each other. In that case, it will send a collision warning message to the user using a Telegram bot via a Wi-Fi network. If the user made accident, then the accident occurred message and location are sent to emergency services based on the vibration sensor and Global Positioning System module.

Keywords Collision warning · SSD MobileNetV1 algorithm · Raspberry Pi · Vibration sensor · GPS · Wi-Fi · Telegram bot

1 Introduction

Road fatalities are increasing daily, placing a heavy load on public health and other relevant organizations [1]. Road accidents and collisions cause the majority of deaths in modern society. The WHO estimates that 1.3 million individuals per year pass away due to road accidents [2]. Additionally, between 20 and 50 million individuals experience non-fatal injuries, many of which leave them crippled [3, 4]. A significant

A. V. Lakshmi (✉) · P. A. K. Goud · G. M. Raj · K. Rahul
Vardhaman College of Engineering, Hyderabad, India
e-mail: vijayalakshmivardhan@gmail.com

G. Rajakumar et al. (eds.), *Intelligent Communication Technologies and Virtual Mobile Networks*, Lecture Notes on Data Engineering and Communications Technologies 171,
https://doi.org/10.1007/978-981-99-1767-9_16

factor in the post-accident survival rates is the time between the event and the arrival of an emergency medical service at the accident site. Death rates between the time of the accident and the arrival of medical help must be reduced to save more lives. In this study, vehicle collision warning and accident detection are performed. Nowadays [5], processing digital images are more common using computer vision. This paper's object detection system was built on an embedded Internet of Things platform. To successfully prevent road accidents in our daily automobiles [6], it is becoming increasingly important to have an active safety system.

The SSDMobileNetV1 algorithm and OpenCV libraries are used in the proposed design. Live video from [7, 8] the Raspberry Pi's 1.3 Camera Rev is fed to the Raspberry Pi 3B+, which uses the SSD MobilenetV1 algorithm. One of the main issues is the detection of objects. When an object is present in an image, the SSD MobileNetV1 algorithm recognizes it and generates bounding boxes indicating its location. After detecting vehicles, it measures and checks the distance between them. When a user's distance is not maintained, the Raspberry Pi uses a Telegram bot to deliver messages alerting them about a potential collision and examines the vibration sensor data to determine whether an accident happened. When an accident occurs, a telegraph bot notifies the emergency services of the GPS position, vehicle information, and buzzer sounds.

2 Related Work

Deep learning methods have developed rapidly in recent years. CNN is the basis for many object detection algorithms [9]. The object detection methods currently used are [10, 11] one-stage and two-stage. The most common type may be a two-stage R-CNN, Fast R-CNN, and Faster R-CNN approach. Regional candidate-based detection [12–14] is mainly used in these methods. While having high detection, these methods are slow. Methods like [15] SSD and [16] YOLO are one-stage. These methods are based on [17] regression-based detection algorithms that can meet real-time requirements and offer faster detection rates. Even though [18], the deep CNN system provides a very reliable and powerful feature representation for object identification and recognition, it is often operationally costly and requires a high-end embedded device to conduct the model evaluation computation. However [19], processing for object detection often needs to be done on an embedded device with limited computing power. Thus, a neural network model with small network size and few calculations is the only one that can manage such a situation. An essential challenge in system development is integrating the embedded system with the object detection network model and having it run with good accuracy and stability. These light-weighted networks include [20] MobileNet, [21] ShuffleNet, [22, 23] MobileNetV2, and [24] EffNet, instances. The optimum networks for [25] embedded devices are the ones that lower their size and characteristics while yet offering comparable accuracy to primary networks. In this study, SSD MobileNetV1 detecting technology is used. The SSD is more straightforward than any other method that provides object detection, removes all pixels or feature samplings throughout the process, and does all network calculations. For each feature map location, Kanimozhi et al. [26] a collection of default boxes with varying aspect ratios and sizes are provided [15]. SSD involves dividing the original image of bounding boxes. Each default box receives an object's presence from the network, which adjusts the box to accommodate the object's shape. Furthermore, the network incorporates predictions from several feature maps with different detection intensities to handle objects of varied sizes. MobileNetV1 [27, 28] uses depth-wise separable convolutions and is built around a simplified design. MobileNetV1 builds a compact deep network and incorporates two hyperparameters to strike a compromise between accuracy and latency effectively. It employs depth-separable convolution to make the traditional network structure simpler. In MobileNetV1, which significantly decreases processing, the depth separable convolution network is implemented in place of the traditional convolution network. The depth-separable convolution is 1/9–1/8 of the conventional one when the channels' input and output picture sizes are equal [29]. The algorithm calculates the separation between the identified vehicles. To forecast future collisions with vehicles, an accurate estimation of the position of the other vehicles is necessary. The algorithm's results classify the vehicle's nature and its separation from the vehicle. Insufficient separation from the vehicle will result in a warning message emerging.

The Raspberry Pi is a device with limited resources [30–33] when it comes to processor speed and deep learning capabilities. The Raspberry Pi was chosen as the SSD algorithm device. Real videos were captured using [34, 35] a Raspberry Pi camera. This method is executed at a high calculation rate and is frame resolution-dependent [36]. Specialized vibration sensors and buzzers [37, 38] are used to detect any problems. In the case of an incident, the GPS receives the accident vehicle's location [39] and sends it. Through a WhatsApp message, this data will be delivered to a mobile number. The circuit's internet connection will be used to receive this message. Information about latitude and longitude values is provided in this message. These values can be used to determine an approximate location for the vehicle.

3 Proposed Methodology

The SSDMobileNetV1 algorithm and OpenCV libraries have been utilized to identify automobiles in the proposed methodology, which uses the design system to detect vehicles in front of them. The Raspberry Pi 3B+ model is used to implement the proposed method, and a Raspberry Pi 1.3 Camera Rev is used to capture the live video. The system performs detecting processes while saving these videos in memory. The SW-420 NC vibration sensor is used to detect whether or not an accident occurred. If an accident occurs, a buzzer module sounds, and the Neo-6 M GPS module subsequently provides position information. The user receives messages from the Telegram bot alerting them to the accident notice. The below Fig. 1 represents the block diagram of collision warning and accident detection.

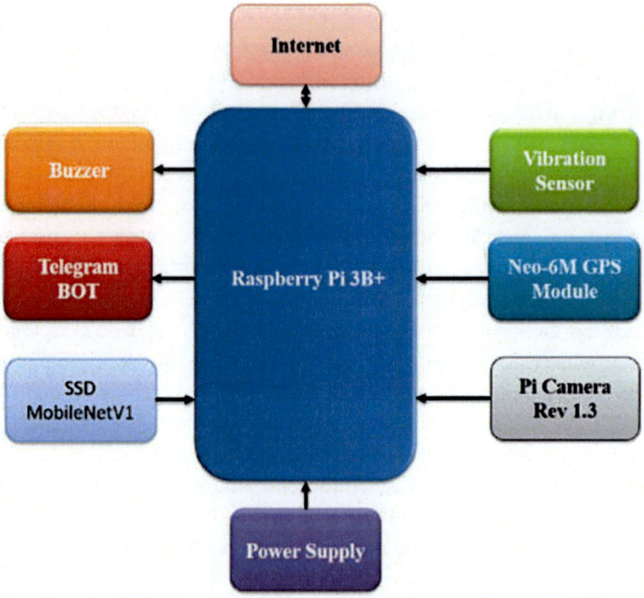

Fig. 1 Block diagram of collision warning and accident detection

3.1 Proposed Algorithm for Object Detection Using SSD MobileNetV1

In the initial step, the algorithm is trained using a vehicle dataset while the vehicles are detected. The vehicles dataset includes the following five classes: bicycle, bike, bus, car, and truck. The SSD MobilenetV1 classifier is used as a pre-trained model to assess whether a vehicle is present or not. The SSD MobilenetV1 model detects vehicles in real time after the classifier has been trained. Real-time video from a Camera Rev 1.3 will be used for input. The SSDMobileNetV1 architecture uses depth-wise separable convolutions to build compact deep neural networks. The video input is taken using OpenCV and split into frames, and then, each frame is converted into a grayscale. The grayscale pixels from the input image are transformed into highlights representing the image's content. The difference between the pixel intensities in the bright and dark areas is used to calculate each feature value. To calculate these attributes, the image is scaled to cover all potential sizes and regions. The object's bounding boxes and related class labels are then determined using the SSD MobileNetV1 model as shown in Fig. 2.

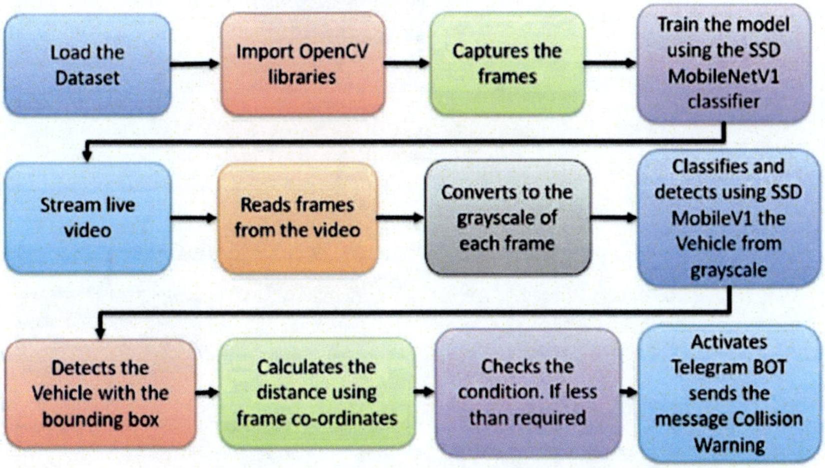

Fig. 2 Flow diagram of collision warning object detection using SSD MobileNetV1

3.2 *Proposed Hardware System Design Using Raspberry Pi*

A significant challenge with vehicles is avoiding collisions and road accidents. In this work, as shown in Fig. 3, using a Raspberry Pi Rev 1.3 Camera, a live video feed is captured and transferred to a Raspberry Pi 3B + for processing and target vehicle recognition. The Raspberry Pi is a device with limited resources when it comes to processor speed and deep learning capabilities. The Raspberry Pi 3B+ was chosen as the SSD MobileNetV1 algorithm device. Object detection is applied to the video stream that the Raspberry Pi camera has captured. In this procedure, the video stream is divided into individual object frames, and each frame is subsequently turned into a grayscale image. The SSD MobileNetV1 model is used to identify and categorize these frames. Bounding boxes are used to display the identified vehicles. Using the coordinates of frames, it determines the distance between two vehicles after detecting the vehicle, as shown in Fig. 5. The cycle continues if the distance is maintained; otherwise, a Raspberry Pi connected to the Wi-Fi network activates the Telegram bot and alerts the user of a potential collision.

Suppose the user ignores the Telegram bot's collision warning message, then the Raspberry Pi receives data from the vibration sensor and processes it to determine the vibration threshold. If the vibration exceeds the threshold, an accident will occur, prompting the buzzer to ring, the GPS module to detect the vehicle's position, and the Telegram bot to send an alarm message to the emergency services telegram, as illustrated in Fig. 6. Hardware circuit of a proposed system design is depicted in Fig. 4.

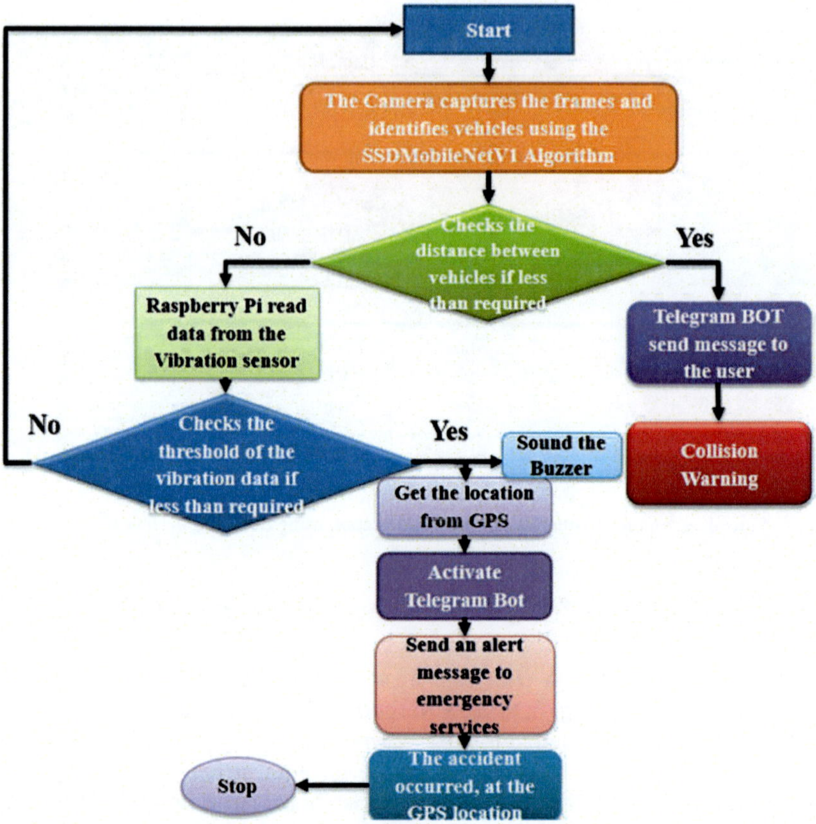

Fig. 3 Flow chart of proposed system design

4 Experimental Results and Analysis

According to Fig. 7, the "tightness" of the predicted bounding boxes to the actual object is represented by the training and validation box losses. The training and validation objectness losses calculate the likelihood that an object exists in the given region of interest. The training and validation classification losses measure whether each predicted bounding box was classified; each box may include either an object class or background.

In detecting the vehicles, the model has achieved 93.1% precision, 99% recall, 99.5% mAP50, and 82% mAP50-95. The model can make a good distinction in all classes of vehicles. Since mAP50 is 99.5%, the model can successfully detect and differentiate unique vehicles.

Fig. 4 Hardware circuit of a proposed system design

Fig. 5 SSD MobilenetV1 detected vehicles using Raspberry Pi VNC viewer

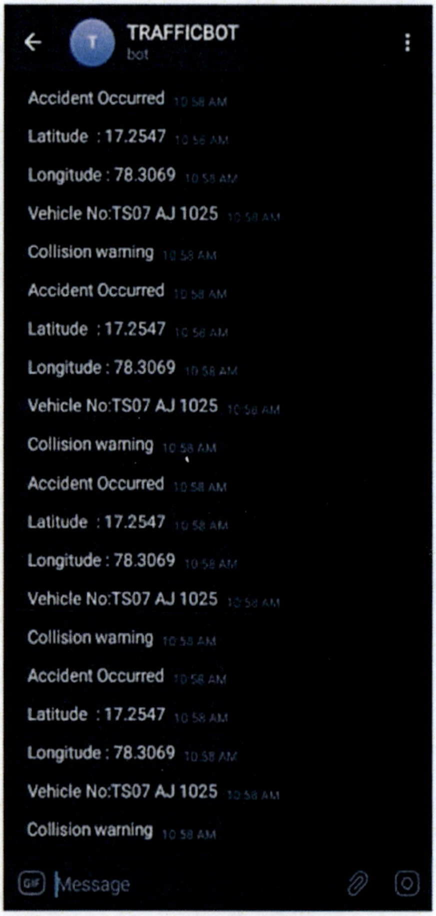

Fig 6. Telegram BOT alert message

4.1 Precision Curve

Precision describes the percentage of accurate predictions within all predictions for a specific class.

$$\text{Precision} = \frac{\text{True Positive}}{\text{True Positive} + \text{False Positive}} \times 100 \tag{1}$$

The precision–confidence graph illustrates it with confidence values on the x-axis and precision values on the y-axis. Confidence is nothing more than the assessment criteria used to get the score. This percentage-based confidence score displays the likelihood that the algorithm will properly recognize the image. Figure 8 illustrates that the precision is confined to 0 and 1. As confidence increased, the blue line

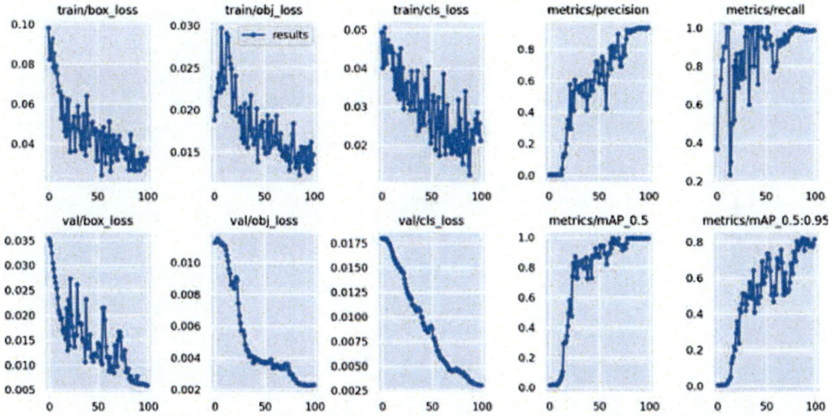

Fig. 7 Loss, precision, and recall graphs of the proposed system during training and validation

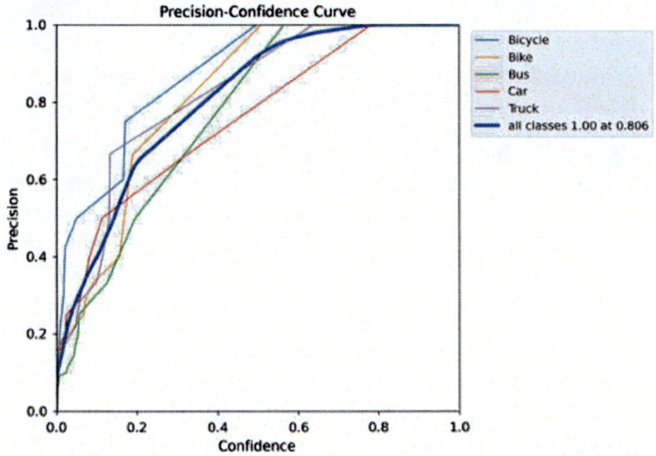

Fig. 8 Precision versus confidence graph

increased. All classes attained precision 1 with a confidence level of 0.806. If the confidence level is set high enough, precision may be arbitrarily good. This precision demonstrates that the system is quite effective at identifying various vehicles.

4.2 Recall Curve

The recall is the percentage of instances from a given class that the model correctly predicted to be members of that class.

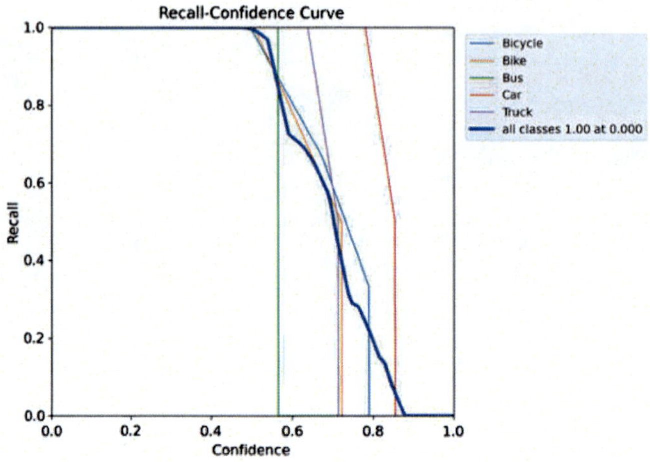

Fig. 9 Recall versus confidence graph

$$\text{Recall} = \frac{\text{True Positive}}{\text{True Positive} + \text{False Negative}} \times 100\%. \quad (2)$$

Recall values are on the y-axis of the recall–confidence graph, while confidence values are on the x-axis. Confidence is nothing more than the assessment criteria used to get the score. This percentage-based confidence score displays the likelihood that the algorithm will properly recognize the image. Figure 9 shows that recall is limited to values between 0 and 1. When confidence rises, the blue line decreases. At confidence 0, all classes obtained recall 1. If the confidence is kept low enough, the recall may be arbitrarily high.

4.3 *Precision–Recall Curve*

Precision–recall can be an excellent indicator of prediction performance when the classes are highly imbalanced. While recall measures the proportion of results that are actually relevant, precision is a measure that determines the relevance of the results in information retrieval. A graph containing precision values on the y-axis and recall values on the x-axis is all that is required to represent a precision–recall curve. In other words, TP/(TP + FN) is present on the y-axis of the precision–recall curve, while TP/(TP + FP) is present on the x-axis. The precision–recall curve shows how precision and recall are traded off at different thresholds. A low false-positive rate is associated with high precision, and a low false-negative rate is related to high recall. High recall and precision are indicated by a high area under the blue curve in Fig. 10. High scores indicate that the classifier produces accurate results with high precision and largely positive results with high recall.

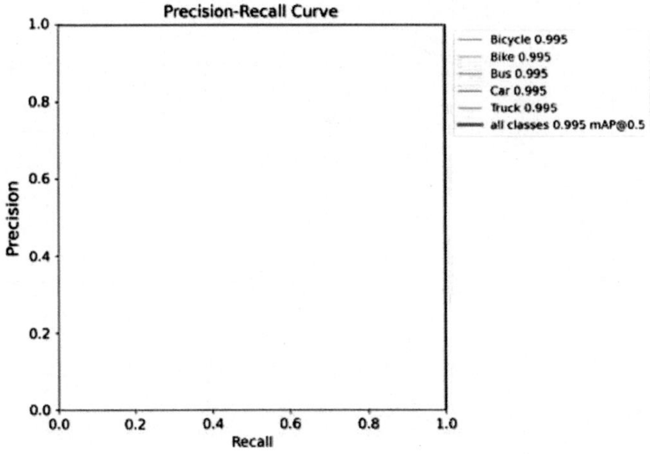

Fig. 10 Precision versus recall graph

4.4 F1-Score Curve

The F1-score, which goes from 1 to 0, may be understood as a harmonic mean of recall and precision. Equal proportions of the F1-score are contributed by both recall and precision.

$$F1\text{ - score} = \frac{2 \times \text{Precision} \times \text{Recall}}{\text{Precision} + \text{Recall}} \times 100\% . \tag{3}$$

The F1-score is beneficial in evaluating the confidence level that balances a given model's precision and recall values. The F1-scores for a certain model may be utilized to create a single value evaluation measure, which may be a credible indicator of the model's overall efficacy.

Figure 11 shows that the F1-score values are on the y-axis, confidence is on the x-axis, and the blue color line indicates the confidence level that maximizes recall and precision is 0.518. In many situations, a higher confidence level is preferred. It could be appropriate to adopt a confidence level of 0.59 for this model, given that the F1 value appears to be about 0.96, which is not too far from the maximum value of 1. It is also confirmed that this would be a good design point by looking at the precision and recall values with a confidence of 0.59. The precision value is still essentially at its maximum value at approximately 0.59, but the recall value begins to decline.

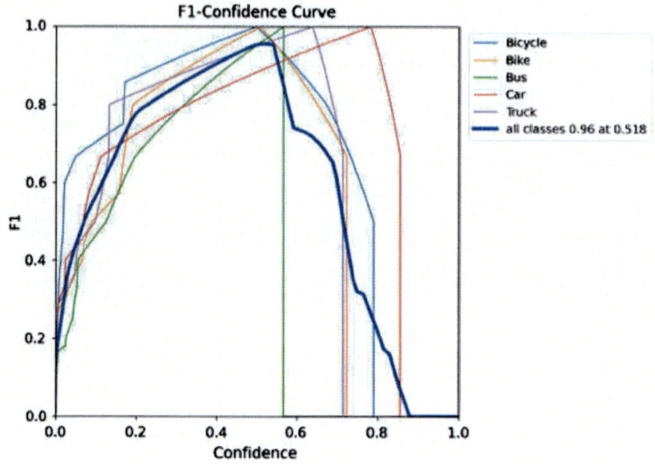

Fig. 11 F1-score versus confidence graph

4.5 Confusion Matrix

An accurate assessment of a classifier's potential is made possible via a confusion matrix. A correctly categorized result is represented by each diagonal element. The confusion matrix's diagonal off-diagonal area shows the results that were improperly classified. Therefore, the confusion matrix for the best classifier will only contain diagonal components, with all other values set to 0. Following the categorizing process, a confusion matrix of actual and predicted values is generated, as shown in Fig. 12.

In Fig. 12, the diagonal elements represent the samples that were successfully predicted. Out of the five samples altogether, five samples were predicted correctly. The total accuracy is 99%.

5 Conclusion

Road accidents are significantly impacted by vehicle collision warning and accident detection using SSD MobilenetV1. The Raspberry Pi, a vibration sensor, and a GPS module are employed to enable accident detection, while the SSD MobileNetV1 platform supports the proposed car collision warning system in this study. The SSD MobileNetV1 model of deep learning techniques is used to develop collision warnings. In this model, vehicles were captured in real-time video using a Rev 1.3 camera and recognized with 99% accuracy, 93.1% precision, 99% recall, and 99.5% mAP50. The model calculates distance based on detection and, if necessary, sends a collision warning message to the Telegram bot. Accident detection is achieved using a

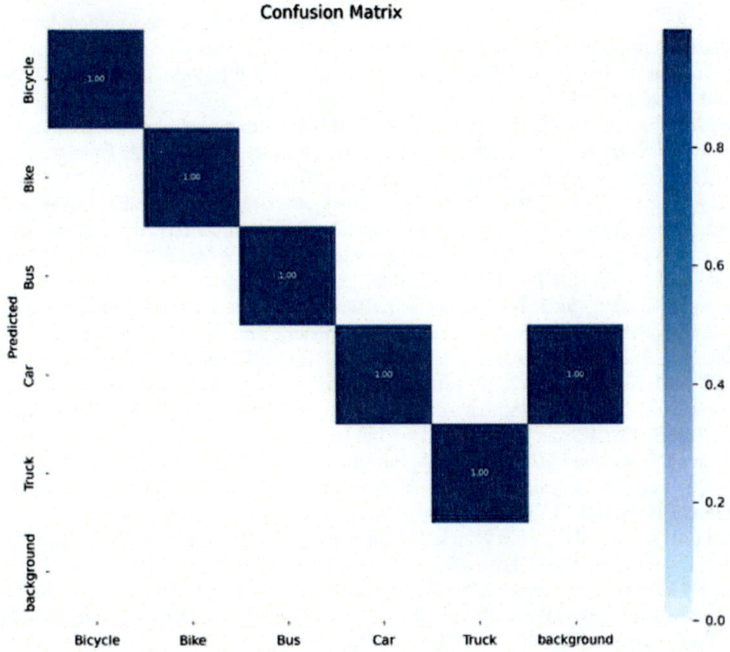

Fig. 12 Confusion matrix

Raspberry Pi. The accident is detected using a vibration sensor, and the vehicle's information, including its longitude and latitude, is sent to the emergency services' Telegram bot. Vehicle collision warning only works when the camera is pointed in the vehicle's forward direction. The future scope of this study may be applied in all directions of the vehicle so that it can determine the distance from any nearby vehicles and provide a collision warning.

References

1. Vaisakhi VS, Surendran D, Prabu T (2018) Smart road accident detection and communication system. In: Proceedings of the 2018 international conference on current trends towards converging technologies, ICCTCT 2018. https://doi.org/10.1109/ICCTCT.2018.8551179
2. Tian D, Zhang C, Duan X, Wang X (2019) An automatic car accident detection method based on cooperative vehicle infrastructure systems. IEEE Access 7:127453–127463. https://doi.org/10.1109/ACCESS.2019.2939532
3. Badhan HS, Oza SK, M. Tech Student (2019) Accident detection using raspberry pi. Int J Res Eng Appl Manage (IJREAM) 5:2. https://doi.org/10.35291/2454-9150.2019.0023
4. Dinesh UV, Johnson J, Sreesankar S, Augustine V, Thomas P, Peter E (2019) Accident alert system using raspberry pi. Int J Inf Secur Softw Eng 5(2):1–23. Accessed 12 Jan 2023. [Online]. https://computers.journalspub.info/index.php?journal=JISSE&page=article&op=view&path%5B%5D=537

5. Ghosh A, Al Mahmud SA, Uday TI, Farid DM (2020) Assistive technology for visually impaired using tensor flow object detection in raspberry pi and coral USB accelerator. In: 2020 IEEE region 10 symposium, TENSYMP 2020, pp 186–189. https://doi.org/10.1109/TENSYMP50017.2020.9230630
6. Song W, Yang Y, Fu M, Qiu F, Wang M (2018) Real-time obstacles detection and status classification for collision warning in a vehicle active safety system. IEEE Trans Intell Transp Syst 19(3):758–773. https://doi.org/10.1109/TITS.2017.2700628
7. Patel R, Dabhi VK, Prajapati HB (2017) A survey on IoT based road traffic surveillance and accident detection system (A smart way to handle traffic and concerned problems). In: 2017 Innovations in power and advanced computing technologies, i-PACT 2017, vol 2017, pp 1–7. https://doi.org/10.1109/IPACT.2017.8245066
8. Ciberlin J, Grbić R, Teslić N, Pilipović M (2019) Object detection and object tracking in front of the vehicle using front view camera. In: 2019 Zooming innovation in consumer technologies conference, ZINC 2019, pp 27–32. https://doi.org/10.1109/ZINC.2019.8769367
9. Luo S, Yu J, Xi Y, Liao X (2022) Aircraft target detection in remote sensing images based on improved YOLOv5. IEEE Access 10:5184–5192. https://doi.org/10.1109/ACCESS.2022.3140876/REFERENCES
10. Carrasco DP, Rashwan HA, Garcia MA, Puig D (2021) T-YOLO: tiny vehicle detection based on YOLO and multi-scale convolutional neural networks. IEEE Access. https://doi.org/10.1109/ACCESS.2021.3137638
11. Dai X (2019) HybridNet: a fast vehicle detection system for autonomous driving. Signal Process Image Commun 70:79–88. https://doi.org/10.1016/J.IMAGE.2018.09.002
12. Girshick R, Donahue J, Darrell T, Malik J, Berkeley UC, Malik J (2014) 1043.0690. In: Proceedings of the IEEE computer society conference on computer vision and pattern recognition, vol 1, p 5000. https://doi.org/10.1109/CVPR.2014.81
13. Ren S, He K, Girshick R, Sun J (2017) Faster R-CNN: towards real-time object detection with region proposal networks. IEEE Trans Pattern Anal Mach Intell 39(6):1137–1149. https://doi.org/10.1109/TPAMI.2016.2577031
14. Deng Z, Sun H, Zhou S, Zhao J, Zou H (2017) Toward fast and accurate vehicle detection in aerial images using coupled region-based convolutional neural networks. IEEE J Sel Top Appl Earth Obs Remote Sens 10(8):3652–3664. https://doi.org/10.1109/JSTARS.2017.2694890
15. Liu W et al (2016) SSD: single shot multibox detector. In: Lecture notes in computer science (including subseries lecture notes in artificial intelligence and lecture notes in bioinformatics), vol 9905 LNCS, pp 21–37. https://doi.org/10.1007/978-3-319-46448-0_2/FIGURES/5
16. Redmon J, Divvala S, Girshick R, Farhadi A (2016) You only look once: Unified, real-time object detection. In: Proceedings of the IEEE computer society conference on computer vision and pattern recognition, vol 2016, pp 779–788. https://doi.org/10.1109/CVPR.2016.91
17. Li H, Zhang S, Kong W (2020) Bilateral counting network for single-image object counting. Visual Comput 36(8):1693–1704. https://doi.org/10.1007/S00371-019-01769-5/METRICS
18. Chiu YC, Tsai CY, da Ruan M, Shen GY, Lee TT (2020) Mobilenet-SSDv2: an improved object detection model for embedded systems. In: 2020 International conference on system science and engineering, ICSSE 2020. https://doi.org/10.1109/ICSSE50014.2020.9219319
19. Othman NA, Aydin I (2018) A new deep learning application based on movidius NCS for embedded object detection and recognition. In: ISMSIT 2018—2nd international symposium on multidisciplinary studies and innovative technologies, proceedings. https://doi.org/10.1109/ISMSIT.2018.8567306
20. Andreetto M et al (2017) MobileNets: efficient convolutional neural networks for mobile vision applications automatic 3-D modeling of textured cultural heritage objects view project MobileNets: efficient convolutional neural networks for mobile vision applications. Accessed 9 Jan 2023. [Online]. https://www.researchgate.net/publication/316184205
21. Zhang X, Zhou X, Lin M, Sun J (2018) ShuffleNet: an extremely efficient convolutional neural network for mobile devices. In: Proceedings of the IEEE computer society conference on computer vision and pattern recognition, pp 6848–6856. https://doi.org/10.1109/CVPR.2018.00716

22. Sandler M, Howard A, Zhu M, Zhmoginov A, Chen LC (2018) MobileNetV2: inverted residuals and linear bottlenecks. In: Proceedings of the IEEE computer society conference on computer vision and pattern recognition, pp 4510–4520. https://doi.org/10.1109/CVPR.2018.00474
23. Vijaya Lakshmi A, Praveen Kumar Goud K, Saikiran Kumar M, Thirupathi V (2022) Real-time face mask detection using MobileNetV2 classifier. In: Smart innovation, systems and technologies, vol 269, pp 63–73. https://doi.org/10.1007/978-981-16-7996-4_6/COVER
24. Freeman I, Roese-Koerner L, Kummert A (2018) Effnet: an efficient structure for convolutional neural networks. In: Proceedings—international conference on image processing, ICIP, pp 6–10. https://doi.org/10.1109/ICIP.2018.8451339
25. Niazi-Razavi M, Savadi A, Noori H (2019) Toward real-time object detection on heterogeneous embedded systems. In: 2019 9th International conference on computer and knowledge engineering, ICCKE 2019, pp 450–454. https://doi.org/10.1109/ICCKE48569.2019.8964764
26. Kanimozhi S, Gayathri G, Mala T (2019) Multiple real-time object identification using single shot multi-box detection. In: ICCIDS 2019—2nd international conference on computational intelligence in data science, proceedings. https://doi.org/10.1109/ICCIDS.2019.8862041
27. Galvez RL, Bandala AA, Dadios EP, Vicerra RRP, Maningo JMZ (2019) Object detection using convolutional neural networks. In: IEEE region 10 annual international conference, proceedings/TENCON, vol. 2018, pp 2023–2027. https://doi.org/10.1109/TENCON.2018.8650517
28. Honmote H, Gadekar S, Katta P, Kulkarni M (2022) Real time object detection and recognition using MobileNet-SSD with OpenCV. Int J Eng Res Technol 11(1). https://doi.org/10.17577/IJERTV11IS010070
29. Gluhakovic M, Herceg M, Popovic M, Kovacevic J (2020) Vehicle detection in the autonomous vehicle environment for potential collision warning. In: 2020 Zooming innovation in consumer technologies conference, ZINC 2020, pp 178–183. https://doi.org/10.1109/ZINC50678.2020.9161791
30. Hegde R, Patel S, Naik RG, Nayak SN, Shivaprakasha KS, Bhandarkar R (2021) Underwater marine life and plastic waste detection using deep learning and raspberry pi. In: Lecture notes in electrical engineering, vol 752 LNEE, pp 263–272. https://doi.org/10.1007/978-981-16-0443-0_22/COVER
31. Khedkar S, Malwatkar GM (2016) Using raspberry Pi and GSM survey on home automation. In: International conference on electrical, electronics, and optimization techniques, ICEEOT 2016, pp 758–761. https://doi.org/10.1109/ICEEOT.2016.7754787
32. Marot J, Bourennane S (2017) Raspberry Pi for image processing education. In: 25th European signal processing conference, EUSIPCO 2017, vol 2017-January, pp 2364–2368. https://doi.org/10.23919/EUSIPCO.2017.8081633
33. Balon B, Simic M (2019) Using raspberry pi computers in education, pp 671–676. https://doi.org/10.23919/MIPRO.2019.8756967
34. Othman NA, Salur MU, Karakose M, Aydin I (2019) An embedded real-time object detection and measurement of its size. In: 2018 International conference on artificial intelligence and data processing, IDAP 2018. https://doi.org/10.1109/IDAP.2018.8620812
35. Asmara RA, Syahputro B, Supriyanto D, Handayani AN (2020) Prediction of traffic density using yolo object detection and implemented in raspberry pi 3b + and intel ncs 2. In: 4th International conference on vocational education and training, ICOVET 2020, pp 391–395. https://doi.org/10.1109/ICOVET50258.2020.9230145
36. Walam SS, Teli SP, Thakur BS, Nevarekar RR, Patil SM (2018) Object detection and seperation using raspberry PI. In: Proceedings of the international conference on inventive communication and computational technologies, ICICCT 2018, pp 214–217. https://doi.org/10.1109/ICICCT.2018.8473068
37. Pavan Manikanta M, Chakrabortty A, Samson M, Akash, M (2022) IoT based accident detection and rescue system. J Positive School Psychol 6(3):6664–6669–6664–6669. Accessed 9 Jan 2023. [Online]. https://www.journalppw.com/index.php/jpsp/article/view/3916
38. Surantha N, Wicaksono WR (2018) Design of smart home security system using object recognition and PIR sensor. Procedia Comput Sci 135:465–472. https://doi.org/10.1016/J.PROCS.2018.08.198

39. Bergonda (2017) IoT based vehicle accident detection and tracking system using GPS modem

A Comprehensive Review on the Development of Pipe Robot

Md. Rawshan Habib, K. M. Monzur Rahaman, Abhishek Vadher, Tahsina Tashrif Shawmee, Md Apu Ahmed, Sibaji Roy, Md Shahnewaz Tanvir, Aditi Ghosh, and Shuva Dasgupta Avi

Abstract The pipe robots are autonomous machines that operate within pipe networks and perform several tasks including surveillance, monitoring, cleaning, rescue, etc., therein. Various elements, such as rust and fractures, could harm pipelines. Consequently, a reliable monitoring system is necessitated to guarantee the security of these pipes. It is difficult for an individual to examine each component of pipes and repair the problems. Pipe robots have been created as a result of such issues. Moreover, the demand for studies on pipe robots has increased in latest days. A variety of pipe robots are usable which include wheel type, screw type, wall-press type, walking type, inchworm type, etc. Relying on its specifications and assessment goal, every model has a unique set of benefits and drawbacks. This paper presents an overall review on the advancement of pipe robots from past few years. Authors believe that ongoing advancements in pipe robot technology are anticipated to be significantly influenced through this paper.

Keywords Pipe robot · Rescue robot · Surveillance robot · In-pipe robot · Review

Md. R. Habib (✉) · A. Vadher
Murdoch University, Murdoch, Australia
e-mail: habib.eee.20@gmail.com

K. M. M. Rahaman
United International University, Dhaka, Bangladesh

T. T. Shawmee · S. D. Avi
Ahsanullah University of Science & Technology, Dhaka, Bangladesh

M. A. Ahmed · S. Roy · A. Ghosh
Chemnitz University of Technology, Chemnitz, Germany

M. S. Tanvir
South Dakota School of Mines and Technology, Rapid City, USA

G. Rajakumar et al. (eds.), *Intelligent Communication Technologies and Virtual Mobile Networks*, Lecture Notes on Data Engineering and Communications Technologies 171,
https://doi.org/10.1007/978-981-99-1767-9_17

1 Introduction

With advancement of new technology, the growth of metropolitan regions is now a constant process. Manufacturing industries are constantly developing in order to provide people with a level of living. This upgrading is placing a lot of strain on the available space on the surface. As a result, the use of underground space for the provision of water, fuel, as well as gas via pipe galleries, which are a collection of various types of pipes, is rising with each day. Oxidation, fractures, and other reasons can cause a lot of damage to such pipelines. To assure the safety of such support systems, a comprehensive monitoring strategy is needed. It is difficult for a person to examine and repair every portion of a pipe [1]. Pipe robots were created as a result of such issues. As a result, pipe robots are becoming a subject of growing interest. Pipe robots are just an interconnected concept that integrates equipment, electrics, and instruments that can travel down the center or outside of a tiny pipe and contain a variety of sensors and instruments to conduct a set of pipe operations underneath the command of a user or software [2].

The use of a pipe surveillance robot has brought forth novel issues in terms of robotics performance, that comprises the robot's propulsion system, which could be self-propelled or aided by a media, as well as its flexibility to changes in diameters. As a result, innovative solutions for autonomous in-pipe monitoring technologies are always searched for. Investigations using the pipe surveillance robot, such as carried out by various researchers, are used in a broad range of research projects. The progress of pipe robots has piqued the interest of specialists. Pipe robots are constructed to examine pipelines. Pipe inspection robot technologies could be divided into two categories: outer-pipe surveillance robot technologies and in-pipe surveillance robotic devices. The outer-pipe surveillance robot systems are in high demand in the market because it can execute inspections without disrupting production process. Such surveillance robot attaches to the pipe's outside surface and can travel across it independently or semi-autonomously, negotiating curves and intersections. In contrast to the in-pipe robotic platform, the outer-pipe surveillance robot scheme is unable to thoroughly examine the inner cylindrical wall [3].

Direct drive systems and differential drive systems are two types of wheel-type robots. Direct drive does have a greater total torque benefit, while differential drive generally takes less motor. In direct drive setup, automobiles take four motors. The harmonic gears transmit the motor's output to the wheels, resulting in a high torque output. Nevertheless, throughout this design, three cameras are necessary, with two of them being used to observe the steering wheels. Walking robots, also known as crawler robots, typically have eight similar pair of legs coupled to a frame [4]. Performance comparison is carried out in [2] to identify the best type of robots and authors of that research found that screw and wall-press types are the ones which provide the most effective performance. Figure 1 depicts the present situation of the pipe robots.

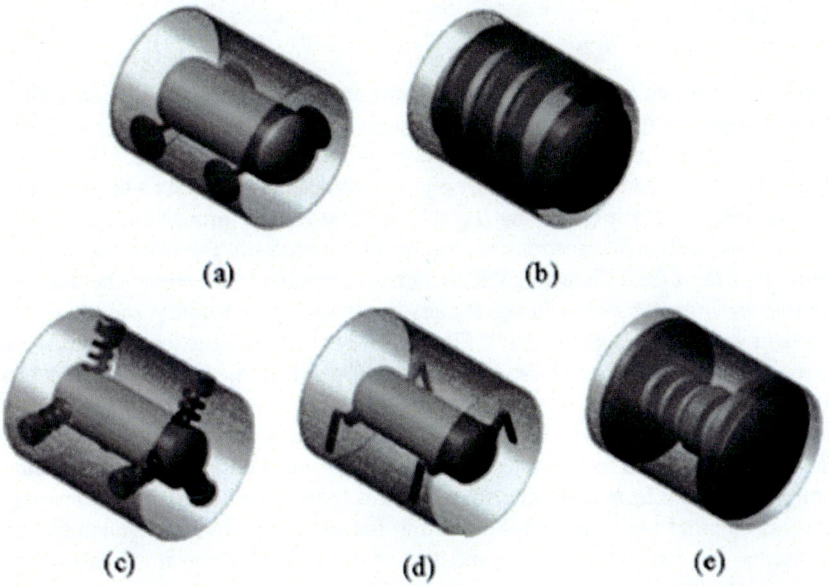

Fig. 1 Different types of pipe robots include **a** wheel type, **b** screw type, **c** wall-press type, **d** walking type, **e** inchworm type [4]

2 Recently Developed Pipe Robots

Pipe robots are mostly made to eliminate human interference from labor-intensive and dangerous work situations. However, robots are occasionally utilized to investigate difficult work locations which are frequently unreachable to people. Since they transport hazardous substances and liquids and frequently have small interior diameters or curves that are unreachable to humans, pipes must be inspected. Robots are required for the checking of certain pipelines due to the complex inner configuration and danger product constraints. The process of developing an in-pipe assessment robot is challenging; thus the developers should consider factors including movement, steering capacity, steering angle, dimension and shape adaptation, etc. flexibility, durability, autonomy, and object tracking online effectiveness on a crooked surface, functioning safely, fabric choice, the nature of the work to be done within the pipe, various actuator, action in a running pipe, bringing back a robot, a user-friendly controller and navigation interface, operating range, quantitative evaluation of pipe defect levels [5].

2.1 Wheel Type

Pipelines represent the most widely used way for moving liquids and gases. They must be regularly inspected to function properly. In order to check pipes, people are required to go hazardous situations. Pipeline robots have been result of this [6]. A pipe robot is created for automatically pipe cracks detection. The six wheel-like feet on this robot's structure ensure that it is able to easily transfer through all orientations. It may make turns at intersections, such as on a T-shaped path. The robot may be used with pipes of various diameters thanks to the adjustable foot length. The device's usefulness is demonstrated in the studies by its ability to identify wide range of clefts in various types of pipes [7]. Figure 2 shows the construction of wheel-type robot. With the use of a self-driving robot equipped with ultrasound detectors and linked through latest technologies, the project, proposed in [8] intends to build and construct a pipeline inspection robot. Additionally, an IP camera is employed for thorough observation, and wireless functionality is included for simple monitoring of the collected information. The advancements made possible by the cutting-edge process required pipeline ultrasonic testing far more attractive. Additionally, this device has the capacity to capture and display the sight of the environments on the workers' monitor screens for reliable monitoring, identification, fast assessment, and evaluation in small, darkened spaces. A unique wheel-type robot is presented in [9] with dual helical angles that can be adjusted. The creature's high adaptation in various pipes' conditions as well as its excellent mobility in bend tubes and T branches has both been tested in investigations. Block diagram of wheeled robot is depicted in Fig. 3.

The architecture and movement management of a wheel-type robot that could check pipes with internal diameters are developed in [6]. Such robot meets movement instability in a bent pipe. In order to solve the existing problems, the suggested robot sets the wheels at various angles, enabling the robot to maintain consistent touch with

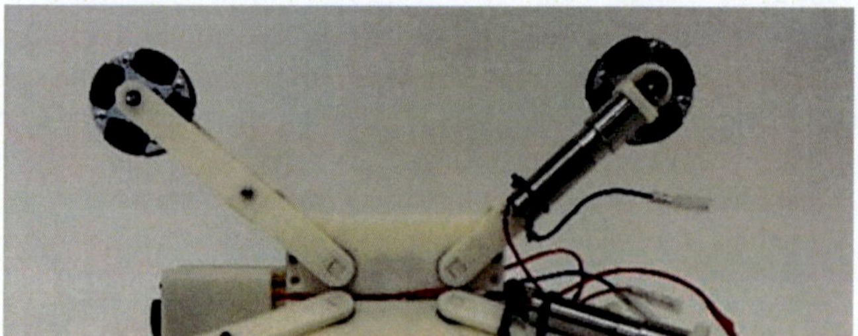

Fig. 2 Design of the robot proposed in [7]

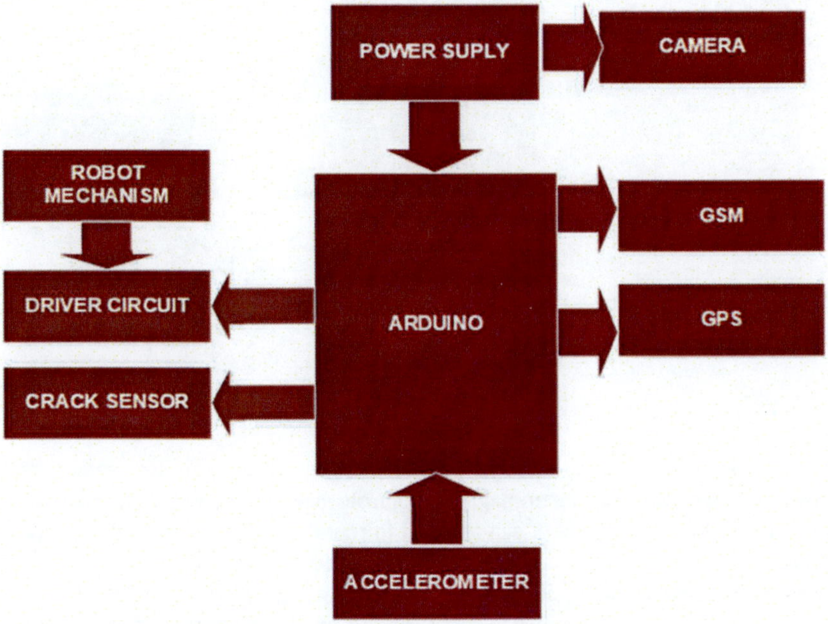

Fig. 3 Schematic representation of the wheel-type robot developed in [8]

pipe's surfaces. The planned robot is being developed with the use of 3D printing. Another robot with several bends is demonstrated in [10]. The active and passive modules make up the robot where every unit includes three possible wheel combinations that are expanded using various mechanisms. The robot is constructed using a proportional integral derivative regulator to match standard velocities. Another device with several movement units is built in [11] with the goal of increasing the flexibility factor of singular movement form pipeline robots. Engineering work has gone into the self-adaptive diameter-changing, wheel movement, as well as rotating mechanisms. The basic architecture of the robot is optimized based on the movement and structural concepts of pipe robots throughout transit. Figure 4 depicts a prototype of wheeled robot.

2.2 *Screw Type*

A sort of efficient smart products for examining pipeline problems is the screw-type robot. Because of the various kinds of pipes and operational environments, pipe robots do have broad variety of design characteristics. Practical or conceptual design methods cannot guarantee the reliability and effectiveness, that raises manufacturing costs and lengthens the production process [12]. A brand-new effective screw-driven

Fig. 4 Wheel-type robot demonstrated in [10]

robot is presented in [13] for pipe status inspection. A suspension system, four drive modules, and four electromagnetic controls make up the three components of the proposed robot. The robot can respond to round and rectangular tubes without changing its construction thanks to a unique movement technique. Construction of screw robot is presented in Fig. 5, whereas working principle is shown in Fig. 6.

A directional pipe surveillance robot relying on screw propulsion is suggested in [14]. It can manage the pitching ratio of the motion as well as avoid potential obstructions. Here, control information is optimized through the use of an optimum control technique. Utilizing MATLAB simulations, the accuracy and effectiveness of the planned robot and controller are confirmed. Another novel robot for navigating bending and branched pipelines is shown in [15]. The robot has been constructed

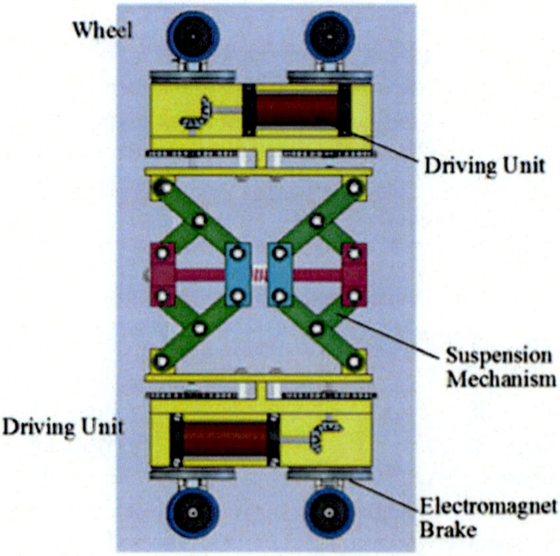

Fig. 5 Novel screw-type robot [13]

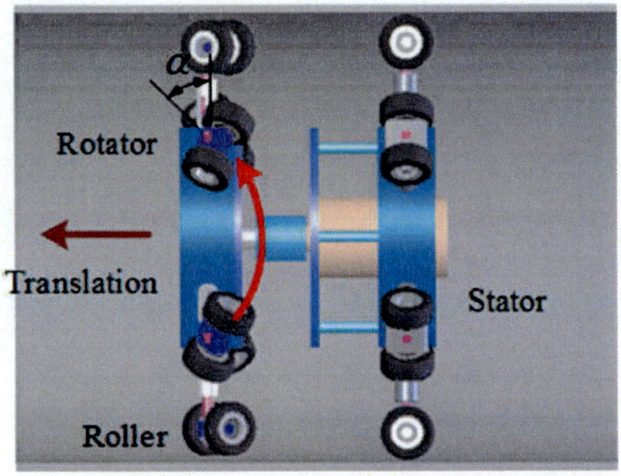

Fig. 6 Principle of screw robot [17]

from three pieces, each of which includes two passive wheels and flexible wings. This allows the robot to go through and maneuver through pipelines that are not just linear but indeed curved and branching. Dual methods are suggested in [16] to reduce the power usage of a unique directional screw pipe robot.

2.3 Wall-Press Type

Robots that are wall-pressed have been built to scale vertical and slanted pipe runs in challenging surroundings. Wall-press feature enables for further pipe mobility. Its structure enables it to go across modifiers, bends, and perhaps even serve as a steering wheel. Considering different sizes of diameters, a robot with several wheels is needed. Nevertheless, a wall-press-type robot just needs one. Operating expenses and management go up as there are more robots. A structure resembling a wall press is constructed in [4]. It is divided into three sections: the camera portion, the front steering unit, and the back operating panel. The benefit of the wall-press type is that its increased diameter may be changed. Front-facing camera part allows for 180-degree unhindered visibility of the inside circumstances. Robot is shielded from obstructions by two driving units. Figures 7 and 8 both depict a prototype of wall-press robot. The creation and execution of an ideal pipe routing method and a moving module to deal with the changeable conditions within metal pipes are the main topics in [18]. Additionally, it provides flexibility to various pipe sizes. The self-reliant feature in the middle of a tube, standard navigation capacity to adjust to in-pipe unbalance, functionality to stay unchanged without falling in pipes, and effective deployment of cleaning supplies are really the significant issues taken into consideration in the concept and deployment. Navigation functionality and driving capabilities are evaluated to ensure that it meets the design parameters on which it is constructed. Focusing on a genuine pipeline scenario, experts developed a straightforward pipe architecture

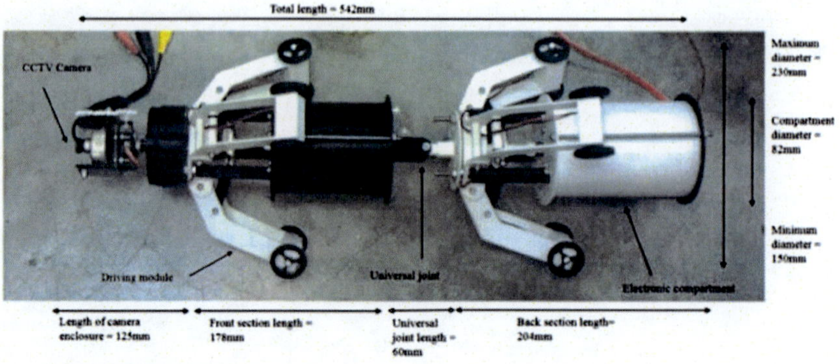

Fig. 7 Wall-press robot [4]

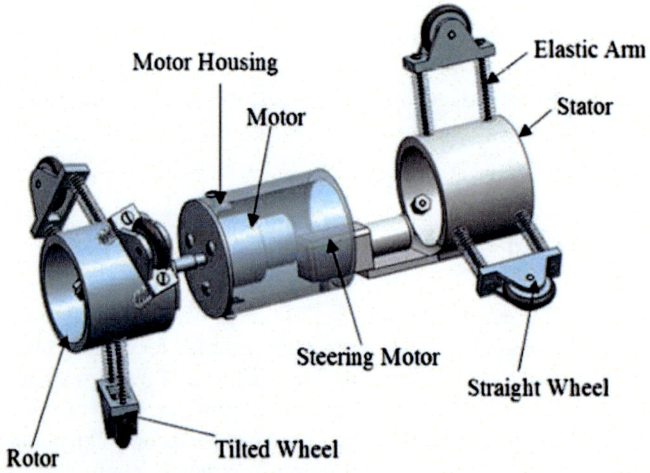

Fig. 8 Prototype of wall-press robot developed in [5]

[19]. To satisfy the design criteria of changeable width and barrier, robot mobility has been examined. An autonomous in-pipe robot concept with a self-adaptive structure for different diameters is created. According to the findings, the developed system has excellent varied diameter characteristics.

2.4 Walking Type

An analysis is done on a walking robot concept in [20] that is made up of three sections joined together by suspension systems. To stay stable while moving within a pipe, the

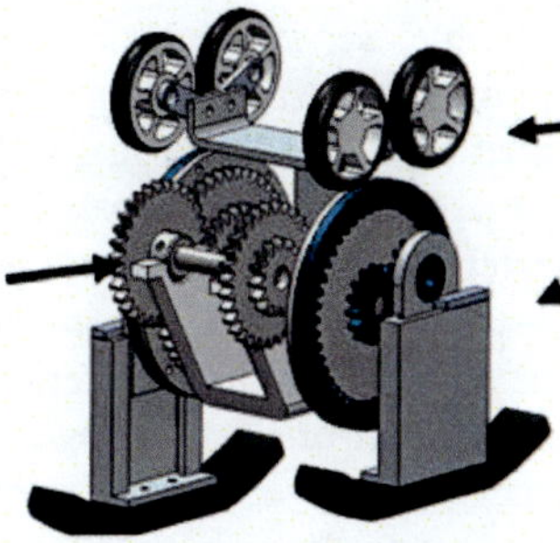

Fig. 9 Experimental prototype for a walking robot [20]

three mechanical components are programmed to crawl using asynchronous movements. It has been done a dynamic study to simulate the variation in driving factor. Additionally, a moving apparatus model is constructed. An experimental prototype for a walking robot is given in Fig. 9. The architecture of a walking robot's state observer is investigated in [21]. The need for this component stems from the idea as detectors do have finite range of precision and therefore are susceptible to buzzing sound generation. Since walking robots employ model-based regulation and need precise knowledge of their present condition, this seems to be particularly troublesome for them. One of the numerous issues that must be resolved before robots can be used in a realistic manner is preparing pathways on the inside ground of a pipe. A method for position control is suggested in [22]. The technique makes utilize a point cloud model of the pipeline, that is simpler to create than the graphics format needed by several of the other movement organizing strategies for walking robots now in operation. Another robot with eight legs is shown in [23] that can navigate inside the pipes. The system prototype of the robot is initially explained, with a focus on its main components. The central regulation work is allocated, managing the entire device behavior and a distributed status, relating to the particular feet, in accordance with the physical platform structure.

2.5 Inchworm Type

Inchworm is another type of pipe robot used for pipe observation. Inchworms can ascend and walk on a variety of complicated structures according to the anatomy and evolutionary habits. It has motivated the creation of several robots that could exhibit comparable skills for a variety of tasks, including the investigation of a complicated environment. Making such machines actually usable with a simple design for enabling continuous adaptation to a complicated landscape, like an outermost layer, is one of the main architectural problems. Scientists offer a fresh idea for a moving robot in [24] that is influenced by inchworms in light of this. The robot is a flexible pipe crawler thanks to a passive foot cap below an electromagnetic leg. According to research observations, the robot can walk on pipes with different curves. The robot may help with closer examination outside of pipes as a novel robotic option, reducing disruption in the transmission of oil and gas. The main objective of the research presented in [25] is to suggest and outline the creation of a robot using smooth sensors that LabVIEW will analyze while it is used in various pipe dimensions. Such research concentrates on how robots may operate efficiently and smoothly in pipelines by suggesting certain key mechanics including gliding features, gripping methods, and twisting units. Figures 10 and 11 show the design architecture of inchworm robot. Inchworm robot progress is discussed in [26], and by applying active shape management of suction cups, a novel adhesion method is experimentally devised and motivated by the activity of inchworms. Three gaits are created, as well as the movement formula is generated.

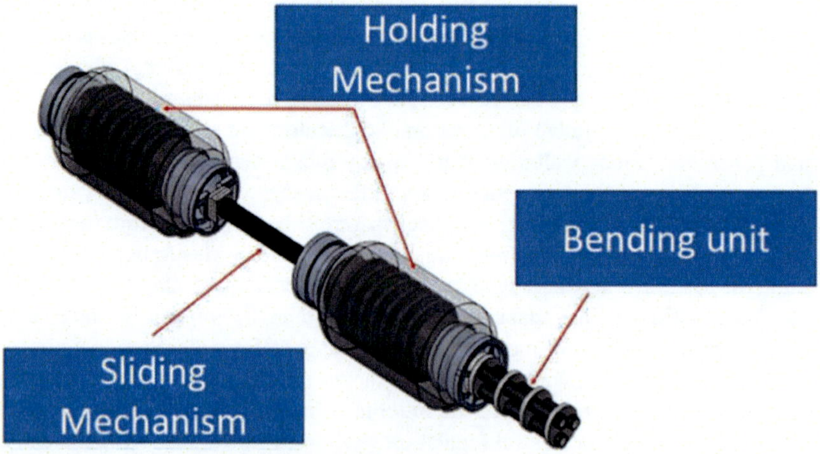

(a) Isometric view.

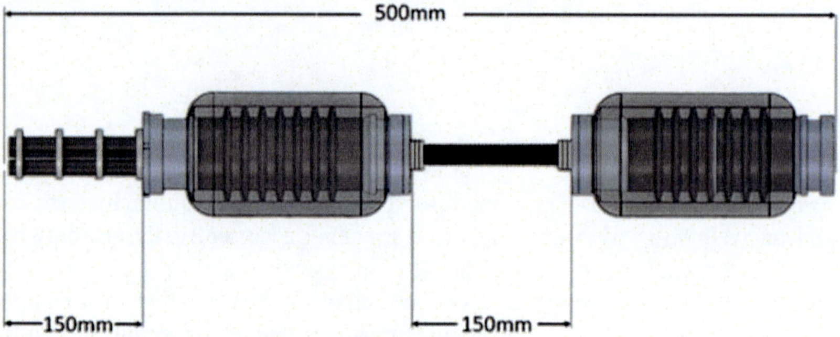

(b) Left view.

Fig. 10 Design of an inchworm robot [24]

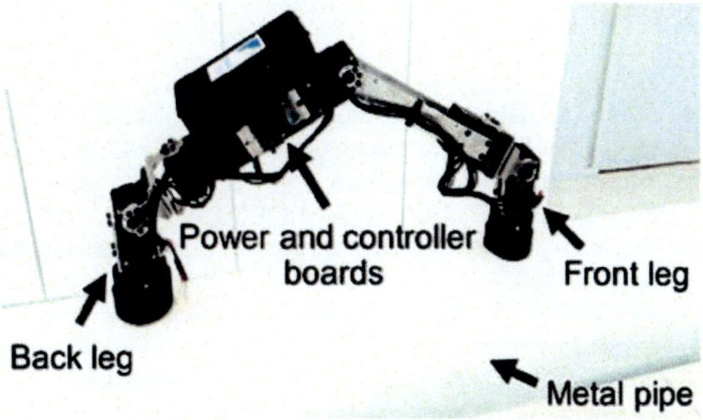

Fig. 11 Prototype of inchworm robot [25]

3 Conclusion

A thorough overview on pipe robot innovation is described in this research. In the topic of pipe robot design and operations, there have been several studies and advancements made. In latest days, a number of pipe robots have been accessible. Some of the most common types include wheel type, wall-press type, walking type, inchworm type, screw type, and electromagnetic force type. Differences among various sorts of pipe robots are provided in Table 1. Though each type of robots has been developed with almost the same objectives, there are still conceptual and functional differences between them. One of most frequent types is the wheel, which has a straightforward architectural style. The operation of a walking-type robot, on the other hand, necessitates the use of a complicated control. Because it is more versatile, a wall-press-type robot seems to have the potential to ascend vertically. The inchworm is a stable worm that may fit into a tiny pipe diameter. The electromagnetic type of robot can only move in one direction, while the screw type can move forward and reverse.

Table 1 Differences among various sorts of pipe robots

Types	Advantages	Disadvantages
Wheel type	• Quick • Simple design • Economical	• Slippery • Not effective in pipes with holes
Screw type	• Flexible • Adaptable to circular and square tube	• Complex design
Wall-press type	• Simple mechanism • Stability without slipping	• Not effective in pipes with holes
Walking type	• Not slippery	• Sometimes causes damage inside the pipe due to the contact with the pipe
Inchworm type	• Flexible in different pipes	• Complex design

References

1. Habib MR, Ahmed K, Reza AS, Mouly FJ, Mahbub HR (2019) An Arduino based robot for pipe surveillance and rescue operation. In: 22nd International conference on computer and information technology. Dhaka, pp 1–5
2. Shao L, Wang Y, Guo B, Chen (2015) A review over state of the art of in-pipe robot. In: Proceedings of 2015 IEEE international conference on mechatronics and automation. Beijing, pp 2180–2185
3. Rashid MZA et al (2020) Modeling of the in-pipe inspection robot: a comprehensive review. Ocean Eng 203:107206
4. Wahed MAA, Arshad MR (2017) Wall-press type pipe inspection robot. In: IEEE 2nd international conference automatic control and intelligent systems. Kota Kinabalu, pp 185–190
5. Nayak A, Pradhan SK (2014) Design of a new in-pipe inspection robot. Procedia Eng 97:2081–2091
6. Elankavi RS et al (2022) Design and motion planning of a wheeled type pipeline inspection robot. J Robot Control 3:415–430
7. Zhang B, Abdulaziz M, Mikoshi K, Lim H (2019) Development of an in-pipe mobile robot for inspecting clefts of pipes. In: IEEE international conference on cybernetics and intelligent systems and IEEE conference on robotics, automation and mechatronics. Bangkok, pp 204–208
8. Mohammed MN et al (2018) Design and development of pipeline inspection robot for crack and corrosion detection. In: IEEE conference on systems, process and control. Melaka, pp 29–32
9. Ye C et al (2015) Development of an in-pipe robot with two steerable driving wheels. In: IEEE international conference on mechatronics and automation. Beijing, pp 1955–1959
10. Min J et al (2014) Development and controller design of wheeled-type pipe inspection robot. In: International conference on advances in computing, communications and informatics. Delhi, pp 789–795
11. Yan F, Gao H, Zhang L, Han Y (2022) Design and motion analysis of multi-motion mode pipeline robot. J Phys Conf Ser 2246:012029
12. Li T et al (2018) Rapid design of a screw drive in-pipe robot based on parameterized simulation technology. SIMULATION 95:659–670
13. Li P, Yang W, Jiang X, Lyu C (2018) Active screw-driven in-pipe robot for inspection. In: IEEE international conference on unmanned systems. Beijing, pp 608–613
14. Tourajizadeh H, Rezaei M (2016) Design and control of a steerable screw in-pipe inspection robot. In: Proceedings of the 4th international conference on robotics and mechatronics. Tehran, pp 98–104
15. Kakogawa A, Nishimura T, Ma S (2014) Development of a screw drive in-pipe robot for passing through bent and branch pipes. In: IEEE ISR 2013. Seoul, pp 1–6
16. Tourajizadeh H, Boomeri V, Rezaei M, Sedigh A (2020) Dynamic optimization of a steerable screw in-pipe inspection robot using HJB and turbine installation. Robotica 38:1–22
17. Li P et al (2018) Design and analysis of a novel active screw-drive pipe robot. Adv Mech Eng 10:1–18
18. Kim YG, Shin DH, Moon JI, An J (2012) Design and implementation of an optimal in-pipe navigation mechanism for a steel pipe cleaning robot. In: 8th International conference on ubiquitous robots and ambient intelligence. Incheon, pp 772–773
19. Ling Z, Chen Y, Tang J, Zheng M (2020) Development and design of self-adaptive variable diameter robot for in-pipe inspection. IOP Conf Series Mater Sci Eng 782:042003
20. YuYu X, Chen Y, Chen MZ, Lam J (2015) Development of a novel in-pipe walking robot. In: IEEE international conference on information and automation. Lijiang, pp 364–368
21. Savin S, Jatsun S, Vorochaeva L (2018) State observer design for a walking in-pipe robot. MATEC Web Conf 161:03012
22. Savin S (2018) RRT-based motion planning for in-pipe walking robots. In: International scientific and technical conference "Dynamics of systems, mechanisms and machines". Omsk, pp 1–6

23. Pfeiffer F, Rossmann T, Loffler K (2002) Control of a tube crawling machine. In: 2nd International conference control of oscillations and chaos. Proceedings. St. Petersburg, pp 586–591
24. Khan MB et al (2020) iCrawl: an inchworm-inspired crawling robot. IEEE Access 8:200655
25. Aliff M et al (2022) Development of pipe inspection robot using soft actuators, microcontroller and LabVIEW. Int J Adv Comput Sci Appl 13:349–354
26. Yuan YY et al (2017) Design and implementation of an inchworm robot. In: International conference on advanced robotics and intelligent systems, Taipei

Design and Analysis of Efficient IoT-Based Pollution Monitoring System in Urban Area

M. Dhivya, N. J. Avinash, P. Bhoomika, K. S. Shanika, Trisha Chatterjee, and V. Preethisree

Abstract As the population is increasing, the environmental resources consumed also started to increase, due to this pollution which becomes a major concern, and it is a day-to-day problem, which is faced by all biological species. Various solutions have been proposed by researchers as countermeasures to control the pollution in past few decades. In this paper, IoT-based pollution monitoring system is presented which encompasses various sensors, which is capable of measuring air pollutants (ammonia, carbon monoxide, smoke, LPG, and methane), temperature, humidity, and sound levels. The microcontroller used is MSP430. The real-time data can be monitored virtually across the world by an application. The application which is used is ThingSpeak, where the level of harmful pollutants can be monitored in the form of tables and graphs.

Keywords Air pollution · Noise pollution · Internet of Things · Sensors · MSP430 · ThingSpeak

1 Introduction

The total population statistics of India as on December 1, 2022, is 1,412,903,336 based on dataset generated by Worldmeter by United Nations data. With ever-increasing pollution, there has been exponential growth of people getting affected due to diseases caused by harmful air pollutants. To make an impact globally, a project is being developed which can help to collect data and that can help to further reinforce probable solution by developing certain product to make it countenance to reduce the pollution and contamination of air.

The sole purpose of this paper is to make an effective change by bringing in a state of acceptance of this product. IoT technology has been used in developing the

M. Dhivya (✉) · N. J. Avinash · P. Bhoomika · K. S. Shanika · T. Chatterjee · V. Preethisree
Department of Electronics and Communication Engineering, New Horizon College of Engineering, Bengaluru, Karnataka, India
e-mail: dhivyamanian@gmail.com

G. Rajakumar et al. (eds.), *Intelligent Communication Technologies and Virtual Mobile Networks*, Lecture Notes on Data Engineering and Communications Technologies 171,
https://doi.org/10.1007/978-981-99-1767-9_18

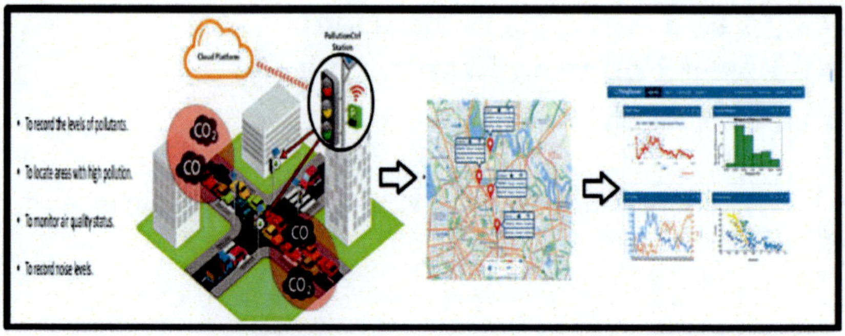

Fig. 1 Graphical abstract of proposed system

product, which can help in getting various data of air containing contaminants like CO_2, sulfur dioxide components gases, nitrogen dioxide gases and so on and so forth which can give us the analysis of how the pollution is affecting the atmosphere, till what threshold value the contamination is taking place so that necessary measures can be devised. Here on, tactfully managing the system with proper acquisition to the tools and approbating the installation in more traffic region can change the world's today's situation. The project consists of hardware part and software part with relatable work to make the best benefit out of it.

The main contribution of this project is to develop a cost-efficient air and noise level monitoring device that senses the various parameters like methane, temperature, smoke, ammonia, carbon monoxide, and humidity and sends the real-time data to the IDE software through the controller, and these values are displayed on the software monitor along with the display. A survey will be conducted to measure the pollutants' contents in each of the four areas. This will help us in acknowledging the factors that actually gives rise to pollution in that particular region. The graphical abstract of the paper is depicted in Fig. 1.

In this project, various types of sensors are used to sense the harmful gases in the surroundings. A Wi-Fi module is used to send these real-time collected data to the ThingSpeak application, where we can display these data using graphs and table representation. This application is an open source; due to this, everyone can monitor these data readings. Most of the respiratory diseases are caused by the air pollution, due to the harmful gases that are released. One example of causing death of the people due to respiratory diseases like bronchitis and asthma, pneumonia, influenza, and whooping cough is given in Fig. 2.

2 Literature Survey

Flores-Cortez et al. [1] constructed a cost-efficient device for pollution monitoring in developing countries. The paper emphasized on developing an IoT station, which

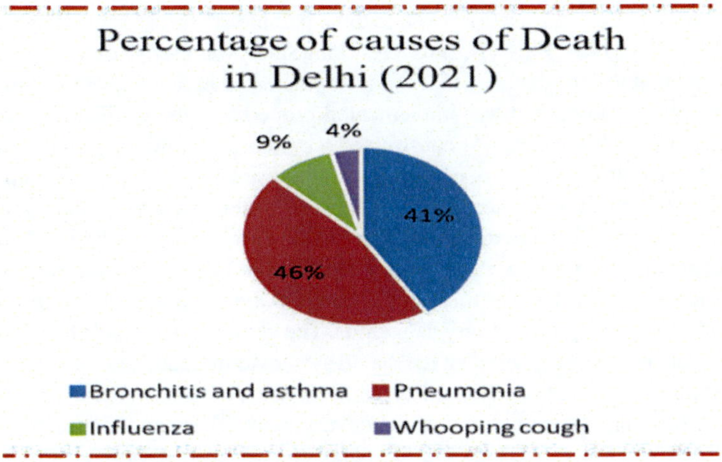

Fig. 2 Percentage of causes of death in Delhi (2021)

monitors PM pollutant level in the environment. The microcontroller used in the project was LoPy ESP32, and the parameters measured were Air (PM2.5, PM10, temperature, and humidity). The tools used were LoPy microcontroller and micro Python language, which made the circuit a low-cost system with high-performance rate. The design was made such that the device sends back three of the values collected by the sensor, every ten minutes as the sensors data will not change drastically for a short time.

Helton et al. [2] constructed a platform to monitor the air quality levels. The paper emphasizes a solution to air pollution. They have covered five important factors given by the WHO, by using cost-efficient hardware components. The microcontroller used in the project was ESP-WROOM-32, and the parameters measured were air pollutants such as PM2.5, PM10, O_3, CO, NO_2, and NH_3. MQTT protocol was used and implemented in the system, in order to send the readings to the application, and it also helps to configure and process the data collected and it stores the information. The tools used to analyze the data graphically was Zabbix.

Bikash et al. [3] developed an IoT-based real-time air level measuring system. The paper depicts a system which is proposed, using dedicated sensors which measure the data remotely from anywhere. The system needed low maintenance, and it had more stability. The microcontroller used in the project was Node MCU, and the parameters measured were temperature, humidity, air (NH_3). The collected data were uploaded to the AWS platform, which can be used to access and monitor the real-time data. The system was designed in a way that it can be used for many applications such as strategic measures and can be used in industries and hospital areas.

Syed et al. [4] proposed a system to measure the air quality using Raspberry Pi4 for indoor purpose. The paper depicted an air quality device integrated between two main networks, IoT and various sensors. This system was a lightweight device, which is easy and convenient to carry the device to different locations. The microprocessor used in the project was Raspberry Pi4, and the parameters measured were air pollutants such as CO, NO_2, H_2, NH_3, and CO_2. ThingSpeak was used as a cloud service to access and monitor the device. MQTT was implemented in the project as a protocol. The security of the system was implemented properly, as the application required authorization. Kalpesh et al. [5] constructed a system for domestic purpose, which detects and monitors leakage of cooking gas for house safety. The paper emphasized a hybrid, low-powered, less cost leakage detection system, which is mainly designed for indoor purpose. The microcontroller used in the project was Arduino UNO, and the parameters measured were air pollutants such as LPG or natural gas, as they mainly focused on leakage of these gases' indoors. Leakage of such harmful gases will lead to high risk, and it is very dangerous. The system is affordable by everyone.

Manglani et al. [6] proposed an IoT-based system to measure and monitor air and sound levels for smart environment. The paper emphasized on developing an efficient, low-cost Arduino system to monitor the environment. The microcontroller used in the project was Arduino UNO, and the parameters measured were air (ammonia, sulfide, and CO_2) and noise levels. The system used very few components to build an easy to use and cheap environmental monitoring system. AQI was taken as an unit for measuring air quality index, and the dB was taken as an unit for measuring the noise level. The four readings each were taken for four days. A buzzer was used to alert purpose; if the readings goes beyond certain value, the buzzer will ring and alerts the surrounding.

Harsh et al. [7] described a framework for air pollution device which can be used by smart cities, based on IoT technology. The paper depicts communication between a cloud platform and an application which is Android based. The microprocessor used in the project was Raspberry Pi, which was used as a gateway to interact with the hardware components. The parameters measured were humidity, temperature, LPG, CO. ThingSpeak was used as a platform to plot graphs received by the sensors. An Android application was built, to monitor the real-time values which are sent by the sensors, and this application needed an account to sign in. To use various applications such as storing and messaging purpose, an API was included in the application. Ajitesh et al. [8] drew an analysis and proposed a design of IoT-based air level measuring device. The paper depicts a portable, real-time pollution monitoring system, which is also cost-efficient. The microcontroller used in the project was Node MCU 1.0 ESP8266 to process the data, and the parameters measured were smoke, CO, PM1, 10, 2.5. The system was also capable of supporting Ruby and Node technologies. ThingSpeak was used to display the values of the collected data using graph representation. Two gas sensors and one particle sensor were used to measure the air quality levels in the surrounding.

Rakib et al. [9] proposed a prediction system for air pollution system. The paper emphasizes a model, which can be monitored by anyone and anywhere in the world, as the data were stored in the cloud. The microcontroller used in this project was Arduino MEGA 2560, and the parameters measured were air (NH_3, CO, PM2.5, temperature, and humidity). They used mathematical calculations such as MAPE and data analysis to predict the pollution. It was a 2 in 1 system, which had prediction system as well as pollution measuring device, which measures air levels. The system had given 90%

accurate readings. Cornelius et al. [10] constructed a pollution tracking system based on IoT. The paper depicts a low-cost, efficient monitoring system, with a module that tracks the location of the system, and it was given to the user. The microprocessor used in the project was Raspberry Pi and Arduino controller was also used, and the parameters measured were air and noise levels. They also used a web server and dashboards for the collecting the real-time data from various sensors.

Literature review depicts a survey of ten papers from 2019 to 2022. Figure 3 shows the analysis of pollution monitoring systems as number of papers published versus various parameters.

As air pollution causes too many airborne diseases, to analyze how every year smoke and air pollutant's density is increasing in air, a survey has been conducted to check how it varied over years. The survey had been conducted for both air and noise pollutants using air and noise sensors. According to survey of 2018, 30% pollution took place because of pollutants from automobile. In year 2020, the contaminants that polluted are as follows—0.31 CO mg/m^3, 1177 NH_3 ug/m^3, 62% humidity. In

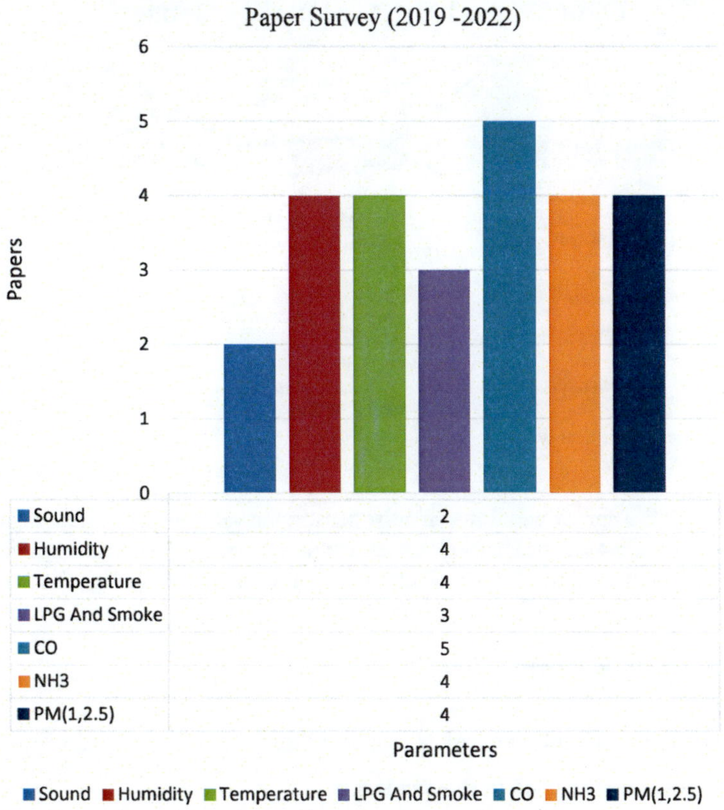

Fig. 3 Paper surveys from 2019 to 2022

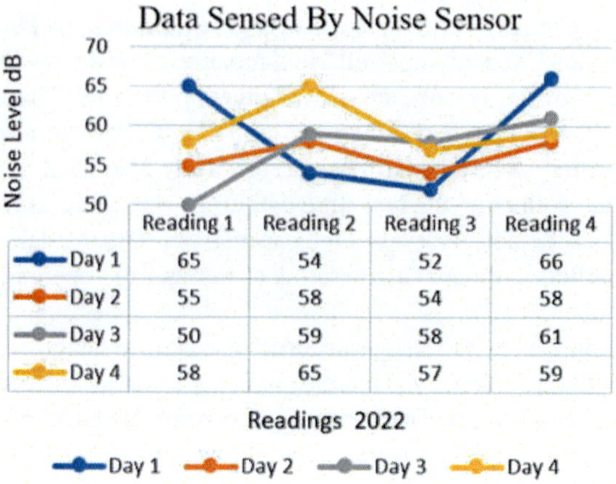

Fig. 4 Survey on noise sensor

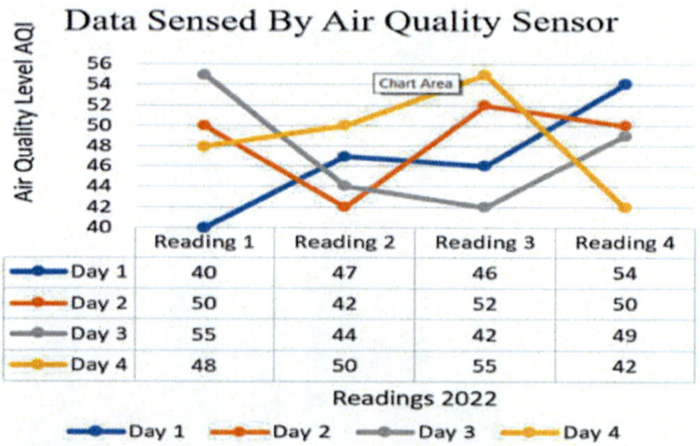

Fig. 5 Survey on air quality levels

the year 2022, a survey was done on noise sensor; the value as per the readings were obtained. Figure 4 shows a survey, which is done by taking four readings of sound levels (dB) for four days. Survey on air quality levels is shown in Fig. 5.

3 Proposed System

Proposed architectural framework is shown in Fig. 6.

The microcontroller used in this project is:

MSP430: It is a microcontroller that offers mixed signal with a very low power, it is 16-bit design that is good and used for wireless industries which uses low-power technology, and it also used mainly in medical applications. The MSP430

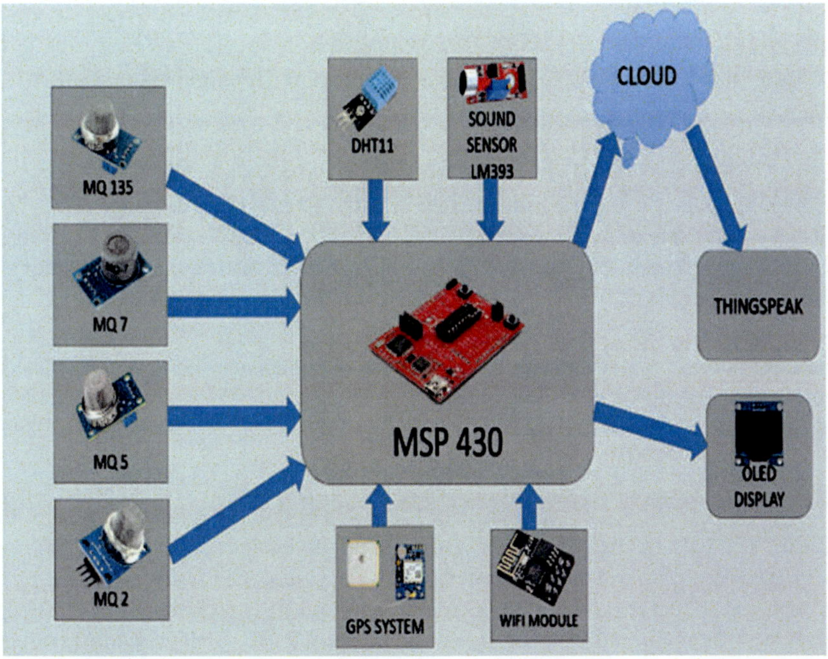

Fig. 6 Proposed architectural framework

microcontroller has twenty main pins. All of these pins play their role to ensure higher practicality of the microcontroller. The MSP430 is C-complier friendly and additionally utilizes low power. It can even effectively wake up from standby mode in less than 5 microseconds. They are excellent for high performance.

During the design process, MSP430 was selected since it is one of the simplest microcontrollers and designed for low-cost, low-power consumption embedded applications.

The software IDE used is Energia IDE:

ENERGIA IDE: Energia is an open-source electronics platform started in 2012 to bring together wiring and Arduino on the same platform. As this Energia IDE is supported on OS, Windows, Mac, and Linux, it is known as a cross-platform tool. It is also a portable as it can be used in different IDEs. The framework of this IDE is implemented and designed in such a way that it is easy to use by all and it is also beginner friendly. Energia together with the Launchpad can be used to develop and implement variety of switches and sensors, by controlling them taking the inputs, and we can also control a variety of other components such as motor, lights, and other physical outputs.

The various sensors used in this project are:

MQ135: It is a sensor that detects methane (CH_4) within the range of 300–10,000 ppm.

MQ2: It is a sensor that detects smoke and LPG within the range of 30–10,000 ppm.

MQ7: It is a sensor that detects carbon monoxide (CO) within the range of 10–500 ppm.

MQ5: It is a sensor that detects and measures the composition of ammonia (NH_3), sulfur (S), benzene (C_6H=6), carbon dioxide (CO_2) within a range of 10–1000 ppm.

LM393: This is a sound sensor that measures the sound within the range of 52–48 db and a frequency of 3 kHz–6 GHz.

DHT11: This is a temperature and humidity sensors that measure temperature within a range of 0–50 °C/±2 °C and humidity within a range of 20–80%/±5%.

The Location of the project can be identified using:

GPS SYSTEM: The GPS system stands for Global Positioning System, it is a navigation device, and based on satellite system, it is used to identify and show the location and time information to a GPS receiver in real time.

The module which is used to connect with cloud is:

ESP01 ESP8266 MODULE: it is a Wi-Fi module-based system which is a SOC. It is a system or a chip which is integrated highly and is designed as a very small chip, in such a way that it provides an internet connection. It is a cheap Wi-Fi-enabled module used for controlling devices over the Internet. It can either be worked with a microcontroller or can be programmed on its own. The ESP8266 comes with factory programmed firmware inside allowing you to control it with standard programmed commands. It is highly flexible, as it is compatible with many controllers and it is a powerful module.

The display which is used for displaying the values is:

OLED DISPLAY: It stands for Organic Light-Emitting Diode and is a solid-state device which is made up of organic thin films that are placed in series between two conductors. They work on the phenomenon of electroluminescence. Due to the presence of organic layers, OLEDs are brighter than LEDs and they also consume less power. They response time is very fast, and compared to LEDs, it has better picture viewing and quality.

The software application used is:

THINGSPEAK: It is an IoT analytics-based service area that allows users to analyze, visualize, and aggregate data streams which are live in the cloud. ThingSpeak is written in Ruby, it is an open-source platform, where everyone can use it, and it allows users to communicate with IoT device. It facilitates logging, retrieval, and data access of data using an API for both the devices and social network websites. We can send live data from sensors to ThingSpeak and we can plot automatically an instant live graph using the data received from the sensors used in this project and find the rising and falling trends form the graph. The device is used to monitor air and sound quality in indoors, but this device can be used to monitor pollution on roadside, and it can be also used to monitor the industries, monitoring activities like shooting, open-air events and other sport matches, monitoring traffic noise in city roads, by changing some of the components to other components according to the area where the device is used. The flowchart of the proposed system is depicted in Fig. 7. The sensors and other components are connected with MSP430, and the data are displayed using OLED. The status of the Wi-Fi module with webserver and the data are displayed on ThingSpeak.

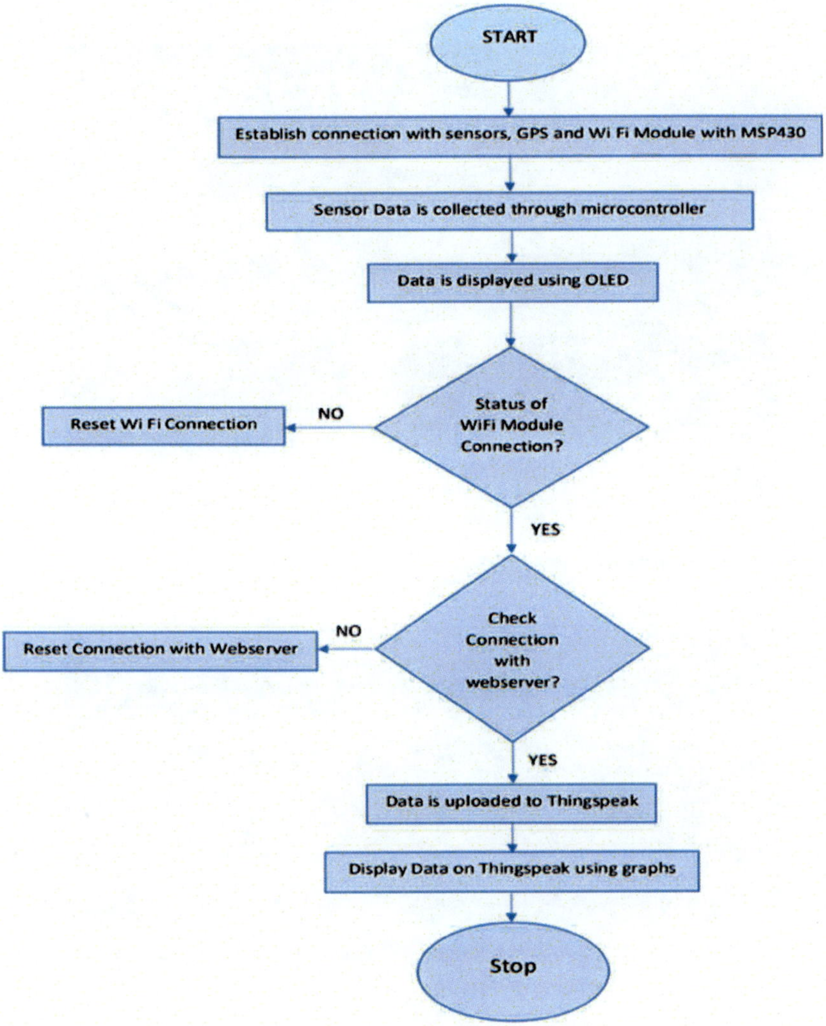

Fig. 7 Flowchart of the proposed model

4 Results and Discussion

Figures 8 and 9 depict the connection of the components. It shows the connection of microcontroller MSP430 with various sensors, such as MQ7, MQ135, MQ5, MQ2 (Table 1).

Figure 10 depicts the methane values in ppm, and these are the values sensed by MQ5 sensor. These values were measured in an open area.

Fig. 8 Hardware connection 1

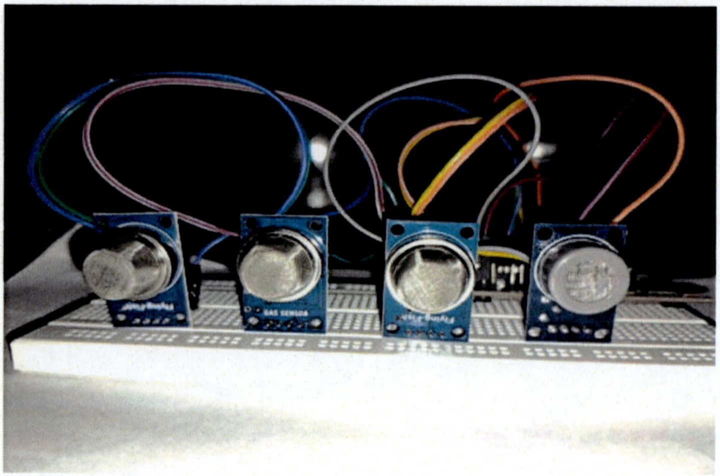

Fig. 9 Hardware connection 2

Table 1 Specifications of the components employed in the project

Name of the component	Specification
MSP430 microcontroller	Low supply voltage range: 1.8–3.6 V Five power-saving modes 16-bit RISC architecture, 62.5-ns instruction Up to 24 capacitive touch-enabled I/O pins
MQ135 sensor	Operating voltage: 2.5–5.0 V Power consumption: 150 mA Detect/measure: NH_3, No_x, CO_2, alcohol, benzene, smoke Typical operating Voltage: 5 V
MQ2 sensor	Operating voltage is +5 V Analog output voltage: 0–5 V Detect/measure: LPG, smoke, methane, butane Preheat duration 20 s
MQ7 sensor	Operating voltage is +5 V Can be used to measure or detect CO Analog output voltage: 0 V–5 V Fast response time Heater consumption about 350 mW
MQ5 sensor	Heater voltage: 5.0 V Power supply: 5 V Detect/**measure**: methane, hydrogen, alcohol Working current: 150 mA
Temperature and humidity sensor (DHT11)	Power supply 3.5–5.5 V Operating current: 0.3 mA (measuring) 60 uA (standby) Temperature range: 0–50 °C/±2 °C Humidity range: 20–80%/±5%
Sound sensor (LM393)	Differential input voltage range equal to maximum rated supply voltage: ±38 V Ambient temperature, TA = 0–70 °C Determines noise within decibel 3 kHz–64 Hz frequency
WI-FI module (ESP01 ESP8266 module)	Wi-Fi direct (P2P), soft-AP 1 MB flash memory 32-bit Tensilica Xtensa LX106 CPU running 80 MHz 3.3 V supply
GPS system (U-Blox neo-6 M)	Power supply voltage: 2.0–3.6 V Antenna gain: 50 dB Antenna type: passive and active antenna

(continued)

Table 1 (continued)

Name of the component	Specification
OLED display	High-resolution at 128 × 64 pixels, 160 degrees viewing angle Lower power consumption: only 0.06 W with normal use Power supply AC 3–5 V, working very well with Arduino Working temperature: −30–70 °C

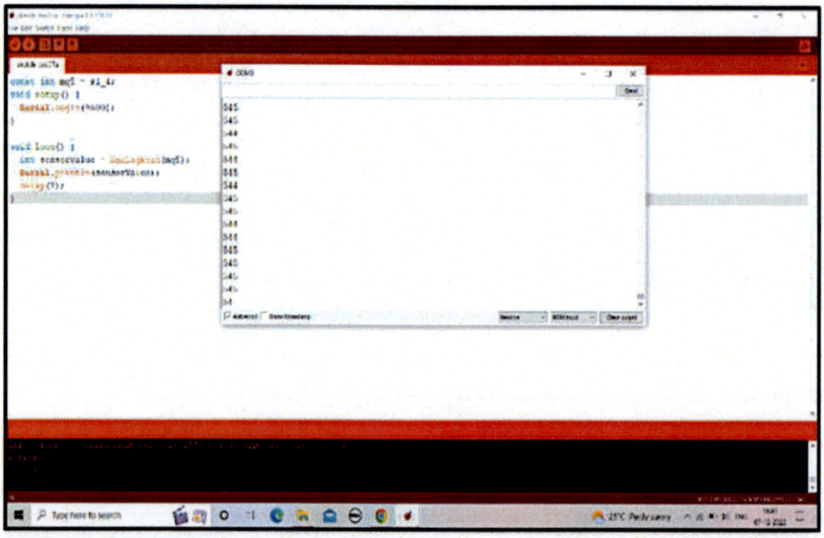

Fig. 10 MQ5 sensor values

Figure 11 depicts the ammonia, sulfur, benzene, and carbon dioxide values in ppm, and these are the values sensed by MQ135 sensor. These values were measured in an open area.

5 Conclusion

In this paper, a general problem faced by the society is taken up and come up with a solution. The proposed system will measure various harmful gases' (ammonia, carbon monoxide, sulfur, smoke, LPG, and methane) sound level, temperature, and humidity. These values are measured, by fixing the proposed system in various locations. The system collects the real-time data and displays it using the OLED display. These values can also be monitored using the ThingSpeak application, where the

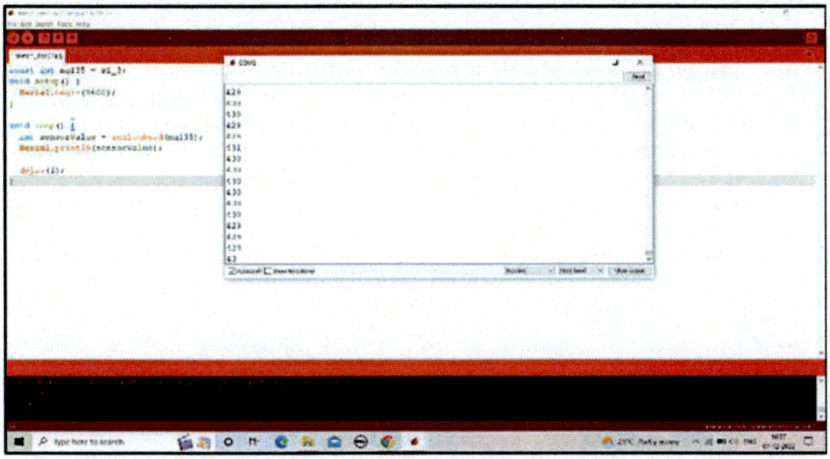

Fig. 11 MQ135 sensor values

values are displayed in the form of graphs. Using GPS module, we have also come up with the location of the device. The project can be further be innovated to detect dust in the areas. Using IOT, it could be much use to detect radioactively, noise levels near nuclear plants and prevent any nuclear disaster from happening, and it can also be modified further to monitor air and sound pollution levels at any place of the world. The battery can also be replaced with solar panels, to reduce the consumption of batteries.

References

1. Flores-Cortez OO, Adalberto Cortez R, Rosa VI (2019) A low-cost IoT system for environmental pollution monitoring in developing countries. In: 2019 MIXDES—26th international conference mixed design of integrated circuits and systems. IEEE, pp 386–389
2. de Medeiros HP, Girão G (2020) An IoT-based air quality monitoring platform. In: IEEE international smart cities conference (ISC2). IEEE, pp 1–6
3. Moharana BK, Anand P, Kumar S, Kodali P (2020) Development of an IoT-based real-time air quality monitoring device. In: 2020 International conference on communication and signal processing (ICCSP). IEEE, pp 191–194
4. Faiazuddin S, Lakshmaiah MV, Alam KT, Ravikiran M (2020) IoT based indoor air quality monitoring system using raspberry Pi4. In: 2020 4th International conference on electronics, communication and aerospace technology (ICECA). IEEE, pp 714–719
5. Gupta K, Gokul Krishna G, Anjali T (2020) An IoT based system for domestic air quality monitoring and cooking gas leak detection for a safer home. In: 2020 International conference on communication and signal processing (ICCSP). IEEE, pp 0705–0709
6. Manglani T, Srivastava A, Kumar A, Sharma R (2022) IoT based air and sound pollution monitoring system for smart environment. In: 2022 International conference on electronics and renewable systems (ICEARS). IEEE, pp 604–607

7. Gupta H, Bhardwaj D, Agrawal H, Tikkiwal VA, Kumar A (2019) An IoT based air pollution monitoring system for smart cities. In: 2019 IEEE international conference on sustainable energy technologies and systems (ICSETS). IEEE, pp 173–177
8. Kumar, Kumari M, Gupta H (2020) Design and analysis of IoT based air quality monitoring system. In: 2020 International conference on power electronics & IoT applications in renewable energy and its control (PARC). IEEE, pp 242–245
9. Rakib M, Haq S, Hossain MI, Rahman T (2022) IoT based air pollution monitoring & prediction system. In: 2022 International conference on innovations in science, engineering and technology (ICISET). IEEE, pp 184–189
10. Cornelius K, Kumar NK, Pradhan S, Patel P, Vinay N (2020) An efficient tracking system for air and sound pollution using IoT. In: 2020 6th International conference on advanced computing and communication systems (ICACCS). IEEE, pp 22–25

Internet of Things in Entrepreneurship: The Intellectual Structure Perspective

Agung Purnomo, Nur Asitah, Elsa Rosyidah, Gusti Pangestu, Meiryani, and Fairuz Iqbal Maulana

Abstract Entrepreneurship and business require Internet of Things support to be able to develop to meet market and consumer needs. This study aims to map the current state of global research and the future trends of IoT development in entrepreneurship studies. Based on the Scopus database, this study uses a bibliometric intellectual structure perspective by analyzing 1,075 scientific publications for half a century. The findings show that the United States Environmental Protection Agency and the United States were the most prolific countries and research institutions studying IoT in entrepreneurship. The findings of this study propose the ISEEEII research theme concept for IoT in entrepreneurship research.

Keywords Bibliometric · Digital business · Entrepreneurship · Internet of Things · Intellectual structure

1 Introduction

The Internet of Things (IoT) idea attempts to demonstrate its primary uses in various aspects of our lives, as well as its fundamental advantages, development patterns, issues, and concerns to be solved [1, 2]. The emergence of a website technological breakthrough that bridges the digital and physical worlds marks the beginning

A. Purnomo (✉)
Entrepreneurship Department, Bina Nusantara University, Jakarta, Indonesia
e-mail: agung.purnomo@binus.ac.id

N. Asitah · E. Rosyidah
Institute for Research and Community Services, Universitas Nahdlatul Ulama Sidoarjo, Sidoarjo, Indonesia

G. Pangestu · F. I. Maulana
Computer Science Department, Bina Nusantara University, Jakarta, Indonesia

Meiryani
Accounting Department, Bina Nusantara University, Jakarta, Indonesia

G. Rajakumar et al. (eds.), *Intelligent Communication Technologies and Virtual Mobile Networks*, Lecture Notes on Data Engineering and Communications Technologies 171,
https://doi.org/10.1007/978-981-99-1767-9_19

of the Industry 4.0 Revolution [3, 4]. The Internet of Things includes tools for researching attitudes, consumer characteristics, spending, and preferences, which impact marketing studies [5]. IoT technology can also track incidents and modifications in the institutional environment that can impact security and raise hazards, resulting in time and cost reductions, dematerialization, quality improvements, and increased productivity for digital businesses and entrepreneurship [6, 7]. The business models of numerous entrepreneurs and businesses are being altered by the Internet of Things (IoT), as is how customers engage with these entities and other stakeholders [8, 9]. This IoT technology can boost competitiveness by converting a company's products and services into digital business and entrepreneurship opportunities in the digital economy era [10].

Entrepreneurship and business require technical support to develop to meet market and consumer needs. Smart homes, smart cities, energy savings, pollution control, smart living, and smart economy are transformations because of IoT [11, 12]. IoT in entrepreneurship many studies state that IoT helps advance entrepreneurial businesses [13–15]. The IoT in entrepreneurship has not gained much attention in the form of a sizable visual map displayed on an international level year after year using information from various published researches. However, no study has not been any research that specifically addressed the substantial correlation between affiliation, scholars, and the influence of academic works.

One strategy for comprehending research is bibliometric analysis, in general. Bibliometrics is a technique for analyzing and measuring academic publications that combine mathematical and statistical methods. Statistical methods known as bibliometrics are used to analyze data on peer-reviewed publications, such as books, magazines, reports, reviews, and conference proceedings. Utilizing bibliometric methodologies, the relationship between the quantitative approaches and the subject of the investigation has been demonstrated frequently [4, 16]. This study proposes a research question, what is the bibliometric intellectual structure of mapping and trends for IoT in entrepreneurship research around the world? Using a bibliometric academic structure perspective, this research aims to map the current state of global research and future development trends of IoT in entrepreneurship studies.

2 Research Methods

This work's bibliometrics of intellectual structure analysis was conducted on a significant literature database. This survey includes significant keywords associated with IoT in entrepreneurship studies to locate and classify similar works in the academic Scopus database. Since academics regard the Scopus database as a reliable source of academic papers, it served as the primary information source for this article.

This research used the terms "Internet of Things" and "entrepreneurship" in the title, abstract, and author keywords to find relevant information in the Scopus database. Data mining was limited to annual data to acquire fully revealed data for an entire year. The search method was used in data mining (TITLE-ABS-KEY

("Internet of Thing*" OR IoT) AND TITLE-ABS-KEY (entrepreneur* OR start-up OR startup OR SMEs OR SME OR "small and medium enterprise*" OR "small and medium-sized enterprise*")) AND PUBYEAR < 2022 as of December 2022. One thousand seventy-five papers over the past 16 years—from 2005 to 2021—were uncovered in this process. At this inquiry stage, the Scopus result information has been extracted in CSV dataset format.

A feature on the Scopus web called to analyze search results displays bibliometric information from certain kinds of literature [10]. Using this technology, we evaluated and visualized the publication performance of researchers, research organizations, and countries. With the help of this tool, authors can also monitor the number of publications made each year and the proportion of publication sources and subject areas [7].

The researcher then used VOSviewer version 1.6.16 to perform co-occurrence analyses on the collected documents. This research uses VOSviewer to create a network of keyword maps conceptual for research themes, combined with a detailed co-occurrence analysis, keyword association analysis, and a fully systematic computation technique [17, 18]. Microsoft Excel was used to construct and tabulate simple tables and descriptive statistics. After that, the investigation's findings were collated and triangulated [12].

3 Result and Discussion

3.1 Organizations with the Highest Productivity in IoT in Entrepreneurship

There were 1,075 articles affiliated with 556 research institutions. The most productive organization in researching IoT in entrepreneurship publications was the Instituto de Desenvolvimento de Novas Tecnologias, Portugal ($n = 10$), as shown in Fig. 1, then followed by the Tampere University, Finland ($n = 8$); Texas A&M University, USA ($n = 8$); Friedrich-Alexander-Universität Erlangen-Nürnberg, Germany; Sungkyunkwan University, South Korea; Fachhochschule Salzburg, Austria; Rheinisch-Westfälische Technische Hochschule Aachen, Germany; Bina Nusantara University, Indonesia ($n = 7$); Chiang Mai University, Thailand; Universität Stuttgart, Germany; Georgia Institute of Technology, USA; Universidade do Minho, Portugal; Tokyo Institute of Technology, Japan ($n = 6$).

Various research institutions have studied IoT research in entrepreneurship. IoT in entrepreneurship research was dominated by affiliates from Germany ($n = 3$), the USA, and Portugal ($n = 2$).

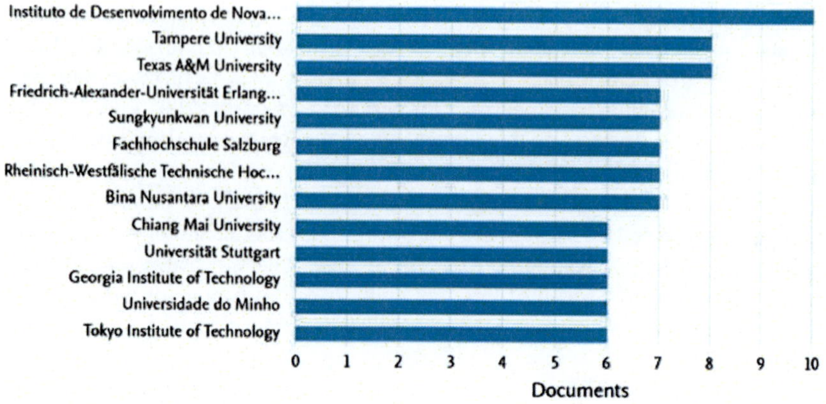

Fig. 1 Affiliation in an organization annual publication IoT in entrepreneurship count

3.2 *Countries with the Highest Productivity of IoT in Entrepreneurship*

These 212 countries were identified to have researched IoT in entrepreneurship. The most productive country in researching IoT in entrepreneurship publications was the USA ($n = 114$), as shown in Fig. 2, then followed by China ($n = 111$); Germany ($n = 85$); India ($n = 82$); Italy ($n = 62$); UK ($n = 57$); South Korea ($n = 45$); Spain ($n = 39$); Japan ($n = 37$); and Australia ($n = 36$).

Developed countries dominated most IoT in entrepreneurship publications. The productive institution supported the most productive countries in the publication of IoT in entrepreneurship, as shown in Fig. 1.

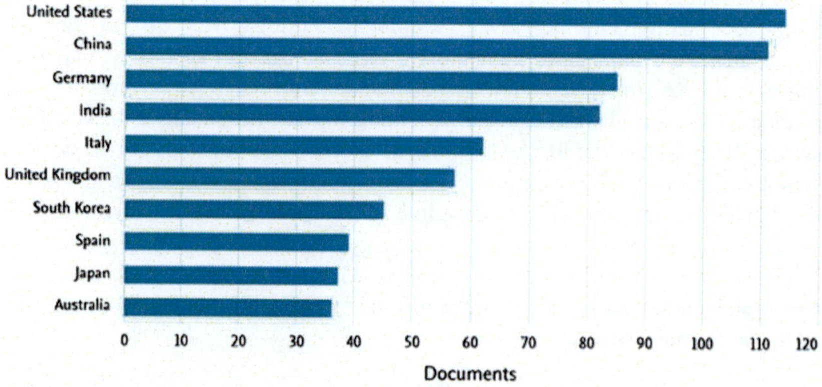

Fig. 2 Country in an annual publication IoT in entrepreneurship count

3.3 Researchers with the Most Individuality in IoT in Entrepreneurship

Two thousand nine hundred and forty-one individual researchers from various affiliations and countries studied IoT in entrepreneurship, as shown in Fig. 3. Agostinho, Carlos from Universidade NOVA de Lisboa, Portugal, and Müller, Julian M. from Privatuniversität Schloss Seeburg, Austria ($n = 9$), had the most papers in the field of IoT in entrepreneurship. Others were working on the case; there were authors with four publications each, Ferreira, José from Universidade NOVA de Lisboa, Portugal ($n = 6$); Sanchez-Sinencio, Edgar from Texas A&M University, USA; Voigt, Kai-Ingo from Universität Erlangen-Nürnberg, Germany; Freire, Mário M. from Universidade da Beira Interior, Portugal; Jardim-Goncalves, Ricardo from Universidade NOVA de Lisboa, Portugal; Mohammadian, Hamid Doost from University of Applied Sciences, Germany; Okada, Kenichi from Tokyo Institute of Technology, Japan; Parks, D. from CDC NIOSH, USA; Samaila, Musa G. from Universidade da Beira Interior, Portugal; Sequeiros, João B.F. from Universidade da Beira Interior, Portugal; and Woschank, Manuel from Montanuniversität Leoben, Austria.

It was clear that the affiliation of Universidade NOVA de Lisboa ($n = 3$) and Universidade da Beira Interior ($n = 3$) and the countries of Portugal ($n = 6$) have produced the most effective individual researchers for the IoT in entrepreneurship study. The researchers with the most publications in the field of IoT in entrepreneurship were Agostinho, Carlos, and Müller, Julian M., who researched the industrial revolution's influence on entrepreneurship [19, 20].

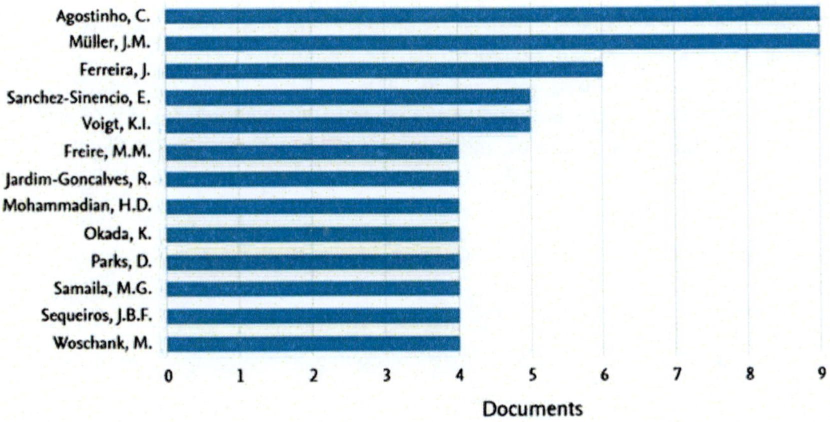

Fig. 3 Most individual IoT in entrepreneurship publication researchers

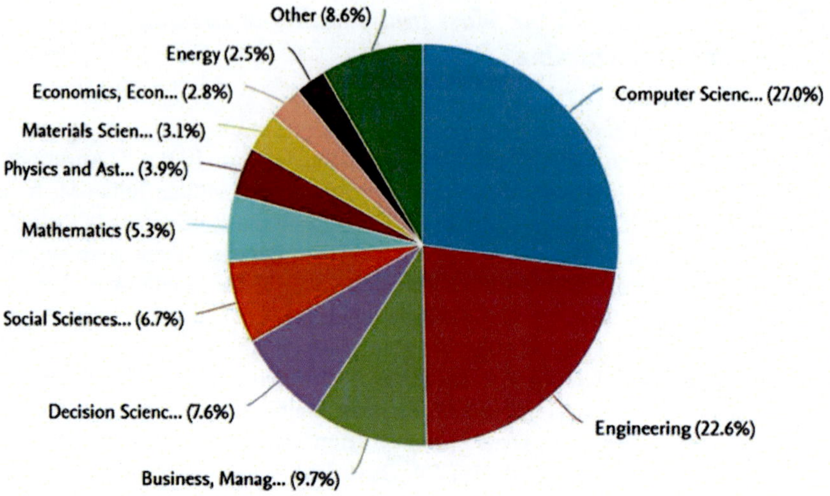

Fig. 4 Subject area frequency of IoT in entrepreneurship research

3.4 The Research Topic Frequency of IoT in Entrepreneurship

Research has been done on IoT in entrepreneurship from various affiliations and countries (see Fig. 4).

Computer science (n = 593, or 27%) has the most global publications in IoT in entrepreneurship research. The following most popular fields were engineering (n = 497, or 22.6%), business, management, and accounting (n = 214, or 9.7%), decision sciences (n = 167, or 7.6%), social sciences (n = 147, or 6.7%), mathematics (n = 117, or 5.3%), physics and astronomy (n = 86, or 3.9%), materials science (n = 69, or 5.3%), economics, econometrics, and finance (n = 62, or 2.6%), and energy (n = 56, or 2.6%). IoT in entrepreneurship study dominated computer science; engineering; business, management, and accounting.

3.5 The Document-Type Frequency of IoT in Entrepreneurship

On a wide range of document types, IoT in entrepreneurship research has been undertaken (see Fig. 5).

The document type with the most worldwide publications in IoT and entrepreneurship research was conference papers (n = 508 or 47.3%) (see Fig. 5). The following most popular choices were article (n = 349 or 32.5%), conference review (n = 96

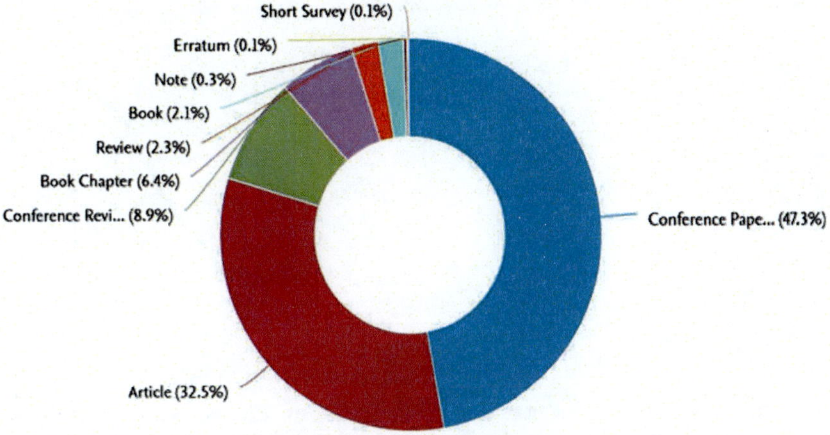

Fig. 5 Document-type frequency of IoT in entrepreneurship research

or 8.9%), book chapter ($n = 69$ or 6.4%), review ($n = 25$ or 2.3%), book ($n = 23$ or 2.1%), note ($n = 3$ or 0.3%), erratum and short survey ($n = 1$ or 0.1%).

3.6 Annual Publications for the IoT in Entrepreneurship

There were more international articles on IoT in entrepreneurship studies published each year. With 244 publications, Fig. 6 shows that 2020 has been the year of publishing's highest peak. According to the Scopus record, scientists have been researching IoT in entrepreneurship since 2005. In general, based on the number of annual publications, IoT in entrepreneurship research shows an upward trend in growth. One thousand seventy-five publication sources facilitated the development of publications on IoT in entrepreneurship.

3.7 Conceptual Research Map

The research theme map was a search for IoT in entrepreneurship research based on previously disclosed keyword correlations. Figure 7 shows the concept of the research theme in IoT in entrepreneurship.

The IoT in entrepreneurship keyword scheme was examined and presented using the VOSViewer application for the research topic map's IoT in entrepreneurship. Six repeats were required for the minimal minimum of keyword-related articles. Out of 295 keywords, 6,916 were determined to satisfy the requirements. Based on

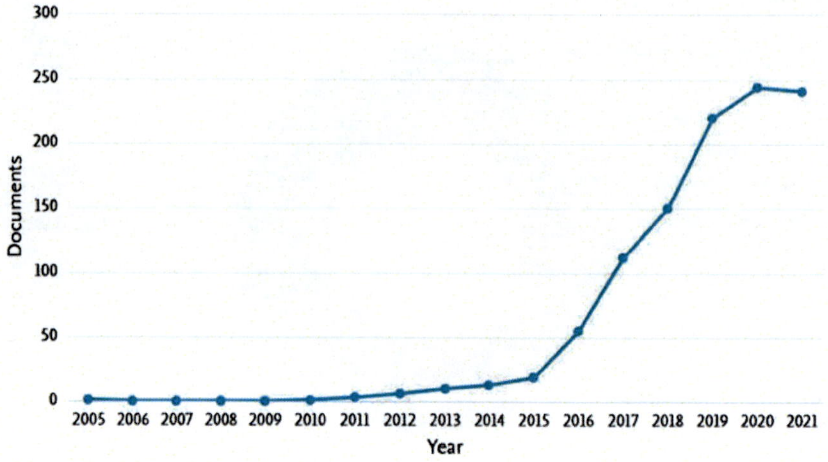

Fig. 6 IoT in entrepreneurship sector's annual publications

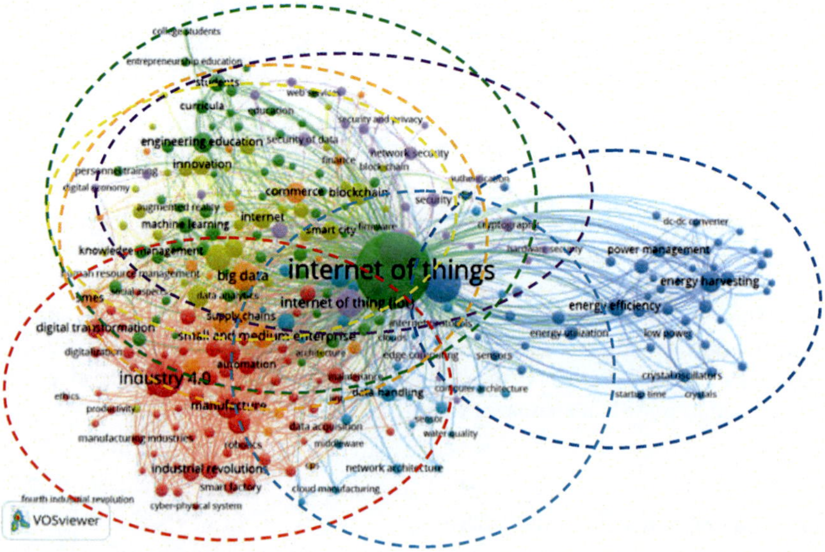

Fig. 7 Map of research themes

research keywords, seven research subject groups—information management, security, energy, entrepreneurship, education, Industry 4.0, and innovation—for global academic publication of IoT in entrepreneurship have been abbreviated and simplified as conceptual of ISEEEII research themes.

1. Information management cluster (light blue). The keyword of cloud, computer-aided manufacture, computer architecture, digital storage, green computing,

information management, middleware, network architecture, and sensor dominated the information management cluster.

2. Security cluster (purple). This cluster of themes focuses on security. These keywords form a cluster that connects to cyber security, cryptography, data privacy, network security, security, security of data, security system, and hardware security was related.
3. Energy cluster (blue). This cluster contains energy themes. The keyword of electric power utilization, electric rectifiers, energy efficiency, energy harvesting, energy management, energy utilization, power management, low power, low power electronics, maximum power point, power management, and ultra-low power appear in this cluster. Energy components dominated this cluster. Numerous of these keywords relate in some way to energy. Energy is a critical resource for various IoT devices, such as sensors, wireless devices, and mobile devices. The topic of energy is significant because it aims to develop technologies that can increase the efficiency and durability of devices. One of the essential roles of energy in the relationship between IoT research and entrepreneurship is a focus on developing innovative and sustainable technologies. Developing efficient and energy-efficient IoT technology will open up new business opportunities, such as IoT devices that can be used in renewable energy, such as solar panels and wind turbines.
4. Entrepreneurship cluster (orange). The keyword of business opportunities, commerce, competitive advantage, crowdsourcing, finance, sales, supply chains, SMEs, and small and medium enterprises dominated the entrepreneurship cluster.
5. Education clusters (green). This cluster of themes focuses on education. The keyword of augmented reality, college students, curricula, education computing, education, e-learning, engineering education, entrepreneurship education, learning systems, personnel training, students, and teaching were often used in this cluster.
6. Industry 4.0 cluster (red). Scientists can find industrial revolution topics in this research cluster. The keyword of Industry 4.0 were 3D printers, cyber-physical systems, digital transformation, digital manufacturing, the fourth industrial revolution, industrial economics, industrial Internet of Things, industrial management, industrial revolutions, industrial robots, Industrie 4.0, Industry 4.0, manufacture, productivity, IoT, robotics, digitalization smart factory, and smart manufacturing.
7. Innovation technology clusters (yellow). This cluster of themes focuses on innovation. The keyword of 5G mobile communication, AI, blockchain, business model innovation, digital technologies, innovation, internet open innovation, smart city, and technology adoption have dominated this cluster.

4 Conclusion

According to data, there have been more worldwide publications on IoT in entrepreneurship, maps, and visual trends each year. The Instituto de Desenvolvimento de Novas Tecnologias and the USA were the most active research institution and countries, with 10 and 114 publications published in the IoT in entrepreneurship publications. The authors with the most publications in the IoT in entrepreneurship journal were Agostinho, Carlos, and Müller, Julian M., with nine articles. Computer science was the area of the subject with the most documents on the IoT in entrepreneurship publications, totaling 593 documents (27%). With 508 papers, conference papers were the most often published type of IoT in entrepreneurship documents (47.3%). 2020 had the most academic publications, with 244 papers on IoT in entrepreneurship studies published globally.

This research has limitations on the database used only from Scopus, so it is possible that some publications relevant to the topic being studied can be overlooked and not included in this study. This research has a limitation covering only 16 years from 2005 to 2021. The study is also limited in understanding the research's context and purpose. In a bibliometric analysis, researchers can only analyze data related to the research topic but cannot understand the context or research objectives in depth.

In this study, the concept of ISEEEII research themes—information management, security, energy, entrepreneurship, education, industry 4.0, and innovation—was utilized to categorize the body of knowledge produced during fifty-seven years of scholarly study in terms of knowledge contributions. Practical field studies are crucial to identify broad backgrounds and disciplines, as well as research gaps, as a result of defining key issues in the IoT in the entrepreneurship industry and a fuller awareness of their requirement. The lack of advanced knowledge, assessment, and analysis in the disciplines will be the subject of new research made possible by all of these. Research on sustainability, technology, and IoT in entrepreneurship's potential to enhance digital business, technopreneurs, and start-up decision-making is everyday.

References

1. Angelova N, Kiryakova G, Yordanova L (2017) The great impact of internet of things on business. Trakia J Sci 15:406–412. https://doi.org/10.15547/tjs.2017.s.01.068
2. Lu Y, Papagiannidis S, Alamanos E (2018) Internet of Things: a systematic review of the business literature from the user and organisational perspectives. Technol Forecast Soc Change 136:285–297. https://doi.org/10.1016/j.techfore.2018.01.022
3. Capetillo A, Abraham Tijerina A, Ramirez R, Galvan JA (2021) Evolution from triple helix into penta helix: the case of Nuevo Leon 4.0 and the push for industry 4.0. Int J Interact Des Manuf 15:597–612. https://doi.org/10.1007/s12008-021-00785-x
4. Maulana FI, Zamahsari GK, Purnomo A (2020) Web design for distance learning Indonesian language BIPA. In: 2020 International conference on information management and technology

(ICIMTech). IEEE, Jakarta, pp 988–991. https://doi.org/10.1109/ICIMTech50083.2020.9211175
5. Sestino A, Prete MI, Piper L, Guido G (2020) Internet of Things and big data as enablers for business digitalization strategies. Technovation 98:102173. https://doi.org/10.1016/j.technovation.2020.102173
6. Dey N, Hassanien AE, Bhatt C, Ashour AS, Satapathy SC (2018) Internet of Things and big data analytics toward next-generation intelligence. Springer International Publishing, Cham. https://doi.org/10.1007/978-3-319-60435-0
7. Purnomo A, Susanti T, Sari AK, Firdaus M, Dewi R (2020) A study of digital entrepreneurship through bibliometric visualizing from 1993 to 2019. In: 2020 International conference on information management and technology (ICIMTech). IEEE, pp 911–915. https://doi.org/10.1109/ICIMTech50083.2020.9211270
8. Langley DJ, van Doorn J, Ng ICL, Stieglitz S, Lazovik A, Boonstra A (2021) The Internet of Everything: Smart things and their impact on business models. J Bus Res 122:853–863. https://doi.org/10.1016/j.jbusres.2019.12.035
9. Lo F-Y, Campos N (2018) Blending Internet-of-Things (IoT) solutions into relationship marketing strategies. Technol Forecast Soc Change 137:10–18. https://doi.org/10.1016/j.techfore.2018.09.029
10. Purnomo A, Susanti T, Rosyidah E, Firdausi N, Idhom M (2022) Digital economy research: thirty-five years insights of retrospective review. Procedia Comput Sci 197:68–75. https://doi.org/10.1016/j.procs.2021.12.119
11. Kumar S, Tiwari P, Zymbler M (2019) Internet of Things is a revolutionary approach for future technology enhancement: a review. J Big Data 6:111. https://doi.org/10.1186/s40537-019-0268-2
12. Purnomo A, Septianto A, Sutiksno DU, Hikmawan MI, Kumalasari RD (2020) Technopreneur publication: a bibliometric analysis (2000–2019). In: 2020 International conference on information management and technology (ICIMTech). IEEE, pp 521–526. https://doi.org/10.1109/ICIMTech50083.2020.9211111
13. Adamseged ME, Grundmann P (2020) Understanding business environments and success factors for emerging bioeconomy enterprises through a comprehensive analytical framework. Sustainability 12:9018. https://doi.org/10.3390/su12219018
14. Moeuf A, Lamouri S, Pellerin R, Tamayo-Giraldo S, Tobon-Valencia E, Eburdy R (2020) Identification of critical success factors, risks and opportunities of Industry 4.0 in SMEs. Int J Prod Res 58:1384–1400. https://doi.org/10.1080/00207543.2019.1636323
15. Nižetić S, Šolić P, López-de-Ipiña González-de-Artaza D, Patrono L (2020) Internet of Things (IoT): Opportunities, issues and challenges towards a smart and sustainable future. J Clean Prod 274:122877. https://doi.org/10.1016/j.jclepro.2020.122877
16. IGI Global: What is Bibliometric?, https://www.igi-global.com/dictionary/bibliometric/49021
17. Ranjbar-Sahraei B, Negenborn RR (2017) Research positioning & trend identification. TU Delft, Walanda
18. van Eck NJ, Waltman L (2010) Software survey: VOSviewer, a computer program for bibliometric mapping. Scientometrics 84:523–538. https://doi.org/10.1007/s11192-009-0146-3
19. Müller JM (2019) Antecedents to digital platform usage in Industry 4.0 by established manufacturers. Sustainability 11(4):1121. https://doi.org/10.3390/su11041121
20. Agostinho C, Lopes F, Ferreira J, Ghimire S, Marques M (2019) A lightweight IoT hub for SME manufacturing industries. In: Enterprise interoperability VIII. Springer, Cham, pp 371–383. https://doi.org/10.1007/978-3-030-13693-2_31

IoT-Based App for Commuter Travel Tracking System: A Review

Md. Sohaib, V. P. M. B. Aarthi, M. Naga Laxmi, G. Govind Shivaji, T. Arun Kumar, and M. Ramesh

Abstract In this paper, the groundwork on commuter-based travel system is studied by considering the several existing techniques in this problem. In the present scenario, IoT plays an inevitable role in all emerging technologies because of the increasing usage of the internet. It aims at smart usage of human time in real-time applications. Transportation has become the essential factor of regular activities. Hence, this paper reviews all the system that can automatically measure the distance travelled by passengers on the bus and collect the distance travelled from the accounts of passengers. Several new systems could be implemented in the comfort zone of passengers only after a thorough review on literature. A few ideas of passenger's convenience like a method of charging tickets are by distance travelled is one of the most innovative ways. The usage of most secure smart card as the prepaid travel card, storing the amount in its internal memory, allowing users to easily board any bus in the area. The idea of incorporating the GPS receiver to calculate position and driving distance avoids relying on the vehicle's built-in distance of metres. To develop an effective system for IoT-based APP for commuter travel tracking system, this paper provides an exhaustive survey for real-time applications and compares the existing systems in various aspects guiding the future researchers.

Keywords Smart card reader · Commuter travelling system · Radio frequency identification tags · Global positioning system receiver

1 Introduction

Indoor and outdoor localization of objects and tracking of moving objects are required for many location-based services. Many technologies including infrared, wireless sensor networks (WSNs), Wi-Fi ultrasound and video camera, are used as the mechanism for localization and tracking in transportation. Several techniques are adopted

Md. Sohaib (✉) · V. P. M. B. Aarthi · M. N. Laxmi · G. G. Shivaji · T. A. Kumar · M. Ramesh
Kalasalingam Academy of Research and Education, Krishanankoil, Srivilliputhur, India
e-mail: mdsohaib09433@gmail.com

G. Rajakumar et al. (eds.), *Intelligent Communication Technologies and Virtual Mobile Networks*, Lecture Notes on Data Engineering and Communications Technologies 171,
https://doi.org/10.1007/978-981-99-1767-9_20

for indoor and outdoor localization procedures. A few like GPS, RTPIS, RFID are most common. Radio Frequency Identification (RFID) is a significant technology for the purpose of tracking of objects using RF signals. IOT as an emerging technology could be adopted for building a comfortable App for passengers for commuter travel tracking system. This paper is organized as follows: Sect. 2 describes the related works on the commuter travelling system. Section 3 discusses the multiple comparisons of the several techniques with respect to overall review on literature according to its broadest methodologies, converging at RFID system with their implementation, limitations, and performance metric analysis. Section 4 converges the challenges with further scope for the researchers. Finally, Sect. 5 concludes the paper.

2 State of the Art

There exist several techniques in literature in the implementation of commuter travel tracking system. A few techniques or devices like GPS receiver, RFID Module, LCD Display are utilized in this. How so, each of the implementation has its own advantages and implementation difficulties in real-time scenario. The state of the art is presented as a selective study that supports the real-time implementation with zero practical difficulties.

A. Design of Real-Time Bus Monitoring and Passenger Information System:

Sinha et al. [1] proposed the main objectives of their research project as the following: RTPIS moving presentation on transportation stops anticipates season of appearance continually, electronic connection point for screen to manage space transportation gradually, to calculate transport plans and RTPIS, the end-user can use a portable application. However, some drawbacks like change in the expected time of arrival (ETA) due to Internet connectivity and security issues like attack by malicious agents.

B. Design of IoT-Based Intelligent Bus Positioning for Smart Campus:

Feng et al. [2] proposed the system that uses RFID technology for transportation. In addition, touch screens at bus stops can calculate the passengers count waiting for each bus, and electronic bus station loading can communicate information like the season in which each bus will appear. How so, a few limitations of the Zigbee wireless communication like reduction in accuracy for long-distance communication and the overall cost of project is not economical.

C. Passenger Assistant for Public Bus Commute using Smartphone:

Verma et al. [3] researched RTPIS shifting presentation on transport stops—predicted season of presence continually are the main objectives of this research study. Electronic transfer from control space to screen connection point, RTPIS, and transport plans can be calculated using a portable application for end-users. Some drawbacks are the speed breakers, and pots may not be detected by the system always and the requirement of Internet connectivity at all the time.

D. UHF-RFID for Vehicle Localization in GPS-Less Environments:

Chen et al. [4] offered a brand new real-time vehicle localization strategy in GPS-less environments, which is based on UHF-RFID. They use a single antenna multi-frequency ranging system on a vehicle. The coordinates of the vehicle can then be determined using the tags reader distant factor. A few drawbacks are due to the presence of antennas which should take care of the system to prevent from damage. UHF-RFID tags cannot generate the omni-directional signal whose pace is independent of the reflecting surface.

E. A passive UHF-RFID for the Indoor Autonomous vehicle:

Digiampaolo and Martinelli [5] proposed the global localization challenge for a mobile robot that can detect the availability of passive RFID tags placed at known positions of the environment's ceiling when moving beneath them was the focus of this article. As a result, a successful localization method based on Kalman filtering to combine RFID readings with odometry data is developed. A few limitations like the cost of the project, requirement of a large detection range (two metres or more) resulting in electromagnetic interference with environment.

F. An Embedded Sensing Network Integrated Monitoring System for Pavement and Traffic:

Xue et al. [6] proposed a prototype to monitor pavement conditions and collect traffic data using the embedded sensing network. The system's two monitoring patterns namely regular monitoring and periodic testing are performed, and the data processing techniques used in each pattern. Few challenges like autonomous, low-cost, and self-powered sensors can be added up in order to minimize the cost along with various environmental and loading conditions.

G. Bumping: Smart phones-based Inertial Navigation Method for Indoor Vehicles:

Tan et al. [7] studied and introduced Bumping, a Bump-Aided Inertial Navigation technique that greatly reduces the two issues. The Bump Matching algorithm, which is at the heart of this technique, uses the position data of the nearby speed bumps to generate helpful references for the INS. The suggested approach uses almost little extra energy and can be easily implemented without the need for infrastructure. Real trials were carried out in parking garages with trees that had various environmental parameters. A smart phone-based Inertial Navigation System (INS) has two significant issues. At first, it's susceptible to mistakes that can develop and add up very quickly over time, potentially making navigation useless. Secondly, since a smart phone can only learn its acceleration, it is difficult for it to infer its initial position and velocity, which form the basis for calculating distant metric, without human input or outside references.

H. GPS Localization Accuracy Classification:

Drawil et al. [8] proposed a Context-Based Approach, a method for estimating localization accuracy includes two steps in this technique. The effectiveness of the suggested approach for categorizing GPS localization accuracy under various measurement settings is shown by real-world comparing studies. Although localization techniques have lately begun to show a noticeable improvement in their localization performance, the measurement conditions in their surroundings still have a considerable impact. Because the measuring process is incongruent, the effect of the measurement conditions on the localization accuracy is in fact an ill-defined problem in and of itself.

I. Error-Cognitive Vehicle Localization in GPS-Less Environments:

Qin et al. [9] researched to attain great location precision; however, high tag density and a high cost are necessary. As a result, it frequently serves as an extra upgrade to GPS and servers. To provide a low-cost, low-density RFID infrastructure in addition to the already presented prototype RFID-based location system.

J. J. RFID-based localization and tracking technologies:

Ni et al. [10] proposed an implementation of RFID-based travel tracking system. These technologies include tag-based (such as LANDMARK), reader-based (such as reverse RFID), transceiver free, and hybrid techniques. To locate the target objects, these systems primarily rely on information about radio signal strength or RSS change that is easily accessible. Although RSS may be cost-effective, multipath and interference are the limitations.

K. Urban Eye—An Outdoor Localization for Public Transport:

Verma et al. [11] created Urban Eye, a route navigation and travel time estimation system for city commuters that are energy-efficient outside. The existing algorithm for localization and navigation shows poor performance when used on public buses in the cities, which is the primary motivation for this work. They develop a device with considerably less power consumption than GPS that just uses inertial sensors.

3 Comparative Analysis

From the thorough study of the literature, we created a diagrammatic representation of different aspects of the commuter travel tracking system. The implementation scenario of the papers is divided into several aspects based on the methodology it follows. Figure 1 consolidates the analysis of entire review of commuter travel tracking system.

This shows three different scenarios RTPIS [1] (Real-Time Passenger Information System), RFID [3, 12] (Radio Frequency Identification) and outdoor localization [6–8, 11]. RTPIS has two sub-types: General RTPIS and RTPIS using smart phone. The

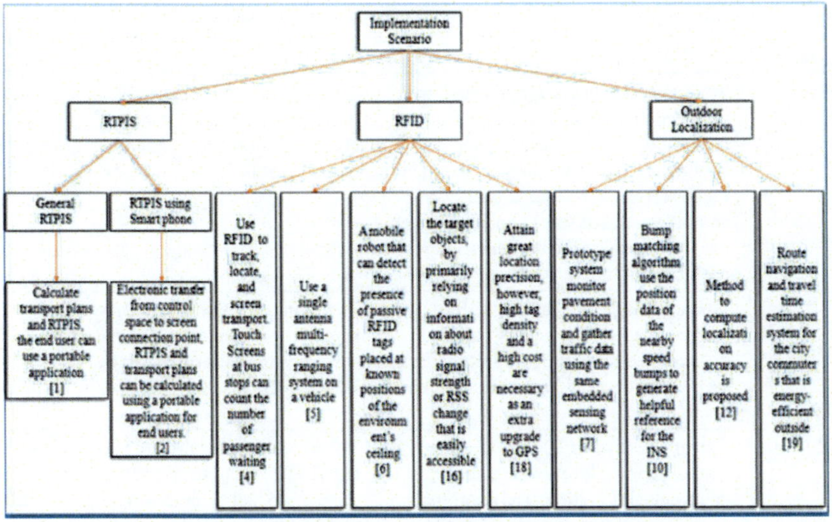

Fig. 1 Comparison on various techniques for commuter travel tracking system

five different techniques under RFID [2, 4, 5, 9, 10] are analysed by their methodology briefly, and some drawbacks are identified. The outdoor localization is also analysed effectively. The cost-effective solution for commuter travel tracking system is highly supported by RFID which can be applied practically in real life usage [13–18]. Table 1 converges the in-depth analysis on RFID-based techniques.

However, the final part of implementation is highly important to measure the performance. Table 2 depicts the comparison of implementation techniques along with the metric, i.e. analysed to measure the performance of the commuter travel tracking system using RFID.

4 Challenges and Future Scope

This paper reviews the entire commuter travel tracking system with the objective of developing a successful real-time cost-effective solution that alters the paper ticketing system using RFID technique as the background system. This analysis on literature is to create a mobile application that will allow users to track their travel positions, as well as their payment of the fare for the distance they cover. Figure 2 depicts that the draft of one of the efficient systems is the progressive work of the authors.

The GPS, RFID reader, and RFID cards, measure the distance metric from the board on point of the passenger; the RFID reader will begin tracking their whereabouts and recording their boarding and alighting times. This proposed cost-effective RFID system utilizes MSP430g2553 with salient features of inbuilt watchdog Timer, UART, LCD control and mainly suits for low power application.

Table 1 Comparison on methodology and drawbacks for RFID-based techniques

Literature	RFID-based methodology	Limitations
[2]	Transportation is tracked, located, and screened by the system using RFID technology. The number of people waiting for each bus can be counted via touch screens at bus stops, and computerized bus station loading systems can share details like the season in which each bus will arrive	Zigbee wireless communication may not be accurate in long-distance communication cases, and the project's overall cost is high
[4]	Usage of single antenna multi-frequency ranging device. The reconstructed distances between the tags and reader can then be used to get the coordinates of the vehicle	The omni-directional signal, whose speed is independent of the reflecting surface, cannot be produced by UHF-RFID tags
[5]	Identify passive RFID tags when moving beneath known locations on an environment's ceiling. The combination of RFID readings and odometry data allows us to create a viable localization approach based on Kalman filtering	The expense of the project is one of its drawbacks. Requires a wide detection range (two metres or more), which could lead to erratic detections due to electromagnetic interference from the surroundings
[10]	These technologies include transceiver-free, reader-based, tag-based (like LANDMARK), tag-based (like reverse RFID), and hybrid approaches. These systems largely rely on readily available information regarding radio signal strength or RSS change to locate the target objects	Despite the potential cost-effectiveness of RSS, multipath and interference pose challenges
[9]	High tag density and a high cost are required in order to achieve high location precision, though. Therefore, it frequently functions as an additional upgrade to servers and GPS	The exact necessity for a low density and low-cost RFID infrastructure

Finally, when they exit the bus, the base fare for the passenger they just travelled will be deducted. The passenger will be able to see the details of the tickets that are charged. This will enable us to eliminate the paper ticketing system and serves as a cost-effective real-time implementation of commuter travel tracking system.

In this perspective, the authors are at the exit phase of developing a system as shown in a process flow model as depicted in Fig. 3. It sends the details regarding distance, time, date, and the fare price to the web applications so that the passenger can track the information about the journey.

Table 2 Comparison on implementation and performance metrics for RFID-based techniques

Literature	Implementation	Performance metrics
[2]	Integrate RFID technology with Zigbee wireless technology	Taking bus of certain examples on road, the effective calculation of dispatching time was measured
[4]	A combination of UHF and RFID technique	Time consumption and computational complexity (Road Test)
[5]	Localization procedure based on Kalman filter is used and is fused with RFID tags with odometry data	Average position error
[10]	Combines the RSS technique along with challenges in multipath and interference	Cost-effective
[9]	Correction based scheme to systematically cognize and correct localization error	Accurate vehicle localization

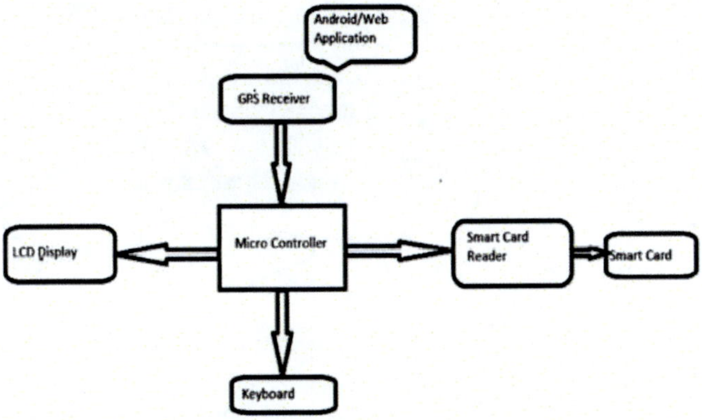

Fig. 2 Proposed cost-effective RFID-based commuter travel tracking system

5 Conclusion

This paper provides an exhaustive review on the existing commuter-based travel system with their preliminaries, state of the art along with the performance comparison to identify a cost-effective solution. In implementation perspective, the practical difficulties of replacing the paper ticketing system by this commuter-based App should be successfully rectified. Implementation of effective IOT App for this problem is still under the progress. The implementation of the efficient low cost IOT-based App for commuter travel tracking system using RFID is the progressive work by the authors.

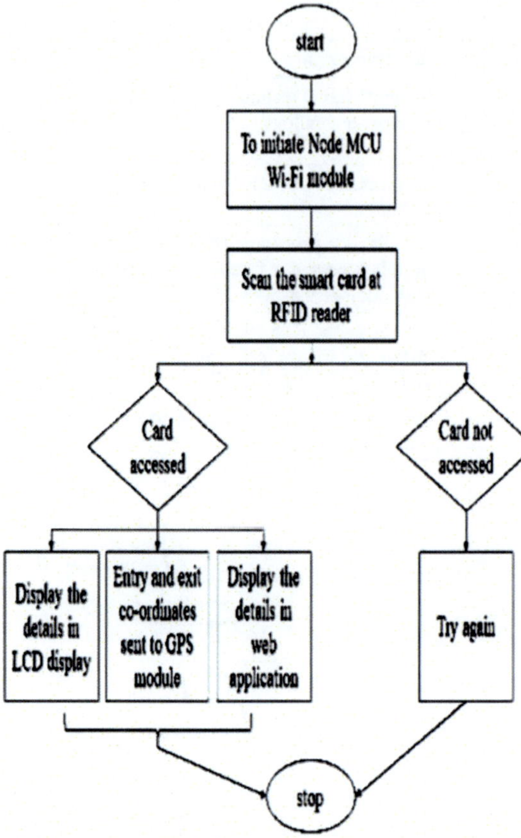

Fig. 3 Proposed flow model for commuter travel tracking system

References

1. Sinha S et al (2013) Implementation of real time bus monitoring and passenger information system. Int J Eng Res Appl 3(2):1624–1627
2. Feng X, Zhang J, Chen J, Wang G, Zhang L, Li R (2018) Design of intelligent bus positioning based on internet of things for smart campus. IEEE Access 6:60005–60015. https://doi.org/10.1109/ACCESS.2018.2874083
3. Verma R et al (2020) A smartphone-based passenger assistant for public bus commute in developing countries. IEEE Trans Comput Soc Syst 7(2):465–476. https://doi.org/10.1109/TCSS.2019.2961390
4. Chen R, Huang X, Zhou Y, Hui Y, Cheng N (2022) UHF-RFID-based real-time vehicle localization in GPS-less environments. IEEE Trans Intell Transp Syst 23(7):9286–9293. https://doi.org/10.1109/TITS.2021.3085824
5. Digiampaolo E, Martinelli F (2012) A passive UHF-RFID system for the localization of an indoor autonomous vehicle. IEEE Trans Industr Electron 59(10):3961–3970. https://doi.org/10.1109/TIE.2011.2173091
6. Xue W, Wang L, Wang D (2015) A prototype integrated monitoring system for pavement and traffic based on an embedded sensing network. IEEE Trans Intell Transp Syst 16(3):1380–1390. https://doi.org/10.1109/TITS.2014.2364253

7. Tan G, Lu M, Jiang F, Chen K, Huang X, Wu J (2014) Bumping: a bump-aided inertial navigation method for indoor vehicles using smart phones. IEEE Trans Parallel Distrib Syst 25(7):1670–1680. https://doi.org/10.1109/TPDS.2013.194
8. Drawil NM, Amar HM, Basir OA (2013) GPS localization accuracy classification: a context-based approach. IEEE Trans Intell Transp Syst 14(1):262–273. https://doi.org/10.1109/TITS.2012.2213815
9. Qin H, Peng Y, Zhang W (2017) Vehicles on RFID: error-cognitive vehicle localization in GPS-less environments. IEEE Trans Veh Technol 66(11):9943–9957. https://doi.org/10.1109/TVT.2017.2739123
10. Ni LM, Zhang D, Souryal MR (2011) RFID-based localization and tracking technologies. IEEE Wirel Commun 18(2):45–51. https://doi.org/10.1109/MWC.2011.5751295
11. Verma R et al (2016) Urban eye: an outdoor localization system for public transport. In: IEEE INFOCOM 2016—The 35th annual IEEE international conference on computer communications, pp 1–9. https://doi.org/10.1109/INFOCOM.2016.7524393
12. Jimoh OD, Ajao LA, Adeleke OO, Kolo SS (2020) A vehicle tracking system using greedy forwarding algorithms for public transportation in urban arterial. IEEE Access 8:191706–191725. https://doi.org/10.1109/ACCESS.2020.3031488
13. Chang T-H, Wang L-S, Chang F-R (2009) A solution to the Ill-conditioned GPS positioning problem in an urban environment. IEEE Trans Intell Transp Syst 10(1):135–145. https://doi.org/10.1109/TITS.2008.2011709
14. Fujii S et al (2011) Cooperative vehicle positioning via V2V communications and onboard sensors. In: 2011 IEEE vehicular technology conference (VTC Fall), pp 1–5. https://doi.org/10.1109/VETECF.2011.6093218
15. Jo K, Chu K, Sunwoo M (2012) Interacting multiple model filter-based sensor fusion of GPS with in-vehicle sensors for real-time vehicle positioning. IEEE Trans Intell Transp Syst 13(1):329–343. https://doi.org/10.1109/TITS.2011.2171033
16. Velaga NR, Quddus MA, Bristow AL, Zheng Y (2012) Map-aided integrity monitoring of a land vehicle navigation system. IEEE Trans Intell Transp Syst 13(2):848–858. https://doi.org/10.1109/TITS.2012.2187196
17. Fouque C, Bonnifait P (2012) Matching raw GPS measurements on a navigable map without computing a global position. IEEE Trans Intell Transp Syst 13(2):887–898. https://doi.org/10.1109/TITS.2012.2186295
18. Guin A (2006) Travel time prediction using a seasonal autoregressive integrated moving average time series model. In: 2006 IEEE intelligent transportation systems conference, pp 493–498. https://doi.org/10.1109/ITSC.2006.1706789
19. Stauffer C, Grimson WEL (1999) Adaptive background mixture models for real-time tracking. In: Proceedings. 1999 IEEE computer society conference on computer vision and pattern recognition (Cat. No PR00149), vol 2, pp 246–252. https://doi.org/10.1109/CVPR.1999.784637

Analysis of Block-Level Prediction-Error Expansion Approach in Data Encryption

Jangam Deepthi and T. Venu Gopal

Abstract Due to the difficulty and inefficiency of the reserved data hiding approach, several encryption domain-based data (reversible) concealing algorithms have drawbacks such as a low embedding rate and poor resolution of the immediately retrieved data. This article will give complete details about different reserved model approaches for data hiding and proposes a different algorithm for data hiding. By taking advantage of the pixel redundancy inside each block, the technique can incorporate confidential data into 2X2 picture blocks. Our proposed mode extends this idea to the encrypted realm (ABPEE-RDHEI). In order to maintain spatial redundancy for data embedding, to further boost the security level, a stream cipher is then applied to the block-permutated image. The proposed ABPEE-RDHEI can produce marked decrypted images with a high embedding rate and good resolution due to the advanced data combining techniques using iteration and data encryption. ABPEE-RDHEI outperforms a number of state-of-the-art approaches, according to experimental findings and analyses. The ability to forecast an image accurately and insert a message into the image with minimal distortion are two essential components in the field of RDHEI. However, it is difficult to increase forecast accuracy because linear regression prediction is susceptible to outliers. To overcome this issue, this work suggests an RDHEI approach based on an adaptive prediction-error label map.

Keywords Data hiding · RDHEI · ABPEE · Reserved room algorithm

1 Introduction

In data encryption, reversible approach in photographs seeks to covertly incorporate the hidden data into the source image. RDH places a strong emphasis on flawless picture recovery and data extraction at the receiver side, in contrast to watermarking,

J. Deepthi (✉) · T. V. Gopal
Department of CSE, JNTUH Hyderabad, Hyderabad, India
e-mail: angamdeepthi121@gmail.com

G. Rajakumar et al. (eds.), *Intelligent Communication Technologies and Virtual Mobile Networks*, Lecture Notes on Data Engineering and Communications Technologies 171,
https://doi.org/10.1007/978-981-99-1767-9_21

which should be resistant to harmful assaults. There are so many data hiding algorithms that are available [1–5]. A no. of potential features from the source picture is retrieved and losslessly compressed in the image compression-based RDH, which is based on. The confidential data is embedded in the designated spare area. Later, new RDH techniques that take use of the image's pixel spatial correlations have been presented (EMD). While HS-based approaches incorporate hidden material into the shifted histograms, DE methods enlarge the difference between two adjacent pixel values to do so [6–10]. Prediction-error expansion is another RDH technique that is often employed (PEE). Utilizing the HS approach, it takes use of the prediction mistake to incorporate the hidden data. Reversible data-hiding techniques in encrypted images are currently attracting a lot of research attention. The data encryption and the data retrieval are both objectives of RDHEI's receiver side work [6, 11–13]. The reversibility attribute of RDHEI allows for a wide range of applications in a variety of crucial situations, including medical picture sharing, legal forensics, cloud storage, and military uses. Users can encrypt their photographs before uploading them to the cloud if they wish to save them there but do not want illegal access. Since there is no communication in this instance, there are no mistakes or attacks [14–19].

2 Related Works

In this part, we examine a few predictors that are often employed in image processing as well as some current VRAE techniques for RDHEI. Many predictors, including the simplified gradient adjusted predictor (SGAP), the 65 gradients adjusted predictor (GAP), the partial differential equations (PDE) predictor, have been suggested recently to estimate the values of pixels in the spatial domain (CBP). GAP and SGAP anticipate the target pixel using the pixels in a half-encircled structure with an irregular form. They work well for anticipating the placement of pixels in raster-scan order. Using the predetermined reference pixels, PDE completes the prediction procedure. Prediction of pixels with a 3X3 area using 25% of the pixels in the host picture, which contains the remaining 75% of the pixels.

2.1 J-VRAE Methods

These are typically first use a stream cipher to encrypt the source picture before embedding the confidential data using various methods including least significant bits (LSBs) flipping and public key modulation. By dividing the encrypted image into a different of non-overlapping blocks. Data retrieval and source photo recovery are made possible by contrasting the smoothness of the decrypted picture blocks. Yun et al. presented an enhancement to Zang's approach. As a result, Yu et al.'s technique [12] with $p = 100$ is a special instance of Zhang's method [19]. The techniques of Kuy [5] compute the complexity of the picture block using a more exact fluctuation measure function and employ the side pole approach to lower the rate of erroneous data retrieval. By flipping the *i*th(1 I 6) LSBs of a group of pixels, Wat et al. [9] insert one bit of the confidential data into the group of pixels (LSBs). By changing a pixel's three LSBs, Li et al.'s technique embeds one bit of the confidential data into a previously chosen pixel. In order to lower the rate of inaccurate data extraction, it replicates the confidential data before embedding. To enhance the resolution of the annotated decrypted image, the approach in [17] flips just the LSBs of fewer pixels using a complex selection. The mistake rate of the extracted confidential data bits is successfully reduced by the adaptive judgment function distributed feature-based local contents of the picture.

2.2 S-VRAE Methods

Many S-VRAE algorithms have been presented to accomplish data extraction and picture recovery independently. In order to include the confidential data, Wu et al. [6] changed the pixel's *i*th ($k > 7$) bit. The stream cipher is used in the procedures in [20, 21], and [17] to encrypt the source picture. A few LSB planes and the fourth bit plane of a stream-cipher-encrypted picture, respectively, are compressed in [21] and [20] to make room for hidden data. Zhang et al. [17] integrate confidential data into the encrypted image's three LSB planes using the pseudo-random sequence modulation approach. The confidential data can be properly and error-free extracted using previous S-VRAE methods, but the restored image can lose data. This is because picture recovery works by looking at the smoothness or local standard deviation of the encrypted image. When the source image has several textures, this might lead to inaccurate findings or image recovery. S-VRAE techniques in [18, 19, 22, 23] encrypt the source picture by permutation in contrast to these S-VRAE methods, which primarily use the stream cipher with the source image's pixel positions unaltered. In [5], the confidential data is encrypted using discrete wavelet transforms. Using DWT, image can be divided in to sub-bands, LL, LH, HL, HH. By employing 1/4 of an source image's pixels to forecast the remaining 3/4 of its pixels, Yi et al. [24] enhanced [5], and HS integrated the confidential data into the permutated prediction-error values. In [17], the source image is first split into a number of sub-bands, and then blocks in the entire image are permutated using a coarse-grained permutation, and then each pixel inside the block is permutated using a fine-grained permutation. The peak spots to incorporate the confidential data are then chosen by HS at random from two pixels in each block. Both confidential data extraction and picture recovery are lossless. The techniques in [25–30] separate the source picture into blocks, encrypt each block using stream encryption and permutation. Then, embeds confidential data into the encrypted picture using two conventional RDH techniques, difference histogram shifting (DHS) and prediction-error histogram shifting (PEHS). The AMBTC compressed picture is encrypted by Niu et al. who also employ HS to insert confidential data into the encrypted data. According to the method in [6], the integer wavelet transformed coefficients are encrypted to produce the encrypted picture. The HS technique is then used to include confidential material into the encrypted coefficients. The embedding rate is still constrained, though [31–33].

3 Methodology

3.1 ABPEE-RDHEI

In Fig. 1, the ABPEE-RDHEI framework is depicted. The image provider initially uses the permutation key Kp to block-level permute the source picture to encrypt it. The image's pixel values are then altered using a stream encipher and the sharing key Ks. The stream decipher is used with Ks at the data hider side. The data hider then uses the ABPEE technique to embed confidential data into the encrypted picture in accordance with the data embedding key Kd. The confidential data and picture content are then further protected using the stream encipher. Although any safe encryption technique can be used in lieu of the stream encipher and decipher in this work, we utilize the stream cipher as an example. The annotated encrypted picture is the encrypted image that contains sensitive information. To meet the needs of various applications, we offer two strategies for the receiver side to restore the source image and extract the hidden data. The first method involves extracting data before picture decryption, whereas the second method involves extracting data after image decryption. The receiver can get the designated decrypted picture immediately by using encryption key (or confidential info). The receiver may simultaneously retrieve the entire source picture and the hidden data using both of two security keys.

In the proposed model, we have three sections, image provider, data hider, and receiver as shown in Fig. 1. In this approach, we are using image partition and

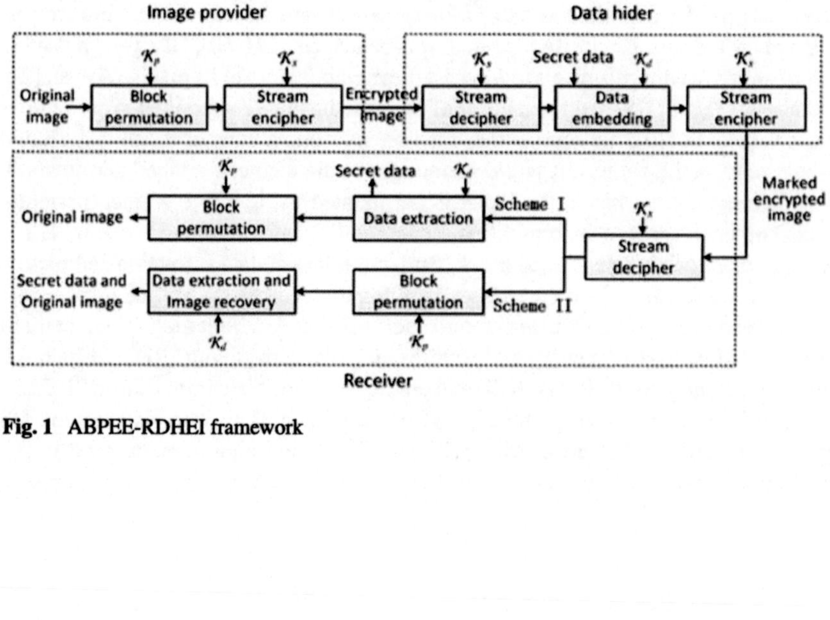

Fig. 1 ABPEE-RDHEI framework

Fig. 2 Image partition and self-contained embedding

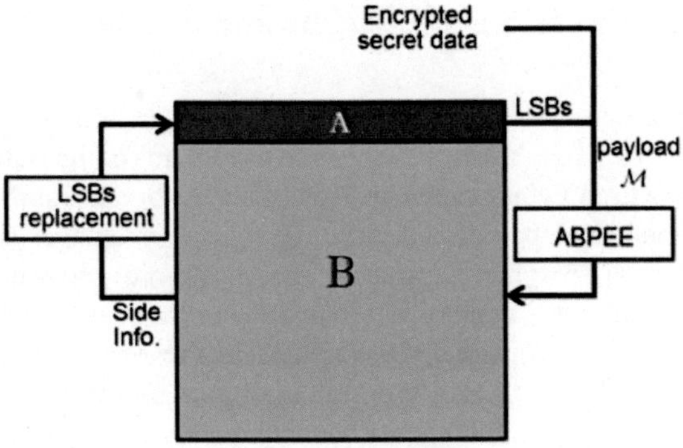

data embedding algorithms. İn this approach, image will be embedded using least significant bits. Figure 2 depicts the ımage partition and self-contained embedding.

3.2 *Pixel Prediction*

The projected value $k(p, q)$ is calculated as the median in its forecast context for each pixel in A.

$$k(p.q) = \frac{k(1) + k(2)}{2} + 1/3 \quad (1)$$

The ordered Gray level values of the pixels that make up the prediction context are k(1), k(2), k(3), and k(4). Note that set U is the home of all of the context pixels for set A. $\tilde{I}(i, j)$ is then utilized in the subsequent equation to work out the hidden data bit within the current group.

$$b = \begin{cases} 0 \text{ If } \sum_k^l \mathrm{e}^{-i\omega t}(k[1]|k[2]) < 1 \\ 10 \text{ If } \sum_k^l \mathrm{e}^{-i\omega t}(k[1]|k[2]) > 1 \end{cases} t. \quad (2)$$

All n pixels in the current group are changed to their respective $k(p, q)$ values if $b = 1$, else the pixels are left alone. A larger number for n increases accuracy but decreases the amount of space that can hold the concealed data.

4 Experimental Results and Findings

We consider three test images from open source. They are Jetplane, Man, and Airplane as shown in Fig. 3. The histograms of the stated figures are shown in Fig. 4. We used Python commands to create histograms of the images. These images were applied proposed algorithm. Histograms of images (Jetplane, Man, and Airplane) before encryption and after encryption are given in Figs. 4 and 5, respectively. Experiments demonstrate our proposed method gives good embedded and PSNR (40 dB). A completely new approach of reversible data concealment after encryption is suggested in [10] and uses both joint and independent methods. The two-stage embedding/decoding procedure, pixel prediction that supported the median context value, and many parity-based embedding for the separate approach are the proposed scheme's biggest advantages. Embedding rate, error rare, and encryption efficiency are the important factors in our model. The PSNR of source picture and the watermarked decrypted version is used to assess and justify the embedding distortions. The drawback is PSNR ratio is a smaller amount, i.e., 30 dB.

Experiment results in our research, the embedding rate of the recommended approach, and its performance are compared to those of related research in 10. Table 1 gives the PSNR factor in terms of embedding rate is given, to the RDHEI techniques already in use. As given in the table, the suggested method frequently achieves noticeably higher embedding rates than the other approaches.

5 Conclusion and Future Scope

In this study, a revolutionary low computing complexity reversible data concealing system for encrypted images is suggested. A stream cipher has completely encrypted the source image's data. Even if a data hider is unaware of the source content, he can add new data to the encrypted image by changing a small portion of the encrypted data. As a result, the data hider may use the extra space that was freed up in the previous stage to simplify the data concealing process. By utilizing all traditional RDH techniques, the recommended method may achieve high performance without losing total privacy for plain photos. By reserving space prior to encryption and pixel prediction, the suggested approach allows us to introduce visible watermarks and some extra data in an encrypted domain in a way that is reversible.

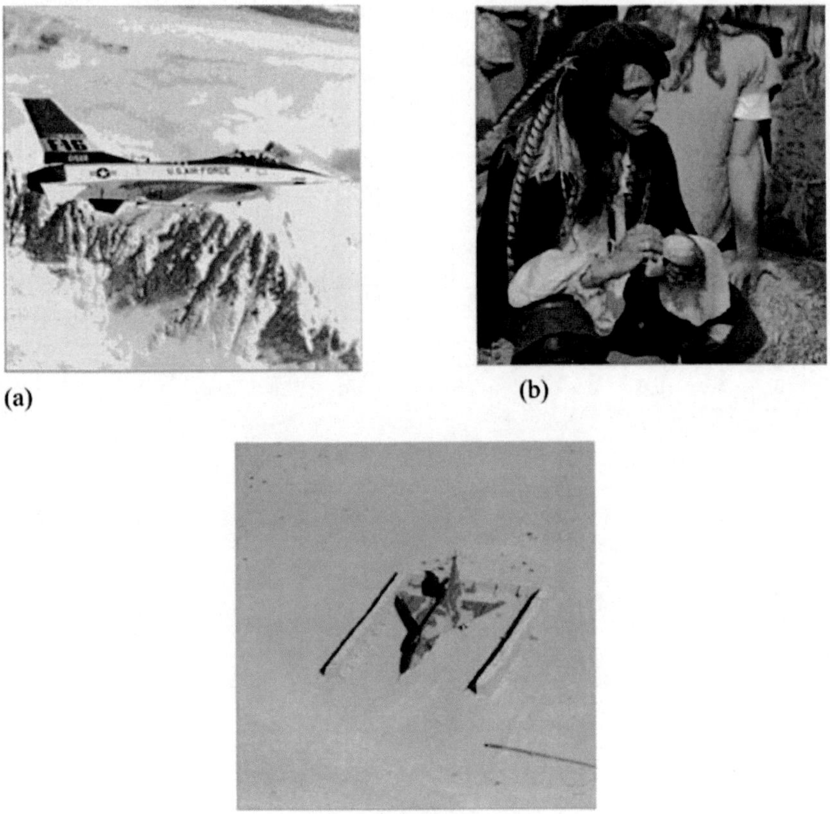

(a) (b) (c)

Fig. 3 Three test images. **a** Jetplane, **b** Man, and **c** Airplane

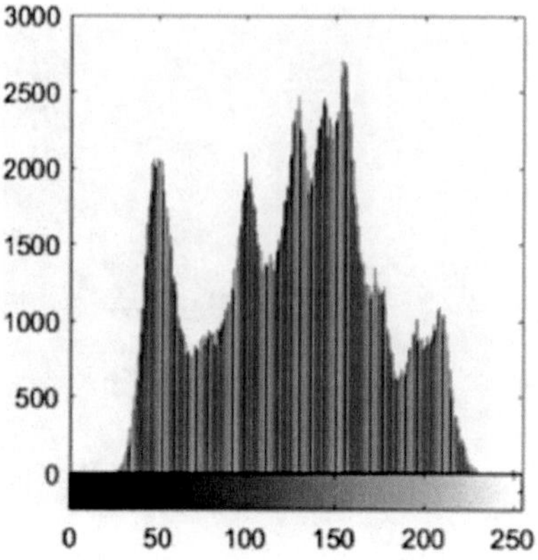

(a)

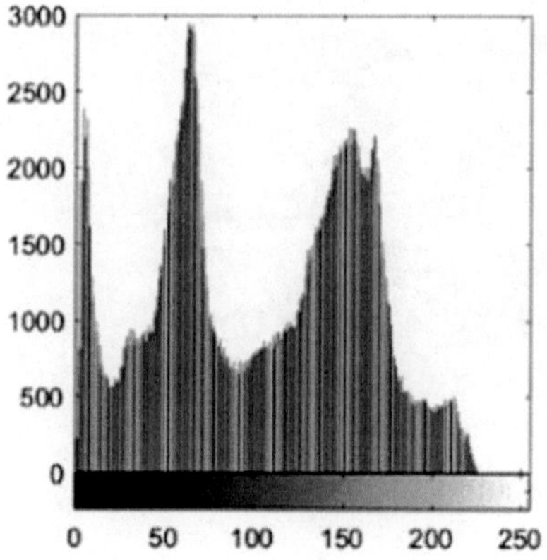

(b)

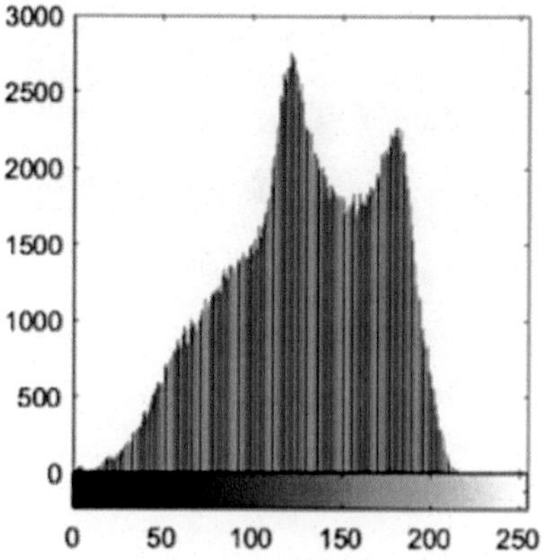

(c)

Fig. 4 **a** Histograms of the Jetplane image, **b** Histograms of the Man image, and **c** Histograms of the Airplane image

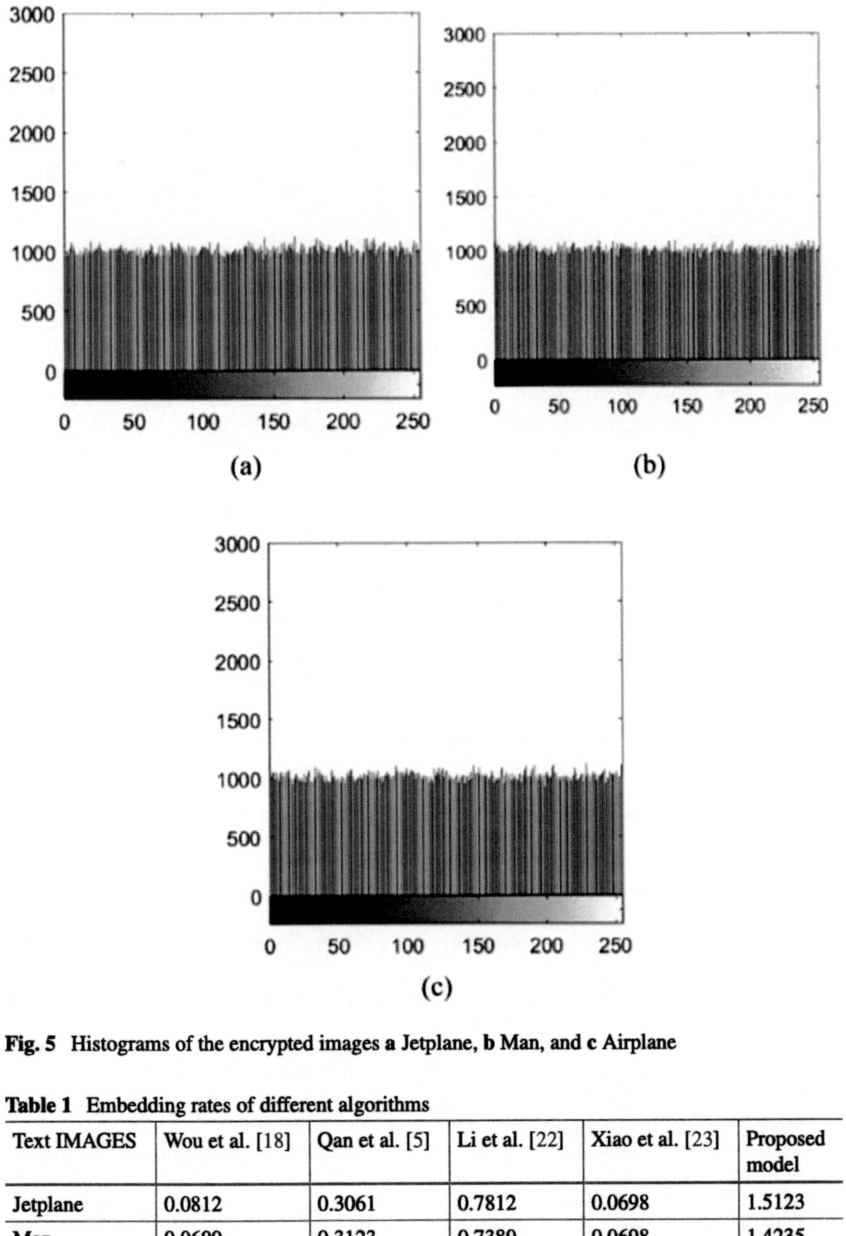

Fig. 5 Histograms of the encrypted images **a** Jetplane, **b** Man, and **c** Airplane

Table 1 Embedding rates of different algorithms

Text IMAGES	Wou et al. [18]	Qan et al. [5]	Li et al. [22]	Xiao et al. [23]	Proposed model
Jetplane	0.0812	0.3061	0.7812	0.0698	1.5123
Man	0.0699	0.3123	0.7389	0.0698	1.4235
Airplane	0.0679	0.2952	0.2136	0.0679	0.2897

References

1. Naha RK, Garg S (2018) Fog computing—Survey of trends, architectures, requirements, and research directions. IEEE Access 6:47980–48009
2. Rimal BP, Van DP, Maier M (2017) Mobile-edge computing vs. centralized cloud computing over a converged FiWi access network. IEEE Trans Netw Serv Manag 14(3):498–513
3. Zhhang X (2011) Reversible data hiding in encrypted images. IEEE Signal Process Lett 18(4):255–258
4. Ahmad I, Pothuganti K (2020) Smart field monitoring using ToxTrac: a cyber-physical system approach in agriculture. In: 2020 International conference on smart electronics and communication (ICOSEC), pp 723–727. https://doi.org/10.1109/ICOSEC49089.2020.9215282
5. Hong W, Chen T, Wu H (2012) An improved reversible data hiding in encrypted images using side match. IEEE Signal Process Lett 19(4):199–202
6. Ramana S, Ramu SC, Bhaskar N, Murthy MVR, Reddy CRK (2022) A three-level gateway protocol for secure M-commerce transactions using encrypted OTP. In: 2022 International conference on applied artificial intelligence and computing (ICAAIC), pp 1408–1416. https://doi.org/10.1109/ICAAIC53929.2022.979290
7. Liao X, Shu C (2015) Reversible data hiding in encrypted images based on absolute mean diference of multiple neighboring pixels. J Vis Commun Image Represent 28:21–27
8. Swathi P (2022) Industry applications of augmented reality and virtual reality. J Environ Impact Manage Policy (JEIMP) 2(2):7–11. ISSN: 2799-113X
9. Wu X, Sun W (2014) High-capacity reversible data hiding in encrypted images by prediction error. Signal Process 104:387–400
10. Qian Z, Zhang X (2016) Reversible data hiding in encrypted image with distributed source encoding. IEEE Trans Circuits Syst Video Technol 26:636–646
11. Xiao D, Xiang Y, Zheng H, Wang Y (2017) Separable reversible data hiding in encrypted image based on pixel value ordering and additive homomorphism. J Vis Commun Image Represent 45:1–10
12. Li M, Xiao D, Zhang Y, Nan H (2015) Reversible data hiding in encrypted images using cross division and additive homomorphism. Signal Process Image Commun 39:234–248
13. Celik MU, Sharma G, Tekalp AM, Saber E (2005) Lossless generalized-LSB data embedding. IEEE Trans Image Process 14(2):253–266
14. Ni Z, Shi Y-Q, Ansari N, Su W (2006) Reversible data hiding. IEEE Trans Circ Syst Video Technol 16(3):354–362
15. Tian J (2003) Reversible data embedding using a difference expansion. IEEE Trans Circuits Syst Video Technol 13(8):890–896
16. Li X, Li B, Yang B, Zeng T (2013) General framework to histogram-shifting-based reversible data hiding. IEEE Trans Image Process 22(6):2181–2191
17. Ou B, Li X, Zhao Y, Ni R, Shi Y (2013) Pairwise prediction-error expansion for efficient reversible data hiding. IEEE Trans Image Process 22(12):5010–5021
18. Dragoi I-C, Coltuc D (2014) Local-prediction-based difference expansion reversible watermarking. IEEE Trans Image Process 23(4):1779–1790
19. Li X, Zhang W, Gui X, Yang B (2015) Efficient reversible data hiding based on multiple histograms modification. IEEE Trans Inf Forens Secur 10(9):2016–2027
20. Ma K, Zhang W, Zhao X et al (2013) Reversible data hiding in encrypted images by reserving room before encryption. IEEE Trans Inf Forensics Secur 8(3):553–562
21. Zhang W, Ma K, Yu N (2014) Reversibility improved data hiding in encrypted images. Signal Process 94:118–127
22. Wang J, Ni J, Zhang X, Shi Y (2017) Rate and distortion optimization for reversible data hiding using multiple histogram shifting. IEEE Trans Cybernetics 47(2):315–326
23. Puech W, Chaumont M, Strauss O (2008) A reversible data hiding method for encrypted images. In: Proceedings of SPIE, vol 6819

24. Zhang X (2012) Separable reversible data hiding in encrypted image. IEEE Trans Inf Forensics Secur 7(2):826–832
25. Zhang X, Long J, Wang Z, Cheng H (2016) Lossless and reversible data hiding in encrypted images with public-key cryptography. IEEE Trans Circuits Syst Video Technol 26(9):1622–1631
26. Xu D, Wang R (2016) Separable and error-free reversible data hiding in encrypted images. Signal Process 123:9–21
27. Huang F, Huang J, Shi YQ (2016) New framework for reversible data hiding in encrypted domain. IEEE Trans Inf Forensics Secur 11(12):2777–2789
28. Puteaux P, Puech W (2018) An efficient MSB Prediction-based method for high-capacity reversible data hiding in encrypted images. IEEE Trans Inf Forensics Secur 13(7):1067–1681
29. Puyang Y, Yin Z, Qian Z (2018) Reversible data hiding in encrypted images with two-MSB prediction. In: IEEE international workshop on information forensics and security (WIFS)
30. Yi S, Zhou Y (2018) Separable and reversible data hiding in encrypted images using parametric binary tree labeling. IEEE Trans Multimedia (99):1–1
31. Bas P, Filler T, Pevný T (2011) Break our steganographic system—the ins and outs of organizing BOSS. In: Proceeding of 13th international conference, pp 59–70. http://dde.binghamton.edu/download/
32. Bas P, Furon T (2017) Image database of BOWS-2. Accessed 20 June 2017. [Online]. http://bows2.ec-lille.fr/
33. Schaefer G, Stich M (2003) UCID: an uncompressed color image database. In: Proceedings of SPIE electronic imaging, storage and retrieval methods and applications for multimedia, vol 5307, pp 472–480. http://vision.doc.ntu.ac.uk/
34. Zhang X (2011) Reversible data hiding in encrypted image. IEEE Signal Process Lett 18(4):255–258
35. Qian Z, Zhang X (2016) Reversible data hiding in encrypted images with distributed source encoding. IEEE Trans Circuits Syst Video Technol 26(4):636–646

Blockchain-Based Implementation on Electronic Know Your Customer (e-KYC)

K. S. Chandraprabha

Abstract The electronic Know Your Customer (e-KYC) is basically a system for all the banking enterprises or for the identity providers to establish and provide a customer identity such as data verification process between agreeing parties. Nowadays, most of the banks are implementing e-KYC on cloud due to its efficient resource consumption and high degree of accessibility and availability of cloud computing. KYC is also a procedure by which banks gather information about their customers' identities and addresses. It is a regulated practice of completing due diligence on clients to verify their identity. This procedure aids in ensuring that clients are genuine and the bank's services are not being abused. The banks are in charge of completing the KYC process. In this paper, we have discussed the demerits of the process for establishing accounts and KYC may be a time-consuming, manual process that is duplicated across institutions. Financial institutions would be able to use blockchain to share KYC information, to improve compliance outcomes, efficiency, and customer service, and to avoid redundancy by gaining experience.

Keywords KYC · Encryption · OCR · Metamask · Ganache · Smart contracts · Blockchain

1 Introduction

1.1 KYC

KYC stands for 'Know Your Customer' in its most comprehensive form. It is a Reserve Bank of India-approved verification technique that allows an organization to authenticate and so verify a customer's validity. Banks, financial institutions, and

K. S. Chandraprabha (✉)
Department of Computer Science and Engineering, Siddaganga Institute of Technology, Tumkur, Karnataka 572103, India
e-mail: chandraprabhaks@sit.ac.in

G. Rajakumar et al. (eds.), *Intelligent Communication Technologies and Virtual Mobile Networks*, Lecture Notes on Data Engineering and Communications Technologies 171,
https://doi.org/10.1007/978-981-99-1767-9_22

brokerages, among others, can use KYC to verify a company's legal status. Validating the names of their beneficial owners and authorized signatories, as well as cross-checking their working address, might be part of this process. The most important reasons for the KYC verification procedure are, it is very successful at guaranteeing that, any financial institutions cannot use any form of money laundering. Money laundering is often carried out without the knowledge of the financial body whose platform is being used. We can establish a single decentralized database that is secure and eliminates data redundancy by leveraging blockchain to implement KYC. When a node in the blockchain network confirms a customer's KYC, it is made available to all nodes and stored in the blockchain database using SHA encryption methods, which improves security. Only verified financial institutions are allowed to act as network nodes and have access to KYC verification.

The blockchain is a permanent digital ledger that can be set up to record every virtual exchange of value in addition to financial transactions. The foundation of a new kind of Internet is blockchain, which enables the distribution of digital information across it while preventing its copying. Blockchain is a collection of incorruptible data records that have been time-stamped and are administered by a network of computers rather than by a single entity. The block's data is all encrypted, and it uses cryptographic techniques to link the blocks together in chains or links.

The blockchain is an unchangeable digital ledger that can be configured to record every virtual transaction of value, not only financial transactions. A new kind of Internet's foundation is blockchain, which enables the spread of digital information online without being copied. Blockchain is a collection of time-stamped, unalterable data records that are controlled by a network of computers rather than by a single device authority. The block's data is all encrypted, and the blocks are connected to one another utilizing a cryptographical technology in a chain or link. The specialty about blockchain includes blockchain networks do not require centralized authority, it is accessible to everyone, since the information is secure and unchangeable, and anybody may see it. Everything created with blockchain technology is transparent, and everyone involved is accountable for their actions. Nobody can be able to change any of the information on it. Due to its immutability, it is risk-free.

1.2 Offline KYC Verification Process

The KYC process procedures are nearly identical when carried out offline as they are when carried out online. One of the main differences between offline and online verification process is that in offline process we need original docs and other application papers. You will be requested to provide your Aadhaar and PAN numbers in this form, so double-check that both numbers are right.

Once you have completed the form, go to your nearest KRA to submit your application in person. With your application form, you will be needed to present proof of your identification as well as verification of your address. In addition to confirming identity, you can be required to provide a biometric scan fingerprints, handprints, and, in rare cases, a picture are all necessary. You will be granted an official application number with which you can trace the progress of your KYC verification after you have submitted your form and completed your biometrics.

1.3 Implementation of KYC Using Blockchain

Currently, every financial institutions (FI) does KYC for the same customer, which is a repetitive procedure that takes additional time due to manual verification, which is a time-consuming process that is inconvenient for customers. This raises the cost while also increasing data redundancy. Many financial institutions use third-party suppliers to verify their clients' KYC, which presents security concerns because third-party vendors have access to consumers' personal information. To address all of these concerns, we need a solution that automates the verification process that is decentralized, transparent, and safe. One such system is blockchain, which provides a global, distributed transactional database with nodes connected through a peer to peer connected network having its own layer of message protocol for the node communication.

By using blockchain to implement KYC, we can create a single decentralized database that is safe and reduces data redundancy. When any node in the blockchain network verifies a customer's KYC, it is offered to all nodes and added to the blockchain database using SHA encryption algorithms, which strengthens security [1, 2]. Only verified financial institutions can function as network nodes and have access to the KYC verification.

2 Literature Survey

In this part, the researchers' contributions to the blockchain implementation on e-KYC are explored.

2.1 KYC Blockchain System Optimization

The disadvantages of traditional manual KYC procedure are less security, more time-consumption, and costlier [1, 3]. The main aspects of blockchain technology to overcome the above disadvantages include immutability, security, and decentralization. Author stated that symmetric AES encryption technique was used to achieve an Ethereum-based optimized KYC blockchain system that also uses and an LZ-based compression method. Individual fields can be given distinct access keys in future editions of this system to provide consumers more control over their personal information. A thorough examination of all encryption and compression algorithms may be carried out in order to examine and arrive at a more efficient general solution in a real-world decentralized environment. The proposed method tackles the current situations such as problem of duplicate registrations in the KYC procedures. Additionally, encryption utilizing the AES method and random key generation are used to provide customers access control over their KYC data while maintaining their privacy.

2.2 A Blockchain-Based Solution to KYC Dilemma

This blockchain-based application seeks to deliver this as an electronic-KYC solution to all financial institutions as 'platform as a service', hence making life easier for end-users. This project presently stores user KYC data in plaintext without encryption [4]. This is because the blockchain's core premise does not prioritize secrecy. We may utilize a centralized key distribution server to keep the symmetric key required for encryption of each and every KYC record in order to issue the keys to banks when they request it. KYC Chain, a distributed blockchain application was demonstrated in this article. KYC Chain is built to meet the demands of today's Know Your Customer procedure. It uses the advantages of the blockchain platform like distributed ledger and immutability [5, 6].

2.3 KYC Process Through Blockchain

The KYC management process is a persistent issue in the banking industry. This is a tedious process because it requires the same procedure to be followed for different institutions for the same customer, resulting in increased costs [7]. Customers will waste time because the process is repeated for each bank or bank with which they wish to do business. In comparison with the typical KYC procedure, this article proposes an alternative that lowers the entire cost [8]. Customers register with a single financial institution, eliminating the redundancy job by avoiding numerous financial institution registrations. Our proposed solution's ultimate efficiency advantage was the twin benefit of lower costs for institutions and a better consumer experience [9].

In [4], the core KYC verification process is only conducted once for every customer, irrespective of the financial institutions in which the customer likes to work. The result of the core KYC verification is securely shared by customers. This system allows for increased efficiency, cost reduction, rich customer experience, and increased transparency. In [6], a novel-based e-KYC scheme called e-KYC Trust Block based on the ciphertext policy attribute-based encryption (CP-ABE) method was discussed. It achieves the binding with the client to deliver trust, security, and privacy compliance. It also delivers privacy preserving and fine-grained access of sensitive transactions which were stored in the blockchain.

In [7], the proposed system allows a customer to open an account at one bank, complete the KYC process there, and generate a hash value using the IPFS network and share it using the blockchain technique. Yuan et al. [10]Wu et al. [11] employed a CP-ABE approach to support data privacy protection and fine-grained sharing in the blockchain system. In these schemes, any changes to the data are recorded on the blockchain and the access policy is enforced to manage the different permissions of access. If there is any key abuse case initiated by any malicious users or authorities, the system provides audit trails to support the traceability of cryptographic operations and transaction activities.

Although user privacy is a critical issue that commercial blockchain solutions aim to solve, the main goal of blockchain is to promote complete transparency. Two distinct smart contract types that provide user privacy protection on commercial blockchains within the parameters of Know Your Customer (KYC) regulations are displayed in a decentralized schema. A user registers and uploads their KYC information to the exploited IPFS store via the public KYC smart contract. The permissioned network blockchain then converts these activities into blockchain transactions. The validity and expiration date of the first user's KYC documents can also be approved or rejected by an admin user using the public KYC smart contract. Future system additions Department of Computer Science and Engineering 11 blockchain implementation on e-KYC 2021–22 are currently being worked on, including the creation of an API that will allow to incorporate more data sources into our architecture and support more intricate procedures for compiling a variety of data about certain entities. The KYC procedures will come into even greater attention in future blockchain-based transactions. The research has demonstrated how modular, all-purpose, and straightforward architectural frameworks may be deployed on top of permissioned blockchains (like Quorum), but with a feel of an open public blockchain (such as through the flavour of solidity-based smart contracts). As a result, installing KYC procedures may become feasible and appropriate for a wide range of decentralized apps [12].

3 Proposed Architecture

3.1 Current KYC Architecture

Each bank conducted its own identification check under the former KYC system, implying that each user was inspected independently by a different institution or government authority. As a result, double-checking each identification from the start is pointless [13].

3.2 Proposed KYC Architecture Using Blockchain

In the proposed architecture, it combines the information from several service providers across the world into a single cryptographical, secure and immutable database using the intrinsic blockchain architecture, by removing the need of a third party to verify the authenticity of the knowledge [2, 4]. To verify his or her identity, our proposed method enables the creation of a system, in which a user only needs to complete the KYC procedure [2, 4] once as shown in Fig. 1. Figure 2 represents the architecture of the proposed work. A user who wants to take out a loan or utilize another service must submit the necessary paperwork to one of the banks as part of the KYC procedure. Personal data is gathered by individual parties (banks, governments, businesses, or consumers themselves) and kept in a decentralized network. If everything is in order, the bank verifies and certifies that KYC has been completed. The blockchain platform, which is accessible to other banks, businesses, and state entities, is where the bank enters the user's data. The KYC procedure is controllable and regulated by all stakeholders. The system will keep track of updates and modifications to user data, and if somebody breaches the rules, everyone will be made aware of it. The second bank accesses the system and verifies the user's identification when a user requests to utilize the services of another bank. Only the user's consent will be used to get access to their data. The user must commence the information exchange procedure by logging in using bitcoin transactions, i.e. by using the private key. To complete the KYC procedure, a user must submit documents to one of the institutions where he desires to take out a loan or use other services. Individual parties gather and store personal data in a decentralized network (banks, government agencies, enterprises, or consumers). The bank double-checks and verifies the completion of KYC if everything is in order. The bank is in charge of entering user information onto the blockchain network, which is accessible to other banks, companies, and government entities. The KYC procedure is within the control of all financial institutions. Changes and updates to user data will be monitored by the system and anybody who breaches the rules will be made aware of it as shown in Fig. 2. When a user requests services from another bank that bank logs into the system and validates the user.

4 Design and Implementation

4.1 Design

The backbone of the work is a website consisting of 2 portals, one for banks and the other for customers. Portal 1 is where banks can sign up from the portal and register themselves for the process. Banks can do the following functionalities such as add customer details to the blockchain after verifying customer details and can modify customer data; the data in the customers' identity card (Aadhaar card/PAN card) is taken as it is using OCR. Bank can send requests to customers using customer

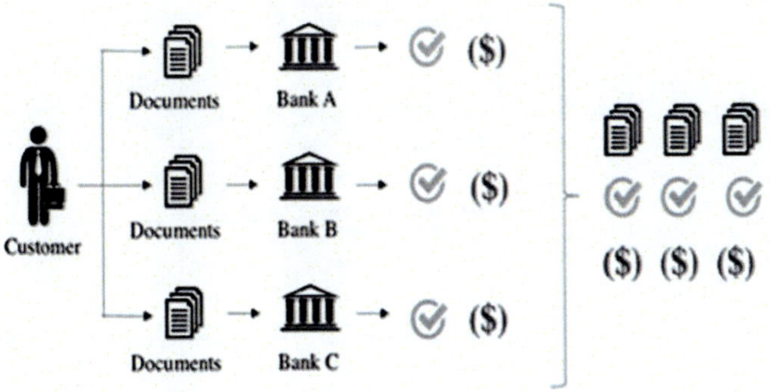

Fig. 1 Block diagram of current KYC system

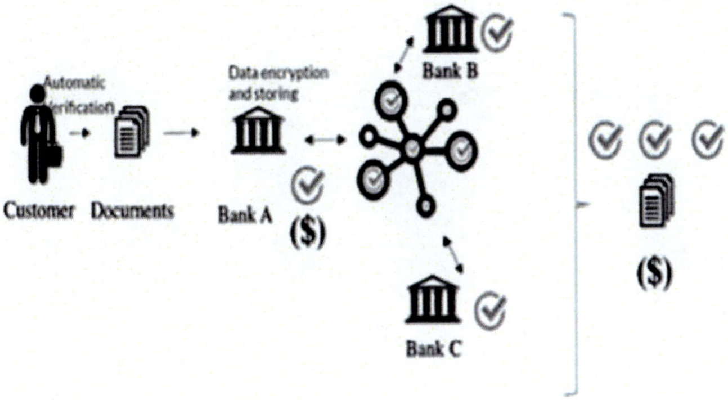

Fig. 2 Block diagram of the proposed KYC system using blockchain

user ID for accessing the details and get permission only if the customer allows it. And in portal 2, customer can sign up through the portal and gets a user ID. The customer should provide any original ID card to bank for the first time and get KYC verification done. From next time onwards, customer can provide a user ID to other banks and allow banks to access verified data by accepting the request sent by the bank to the customer through the portal. This process will ensure that customer data will be controlled only by the customer [14].

Use Case Diagram:

One of the unified modelling languages, the use case diagram, portrays actors, or users, together with their responsibilities and roles in relation to the system. Users and use cases involved in the system are shown in the use case diagram. Use cases are the activities or jobs carried out by the user when utilizing the software. Use case diagrams are used to illustrate the functional relationships between systems and actors graphically. Figure 3 shows how the system's user roles are depicted.

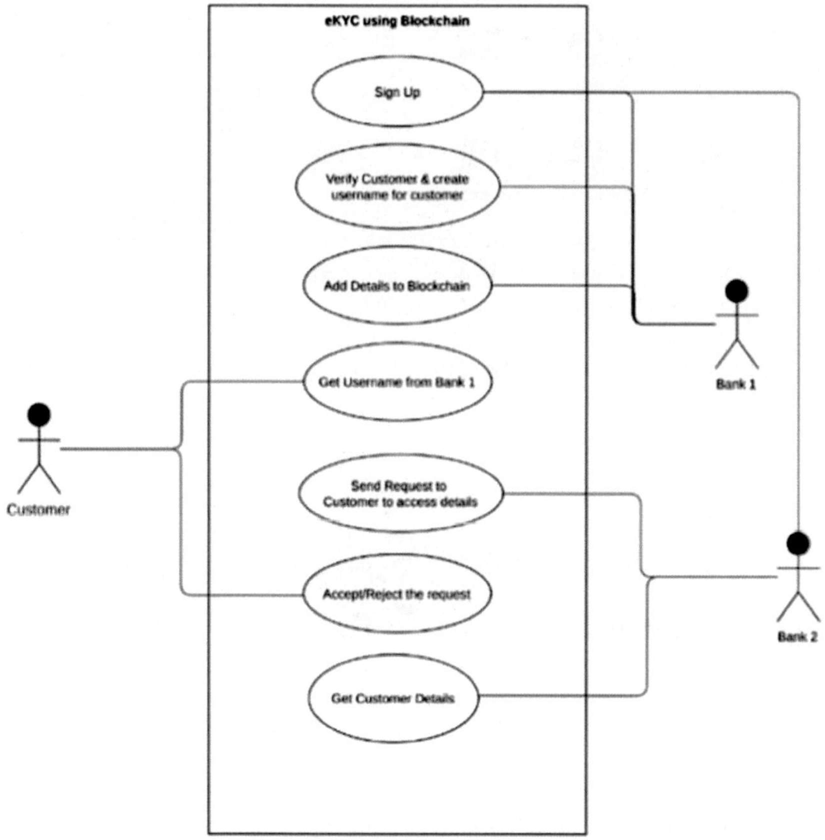

Fig. 3 Use case diagram

There are three different user types in the system such as Customer, Bank1 and Bank2.

Use case 1: Sign up.

Primary actor: Bank1.

Description: If a bank uses the application for the first time, they must first sign up for the system.

Use case 2: Verify customer and create username to customer.

Primary actor: Bank1.

Description: here, the Bank1 will verify the customers details whether customer is a genuine and create a username and temporary password for the customer.

Use case 3: Add details to blockchain.

Primary actor: Bank1.

Description: After verifying the customer details Bank1 will add the details to the blockchain.

Use case 4: Get username from Bank1.

Primary actor: Customer.

Description: After Bank1 adds customer details to blockchain, it gives customer username and temporary password to sign in in customer portal where customer can change password and access his account and other features in it.

Use case 5: Send request to customer to access details.

Primary actor: Bank2.

Description: Now if Bank2 wants to access the details of customer and verify customer, then Bank2 needs to send a request to customer using customer's username.

Use case 6: Accept/reject request.

Primary actor: Customer.

Description: After request sent by the Bank2 to customer now customer can reject or accept the request based on whether customer wants to give access to Bank2 or not.

Use case 7: Get customer details.

Primary actor: Bank2.

Description: If the request is accepted by the customer, then Bank2 can access the details of the customer.

The two different kinds of interaction diagrams are collaboration and sequence diagrams. Sequence diagrams are used to illustrate the communication that occurs between the many items that are listed in our project. Above diagram demonstrates how messages are sent between processes. As shown in Fig. 4, we totally have five objects those are Customer, Portal, Bank1, Blockchain, and Bank2. here as shown in the diagram banks will sign up in portal. Then suppose first time customer wants to open account with Bank1, then he will provide his details like Aadhaar and PAN card to Bank1 for KYC verification. Then, Bank1 will verify and add details to blockchain if he or she is a genuine user and provide username and temporary password to customer for further use. Customer can change the password in the customer portal. Now, if a customer wants to open an account in Bank2, then he just needs to provide his username to Bank2 using which Bank2 will send a request to view the user details. Now a customer can accept or reject the request sent by the Bank2 in the portal if he or she accepts the details, then Bank2 can view customer details and proceed.

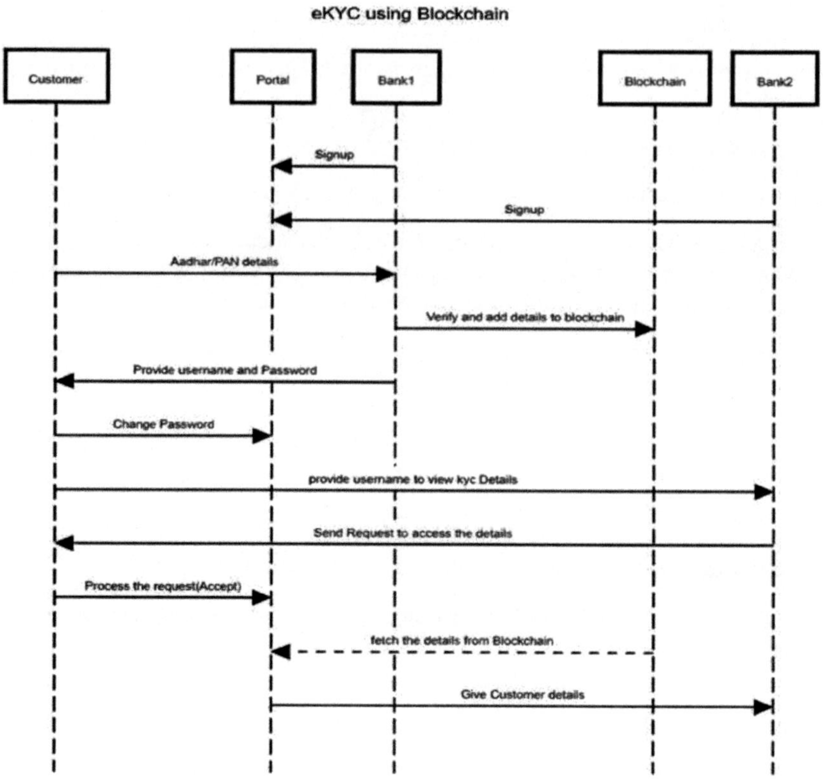

Fig. 4 Sequence diagram of e-KYC

4.2 Implementation

The algorithm used to carry out the work is written in steps below:

- Create a local blockchain network using Ganache and create an Ethereum workspace named KYC.
- Install metamask extension in chrome.
- Write solidity contract named KYC with all the variables and methods to perform necessary functions of the application.
- Compile the contract and deploy it to Ganache using Truffle. This will provide Kyc.json file which can be used in our React.js app to interact with the contract.
- Create a react app with all necessary libraries like Web3, solc required to develop a Dapp and import Kyc.json file into the react project.
- Connect the Ganache with metamask and import the accounts from Ganache to metamask to perform necessary transactions.
- Each function call is considered as a transaction and appropriate gas fees is required to perform a transaction.

- Each transaction is recorded, and blocks are created in Ganache.

The following points give the brief idea of functionalities implemented.

A new digital account of bank can be created by creating a username and password for the bank, and it is verified and added. Smart contracts used is bank—to store bank details like name, password and to perform functions like AddBank to add bank, RemoveBank to remove the bank. A digital account of bank is mandatory for processes like adding, verifying customer, and accessing customer details, due to which customer's details are not in wrong hands. Customer must create his digital account before visiting to bank for KYC verification [9, 15, 16]. If a new customer visits the bank, both username and password are mandatory for account creation. Username is taken so that it is easy to remember instead of public address which is long and password is for authentication.

Customer who has not verified his details in any bank needs to provide his username and physical documents to the bank, now bank can do new KYC process also it can modify old KYC. Using OCR technique, the details are scanned and manual operations are skipped. These details are verified and added to blockchain. Smart contracts used to store and retrieve customer data and functions like add customer. If the customer has already verified his KYC, then he must provide his username to bank. Bank must send the request to customer using the username and wait for customer's response. Customer can accept/reject the request in the interface provided to him based on which bank can access customer's data. In this process, bank can access data only when customer allows the bank, so that customer's privacy is only controlled by customer [8, 17].

Smart contract is used in KYC to perform functions like send Request, approve Request for accepting or rejecting the request. Authentication is done using blockchain, and local blockchain network Ganache is used. Figure 5 shows the snapshot of sign in process [2, 18].

We can access Ganache from the browser by integrating it with an Ethereum browser wallet called metamask as shown in Fig. 6.

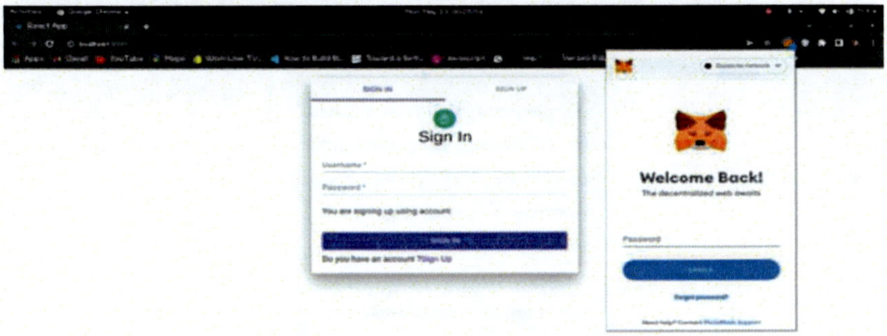

Fig. 5 Snapshot of sign in

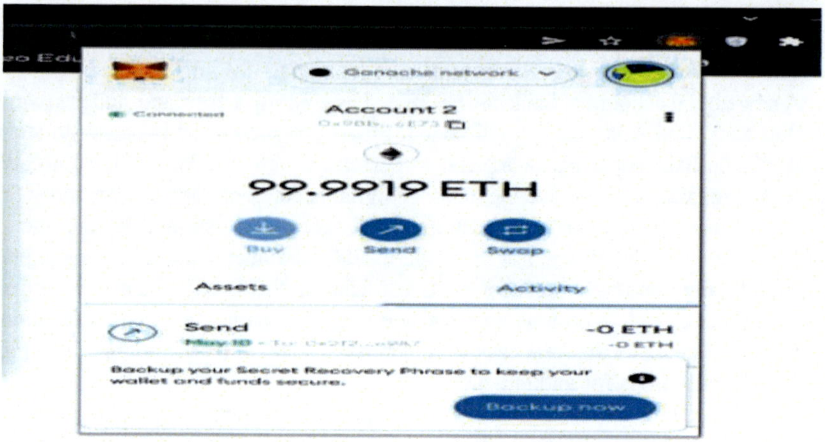

Fig. 6 Snapshot of metamask wallet

Contracts for bank login signup, customer login signup, add customer, remove customer, view customer, and modify customer details are written in solidity and deployed in Ganache. Contracts are deployed in Ganache using Truffle. Then, the blockchain is connected. Figure 7 shows the snapshot of add customer page.

We can view the customer details in the view customer page as shown in Fig. 8.

As shown in Fig. 9, customer details can be modified here using customer's username if and only if the bank has access to that customer details.

As shown in Fig. 10, here banks can send a request to customers to get access to their details. Banks can only view customer's details after request is accepted by the customer.

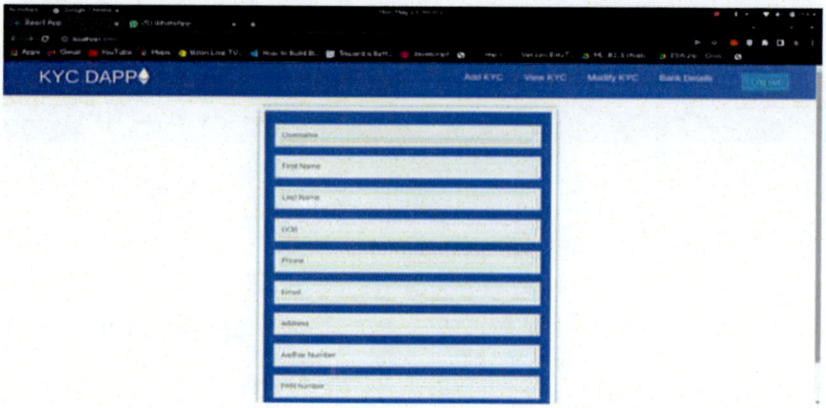

Fig. 7 Snapshot of add customer page

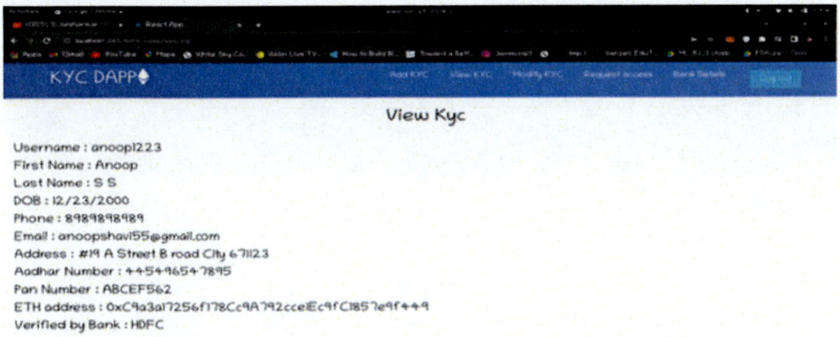

Fig. 8 Snapshot of view customer page

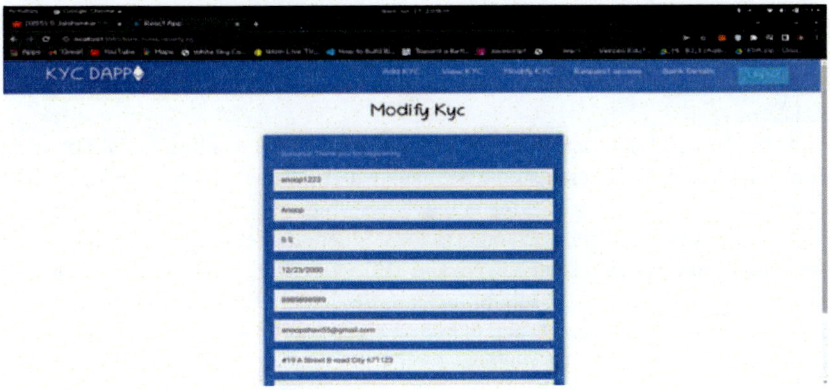

Fig. 9 Snapshot of modify customer page

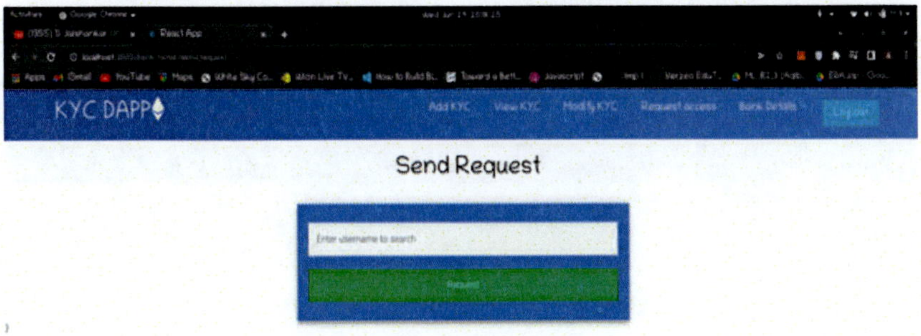

Figure10 Snapshot of send request to customer page

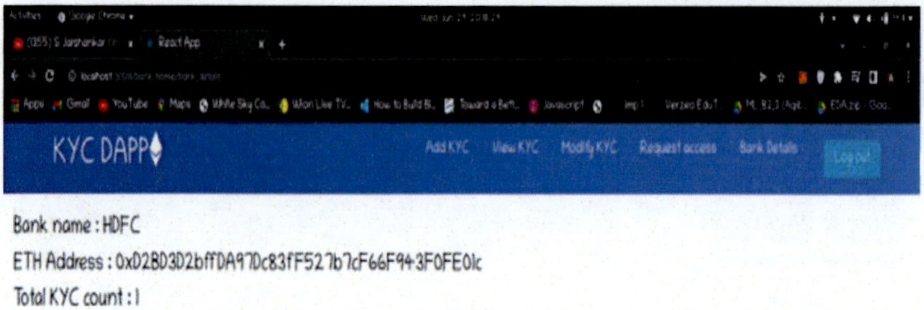

Figure11 Snapshot of bank details page

As shown in Fig. 11, in this page, bank details like bank name, Ethereum address of bank, and total KYC count of that bank can be seen.

As shown in Fig. 12, customer sign in page can be seen where customer needs to sign in using username and password provided by the first bank that customer registered.

As shown in Fig. 13, customer details like name, date of birth, mobile number, Aadhaar number, PAN number, and the bank which verified customer details can be seen.

As shown in Fig. 14, here accounts in local Ethereum network can be seen. In our work, we need all users, i.e. customers and banks to be in Ethereum network through which they create their accounts in our website having their individual Ethereum accounts. We can see various accounts showing its public address.

As shown in Fig. 15, transaction details like among which two accounts the transaction is happening can be seen. The transactions between various accounts will be happening in our work like between customer and banks. Here, we can also see gas used for each transaction.

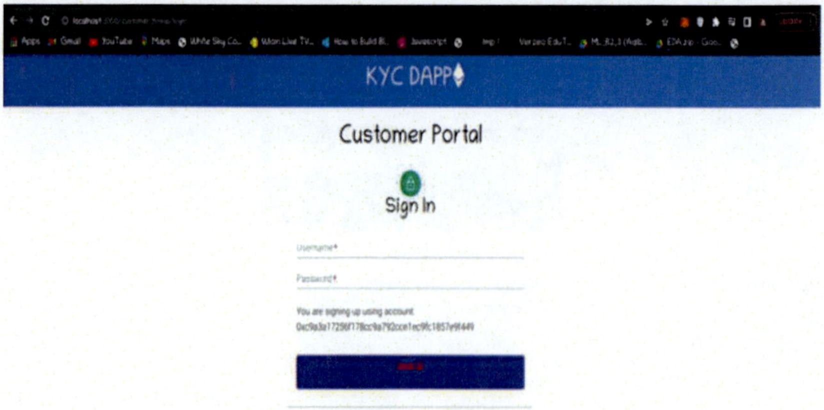

Fig. 12 Snapshot of customer sign in page

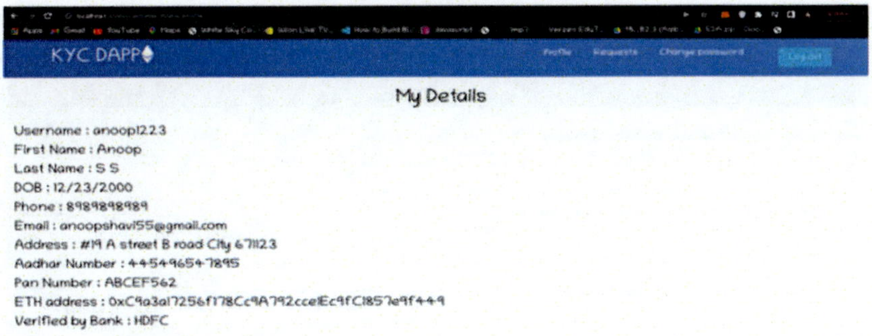

Fig. 13 Snapshot of customer profile page

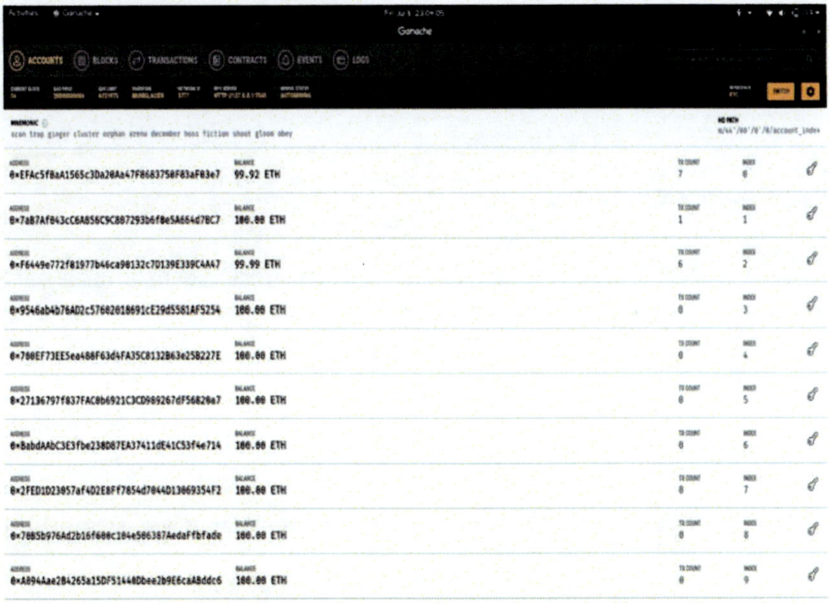

Fig. 14 Snapshots of accounts in local blockchain network (Ganache)

5 Conclusion

The traditional method for KYC verification is manual and long process which requires submitting the details every time for different banks and sometimes need to submit the personal details to third-party vendors which is of great security concern. Since advance technology and fast computing computers are available, we can make use of blockchain technologies to implement KYC system. Using the OCR we can automate the verification process, and using blockchain we can build a decentralized system, with a very high security of the user data and reducing the redundancy of

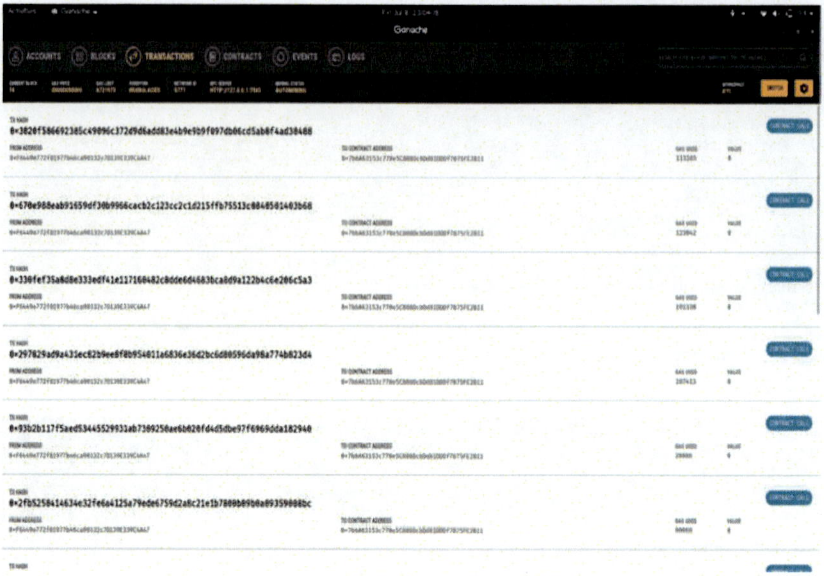

Fig. 15 Snapshot of transactions

verification of KYC in every bank. The proposed system can save time, money, and repetitive work during the KYC process when someone tries to open an account at multiple banks.

References

1. Zyskind G, Nathan O (2015) Decentralizing privacy: using blockchain to protect personal data. In: IEEE security and privacy workshops, San Jose, CA
2. Yuan R, Xia YB, Chen HB, Zang BY, Xie J (2018) Shadoweth: private smart contract on public blockchain. J Comput Sci Technol 33:542–556
3. Ayyasamy S (2022) Metadata securing approach on ubiquitous computing devices with an optimized blockchain model. J Ubiquitous Comput Commun Technol 4(2):57–67
4. Parra Moyano J, Ross O (2017) KYC optimization using distributed ledger technology. Bus Inf Syst Eng 59:411–423
5. Know your customer (KYC) laws by country, [online] https://en.wikipedia.org/wiki/Know_your_customer
6. Fugkeaw S (2022) Enabling trust and privacy preserving e-KYC system using blockchain. IEEE, pp 49028–49039. Electronic ISSN: 2169-3536
7. Mamun AA, Al Mamun A, Hasan SR, Hasan SR, Bhuiyan MS, Bhuiyan MS et al (2020) Secure and transparent KYC for banking system using IPFS and blockchain technology. In: Proceeding of IEEE region symposium (TENSYMP), pp 348–351
8. Kapsoulis N, Psychas A, Palaiokrassas G, Marinakis A, Litke A, Varvarigou T (2020) Know your customer (KYC) implementation with smart contracts on a privacy-oriented decentralized architecture. Future Internet 12(41):1–13

9. Ullah N, Al-Dhlan KA, Al-Rahmi WM (2021) KYC optimization by blockchain based hyperledger fabric network. In: International conference on advanced electronic materials, computers and software engineering (AEMCSE), pp 1294–1299
10. Yuan C, Xu M, Si X, Li B (2017) Blockchain with accountable CP-ABE: how to effectively protect the electronic documents. In: Proceedings of IEEE 23rd international conference on parallel and distributed systems (ICPADS), pp 800–803
11. Wu A, Zhang Y, Zheng X, Guo R, Zhao Q, Zheng D (2019) Efficient and privacy-preserving traceable attribute-based encryption in blockchain. Ann Telecommun 74(7):401–411
12. Kapsoulis N, Psychas A, Palaiokrassas G (2020) Implementation with smart contracts on a privacy oriented decentralized architecture
13. Zhong Y, Zhou M, Li J, Chen J, Liu Y, Zhao Y et al (2021) Distributed blockchain-based authentication and authorization protocol for smart grid. Wirel Commun Mobile Comput 2021:1–15
14. Shbair WM, Steichen M, François J (2018) Blockchain orchestration and experimentation framework: a case study of KYC. In: International workshop on managing and managed by blockchain (Man Block), pp 23–25
15. Pic M, Mahfoudi G, Trabelsi A (2019) Remote KYC: attacks and counter-measures. In: Proceeding of European intelligence and security informatics conference (EISIC). IEEE, pp 126–129
16. Norvill R, Steichen M, Shbair WM, State R (2019) Demo: blockchain for the simplification and automation of KYC result sharing. In: Proceedings of International conference on blockchain and cryptocurrency (ICBC), pp 9–10
17. Know Your Customer: Quick Reference Guide Understanding Global KYC differences by PWC (2016) [online] https://www.pwc.com/gx/en/financial-services/publications/assets/pwc-anti-money-laundering-know-your-customer-quick-reference-guide.pdf
18. Lim SY, Fotsing PT, Almasri A, Musa O, Kiah MLM, Ang TF et al (2018) Blockchain technology the identity management and authentication service disruptor: a survey. Int J Adv Sci Eng Inf Tech 8:1735–1745

Cryptography Using GPGPU

Swati Jadhav, Uttkarsh Patel, Atharv Natu, Bhavin Patil, and Sneha Palwe

Abstract Today, with an ever-increasing number of computer users, the number of cyberattacks to steal data and invade privacy is of utmost importance. A group of applications uses the Advanced Encryption Standard (AES) to encrypt data for security reasons. This mainly concerns enterprises and businesses, which ultimately handle user data. But many implementations of the AES algorithm consume large amounts of CPU horsepower and are not up to the mark in terms of throughput. To tackle this problem, the proposed system makes use of GPUs, which are targeted for parallel applications. These enable parallel operations to be performed much faster than the CPU, ultimately increasing throughput and reducing resource consumption to some extent. The vital aspect of this approach is the speedup that is achieved due to massive parallelism. This research aims to implement AES encryption and decryption using CUDA and benchmark it on various compute devices.

Keywords Cyberattacks · AES · CPU · Throughput · GPUs · SIMD · Parallel · Massive parallelism · CUDA · Compute devices

S. Jadhav (✉) · U. Patel · A. Natu · B. Patil · S. Palwe
Vishwakarma Institute of Technology, Pune, Maharashtra, India
e-mail: swati.jadhav@vit.edu

U. Patel
e-mail: patel.uttkarsh21@vit.edu

A. Natu
e-mail: atharv.natu21@vit.edu

B. Patil
e-mail: bhavin.patil21@vit.edu

S. Palwe
e-mail: palwe.sneha21@vit.edu

G. Rajakumar et al. (eds.), *Intelligent Communication Technologies and Virtual Mobile Networks*, Lecture Notes on Data Engineering and Communications Technologies 171,
https://doi.org/10.1007/978-981-99-1767-9_23

1 Introduction

Cybersecurity is critical these days, with cybercrime and cyberattacks carried out by malicious users and hackers on the rise. To protect our data, encryption algorithms are used, like Advanced Encryption Standard (AES), which was developed to tackle security problems found in Data Encryption Standard (DES) in 2001 by the National Institute of Standards and Technology (NIST). AES is a block cipher-based encryption technique. Even after 21 years, AES still withstands all cyberattacks and is also considered to be arguably the most used encryption algorithm. The widespread use of AES led to the development of many optimized implementations for a variety of CPU architectures.

Traditionally, graphical processing units (GPUs) are used by enthusiasts or developers for game playing or development, respectively, video rendering, and all operations that require a considerably large amount of video memory and dedicated processing. GPU architectures work on single instruction multiple data (SIMD) which allows them to execute the same instruction on multiple data streams in parallel fashion. This is also known as "massively accelerated parallelism" [1–5].

The main motive behind adapting this research topic is to enhance data security using encryption algorithms that allow for minimal power consumption, high throughput, and low latency. This includes exploring the field of computation to extract all the benefits we gain from general-purpose computing. Encrypting large files on a CPU tends to take a lot of time as they sequentially perform each calculation, while offloading this computation to a GPU drastically reduces the time taken as it parallelly performs the same calculations [5–10]. This means that many similar calculations are performed concurrently, resulting in a faster result. When GPUs are used to perform general tasks rather than video processing, they are known as general purpose graphical processing units (GPGPUs). GPGPUs are used for tasks that are generally performed by CPUs, such as mathematical equations and cryptography, as well as to create cryptocurrency. These GPGPUs are accessible by making use of parallel platforms like OpenCL or CUDA [10–15]. The proposed project makes use of Compute Unified Device Architecture (CUDA). It is a NVIDIA-exclusive technology that will be available on select NVIDIA compute devices. This compatibility can be checked on NVIDIA's official website.

The proposed study seeks to show the potential speedup and advantage of using a GPU to encrypt files using the AES algorithm. Despite making a significant improvement in performance, this speedup is not directly beneficial to end users. Large corporations can truly harness this power as they have to continuously encrypt a large number of files while being confined by time. As a consequence, end users benefit indirectly since it takes less time to respond to their requests. This technique not only saves a lot of time, but also power if the resources are used efficiently. This will save money not only by lowering electricity consumption but also by lowering the cost of cooling the machines. The applications of using GPUs for general-purpose workloads are limitless; encryption is just one of the many others.

2 Related Work

Survey of related works is shown in Table 1.

3 Proposed Work

(A) AES Algorithm

The AES block cipher works with 128 bits, or 16 bytes, of input data at a time. The substitution-permutation network principle is used in AES, which is an iterative algorithm (Fig. 1). The total number of rounds needed for the encryption or decryption process is determined by the size of the cryptographic key employed. AES's key length and number of rounds is shown in Table 2.

The input is represented as a 4 × 4 bytes grid or matrix in a column major arrangement, in contrast to traditional row major arrangement followed in system programming. The below equation shows the AES 16-byte matrix of 4 rows and 4 columns, which will be mapped as an array for converting plaintext to ciphertext.

$$\begin{bmatrix} b0 & b4 & b8 & b12 \\ b1 & b5 & b9 & b13 \\ b2 & b6 & b10 & b14 \\ b3 & b7 & b11 & b15 \end{bmatrix}$$

Round is composed of several processing steps, including substitution, transposition, and mixing of the input plain text to generate the final output of cipher text. Each round is divided into 4 steps—

(1) SubBytes—Input 16 bytes are substituted by looking up the S-Box.
(2) ShiftRows—Each of the four rows of the matrix is shifted to the left.
(3) MixColumns—Each column of four bytes is transformed using a special mathematical function. This operation is not performed for the last round.
(4) Add Round Key—The 128 bits of the matrix are XORed to the 128 bits of the round key. If the current round is the last, then the output is the ciphertext.

(B) CUDA Implementation

The proposed project consists of a parallel implementation for

- AES-128-bit encryption
- AES-192-bit encryption
- AES-256-bit encryption.

The proposed implementation is developed using Compute Unified Device Architecture (CUDA). It is a parallel computing platform and programming model created by NVIDIA for general computing on NVIDIA GPUs only. CUDA is not just an API,

Table 1 Literature survey

Authors	Year	Parallel platform	Research
Tezcan [16]	2021	CUDA	• Aims to use a single GPU as a cryptographic coprocessor, with optimizations that can help with cryptanalysis • Optimized version of AES providing way faster and higher throughput levels than that of CPU AES instructions • Benchmarked on multiple devices, from one of the entry-level GPUs (NVIDIA GeForce MX 250) to one of the most powerful GPUs (NVIDIA GeForce RTX 2070)
Sanida et al. [17]	2020	OpenCL	• Demonstrate the AES algorithm's implementation in the CTR and XTS parallel operating modes • Portable implementation with key length options of AES-128, AES-192, and AES-256 • File encryption and decryption from range of sizes starting with 512B to 8 MB • Throughput achieved is 12.53 and 14.71 Gbps for XTS and CTR, respectively
Bharadwaj et al. [18]	2021	CUDA	• GPU-accelerated implementation of an image encryption algorithm • To encrypt and decrypt the images, a customized XOR cipher was used with an encryption pad generated using the shared secret key and initialization vectors
Wang and Chu [19]	2019	CUDA	• Benchmarking approach for the GPU-based AES algorithm • Adapts the electronic code book (ECB) cipher mode for cryptographic transformation and the T-box scheme for the purpose of fast lookups • Follows a single thread per state granularity that basically means that every thread is mapped to an AES state for scheduling threads on the GPU • Makes efficient use of GPU hardware in terms of memory allocation and register allocations to increase the overall efficiency and throughput of the operation
Inampudi et al. [20]	2018	OpenCL	• Parallel OpenCL implementation of AES-256-bit algorithm • By using 256 work items, which simultaneously assign the elements to various threads and GPU cores for execution, data parallelism is achieved • Testing done on AMD Radeon 8550 M and 8570G GPU • Compared to sequential implementation, 1024 work items were sped up by 99.8%

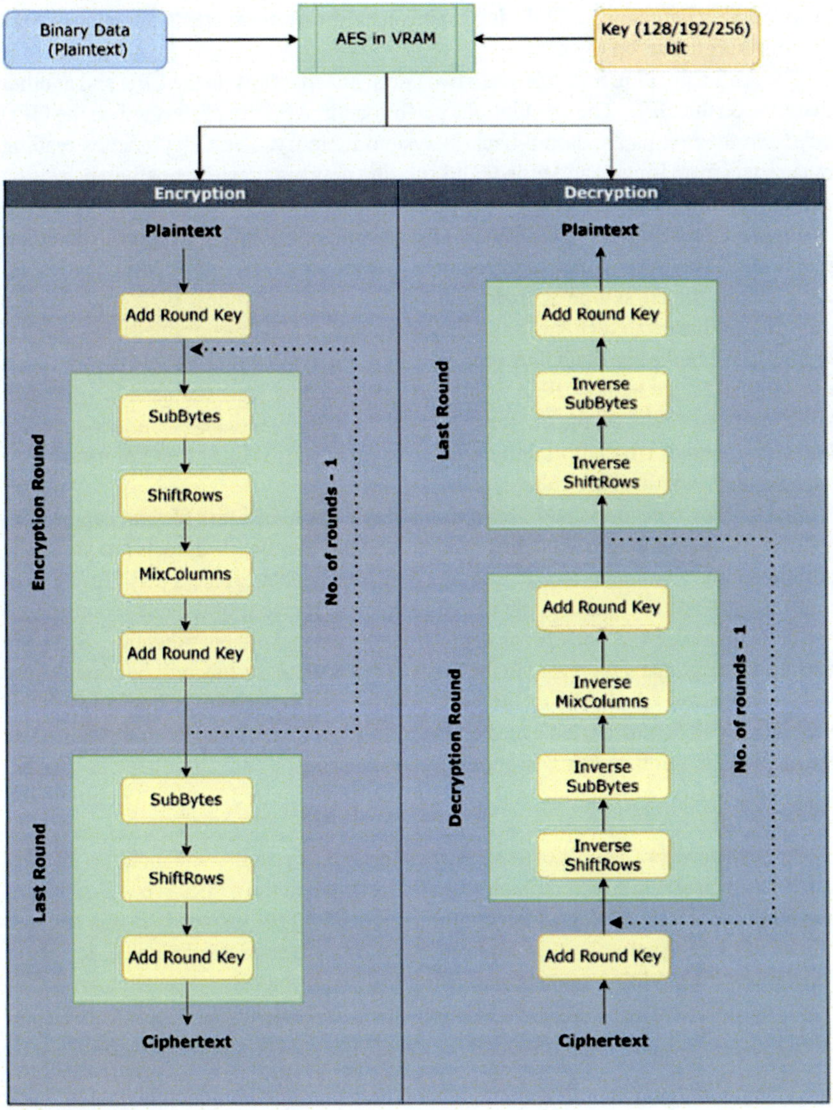

Fig. 1 AES algorithm

Table 2 AES's key length and number of rounds

Key length	Number of rounds
AES-128	10
AES-192	12
AES-256	14

programming language, or SDK. It is mainly a hardware component, allowing drivers and libraries to run on top of it.

The AES algorithm is divided into two parts: one that runs on the CPU and another that runs on the GPU. The calculations performed in AES are performed on the GPU side, and the results are stored back in system memory. The CPU handles reading binary data from images and videos and creating new binary streams after encryption and decryption by the GPU.

Figure 2 presents the NVIDIA CUDA Compiler (NVCC) trajectory. It includes transformations from source to executable that are executed on the compute device. As a result, the.cu source will be divided into host and guest parts and execute on different hardware. This is called hybrid code, which then implements parallel working. We make use of CUDA specifiers like **__global**, which denote that the code runs on the device and is called from the host. The next specifier used is **__device**, which runs and calls on the device itself. Along with this, there is another specifier known as **__host**, which runs and makes calls on host, just like other library APIs or user-defined functions are called.

The key for encryption and decryption will be stored in a text file and can be 128, 192, or 256 bits in length. The binary data, key, and number of threads are passed as command line arguments to the program. The binary data can be a text file, a video, or an image to be encrypted given as a relative path. Number of threads is considered for benchmarking the performance of GPUs to measure their potential. After the execution starts, the data stored in the form of blocks and the key will be copied from system memory, i.e., RAM to the GPU Video RAM (VRAM), using arrays. The operations will be performed based on the number of rounds which is determined by the key length. After encryption and decryption operations are completed in VRAM, the results will be copied back to RAM, and the time required for the calculation will be displayed.

Figure 3 depicts a sample code snippet using CUDA specifiers. It includes the byte substitution process, which replaces the state array with the respective S-Box values and addition of a round key which comprises Binary XOR operation. In this manner, all the AES encryption and decryption operations are implemented as C language functions with modified extensibility using CUDA to implement parallelism.

4 Result Analysis

(A) Evaluation Environment

For the purpose of evaluating the performance of our proposed algorithm, we have used a set of hardware and software components specified in Table 3.

Figure 4 demonstrates the use of our implementation. All of the samples are bitmap images having.bmp file extension. After the program has finished the execution, in the root directory of the application, 2 bitmap images will be generated, **EncryptedImage.bmp** and **DecryptedImage.bmp**. Above figure is a screenshot

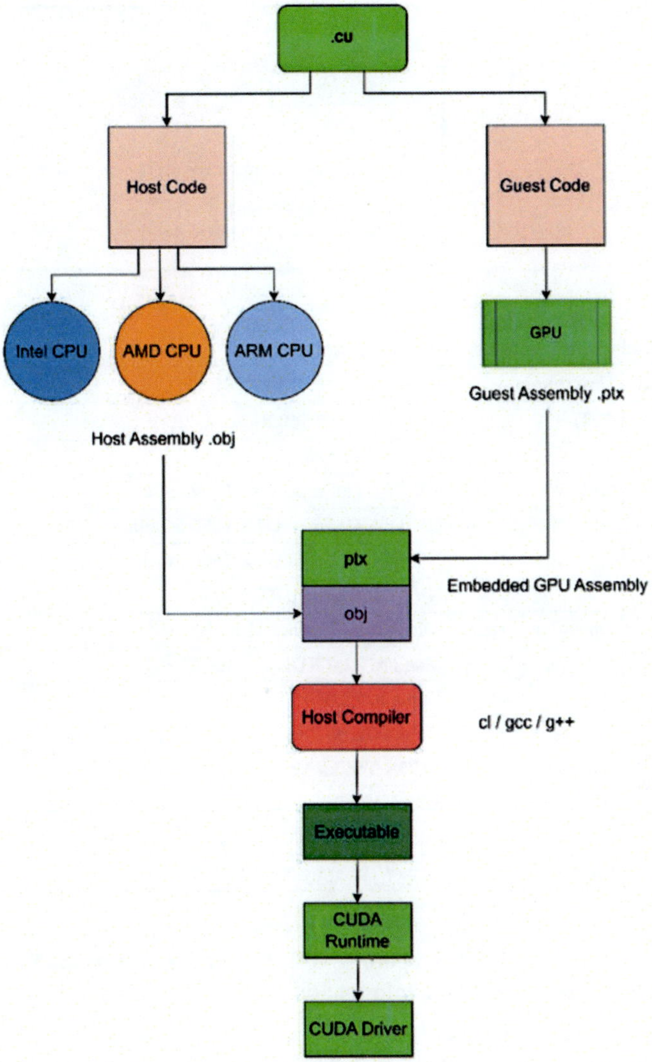

Fig. 2 NVCC trajectory

which combines both the specified files. It has two parts, the left part of the image contains the encrypted file, which the default image application is unable to open, and on the right side you can see the decrypted image.

Figure 5 describes the CUDA information about the computer system you are using to run the program. This uses the APIs from **cuda.h**, which includes **cudaGetDeviceCount()** and **cudaGetDeviceProperties()**. The summary lists out different parameters like—

Fig. 3 Sample code

```
__device__ void sub_bytes(BYTE state[], BYTE sbox[])
{
   for (int i = 0; i < LENGTH; i++)
      state[i] = sbox[state[i]];
}


__device__ void add_round_key(BYTE state[], BYTE round_key[])
{
   for (int i = 0; i < LENGTH; i++)
      state[i] = state[i] ^ round_key[i];
}
```

Table 3 Hardware and software technical specifications

Component	Description	Name
Hardware	CPU	AMD Ryzen 5 5600H
		Intel i7-12700 K
	GPU	NVIDIA GeForce MX 450 Mobile
		NVIDIA GeForce GTX 1650 Mobile
		NVIDIA GeForce GTX 1660 Super
		NVIDIA GeForce RTX 3060
	Operating system	Microsoft Windows 11 64-bit
		Manjaro 22.0 KDE Plasma
		Kubuntu 22.04 LTS
	Drivers	Nvidia Game Ready Driver 528.24
Software	Toolkit	CUDA Toolkit Version 12.0
	IDE/text editor	Microsoft Visual Studio 2022, Microsoft Visual Studio Code
	Compiler	Microsoft CL, GNU GCC
	Library	Helper Timer by Nvidia

(1) Total Number of CUDA Supporting GPU Device/Devices on the System
(2) CUDA Driver and Runtime Information

   a. CUDA Driver Version
   b. CUDA Runtime Version

(3) GPU Device General Information

   a. GPU Device Number
   b. GPU Device Name
   c. GPU Device Compute Capability
   d. GPU Device Clock Rate
   e. GPU Device Type—Integrated or Discrete

(4) GPU Device Memory Information

   a. GPU Device Total Memory

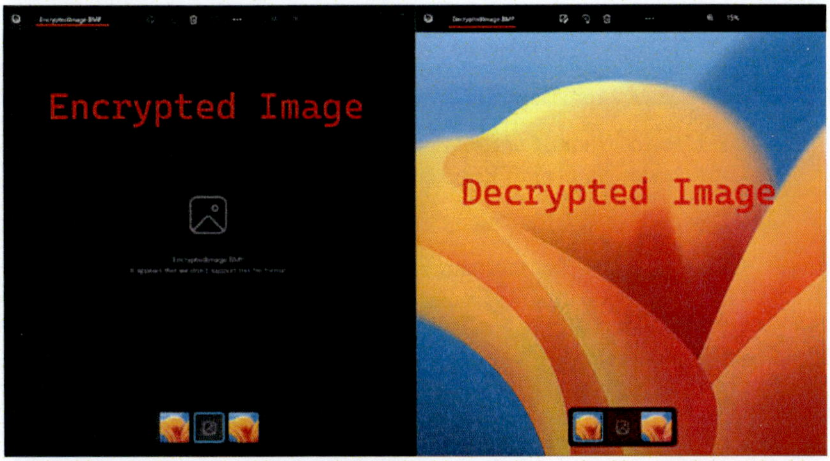

Fig. 4 Image encryption and decryption

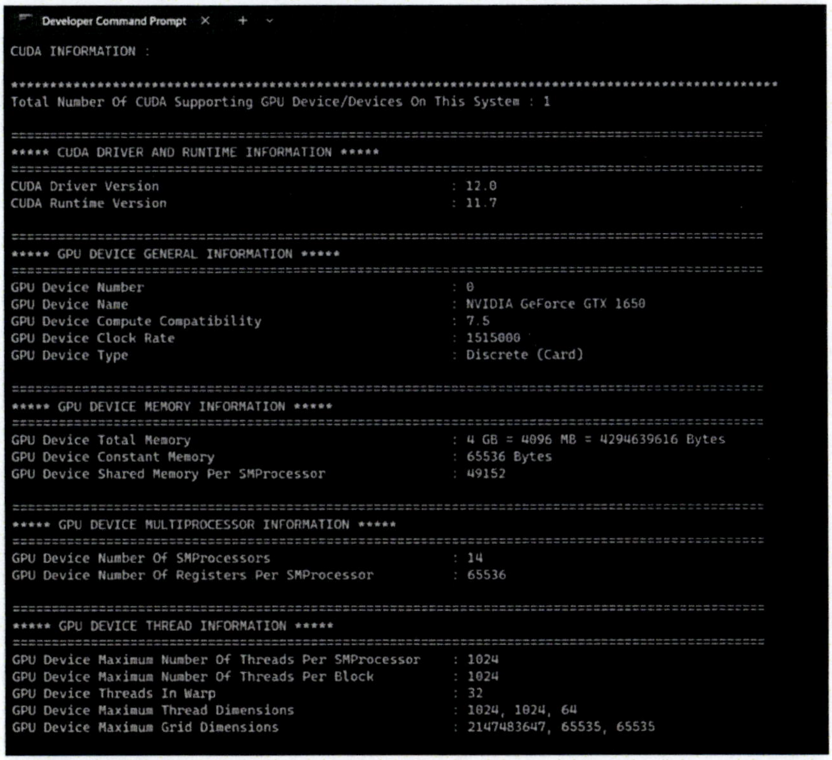

Fig. 5 CUDA device properties

b. GPU Device Constant Memory
c. GPU Device Shared Memory per SMProcessor

(5) GPU Device Multiprocessor Information

a. GPU Device Number of SMProcessors
b. GPU Device Number of Registers per SMProcessor

(6) GPU Device Thread Information

a. GPU Device Maximum Number of Threads Per SMProcessor
b. GPU Device Maximum Number of Threads Per Block
c. GPU Device Threads in Warp
d. GPU Device Maximum Thread Dimensions
e. GPU Device Maximum Grid Dimensions

(7) GPU Device Driver Information

a. Error Correcting Code (ECC) Support—Enabled/Disabled
b. GPU Device CUDA Driver Mode—Tesla Compute Cluster(TCC)/Windows Display Driver Model (WDDM).

(B) **Evaluation Result**

While comparing results with existing implementations, the proposed system includes 2 different performance benchmarks, one which compares the performance obtained on different CPUs and GPUs, thus specifying the need to use parallel computing and the second compares compute capability of different GPUs.

The calculation of the time taken for performing operations on the binary data is done using the "**helper_timer**" library offered by NVIDIA. This is achieved using the set of APIs—

a. sdkCreateTimer()—To create a timer pointer of type StopWatchInterface
b. sdkStartTimer()—To start the timer
c. sdkStopTimer()—To stop the timer
d. sdkGetTimerValue()—To get the timer value after the timer is stopped
e. sdkDeleteTimer()—To free the timer pointer.

Figure 6 depicts the program results obtained using 2048 threads. The time required to encrypt and decrypt the images is calculated and displayed in seconds. The number of threads passed to the application is modifiable and is passed as command line arguments to the program.

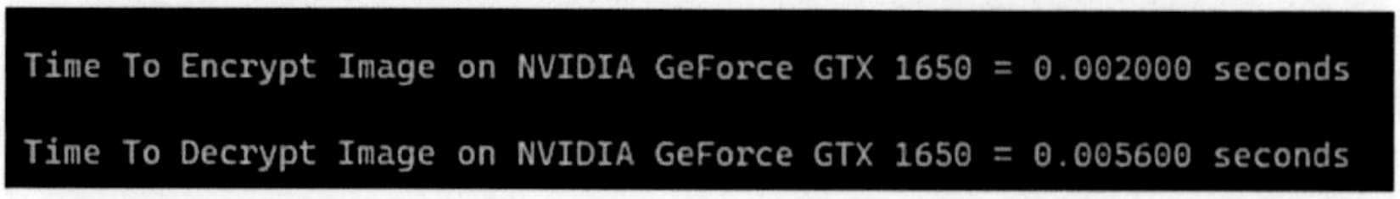

Fig. 6 Program execution using 2048 threads

Table 4 CPU and GPU performance comparison

Device	Sample size	Encryption time	Decryption time
AMD Ryzen 5 5600H	800 Kb	0.093	0.149
	3 Mb	0.373	0.588
	7 Mb	0.841	1.312
	50 Mb	6.185	9.847
	100 Mb	11.072	16.378
Intel i7-12700 K	800 Kb	0.071	0.097
	3 Mb	0.274	0.387
	7 Mb	0.598	0.876
	50 Mb	4.319	6.364
	100 Mb	7.542	11.1
Nvidia GeForce GTX 1650	800 Kb	0.0019	0.0053
	3 Mb	0.0018	0.0043
	7 Mb	0.0018	0.0042
	50 Mb	0.0017	0.0039
	100 Mb	0.0016	0.0036
Nvidia GeForce RTX 3060	800 Kb	0.0012	0.0044
	3 Mb	0.0012	0.0036
	7 Mb	0.0011	0.0035
	50 Mb	0.0011	0.0033
	100 Mb	0.0009	0.0029

Table 4 shows the different time values required to perform the encryption and decryption on various CPUs and GPUs.

a. Column 1—Represents the device on which the program is tested.
b. Column 2—Specifies the sample size. Samples are the bitmap images used for testing.
c. Column 3—Time required to encrypt the data, represented in seconds.
d. Column 4—Time required to decrypt the data, represented in seconds.

Figures 7 and 8 portray the time required for encryption and decryption on CPU and GPU for different sample sizes. According to the results specified in Table 4, as the size of input data increases the GPU takes less time to perform AES operations.

Table 5 depicts the different time values required to perform the encryption and decryption on various GPUs with variable threads. First column represents the name of the GPU, second column specifies the number of threads tested on that GPU. The third and fourth columns state the time required for encryption and decryption measured in seconds, respectively. This data is dynamic as the values can change over different runs. But overall, it gives the idea of performance capabilities of different NVIDIA GPUs.

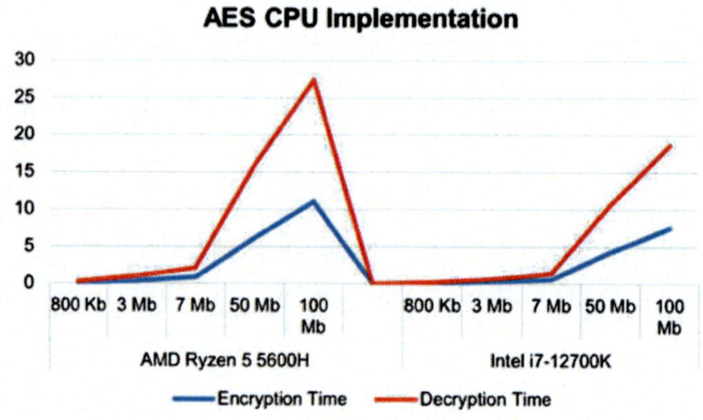

Fig. 7 CPU performance comparison

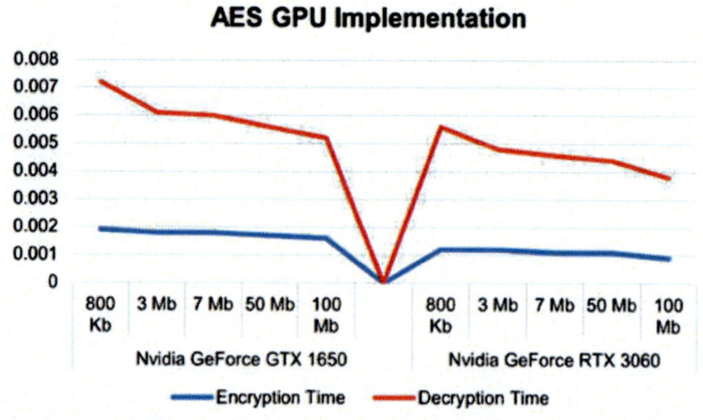

Fig. 8 GPU performance comparison

Table 5 GPU benchmarks

GPU	Threads	Encryption time	Decryption time
MX 450	1024	0.029	0.0284
	2048	0.0022	0.0124
	4096	0.0021	0.0095
GTX 1650	1024	0.0222	0.0122
	2048	0.0024	0.0056
	4096	0.0021	0.0053
GTX 1660 SUPER	1024	0.0145	0.0121
	2048	0.0015	0.0046
	4096	0.0014	0.0045
RTX 3060	1024	0.0115	0.0101
	2048	0.0008	0.003
	4096	0.0007	0.0028

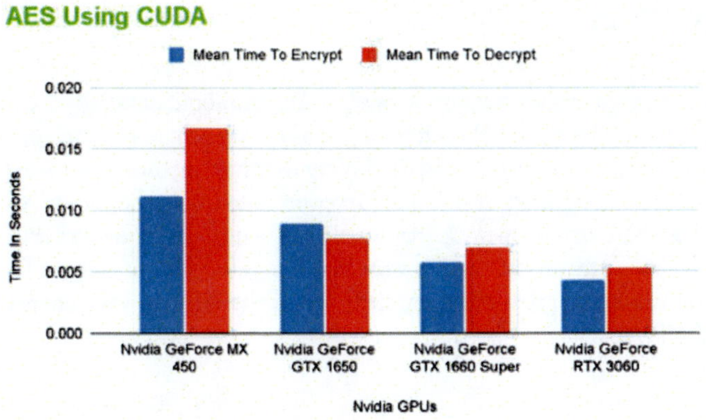

Fig. 9 GPU benchmarks

Figure 9 is used to represent the time and speedup factor by visualizing it. From the results, we can clearly see that, the more powerful the GPU, the less time required to complete the task. Here, the number of threads is also a crucial factor while determining the best GPU. For our testing, NVIDIA RTX 3060 was the best performer.

From the results, we can say that using CUDA extensively saves time and increases the throughput. This can be useful in hash algorithms as well, which can then be implemented in blockchain technology which will compute the hash of the block a lot faster. CUDA can also be used to conserve the amount of energy and power required to maintain the blockchain network. Basically, it will save time, resources and computational cost will be reduced to a great extent.

5 Conclusion

We proposed a method to parallelize the encryption and decryption processes in order to overcome the issue of high resource consumption in the traditional implementation of AES that would run on the CPU. We designed and implemented the AES encryption and decryption algorithm, which works on 128-bit, 192-bit, and 256-bit key sizes, to run on GPUs using CUDA, thereby reducing power consumption and increasing efficiency. This method provided a significant speedup over the CPU, providing high speed. This may change the way traditional resources are used, as these implementations can be used to encrypt binary data in all forms, including images and videos, as well as full disk encryption like Microsoft BitLocker. This process would require extreme fine-tuning to make such implementations a standard for other security techniques.

6 Future Scope

Currently, the proposed system presents a parallel implementation of the AES algorithm that can only be run on NVIDIA GPUs, as the presented research uses CUDA. This limits portability of testing and deploying to infrastructure using AMD or Intel GPUs, may it be integrated or discrete. To overcome this issue, we would need to develop a codebase using OpenCL that would allow us to cover every GPU and CPU device. But there are various parameters that are yet to be considered to optimize the algorithm to make proper and efficient use of the GPU to save energy and still produce similar results.

References

1. Yuan Y, He Z, Gong Z, Qiu W (2014) Acceleration of AES encryption with OpenCL. In: 2014 Ninth Asia joint conference on information security, pp 64–70
2. Jaiswal M, Kumari R, Singh I (2018) Analysis and implementation of parallel AES algorithm based on T-table using CUDA on the multicore GPU. IJCRT 6(1). ISSN: 2320-2882
3. Ma J, Chen X, Xu R, Shi J (2017) Implementation and evaluation of different parallel designs of AES using CUDA. In: 2017 IEEE second international conference on data science in cyberspace (DSC), pp 606–614
4. Abdelrahman AA, Fouad MM, Dahshan H, Mousa AM (2017) High performance CUDA AES implementation: a quantitative performance analysis approach. In: 2017 Computing conference, pp 1077–1085
5. An S, Seo SC (2020) Highly efficient implementation of block ciphers on graphic processing units for massively large data. NATO Adv Sci Inst Ser E Appl Sci 10(11):3711
6. Biryukov A, Großschädl J (2012) Cryptanalysis of the full AES using GPU-like special-purpose hardware. Fund Inform 114(3–4):221–237
7. Li Q, Zhong C, Zhao K, Mei X, Chu X (2012) Implementation and analysis of AES encryption on GPU. In: 2012 IEEE 14th international conference on high performance computing and communication & 2012 IEEE 9th international conference on embedded software and systems, pp 843–848
8. Iwai K, Kurokawa T, Nisikawa N (2010) AES encryption implementation on CUDA GPU and its analysis. In: 2010 First international conference on networking and computing, pp 209–214
9. Mei C, Jiang H, Jenness J (2010) CUDA-based AES parallelization with fine-tuned GPU memory utilization. In: 2010 IEEE international symposium on parallel & distributed processing, workshops and Phd forum (IPDPSW), pp 1–7
10. Jadhav S, Vanjale SB, Mane PB (2014) Illegal access point detection using clock skews method in wireless LAN. In: 2014 International conference on computing for sustainable global development (INDIACom). https://doi.org/10.1109/indiacom.2014.6828057
11. Stallings W (2022) Cryptography and network security: principles and practice. Pearson.
12. Daemen J, Rijmen V (2013) The design of Rijndael: AES—the advanced encryption standard. Springer Science & Business Media
13. Sanders J, Kandrot E (2010) CUDA by example: an introduction to general-purpose GPU programming. Addison-Wesley Professional
14. Wilt N (2020) The Cuda handbook: a comprehensive guide to Gpu Programming. Addison-Wesley Professional
15. Nguyen H, NVIDIA Corporation (2008) GPU gems 3. Addison-Wesley. https://developer.nvidia.com/gpugems/gpugems3/part-vi-gpu-computing/chapter-36-aes-encryption-and-decryption-gpu

16. Tezcan C (2021) Optimization of advanced encryption standard on graphics processing units. IEEE Access 9:67315–67326
17. Sanida T, Sideris A, Dasygenis M (2020) Accelerating the AES algorithm using OpenCL. In: 2020 9th International conference on modern circuits and systems technologies (MOCAST), pp 1–4
18. Bharadwaj B, Saira Banu J, Madiajagan M, Ghalib MR, Castillo O, Shankar A (2021) GPU-accelerated implementation of a genetically optimized image encryption algorithm. Soft Comput 25(22):14413–14428
19. Wang C, Chu X (2019) GPU accelerated AES algorithm. arXiv [cs.DC]. arXiv. http://arxiv.org/abs/1902.05234
20. Inampudi GR, Shyamala K, Ramachandram S (2018) Parallel implementation of cryptographic algorithm: AES using OpenCL on GPUs. In: 2018 2nd International conference on inventive systems and control (ICISC), pp 984–988

Identifying and Predicting Sinkhole Attacks for Low-Power and Lossy IoT Networks

Animesh Giri, Abhishek Goyal, Arpit Kogta, Priyansh Jain, and Pihoo Verma

Abstract Routing protocols in the Internet of Things (IoT) are quite vulnerable to attacks by design. There are various types of the routing attacks like the blackhole attack, selective forwarding, sinkhole attack, wormhole attack, and decreased rank attack. Other attacks such as version number modification, hello flooding attack, and Sybil attack are considered resource depletion attacks. These attacks can be designed in a way to corrupt information, reduce bandwidth, and threaten the integrity of the network which is why it is critical to identify the attacks and avoid as much damage as possible to the network. In this paper, we will be focusing on some of the attacks mentioned above. We have proposed a methodology to detect and predict IoT attacks. We propose different machine learning algorithms and conclude which algorithm is the most accurate in detecting every attack. The paper primarily focuses on dataset creation, followed by implementing various machine learning algorithms and identifying if a network is under attack and then what attack is taking place in that scenario.

Keywords Internet of Things (IoT) · Routing protocol · RPL attacks · Low-power lossy networks · Contiki OS · Cooja simulator · Sinkhole attack · Isolation attack

A. Giri
PES University, PES University Electronic City Campus, Hosur Rd, Konappana Agrahara, Electronic City, Bengaluru, Karnataka 560100, India
e-mail: animeshgiri@pes.edu

A. Goyal (✉) · A. Kogta · P. Jain · P. Verma
Department of Computer Science, PES University, PES University Electronic City Campus, Hosur Rd, Konappana Agrahara, Electronic City, Bengaluru, Karnataka 560100, India
e-mail: abhishek333goyal@gmail.com

G. Rajakumar et al. (eds.), *Intelligent Communication Technologies and Virtual Mobile Networks*, Lecture Notes on Data Engineering and Communications Technologies 171,
https://doi.org/10.1007/978-981-99-1767-9_24

1 Introduction

IoT, or the Internet of Things, refers to a network of interconnected devices, software services, and machines that plays a significant role in modern society by enabling energy-efficient automation and improving the quality of life.

Of late, there has been an increasing interest in IoT domains such as transportation, healthcare, and education in the industry. However, due to their ad-hoc and resource-constrained nature, IoT systems are susceptible to cyberattacks. To make certain the security of IoT systems, continuous monitoring and unexpected conditions, safeguard sensitive data, maintain continuity, and minimize potential losses. The abundance of sensing and network data produced by IoT devices and systems can be effectively analyzed using big data and machine learning methods for continuous monitoring and security of IoT systems.

1.1 Introduction to IoT LLN

The primary requirements of most of these setups are low-power and efficient networks. LLNs or low-powered lossy networks provide IoT networks provide for resource-constrained and size-constrained IoT devices to be interconnected in a hassle-free manner.

1.2 RPL in LLN

Proposed by the IETF for lossy IoT and low-powered networks, for low-powered and lossy networks (RPL), the IPv6 routing protocol provides IoT networks with structured and loop-free IPv6 routing. The implementation of RPL, however, is dependent on a highly modifiable objective function, whose parameters depend totally on the type of IoT application it is being used for. RPL, however, is vulnerable to a number of attacks, especially those targeting data transmission through the network. One such attack is the sinkhole attack.

1.3 Attacks in RPL

An asset refers to a crucial system resource, including information, processes, or physical resources. The loss or corruption of an asset negatively affects the system, particularly in the control plane where the asset encompasses information about the network, processes for managing and manipulating data, and the physical devices storing and processing the data. Losing or corrupting these assets can harm the control plane of the network.

Network security attacks are unlawful attempts to damage, alter, or steal sensitive data from private, public, or corporate IT assets. More businesses are allowing employees to access data via mobile devices, making networks more susceptible to data theft or network or data erasure entirely. An attempt to maliciously breach network security is hence called an attack.

RPL was created for the 6LoWPAN protocol in order to reduce the amount of power used when communicating across several intricate IoT devices. The smooth operation of the RPL protocol is essential since cost-effectiveness depends on the efficient use of energy. However, due to the RPL protocol's intrinsic complexity and the inadequate security of 6LoWPAN devices, it is susceptible to vulnerabilities from both inside and outside the network.

Various attacks such as sinkhole attacks, Sybil attacks, selective forwarding attacks, blackhole attacks, hello flooding attacks, wormhole attacks, rank attacks, and version number attacks have occurred while using RPL routing protocol in IoT.

1.4 Overview of a Sinkhole Attack in RPL

A sinkhole, in its simplest terms, is an IoT node having the intention to capture a lot of traffic flowing through the network, with the sole intent to drop data packets and reduce the end-to-end communication throughput.

1.5 Focus of the Paper

This paper mainly focuses on the following:

- To simulate sinkhole attacks by making appropriate code changes
- To predict a best-fit ML model for predicting sinkhole attacks in RP.

2 Related Work

In [1], an IDS system called CHA-IDS, which uses a compression header analyzer, has been developed to detect sinkhole attacks, hello flooding attacks, and wormhole attacks in RPL networks. The system uses 77 features and compares the performance of different machine learning classifiers, such as MLP, SVM, J48, NB, logistic, and RF. After evaluating the results, it appeared that J48 performed the best in this specific scenario. However, the study only examined one network topology and a small network of 8 nodes, so further research is needed to fully evaluate the potential of this ML-based IDS approach for IoT networks.

In [2], to detect RPL routing attacks, the authors utilized a deep learning technique with five hidden layers. They created datasets for decreased rank, hello flooding, and version number attacks using various network topologies. In the end, the results measured by F1-score were 94.7

In [3], Meysam proposed to design an anomaly based and distributed IDS based on RPL using a genetic algorithm as a feature selection and then to evaluate the same algorithm's accuracy using multiple experiments. The authors cover the multiple stages like data collection, preprocessing, i.e., removing useless values and making the data meaningful, making a compute classifier, applying the same, and finally, running the model for accuracy.

In [4], the authors warned that there might be certain RPL attacks which when combined can have a much more destructive tendency, more particular, the article focuses on how CloneID and sinkhole attacks together are more detrimental to network performance by raising malicious traffic and the network's typical power consumption. Above this, the authors suggested detection mechanisms for hybrid attacks over standalone attacks. Raza et al. [5] have developed SVELTE, a combination intrusion detection system that targets RPL routing attacks, such as sinkhole and selective forwarding attacks. The goal of SVELTE is to find a balance between the cost of storing signatures and the cost of computing anomalies. In SVELTE, the main operations and processes are handled by the border router, while the network's nodes host lightweight agents that send data to the border router for analysis. The simulation results indicated that SVELTE can detect all malicious nodes, but the true positive rate is not perfect and there were some false alarms during the detection process. In [6], the authors have designed an anomaly based distributed IDS architecture where detection agents are learned using algorithms such as SVM, Nave Bayes classifier, and decision trees, using the genetic algorithm as a feature selection mechanism to improve the accuracy of their classifier. They optimized the decision trees using the pruning method and genetic algorithm using selected appropriate features. The best-case accuracy was 99.35 percent.

3 Proposed Methodology

Our methodology to detect an attack scenario consists of several stages which are

1. Identifying commonly occurring attacks in RPL
2. Simulation of an attack in normal and attacked scenario
3. Capturing the radio messages transmission between the motes
4. Converting the packet capture to CSV using wireshark
5. Making raw data meaningful to generate it into meaningful labeled datasets
6. Statistical analysis of meaningful data
7. Analyzing the datasets performing different machine learning algorithms.

For attack simulation, various options were available such as NS3, Netsim, and Cooja but Cooja simulator was taken into use due to the functional GUI interface and

Contiki operating system (open source code). We made necessary code changes to simulate the behavior of a sinkhole attack for a group of sensor motes. A client–server architecture was followed in the DODAG wherein the root mote was compiled using udp-server.c file in the Contiki/example/ipv6/rpl-udp directory and the rest of the motes were compiled using UDP-client.c of the same directory. The code changes are explained as follows in Table 1 (Fig. 1).

The network with the z-motes is simulated for a certain period of time, and the Cooja simulator captures radio packet transmissions between the motes which are further used to generate the datasets. The .pcap file generated is processed into a comma-separated file making the use of wireshark packet analyzer, which contains rows with features such as source, destination, time and type of packet, protocol, and packet length for each individual packet. The dataset's features, however, are not suitable to make any inferences about the network and its malicious behavior. Hence, the raw dataset is preprocessed into a meaningful dataset. This is done by segmenting the raw data into one-second frames and computing the following values for 1-s frame duration:

1. Source Mote
2. Destination Mote
3. Packet Count
4. Source Mote Ratio
5. Destination Mote Ratio
6. Source Mote Duration
7. Destination Mote Duration
8. Total Packet Duration
9. Total Packet Length
10. Source Packet Ratio
11. Destination Packet Ratio
12. DIO Message Count
13. DIS Message Count
14. DAO Message Count
15. Other Message Count.

The input dataset was obtained and labeled as 0 or 1 based on whether the dataset is of the attacking or normal scenario. The final dataset contains the labeling accordingly. Then, for every 1-s duration post 60 s, the various parameters between every source and destination mote were obtained. During the simulation, it was observed that a system consisting of 10–12 nodes fully formed the DODAG structure after the 30th second. Due to the nature of RPL, when DODAG is occurring, devices will send DIO, DAO, and DAO-ACK messages to each other, and packet traffic will be different from the traffic after DODAG occurs. In order to prevent this difference from being learned by the machine, the data after the 60th second of the raw dataset was taken and the new dataset is created. The datasets created for each attack and classified as vulnerable-normal have become ready to be compared with different machine learning algorithms. Several machine learning algorithms were applied to predict

Table 1 Code changes for sinkhole attack

Location	Replace/Add	With
contiki-3.0/core/net/rpl/rpl-private.h	(Addition)	#defineRPI_CONF_MIN_HOPRANKINC
contiki-3.0/core/net/rpl/rpl-private.h	#defineRPL_MAX_RANKIC(7*RPL_MAX _RANKIC)	#define RPL_MAX_RANKIC 0
contiki-3.0/core/net/rpl/rpl-private.h	#define INFINITE_RANK 0xffff	#define INFINITE_RANK 256
contiki-3.0/core/net/rpl/rpl-timers.c	(Remove the line)rpl_recalculate_ranks();	
contiki-3.0/core/net/ipv6	UIP_STAT(+ + uip_stat.ip.forwarded);	UIP_STAT(+ + uip_stat.ip.drop);
contiki-3.0/core/net/ipv6	goto send;	goto drop;

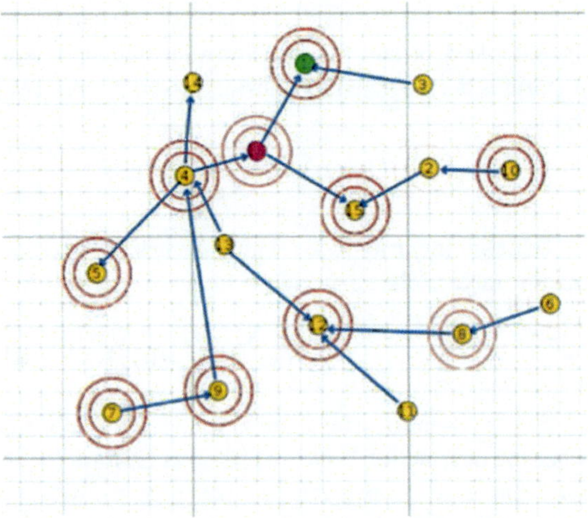

Fig. 1 Sinkhole attack simulation

the attacks such as logistic regression, decision trees, random forest classification, Naive Bayes, KNN, and SVM. The following has been explained below:

1. Logistic regression—Before explaining logistic regression, it is important to first discuss linear regression. Logistic regression is used to classify a dataset, which includes a dependent variable (y) and an independent variable (x), using a linear equation [10].

$$y_i = \beta_0 + \beta_1 x_i + e_i \tag{1}$$

For more than one linear variable, the equation would look like

$$y_i = \beta_0 + \beta_1 x_1 + \beta_1 x_2 + \beta_1 x_3 + \cdots + \beta_1 x_n + e_i \tag{2}$$

The probability of having y independent variable will be as shown in Eq. 3.

$$P(y) = \frac{1}{1 + \mathrm{e}^{-(\beta_0 + \beta_1 x_1 + \beta_1 x_2 + \beta_1 x_3 + \cdots + \beta_n x_n + \varepsilon_i)}} \tag{3}$$

2. Decision tree—In decision trees, firstly, the strongest feature in the dataset is determined, and then one of the branches is selected. This procedure is repeated until the searched classification target is reached [10]. The selected independent variables are defined as in Eq. 4.

$$\sigma = \langle i; t_k \rangle \tag{4}$$

The selection of the optimal threshold plays a crucial role in determining the performance of the tree, as it affects its structure. The goal is to minimize residual impurities by making the fewest possible divisions, resulting in a short path between the sample data and the classification outcome. Additionally, a measure of total impurities is calculated by considering both branches [10].

$$I(D, \sigma) = \frac{N\text{left}}{N_D} I(D\text{left}) + \frac{N\text{right}}{N_D} I(D\text{right}) \tag{5}$$

The Gini impurity index which measures the probability of misclassification if a random label is chosen using the branch's probability distribution is defined in Eq. 6.

$$I_{\text{Gini}} = \sum p(i|j)(1 - p(i|j)) \tag{6}$$

The cross-entropy measure is founded on information theory, it is at its highest when the distribution between classes is uniform (which is one of the worst scenarios for decision trees, as it implies that there are still many decision steps before final classification) and samples from one class are present in a split, while the values are null. [10] It is defined in the equation in Eq. 7.

$$I_{\text{Cross - Entropi}}(j) = -\sum_i p(i|j) \log p(i|j) \tag{7}$$

3. Random forest—A random forest is a collection of decision trees created from random samples with a different approach to splitting a node: instead of looking for the optimal option, a random subset of features (for each tree) is used to find the threshold that best separates the data. This results in many weaker decision trees, each producing a different estimate. There are two ways to interpret these results; the more common approach is based on majority voting (the class with the most votes is considered correct). Although they differ in theory, the probability mean of a trained random forest is unlikely to be vastly different from most of the predictions (otherwise there would be different fixed points); as a result, the two methods often produce similar results [10]. The concept of feature importance can also be applied to random forests as in Eq. 8 by calculating the average of all trees in the forest:

$$\text{Importance}(x_i) = \frac{1}{N_{\text{Trees}}} \sum \sum \frac{N_k}{N} \Delta I x_i \tag{8}$$

4. Naive Bayes—Naive Bayes is a family of powerful and easy-to-train classifiers that determine the probability of an outcome given a set of conditions using Bayes' theorem. [10] The Bayes theorem is defined in Eq. 8.

$$P(A|B) = \frac{P(B|A)P(A)}{P(B)} \tag{9}$$

5. KNN—In KNN, the nearest neighbors of each point are calculated by simple majority vote. Usually, this is the majority Euclidean distance.

$$|AB| = \sqrt{(x_2 - x_1)^2 + (y_2 - y_1)^2} \tag{10}$$

6. Artificial neural networks—It is capable of performing highly intricate classifications and very powerful algorithms. However, the training time for this algorithm is longer than for other algorithms and requires more processing power. They generally consist of three types of different layers—an input layer, a hidden layer, and an output layer as shown in fig x. Every neuron is connected to all neurons of the next layer with three weight matrices.

$$W_{ij}, H_{ij}, \text{ and } M_{ij}$$

$$\begin{cases} z_j^{\text{Input}} = W_{0j}X_0 + W_{1j}X_1 + \cdots + W_{nj}X_n = \sum_i W_{ij}X_i \\ z_j^{\text{Output}} = f_a^{\text{Hidden Layer}}\left(z_j^{\text{Input}} + b_j^{\text{Hidden Layer}}\right) \end{cases} \tag{11}$$

Likewise, the network output is calculated as in Eq. (12):

$$\begin{cases} y_k^{\text{Input}} = h_{0k}z_0^{\text{Output}} + h_{1k}z_1^{\text{Output}} + \cdots + h_{pk}z_p^{\text{Output}} = \sum_i h_{jk}z_j^{\text{Output}} \\ y_k^{\text{Output}} = f_a^{5\text{Output}}\left(y_k^{\text{Input}} + b_k^{\text{Output}}\right) \end{cases} \tag{12}$$

An error (loss) function is defined as the total square of error and this process is applied to all layers (Eq. 13).

$$L = \frac{1}{2}\sum_n \left\| \bar{y}_n^{\text{Predicted}} - \bar{y}_n^{\text{Target}} \right\|^2 \tag{13}$$

This function depends on all variables (weights and Bayesian values), then a backward calculation is performed. This process is the calculation of how much the change is, that is, the derivative function. For a single layer, this function is as in the equation. This equation is extended to all layers [10] (Figs. 2 and 3).

$$\frac{\partial L}{\partial h_{jk}} = \sum_n \frac{\partial f_a^{\text{Output}}}{\partial y_k^{\text{Input}}} \frac{\partial y_k^{\text{Input}}}{\partial h_{jk}} = \sum_n \partial_k z_j^{\text{Output}} \frac{\partial f_a^{\text{Output}}}{\partial y_k^{\text{Input}}} = \sum_n a_k z_j^{\text{Output}} \tag{14}$$

Here, the alpha term (proportional to the error delta) propagates back from the output layer to the hidden one. If there are many hidden layers, this procedure is

Fig. 2 Preprocessing of the dataset

→ input filname1="Attacker_meaningful+out"
→ input filename2="Normal_meaningful_out"
→ generate df1=comma separated value ← filename1
→ generate df2=comma separated value ← filename2
→ make both data frames of equal-length rows
→ concat both df1 and df2
→ x←slice 3-16 columns (important features)
→ y ← slice 16-17 column (label)
→ split dataset into 80% training and 20 % testing
→ normalize training and testing dataset

Fig. 3 Applying ML models in preprocessed dataset

→ Script for **Machine Learning classification**

- start_time←obtain initial time
- de←create an object
- train the data
- end_time←obtain end time
- MLCduration←calculate the duration (end_time-start_time)
- y_pred←predict data
- cm_mlc←create confusion matrix
- acc_mlc← calculate accuracy

repeated iteratively up to the first layer. The algorithm adopts the gradient descent method; so it iteratively updates the weights until convergence. Thus, machine learning is performed with the data in hand [10]. We used Python Script to implement and run a machine learning algorithm. The pseudo-code of the same has been mentioned below.

4 Results and Discussion

Cooja simulator has been employed to simulate the attacks. The top few attacks that are thought to be the most susceptible were duplicated.

Additionally, datasets are produced by producing the simulations of such attacks. Packets of legitimate simulations are mingled with malicious simulations to create a balanced dataset. The number of nodes in the simulations have never surpassed 12–15 due to hardware limitations.

Several approaches to data analysis and attack prediction were investigated for various attack types using the data gathered during various simulation scenarios [10] (Fig. 4; Table 2).

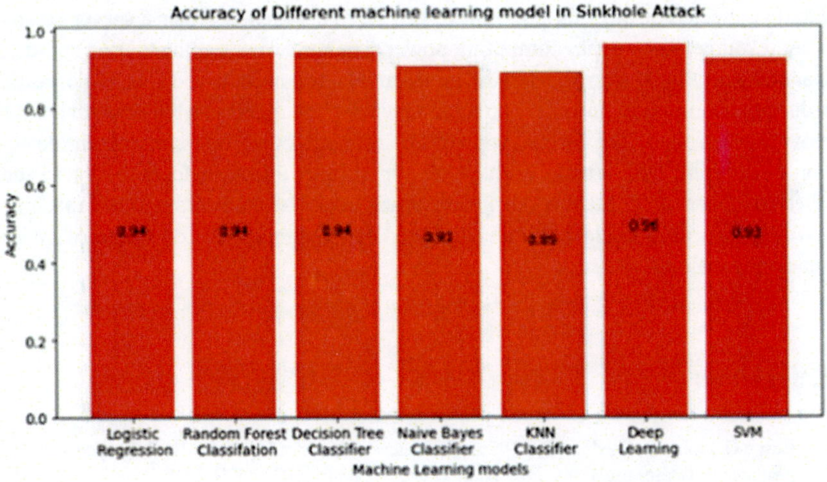

Fig. 4 Comparison of various machine learning algorithms for predicting sinkhole attack

Table 2 Comparison of various machine learning algorithms for predicting sinkhole attack

Name of ML model	Accuracy (%)
Random forest classifier	92.59
Support vector machines (SVM)	92.59
Logistic regression classifier	94.44
Decision tree classifier	94.44
k-nearest neighbors (KNN)	90.44
Gaussian Naive Bayes classifier	88
Deep learning model	96.9

The best accuracy was recorded for artificial neural network algorithm. However, the training time in the artificial neural network alone is longer than in other algorithms. It requires more processing. Hence, it is not a very feasible approach for detection. Other approaches like logistic regression and decision tree can be used since they require much less training time and processing scenario and are more feasible for attack detection for an IDS in an RPL network scenario [10].

5 Conclusion and Future Work

Six of the most common machine learning algorithms like KNN, decision tree, logistic regression, random forest, and SVM were applied to the sinkhole attack dataset, and ANN gave the best accuracy at around 96.9. Our solution in practice is supposed to work on a device with a good amount of computing power. So the data is collected at the sink and sent to this device which is either a server or any other resource with certain computing power. The next work can be to try a solution that can partially be computed in the motes and later be sent to the sink. This would reduce the computing power required and increase the speed at which the malicious mote can be found. The future scope of this research could be to implement detection of different RPL attacks such as version number attack, blackhole attack, and selective forwarding attacks. Also each attack could be analyzed and the machine learning algorithm could be more improved and optimized for coverage of major attack mitigation.

References

1. Eason G, Noble B, Sneddon IN (1955) On certain integrals of Lipschitz-Hankel type involving products of Bessel functions. Phil Trans Roy Soc London A247:529–551
2. Maxwell JC (1892) A treatise on electricity and magnetism, vol 2, 3rd edn. Clarendon, Oxford, pp 68–73
3. Yadollahzadeh Tabari M, Mataji Z (2021) Detecting sinkhole attack in rpl-based internet of things routing protocol. J AI Data Min 9(1):73–85. https://doi.org/10.22044/jadm.2020.9253.2060
4. Mirshahjafari SM, Ghahfarokhi BS (2019) Sinkhole+ CloneID: a hybrid attack on RPL performance and detection method. Inf Secur J Glob Persp 28(4–5):107–119. https://doi.org/10.1080/19393555.2019.1658829
5. Raza S et al (2013) SVELTE: real-time intrusion detection in the internet of things. Ad Hoc Netw 11(8):2661–274. , Elsevier BV. https://doi.org/10.1016/j.adhoc.2013.04.014
6. Deep learning for detection of routing attacks in the internet of things. Int J Comput Intell Syst
7. Algahtani F, Tryfonas T, Oikonomou G (2021) A reference implementation for RPL attacks using Contiki-NG and COOJA. In: 2021 17th International conference on distributed computing in sensor systems (DCOSS), Pafos, Cyprus, pp 280–286. https://doi.org/10.1109/DCOSS52077.2021.00053
8. Winter T, Thubert P, Brandt A, Hui J, Kelsey R, Levis P, Alexander R (2012) RPL: IPv6 routing protocol for low-power and lossy networks. Internet Engineering Task Force
9. Mayzaud A, Badonnel R, Chrisment I (2016) A taxonomy of attacks in RPL-based internet. Int J Netw Secur ACEEE Div Eng Netw, pp 459–473. https://hal.inria.fr/hal-01207859/document
10. Bonaccorso G (2017) Machine learning algorithms. Packt Pub- lishing Ltd., Birmingham
11. Verma A, Ranga V (2019) Addressing flooding attacks in IPv6-based low power and lossy networks. In: Proceedings of IEEE region conference (TENCON), pp 552–557. https://doi.org/10.1109/TENCON.2019.8929409
12. Ahmed HI, Nasr AA, Abdel-Mageid S, Aslan HK (2019) A survey of IoT security threats and defenses. Int J Adv Comput Res 9(45):325–350. https://doi.org/10.19101/IJACR.2019.940088
13. Hamid MA, Rashid M, Hong CS (2006) Routing security in sensor network: Hello flood attack and defense. In: Proceeding of IEEE ICNEWS, pp 2–4
14. Kasinathan P, Pastrone C, Spirito MA, Vinkovits M (2013) Denialof- service detection in 6LoWPAN based Internet of Things. In: Proceedings of IEEE 9th international conference on wireless and mobile computing, networking and communications (WiMob), pp 600–607

Innovative Online Ticketing Model on an Intelligent Public Blockchains

S. C. Prabanand, M. S. Thanabal, A. Durai Murugan, P. Mahalakshmi, S. Tarunbalaji, and T. Sriharish

Abstract Online Ticketing systems face the issue of eliminating ticket counterfeit and scalping while ensuring privacy protection and information openness. Another concern is ticketing fraud, when duplicate tickets allow unauthorised entry and cost hosts money. An Ethereum-based ticketing DApp would fix all the issues. Ticket holders may easily sell their tickets using the DApp. Ticket numbers are set, and each ticket is owned by a concert goer. When paying using an Ethereum wallet, the consumer gains ownership of the tickets. The ownership of tickets cannot be modified once changed, thereby preventing ticket fraud. When tickets are purchased, they become the purchasers' property. This paper proposes a hybrid online ticketing model based on the blockchain to address these problems. It makes need for blockchain technology to make ticketing data transparent and utilises encryption techniques to keep

S. C. Prabanand (✉) · S. Tarunbalaji
Department of Artificial Intelligence and Data Science, Bannari Amman Institute of Technology, Sathyamangalam, Tamil Nadu, India
e-mail: prabanandsc@bitsathy.ac.in

S. Tarunbalaji
e-mail: tarunbalajai.ad21@bitsathy.ac.in

M. S. Thanabal
Department of Computer Science and Engineering, PSNA College of Engineering and Technology, Dindigul, Tamil Nadu, India
e-mail: msthanabal@psnacet.edu.in

A. D. Murugan
Department of Computer Science and Business Systems, M. Kumarasamy College of Engineering, Karur, Tamil Nadu, India
e-mail: duraimurugana.csbs@mkce.ac.in

P. Mahalakshmi
Department of Computer Science and Engineering, Solamalai College of Engineering, Madurai, Tamil Nadu, India

T. Sriharish
Department of Computer Science and Engineering, Bannari Amman Institute of Technology, Sathyamangalam, Tamil Nadu, India
e-mail: sriharish.cs21@bitsathy.ac.in

G. Rajakumar et al. (eds.), *Intelligent Communication Technologies and Virtual Mobile Networks*, Lecture Notes on Data Engineering and Communications Technologies 171,
https://doi.org/10.1007/978-981-99-1767-9_25

user information private. It also employs the use of digital signature advanced technologies to assure ticket validity and includes a revolutionary ticket approval process to avoid digital piracy. An assessment of the system concludes with a description and analysis of the tests carried out during the installation of the system.

Keywords Cryptography · Smart contracts · Consensus · DApp

1 Introduction

Compared to conventional paper tickets, the rising usage of electronic ticketing systems has increased ticketing efficiency, decreased administrative costs, and lowered the cost of storing ticketing information [1]. Although there are still a number of issues to overcome, centralised ticketing systems lack transparency, consumer rights protection, data protection and privacy, and traceable legers. The decentralised nature of blockchain technology is used to construct ticketing systems to address these issues [2]. Autonomous online mechanisms for issuing and collecting tickets, however, have their own set of issues, including inefficiency, scalping of tickets, and assuring data security and privacy. To address these issues, this paper proposes a blockchain-based hybrid ticketing that is both transparent and traceable. Concerts, plays, and other large-audience productions are the exclusive subject of this paper. Flight, rail, bus, fast transit, and cinema tickets will not be mentioned in this article.

Electronic ticketing systems have not only made it easier to acquire and sell tickets, but they've also made managing and replacing lost or stolen tickets a thing of the past [1]. However, contemporary ticketing systems are plagued with flaws. Clients must have faith in centralised systems before they can use them. In centralised systems, a trustworthy third party generally supplies a central controller for data processing and storage, and delivering facility. Centralised computerised ticketing systems have these qualities. When it comes, for example, organisers only have access to ticket sales data from the firm that sold them. This information cannot be confirmed since it is difficult for them to verify its veracity. Consumers are also affected by this problem. A customer's ticket purchasing details cannot be verified. As a result, it is impossible for customers to determine whether or not the tickets they purchased are legitimate. For the centralised system to work, everyone involved has to have complete faith in the central controller. Second, ticketing systems must ensure that their customers' personal information is protected. Customer records, for example, may be helpful in the study of consumer behaviour. As a result, any use of the customer's private data must be approved by the consumer. Consumers, on the other hand, have a hard time establishing confidence because the centralised system is a mystery [22, 23]. Another critical concern is the safety of the ticketing systems [14–17]. There is just one location for data storage, maintenance, and processing in a central system. It's more susceptible than a decentralised system. Detecting data manipulation in a centralised database is tough. To add insult to injury, existing ticketing systems are plagued by ticket reselling since the tickets may be resold for several times their original price. There is uncertainty in the ticket market due to this kind of resale since no one gains but the intermediate ticket dealers [6, 8, 12].

There have been significant advances in decentralised solutions in order to address these issues. The use of blockchain technology allowed these solutions to address some of the drawbacks of traditional ticketing systems. There are still unresolved issues in the research literature. Research on blockchain-based ticketing systems is reviewed in this paper, as well as the issues that haven't been addressed in the literature [25, 26].

2 Problem Definition

There are a number of issues with online ticketing systems, as seen in Fig. 1, which is discussed in this paper.

In this paper we proposed to design an online ticketing system that overcomes the issues shown in Fig. 1. Safeguarding users (including consumers and organisers) safety while maintaining needed openness to varied user groups. Follow the steps below to achieve these goals.

1. How to create an unforgeable system of tickets?

Fraud tickets are the most common and serious issue facing any ticketing system. The objective is to create traceable tickets with unforgeable proof of authenticity. The user may check their ticket's details and establish its legitimacy.

2. To defend consumer rights, how to assure data transparency?

Transparency and tracking of tickets protect users' rights. Except for the centre controller, centralised ticketing systems are a mystery. Consumers and staff should be able to track and monitor public information such as times, locations, ticket and revocation agreements. This prohibits unilateral modification of public ticket information and protects public information records that may be utilised to safeguard

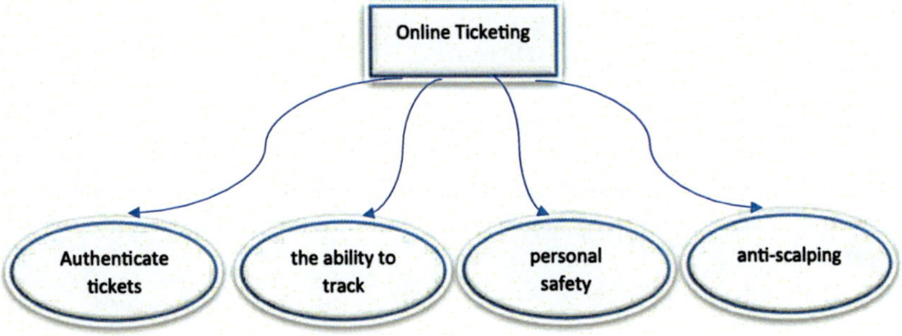

Fig. 1 The description of the issue

customers' rights and interests when a dispute arises. Similarly, all organisers should oversee ticket sales. It is verifiable and tamper-proof since all nodes execute each transaction.

3. How can you safeguard users' privacy?

User privacy is as important as availability of ticketing system information. A ticketing system should consider and protect the customer's identity. Less so, the ticket sales information. Consumers and participants should be segregated to address the privacy problem.

The ticketing system must not just that protect the consumer's genuine identity and also the tracking of their ticketing activity while making the ticket information clear and traceable. Encrypting ticket and purchaser identifying information helps preserve customers' privacy in a transparent ticketing system.

Private data, such as ticket sales data, should only be shared between attendees and not be available to the outside world. The answer to this challenge is to use several blockchain technologies to satisfy various criteria.

4. Scalping of tickets

Scalping of tickets (reselling) is a perennial issue in ticketing. Unlawful ticket resale may devastate and the ticket market. A verified ticket checking method before audience entry might avoid it. Ticket scalping may be prevented by requiring legitimate identification, such as identity cards, driving licences, or facial identity. Regardless, these approaches connect sold tickets to real-world identities and track purchases, compromising customer privacy. For this reason, a novel approach should be investigated.

3 Literature Review

This segment reviews and critiques alternative possibilities while introducing blockchain and other comparable technologies.

3.1 Bitcoin

An evaluation of different blockchain types (Hyperledger Fabric and Ethereum) and different blockchain platforms (Ethereum and Hyperledger Fabric) is presented in this section. Previous study on blockchain-based ticketing systems is discussed in this section.

3.1.1 Blockchain Origin

Bitcoin: A Peer-to-Peer Electronic Cash System was published by Nakamoto Satoshi in 2008. Nakamoto proposed a new decentralised peer-to-peer E-currency system that does not need third-party verification of payment transactions. A network transaction is validated before it is approved. The network checks the submitted transaction's history to see whether it's been spent. All legitimate transactions are hashed into a Merkle Tree [3]. Any further transactions from the same payer will be rejected. A blockchain is built using the previous block's hash. New blocks are delivered to all network nodes for verification and then added to a growing chain. It receives the lengthiest verified chain and attempts to extend it. A transaction recorded in a block cannot be changed. Faking a transaction involves regenerating it and all following blocks to catch up. A bulk of the network's processing capacity is required to build blocks via proof-of-work. Instead, Nakamoto (2008) built an incentive system within their network topology. A network node that produces a very new block gets rewarded. As a consequence, an attacker who controls the network's computing power owns all future coins. Being honest benefits the attacker.

3.1.2 Proof-Of-Work

A proof-of-work technique assures that fresh block creation decisions are based on computer power. Here we explore the Bitcoin blockchain's proof-of-work and block composition. Dai (1998) introduced the notion in his white paper b-money, but did not fully explain the implementation. H. Finney implemented the author's idea by solving a difficult hash problem in a network.

So, while trying to construct a new block, a node must keep generating nonces until it finds one that starts the block's hash value (made of the previous block's hash, transaction Merkle tree's root hash, and the nonce). The node with the most processing power will be the one to discover the nonce first using this method. To proceed, the node will broadcast the nonce and the freshly generated block to the network. The other nodes will verify the nonce upon getting a new block. They will abandon the existing block in order to include the new one into the main chain. Next, we'll try to find a different nonce that satisfies the requirements for the longer chain. Proof-of-work complexity governs the distance between blocks. The longer it takes for nodes to find the nonce, the more difficult the proof-of-work gets. A transaction can't be changed after it's recorded in a block. Figure 2 illustrates, in accordance with Nakamoto, a block on the Bitcoin blockchain.

There is a Merkle Tree and a Block Header [2] in the Bitcoin blockchain, as seen in Fig. 2. All three parts of the Block Header are included in this header: the Merkle Tree's initial hash as well as the previous hash [2]. The hash of the prior block is used to establish a relationship with the hash of the block before it. Proof-of-work is met by the nonce, which is a discovered nonce. A Merkle Tree's root is the Root Hash. Merkle trees are the result of all transactions in a block.

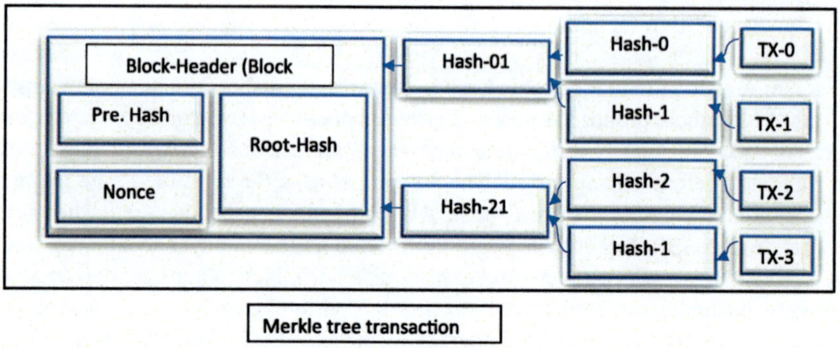

Fig. 2 Bitcoin Blockchain block composition (Nakamoto 2008)

3.1.3 Ticketing System Based on Blockchain Technology

The previous sections presented some blockchain-based ticketing research and literature. This section covers various studies and evaluates the remedies proposed as well as the issues that remain unresolved.

Existing blockchain-based ticketing systems fall into two groups. The permissioned blockchain is one kind, while the public blockchain is another [2] devised a solution to handle the problems of double-selling, false tickets, invalid ticket resale, and multiple organiser trust. The organisers hold the nodes in their solution. Using Hyperledger Fabric's permissioned blockchain technology, all transactions, including revenue distributions, are stored and processed in real time. Furthermore, all permissioned blockchain transactions are viewable and controlled by all nodes, ensuring revenue distribution process. In order to ensure payment security and ticket validity, Becker [2] created a smart contract-based mobile ticketing model. An app on your mobile device may save and display this QR code. It would be signed twice, once by the host to verify ticket validity, and once by the consumer upon admission to complete payment and prevent ticket theft. They also use a smart contract with the host's and the ticketing agency's digital signatures to preserve sales plan data and validate both parties' approval. Created permissioned blockchain smart contracts for tickets. In 2020, a permissioned blockchain-based ticketing system was developed to stop agents from utilising macros to book large blocks of tickets in bulk. Their approach is based on Hyperledger Fabric [15]. Only permitted users may join their own blockchain network, preventing customers from buying large tickets.

Each of the following remedies addressed a distinct problem. But they are flawed. Note that data in an authenticated blockchain network is only available to its nodes. Non-nodes cannot see what goes on within the permissioned blockchain [7] Participants in these permissioned blockchain-based ticketing systems are not nodes. Not everyone can host a network node. The consumer's right to know is therefore not guaranteed. Consumers may be unaware of network data changes. Conflicts do not safeguard consumers' rights and interests [3] did not mention scalping. A scalper may sell a ticket by giving the QR code to another individual. Ko et al. But scalpers may still acquire tickets and resell them outside the system at a greater price. Determining payment upon system arrival also creates considerable risks (2019). Bad actors may buy tickets but not show up. It costs the hosts a lot.

4 Architectural Design

For the sake of addressing the research issues previously stated, outlines the general architecture of the Blockchain-Based Hybrid Ticketing model. Explanatory diagrams and conceptual models of the system's structure are shown in a demonstration to demonstrate how the system works.

4.1 Overall Architecture

Figure 3 illustrates the overall architecture of the Blockchain-Based Hybrid Ticketing model.

Figure 3 Permissioned Blockchain, Public Blockchain, and the Online Ticket Verification System.

The design was created to use several blockchains to manage various data types in order to meet various openness and privacy requirements [4]. Component 2: Methods in the Public Blockchain Information that Must be Supervised by a Protected Body, such as evidence or contracts that cannot be altered, is encrypted and handled on the public blockchain. It manages information that must be transmitted throughout the many players. Rather of storing data directly on the public blockchain component, this work used [5] as a decentralised storage option. As a consequence, the public blockchain now only stores IPFS file addresses. The Web Service interface to the ticketing system will manage interactions between system parts [7] Confidential information is not processed by the Web Service. King and Nadal [18] Decrypting

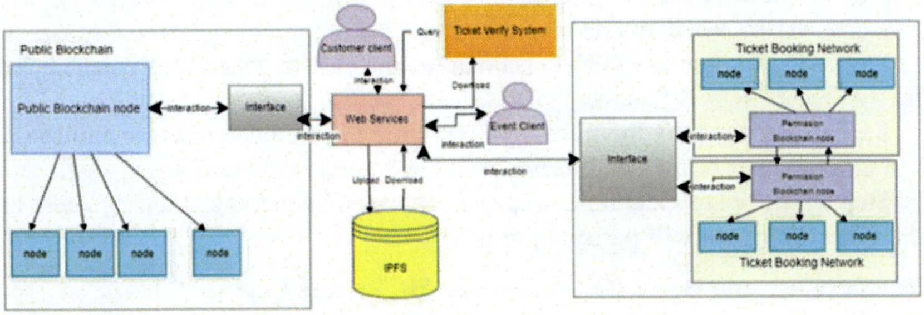

Fig. 3 A blockchain-based hybrid ticketing model architecture

sensitive information before sending it to a Web Service is a must. Also, the Web Service participates in the permissioned blockchain as a node. The Consumer Client allows consumers to purchase tickets, have their tickets authenticated, and request refunds. This service also protects users' sensitive data, such as their private key. Using the Client API, participants may register and connect to participating permissioned blockchain-based networks. The Client may also query data and the member's private details, such as network identification. It is intended to download and validate all tickets sold at the moment of entry [4].

4.1.1 Unforgeable Ticket Solution

In problem definition outlined a study aim to develop a ticketing system that eliminates bogus tickets. To accomplish this purpose, each ticket may be traceable, tamper-resistant, and carry irrefutable proof of authenticity. The traceability should enable consumers to check ticket details after purchase. This is required to verify the ticket's legitimacy. To verify the ticket's legitimacy, the information on the ticket must also be tamper-proof. It is possible to manufacture a fake ticket by forging information from a legitimate one. There are three questions that need to be asked to begin fixing these research issues [22].

This paper utilises the blockchain to handle and keep a record of each and every ticket sale. As a result, the blockchain will keep every node-generated block, allowing any transaction to be accessed and validated. Tickets being a unique commodity, all ticket transactions will be recorded on the blockchain, allowing for ticket record traceability. As discussed in literature review, blockchain may also address issue b the blockchain records transactions forever once they are placed into a block. It costs half the total processing power of all nodes for the moderator to construct a newest branch longer and faster than the public blockchain's original block chain [24].

To address question c, to prove ownership, the hash should be not modifiable and unique. If each ticket has a digital signature, it may assist keep track of here is how to create a ticket signature.

Step 1: Create a public/private key combination for each to preserve privacy and confirm identification.

Step 2: The ticketing model uses the public key, which is made available by the ticketing system, which in turn uses the private key.

Step 3: In response to a fresh ticket request, our system generates a unique ticket ID.

Step 4: Ask to verify the ticket by encrypting the provided ticket identification and the consumer ticket identification.

Step 5: The ticket's legitimacy may be confirmed by looking at the signature on the ticket (as evidence of purchase) as in Fig. 4.

Proof = Encrypt (Ticket ID + Consumer ID, Private Key)

The proof is used to establish the legitimacy of a ticket.

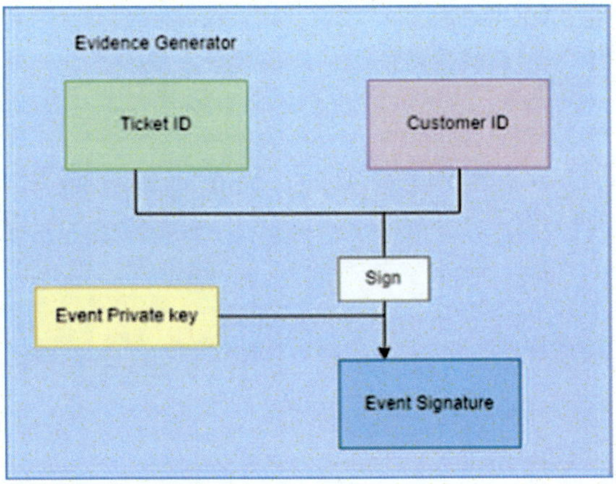

Fig. 4 A ticket's proof of purchase

Encrypt (A, K): Using key K, encrypt information A.

Ticket ID: A ticket's unique identifier.

Consumer ID: A consumer's unique identifier.

4.1.2 Solution to Transparency and Privacy

Mentioned in problem definition, the centralised system is an impenetrable black box for users, making consumer protection impossible. Consumers have no control over the system; therefore, the owner may update data without their awareness. As a consequence, customers struggle to gather proof to safeguard their rights. Because the product owner may change the data stored in the system, it is hard for consumers to keep their rights. Thus, customer engagement is entirely contingent on system owner confidence. To address this issue, this study proposes a decentralised blockchain-based ticketing system. Like centralised systems, a blockchain-based decentralised system cannot edit or erase data once it is recorded on the blockchain [16]. Every node also contains all blockchain transaction records. So long as a user is or has access to a blockchain node, they may query all transaction data, maintaining system transparency [13].

The next category is the manager's private and shared property, including the private key. This category's access control should enable transparency among all organisers and complete data security. This issue may be solved by encrypting all data and distributing the decryption procedure across all organisers. But the necessity is not simply storage, but also transactional usage of the data. All of these transactions should be visible to all involved organisers. But these transactions should be hidden from others. The content of these transactions should be encrypted if we use a public blockchain to complete them. Even if it's feasible, the public blockchain won't be able to carry out the transaction until the material is encrypted. Encryption and decryption

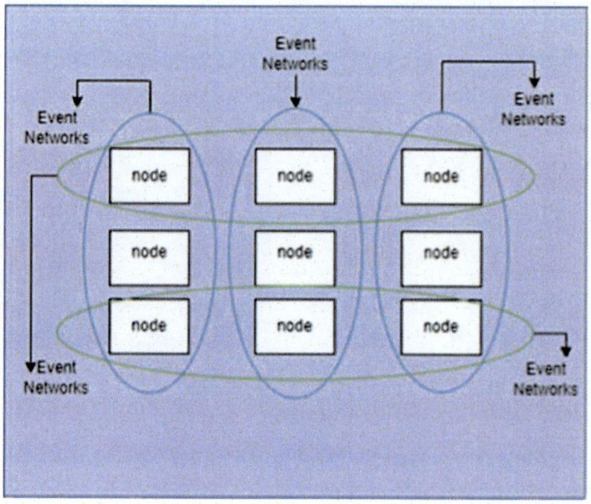

Fig. 5 The distributed ledger technology-based ticketing system

techniques use on-chain computer resources, which is costly for public blockchains. Given these constraints, employing a permissioned blockchain to store data and conduct transactions makes more sense. To become a node on the public chain, you must be part of the permissioned group [10]. In terms of ticketing, only the organisers may join the permissioned blockchain-based ticketing network. This ensures that all organisers have access to the confidential information and transactions. This ticketing system built on a private blockchain cannot be penetrated by anybody who is not authorised to do so, guaranteeing complete privacy. This image is a representation of the whole ticketing network based on permissioned blockchain technology in Fig. 5.

Aside from the aforesaid, data and transactions belonging to customers and organisers should also be examined. For example, ticket sales information should only be shared with the organisers. These latter details, along with related transactions like the creation of ticket record data, need to be recorded on the public blockchain. Ticket purchaser identities and other personal data should be encrypted before being processed or stored. When it comes to the latter, the details ought to stay inside the circle of organisers and not be broadcast. This information is subject to the same openness and privacy rules as the organisers and their shared property. Decentralised blockchains should store and process this data [9].

4.2 Blockchain-Powered Online Ticketing in Hybrid

With the use of these diagrams, this research presents a conceptual model of a Hybrid Blockchain-Based Ticketing System.

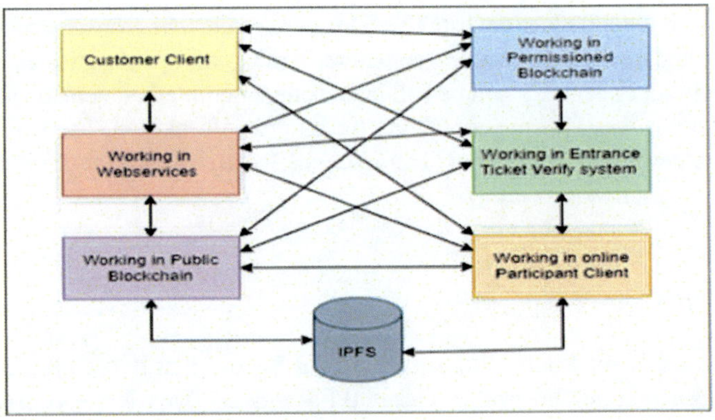

Fig. 6 Online ticketing system with hybrid blockchain

Figure 6 depicts the class diagram's conceptual structure. Each class in the diagram represents a system component. A class's colour correlates to the component it interacts with. Each line shows a class connection. This paper elaborates on each class's function in the following parts [11].

4.2.1 Customer Client Component Model Classes

Customer Ticket Record handles the Consumer Client component's ticket data structure. This includes the ticket's unique identifier, its record, its IPFS address, and its identity hash. Every single ticket is one of a kind. Authorised Group for Viewing Ticket Blockchain Record (The Permissioned Blockchain Ticket Record class will be introduced later). Ticket object's IPFS address (The Ticket Class will be introduced later). An incomplete identification number of the ticket holder is "salted" in this case. It re-creates a ticket's owner's ID.

The controller class for Customers and Clients. Five methods and an object of the consumer class Using the Request Create Ticket method, one may enquire about the availability of a ticket using the Web Service component of the ticketing system. The client's private key is used to send revocation requests to the service provider's server. An instance of the Consumer Ticket Record class may be queried using the Query Ticket function. Using the Validate Ticket Legitimacy method, a customer is required to verify the legitimacy of a ticket. A new instance of the consumer class is created [15].

4.2.2 IPFS Component Class

The Ticket Class specifies the IPFS ticket data structure. In addition to the IPFS address, the Ticket ID contains the Issue Date (Consumer ID). Each ticket is unique.

Each ticket buyer's ID is different. The ID is connected. The Consumer Public Key is used to validate the ticket buyer's signature. The signature authenticates the ticket. The ticket holder's identity will be validated using encrypted ciphertext. The signed IPFS address is the IPFS file containing the party's information. The IPFS component's ticket revocation agreement file's address. The issue date is the ticket's issuance date.

4.2.3 Public Blockchain Class

A model class is the Public Blockchain Ticket Record that defines the ticket data record data structure. Encrypted Ticket IPFS address–This is the address of the encrypted ticket. Each Ticket Class object has its own ID. Ticket IPFS Encrypted Ciphertext is the ciphertext of a ticket file encrypted by public key. ESET IPFS is a digital signature. It's created by encrypting an IPFS ticket with private key.

The information architecture of the Public Blockchain module is reflected in the Public Blockchain Record class of the corresponding model. They are a valid photo ID, a signed IPFS address, and an IPFS address. Each Class object has a unique ID. The IPFS address property specifies the location of IPFS file containing data such as location, entry time, and description a ciphertext of the IPFS address. It acts as signature to verify the IPFS address [19].

4.2.4 Web Services Classes

The Class is a model class that governs the Web application component's data structure. There's also a public address, a signed Ticket Agreement Address, and a Permissioned Blockchain Identity. One for each. The public key of a key pair. It encrypts data and verifies signatures. A Public Blockchain Record object. It enables Web Service communication with the permissioned blockchain network. The Web Service's Public Blockchain Identification is used to public blockchain data uploading and querying. The IPFS component's Contract for Purchase of Tickets, Signed file's private key encrypted address. The IPFS component's ticket revoke agreement file address.

Data from the Consumer Client component is used to build the Ticket Class data. It contains the purchaser's unique identifier, the public key, and the ticket's owner's unique identifier and hash value. The consumer wants the ticket. It uploads them. It's a salted text hash value that identifies the prospective ticket owner. This is an IPFS address with a Ticket ID. The consumer and private keys sign it. This is seen in Fig. 7 Unique ticket ID (ID).

Signed Ticket Agreement = Cpri(Epri(IPFS address of the Agreement File) + Ticket ID).

Epri(A): Encrypting A using the private key.

Cpri(A): Encrypting A using Consumer's private key.

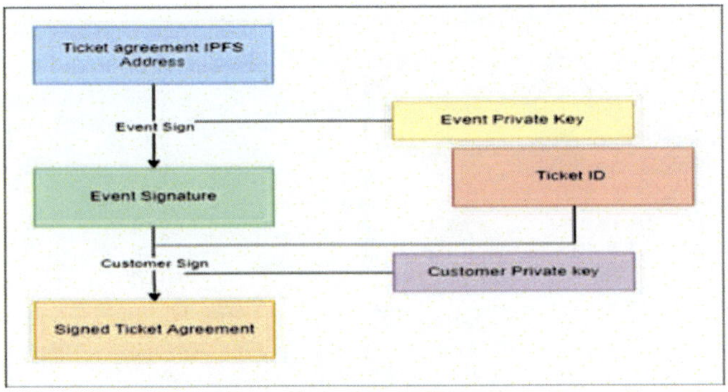

Fig. 7 Creation of a Ticket Agreement with a Signed Contract

Ticket Class has two methods: Make a payment and get your ticket via check. Financial Transaction Conducted Via Check option verifies ticket payment. As stated in the problem statement, this research does not offer a complete payment processing and verification solution. A Ticket Class object is created utilising Consumer-generated input and persistent information. You are placed in the Ticket Class.

Generate Ticket ID is a controller class with the same name function. It creates an ID for a Ticket Class object. An ID is given.

Encrypt Ticket Owner Identity method in Controller class. Class is the public key. With this function you may encrypt your ticket owner's identity using your public key.

The IPFS Uploader Class contains a ticket property and an upload method. Ticket object class. Upload to IPFS uploads a Ticket Class object as a file to IPFS. The IPFS address is returned.

A controller class property of the Generate Public Blockchain Ticket Record Class is the IPFS address. The location on IPFS where the ticket file for an item of type IPFS Uploader Class may be found. A Public Blockchain Ticket Record Class object is generated by this procedure [21].

4.2.5 Client Class

The Participant Class keeps a record of an identifier in the Client component. A property-rich blockchain node. It's a thing. The Permissioned Blockchain Nodes list includes node addresses.

A user record in the Client component is described by the User Class. It has user ID, procedures, and add/remove properties. Each user is identified by their User ID property. An array of Participant class objects identifies a user's permissioned blockchain networks. Creates a Participant class object. Removes a Participant class object from a variable.

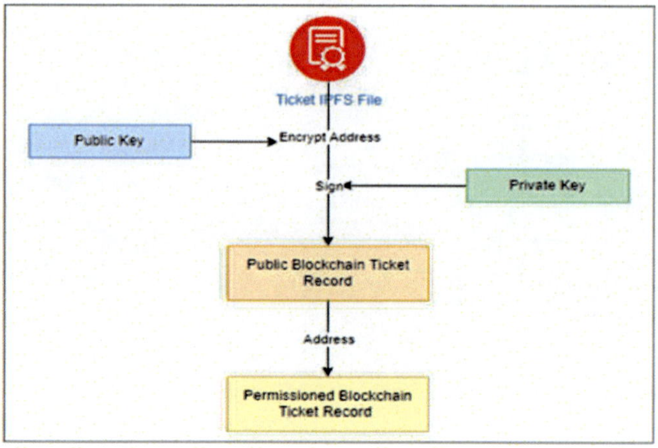

Fig. 8 Shows a novel blockchain-based booking system

As indicated above, a ticket has numerous components. This paper shows how the envisaged system ensures component ticket record consistency. Figure 8 depicts the proposed system's ticket record.

An IPFS file, a Public Blockchain Ticket Record, and a Permissioned Blockchain Ticket Record store a ticket. It's stored in an IPFS file. To connect a public blockchain transaction to an IPFS ticket file, the IPFS file address is encrypted and signed. The public blockchain transaction's address is retained. By linking these three ticket records together, the intended system ensures ticket record consistency.

4.2.6 The Method of Making Tickets

How a new ticket is made in the future system is detailed here. Tickets are generated by a collaborative effort between the Consumer Client, the Web Service, the Permissioned Blockchain, the Public Blockchain, the IPFS, and the Consumer User as an actor. To produce a ticket, the following describes how the desired system interacts with these parts.

Step 1: The Consumer Client utilises the Verify Ticket Authenticity operation to make sure that a ticket is legitimate (in the Consumer Client class). The Query Ticket method of the Consumer Client class is used to send a request to the Web Service component with a ticket ID and an event ID.

Step 2: After receiving the request, the Web Service then invokes the Check If Ticket Is Revoked procedure (in the Check If Ticket Is Revoked class).

(a) To find out more about a specific event, you may use the Query Event method and its associated class object by entering the event's ID (in the Query Event class).

(b) If the ticket has been revoked, you can find out by using the Permissioned Blockchain component's Check If Ticket Is Revoked method and passing in the ticket's ID.
(c) Inform the Web Service component, which notifies the Consumer Client component, which notifies the Consumer User, if the ticket is revoked. Just tell the Web Service component that the ticket is invalid and go on. This is a continuation of the Consumer Client.

Step 3: The Public Blockchain Ticket Record class object is obtained through a consumer client query to the smart contract storing the target ticket record.

Step 4: The third step involves the Consumer Client checking the class object of the Public Blockchain Ticket Record acquired in Step 2.

(a) Invoke the Query Event Information method from the Web Service component to get the event's public key.
(b) Public Blockchain Ticket Record class object's encrypted Ticket IPFS address. Tell the customer that the ticket is false if the information doesn't match. The next step is to see whether they are the same.

Step 5: The Consumer Client component makes a request to the Web Service component to decrypt the Encrypted Ticket IPFS address of an object of type Public Blockchain Ticket Record.

(b) When an IPFS ticket file is used, the Web Service component establishes a connection to the appropriate event network and calls the Decrypt method on the Permissioned Blockchain component to reveal the encrypted address.

Step 6: An object of type Ticket is stored in an IPFS ticket file, which the Consumer Client component downloads in Step 5.

Step 7: The event signature of the Ticket Class object is verified by the Consumer Client module. The event signature may be unlocked by using the event public key.

Object of the Ticket Class with the decrypted data replaced in its content. If they don't match, you must inform the buyer that the ticket is fraudulent. Continue if they are consistent.

Step 8: The Consumer Client module performs a cross-reference check using the ticket's partial identifier and the Ticket Class object's ticket identifying evidence.

4.2.7 Process for Verifying the Authenticity of a Ticket

This section describes how consumers may authenticate their tickets. The ticket authenticity verification method involves five components and an actor: Consumer Client, Web Service, Permissioned Blockchain, Public Blockchain, IPFS, in addition to the role of the consumer user. In order to determine whether a ticket is legitimate, the following steps must be taken.

Step 1: The Arrival Ticket Verify System employs the Download technique to retrieve tickets (in the Arrival Ticket Verify class). Using an event network participant identity, the Arrival Ticket Verify System component retrieves the Tickets List class object from the Permissioned Blockchain component.

Step 2: Using the Public Ticket Address, the Arrival Ticket Verify System accesses the warrant ticket data from the Public Blockchain component's smart contract. Get a Copy of a Ticket's Blockchain Transaction History.

This part ensures the provided Public Blockchain Ticket Record class object has the correct event signature. The event signature may be unlocked by using the event public key.

(b) Check the decrypted data against the IPFS address that was returned in the encrypted ticket. Otherwise, please have the ticket number printed. The next step is to see whether they are the same.

Step 5: The part of the Ticket Verification System that deals with arriving passengers requests the IPFS ticket file at the specified return address. Ownership Identity Verification through Permissioned Blockchain's Decrypt Method and the Ticket Class Object (in the Decrypt by Event Private Key class).

In order to securely store the Ticket and the encrypted Ticket Owner Identity Proof, the Arrival Ticket Verify System creates an object of the Arrival Ticket Record class with the characteristics Ticket Record and Ticket Owner Identity Ciphertext.

Checking arrival tickets:

One module of the Arrival Ticket Verification System scans the QR code and the owner's partial identifying number to determine their identity and the salting information for the ticket.

When a passenger arrives, the Verify Ticket system function is used to begin checking tickets. Ticket Owner Identity Hash Value is generated by hashing the salted text by the Arrival Ticket Verify System component.

As soon as a new Owner Identity Hash Value is generated, the Arrival Ticket Verify System component checks its local storage to see whether it already has the value. Recognise a pass message and open the door for the ticket bearer. If the answer is no, then the ticket holder will be denied access.

5 Conclusion

This paper proposes a hybrid blockchain-based ticketing system that avoids scalping, counterfeit tickets, and privacy protection. In terms of transparency, privacy, and information exchange, the ticketing system is complex. So, data from the system should be categorised by user groups. And data should be housed on separate infrastructures. It has permissioned and public blockchains. The permissioned blockchain stores private ticketing data that should only be shared by attendees. The public blockchain stores public data like ticket purchase agreements and public details. Tickets may also be monitored via the public blockchain. The outcome is fixed. The Public Blockchain Ticket Record might safeguard consumer rights in the case of a dispute. As an alternative to maintaining full ticket details on the public blockchain, this approach makes use of IPFS files [4] and simply saves addresses to these files.

This study also creates a digital signature system for each ticket to guarantee its genuineness and check its validity with public records. Each ticket's private information, such as name and address, is encrypted to prevent unauthorised access. Permissioned and public blockchains are linked through a web service component to ensure data integrity. All sensitive information is secured before being uploaded using the web service module. Tickets are checked for scalping, and personal information is kept secure. The proposed solution will increase the difficulty, as well as the potential consequences, of ticket scalping, while also protecting the identity of the individual doing the purchasing. The ticket's owner is validated using the hash value of the salted partial consumer real ID number. To build a fresh identity evidence, consumers keep salting information until they submit a partial real identification number during ticket verification. The system saves just the ticket owner's identity verification data, and the consumer's genuine identity cannot be identified without the salting information. Abstract system design technology does not restrict it in any way.

5.1 Future Work

The issues are addressed by the adaptable ticketing system based on the blockchain technology. Some issues will need to be addressed in future study.

Ticket generating concurrency must be improved. Due to the blockchain-based storage structure, high concurrency ticket production is not possible. Reduce the cost of data storage. The system's particular implementation may be analysed. System architecture is abstract. Future research might compare and analyse specific implementation strategies to determine the best fit for various application situations. Experiments must be done under realistic conditions. In a real application, various components and nodes are deployed. They communicate remotely. The delay induced by the Internet connection might considerably alter the findings. It is necessary to examine the whole ticket creation life cycle. Due to budget and time constraints, this research did not test the whole ticket generating process. Larger data sets are needed to properly portray real-world circumstances. To shield your work from outside influence, use a cloud-based virtual machine or a cluster of computers.

References

1. Aventus Protocal Foundation (2018) A blockchain-based ticketing protocol [White paper]. Retrieved 18 Mar 2019, from Aventus: https://3.9.12.207/wp-content/uploads/2019/03/Whitepaper.pdf
2. Becker G (2008) Merkle signature schemes, merkle trees and their cryptanalysis. Ruhr-University Bochum, Tech. Rep.
3. Bell J (2005) Ticket scalping: same old problem with a brand-new twist. Loy Cons L Rev 18:435

4. Benet J (2014) Ipfs-content addressed, versioned, p2p file system. arXiv preprint arXiv:1407.3561
5. Buterin V (2016) Ethereum platform review: opportunities and challenges for private and consortium blockchains. Retrieved 2 Mar 2021, from internet: http://www.smallake.kr/wp-content/uploads/2016/06/314477721-Ethereum-Platform-Review-Opportunities-and-Challenges-for-Private-and-Consortium-Blockchains.pdf
6. Cachin C (2016) Architecture of the hyperledger blockchain fabric. In: Workshop on distributed cryptocurrencies and consensus ledgers, vol 310, no 4
7. Castro M, Liskov B (1999) Practical Byzantine fault tolerance. In: OSDI, vol 99, no 1999, pp 173–186
8. Cha SC, Peng WC, Hsu TY, Chang CL, Li SW (2018) A blockchain-based privacy preserving ticketing service. In: 2018 IEEE 7th global conference on consumer electronics (GCCE). IEEE, pp 585–587
9. Chung K, Yoo H, Choe D, Jung H (2019) Blockchain network-based topic mining process for cognitive manufacturing. Wireless Pers Commun 105(2):583–597
10. Diffie W, Hellman M (1976) New directions in cryptography. IEEE Trans Inf Theory 22(6):644–654
11. Eastlake D, Jones P (2001) US secure hash algorithm 1 (SHA1). Retrieved 2 Mar 2021, from internet: https://www.hjp.at/(st_a)/doc/rfc/rfc3174.html
12. Finney H (2004) Rpow-reusable proofs of work. Retrieved 2 Mar 2021, from internet: https://cryptome.rg/rpow.htm
13. Finžgar L, Trebar M (2011). Use of NFC and QR code identification in an electronic ticket system for public transport. In: SoftCOM 2011, 19th international conference on software, telecommunications and computer networks. IEEE, pp 1–6
14. GET Foundation Team (2017) Guaranteed arrival token smart ticketing protocol [White paper]. Retrieved 18 Mar 2019, from GUTS Tickets: https://guts.tickets/files/GET-Whitepaper-GUTS-Tickets-latest.pdf
15. Göbel J, Krzesinski AE (2017) Increased block size and Bitcoin blockchain dynamics. In: 2017 27th International telecommunication networks and applications conference (ITNAC). IEEE, pp 1–6
16. Isaksson C, Elmgren G (2018) A ticket to blockchains. Retrieved 2 Mar 2021, from internet: https://www.diva-portal.org/smash/get/diva2:1282090/FULLTEXT02
17. Jeppsson A, Olsson O (2017) Blockchains as a solution for traceability and transparency. Retrieved 2 Mar 2021, from internet: https://lup.lub.lu.se/student-papers/search/publication/8919957
18. King S, Nadal S (2012) Ppcoin: Peer-to-peer crypto-currency with proof-of-stake. self-published paper, August, 19, 1. Retrieved 2 Mar 2021, from internet: https://www.chainwhy.com/upload/default/20180619/126a057fef926dc286accb372da46955.pdf
19. Ko DH, Choi HK, Kim KS (2020) A design and implementation of macro pr ion ticket booking system using blockchain. In: Proceedings of the 2020, the 6th international conference on e-business and applications, pp 95–98
20. J. Rajalakshmi, Ranjani SS, Sugitha G, Prabanand SC (2022) Electroencephalogram data analysed through the lens of machine learning to detect signs of epilepsy. In: 2022 4th International conference on inventive research in computing applications (ICIRCA), Coimbatore, India, pp 898–902. https://doi.org/10.1109/ICIRCA54612.2022.9985641
21. Kuznetsov A, Svatovskij I, Kiyan N, Pushkar'ov A (2017) Code-based public-key cryptosystems for the post-quantum period. In: 2017 4th International scientific-practical conference problems of infocommunications. Science and Technology (PIC S&T). IEEE, pp 125–130
22. Lamport L (1983) The weak Byzantine general's problem. J ACM (JACM) 30(3):668–676
23. Larimer D (2014) Delegated proof-of-stake (dpos). Bitshare whitepaper, 81, 85. Retrieved 2 Mar 2021, from internet: https://whitepaper.io/document/388/bitshares-whitepaper

24. Leach P, Mealling M, Salz R (2005) A universally unique identifier (uuid) urn namespace. Retrieved 2 Mar 2021, from internet: https://www.hjp.at/doc/rfc/rfc4122.html
25. Leslie L (1998) The part-time parliament. ACM Trans Comput Syst 16(2):133–169
26. Lin KP, Chang YW, Wei ZH, Shen CY, Chang MY (2019) A smart contract-based mobile ticketing system with multi-signature and blockchain. In: 2019 IEEE 8th global conference on consumer electronics (GCCE). IEEE, pp 231–232

Machine Learning for a Payment Security Evaluation System for Mobile Networks

Solleti Ramana, N. Bhaskar, M. V. Ramana Murthy, and M. Raghavender Sharma

Abstract Consumers can now access services using a wide range of handheld devices due to the recent explosive expansion of various mobile network payment gateways. The method used to assess the security of mobile network payment gateways covers complex, complex situations such as malware detection, two-factor authentication, and fraudulent payment system detection. To help with mobile transaction fraud detection, virus detection, and authentication issues, this work recommends using the Secure Mobile Electronic Payment Framework Enhanced by Machine Learning.

Keywords Transfer learning · Machine learning · Naive Bayes · DenseNet · Cloud security

1 Introduction

The massive development of wireless networks is generating a record-high rise in mobile devices. However, smartphone users are all still seriously challenged by financial malware, malware, spyware, and malicious applications. Modern smartphones may be used for e-commerce, personal transactions, and social media beyond simple communication. The methods now being used for malware detection are usually split into three divisions, with many practical approaches having been developed. Network-level behavior detection, dynamic system-level behavior analysis, and static analysis. Techniques for identifying malicious apps usually use static

S. Ramana (✉)
Department of Computer Science, Bhavan's Vivekananda College, Saini-Kpuri, Secunderabad, Telangana, India
e-mail: ramana_msc008@yahoo.co.in

N. Bhaskar
Department of CSE, CBIT, Hyderabad, Telangana, India

M. V. R. Murthy · M. R. Sharma
Department of Statistics, Osmania University, Hyderabad, India

G. Rajakumar et al. (eds.), *Intelligent Communication Technologies and Virtual Mobile Networks*, Lecture Notes on Data Engineering and Communications Technologies 171,
https://doi.org/10.1007/978-981-99-1767-9_26

analysis and dynamic system approaches [1]. Static analysis, based on semantic signatures and focused on looking at code samples without running them, employs reverse-engineering techniques to analyze programmers' code. To determine if an app displays malicious activity, it collects static data from malware attacks, such as binary sequences, callback sequences, authorization request sequences, and any other static information.

Although masking methods can more readily circumvent these strategies, they are more resilient to changes in malware development [2]. The dynamic system-level evaluation concentrates on gathering accurate data from the execution traces of malicious applications. Dynamic analysis has the disadvantage that it requires the applications to be executed in a controlled environment, which might also affect malware behavior, making it challenging to identify the specific programming location where such activity happened. We initially collect Mobile malware activity during the initial minutes for use as training and testing data [3] to circumvent these issues. Then, we detect harmful activity using various well-known machine learning models, including SVM, C4.5, and Naive Bayes. The study's results suggest that network traces and the selected criteria may be used to identify some malicious behaviors. [4]. They also highlight the limitations of traditional machine learning approaches in dealing with data sets.

2 Related Work

Researchers work hard to accurately pinpoint malicious network activity and identify the related malicious software. One research thoroughly analyzes the Internet infrastructure for mobile virus detection. The links between the network and user activity are examined in other studies, which is a typical method for identifying anomalies [5]. It offers a technique for clustering behavioral malware just at the network level. Several studies have also looked at using machine learning to detect malicious apps. Fortunately, there are excellent and efficient solutions to safeguard the information on your mobile device [6]. Data storage preservation and transfer on smartphones, laptops, and other mobile devices are called mobile Safety. The phrase "mobile security" refers to a broad category that covers anything from malware protection to smartphones.

Furthermore, it decreases the chance that mobile devices will escape theft, unauthorized or accidental loss, and their data will be secured. "Contactless payments" refers to a customer's transaction using their smartphone. This device converts the image data from standard and botnet network traffic into a robust convolutional neural network called DenseNet [7]. The effectiveness of each other's simulation may be increased by the outcomes of transfer learning from 33.41% to 99.98%. Afterward, a security strategy construct based on learning is dynamically recommended to be transferred using the security strategy update approach. The outcomes of the security plan formulation and network performance simulation confirm the viability and efficacy of the suggested security feature [8].

The methodology allows you to take advantage of the data gathered from numerous IoT devices but only includes some. The experimental results demonstrated that, compared to the basic in-depth training technique, the proposed DTL model significantly increases IoT attack precision [9].

3 Machine Learning Smart Payment Framework

The framework necessary for the efficient operation of companies that are engaged in the development of mobile handsets is offered by the mobile infrastructure. The Global System for Mobile and 2-Generation, one of the most widely used mobile networks, was established with various goals. Globe roaming, increased voice, quicker data speeds, fraud protection, and collecting data [10]. The development of payment options significantly impacts the standard of living. New payment methods are now presenting future possibilities and challenges. Contactless payment became increasingly complicated as the process became more accessible and quicker, especially in dealers where throughput was the most crucial component.

The area covered by an emitter is shown in Fig. 1 as the primary physical component of a cellular network. In its simplest form, a mobile phone communicates with an access point through radios. A base station connects and manages multiple base stations [11]. Every wireless connection can serve as the communication path between a base station and its controllers.

The mobile switching center was interconnected with several base station controllers and downstream. Switch or Hub, connected through wired networks like the Public Switched Telephone Network (PSTN), handles accounting data. Some applications require signing in through the home network. The web and the Mobile

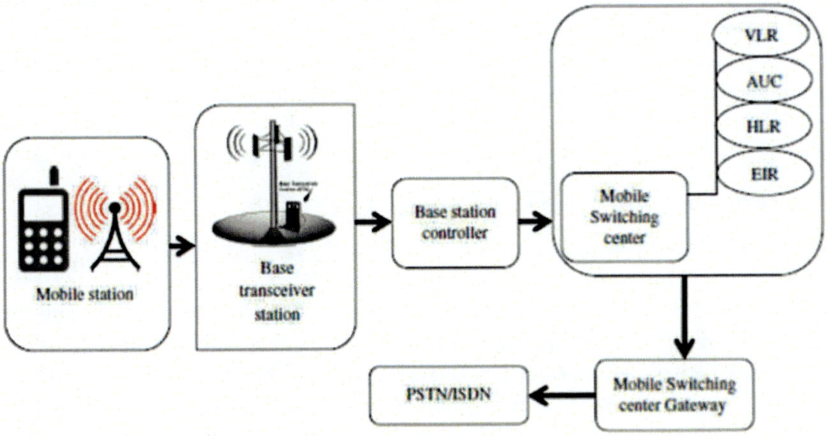

Fig. 1 An overview of a mobile system's essential components

Table 1 The dataset's top 10 families

Sl. no.	Family name	Number
1	Plankton	626
2	DroidKungFu	668
3	Fake Installer	927
4	Icono system	154
5	GinMaster	340
6	Opfake	615
7	K min	155
8	Geinimi	94
9	Droid Dream	83
10	SMS reg	42

Switching Center provide a direct connection. The majority of the database comprises the user's phone number, the services provided, and the secret key on a mobile device listed in the home location register [12]. The home location register, a database, is updated by the mobile switching center with user information.

4 Analysis of Imbalanced Issues and the Collection of Mobile Malware Traffic

4.1 Malware Sample Data Set for Android

The malicious traffic data was collected to identify the project's 5560 malware samples. The malware samples in question were collected between August 2021 and November 2022.

Table 1 gives an overview of the top 10 malware families in the data collection, including those currently being aggressively distributed in the app store [13].

4.2 Platform for Automated Traffic Collection

We created helpful traffic creation and monitoring software to detect malware traffic traces. The foundation platform, the traffic generator, the traffic collector, and the networking proxy/firewall are displayed in Fig. 2, together with their specific roles. The foundation platforms offer a regulated setting for simulating Android and its essential features, including creating simulators and installing apps [14]. The traffic generator, designed to set up and run malware samples to generate traffic traces automatically, comprises two components: the automated traffic generator and the

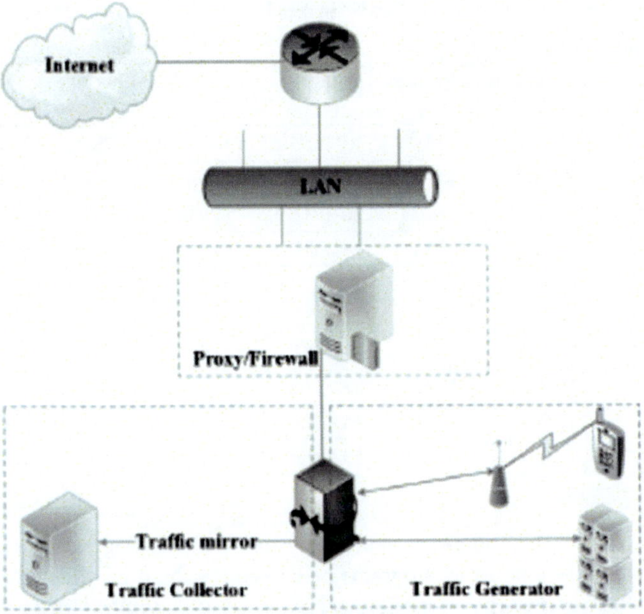

Fig. 2 Platform for creating and collecting mobile traffic

malware execution controller. The previous module is required for a specific malware sample's deployment and activation. In contrast, the later module is developed to preserve incoming and passing data via TCP dump automatically [15].

4.3 *Collection of Mobile Traffic*

Using the network collection platform, within the first 5 min, we collect the network traffic of 5560 malware samples.

To increase traffic quality, a sizable portion of useless traffic, including SSDP, DHCP, ARP, NBNS, IGMP, and SMB packets, is filtered out of the received raw data. Table 2 provides an overview of the most critical data from the traffic collected for the top 10 families [16].

4.4 *Recognizing Malicious Traffic Flows*

We develop a method to transform mobile traffic packets into traffic flows. The TCP dump data packets must be converted into flows to detect malicious flows. According

Table 2 Data about the top 10 families' traffic in detail

Name of family	Data outbound			Data inbound		
	Number of traffic	Average	Bytes	Number	Average	Bytes
DroidKungFu	81,008	223.98	9,072,298	169,299	2752.05	213,138,982
Plankton	57,765	347.85	10,075,699	49,755	763.16	18,935,688
Fake installer	2672	246.73	2670 329 034	2938	1578.68	2,311,118
Opfake	2298	336.19	382,654	2218	1036.90	1,152,088
Base Bridge	6738	377.76	1,277,562	4812	881.95	2,117,626
Iconosys	688	312.94	108,115	467	608.08	142,833
K min	3958	291.5	575,357	3622	1152.60	2,085,299
Fake doc	19,955	250.35	2,475,638	23,451	1705.69	19,988,415
Geinimi	1138	288.9	165,237	835	506.03	210,344
SMS reg	2382	244.92	292,572	3496	1977.52	3,458,448

to research, over 80% of malware applications use the repackaging strategy to hide their true nature as innocuous programs by embedding harmful code into innocent ones [17]. So, the majority of malicious apps pose as "harmless" ones. Naturally, both good and bad traffic data will be included in the traffic that has been collected. Both good and bad traffic data will be included in the traffic that has been gathered. The transformed flows help locate "pure" malicious traffic frequently provided by malicious apps.

Based on extensive data analysis, we discover that each malware sample creates 4–5 flows, 1–2 of which are malicious, on average, during the first 5 min. In contrast, a good application contains 8.02 flows [18]. Consider such a condition when five other Android devices are safe, but one Android device gets infected with a particular malware sample. Ninety-five secure applications are pre-installed on each smartphone. In a typical network context, just one out of every 4565 flow pattern average IR valuation is malicious.

4.5 Analysis and Extraction of Flow Features

We describe how the amount of downlink traffic differs from the amount of uplink traffic using a graph where the x-axis represents downlink traffic and the y-axis uplink traffic. The red line indicates that $x = y$, suggesting that the uplink and downlink traffic rates are equal in this case. By comparing the proportion of uplink traffic to downlink traffic in terms of packets, it can be shown that most malware families have packet quantity ratios higher than the threshold [19]. The study demonstrates that uplink traffic successfully and efficiently decreases downlink traffic and that most

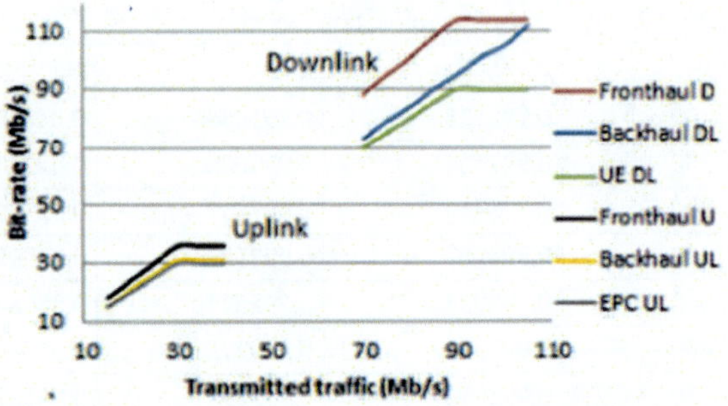

Fig. 3 The top 10 families' downlink to uplink traffic amount ratio

harmful applications upload data more often than helpful ones. The top 10 families' downlink to uplink traffic amount ratio is presented in Fig. 3.

5 Experimental Results

Using virtual mobile simulation software, simulation research is carried out. The performance, security, cost, and accuracy ratios were all used to show how dependable the suggested technique was. Several security procedures have been put in place to ensure mobile payment security. Security issues with contactless banking include malware detection, multi-factor authentication, safety, fraud detection, and data intrusion prevention. Figure 4 provides the analysis of the ecosystem's payment chain security.

Each member of the digital payments system should be able to provide proof that they have implemented the necessary security protocols. Are included cloud-based, token service providers, and the mobile payment system. The parties might be formed separately from the end-to-end safety check, with each component individually checked. Therefore, the many parties must collaborate to integrate and provide a service while enhancing health. It is essential to test the protection of all participants from threats, particularly privileged access for network administrators.

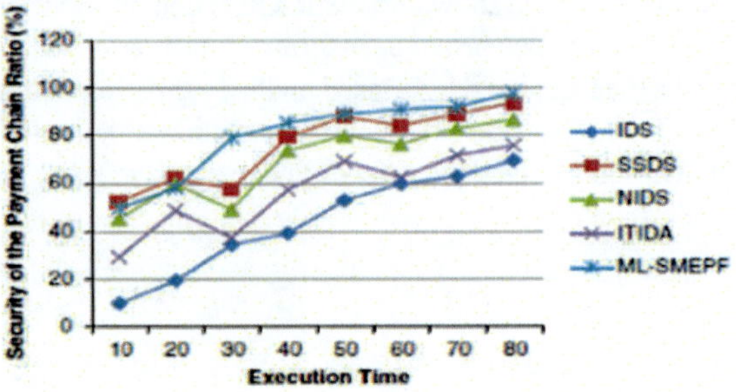

Fig. 4 Analysis of the ecosystem's payment chain security

Table 3 Accuracy of classification models

IR	Bayes Net	SVM	Naive Bayes	C4.5
100	0.995528	0.995454	0.818124	0.999568
200	0.997534	0.997008	0.422445	0.999636
300	0.998256	0.997457	0.521018	0.996878
400	0.998257	0.998183	0.589985	0.999854
500	0.998983	0.998544	0.856845	0.999565
600	0.998984	0.999273	0.956425	0.999781
700	0.998618	0.999271	0.845233	0.999780

The unbalanced mobile traffic flow feature data set is divided into seven groups to be taken into consideration to fix the problem with the IR values in the anomalous detection problem. 100, 200, 300, 400, 500, 600, and 700 are the IR values' equivalents. The ratios of positive and negative samples across various network topologies are represented in these graphs. The majority of the classifiers in the ACC findings perform well. Except for the Naïve Bayes classifiers, these classifiers' accuracy rates may exceed 99.9%. The AUC results, however, demonstrate that when the IR value rises, the accuracy of every classifier falls as in Table 3.

6 Conclusion

Modern technological technology is more dependent on mobile payment systems in various settings as an alternative payment method. Adaptability, perceived utility, robustness, comfort, and mobility are a few advantages of this measurer examining the traffic flow statistics and comments. We recognize the data imbalance problem that negatively impacts the Android malicious flow detection secure statistics. The results show that most frequent classifiers work rather well, demonstrating the effectiveness of machine learning methods for spotting fraudulent mobile traffic. When the amount of imbalance and the IR value increase, the accuracy of pleasant classifiers rapidly decreases. By comparing it to other classifiers using the same traffic flow data set, we show that the IDGC model is significantly more stable. The IDGC's classification performance in terms of AUC and GM stays within the range of 0.8 and 1.0 as the IR value rises. Based on the foundation of machine learning technology, the paper evaluates the safe electronic payment system of mobile payment technology. It also summarizes the safety concerns and security risks involved with each object.

References

1. Gangadevi G, Jayakumar C (2022) Review of deep learning architectures used for identification and classification of plant leaf diseases. In: Artificial intelligent techniques for wireless communication and networking. Wiley: Hoboken, NJ, USA, pp 75–90
2. Khandokar I, Hasan M, Ernawan F, Islam S, Kabir M (2021) Handwritten character recognition using convolutional neural network. J Phys Conf Ser 1918:042152
3. Dora S, Terrett OMOM, Sánchez-Rodríguez C (2022) Plant–microbe interactions in the apoplast: communication at the plant cell wall. Plant Cell 34:1532–1550
4. Scharr H, Pridmore TP, Tsaftaris SA (2017) Computer vision problems in plant phenotyping, CVPPP 2017–Introduction to the CVPPP 2017 workshop papers. In: Proceedings of the IEEE international conference on computer visionworkshops, Venice, Italy, 22–29 Oct 2017, pp 2020–2021
5. Al Farid F, Hashim N, Abdullah J, Bhuiyan MR, Shahida Mohd Isa WN, Uddin J, Haque MA, Husen MN (2022) A structured and methodological review on vision-based hand gesture recognition system. J Imag 8(6):153
6. IIslam MS, Sultana S, Kumar Roy U, Al Mahmud J ()2020 A review on video classification with methods, findings, performance, challenges, limitations and future work. J Ilm Tek Elektro Komput Dan Inform (JITEKI) 6:47–57
7. Islam MS, Sultana S, Roy UK, Al Mahmud J, Jahidul S (2021) HARC-new hybrid method with hierarchical attention based bidirectional recurrent neural network with dilated convolutional neural network to recognize multilabel emotions from text. J Ilm Tek Elektro Komput Dan Inform (JITEKI) 7:142–153
8. Islam S, Ab Ghani N, Ahmed M (2020) A review on recent advances in deep learning for sentiment analysis: performances, challenges and limitations. J. Ilm. Tek. Elektro Comput. Dan Inform. (JITEKI) 9(7):3775–3783
9. Islam M, Ghani NA (2022) A novel BiGRUBiLSTM model for multilevel sentiment analysis using deep neural network with BiGRU-BiLSTM. In: Recent trends in mechatronics towards industry 4.0. Springer: Berlin/Heidelberg, Germany, pp 403–414
10. Islam MS, Sultana S, Islam MJ (2021) New hybrid deep learning method to recognize human action from video. J Ilm Tek Elektro Komput Dan Inform 7:306–313
11. Islam MN, Ahmed F, Ahammed MT, Rashid M, Bari BS (2022) Rice disease identification through leaf image and IoT based smart rice field monitoring system. In: Enabling Industry 4.0 through advances in mechatronics. Springer, Berlin/Heidelberg, Germany, pp 529–539
12. Bari BS, Islam MN, Rashid M, Hasan MJ, Razman MA, Musa RM, Ab Nasir AF, Majeed AP (2021) A real-time approach of diagnosing rice leaf disease using deep learning-based faster R-CNN framework. PeerJ Comput Sci 7:e432
13. Wijesinha-Bettoni R, Mouillé B (2019) The contribution of potatoes to global food security, nutrition, and healthy diets. Am J Potato Res 96:139–149
14. Grieneisen ML, Aegerter BJ, Scott Stoddard C, Zhang M (2018) Yield and fruit quality of grafted tomatoes, and their potential for soil fumigant use reduction. A meta-analysis. Agron Sustain Dev 38:29
15. Meng F, Li Y, Li S, Chen H, Shao Z, Jian Y, Mao Y, Liu L, Wang Q (2022) Carotenoid biofortification in tomato products along the whole agro-food chain from field to fork. Trends Food Sci Technol 124:296–308
16. Savary S, Ficke A, Aubert JN, Hollier C (2012) Crop losses due to diseases and their implications for global food production losses and food security. Food Secur 4:519–537

17. Islam M, Dinh A, Wahid K, Bhowmik P (2017) Detection of potato diseases using image segmentation and multiclass support vector machine. In: Proceedings of the 2017 IEEE 30th Canadian conference on electrical and computer engineering (CCECE), Windsor, ON, Canada, 30 Apr–3 May 2017, pp 1–4
18. Azlah MA, Chua LS, Rahmad FR, Abdullah FI, Wan Alwi SR (2019) Review on techniques for plant leaf classification and recognition. Computers 8(4):77
19. Ramya V, Lydia MA (2016) Leaf disease detection and classification using neural networks. Int J Adv Res Comput Commun Eng 5:207–210

Secure and Efficient Routing Mechanism for Healthcare Networks

Animesh Giri, B. V. Balaji, Bhoomika P. Bhavimath, V. Durgalakshmi, and B. Rahul

Abstract The demand for healthcare services is increasing with the growing population. Integration of the Internet of Things (IoT) is a key move to cater to this demand. Routing protocols form a key aspect of a network. The three principal elements of routing protocols deployed in the healthcare sector include compliance with the constrained nature of devices, capacity to manage heterogeneous traffic, and data security. In this study, the RPL routing protocol is being explored. This paper compares the usage of Multiple RPL Instances with Multi Sink against Single RPL Instance with Single Sink over Latency, Energy consumption, Control traffic overhead, and Packet Delivery Ratio in Cooja Simulator. The results obtained illustrate that the Multiple Instance and Multi Sink approach performs better, thereby indicating its ability to handle heterogeneity. Further, cryptographic mechanisms are added to ensure confidentiality. Additionally, data from the sinks are transferred to a backend server hosting a website.

Keywords Internet of things · Routing protocol · Heterogeneous traffic · Multiple instance · Cooja simulator

1 Introduction

In this period of digital transformation, the "Internet of Things" (IoT) is one of the main driver technologies. The IoT devices agree with the IEEE 802.15.4 standard making them constrained by nature. Such wireless nodes form a network termed "Low-Power Wireless Personal Area Networks" (LoWPANs). Thus, the conventional Internet Protocol (IP) cannot be implemented directly. The "Internet Engineering Task Force" (IETF) thus developed a solution to make Internet Protocol Version 6 (IPv6) to communicate with LoWPANs with strategies like compression of IPv6 headers and addition of an adaptation level/layer between IP and Data Link

A. Giri (✉) · B. V. Balaji · B. P. Bhavimath · V. Durgalakshmi · B. Rahul
Department of Computer Science and Engineering, PES University, Bangalore, India
e-mail: Animeshgiri@pes.edu

G. Rajakumar et al. (eds.), *Intelligent Communication Technologies and Virtual Mobile Networks*, Lecture Notes on Data Engineering and Communications Technologies 171,
https://doi.org/10.1007/978-981-99-1767-9_27

layer [1]. "Low-Power and Lossy Networks" (LLNs) consist of resource-constrained nodes connected through unstable links with high packet loss. One of the major considerations in such scenarios is the building of a suitable routing protocol. The protocol must deal with the less energy and processing power of nodes along with providing reliability given the lossy nature of the network [2]. The IPv6 Routing Protocol for LLNs (RPL) is a routing protocol explicitly designed for LLNs by the ROLL Working Group in the IETF. It is a proactive; "distance-vector" protocol that constructs a graph termed "Destination Oriented Directed Acyclic Graph" (DODAG) with no cycles [3]. Most healthcare organizations are using IoT technology to transform into "Smart Healthcare" with a focus on personalized care. This inclusion allows to collect, analyze, and transfer data about patients and the hospital environment efficiently to a central database for better decision-making [4, 5]. It is observed that sensors on patients measuring vitals represent critical data with higher sending intervals. While static sensors measuring environmental conditions have a smaller sending rate. It is thus observed that different sensors within the healthcare environment generate different types of traffic that need to be handled in a varied manner. In summary, IoT devices in the healthcare environment generate heterogeneous traffic.

2 Background

2.1 RPL Overview

RPL is a reliable routing solution that aims to satisfy the needs and requirements of constrained nodes and lossy networks. It is a "distance-vector" protocol that splits the given network topology of nodes/sensors into one or more instances. Each instance consists of at least one tree-like structure termed DODAG. All sensors in the graph are directed to a single node called the DODAG root/sink. A DODAG is uniquely recognized using DODAG ID and RPL Instance ID. Every RPL instance is linked to a single Objective Function (OF) that helps in the selection and optimization of paths within it [6, 7]. The OF helps a node to take primary decisions such as creating a potential parent list, choosing a parent over the others with an additional metric, choosing a DODAG to join, and rank calculation [8]. Figure 1 illustrates the route selection process followed by RPL.

2.2 Heterogeneous Traffic and RPL

In a heterogeneous traffic environment, each type of traffic requires different Quality of Service (QoS) parameters [9]. An RPL instance is a collection of one or more DODAGs all operating with the same OF. Thus, all nodes in an instance are said to have one objective (in terms of how to optimize their routes). A heterogeneous traffic

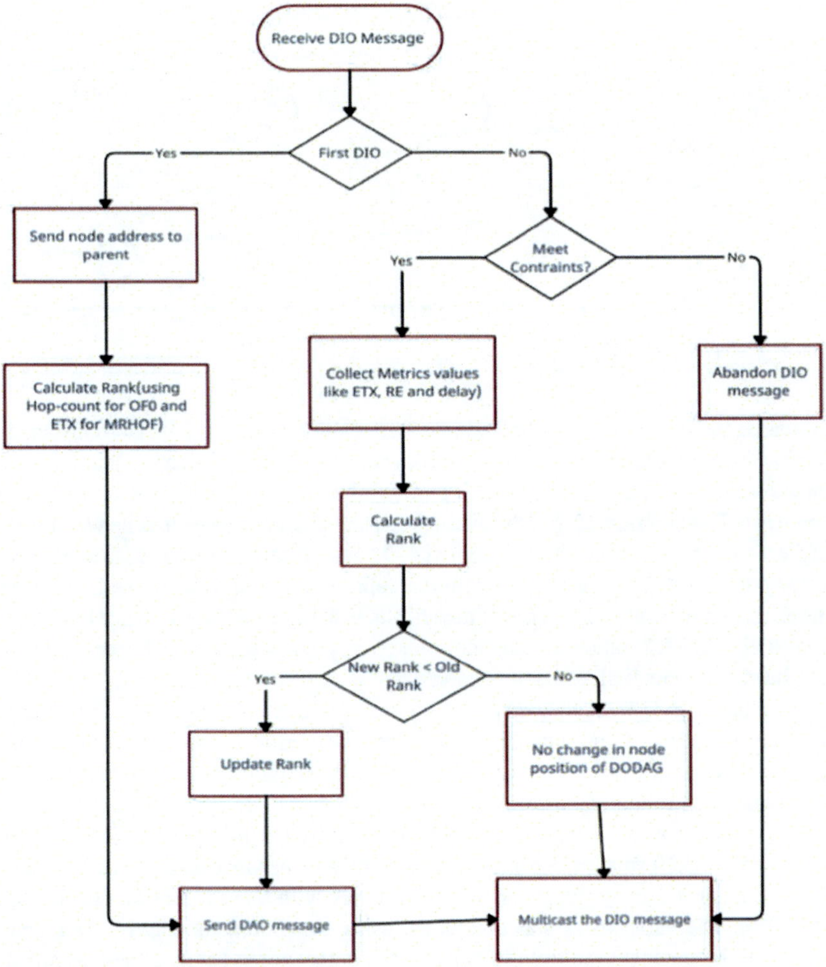

Fig. 1 Flow chart representing route creation

context would need multiple routing objectives. Thus, forming multiple instances is a way to handle this requirement in RPL [10, 11]. Figure 2 shows the formation of multiple instances.

3 Related Work

The authors in [12] built a network of 60 nodes generating critical and regular/periodic type data traffic. Sending intervals and packet sizes were modified for each. They

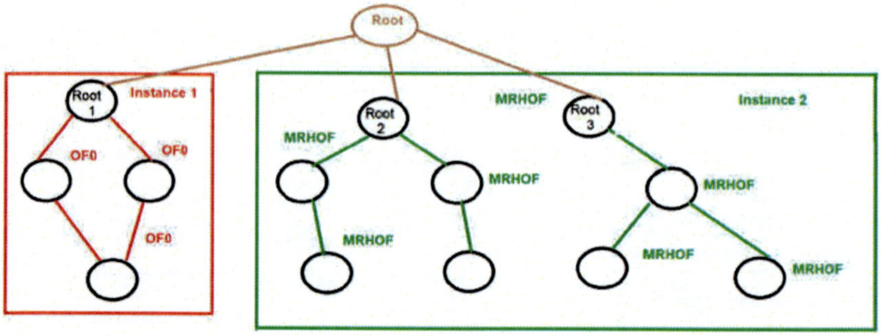

Fig. 2 Multiple Instances

evaluated the performance of both single and multiple instance RPL (two instances in total) against average PDR, average convergence time, and average latency with three reception ratios (100, 85, 70%). The simulations were conducted in the Cooja simulator. It was observed that the average routing tree convergence time was a little higher in multiple instance by a margin of 4–5 s at all reception ratios. This is because every node must form two DAGs with the sink node as opposed to one in a single instance. Additionally, both types of data traffic had improved performance in latency and PDR at all RX ratios. Parameters like energy consumption and control traffic overhead were not explored in this paper.

Al-Abdi et al. [13] considered an IoT system deployed on a floor with four wards with three kinds of traffic, namely critical, medium-critical, and periodic. Sending intervals and packet sizes were modified for each. Requirements included minimal delay and high reliability for critical data, considerable delay for medium-critical data, and least priority in terms of delay and reliability for periodic data. The authors evaluated the performance of both single and multiple instance RPL (three instances in total) against average PDR and average latency with three reception ratios (100, 85, 70%). The experiment was conducted using the Cooja simulator. The results obtained illustrate that multiple RPL instances outperformed at all reception ratios. Considering the RX ratio of 100%, an overall latency of 21.2 s was obtained with single instance. In contrast, with multiple instances, an average latency of 0.04 s was observed. Similarly, the average PDR was 28% with single instance and 93% with multiple instance RPL. However, only two metrics were evaluated against single instance RPL.

Mardini et al. [10] demonstrated the performance of multiple instances in a 6G/IoE-based sample health monitoring system. The system is assumed to be employed at King Abdullah University Hospital (KAUH), Jordan. Sensors were deployed on every floor of the hospital, each segregated into departments such as labs and the Intensive Care Unit (ICU). Thus, each floor would generate different traffic types, namely high-critical, medium-critical, and periodic. Sending intervals and packet sizes were different for each type. Requirements included minimal delay and high reliability for critical data (data from ICU), considerable delay for medium-critical data, and the least priority in terms of delay and reliability for periodic data (data from general wards). Three floors (1, 9, 11) were simulated in the Cooja simulator. Average PDR and average latency were tested for three reception ratios (100, 85, 70%). Latency was improved in multiple instance RPL at all RX values. Further, PDR has also improved in most cases except on the eleventh floor at 100% and 85% reception ratios. Other metrics such as convergence time and energy consumption were not compared.

Zaatouri et al. [14] have evaluated the performance of multi-sink in a network topology of randomly spaced 35 sender nodes. Cooja Simulator was used to contrast the performance of 1 sink, 5 sinks, and 10 sinks against PDR, energy consumption, number of lost packets, and throughput. Additionally, the results were compared with all nodes running with OF0 and MRHOF separately. The results obtained proved that the multi-sink approach improved the performance in all four aspects. Here every node tries to form a DODAG with the nearest sink thereby decreasing the number of packets lost, energy consumption, and increasing PDR and throughput. Further, MRHOF (ETX) outperformed OF0 (HC) in all cases.

The limitations of the existing work in this area include:

- Testing of the Multiple Instance and Multi-sink approach together in a network generating heterogeneous traffic
- Relaying sensory data outside the Cooja simulator to a backend server

This paper aims to address both these limitations with a new network topology and with the integration of COAP and SSE service.

4 Proposed Approach

In this paper, we have assumed an IoT system to be deployed on a hospital floor with three departments, ICU, a Special ward, and a laboratory. Sensors are placed on patients in Special wards and ICUs to closely track their health statistics while sensors are placed on walls of laboratories to get environmental statistics. Information from the ICU and Special ward sections are regarded as critical traffic and medium-critical traffic, respectively. Data from labs are classified as periodic traffic (to be sent after a fixed period) considering it measures values such as temperature and humidity of the room (Fig. 3).

A multiple instance approach is proposed to route traffic efficiently in this scenario. This is because, a single OF is used when all the nodes have the same Quality of Service (QoS) requirement while in a heterogeneous environment, a Multiple Instance approach (DODAGs running on different OFs to create routes) is to be deployed [5]. Data from ICU and Special ward is considered as one instance while traffic from labs forms the second instance. This is because both critical and medium-critical data share characteristics such as high sending intervals and less delay. While

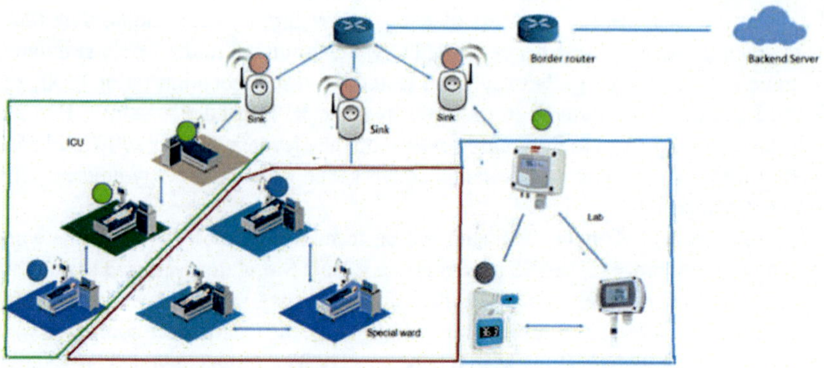

Fig. 3 High-level design of the proposed mechanism

periodic data is sent after a large fixed interval and is delay tolerant. Here heterogencity is only defined based on the QoS parameters requirement. Further, a multi-sink approach is proposed. Each department/traffic type is provided with a separate sink to relay information as opposed to having one single sink for all the departments/traffic types. Thus, three DODAGs would be formed in total with two of them being part of one instance (critical traffic) and the other forming a separate instance (periodic traffic). Figure 4 illustrates the proposed mechanism. This paper aims to compare and contrast the single instance and single sink approach with the multiple instance and multi-sink [with 2 and 3 sinks] approach for the proposed architecture. Further, the communication between the Constrained Application Protocol (COAP) server and the backend server is illustrated in Fig. 5.

Each sink (act as COAP client) buffers the data received from the sender nodes and pushes it periodically to the COAP server as a COAP request. The backend server

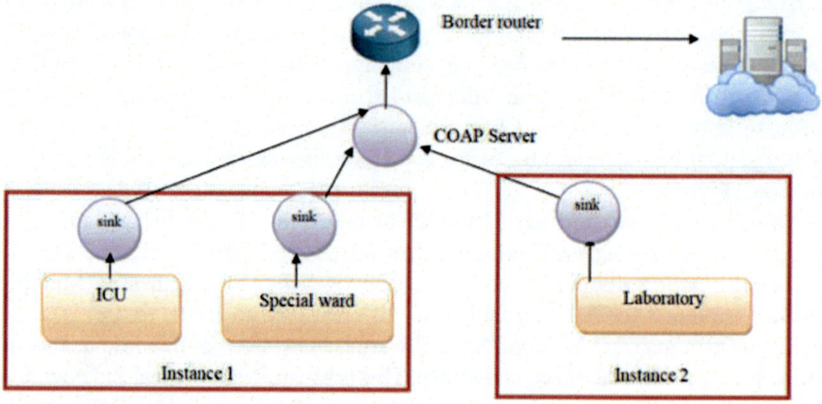

Fig. 4 Proposed approach to handle heterogeneous traffic

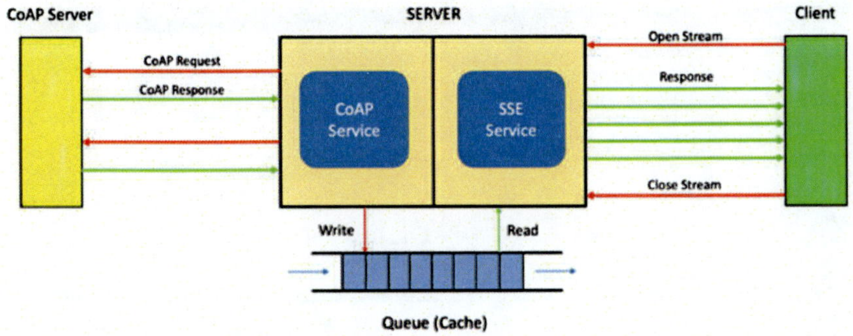

Fig. 5 COAP-proxy-backend server communication

(indicated as the client in Fig. 5) establishes an Server-Sent-Events (SSE) connection with the Proxy server (shown as the server in the Fig. 5). Thus, the client/ backend server continues to listen to this connection. The server periodically sends a COAP request to the COAP server. The entire queue (maintained by the COAP server) is sent back as a COAP response to the proxy server. On receiving data, the Server/ Proxy server sends it to the client/Backend server via the SSE connection to update the website.

COAP is a lightweight application protocol (majorly following a client–server architecture) running on User Datagram Protocol (UDP). Any COAP client and server communicate via request and response messages. Request messages are of four types GET, POST, PUT, DELETE similar to Hyper-Text Transfer Protocol (HTTP). Figure 6 displays a sample communication.

Certain measures have been taken to ensure data security within the simulation alone (majorly confidentiality tackled). Contiki-NG provides link-layer security for IEEE 802.15.4 Time Slotted Channel Hoping (TSCH). Link-layer security encrypts frames as it travels across the medium using the AES-128-bit algorithm (Advanced Encryption Standard). Suitable changes have been made to the configuration files to deploy and activate this feature at all sender nodes (at sink nodes, data gets decrypted). This feature maintains 2 cryptographic keys for data transfer and control/management

Fig. 6 COAP communication between server and client

```
Client              Server
   |                  |
   |  CON [0xbc90]    |
   | GET /temperature |
   |  (Token 0x71)    |
   +----------------->|
   |                  |
   |  ACK [0xbc90]    |
   |  2.05 Content    |
   |  (Token 0x71)    |
   |    "22.5 C"      |
   |<-----------------+
```

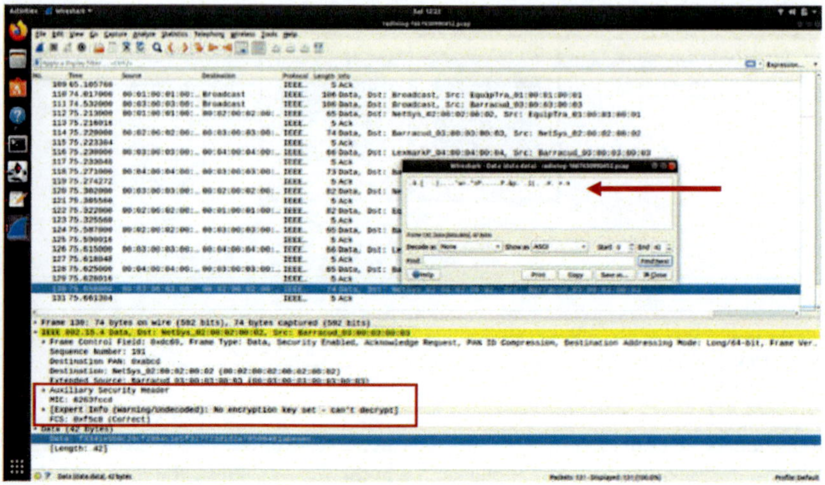

Fig. 7 Packet encrypted with AES-128 algorithm

messages, respectively. Further Datagram Transport Layer Security (DTLS) is used in this project. It is a security layer built over UDP that has both cryptographic and authentication mechanisms all resolved within a short run of time thereby not adding to latency.

Figure 7 shows all the packets transferred during a simulation run (pcap file loaded in Wireshark). The dialogue box (marked with an arrow) illustrates the data/payload in the packet. It is seen that it is encrypted successfully and the cipher text is shown in place of plain text. Further, the Auxiliary Security Header is included as part of the frame (highlighted with a box) with information such as Message Integrity Code (MIC) and Frame Check Sequence (FCS) checksum. Thus, the inclusion of Link-layer security and DTLS protocol ensures confidentiality of data is maintained as it travels from sender to sink nodes.

5 Simulation

Critical, medium-critical, and periodic traffic is distinguished based on sending intervals and payload sizes. Critical data from ICU would be sent regularly at short intervals. While medium-critical data would also be sent regularly. Periodic data will be sent after large intervals [12]. Further, the payload length will be in the order critical, medium-critical, and periodic from smallest to largest. Tables 1, 2 and Fig. 8 highlight these values. The two approaches (Single Instance with Single Sink and multi-instance with multi-sink) were compared against four metrics, Packet Delivery Ratio (PDR), Latency, Average Energy consumption, and Control traffic overhead.

Table 1 Sending Intervals of sensor nodes at different departments

Critical traffic	Medium-critical traffic	Periodic traffic
15 s	30 s	200 s

Table 2 Simulation environment parameters.

Parameters	Values
OS	Contiki-NG
Simulator	Cooja
Communication protocols	IEEE 802.15.4, 6LoWPAN, RPL, IPv6,UDP
Radio medium	Unit Disk Graph Medium: with distance loss
Mote type	Cooja mote
Total number of sender nodes	38
Node positioning	Linear, eclipse
Transmission ratio	100%
Transmission range	50 m
Interference range	100 m
Simulation period	45 min
Squared area	360 m × 360 m
Topology	Grid topology
TX ratio	100%
RX ratio	80%, 85%, 90%, 95%, 100%
Speed	No limit
OF	Single instance—OF0, Multiple instance: OF0 (periodic), and MRHOF (critical and medium-critical data)

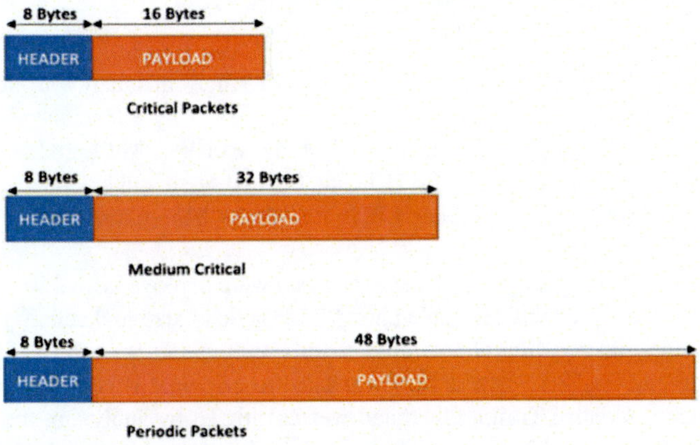

Fig. 8 Payload length for UDP packet

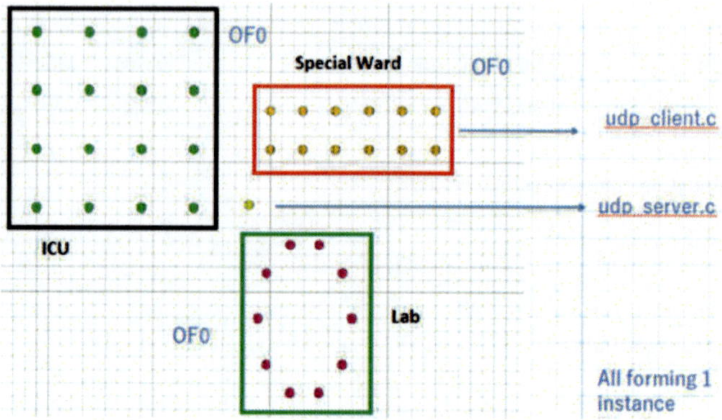

Fig. 9 Topology for single instance with single sink

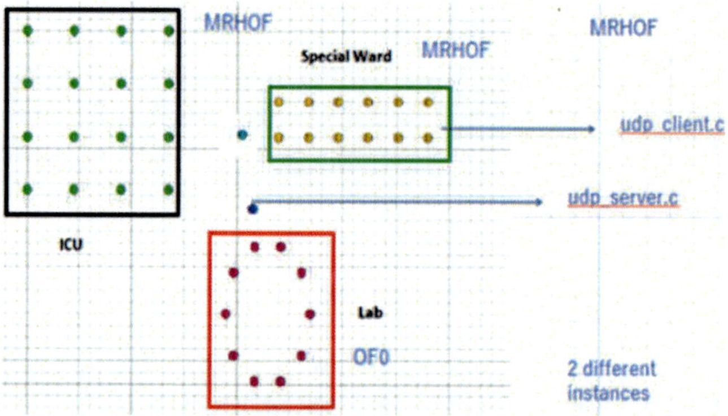

Fig. 10 Topology for multiple instance with multi (2) sinks

Figures 9, 10 and 11 display the topology for single instance with single sink and multiple instance with multiple sinks approaches. Single Instance with Single Sink approach has a single sink/collector for all departments. Multiple Instance RPL with 2 sinks, has a sink common for ICU and Special ward and a dedicated sink for Lab sensors while Multiple Instance RPL with 3 sinks, has a dedicated sink for each department.

These three topologies (with 38 sender nodes) were deployed in a 360 × 360 grid in the Cooja simulator with suitable changes in payload size and sending intervals and run for 45 min at different reception ratios (100%, 95%, 90%, 85%, 80%). The in-built Simulation Java Script Editor was used to calculate Packet Delivery Ratio and Latency [1]. Control Traffic Overhead was calculated by analyzing the resultant pcap file in Wireshark. The energy consumption of each node was calculated using the "Energest module" in Contiki-NG. This library helps to track the components of

[1] https://github.com/DURGALAKSHMI0108/Multiple-Instance-RPL-with-Multiple-Sink.git

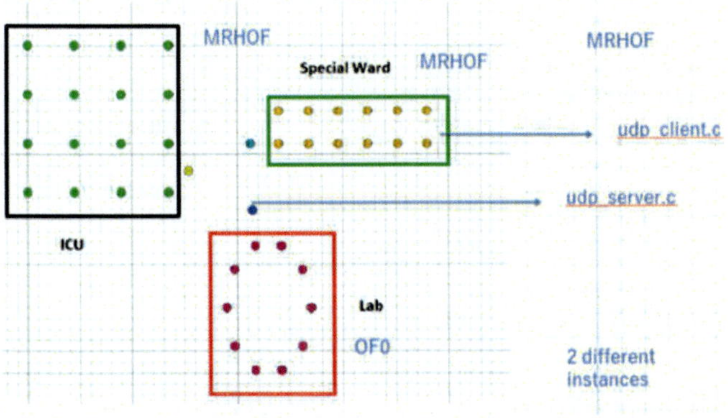

Fig. 11 Topology for multiple instance with multi (3) sinks

the node and prints those details as part of the mote output. The log file (indicating simulation output) was given as input to a python script written to calculate the energy consumption with the outputted values (See footnote 1).

6 Results and Discussion

6.1 Packet Delivery Ratio (PDR)

The PDR values are drastically improved in the proposed approach for all types of traffic (at all reception ratios). For critical traffic, Single Instance and Single Sink approach provides lesser values compared to Multiple Instance + Multi Sink [3 sinks] approach at all RX values. This difference becomes large at RX ratios below 80%. At an RX ratio of 50%, Single Instance and Single Sink approach provide only 83.6734% while this was improved to 99.6817% in the Multiple Instance + Multi Sink approach. For medium-critical traffic, the Single Instance and Single Sink approach gives only 98.84% and 47.4137% at 85 and 50% RX ratios while these values are increased to 100% and 100% in the Multiple Instance + Multi Sink approach. Similarly, value jumps are seen at the other reception ratios. For periodic traffic, Single Instance and Single Sink approach present 0 at 50% RX ratios. While this value is increased to 97.0588% in the Multiple Instance + Multi Sink approach, the rest of the values are 100 in both approaches for periodic traffic. Generally, at a 50% RX ratio, the PDR values will be considerably low because of the low reception ability of the sink node thereby leading to high data loss. In the Single Instance and Single Sink approach, all kinds of data traffic packets are aimed at one node making it congested. Critical packets are sent at a 4 ppm rate and medium-critical

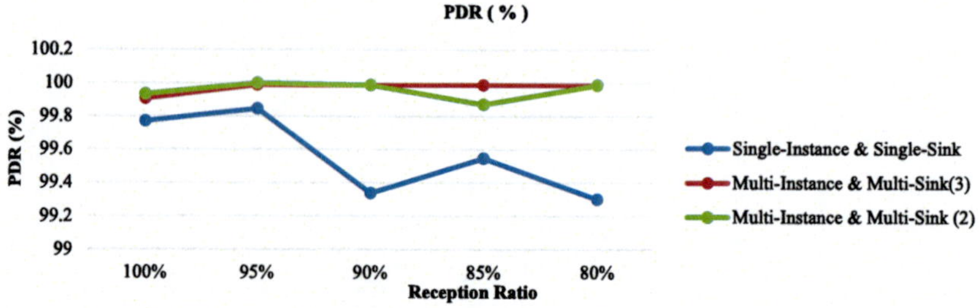

Fig. 12 Results of PDR values obtained

with a 2 ppm rate. Periodic packets are sent at a 0.33 ppm rate thereby around 18–19 packets are sent in the entire simulation. With a single congested sink, none of the periodic packets get a chance to reach the sink node making the PDR 0%. In contrast, in the Multiple Instance + Multi Sink approach, each kind of traffic has its sink node. Therefore, periodic traffic gets a better PDR value of 97.0588% at a 50% RX ratio. The reduction in congestion is one of the main reasons for the improvement in PDR in all traffic kinds. Another reason for the change in PDR for critical and medium-critical traffic is due to the use of ETX (in Multiple Instance + Multi Sink). ETX considers the quality of the communication links and can rapidly respond to dynamic network conditions, a feature that is missing in OF0. This feature helps us to meet high-reliability requirements for critical and medium-critical traffic. Further, the variation in PDR in Single Instance and Single Sink approach with RX values is large compared to the Multiple Instance + Multi Sink method thereby making it a better choice in realistic scenarios where RX ratios are less than 100%. The reason for the small differences between Single Instance and Single Sink and Multiple Instance + Multi Sink approach at higher RX ratios could be attributed to the small number of nodes in the network. The same experiment was conducted for Multiple Instance + Multi Sink (with 2 sinks) and the results are shown in Fig. 12. It is seen that Multiple Instance + Multi Sink (3 sinks) outperforms all the approaches.

6.2 *Average Latency*

Figure 13 displays the overall latency results obtained. Here latency is calculated as the delay of every packet received to the total number of packets received irrespective of the traffic type. The Multiple Instance + Multi Sink [with 3 sinks] approach gives better performance at almost all RX ratios. The Multi Sink approach gives lower latency due to the presence of a closely placed dedicated sinks. Further, it is proved that ETX has a lesser hop count than OF0 thereby giving way for lesser delay [15]. Furthermore, as the RX ratios decrease, latency values increase in both methods. But the variation is reduced in Multiple Instance + Multi Sink approach. This is

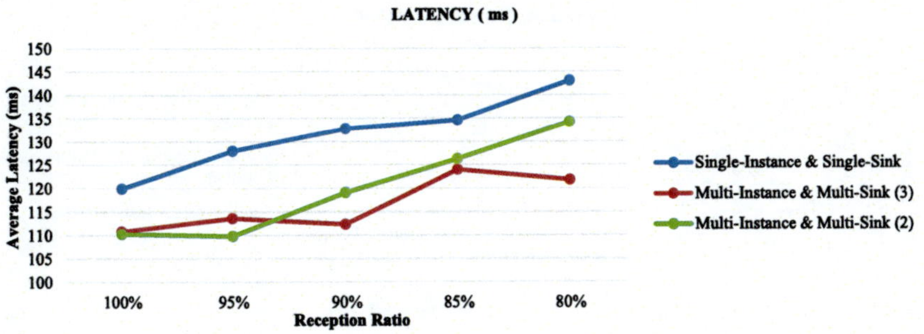

Fig. 13 Results of average latency values obtained

because MRHOF was used for critical and medium-critical traffic and it considers data related to link quality such as ETX while forming routes.

6.3 Average Energy Consumption

Figure 14 displays the results obtained. The average energy consumption of all nodes is the least in our proposed approach (Multiple Instance + Multi Sink [3 sinks]) at all RX ratios. This is attributed to the presence of dedicated sinks. This avoids normal sender nodes being relay nodes (nodes that transmit information of other nodes to the sink), thereby reducing energy consumption. Further, the energy consumption of MRHOF is lesser than OF0 because it measures link quality while forming routes thereby creating better routes, allowing lesser re-transmissions, and saving energy [16].

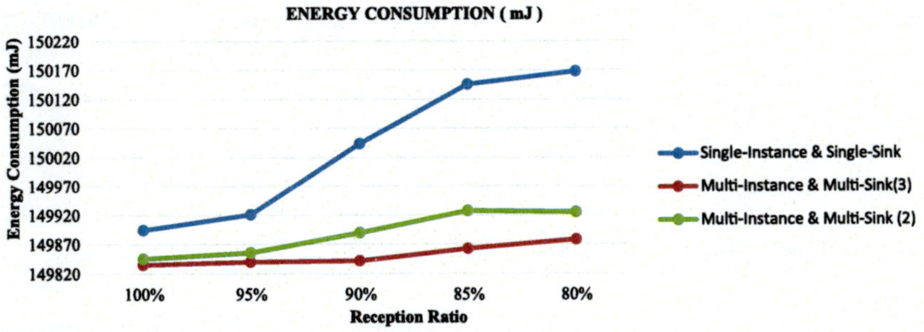

Fig. 14 Results of Average energy consumption values obtained

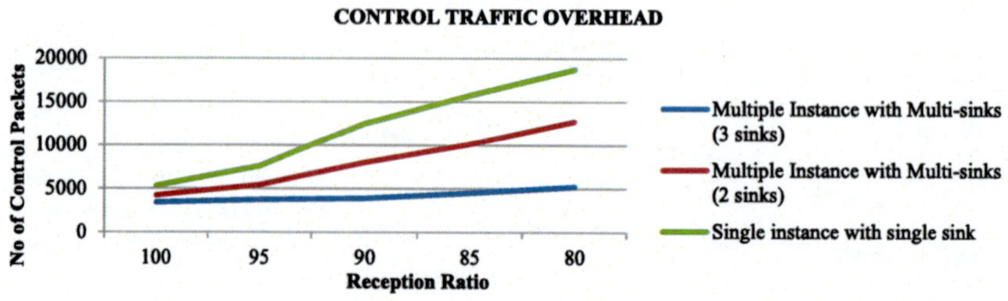

Fig. 15 Results of control traffic overhead values obtained

6.4 Control Traffic Overhead

Figure 15 displays the results obtained. Control traffic overhead is calculated as the total number of control messages transmitted throughout the simulation time to form and maintain routes. It is observed that Multiple Instance + Multi Sink [3 sinks] has lesser control traffic overhead among all the approaches at all RX ratios. Further, as the reception levels decrease, there is an increase in the control messages passed, but the variation is less in our proposed approach. The effects of higher overhead include higher network traffic and energy drain.

6.5 Foren6

Foren6 is a visualization and diagnosis tool specifically designed for 6LoWPAN networks. The simulation pcap file is processed to display the DODAG formed, its parameters (DODAG ID, Instance ID, Objective function, Node rank). Figures 16 and 17 show the DODAG formation at a 50% RX ratio with Single Instance with Single Sink and Multiple Instance with 3 sinks, respectively. It is observed that many nodes are not connected at all in the Single Instance and Single Sink approach while in the proposed approach none of the nodes are left out. Thus, the proposed approach ensures reliability even at an RX ratio as low as 50%.

7 Conclusion and Future Scope

In summary, our project proposes the use of Multiple Instance RPL to handle heterogeneous traffic within a single network. The instances were formed based on variations in sending interval, QoS requirements, and payload sizes. Further, a multi-sink [3 sinks–one for each department] approach was also chosen to avoid hotspots within the network. The proposed approach was compared with Multiple Instance + Multi

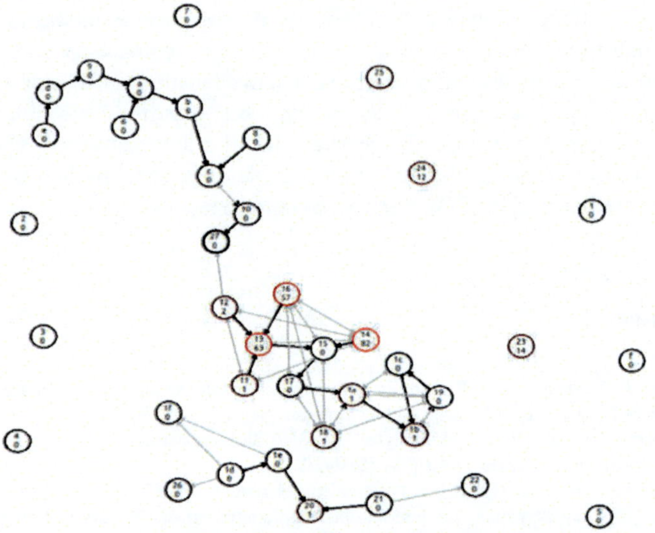

Fig. 16 Foren6 visualization of single ınstance with single sink approach

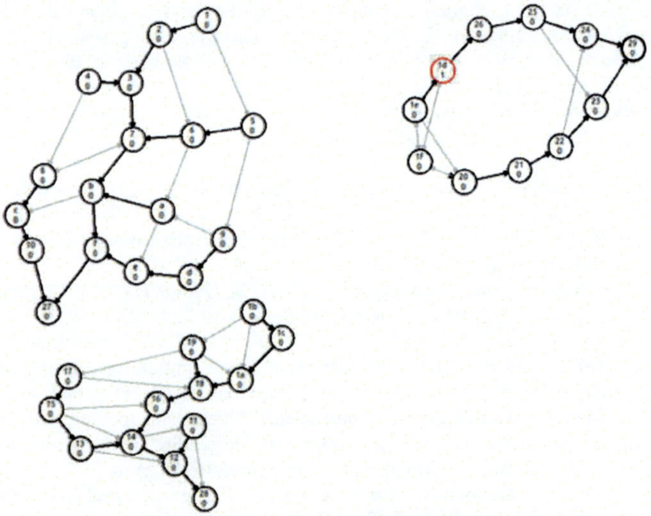

Fig. 17 Foren6 visualization of multiple ınstance with 3 sinks approach

Sink [2 sinks] and Single Instance with Single Sink approach in the Cooja Simulator. Results showed that the proposed approach is better at all reception ratios in terms of PDR, Latency, Control Traffic Overhead, and Energy consumption. As part of future implementation, the proposed routing mechanism can be tested with a topology with multiple departments and floors thereby adding more nodes to the network. The routing mechanism could be altered by making the number of instances dynamic rather than pre-defined. The Multiple Instance approach could be extended by proposing an improved and optimized Objective function and Trickle algorithm. Additional security mechanisms (such as Firewalls, Intrusion Detection Systems) can be incorporated to ensure all security principles are maintained. Our project aimed at securing data transfer within the simulation bounds, further, attacks, and mitigation mechanisms on COAP and SSE could be tackled.

References

1. Kumar V, Tiwari S (2012). Routing in IPv6 over low-power wireless personal area networks (6LoWPAN): a survey. J Comput Netw Commun
2. Martocci J, De Mil P, Riou N, Vermeylen W (2010) Building automation routing requirements in low-power and lossy networks (No. rfc5867)
3. Witwit AJ, Idrees AK (2018) A comprehensive review for RPL routing protocol in low power and lossy networks. In: International conference on new trends in information and communications technology applications. Springer, Cham, pp 50–66
4. Zhao W, Luo X, Qiu T (2017) Smart healthcare. Appl Sci 7(11):1176
5. Kodali RK, Swamy G, Lakshmi B (2015) An implementation of IoT for healthcare. In: 2015 IEEE recent advances in intelligent computational systems (RAICS), pp 411–416
6. Winter T (2011) RPL: IPv6 routing protocol for low power and lossy networks. RFC6550
7. Abdel Hakeem SA, Hady AA, Kim H (2019) RPL routing protocol performance in smart grid applications based wireless sensors: experimental and simulated analysis. Electronics 8(2):186
8. Lamaazi, H., Benamar, N. and Jara, A.J., 2016, May. Study of the impact of designed objective function on the RPL-based routing protocol. In International Symposium on Ubiquitous Networking (pp. 67–80). Springer, Singapore.
9. Kim HS, Ko J, Culler DE, Paek J (2017) Challenging the IPv6 routing protocol for low-power and lossy networks (RPL): a survey. IEEE Commun Surv Tutor 19(4):2502–2525
10. Mardini W, Aljawarneh S, Al-Abdi A (2021) Using multiple RPL instances to enhance the performance of new 6G and Internet of Everything (6G/IoE)-based healthcare monitoring systems. Mob Netw Appl 26(3):952–968
11. Nassar J, Berthomé M, Dubrulle J, Gouvy N, Mitton N, Quoitin B (2018) Multiple instances QoS routing in RPL: application to smart grids. Sensors 18(8):2472
12. Banh M, Mac H, Nguyen N, Phung KH, Thanh NH, Steenhaut K (2015) Performance evaluation of multiple RPL routing tree instances for Internet of Things applications. In: 2015 international conference on advanced technologies for communications (ATC). IEEE, pp 206–211
13. Al-Abdi A, Mardini W, Aljawarneh S, Mohammed T (2019) Using multiple RPL instances for enhancing the performance of IoT-based systems. In: Proceedings of the second international conference on data science, e-learning and information systems, pp 1–5
14. Zaatouri I, Alyaoui N, Guiloufi AB, Kachouri A (2017) Performance evaluation of RPL objective functions for multi-sink. In: 2017 18th International conference on sciences and techniques of automatic control and computer engineering (STA). IEEE, pp 661–665
15. Yassein MB, Flefil A, Krstic D, Khamayseh Y, Mardini W (2019) Performance evaluation of RPL in high density networks for internet of things (IoT). In: Proceedings of the 2019 8th international conference on software and information engineering, pp 183–187
16. Dalal HN, Soni NV, Razaque A (2016) Header encryption of IEEE802. 15.4. In: 2016 IEEE long island systems, applications and technology conference (LISAT). IEEE, pp 1–6

Secured Ticketless Booking System to Monuments and Museums Using Cryptography

Kuldeep Vayadande, Ankur Raut, Roshita Bhonsle, Vithika Pungliya, Atharva Purohit, and Varad Ingale

Abstract The background of our prototype—'Safar', is to develop a web-based application based on the effectiveness of QR codes for authenticating an E-ticket. The aim is to provide a simple and sophisticated system to generate an E-ticket, it will be able to increase ticket sales. The proposed system will allow the user to book a ticket for a particular monument on a particular day. After booking a ticket, the user is directed to the payment portal where the money transaction is carried out using the RazorPay testing API. If the payment is successful, an E-ticket containing the QR is generated which is sent to the email of the tourist. The tourist then scans this QR code at the monument to gain entry into the monument. The proposed website also contains an admin portal where the live count of visitors with their credentials and the history of tourists for that monument is stored. All the user data will be encrypted using a novel cryptographic approach before it is saved to the database, to provide added security. As tourism accounts for 7% of India's GDP, the proposed system aims to attract tourists by giving them an all-round, complete experience. It not only reduces the time to wait for a long duration in queues but also makes the process of visiting a monument seamless and hassle free. The proposed system is developed using React, HTML, CSS, JavaScript, and Django. It will give a boost to the heritage and culture of our country and also ensure a greener and cleaner economy.

K. Vayadande (✉) · A. Raut · R. Bhonsle · V. Pungliya · A. Purohit · V. Ingale
Vishwakarma Institute of Technology, Pune, India
e-mail: kuldeep.vayadande@gmail.com

A. Raut
e-mail: ankur.raut20@vit.edu

R. Bhonsle
e-mail: roshita.bhonsle20@vit.edu

V. Pungliya
e-mail: vithika.pungliya20@vit.edu

A. Purohit
e-mail: atharva.purohit20@vit.edu

V. Ingale
e-mail: varad.ingale20@vit.edu

G. Rajakumar et al. (eds.), *Intelligent Communication Technologies and Virtual Mobile Networks*, Lecture Notes on Data Engineering and Communications Technologies 171,
https://doi.org/10.1007/978-981-99-1767-9_28

Keywords Booking system · Cryptography · Django · E-ticketing · JavaScript · QR code · RSA algorithm

1 Introduction

Even if the idea of E-ticketing is widely used in the travel industry, it is challenging to put into practice in a setting where the entire company strategy is built around flexible travel and tourism. The daily verification of identification documents for 22 million visitors requires significant human resources, but all of this can be digitally streamlined by connecting the location of the app with the information card database, with local museums generating a major portion of this income. With the use of digitalization, technology is becoming increasingly important in today's fast-paced environment. E-ticketing systems will be able to integrate effective verification mechanisms and also be time and cost-effective. The proposed work aims to help individuals by offering a simple, user-friendly layout by providing a simple and secure way for museums to verify visitors. It also does away with the need for human resources to check tickets, which is another benefit. With the present trend heading toward digitalization, smart ticketing systems are imperative. Ticketing systems are projected as Museums and Heritage Sites in recent plans for smart cities.

With technology taking over the world, the E-ticketing system is in tourism as a 'paperless' revolution. The process of booking tickets is getting better every day. There is no more need for people to be in long queues. United Airlines was the first airline to experiment with E-ticketing in 1984, but it took another six years for the process to be put into practice. But once it got going, the world's travel industry recognized the potential for a significant change in ticketing methods. There are various ticketing media and are categorized into:

- Cash
- Tokens
- Paper Tickets
- Magnetic Strip Ticket
- Contact-based Smart Cards
- Contactless Card.

In this work, an E-ticketing system with the help of a QR system is proposed. Through this system people can generate tickets to monuments and museums all over the country. Through this approach, the booking system is more centralized and real time. Some of the advantage QR code offers are:

- An inexpensive printer can be used to print QR barcodes and placed on objects as compared with printing RFID tags which requires a special purpose device to do that.
- Ability to store large amounts of information as compared to NFC and RFID.
- It performs high-speed encryption and decryption with good reliability.

- By using Smartphone's QR scanner, data entry can be done without using the phone's numeric keypad. This is suitable for the transfer of data from the physical world into electronic devices.

E-ticketing is a widely used idea in the travel industry, but it is challenging to put into practice in a setting where the entire business strategy is built around an 'open access' system for flexible travel and tourism. A system is proposed for generating real time QR code for various monuments, this paper describes the working of the system and briefs over the features. It also talks about the models:

1. Admin module
2. User module.

2 Literature Review

The design and development of a ticketless mobile web-based travel booking system that can be delivered by QR code are covered in this study [1]. The website includes timetables for all company buses as well as an online booking system that enables users to purchase tickets electronically using a valid debit or credit card and a mobile device with Internet connection. The web application is a mobile-based platform that enables potential consumers to request corporate information and make reservations over the web and through Smartphones in particular. The client side architecture was built using JavaScript, HTML5, Cascading Style Sheets, and jQuery. For server-side scripting, PHP was used.

The objective of [2] was to set up an online system for frequent passengers of Nagpur city to buy bus passes and tickets. The information about commuters is essentially stored in a database by program. The system developed in this features two login pages, one for the user and one for the administrator. This is used to create tickets, create bus passes, and renew a user's current bus pass. Every passenger who registers for the first time will receive an one time password (OTP), which will be delivered to the user's email address.

The advantages and disadvantages of E-ticketing in public transportation are discussed in paper [3], although E-ticketing is a common practice in the airline sector, and it is challenging to put into practice in a setting where the entire business model is built around an 'open access' system for flexible travel. Here is a case study of the effective E-ticketing implementation technique used by Deutsche Bahn Mobility, the largest logistical network in the world. Describe E-ticketing tactics in the transportation sector and briefly describe the differences between the trains and aircraft sectors [4].

In [5], the authors propose a model for secure E-tickets which include a QR code containing user-encrypted data. The TLS protocol, which assures secure communication, is used by the user to book a ticket on an application that contacts a web-sales service [6]. A symmetric cryptosystem is used to ensure that another user cannot copy/steal the access to the event. Another layer of security is added to the model and implements hashing of the encrypted token to prevent forgery of the e-ticket. The user can scan his QR code, enter the password and if the ticket is valid, the user is admitted.

The necessity of data security has grown, particularly for Web applications and distributed databases [5]. Using cryptographic methods, whose advancement has been a continual worry, is one way. These algorithms' growing complexity necessitates longer execution durations, which lowers application performance. This study compares the execution timings of three asymmetric key algorithms—RSA, El Gamal, and ECIES—based on the size of the encryption/decryption keys. A benchmark utilizing Java APIs and an application for evaluating the algorithms on a test database were developed for this algorithms comparison [7, 8].

In [9], authors discussed encryption and decryption, everything is quickly becoming completely dependent on Internet communication in the current, digital world. Our lives are made simpler by using the Internet effectively. In the modern day, there are significant security dangers and difficulties with the information that is disclosed online. The answer to protecting data from various security dangers is cryptography. It is evident that stronger encryption technology is needed to increase the security of communication systems. Authors mostly concentrate on data encryption and decryption methods for a superior cryptosystem, and they have put forth a fresh strategy that guarantees improved performance in contrast to cutting-edge approaches. In this work, the plain text is converted into ciphertext after a special key is created using random letters [10].

In [11], authors discussed Cloud, although the cloud-based system is safe and reliable, data can corrupt without the cloud service provider's notice. Better but requiring more storage is the blockchain. In this research, authors developed a key management technique to protect exam storage using public key cryptography; it makes use of the RSA algorithm to give distinctive features of data integrity, secrecy, and authentication. It may be applied to cloud-based systems or centralized server-based systems. The suggested system in this work offers security without extending encoding or decoding times or growing file sizes [12].

3 Methodology

This Section presents the methodology for the proposed system (Fig. 1) that allows users to book tickets to monuments/museums online.

3.1 Tools

Using HTML, CSS, JS, and React for the Frontend and Django for the backend, to provide users a seamless experience to book their tickets. The booking details and information entered by the users were stored in an SQLite database.

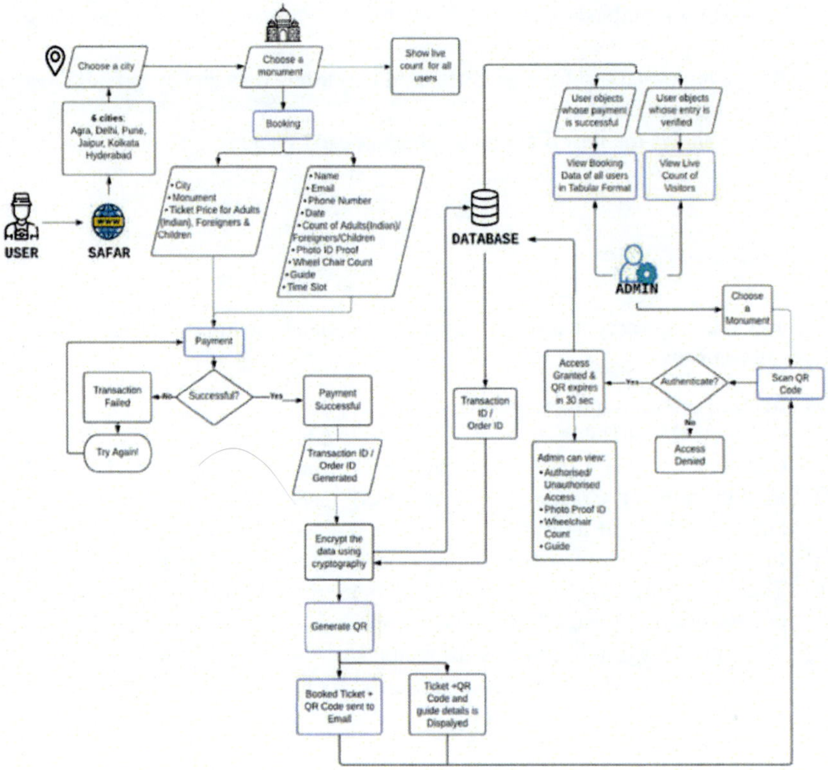

Fig. 1 System architecture

3.1.1 Frontend

HTML, CSS. and JavaScript is used for creating the frontend of the webpages. Also, React is used which is an open-to-use JS library that provides all the necessary tools to build an interactive user interface.

3.1.2 Django

Django is a web framework that runs on the server. It's a free, open-source Python web framework for quick construction, easy maintenance. A group of elements needed to construct an application makes up a web application framework. The Django framework's primary goal is to make it easier for developers to concentrate on the new application components. It handles many of the low-level tasks associated with web development, allowing users to concentrate on creating the components required for their application.

3.1.3 SQLite Database

SQLite is a relational database which is used for storing data and by default Django is used for SQLite Database.

Figure 1 shows the flow diagram for the proposed system.

3.2 RSA Algorithm

The user's data is encrypted before storing to the database to provide an additional layer of security.

(1) Select 3 prime numbers randomly—$p1, p2, p3$.
(2) Calculate the modulus (n) by multiplying $p1 * p2 * p3$.
(3) Calculate the totient '$\Phi(n)$', such that $\Phi(n) = (p1 - 1) * (p2 - 1) * (p3 - 1)$
(4) Find the number 'e' such that 'e' and '$\Phi(n)$' have 1 as the only common factor and 'e' is co-prime to $\Phi(n)$
(5) Find a number 'd' such that d is the modulo inverse of 'e' and '$\Phi(n)$'.
(6) Thus, the public key obtained is $\langle e, n \rangle$ is used to encrypt the data, while the private key $\langle d, n \rangle$ is used to decrypt the data.
(7) The ciphertext generated using the public key is done by finding the modulo of each character in the string to the power 'e' by n.
(8) The ciphertext is decrypted by finding the modulo of each character in the ciphertext to the power of 'd' by n.

To further increase the security, Python's cryptography package is used. The cryptography package's fernet module has built-in functions for generating keys, converting plaintext to ciphertext, and recovering plaintext from ciphertext using the encrypt and decrypt methods, respectively. In the proposed work, the message is encrypted using the above algorithm and a cipher text is generated. The 'd' and modulus 'n' generated are also encrypted using the fernet module.

3.3 Working/Features

Safar contains two main modules whose working and functionalities are discussed below.

User Module

As discussed previously, Safar features an interactive user interface to allow users to book their tickets online without any hassle. The user can access the 'Safar' website in order to book their tickets. The user is then presented with a choice of 6 of the main cities in India–namely–Agra, Delhi, Jaipur, Pune, Kolkata, and Hyderabad. On choosing one of these cities, the user is then directed to a page featuring 3 of the most popular monuments or museums in the chosen city.

Once the user chooses a monument/museum to visit, he/she is directed to the booking page.

The Booking Page consists of a user form containing the following fields to be entered by the user–Name, Contact Number, Email ID, Date of visit, Number of Children, Number of Adults, and Number of Foreign Nationals (if any). It also contains some additional information about the ticket prices, the contact details and the address of the museum/monument and also provides a link to Google's Art and Culture 360° view of the monument/museum. The user must also enter his/her ID proof details such as his/her Aadhar No. or PAN No. or Passport No. for authentication and identification purposes.

This along with the information about the city and monument chosen and the ticket price is saved to the database. The user is then directed to a page where the user can make the payment. If the payment is successful, a QR is generated using the user's order_id and displayed on the screen along with the ticket. This is also sent to the user's email.

If the payment is unsuccessful, a message 'Transaction failed' is displayed and the user can make the payment again. After a successful payment and booking, the ticket with the relevant details will be displayed on the screen as well as it will be mailed to the Email ID provided by the user in the Booking Form. The E-ticket will contain all the relevant information about the booking along with a QR code which the user will need to scan to verify his/her booking before entering the monument/museum.

Admin Module

The admin, the other two features are available to the admin. The first one is to view the booking details of all the users who have made the payment and second one is to view the daily live count of all the visitors whose entries have been verified. To access the Admin Module, the admin is required to login using the username and password provided to him/her. After successfully logging in, he/she is directed to the page for QR scanning. The admin can select the monument/museum whose QR code has to be scanned from the given drop-down list. If the QR code is verified a green box is created around the QR code indicating that the verification has been successful and the tourist can be allowed to proceed inside the monument/museum. However, if the QR code fails verification, a red box will be created around the QR code, indicating that the tourist shall be denied entry inside the museum/monument. The Admin Module has two other features that allow the admin to view all the booked records for the current day. Users can also sort these records to view the specific records for a given city and whether the tourist has visited the monument or not. The admin can also view the Live Daily visitor count of the tourists sorted according to the Monuments/Museums displayed using an elegant dashboard.

QR Scanning Feature

The user can enter the monument by showing the E-ticket to the admin who scans the QR code and verifies the booking. The QR code contains the encrypted 'order_id' of the visitor, which is then decrypted using the OpenCV library and programming logic in the backend. The decrypted 'order_id' is checked against the database to verify the booking of a ticket. The time of scan of the QR code is noted and updated to the database. The QR code remains valid for 30 s after scanned at the entry by the Admin. After this time elapses, the user will not be granted entry inside the museum/monument. This time window is introduced to expire the QR code on the E-ticket after a single use and will prevent tourists from entering the monument/museum multiple times using a single ticket.

Payment Feature

For the payment feature, RazorPay Testing API is used to introduce a payment gateway in the proposed system.

4 Results and Discussion

4.1 User

The user can access the website through the link and will be prompted to the home page where they will be able to select a city for which they want to book a ticket. This can be seen in Fig. 2.

After the user has selected the city, they will be redirected to the locations page where the monuments in the city can be selected. The page also contains general

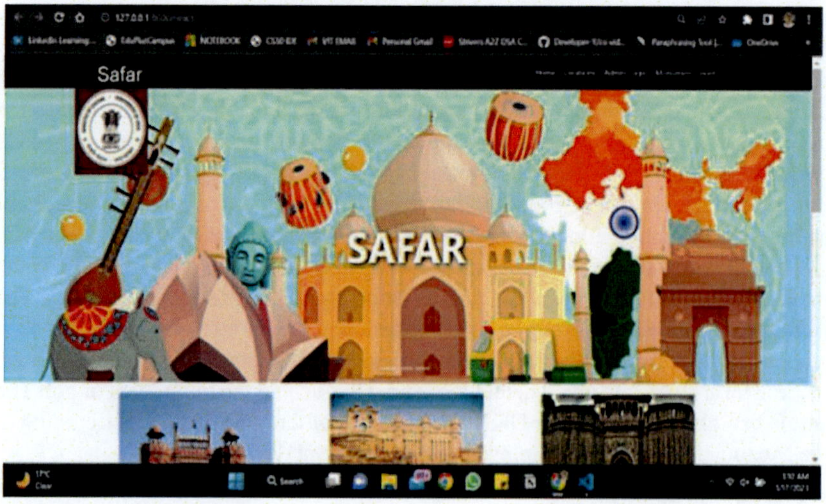

Fig. 2 The homepage of Safar

information about the city and its culture to enhance the user experience. This can be seen in Fig. 3.

Once the user selects the monument, it will be redirected to the booking page as in Fig. 4, where the user can enter the required information and book their tickets.

After all the steps, the user will be redirected to the payment portal where payment can be made using either card or UPI as seen in Fig. 5. Once the payment is done a

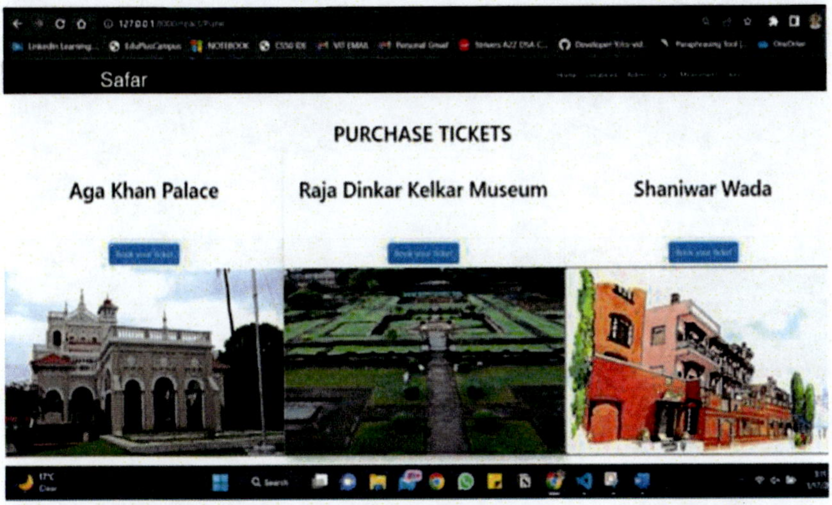

Fig. 3 Choice of monuments/museum provided by user

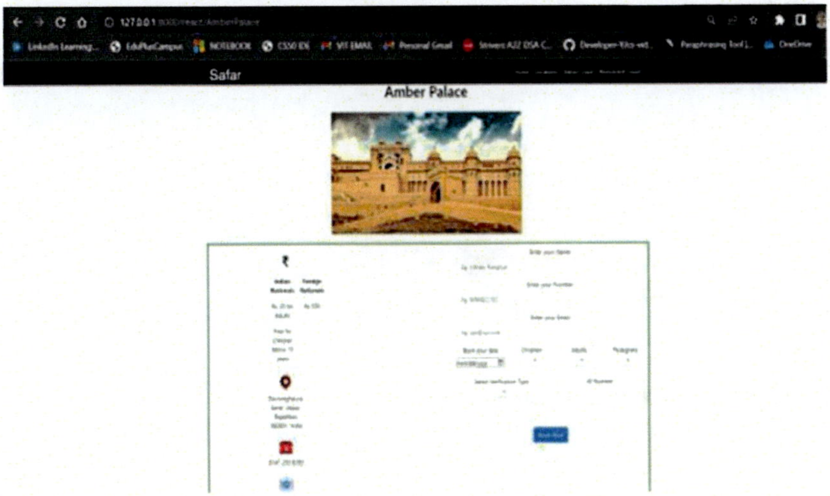

Fig. 4 Booking page for a chosen monument/museum

ticket with a QR code is generated and sent to the registered email id entered. This can be seen in Fig. 6.

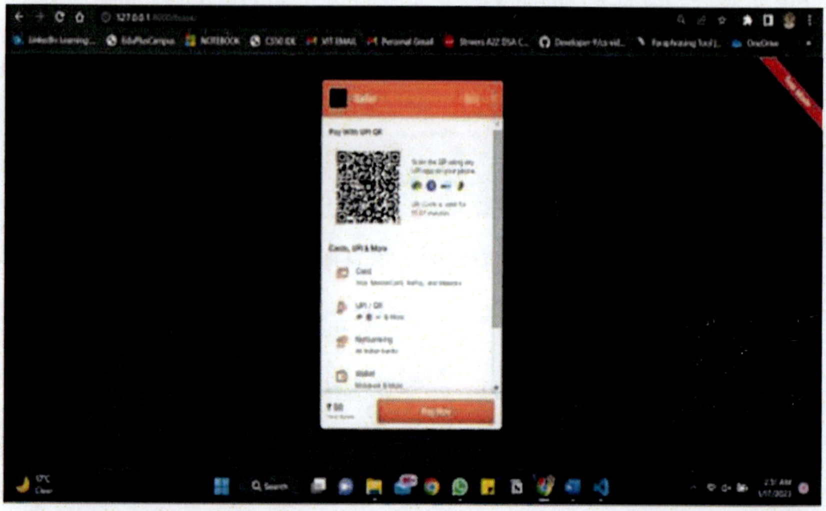

Fig. 5 Payment page

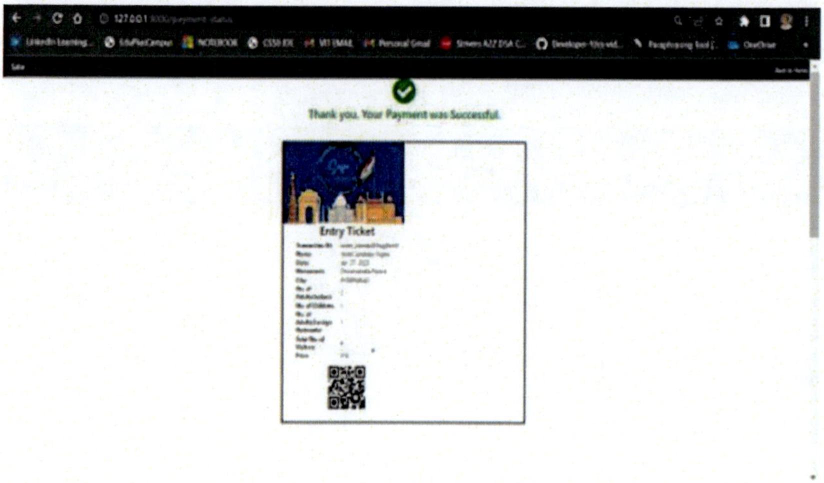

Fig. 6 Booked E-ticket displayed on the screen

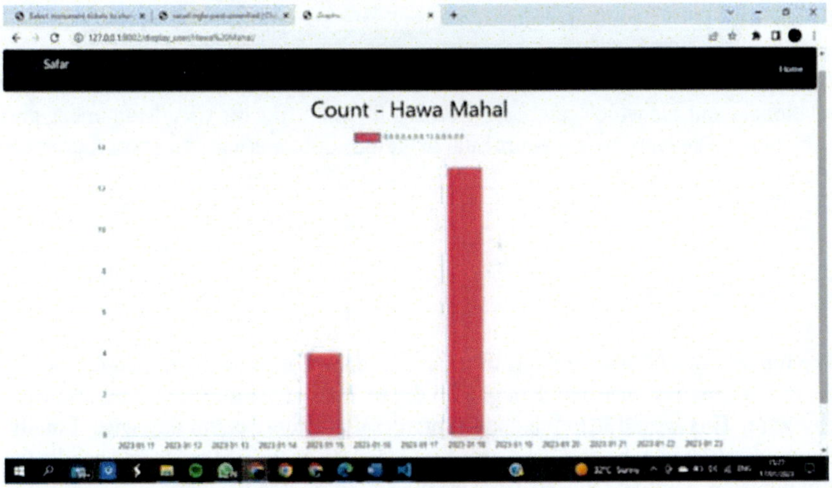

Fig. 7 Total counts of booking for the selected monument/museum for all days

4.2 Admin

To enter into the admin panel, the admin first has to log in. After successfully logging into the admin panel, the admin can view the scanning page wherein the admin has to choose the monument from the drop-down list and click on the scan icon.

Next, the user has to place the QR code, sent on his/her email, in front of the camera to get it scanned. Now if the ticket is valid, then the access is granted and is depicted using a green box around the QR code. Once the access is granted, the QR is valid for another 30 s after which the QR code expires and the access is denied.

The admin dashboard has a ticket page where the daily records of all the monuments are displayed and the admin can filter the records according to the city and verified status. The admin dashboard has a count page where the live count of all the monuments is displayed and by clicking on a particular monument, the whole visitor's history till the date of the monument can be viewed by the admin and the admin can filter the records according to the date. The admin dashboard can be seen in Fig. 7.

5 Conclusion

The proposed system is a prototype to an E-Tourism Booking System for monuments and museums. The web stack which was used for the prototype is React, Html, CSS, and Django. The moveable camera is responsible for scanning the QR code and the access message of Access Denied or Access Granted, as well as the count of people.

By using the RSA Cryptographic Algorithm for the encryption and decryption of data stored in the database, an additional degree of protection is added to the proposed system. The admin side is responsible for collecting all the data and maintains a relational database which can be queried and filtered using the City, Monument, and Date Filter. The website is instrumental in various domains and can boost tourism in India.

6 Future Scope

The future scope of this system is to collect the data of the number of people visiting the monument day wise and then predict the number of visitors for the monument's day wise. This would help the user to predict the visitors count day wise. Finally, this application may be deployed on a cloud to manage the large data and handle the simultaneous booking requests from users in an error-free manner.

References

1. Sunday D, Aliyu WA (2016) Design of mobile-based travel e-ticketing using QR-code. In: Proceedings of the second annual research conference of Federal University Lafia, Nigeria, p 241
2. Gupta A, Iram B, Samrit B, Dhage C, Khan N (2018) Online facility of ticket booking and generating bus pass using qr code. Int Res J Eng Technol 5:03–51
3. Sahdev N, Kumar S. e-ticketing strategy and implementation
4. Pasyeka M, Malitchuk A (2020) Software implementation of movie ticket booking system. In: 2020 IEEE international conference on problems of infocommunications. science and technology (PIC S&T), Kharkiv, Ukraine, pp 53–56. https://doi.org/10.1109/PICST51311.2020.9468015
5. Satpalkar T, Salian S, Stephen S, Shaikh S (2016) Smart city parking: a QR code based approach. Int J Eng Res Technol (IJERT) 5(2):2278–0181
6. Conde-Lagoa D, Costa-Montenegro E, González-Castaño FJ, Gil-Castiñeira F (2010) Secure eTickets based on QR-codes with user-encrypted content. In: 2010 digest of technical papers international conference on consumer electronics (ICCE), pp 257–258. https://doi.org/10.1109/ICCE.2010.5418880
7. Qadir AM, Varol N (2019) A review paper on cryptography. In: 2019 7th international symposium on digital forensics and security (ISDFS), pp 1–6. https://doi.org/10.1109/ISDFS.2019.8757514
8. Galla LK, Koganti VS, Nuthalapati N (2016) Implementation of RSA. In: 2016 international conference on control, instrumentation, communication and computational technologies (ICCICCT), Kumaracoil, India, pp 81–87. https://doi.org/10.1109/ICCICCT.2016.7987922
9. Kumar S, Mahotsav RS (2015) E-ticketing of all ticketed monuments. Survey plans by ASI through ISRO, 31 Dec 2015
10. Imam R, Areeb QM, Alturki A, Anwer F (2021) Systematic and critical review of RSA based public key cryptographic schemes: past and present status. IEEE Access 9:155949–155976. https://doi.org/10.1109/ACCESS.2021.3129224

11. Masud KI, Hasan MR, Hoque MM, Nath UD, Rahman MO (2022) A new approach of cryptography for data encryption and decryption. In: 2022 5th international conference on computing and informatics (ICCI), pp 234–239. https://doi.org/10.1109/ICCI54321.2022.9756078
12. Hamza A, Kumar B (2020) A review paper on DES, AES, RSA encryption standards. In: 2020 9th international conference system modeling and advancement in research trends (SMART), Moradabad, India, pp 333–338.https://doi.org/10.1109/SMART50582.2020.9336800

Security Enhanced RFID-Based Digital Locking System in Home Automation Using IoT Framework

M. Dhivya, A. B. Gurulakshmi, G. Rajesh, and Sanjeev Sharma

Abstract In modern era, smart home automation system provides sophistication and comfort in day-to-day life. Secure, scalable, efficient authenticated protocols are designed for indoor localization, real-time tracking, and flexibility of applications in home automation. In this paper, a novel, secure, and cost effective IoT framework is implemented using Raspberry Pi 3 B+ controller and sensor network. The proposed model is an advanced access control system encompassing face recognition, biometric authentication that provides decentralization of information in the home environment including protection of the PET animals. The proposed model is analysed for three different scenarios and an exhaustive theoretical and extensive experimental analysis is carried to prove the significance of the proposed model.

Keywords Home automation · Radio frequency identification (RFID) · Internet of things (IoT) · Digital locking system · Face recognition · Biometric authentication

1 Introduction

Automation and embedded systems market has become a trend to design various cyber physical systems incorporating IoT and IIoT for aerospace, military systems, consumer electronics, communications, automotive, industrial controls, and other sectors, including smart cities. One vital aspect in day-to-day application is Smart Home automation. Smart Home automation has extended swiftly in terms of access to network technology and enhancement of systematic analytics. People can control and secure their houses conveniently by using IoT and RFID technology. The control strategies in designing a home includes durability, ease of control/convenience, and assist with security and energy savings of home.

M. Dhivya (✉) · A. B. Gurulakshmi · G. Rajesh · S. Sharma
Department of Electronics and Communication Engineering, New Horizon College of Engineering, Bengaluru, Karnataka, India
e-mail: dhivyamanian@gmail.com

G. Rajakumar et al. (eds.), *Intelligent Communication Technologies and Virtual Mobile Networks*, Lecture Notes on Data Engineering and Communications Technologies 171,
https://doi.org/10.1007/978-981-99-1767-9_29

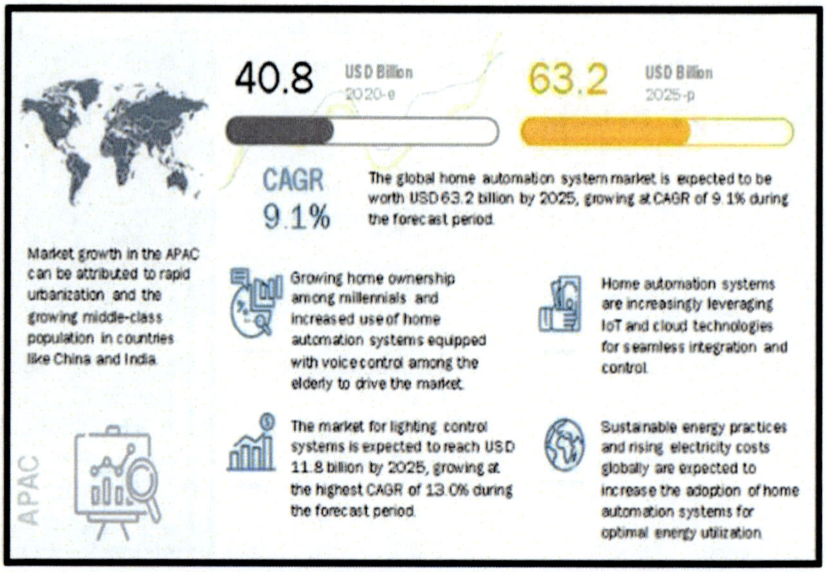

Fig. 1 Statistics of home automation system

According to Markets and Markets [1], the home automation system market is projected to grow from USD 40.8 billion in 2020 to USD 63.2 billion by 2025; it is expected to grow at a Compound Annual Growth Rate (CAGR) of 9.1% from 2020 to 2025. The ease of use and affordability of home automation systems drives the growth of home automation system market from sophisticated HAS to DIY HAS.

The statistics of home automation systems and RFID technology are depicted in Figs. 1 and 2. The objective of this paper is to design a Security Enhanced RFID-Based Digital Locking System In Home Automation Using IoT Framework. The rest of the paper is organized as follows: Sect. 2 discusses the related work in home automation system. In Sect. 3, the proposed framework is discussed. Description of the proposed model and the flowchart is depicted in Figs. 4 and 5. Section 4 presents the results and discussion. Section 5 elaborates on the conclusion and future directions.

2 Theoretical Background

Home automation is an inevitable topic in consumer electronics and intelligent control and computational systems. Home automation dates back to late 1960's in literature. Thring [2] has stated the inventive skills of engineer for satisfactory lifestyle of men and women. In early eighties, standardization of the Home Bus System was implemented by Murata et al. [3], for effective realization of

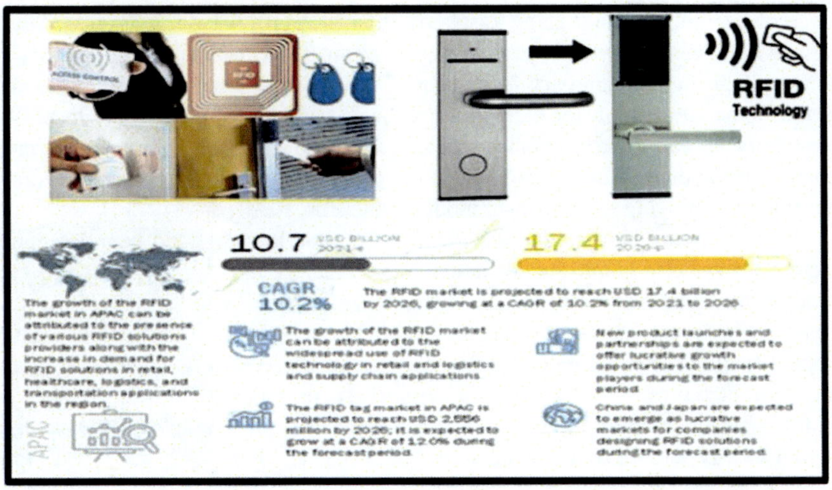

Fig. 2 Statistics of RFID automation systems

home automation system. In late eighties, conformance test system, development tools, microcomputer-based intelligent sensors and various revised protocols were devised for home automation system. Reliability of home automation system, simulation models for home automation employing artificial intelligence, self-optimization-based systems, spectrum technologies, comprehensive description of system encompassing features, architecture, control algorithms etc., were discussed during 1990–2000 [4, 5].

Home automation system employing java-based stand-alone embedded system board integrated into a PC-based server at home was implemented in 2004 by Al-Ali and Al-Rousan [6]. The authors developed the system incorporating password protection to block unauthorized users from using the home appliances. Zigbee-based wireless sensor network, virtual reality–3D virtual world, routing requirements were garnering significance during 2005–2010.

Zigbee-based home automation by Gill et al. [7] incorporated a light switch, radiator valve, safety sensor, and ZigBee remote control. The model developed is a virtual home security system with feasibility and effectiveness. X10, Z-Wave, ZigBee, INSTEON, and EnOcean home automation technologies were surveyed by Withanage et al. [8]. Since the inception of home automation, for four decades it had significant increase in the usage of technologies enabling to address the barriers like security, connectivity of Internet, protocols, and smart communication.

Raspberry Pi-based home automation gained significance in literature from 2014 onwards. As per google scholar reports, nearly 300 articles are published in the smart home automation based on Raspberry Pi from 2014 to 2021. Jain et al. [9] authored the paper on smart home automation through E-mail in 2014. Python programming environment was used with Raspberry Pi for implementation of the proposed algorithm. Vladimir Vujović and Mirjana Maksimović implemented Sensor Web node as a part of Internet of Things using Raspberry Pi in 2015 [10]. Vladimir developed the testbed with low energy and resource consumption. They also elaborated the usage of fuzzy logic in their approach for effective flexibility and scalability of the project. Smart home automation for elderly people was designed using python, OpenCV, Raspberry Pi, and android application by Vaidya [11]. All the equipment's were controlled by voice of the user from the mobile phone. Advancements in automation developed employing loT of sensor modules like PIR, RFID, etc. [12, 13].

In recent years, blockchain technologies, machine learning, and deep learning techniques are also employed for home automation. Popa et al. [14] developed a modular platform used for collecting, aggregating, and storing of the information of smart environment. Parvaresh and Faghani [15] has emphasized the need of information routing for effective data delivery in automation. Thus considering a brief survey of literature, we have proposed a model incorporating RFID with Raspberry Pi for home automation. In this model, the user recognition is employed enhancing digital lock system for both the inmates of the house and Pets.

3 Proposed Architecture

3.1 Problem Statement

Locking systems commonly employed has physical key and lock. Physical keys are unique for the lock and there can be fixed number of possible key 'codes' or blend of diverse inner elements that can unlock it. Physical keys provide safety and control access authorization. If keys are out-of-place (or) missing (or) pilfered then entire locking mechanism to be revamped. Relatively depending on a key to turn, at present prevalent electronic locks are furnished with motors concealed within the door itself. Large firms go with two factor authentication for hardware keys even-though it is susceptible to get lost and it's a burden to have numerous keys for different doors.

Alternative measures for physical keys are smart door lock with keyless entry, keypad entry, smart card entry through Radio Frequency Identification (RFID). The device can pair with more than one card. Nevertheless, they are susceptible to theft and loss. And carrying a key defeats the purpose of not carrying one. Face recognition is proposed as a replacement for existing systems to overcome all these problems.

3.2 Proposed Solution

To tackle the above issues, we propose to replace physical keys with an equipment that contains digital information like fingerprints and face recognition which allows the user to unlock the lock in multiple ways. For example, you can still open the lock with face recognition even if your fingerprint is not working. The equipment

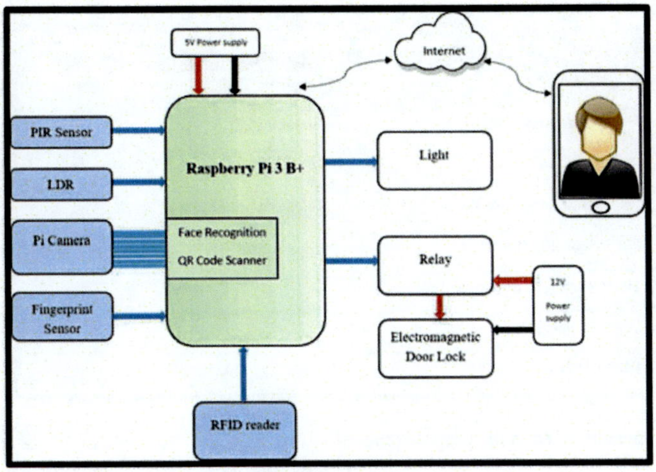

Fig. 3 Proposed architecture of digital locking system

consists of a PIR sensor to detect any motion in front of the door, LDR to measure the light intensity, Camera module to monitor the surroundings, recognize faces and scan barcodes, fingerprint scanner for user authentication, RFID reader to read the RFID tags and a processor to process all the digital information, electromagnetic lock for opening and closing the door and a processor to process all the digital information obtained by all the modules and unlock the electromagnetic lock.

The proposed model also includes the provision to control the 'IN' and 'OUT' passage of the pets using RFID Reader. Proposed Architecture of Digital Locking System is shown in Fig. 3. It reads the tag present in the pet's collar for authentication. The Face recognition system and Fingerprint scanners are used for user authentication using which the user can unlock the lock using multiple ways. The camera module is also used as a Barcode scanner, which scans the tracking ID present in the couriers and receive the packages from the personnel. The hardware and software components used in the project are listed in Fig. 4.

3.3 Description of the Proposed Model

- A live person generally emits infrared energy and when a person approaches the operating range of the PIR sensor, PIR sensor detects the change in temperature and send the equivalent voltage (sensing signal) to the Raspberry Pi. The Raspberry Pi checks for the light intensity using the LDR sensor. During daytime the light intensity is more. Hence, the resistance offered by the LDR is very low and current flows through the LDR and ground and low voltage is passed to the Raspberry Pi. Similarly, in night time the resistance of LDR is high compared to other

Components	Software
Raspberry Pi 3 B+	Raspbian OS (4.19)
Pi Camera (module V2)	OpenCV (4.2)
EM-18 RFID module	Python 3.8.2
Fingerprint Sensor (R-307)	Telegram bot API
PIR Sensor (HC-SR501)	
Electromagnetic door lock.	
LDR	
Power Supply	

Fig. 4 Components used in the proposed model

part of the circuit. As a result, the current flows through the low resistance lane such that high voltage is passed to the Raspberry Pi. Whenever the signal is high Raspberry Pi turns on the light.

- It activates the camera module and the fingerprint scanner for user authentication. The user can unlock the door in multiple ways. The camera module is used for face recognition which uses Local Binary Pattern Histogram (LBPH) algorithm. Fingerprint scanner is used as an alternative way to unlock the door.
- If the user's face is recognized by the algorithm or fingerprint matches the saved data in the database, Raspberry Pi sends the logical high command to the electromagnetic door lock.
- The electromagnetic door lock unlocks the door when it receives a high voltage input. Raspberry pi automatically locks the main door after a fixed time delay if there is no further movement within the PIR operating range.
- In contrast, if the person's face is not recognized by the algorithm, an image of the person is captured and sent to the owner through Internet. This is done using the Telegram Bot API. Along with the image, Raspberry Pi sends two inline keyboards such that the owner can send the call back message whether to unlock the door or not for the user.
- Apart from these, we have designed a pet door which lets the pet come and go as they please. A RFID reader is places on top of the pet door such that it reads the RFID tag on the pet's collar. Raspberry Pi opens or unlocks the pet door whenever it detects the RFID tag. This feature also restricts all other stray animals from entering the house.
- Our system can also receive the packages when there is no one in the house. The Raspberry Pi can read the barcode of the package which contains the tracking id using the camera module. If the tracking id matches your courier tracking id, it unlocks the pet door such that the package can be placed inside the house.

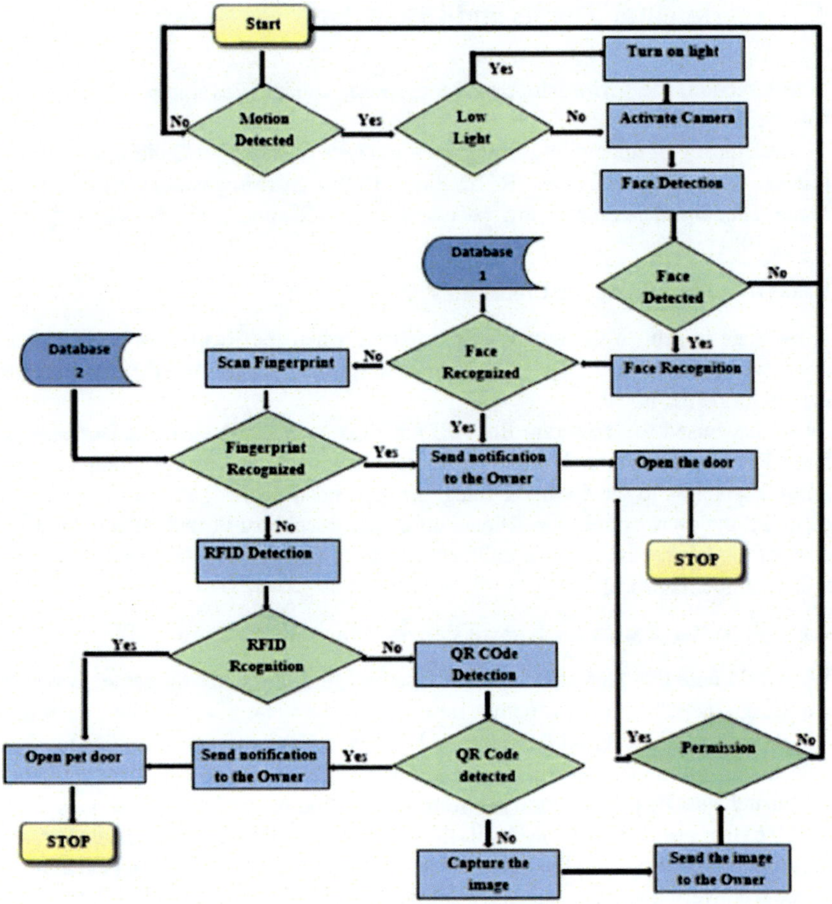

Fig. 5 Flowchart of the proposed model

3.4 Flowchart of the Proposed Model

The flowchart of the proposed model is depicted in Fig. 5. Database 1 and Database 2 is fed to fingerprint recognition system and face recognition system. When motion is detected in camera module, it is checked with the existing database.

Database 1 and Database 2 is fed to fingerprint recognition system and face recognition system. When motion is detected in camera module, it is checked with the existing database. If the match is obtained notification is sent to the owner to open the door. Else, finger print is obtained through biometric fingerprint scanner and if the match is obtained the owner will open the door. For the pets, the pet door is accessed through RFID.

4 Experimental Results and Discussion

In this section, the performance of the proposed model is evaluated via hardware software integration.

The efficacy of the model is tested as four cases as shown in Table 1. The experimental setup of the enhanced RFID-Based Digital Locking system is depicted in Fig. 6. The sensed parameters are processed through Raspberry Pi 3 B+ and Raspbian OS (4.19).

Case (i) Authorization for Unknown Person

A message is sent to the owner's Smartphone when the Raspberry Pi detects an unknown person (Fig. 7). The owner can take further decision whether to allow the person inside the house.

We have used the Telegram Bot API which is an HTTP-based interface used to build Bots for Telegram. It will send the unknown person's image along with the inline keyboards to the Owner's Telegram app and wait for the response from the owner. If the owner clicks the Allow button, the Raspberry Pi will unlock the main door of the house and automatically close after a fixed time delay else the system will continue its process.

Case (ii) Authorization for Known Person—Face Recognition

We should note that face detection and face recognition are not the same. Since we are using OpenCV, it comes with a lot of pre-trained classifiers for face detection and its data is stored in XML files. We have used Haar cascade classifier for face detection in our model. We will import the Opencv library which is used for computer vision and NumPy python library to store the location of the detected faces. We will load the Haar cascade classifier XML file which contains all the trained data for face detection. Start capturing the video using the Raspberry Pi camera and each video frame is converted into grey scale. The algorithm finds faces in the image. If the faces are found, the function returns the position of the face. The facial images cropped and converted to grey scale are used to train the algorithm. The Raspberry Pi will capture 50 images of each people we want to recognize and a database is created with

Table 1 Different analysis of the proposed model

	Description	Inference
Case 1	Unknown person	Telegram API BOT is used to permit access (or) deny the person enter the house
Case 2	Face recognition • Known person	• Permitted access
Case 3	Biometric access • Known person	• Permitted access
Case 4	RFID recognition and bar code detection	• Permitted access for pets and to receive courier

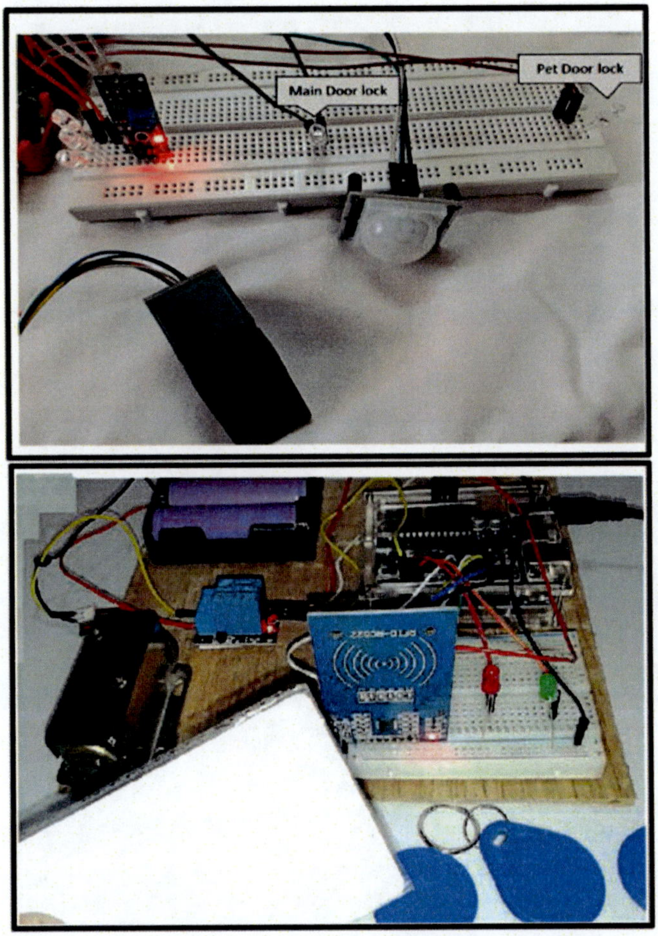

Fig. 6 Experimental setup of security enhanced RFID-based digital locking system in home automation

that facial images where images of same person is given same ID. As a result, the algorithm use this information and recognize the input image. The original images are converted into intermediate images which describes the images in better way, by highlighting the facial characteristics. The LBP algorithm uses sliding window concept considering the radius and neighbours of each pixel. A part of the image is considered as a window of 3×2 pixels and represented as a matrix containing intensity of pixels (0–255).

The central value of the matrix is set as a threshold and based on the threshold value, the values of the 8 neighbouring pixels is defined with new value. If the pixel value is greater or equal to the threshold, the pixel value is considered as 1 and if pixel value is less than the threshold, the pixel value is considered as 0. As a

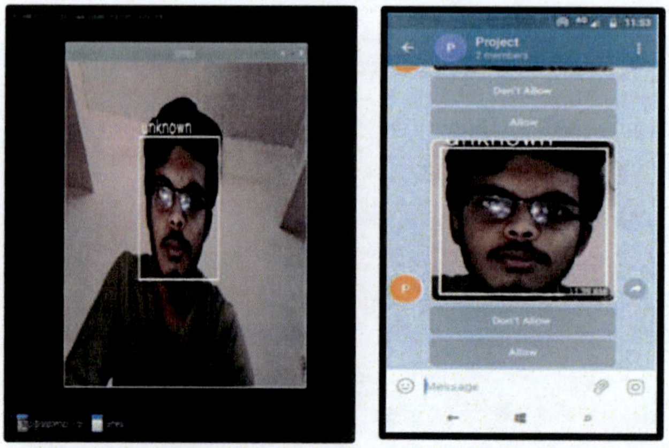

a. Unknown user is trying for access to enter to house **b. Image sent to owner through Telegram Bot API**

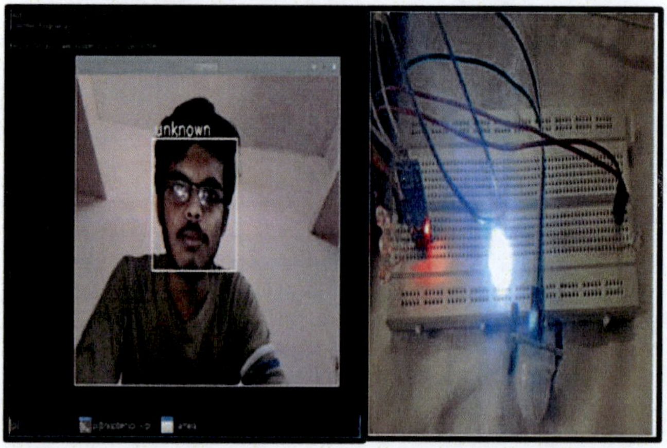

c. Owner Provides Access for Entry **d. Door is unlocked**

Fig. 7 Authorization for unknown person

result, we will get an 8-bit binary number when we traverse the matrix in clock-wise direction. The 8-bit binary number is converted to decimal and set it as the central value of the matrix. At the end of the LBP procedure, a new image is created which represents the characteristics of the image better than the original image. Using the arguments of the position of the face, the image is divided into multiple grids and a histogram of each region is extracted. Since the image is in grey scale, each histogram will contain values in the range of 0–255 which represents the occurrences of each pixel. Each histogram is concatenated to create a bigger histogram and the final histogram represents the characteristics of the original image. At this movement, the algorithm is completely trained and each image in the training dataset are converted

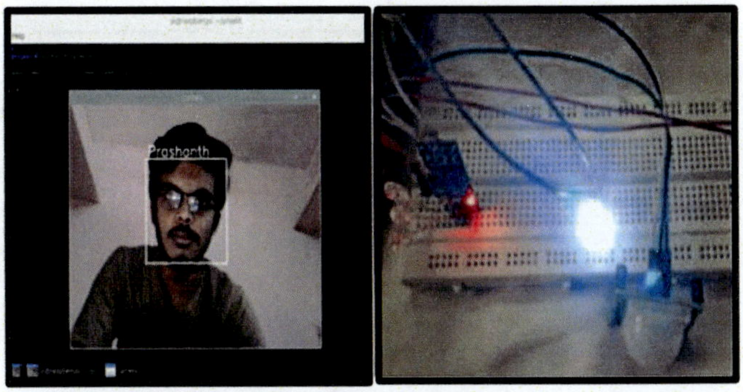

a. User Identified Recognition b. Lock is Opened using Face

Fig. 8 Authorization for known person

into histograms and its data is stored in a YML file. While recognizing a new facial image, all the above steps are repeated to create a histogram of the image. The new histogram is compared with the histograms in the dataset and the ID of the image with the closest histogram is returned. If there is no closest histogram, the algorithm will return -1. The LPBH algorithm is trained using fifty images of each person for face recognition. The training was successful and the output is shown in Fig. 8.

Case (iii) Authorization for Known Person—Biometrics Recognition

Although we have designed a face recognition system, it may fail to recognize due to low illumination or image quality. Therefore, we have biometric authentication as an alternative way to unlock the door (Fig. 9). The fingerprints of each people we want to give access is stored in the database. When the fingerprint scanner detects a finger, it will convert the fingerprint into a unique code and compares it with the unique codes stored in the database and will unlock the electromagnetic lock if it matches. Fingerprint sensor is interfaced with the Raspberry Pi and can be used as an access control system along with face recognition. A HTTP-based interface, Telegram Bot API in Raspberry Pi is made possible using TeleBot, a python package. As a result, the Raspberry Pi can chat with the House owner.

Case (iv) Authorization for Receiving Courier in Absence of Owner

Barcode is a method of representing data in visual form which can be scanned by optical scanners and also from Camera module (Fig. 10). We have used a Raspberry Pi camera module to read the barcode present in the packages. All the packages have a unique tracking number which is printed on the shipping label as a bar code. Tracking numbers are used to locate the time sensitive deliveries. In our project, the system will scan for a barcode in the screen. If the barcode data is same as the data stored in the database, the system will unlock the Pet door so that the package can be placed inside the house in the courier tray. This is shown in Fig. 10.

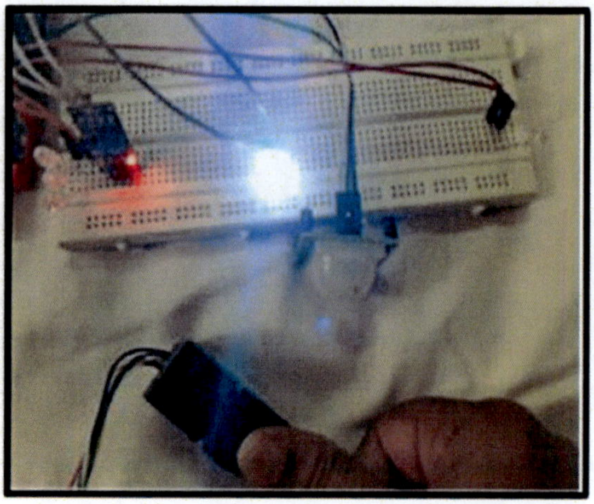

Fig. 9 Lock opened using fingerprint

5 Conclusion

The proposed system allows to unlock the door remotely without physical user intervention. It is more convenient than the mechanical keys. This paper presents a novel Internet of Things-based access control system which can detect user, fetch user Id, analyse and verify, send notifications, and process according to the response which integrates home security with home automation. Since our system is built on wireless network sensors, it is cost efficient, easily installable system. The intruder alert feature enhances the security of the system. The system consumes minimal power and automatically unlocks and locks the door. It is a low-cost access control system based on face recognition, making home automation more secure. Practically this system can be implemented in any place where security, automation or remote controlling is required. Some of the functionality that this system needs to be further developed is to replace the face recognition system with the Face ID which enables secured authentication. We can use this system instead of CCTV to boost security. We can have real-time access to the criminal database to track the criminals.

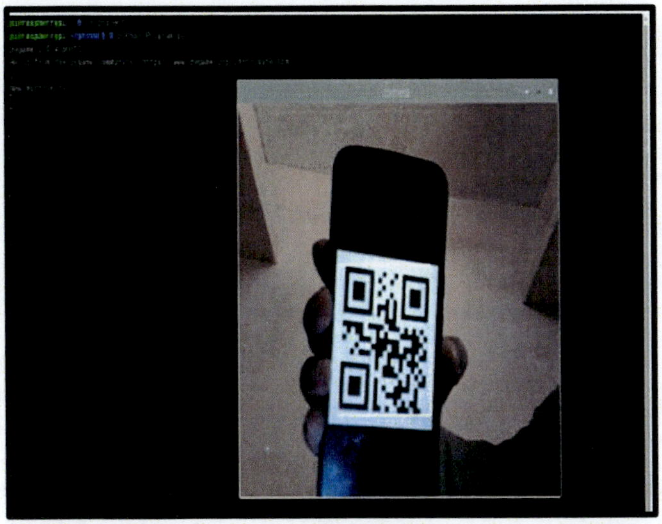

a. Barcode Scanner

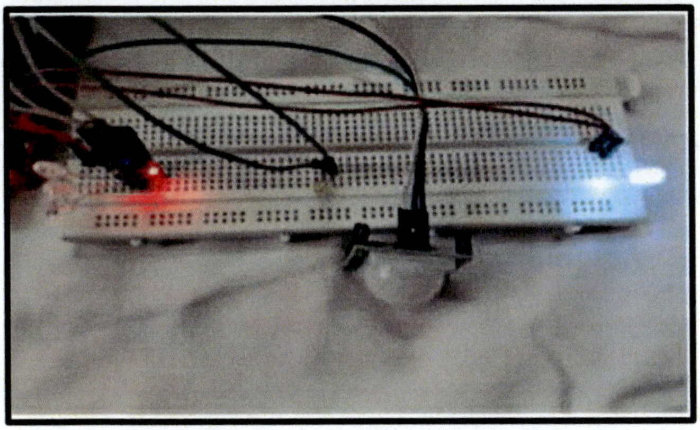

b. Pet Door Lock Opens after authentication

Fig. 10 Authorization for receiving courier in absence of owner

References

1. Markets and markets [online]. https://www.marketsandmarkets.com/Market-Reports/home-automation-control-systems-market-469.html. Accessed 21 July 2022
2. Thring MW (1968) Automation in the home. Electron Power 14(11):440–441
3. Murata M, Namekawa T, Hamabe R (1983) A proposal for standardization of home bus system for home automation. IEEE Trans Consum Electron CE-29(4):524–530
4. Stauffer HB (1991) Smart enabling system for home automation. IEEE Trans Consum Electron 37(2):xxix–xxxv
5. Zainzinger HJ (1998) An artificial intelligence-based tool for home automation using MATLAB. In: Proceedings tenth IEEE international conference on tools with artificial

intelligence (cat. no. 98CH36294), Taipei, Taiwan, pp 256–261
6. Al-Ali AR, Al-Rousan M (2004) Java-based home automation system. IEEE Trans Consum Electron 50(2):498–504
7. Gill K, Yang S-H, Yao F, Lu X (2009) A zigbee-based home automation system. IEEE Trans Consum Electron 55(2):422–430
8. Withanage C, Ashok R, Yuen C, Otto K (2014) A comparison of the popular home automation technologies. In: Proceedings of IEEE innovative smart grid technologies—Asia (ISGT ASIA), Kuala Lumpur, Malaysia, pp 600–605
9. Jain S, Vaibhav A, Goyal L (2014) Raspberry Pi based interactive home automation system through E-mail. In: Proceedings of international conference on reliability optimization and information technology (ICROIT), Faridabad, India, pp 277–280
10. Vujović V, Maksimović M (2015) Raspberry Pi as a sensor web node for home automation. Comput Electr Eng 44:153–171
11. Vaidya B, Patel A, Panchal A, Mehta R, Mehta K, Vaghasiya P (2017) Smart home automation with a unique door monitoring system for old age people using python, OpenCV, Android and Raspberry pi. In: Proceedings of international conference on intelligent computing and control systems (ICICCS), Madurai, India, pp 82–86
12. Ishak SN, Malik NNNA, Latiff NMA, Ghazali NE, Baharudin MA (2017) Smart home garden irrigation system using Raspberry Pi. In: Proceedings of IEEE 13th Malaysia international conference on communications (MICC), Johor Bahru, Malaysia, pp 101–106
13. Román V, Ordieres-Meré J (2018) WiP IoT blockchain technologies for smart sensors based on Raspberry Pi. In: Proceedings of IEEE 11th conference on service-oriented computing and applications (SOCA), Paris, France, pp 216–220
14. Popa D, Pop F, Serbanescu C, Castiglinone A (2020) Deep learning model for home automation and energy reduction in a smart home environment platform. Neural Comput Appl 31:1317–1337
15. Parvaresh F, Faghani F (2021) A new routing algorithm based on 2-hop neighbourhood information scheme for total power minimization in wireless ad hoc networks. Int J Syst Control Commun 12(2):91–115

Towards a More Secure and Transparent Crowdfunding Ecosystem Using Blockchain

M. R. Sumalatha, Rozen Berg, B. S. Sandeep, and M. Tharunraj

Abstract A recent development in industry and a cutting-edge means of generating funds is blockchain-based crowdfunding. While it has similarities with conventional crowdfunding, it has some unique characteristics of its own. In the light of this, the success criteria that affect how traditional crowdfunding turns out might have a different impact on crowdfunding enabled by blockchain. It is currently unknown what distinguishes the success criteria for blockchain-based fundraising efforts from those for conventional crowdfunding, given the fact that the amount of these projects has grown significantly over the past several years. Such information is necessary for organisation to properly organise their blockchain-based fundraising operations and to help potential investors find the important characteristics and motivators of outstanding initiatives. Regulators and market players would also benefit from understanding how the present regulatory framework applies to blockchain-based crowdfunding. Due to the specific properties of blockchain-based crowdfunding, legal structures may need to be interpreted in order for law to be successfully enforced. In order to fill this information vacuum, a variety of significant literature on the success determinants for both conventional and blockchain-based crowdfunding was looked upon. The results of this literature evaluation offer recommendations for the direction of future research and development. The study has helped to clarify the differences and parallels between traditional crowdfunding and blockchain-based crowdfunding.

Keywords Blockchain · Crowdfunding · Ethereum · InterPlanetary file system (IPFS) · Smart contract

1 Introduction

Cyber-risks are one of the most difficult problems, the technology and information industry has recently had to cope with since they have the potential to pose a major

M. R. Sumalatha (✉) · R. Berg · B. S. Sandeep · M. Tharunraj
Anna University, Chennai, India
e-mail: sumalatha@annauniv.edu

G. Rajakumar et al. (eds.), *Intelligent Communication Technologies and Virtual Mobile Networks*, Lecture Notes on Data Engineering and Communications Technologies 171,
https://doi.org/10.1007/978-981-99-1767-9_30

threat to consumers, companies, and organisations. Now that the Internet of things is routinely used, network, wireless network, and information technology, IT systems, and any sensitive data they could contain are extremely exposed. Shifting the risk to the outside entities may be more beneficial for the organisations, even though there are approaches that may help firms discover possible cybersecurity problems and improve their cybersecurity. The practice of transferring the financial risk connected to network and computer accidents to a third network provider known as the insurer is known as cyber-insurance.

Crowdfunding is a method of getting funds from many of different people or businesses [1]. Investors may do this and make money by giving to any cause they are passionate about the benefit of the effort's success. There are currently a large number of sites for crowdfunding that take big contributions from investors in return for unreliable guarantees. Blockchain-powered crowdfunding modifies the conventional approach to handling business money. Before such a person can pitch their concept to prospective clients or collaborators in order to collect money to begin a corporation, they frequently need to develop a plan, data, and models. These funding sources included lenders, seed funding, and venture capital.

In 2022, more seasoned businesses also began investigating the novel financing solutions. Due to the complex and evolving nature of the phenomena, there is a lack of information on the distinctions between crowdfunding that use traditional techniques and those that use blockchain technologies. While traditional and blockchain technology crowdfunding have certain similarities, there are also many areas where they diverge. Because of this, blockchain-based crowdfunding [2] could not be affected by the success criteria that determine the outcome of traditional crowdfunding. The success variables for Ethereum-based initiatives are not as well understood as they should be as compared to traditional fundraising techniques.

1.1 Research Objectives and Contribution

The research objectives of this work are

- Comprehension of the situation of crowdfunding today and the difficulties it faces, including as fraud and a lack of transparency.
- Investigation on how blockchain technology may increase crowdfunding's integrity by increasing transparency and lowering the risk of fraud.
- Creating and implementing a blockchain-based crowdfunding platform, and assessing how well it protects crowdfunding process integrity.
- Scalability and viability of the blockchain-based crowdfunding platform, as well as any implementation issues or restrictions.

The study can show how security and transparency may be improved in crowdfunding by utilising blockchain technology, boosting platform confidence, and drawing in additional investors. The study may also offer a fresh framework for crowdfunding platforms that other businesses may use, advancing the expansion and growth of the crowdfunding sector.

2 Related Work

2.1 *Blockchain*

The capacity of blockchain-based to offer security and transparency has made it extremely popular. A distributed ledger is used by blockchain to store transaction histories. In this scenario, record changes are made by consensus, requiring consent from every node, and instead of having separate versions of the same information, all network members share the very same information. Since data is stored throughout a computer network rather than on a single server, blockchain also offers improved security [3]. Blockchain has the capacity to eliminate errors and spot fraudulent activity since it may generate a public ledger among numerous untrusted parties. By offering a complete history record, a blockchain library can autonomously check the legitimacy of clients, policies, and transactions [4].

2.2 *Smart Contract*

The blockchain's cryptographic protocol technology makes it possible to define the business logic supporting transactions, from determining who is the rightful owner of an object to putting in place intricate self-enforcing regulations. Only two benefits that financial intermediaries and cryptocurrency may offer are the smooth transfer of assets and the ease with which claims may be processed. The Ethereum blockchain currently offers the most support for the development of smart contracts [5]. The Ethereum Smart Engine, a straightforward stack-based Turing perfect 128-bit virtual processor, executes smart contracts. Solidity, a well-known programming language, offers a growing ecosystem for creating smart agreements.

2.3 *Decentralised Autonomous Organisation*

Virtual businesses that are run by autonomous payment systems are known as decentralised autonomous organisations (DAO). According to the usage of the Ethereum platform, an agreement method is used to preserve the state of the organisation, and agreements are used to handle transactions, free cash flow, regulations, and privileges inside the institution [6]. Members can exchange ether and use smart contracts to communicate once a DAO code has been made public. DAOs typically convert tokens tied to the recipient account into ether. The number of tokens issued is inversely connected with the volume of ether transferred, and they are utilised to grant various permissions.

2.4 Crowdfunding Mechanisms

Through a process known as equity crowdfunding, big groups of investors can contribute to new businesses or people in exchange for shares. A wide range of for-profit enterprises, including programmes for peer social entrepreneurship, travel, medical expenditures, and artistic and creative endeavours, have utilised crowdfunding. Fund raising affects a company or person getting assistance through a big series of tiny significant contribution from the a large number of customers, in comparison with the conventional form of raising funds, that also involves a corporation receiving a limited amount of fairly sizable investments from the a specific subset of lenders [7]. Thus, crowdsourcing is a hybrid of social network involvement and economic ventures in which members act as investors [8].

2.5 Limitations and Issues with Existing Models

One of the main problems with the current blockchain-based crowdfunding methods is their inability to grow. A blockchain network can only handle a certain amount of transactions at once, which can result in delays and rising transaction prices as the network's user base expands. Because different blockchain networks adhere to various protocols and standards, interoperability is a problem for existing models. As a result, combining many blockchain platform into an unified crowdfunding site may provide difficulties. The adoption of blockchain technology may make it more difficult to abide by current laws in the tightly regulated field of crowdfunding. Additionally, investors and site operators may experience uncertainty due to the fact that the regulatory environment for crowdfunding models based on blockchain is still developing. Blockchain-based crowdfunding models still have dangers despite the benefits of blockchain technology for maintaining security. For instance, smart contracts, which are a common component of blockchain-based crowdfunding sites, may have security flaws that attackers might take advantage. Last but not least, existing blockchain-based crowdfunding methods have a problem with user adoption. Numerous investors and business owners might not be aware with blockchain technology or comprehend how it might help them during the crowdfunding process.

3 System Model

Vendor, customer, auditor, and insurer are the four parties involved in this paradigm as shown in Fig. 1. The following is a description of these things:

- In order to shift the danger of having a fresh vulnerability exploited to the insurer, a vendor must request life insurance for its computer security. The vendor might be a producer of equipment or software, a manufacturer of applications, a provider of IT services, etc. The vendor wants to expand its market share by providing insurance services to clients. In order to find new vulnerabilities, a vendor cannot reduce security spending, hence the framework must ensure this does not happen.
- A client is the user of a cyber-final product. The buyer expects that the item will have the necessary level of protection and will be safe from cyberattacks when they acquire it from the vendor. In order to be paid in the event that an attacker discovers a new weakness, clients opt to purchase computer security for the botnet [9].
- The task of assessing a product's cybersecurity falls to an auditor. The accountants are also charge of evaluating the validity of the customers' indemnification demands. It is important to design the system to prevent cooperation between the inspectors, patrons, and vendors. An insurance provider is a business that, in

Fig. 1 Crowdfunding system architecture

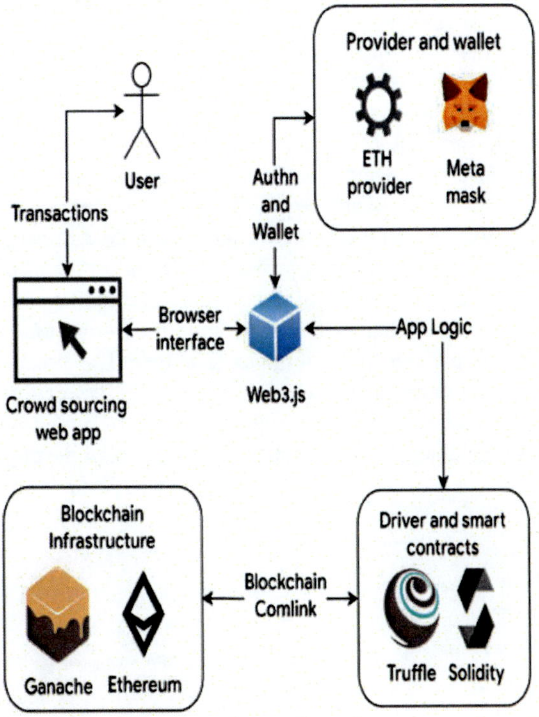

return for a premium, takes the risk posed by a cyber-critic's protection against recently found flaws.

3.1 Design Objectives

The following reasons led to the creation of a cyber-product insurance:

- Insurance for cyber-products: The presence of cyber-product insurance increases the clients' trust in the security of cyber-products and covers the expense of cyberattacks. The vendor's market share grows as a result.
- Risk sharing: Since it might be difficult to predict the likelihood that a product's new vulnerability will be exploited, insurers may not be eager to take on the entire risk and engage in such a market [10]. However, the low-risk tolerance insurers are encouraged to participate in such a process by creating a platform for crowdfunded insurance and risk pooling.
- Increase Security: The method should include measures to increase the security of the cyber-product.

4 Crowdfunding Framework

The problem with the existing system is that firms charge both donors and users hefty fees. There is no tracking of money transactions, contact among investors and users, or openness in the project's development. Trust is the main problem to financing with reputable businesses. The donor guarantee policy is not available from these companies. Absurdly high fees, lack of transparency, there is no track the recordings and no donor assurance policy. The suggested approach states that when campaign creators submit their creative projects in the campaign, interested parties would pay money to the proposed project. It differs from earlier crowdsourcing in that all the funds are presently held in cryptocurrencies like ether [11–15].

The blockchain will monitor and store every single ether currency. The blockchain, in comparison, is an immutable ledger. The donor has authority over the funds. The requests clearance module gives the donor total control over the funds they contributed. Only if the authors' plan is accepted by at least half of the investors. Giving control over invested cash is a part of developing trust. Belief, command of finances, donor guarantee policy: no charges. All actions are tracked, and money is held in a safe place.

4.1 Crowdfunding

That once sensitive data has been broadcast in the blockchain, the vendor starts the auction. The system conducts an encased auction for this. This is due to the fact that insurers put offers in an encased auction depending on their measured principles rather than the auction's present status. Due to the fact that it is a blockchain-based smart contract, users are certain that the seller also is not exploiting the competitive process.

Insurers engage in the encased auction by indicating the quantity of insurance they would be obligated to purchase and the associated desired premium for their obligation. The beneficiaries will be selected after the auction based on the entire indemnification for the requested insurance. The winners are then given the tokens via the smart contract in return for executing their obligations. The insurers will be compensated for their services by delivering these credentials at the end of the insurance term. The insurers may also sell their tokens when in an insurance contract. This boosts the profitability of the insurance sector and attracts additional insurers.

4.2 Algorithms

4.2.1 Campaign Contract Algorithm

The algorithm1 is a contract algorithm for a blockchain-based crowdfunding campaign. The algorithm outlines how requests for funding from a campaign recipient are handled, as well as how donations from contributors are gathered and managed. A number of functions are defined by the algorithm, including "Contribute," "Request," "ApproveRequest," and "FinalizeRequest." If a donation fulfils the minimum contribution criterion, the "Contribute" function designates the contributor as an approver and increases the count of approvers. The "Request" method adds a new request to a list of requests by initialising a new variable of type "appeal." A vote on a specific request can be cast by approved contributors using the "approveRequest" method. The request is authorised, and the count of approvals is increased if the account is designated as an approver. If the number of approvals equals or surpasses half of the total number of approvers, the "finalizeRequest" function sends the cash to the beneficiary.

Algorithm 1 Campaign Contract Algorithm

```
Require: current hash, nodeID, transaction key 0: struct appeal
0: string account;
0: uint val;
0: address receiver;
0: bool finish;
0: uint approvalCount;
0: mapping(addess == bool) approvals; 0: function Contribute
0:          if contribution == minimumContributionRequirement then
0:             mark contributor as approver 0:    increment approvercount
0:          end if
0: end function
0: function Request(account, val, receiver address) 0:initialize new variable of
type appeal

0:          add new Request to to requests list 0: end function
0: function approveRequest(index)
0:          appeal=rappeal[index]
0:          if approvers[account]=true then
0:             appeal.approvals[account]=true
0:             appeal.approvalCount++
0:          end if
0: end function
0: function finalizeRequest(index)
0:          appeal=requests[index]
0:          if appeal.approvalCount == (approversCount/2) then
0:             transfer appeal amount to recipient 0:
               appeal.complete=true
0:          end if
0: end function=0
```

4.2.2 Verifier Algorithm

The algorithm2 explains a verifier that is employed in a blockchain-based system. The three primary operations of the algorithm are verify, VerifyC, and Checker. Input and proof are the two inputs required by the verify function. The contributing credentials and the verifying key are first initialised (v_k). The next step is to determine if the length of the input plus one equals the length of the minimal contributions in the verifying key. The function returns true if this criterion is satisfied; otherwise, it returns false. A struct of data with the property d is the input for the VerifyC function. This function creates an elliptic curve using input d and obtains the blockchain address. The helperFunction is then invoked using the input and evidence as inputs together with the input values from d. The Boolean mapping value is indicated as true if the function's output equals zero. The blockchain address is sent to the Checker function. The Boolean mapping value is then verified. The method returns 1 if the value is true and 0 otherwise.

4.3 Indemnity

The insurers will get their obliged values and tokens back if no weakness was used within the insurance policy's duration. Before utilising a new risk, users must alert the vendor or assessor. Following data verification and blockchain registration, the auditor uses the reimbursement feature to request compensation from the fundraising stage. The insurers will share the cost of the indemnification in an equitable manner. The claiming procedure is recorded on the blockchain, has a negative effect on the company's reputation, and lowers trust in the vendor's security. As a result, fewer underwriters are willing to cover items, which drives up the insurance rate.

4.4 Advantages of Proposed Framework

- Given that it is impossible to predict the risk connected with the cyber-product, an extensive indemnity request could not be granted by the insurers. The insurers mitigate the effect of a loss by utilising a hazard insurer crowdsourcing strategy, nevertheless. This enables companies with different risk appetites to select their desired degree of risk.
- Blockchain makes it possible for insurers to track the progress of the insureds by keeping accurate records of indemnification and claims. Although it can be difficult for insurers to estimate the likelihood that a new vulnerability would be used, the seller's track record in dealing with security-related problems has aided insurers in making wiser investment choices. This kind of design also uses the image of the organisations who provide the service.

Algorithm 2 Verifier Algorithm

```
0: function verify(input,proof)
0:     initialize contributor credentials 0: initialize verifying key as vk
0:     if input.length+1 == vk.minimumcontributors.length then
0:       return true
0:     else
0:       return false
0:     end if
0: end function
0: function VerifyC(struct data d)
0:     fetch and initialize blockchain address 0: construct elliptical
curve input d
0:     take input values from d
0:     if helperFunction(input,proof)==0 then 0: mark
boolean mapping value as true 0: end if
0: end function
0: function Checker(blockchain address) 0: if boolean mapping value is true then
0:       return 1
0:     else
0:       return 0
0:     end if
0: end function=0
```

- Under conventional insurance models, it is difficult, if not inconceivable, for carriers to shift the risk of an insurance products to a third party. In contrast, insurers get paid in tokens thanks to the use of a smart contract. The insurers are able to accomplish this by trading tokens with other insurers.
- When the agreement expires, the insurer will be reimbursed for their obligations. The entities have faith in the veracity of the smart contract since it operates on a blockchain network and is backed by a resolution procedure. The blockchain's architecture also eliminates intermediaries. The entities' services are improved.

4.5 Smart Contract

The encased auction for financing presents a challenge due to the transparency of blockchain transactions. To part of the responsibility from the auction winners, the bidders should transmit their bid amounts to the consensus mechanism during the bidding process. The fundamental reason why this is challenging is because everyone may see the bids that are recorded in the blockchain. As a consequence, bidders modify their proposals accordingly and do not submit genuine bids, which costs the seller money. Consider the scenario where a legitimate appraisal of an insurance cover by an insurer is \$2 premium. The insurer will reduce its bid to \$2.99 in hopes of winning the auction if it discovers that now the lowest desired premium is \$3.

The auction procedure is split into the revealing and bidding phases in order to address this problem. To protect the bid values, the commitment approach is employed. A commitment strategy has elements of binding and concealment. While concealing requires that a promise contain no details about the value being committed, binding property makes sure that a pledge cannot be extended to some other value. The approach employed the Ethereum-SHA3 hash algorithm (Keccak256) as that of the commit mechanism as shown in Eq. 1.

$$E[u] = b \cdot \frac{R - b)^{(N-Y)}}{R} \tag{1}$$

4.6 Realisation of Security

Decentralised ledger technology, cryptographic algorithms, and consensus procedures are used to achieve high security in blockchain-based crowdfunding integrity. These characteristics guarantee the transactions' security, transparency, and immutability. Blockchain's decentralised structure reduces the need for middlemen, lowering the likelihood of fraud, and cyberattacks. Data cannot be changed or removed after it has been stored on the blockchain thanks to cryptographic algorithms and consensus procedures. This results in a lasting and verifiable record of transactions, preserving the legitimacy of crowdfunding campaigns, and fostering participant trust. Additionally, the blockchain's smart contract functionality can automate the funding and disbursement of funds, lowering the possibility of fraud or errors.

4.7 Achievement of Reliability

Using blockchain technology, reliability in crowdfunding integrity is made possible by a number of important qualities.

- Decentralisation: By doing away with middlemen, blockchain lowers the possibility of single points of failure and makes sure the network is always accessible, even if some nodes are down.
- Transparency: The blockchain creates a transparent and accountable system because all transactions are visible to all participants.
- Immutability: The cryptographic methods and consensus procedures employed in blockchain technology ensure that once a transaction is recorded, it cannot be changed or removed. This is known as immutable record-keeping. This ensures the dependability of the system and gives a permanent record of all activities.
- Automation: Automation of smart contracts can be used to reduce the risk of fraud and human mistake by automating the funding and disbursement of cash.

4.8 Monitoring and Storage of Ethers Accurately

Smart contracts and decentralised ledger technology allow for reliable monitoring and storage of ethers in blockchain-based crowdfunding.

- Smart Contracts: These agreements can be set up to automatically track and hold ethers, guaranteeing that the funds are safe and secure. The contracts can impose particular guidelines and requirements, such as establishing a campaign deadline, releasing funds only when specific requirements are satisfied, and allocating the cash to the proper parties.
- Decentralised Ledger: The blockchain's decentralised ledger offers a safe, visible record of every transaction, guaranteeing that the money is properly monitored and kept. A network of nodes maintains the ledger, making it challenging for a single organisation to change or modify the data.
- Security: Blockchain employs cryptographic techniques to encrypt transactions and guard against unwanted access, making sure that the money kept in the smart contract is safe and can only be accessed by authorised parties.

4.9 Suitability for Future Research and Real-Time Implementation

Blockchain offers a safe and open ledger for recording transactions, preserving the legitimacy of the crowdsourcing procedure. Decentralised networks that record transactions make it more challenging for hackers to influence the system. The blockchain-based crowdfunding may automate many traditional crowdfunding operations, such as recording donations, confirming terms, and allocating cash, by using smart contracts. By doing this, the likelihood of human mistake and potential fraud is decreased. The stakeholders' confidence and responsibility may rise as a result of blockchain's openness. A crowdfunding campaign's development may be easily followed by contributors, and the project's funders can be held responsible for spending the money as intended. Blockchain has the potential to increase accessibility of crowdfunding by enabling cross-border transactions and lowering entrance barriers for smaller investors.

5 Results

5.1 IPFS Performance

Blockchain can be combined with the decentralised file storage system known as InterPlanetary File System (IPFS) to enhance the efficiency of crowdfunding campaigns.

- Decentralised Storage: IPFS is a distributed, decentralised file storage system that makes it possible to store and retrieve huge data quickly and effectively. In particular for crowdfunding projects that call for the storing of huge volumes of data, like video or photographs, this helps to lessen the stress on the blockchain network and enhance performance.
- Increased Security: IPFS uses cryptographic hashing to defend against manipulation and guarantee the integrity of the data stored, thus strengthening the security of the crowdsourcing campaign.
- Improved User Experience: By using IPFS, users may access crowdfunding campaign content fast and effectively without waiting for it to load from a central server. This helps to enhance user experience generally and raise campaign involvement.

IPFS is used as the study's privacy data database. To evaluate the storage, Fig. 6 shows the IPFS uploading and receive times over various file sizes. The test environment uses Ubuntu LTS 20.4 version and go-ipfs v0.16.0 as the IPFS version. As noted in the illustration, as file sizes increase, so do IPFS's transfer files times, with the data allow often taking more time than the download speeds. The transfer files speeds are 43.54 and 51.27 MB/s, correspondingly, when the storage file exceeds 50 MB, which can accommodate IoT demand.

5.2 *Algorithm Efficiency*

In this work, an elliptic curve cryptography smart contract using the Go programming language is built. The encryption and decryption timings of the technique for various key lengths are shown in Fig. 4. The execution durations of three different approaches for varied key lengths are shown in Fig. 5. The chart shows that, given the same key length restriction, the encryption time is less time than the decrypting time, and that the time needed for cyphertext plus block cypher is longer than the time needed for cyphertext plus constants but less time than both. The duration of cyphertext + block cypher is 538.61 ms when the key length is 1024 bits, which may meet the blockchain requirement.

5.3 *Meta Mask Integration*

Figures 2 and 3 show us the creation of funding initiative and creation of account using Meta Mask. Figure 4 depicts the login screen for the Meta Mask user to chrome and work as an individual user. The Meta Mask login acts as a private network where multiple users can be created over a network and they can contribute to the blockchain network for the crowdfunding contributions. The final project of all crowdfunding is shown in Fig. 5.

Fig. 2 Creation of a funding initiative

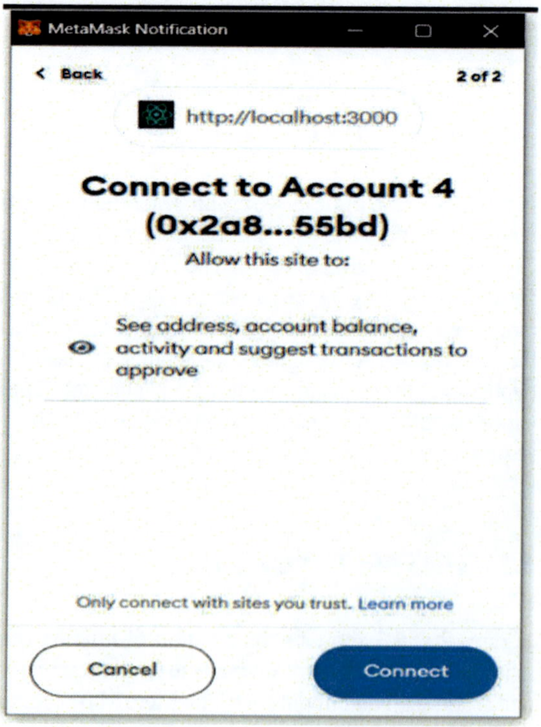

Fig. 3 Configuring with meta mask

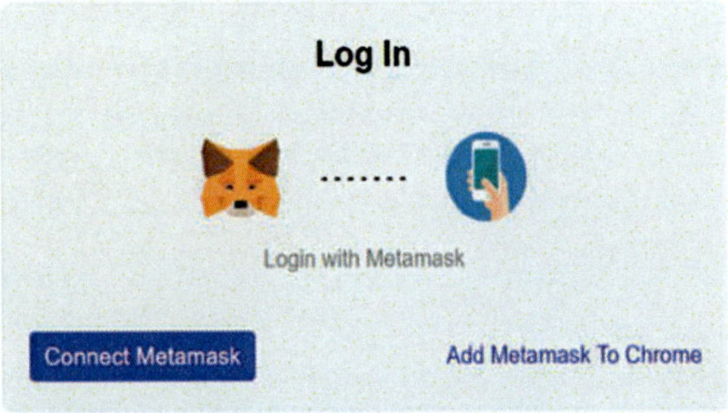

Fig. 4 Logging in using Meta Mask to the particular crowdfund

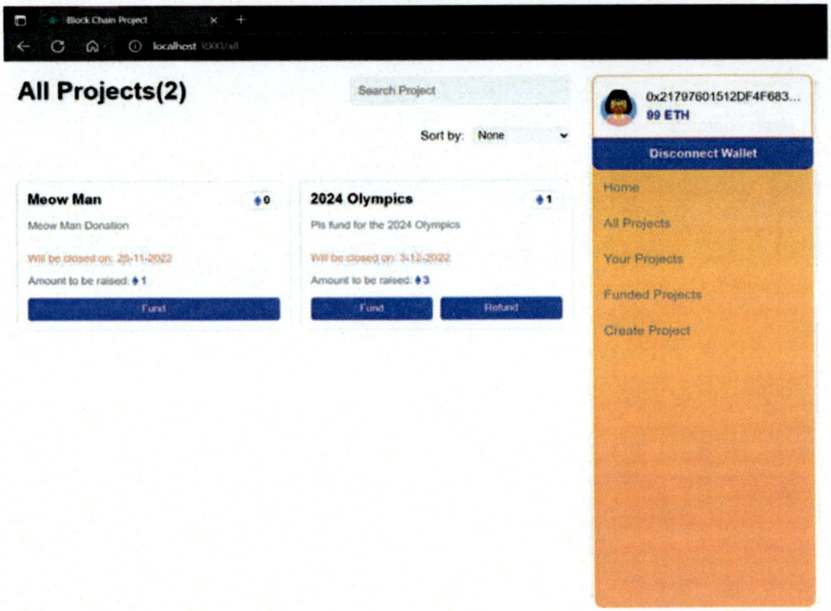

Fig. 5 Status of all crowdfunding initiatives

5.4 Gas Fees

Table 1 represents the gas fees for various operations over the blockchain network. Digital identity smart contract deployment charges the highest gas fee and donate operation require minimum gas fee. Since the digital identity verification is a complex operation, it requires more gas fee compared to other operations. The results of logging in using Meta Mask to the particular crowdfund are provided in Figs. 6 and 7

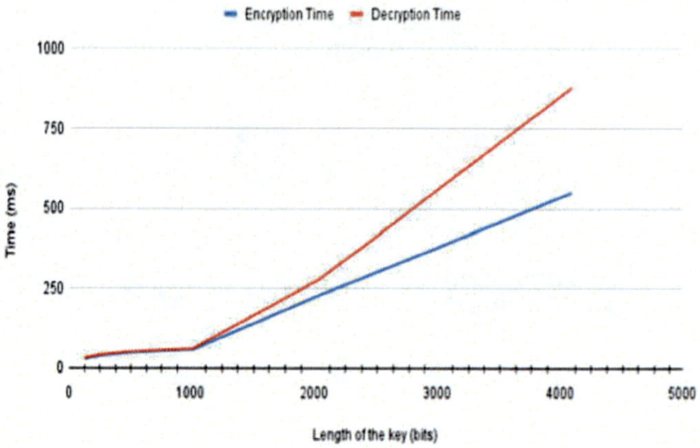

Fig. 6 Logging in using Meta Mask to the particular crowdfund

Table 1 Gas fees associated with smart contracts functionalities

Module	GAS FEE
Smart contract deployment	0.0043542
Function to update keys	0.0035414
Function to update users	0.0023781
Function to update passwords	0.0081923
Function for verification	0.0014234
Function for suspension	0.0076521
Function for creation of campaign	0.0023124
Cost of donation	0.0043212

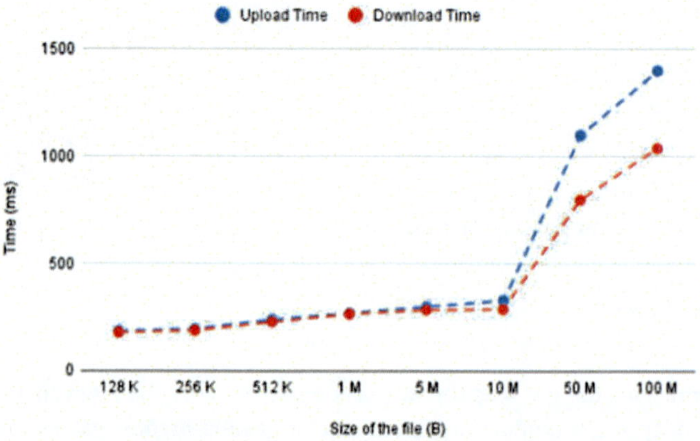

Fig. 7 Upload and dowload time of the nodes

6 Conclusion

In this paper, a novel paradigm for using blockchain technology to guarantee a digital good is provided. To spread the risk involved with insurance, crowdsourcing through secure blockchain platform is proposed and implemented. The advantages of the architecture proposed are discussed and examined the insurers' bid strategy. The execution of an encased auctioneer on the blockchain and proposed a way to safeguard bid values during in the tender process is looked with standard Ethereum tabulation.

Future study should concentrate on creating a thorough knowledge of the criteria for success for ledger crowdsourcing and how they vary from those for conventional crowdfunding. This can entail performing further research on the drivers of investors and business owners as well as examining the legal environment around blockchain-based crowdfunding and how it affects the outcomes of these projects. Examining how established finance patterns could be disrupted by blockchain-based crowdfunding and the effects this would have on the whole economy are the potential for future research.

References

1. Sanjaya YPA, Akhyar MA (2022) Blockchain and smart contract applications can be a support for MSME supply chain finance based on sharia crowdfunding. Blockchain Front Technol 2(1):44–49
2. Aprialim F, Paundu AW (2021) Penerapan Blockchain dengan Integrasi smart contract pada Sistem crowdfunding. Jurnal RESTI (Rekayasa Sistem Dan Teknologi Informasi) 5(1):148–154
3. Cali U, Halden U, Dynge MF, Bukvic-Schaefer AS (2021) Blockchain-enabled equity crowdfunding for energy storage investments. In: 2021 international conference on smart energy systems and technologies (SEST). IEEE, pp 1–6
4. Chang SY, Bathen L, Di Troia F, Austin TH, Nelson AJ. Silicon valley cybersecurity conference
5. Sevindik V (2021) Blockchain based resource tokenization for crowdfunding of wireless network investment. In: 2021 11th IFIP international conference on new technologies, mobility and security (NTMS). IEEE, pp 1–5
6. Li Z (2021) Remolding of the supply chain development mode based on the block chain technology. In: 2021 international conference on computer, blockchain and financial development (CBFD). IEEE, pp 392–395
7. Hegde B, Ravishankar B, Appaiah M (2020) Agricultural supply chain management using blockchain technology. In: 2020 international conference on mainstreaming block chain implementation (ICOMBI). IEEE, pp 1–4
8. Zhu H, Zhou ZZ (2016) Analysis and outlook of applications of blockchain technology to equity crowdfunding in China. Fin Innov 2(1):1–11
9. Boakye EA, Zhao H, Ahia BNK (2022) Emerging research on blockchain technology in finance; a conveyed evidence of bibliometric-based evaluations. J High Technol Manage Res 33(2):100437
10. Velmurugan K, Yuvaraj S, Sumalatha MR, Palivela L (2022) Supply chain financial risk management using business intelligence. Available at SSRN 4128710
11. Desabathina NVM, Merugu S, Gunjan VK, Kumar BS (2022) Agricultural crowdfunding through blockchain. In: ICDSMLA 2020. Springer, Singapore, pp 1643–1650

12. Elangovan P, Sumalatha MR (2021) Weight based deduplication for minimizing data replication in public cloud storage. J Sci Industr Res 80(03):260–269
13. Wehnert P, Beckmann M (2021) Crowdfunding for a sustainable future: a systematic literature review. IEEE Trans Eng Manage
14. Shelke P, Zanjal S, Patil R, Desai D, Chavan H, Kulkarni V (2022) Blockchain technology based crowdfunding using smart contracts. In: 2022 international conference on augmented intelligence and sustainable systems (ICAISS). IEEE, pp 939–943
15. Saranya S, Muvvala SP, Chauhan V, Satwik R (2022) Crowdfunding charity platform using blockchain. In: 2022 international conference on inventive computation technologies (ICICT). IEEE, pp 1–7

A Brief Analysis on Video Compression and Deduplication Models Using Video Steganography with Optimised Feature Extraction Techniques for Integrity Preservation

S. K. Sameerunnisa and J. Jabez

Abstract Increased Internet use has brought about a greater focus on security in recent years. The daily rate of data exchange is increasing as Internet usage grows. Hackers might potentially compromise the daily flow of information. Videos feature a wide variety of sceneries and objects, some of which are static, some of which are constantly moving, some of which are sparsely populated, and some of which undergo complex, non-repetitive motions. The Dynamic Texture (DT) in particular presents a number of difficult challenges due to its constantly shifting appearance and motion. With cloud computing, users have unprecedented flexibility in terms of video data storage, sharing, and accessibility. It has also contributed to the explosive growth of digital data. In order to keep up with this speedy expansion, video deduplication has become a major strategy for Cloud Storage Providers (CSPs) by enabling them to efficiently delete duplicate data from their storage facilities. This research proposes a new compression plot that can safely deduplicate videos stored in the cloud. To efficiently transmit and store digital video files over a network and on computer discs, video compression techniques are used to solve the problem of reducing and deleting superfluous video data. Video steganography, in which information is concealed within video frames, is used in this research. The video compression and deduplication techniques and feature extraction techniques for video steganography for integrity preservation are analysed in this research.

Keywords Video compression · Video deduplication · Video steganography · Optimization · Video feature extraction

S. K. Sameerunnisa (✉) · J. Jabez
Department of Computer Science and Engineering, Satyabhama Institute of Science and Technology, Jeppiaar Nagar, Rajiv Gandhi Salai, Chennai, Tamil Nadu 600119, India
e-mail: sameerask.25@gmail.com

G. Rajakumar et al. (eds.), *Intelligent Communication Technologies and Virtual Mobile Networks*, Lecture Notes on Data Engineering and Communications Technologies 171,
https://doi.org/10.1007/978-981-99-1767-9_31

1 Introduction

Images for a digital video are taken at regular intervals. Digital samples of the images are used to represent the visual data present at each pixel in the image over its full range of time [1]. The values of the three primary colours in the RGB colour space can be used to express visual information at each sample point [2]. Both frames and fields can be used to sample a video signal. One frame is sampled at a time in progressive video. In an interlaced video, just half of the frame is recorded at any given time instant. An increasing variety of video codec technical standards and intelligent algorithms are available, making it feasible to store and transmit video in digital form, making video compression a crucial enabler for these applications. New algorithms and the ever-increasing processing capability of inexpensive integrated circuits like digital media processors are driving changes in compression standards [3].

Compression standards vary, as do implementations of those standards, with the key differences stemming from varying emphasis on secondary needs [4]. Video compression, in the broadest sense, is a technology for compressing video signals that seeks to preserve the original quality under a variety of constraints, such as those imposed by storage space, latency, or processing capacity. It uses redundant data between consecutive frames to cut down on the amount of processing power needed to store the data. It is common practice in the design of data compression systems and make sacrifices in the areas of performance, speed, efficiency, and resource/energy usage [5].

Steganography is crucial to the realm of information security. Pictures and videos are frequently used to conceal information. To practise visual cryptography, an image can be considered and use LSB audio steganography to hide information in its colour channels. Video encryption has the advantage of being able to conceal a sizable quantity of information within a continuously evolving sequence of images [6]. When data encryption and steganography are used together, a multimedia file can be used as a covert communication channel. The information is concealed in the angular and linear motion of the objects. Images, audio, and video files are collections of bits that can be decoded into their respective data types [7]. Excess bits and blank space in the file might be used to store new information or to conceal older information.

Due to the high volume and inherent redundancy of the data produced by IoT devices, both communication capacity and storage space are quickly depleted. As a result, data deduplication has become increasingly popular as a means to dispose of duplicate information [8]. It is a difficulty, however, to ensure the privacy of users' information because the data acquired in IoT are often highly personal and sensitive. When privacy is at stake, the user can select a secure policy and a personal model that best suits their needs. Some devices under the personal model can only offer services, such as the sharing of data acquired by environment sensors, to other devices in a close enough network that they already have a common public key [9], such as devices belonging to the same user.

Thus, encryption of data prior to storage in the cloud is essential and also duplicate removal is required. In many research models, data encryption is a useful tool for protecting individual privacy in videos [10]. While in transit from the peripheral device to the central structure, the sensor gathered video data is encrypted to ensure that even if it is intercepted by another unauthorised device, such as a malicious one, on the same wireless physical media, the interceptor will be unable to read the original data [11]. A downside of this approach is that encryption makes deduplication by the cloud server more difficult to do, since the cypher text will be unique even if the underlying plaintext is similar [12].

A video comparison system is essential to the deduplication framework. The video signature is a concise representation of the video that is used for fast video comparison. Different methods have been offered in the past [13]. Ordinal signature was selected from the presented strategies for the development of video signatures. Because ordinal signature depends on the colour and the spatial–temporal distribution of intensities of video, it is insensitive to differences in the resolution and display formats of the compared videos [14].

Deduplication can reduce the amount of space needed to store this increasingly centralised web data. Deduplication is a process that uses pointers to the original copy of duplicate data to get rid of unnecessary copies of the data [15]. In light of the growing centralization of web data and the widespread redundancy of that data, deduplication offers a promising method for optimising storage space. It is crucial to differentiate between managed and unmanaged redundancy from a storage perspective [15]. When the storage system knows about the duplicates and replicates for a reason, it can be considered that redundancy is managed. Unmanaged redundancy, on the other hand, results in the storage system being oblivious to the existence of the copies and does not explicitly contribute to better characteristics of the system [16]. Problems with data management arise when redundancy in large-scale storage systems is not properly controlled [13].

The term data hiding describes the procedure by which information is secreted in a cover media. Data concealing typically causes some distortion to the cover media and prevents the media from being converted back to its original state [17]. However, reverting tagged media to unmodified media is essential in a variety of contexts, including medical diagnosis, military imagery, remote sensing image analysis, legal certification and proof, and other areas [18]. Even now, reversible video steganography techniques are put to use in a wide range of applications. The primary goal of steganalysis is to reveal hidden information in a given object. Digital media files such as video, audio, and photographs are perfect for using with steganography to conceal a hidden message. Measurable descriptors make it hard to demonstrate certain empirical spreads properly and also greatly muddle the detection of embedding changes [19].

2 Literature Survey

The use of augmented and virtual reality (AR and VR) applications has attracted increasing attention in recent years. It is not uncommon for augmented and virtual reality applications to make use of 360°, or omnidirectional, video. In response to market demand, a new edition of HEVC was recently released, and it contains a number of SEI messages that make it possible to transport omnidirectional video using High-Efficiency Video Coding (HEVC). On the other hand, 360° video compression efficiency still has room for development. To combat this issue, the ITU-T VCEG and ISO/IEC MPEG formed the Joint Video Exploration Team (JVET) to study 360° video coding technologies like projection formats, pre- and post-processing methods, and 360° video coding tools analysed by Ye et al. [1]. Recently, the ITU-T VCEG and ISO/IEC MPEG announced a joint CfP on video compression technologies beyond HEVC, and one of the categories was 360° video. We received 12 replies to our call for proposals for 360° videos. This article discusses 360° video and virtual reality communication methods.

HEVC is a new generation of video coding that features both better compression performance and greater coding complexity. Therefore, the current trend in video production is to lessen perceptual redundancy in order to improve compression. In this research, Zhu et al. [2] expanded and optimised the HEVC by combining a video saliency method based on an attention mechanism with a video compression strategy based on perceptual priority. This research provides a convolutional neural network-based spatial saliency method and a motion vector-based temporal saliency algorithm for identifying important parts of videos. To finish off the saliency map of the input video, the saliency algorithm can merge the motion estimation of each block during the HEVC compression using a convolutional neural network and carry out adaptive dynamic fusion. This study presents a more adaptable approach to QP selection within the context of the perceptual priority video compression algorithm, wherein the QP is chosen in relation to the saliency value of the CU. The author also presented a new rate-distortion optimization approach that uses the current block's saliency characteristic in conjunction with the standard way of calculating rate-distortion to better direct bit allocation and realise the desired effect of perception priority.

Two important new generations of video coding standards have been produced in the 17 years since the initial version of the now-dominant H.264/Moving Picture Experts Group-4 (MPEG-4) Advanced Video Coding (AVC) standard was finalised in 2003. High-Efficiency Video Coding (HEVC) is one of these, along with the VVC (Adaptive Video Coding) standard. The final version of HEVC was released in 2013, maintaining the ten-year cycle time established by AVC's predecessor. The VVC project, which was completed in July 2020, had a shorter cycle by three years and reduced bit rate by nearly 50% compared to its predecessor HEVC. Bross et al. [3] provide a concise overview of post-AVC advances in the standardisation of video coding. The initial version of VVC is reviewed in detail, alongside comparisons to HEVC. As with earlier iterations of development, VVC explains not only the latest innovations in hybrid video compression but also the extensive flexibility of the application domain emphasised by the name. VVC includes capabilities to handle computer-generated/screen video, high-dynamic range information, multilayer and multi-view encoding, and support for immersive media like 360° video, in addition to the more common standard- and high-definition camera-captured content codings.

As a result of the high bit-rate versus video quality trade-off and the complicated perceptual masking that affects performance, highly textured video footage is difficult to compress. This highlights the need for, but scarcity of, test datasets that encompass a wide variety of texture types for use in evaluating codecs. Katsenou et al. [4] presented a data set of synthetic video textures (BVI-SynTex) created in a CGI environment, which may be used to investigate the characteristics of video textures. It has 196 sequences organised into three groups by texture type and the ability to make numerous variations of the same scene using varied video parameters. As a result, it serves as a malleable starting point for investigating how different texture types and characteristics affect both video compression and the impression of video quality. BVI-SynTex is thoroughly tested and compared to genuine video content with a similar texture. The results demonstrate that BVI-SynTex yields encoding statistics that are equivalent to real-world video data sets, and that it also displays comparable coverage in the spatial and temporal domains. In order to conduct a subjective judgments of compression using the MPEG HEVC codec, a subset of the BVI-SynTex data set was chosen. The outcomes illustrate the effect of the content parameters on compression effectiveness and on the experienced quality.

Data deduplication is a powerful tool for preserving storage and bandwidth within the cloud environment. Yet, it is not easy to avoid adding layers of redundancy in the form of more storage or communication because of safety concerns. In spite of the many efforts made to find a unified cryptographic framework for safe data duplication, issues still exist between data privacy, security, and authority. Specifically, despite the fact that convergent encryption makes the data unintelligible, a violent dictionary attack is still effective because the entire pseudorandom process is largely dependent on plaintexts. Similar issues may arise with data ownership, as the hash value used to determine who has download access. Mi et al. [5] introduced a threshold blind signature-based data deduplication methodology in this model to do away with these issues. The outsourced file and deduplication label are computationally distinguishable from random strings with the use of several key servers. To verify that users' claims of ownership are genuine, the author deployed the boom filter as a proof of ownership mechanism. As a result, the attacker is unable to use the stolen tag to gain unauthorised access to the entire file. The main contribution of this research is the introduction of a secret sharing mechanism based on the Chinese Remainder Theorem for hiding signature keys, which strikes a balance between cloud and client security concerns by using homomorphism calculation to aggregate and generate partial signature tags. Limitations of existing models are listed in Table 1.

The study of how to combine information from different databases is known as heterogeneous database integration. There are three key obstacles that make the heterogeneity problem difficult to tackle when integrating disparate databases of the same area. Heterogeneity in meaning, syntax, and structure are the issues. Deduplication techniques, data warehousing, and information retrieval (IR) search techniques are all examples of traditional heterogeneous database integration strategies, but none of these can solve the problem of database integration on its own. The only plausible explanation is that they are ineffective in addressing issues of semantic heterogeneity.

Table 1 Limitations of existing models

Author name	Year of publication	Manuscript title	Proposed model	Remarks
Ye et al. [1]	2020	Omnidirectional 360° video coding technology in responses to the joint call for proposals on video compression with capability beyond HEVC	The ITU-T VCEG and ISO/IEC MPEG formed the joint video exploration team (JVET) to study 360° video coding technologies like projection formats, pre- and post-processing methods, and 360° video coding tools are analysed	High time complexity
Zhu et al. [2]	2020	High-definition video compression system based on perception guidance of salient information of a convolutional neural network and HEVC compression domain	The author expanded and optimised the HEVC by combining a video saliency method based on an attention mechanism with a video compression strategy based on perceptual priority. This research provides a convolutional neural network-based spatial saliency method and a motion vector-based temporal saliency algorithm for identifying important parts of videos	More memory usage

(continued)

Table 1 (continued)

Author name	Year of publication	Manuscript title	Proposed model	Remarks
Bross et al. [3]	2021	Developments in international video coding standardization after AVC, with an overview of versatile video coding (VVC)	The author provided a concise overview of post-AVC advances in the standardisation of video coding. The initial version of VVC is reviewed in detail, alongside comparisons to HEVC. As with earlier iterations of development, VVC explains not only the latest innovations in hybrid video compression but also the extensive flexibility of the application domain emphasised by the name	Unsatisfactory compression levels
Katsenou et al. [4]	2021	BVI-SynTex: a synthetic video texture dataset for video compression and quality assessment	The author presented a data set of synthetic video textures (BVI-SynTex) created in a CGI environment, which may be used to investigate the characteristics of video textures. It has 196 sequences organised into three groups by texture type and the ability to make numerous variations of the same scene using varied video parameters	More complex operations

(continued)

Table 1 (continued)

Author name	Year of publication	Manuscript title	Proposed model	Remarks
Mi et al. [5]	2020	Secure data de-duplication based on threshold blind signature and bloom filter in internet of things	The author introduced a threshold blind signature-based data deduplication methodology in this model to do away with these issues. The outsourced file and deduplication label are computationally distinguishable from random strings with the use of several key servers. To verify that users' claims of ownership are genuine, the author deployed the boom filter as a proof of ownership mechanism	Applicable only on small videos

In this research, Asfand-E-Yar et al. [6] experimented with and talked about a model for ontologies on the semantic web that is grounded in the way queries are executed. There are two stages to ontology modelling; the first involves adapting the database rules to the ontology rules in order to arrive at an abstract ontology model. Second, we need to add more specifics to the abstract ontology model based on the information already present in the databases. Using this technology, similar SPQRAL searches may be applied on database data with ease. In order to access data that is similar in meaning to another set of data, the Jena API is called upon.

In this paper, Xu et al. [7] offered a commutative encryption and data concealing technique that performs well for HEVC videos. Due to the commutative characteristic, it is possible to encipher a steganographic video without affecting the hidden signal, or to execute steganography on an encrypted video while maintaining flawless decryption. Security is maintained against unauthorised viewing by encrypting the HEVC standard's syntactic parts, such as the sign of quantized transformation coefficient (QTC), the sign of motion vector difference (MVD), and the intra-prediction mode (IPM). Due to the usage of a tailored QTC modification mechanism, the visual distortion brought on by data concealing is minimal when using data embedding. Also, the proposed framework will generate a HEVC-compliant bitstream and allow for hidden information extraction in both the encrypted and decrypted domains.

Confidential information can be concealed in digital carriers using information hiding techniques. The presence of sensitive information within digital carriers is often concealed, making it harder for malevolent users to find. Digital media might take the form of audio files, photo files, or video files. Steganography and watermarking are two methods of concealing data. In order to safeguard intellectual property, watermarking is employed in this research by Ko et al. [8]. Steganography is employed to shield the transfer of sensitive data from prying eyes. In steganography, the primary considerations are always concealment and capacity. In this study, the author suggested a magic cube- and modulus-based steganography scheme for photographs. The bits of the partially revealed secret message are counted and then mapped to a cube. The cover art features a modulus operation that encodes the map's coordinates there. With a payload size of only 3 bits per pixel, we are able to produce stego-images with a quality of over 42 dB.

When sending sensitive information via the Internet, security should be the top priority. Therefore, steganography is utilised to protect this information during transmission. Steganography is the practice of hiding information in other media, be it a digital image, video, audio file, written document, or any other type of carrier. Existing picture steganography techniques function as follows: a cover image is assigned, and the secret message is embedded within it via alterations to the image's pixels, yielding a stego-image. This alteration makes it possible for steganography analysis algorithms to read the hidden message. Therefore, Saad et al. [9] proposed a concept of coverless data concealing as a means of resolving this issue. It is important to note that coverless does not imply that the secret message will be sent without a cover file or that the cover file will be removed after transmission. Instead, a cover file or secret message mapping will be generated to conceal the message. In this work, the author offered a new, highly secure coverless image steganography approach that utilises optical mark recognition (OMR) and rule-based machine learning (RBML).

High-capacity picture steganography is a method for protecting private information such as photographs and fingerprints by disguising the hidden image within the cover image. While traditional approaches prioritise secrecy during transmission, they expose users to potential privacy breaches following the successful restoration of hidden photos. Multitask Identity-Aware Image Steganography (MIAIS) is a framework proposed by Cui et al. [10] to directly recognise container images without having to restore secret images. The primary challenge of direct recognition is in the need to both conceal the identities of the secret images within the container images and make the container photos appear similar to the cover images. To this end, the author implemented a minimax optimisation to handle the contrasting factors and introduced a basic content loss to protect the identity information. The author showed that the robustness findings generalise to other cover datasets. The author provided a multitask framework and add an optional restoration network to this method to allow for adaptability when restoring hidden images. Traditional models and limitations are given in Table 2.

3 Proposed Model

In this research, duplicate-free storage architecture models, which allows for streamlined, centralised management of terminals' aggregated data models are analysed. The difficulty lies in achieving remote server-side deduplication that protects user privacy. If users want to keep computations secret, they need to do them in an encrypted format. As a result, a deduplication strategy with compression models to cut down on storage needs and encryption mechanisms to safeguard individual

Table 2 Traditional models and limitations

Author name	Year of publication	Manuscript title	Proposed model	Remarks
Asfand-E-Yar et al. [6]	2020	Semantic integration of heterogeneous databases of same domain using ontology	The author experimented with and talked about a model for ontologies on the semantic web that is grounded in the way queries are executed. There are two stages to ontology modelling; the first involves adapting the database rules to the ontology rules in order to arrive at an abstract ontology model	Ontologies models are applicable to same domain that need to go for heterogeneous
Xu et al. [7]	2019	Commutative encryption and data hiding in HEVC Video Compression	The author offered a commutative encryption and data concealing technique that performs well for HEVC videos. Due to the commutative characteristic, it is possible to encipher a steganographic video without affecting the hidden signal, or to execute steganography on an encrypted video while maintaining flawless decryption	The key generation model need to be more complex for enhancing the security

(continued)

Table 2 (continued)

Author name	Year of publication	Manuscript title	Proposed model	Remarks
Ko et al. [8]	2022	Efficient cost-reduced with high-quality image of imperceptible steganography using modulo and magic cube	Digital media might take the form of audio files, photo files, or video files. Steganography and watermarking are two methods of concealing data. In order to safeguard intellectual property, watermarking is employed in this research	The magic cube model is complex and reduces the performance levels
Saad et al. [9]	2021	Coverless image steganography based on optical mark recognition and machine learning	The author proposed a concept of coverless data concealing as a means of resolving this issue. It is important to note that coverless does not imply that the secret message will be sent without a cover file or that the cover file will be removed after transmission. Instead, a cover file or secret message mapping will be generated to conceal the message	The complex operations can be reduced using light weight operations

(continued)

confidentiality is analysed. Web searching, target recognition, surveillance, criminal detection, and other computer vision applications all face the challenge of recognising object sequences from videos. The goal is to create a powerful video retrieval system that makes advantage of contextual information such as colour, texture, shape, motion, and embedded visual text. Usefulness of the content model varies across domains and audiences depending on the sorts of descriptors used and the level of detail in the description scheme. This raises the problem of the content model being incompatible across different systems, software, suppliers, and intellectual property. The model's syntactic and semantic content descriptions must be fused in

Table 2 (continued)

Author name	Year of publication	Manuscript title	Proposed model	Remarks
Cui et al. [10]	2021	Multitask identity-aware image steganography via minimax optimization	While traditional approaches prioritise secrecy during transmission, they expose users to potential privacy breaches following the successful restoration of hidden photos. Multitask identity-aware image steganography (MIAIS) is a framework proposed to directly recognise container images without having to restore secret images	The time complexity levels can be reduced by reducing the round operations

this way, revealing an organic relationship between the content's structural elements and its concepts. The semantic gap refers to the discrepancy between what a human reader understands to be the content's meaning and what can be recovered through automated means of indexing.

To create a memory-friendly video compression model that makes use of frame deduplication for use with real-time database management is required. Using a robust steganography model for video encryption and decoding to safeguard sensitive information is strongly required for maintaining security. The purpose of this research is to analyse the limitations of existing models and suggest to develop a powerful framework for a novel video steganography feature extraction approach that makes use of regions of motion. The purpose of this survey is to provide suggestions to develop a methodology for optimising video feature extraction that will significantly improve data security.

4 Conclusion

The limited storage and processing capabilities of most end-terminals in a network setting necessitates that users typically outsource data from local to cloud-based computing. In order to keep up with the exponential growth of the data generated by IoT terminals, data deduplication must be implemented in data centres to save on both transmission and archival costs. By focusing on the computation of frame near distance, an effective video compression method based on frames difference methods can be developed. The selection of significant frames, in which comparable frames are eliminated, relies on a number of criteria. Preprocessing, frame extraction, frame selection, frame reordering, two-dimensional (2D) wavelet transform, and finally video production are all processes in the system for deduplication and to secure the data. The compressed video that is output is of a high quality, has a high performance, and is compressed to a fixed ratio. By eliminating redundant data, data deduplication is a potent tool for lowering the footprint of large-scale storage infrastructures. Another intriguing area of study is the development of methods for efficiently and scalably dealing with false positives during video comparison. Steganography refers to the practise of secretly encoding information into other data. Network steganography, which involves transferring data with the use of a network protocol, is another method for concealing information. This method of data protection centres on encrypting and cloaking sensitive information within a multimedia file. The process of video steganography consists of two steps: the extraction of video files and the insertion of a hidden message. In the simple scenario of zero motion between frames, it is straightforward for the encoder to effectively predict the current frame as a duplicate of the prediction frame, and this is the foundation upon which motion estimation is built. This research presents a brief survey on the video deduplication, compression, and steganography models that are used to secure the data and to effectively utilise the memory to store the video content. This research helps numerous researchers to identify the limitations of the existing models.

References

1. Ye Y, Boyce JM, Hanhart P (2020) Omnidirectional 360° video coding technology in responses to the joint call for proposals on video compression with capability beyond HEVC. IEEE Trans Circuits Syst Video Technol 30(5):1241–1252. https://doi.org/10.1109/TCSVT.2019.2953827
2. Zhu S, Liu C, Xu Z (2020) High-definition video compression system based on perception guidance of salient information of a convolutional neural network and HEVC compression domain. IEEE Trans Circuits Syst Video Technol 30(7):1946–1959. https://doi.org/10.1109/TCSVT.2019.2911396
3. Bross B, Chen J, Ohm J-R, Sullivan GJ, Wang Y-K (2021) Developments in international video coding standardization after AVC, with an overview of versatile video coding (VVC). Proc IEEE 109(9):1463–1493. https://doi.org/10.1109/JPROC.2020.3043399
4. Katsenou AV, Dimitrov G, Ma D, Bull DR (2021) BVI-SynTex: a synthetic video texture dataset for video compression and quality assessment. IEEE Trans Multimedia 23:26–38. https://doi.org/10.1109/TMM.2020.2976591
5. Mi A, Li Y, Darong H, Wei T, Zou Q (2020) Secure data de-duplication based on threshold blind signature and bloom filter in internet of things. IEEE Access 8:167113–167122. https://doi.org/10.1109/ACCESS.2020.3023750
6. Asfand-E-Yar M, Ali R (2020) Semantic integration of heterogeneous databases of same domain using ontology. IEEE Access 8:77903–77919. https://doi.org/10.1109/ACCESS.2020.2988685

7. Xu D (2019) Commutative encryption and data hiding in HEVC video compression. IEEE Access 7:66028–66041. https://doi.org/10.1109/ACCESS.2019.2916484
8. Ko H-J, Huang C-T, Tseng H-W, Wang S-J (2022) Efficient cost-reduced with high-quality image of imperceptible steganography using modulo and magic cube. IEEE Access 10:67686–67693. https://doi.org/10.1109/ACCESS.2022.3185120
9. Saad AHS, Mohamed MS, Hafez EH (2021) Coverless image steganography based on optical mark recognition and machine learning. IEEE Access 9:16522–16531. https://doi.org/10.1109/ACCESS.2021.3050737
10. Cui J et al (2021) Multitask identity-aware image steganography via minimax optimization. IEEE Trans Image Process 30:8567–8579. https://doi.org/10.1109/TIP.2021.3107999
11. Ntalianis K, Tsapatsoulis N (2016) Remote authentication via biometrics: a robust video-object steganographic mechanism over wireless networks. IEEE Trans Emerg Top Comput 4(1):156–174. https://doi.org/10.1109/TETC.2015.2400135
12. Wahab OFA, Khalaf AAM, Hussein AI, Hamed HFA (2021) Hiding data using efficient combination of RSA cryptography, and compression steganography techniques. IEEE Access 9:31805–31815. https://doi.org/10.1109/ACCESS.2021.3060317
13. Zhai L, Wang L, Ren Y (2020) Universal detection of video steganography in multiple domains based on the consistency of motion vectors. IEEE Trans Inf For Secur 15:1762–1777. https://doi.org/10.1109/TIFS.2019.2949428
14. Liu Y, Ni J, Zhang W, Huang J (2022) A novel video steganographic scheme incorporating the consistency degree of motion vectors. IEEE Trans Circuits Syst Video Technol 32(7):4905–4910. https://doi.org/10.1109/TCSVT.2021.3135384
15. Liu S, Liu B, Hu Y, Zhao X (2021) Non-degraded adaptive HEVC steganography by advanced motion vector prediction. IEEE Signal Process Lett 28:1843–1847. https://doi.org/10.1109/LSP.2021.3111565
16. Li Q, Chen S, Tan S, Li B, Huang J (2022) One-class double compression detection of advanced videos based on simple Gaussian distribution model. IEEE Trans Circuits Syst Video Technol 32(4):2496–2500. https://doi.org/10.1109/TCSVT.2021.3069254
17. Lu J, Zhou G, Yang C, Li Z, Lan M (2019) Steganalysis of content-adaptive steganography based on massive datasets pre-classification and feature selection. IEEE Access 7:21702–21711. https://doi.org/10.1109/ACCESS.2019.2896781
18. Ma Y, Luo X, Li X, Bao Z, Zhang Y (2019) Selection of rich model steganalysis features based on decision rough set α-positive region reduction. IEEE Trans Circuits Syst Video Technol 29(2):336–350. https://doi.org/10.1109/TCSVT.2018.2799243
19. Xia C, Guan Q, Zhao X, Wu K (2020) Improved JPEG phase-aware steganalysis features using multiple filter sizes and difference images. IEEE Trans Circuits Syst Video Technol 30(11):4100–4113. https://doi.org/10.1109/TCSVT.2019.2954041

A Study on Various Techniques of Two-Dimensional Bin Packing Problem

U. Prabu

Abstract In real-world circumstances, like as logistics, it may be highly valuable to consider all of the features of the products and the container, such as value, weight, and form, for better optimum solutions. Two well-known optimization problems involving placing products in a container are the packing problem and the Knapsack Problem. Knapsack Problem is an optimization problem in the sense that it finds the best solution with maximum profits among a group of objects with varying weights and values. Packing problems are a sort of optimization problem in which objects are packed into containers. The combination of these two issues (called the Joint Problem) defines the set of things that must be placed in a bin such that the greatest amount of space in the bin is used by obtaining the maximum profits. The geometric forms associated with the objects and the container in the packing problem were combined with the value and weight conceptions of the Knapsack Problem. The Joint Problem's purpose is to maximize the total value sum of all things in the container while remaining within the container's maximum weight constraint and avoiding geometric crossings. In this article, a detailed study has been performed on various techniques of two-dimensional bin packing problem. The main categorization of mathematical models or algorithms, as well as the purpose of usage, are also studied and listed so that it will be useful for the researchers to have an insight on the working of each technique.

Keywords Optimization problems · Knapsack problems · Packing problems · Joint problems

U. Prabu (✉)
Department of Computer Science and Engineering, Velagapudi Ramakrishna Siddhartha Engineering College, Vijayawada, India
e-mail: uprabu28@gmail.com

G. Rajakumar et al. (eds.), *Intelligent Communication Technologies and Virtual Mobile Networks*, Lecture Notes on Data Engineering and Communications Technologies 171,
https://doi.org/10.1007/978-981-99-1767-9_32

1 Introduction

The packing bin problem is an optimization problem. In this problem, a set of objects need to be placed in a bin such that the maximum space of the bin is utilized. It is similar to a knapsack technique in which the objects are kept in a knapsack bag without exceeding the weight of the bag. Similarly, the bin packing problem also exists and it needs to be optimized. A two-step technique, i.e., positions making and covering them can be used to get the optimal answers for the two-dimensional (2D) packing bin problem [1]. By integrating the crow search algorithm (CSA) and a type of genetic algorithm (GA), the 2D bin packing issue is addressed in [2].

Generally, in problems bins are considered in rectangular shape but in this problem, the shape of the bin is not limited to a specific shape, and the placeable items also can have different shapes and objects have sizes and fragility [3]. The problem here is limited to two-dimensional that is the bin and the objects are in two-dimensional geometric shapes, but the shape is not limited to a rectangle. For this bin packing problem, as in knapsack weights are included in the bin and the objects, and both problems are considered to be combined. The issue can be known as a 2D irregular shape packing bin problem with weights. The objects placed should not overlap in this problem. The goal is to reduce the complete cost of the bins used for packaging the rectangular objects, which come in a variety of sizes and prices [4].

The problem is solved using the genetic algorithm such as generating the initial population from the given inputs and then generating the offspring from the selected ones and obtaining the solutions for every set and the set with the best outcome is considered as the optimal one. The best is obtained from the maximize function which is generated using the limitations of the problem such as weights and the values of objects and the weight of the bin. The greedy strategy with edge-occupying action is used by the ALNS algorithm [5].

A binary value is considered for every bin such that if it is placed or not that is 1 if it is placed else 0. From the maximum value of the bin obtained the best solution is considered from the rest. An effective variable neighborhood search (VNS) metaheuristic produces first answers using a straightforward greedy heuristic [6].

The objective of this article is

1. To review the state-of-the-art techniques of two-dimensional bin packing problem.
2. To list out the purpose and classification of mathematical model and algorithms used for two-dimensional bin packing problem.

2 Review on Two-Dimensional Bin Packing Problem Techniques

2.1 A Two-Stage Methodology to Obtain Optimal Solutions

Cid-Garcia et al. [1] proposed a two-stage methodology for obtaining optimal solutions. The well-known NP-hard combinatorial problem seeks to fit as many objects (small rectangles) as feasible into as few bins as possible (larger rectangles). The initial step in the locations and covering stage is to establish a list of appropriate placements for every object using a pseudo-polynomial algorithm, showing the numerous ways to pack the item into the bin. In the next step, a novel way for finding the best non-overlapping solution, the set-covering formulation utilized item configuration for each bin, and to back it up, three sets of legitimate inequalities are employed. The problem and covering approach's experimental results are to be better than other techniques.

2.2 A Crow Search-Based Genetic Algorithm

Amri et al. [2] discussed briefly the bin packing problem in two-dimensional. It is comprised of packing that is not intersecting geometrically. A group of rectangle-shaped objects of varying sizes is placed into the least number of rectangle-shaped containers, known as "bins," of comparable proportions. For solving the 2D BPP, a crow search technique is used on basis of genetic operators. A problem-solving hybrid algorithm incorporates two bio-inspired techniques, the genetic algorithm (GA) and the crow search algorithm (CSA). Cutting and packing (C&P) have real-world applications from many domains.

2.3 A Novel Algorithm with Fragile Objects

Laabadi et al. [3] introduced a novel algorithm. The Bin Packing (BP) challenge is to discover the least quantity of bins required for loading objects optimally for a given set of items. The things are large and delicate. The frequency allocation problem is used as an analogy for the fragile object bin packing problem. The BP problem is considered differently. The BP problem affects a variety of applications, including industry and logistics.

2.4 Evolutionary Hyper-Heuristics for Tackling Bi-Objective Problems

Gomez et al. [7] introduced Tackling bi-objective 2D Bins. A multi-objective evolutionary learning mechanism is used in this method. An extensive experimental study was carried out using a large number of 2D BP issues. A 2D BP issue exists with irregular pieces that are convex and nonconvex as well as single-size bins. For fitting a finite large quantity of items into a small number of containers, the suggested framework includes three well-known multi-objective evolutionary algorithms.

2.5 Binary Crow Search Algorithm

Soukaina et al. [5] proposed a strongly sensed problem of optimization. Different metaheuristics are used to solve the 2D BPP or 2D packing bin problem. The 2D BPP issue can be solved using a variety of metaheuristics, however, swarm intelligence methods are rarely used to tackle it. The crow search algorithm (CSA), a newly evolved method based on the ideas of swarm intelligence, has been effectively used to solve continuous optimization issues. To make the CSA suitable for solving the 2D BPP, a sigmoid transformation is used to binarize it.

2.6 Adaptive Large Neighborhood Search

It outlines substantial changes made to the algorithm to finely increase the quality of packaging and suggests the adaptable large neighborhood searching (ALNS) technique [6]. To provide an initial solution, ALNS first calls GACOA. The next step is a regional or local neighborhood searching period, which repetitively perplexes the present solution by picking any two bins used at random. The procedure is somewhat almost equal to that suggested by Okano and Osogami [8] and adopts the topic of He et al. [9] (examining a one-circle packing problem instance), in which every circular item is estimated as a square in the way to find open goods locations to avoid local minimal by exchanging two circular items of the same size or two circles chosen at random.

2.7 Fitness Dependent Adaptive Optimizer

El-Ashmawi et al. [10] proposed an Adaptive Fitness Dependent Optimizer (AFDO) technique for solving the 1D container packaging issue, and the technique is developed by usage of a lately improved swarm optimization and fitness-based optimizer, and a modified firstly fit (FF) intuitive method is used to generate a workable initial population. The performance outcomes of the suggested design are also differentiated from those of PSO, CSA, and Jaya algorithms, and it is clear from these comparisons that the suggested design can find an answer to the robust NP-hard problem.

2.8 A Variable Neighborhood Search Methodology

Luiz et al. [4] proposed a metaheuristic approach which is a variable neighborhood search (VNS) algorithm for solving during last destination delivery to the small warehouses in big city areas, the compatible categories included the packing bin problem, and a difficult optimal technique occurs. To reduce the number of bins necessary, the optimal assignment of delivery of various items to a homogenous fleet of capable vehicles must be found. The suggested VNS metaheuristic uses a straightforward greedy heuristic to produce the first solutions. According to computational studies, the utilized VNS methodology can successfully achieve the answer for the compatible categories including the packing bin problem in the relatively least amount of computational time.

2.9 A Variable Bin Size and Cost-Oriented Methodology

Pisinger et al. [11] discussed the test of squeezing an assortment of rectangular items into an assortment of rectangular receptacles is known as the two-layered variable estimated container pressing issue (2DVSBPP). The objective is to lessen the expense of the receptacles used, and some container types might have variable costs or costs that are not proportionate to the size of the canister. Such issues have different modern applications, including the cutting of metal plates and wood, where more modest bits of wood might be more reasonable than greater ones.

2.9.1 Multi-Dimensional Bin Packing Problem with Guillotine Constraints

Amossen et al. [12] addresses choosing whether a gathering of multi-layered rectangular boxes can be stuffed symmetrically into a rectangular receptacle as a decision issue. There are numerous scenarios when organizing boxes into bins is a difficulty. For example, in business, one goal is to save travel and capacity costs by pressing however many things as doable per space unit.

2.9.2 An Improved Optimization Algorithm for Container Load Balancing

Shunzhi et al. [13] devised a novel method for loading several items into one container using multi-batches. The method employs a "depth priority" technique to find a suitable space using the concepts of "plane" and "block." The algorithm also allows objects to spin in any direction while ensuring effective space consumption and improving placement stability. When the last group packing is inappropriate, the optimum algorithm is expected to withdraw. The algorithm's effectiveness in solving such situations is demonstrated by experimental findings.

2.9.3 Heuristics for Container Loading Problem

Pisinger et al. [14] suggested a novel heuristic based on the wall-building technique, which divides the issue into several tiers, which are then divided into many strips. The loading of a strip is best expressed and solved as a Knapsack problem with a volume equal to the width or height of the container [15–17]. The depth of a layer, as well as the thickness of each strip, are determined using a branch-and-bound technique in which only a subset of branches is investigated.

3 Analysis of Two-Dimensional Bin Packing Problem Techniques

The algorithms or techniques are used to load the goods in a container. The various dimensions of the container are taken into consideration. The items are stored for 1D bin packing as well as 2D bin packing. Table 1 provides an overview of techniques for various bin packing problems. Table 2 provides the usage and classification of algorithms used in the bin packing problem.

4 Conclusion

The various packing problems are examined. In some articles, along with the packing problem, knapsack technique is included to obtain the maximum profits in an optimal solution. The various topics tackled over the period are also noted. This provides a clear understanding for investigating the packing issues and delivering a better solution. It is also intended to work on three-dimensional packing in the future.

Table 1 Overview of techniques of various bin packing problems

Research work	Addressed issue	Compared with	Results
[1]	Position and covering	NP-Hard	The items are stored in bins with possible positions and optimal packing
[2]	2D bin packing problem	Binary particle swarm optimization (BPSO), standard GA	Bin usage is optimized compared to GA and BPSO
[3]	Bin packing problem	Frequency allocation problem	Results are better by following the first fit algorithm
[7]	Tackling bi-objective	Hyper-heuristics	Better solutions generated
[5]	Fixed orientation	Crow search algorithm	To fix the bin packing issues in two dimensions
[10]	Optimization of 1D-BPP	PSOA	Performance is better than other algorithms by obtaining the smallest fitness values and advantage in terms of speed of execution
[4]	Compatible categories	Item-by-item bin packing formulation	The suggested approach can successfully resolve the BPCC with very fast CPU times
[11]	Bin sizes and costs	2DVSBPP	Reduce the overall cost of the bins that will be used to pack the rectangles
[12]	Orthogonal packing	NP-hard	Reduce the price of storage and transportation and maximize the utilization of items The optimal rate of a container is more
[13]	3D bin packing problem	New algorithm based on depth priority strategy [18]	Results are better by following the first fit algorithm
[14]	3D bin packing problem	Wall building approach	Having the capacity to fill larger-sized instances to a volume greater than 95%

Table 2 Usage and classification of algorithms used in bin packing problem

Research work	Mathematical model/algorithm	Purpose of usage	Classification
[1]	Heuristic algorithm	To discover a more expedient solution to an issue when traditional approaches are unsuccessful in doing so	–
[2]	Crow search algorithm	For bin packing	Binary classification
[3]	First fit algorithm	It allocates the empty blocks of space for items among all bins	Bin packing
[7]	Hyper-heuristic algorithm	To find excellent solutions to problems	Multi-objective optimization
[6]	Fitness-dependent optimizer (FDO)	To develop ALDO for solving the BPP	Stochastic optimization technique
[4]	Variable neighborhood search (VNS)	To solve the BPCC	Metaheuristic optimization
[5]	Crow search algorithm Binary crow search algorithm	To solve continuous usage issue To adapt CSA to 2D BBP	Dynamic programming Binary classification
[10]	Fitness-dependent optimizer (FDO) First fit heuristic approach	To develop ALDO for solving the BPP To generate a feasible initial population	Stochastic optimization technique Bin packing
[11]	Branch-and-price algorithm	This generalized to handle the 2DVSBPP	Linear programming
[12]	Guillotine algorithm	Maximize the total area, minimize the amount of waste	Guillotine partition
[13]	Space cut technology Goods placement method	Obtained a good result while using this technique with a loading optimization more than 81%	Heuristic algorithm for multi-cluster container stacking issue
[14]	Space cut technology	Obtained a good result while using this technique with a loading optimization of more than 81%	Heuristic algorithm

References

1. Cid-Garcia NM, Rios-Solis YA (2020) Positions and covering: a two-stage methodology to obtain optimal solutions for the 2D-bin packing problem. PLoS ONE 15(4):e0229358
2. Laabadi S et al (2019) A crow search-based genetic algorithm for solving two-dimensional bin packing problem. In: KI 2019: advances in artificial intelligence: 42nd German conference on

AI, Kassel, Germany, 23–26 Sept 2019, proceedings 42. Springer International Publishing
3. Laabadi S (2016) A new algorithm for the bin-packing problem with fragile objects. In: 2016 3rd international conference on logistics operations management (GOL). IEEE
4. Santos LFOM et al (2019) A variable neighborhood search algorithm for the bin packing problem with compatible categories. Exp Syst Appl 124:209–225
5. Laabadi S et al (2020) A binary crow search algorithm for solving two-dimensional bin packing problem with fixed orientation. Procedia Comput Sci 167:809–818
6. Zeng J, Zhang X (2021) An adaptive large neighborhood search for single-machine batch processing scheduling with 2-D rectangular bin-packing constraints. IEEE Trans Reliab 71(1):139–148
7. Gomez JC, Terashima-Marín H (2018) Evolutionary hyper-heuristics for tackling bi-objective 2D bin packing problems. Gen Program Evolvable Mach 19:151–181
8. Osogami T, Okano H (2003) Local search algorithms for the bin packing problem and their relationships to various construction heuristics. J Heuristics 9(1):29–49
9. K, Tole K, Ni F, Yuan Y, Liao L (2021) Adaptive large neighborhood search for solving the circle bin packing problem. Comput Oper Res 127:105140
10. Abdul-Minaam DS et al (2020) An adaptive fitness-dependent optimizer for the one-dimensional bin packing problem. IEEE Access 8:97959–97974
11. Pisinger D, Sigurd M (2005) The two-dimensional bin packing problem with variable bin sizes and costs. Discret Optim 2(2):154–167
12. Amossen RR, Pisinger D (2010) Multi-dimensional bin packing problems with guillotine constraints. Comput Oper Res 37(11):1999–2006
13. Shunzhi Z, Wenxing H (2009) An improved optimization algorithm for the container loading problem. In: 2009 WRI world congress on software engineering, vol 2. IEEE
14. Pisinger D (2002) Heuristics for the container loading problem. Eur J Oper Res 141(2):382–392
15. George JA, Robinson DF (1980) A heuristic for packing boxes into a container. Comput Oper Res 7(3):147–156
16. Ibaraki T (1987) Enumerative approaches to combinatorial optimization. Ann Oper Res 11
17. Chen CS, Lee S-M, Shen QS (1995) An analytical model for the container loading problem. Euro J Oper Res 80(1):68–76
18. Zhai Y, Sun X (2007) A heuristic algorithm for three-dimensional container loading problem with non-identical items. J-Shanghai Jiaotong Univ-Chin Edn 41(8):1244

Advances in Computer-Aided Diagnosis of Developmental Delay in Children Using Bioengineering Systems: A New Math Model and Algorithm

Igor Val Danilov

Abstract Assessing cognitive abilities in children is one of the contemporary challenges. The classical psychometric approach for assessing is based on a questionnaire of behavioral markers. It does not offer great prospects for developing a possible computer-aided assessment. Recent studies proposed a new approach that provides a computerized diagnosis based on evaluating shared intentionality in mother–child dyads. This new approach does not employ behavior markers. However, the proposed high-tech method is limited by the applied mathematical model described in probabilistic terms. This theoretical study develops the computer-aided approach, shaping the concept design for translational research. The new high-tech method also emulates the mother-newborn communication model by employing human–computer interaction to detect shared intentionality in mother–child dyads. Its novelty lies in introducing the mean baseline value into computations that forms the new mathematical model and algorithm, reducing the shortcomings of the former computerized method.

Keywords Assessing cognition · Communication technologies · Communication model · Human–computer interaction · Shared intentionality

1 Introduction

Assessing cognitive abilities in preverbal children is one of the crucial contemporary challenges. About 15 percent of the world's population has a significant physical or mental disability [1]. The economic impact of mental illness amounts to 600 billion euros in the European Union, or more than 4% of GDP [1]. Until now, the psychometric behavior marker questionnaire is the only generally accepted classical diagnostic method for estimating cognitive development trajectory in young children [2, 3]. However, there are several limitations of this classical assessing

I. V. Danilov (✉)
Academic Center for Coherent Intelligence (Consortium), Riga, Latvia
e-mail: igor_val.danilov@acci.center

G. Rajakumar et al. (eds.), *Intelligent Communication Technologies and Virtual Mobile Networks*, Lecture Notes on Data Engineering and Communications Technologies 171,
https://doi.org/10.1007/978-981-99-1767-9_33

method: some of the social communication features of cognitive delay are not yet fully present, no specific period exists for the manifestation of "marker" skills in a child, and ongoing development bears an extent of uncertainty in developmental diagnosis [2, 3]. The behavior markers approach in diagnosing requires the competence of caregivers in reporting and the level of experience of the professionals to recognize them from the parent's rapport [3]. The continuous monitoring of the child's cognitive development trajectory based on this psychometric approach demands the parent's persistence in self-education with their opportunities in time management (for providing regular monitoring and appointments with specialists). This behavioral marker method becomes even more problematic in a case of a multicultural environment when parents cannot relevantly explain a child's divergent behavior to specialists for disclosing behavior markers because of social, cultural, and linguistic differences. Even more, the behavior markers approach is relative to the situation in which the child is being. Further, similar behavior patterns may have different meanings in different cultures; lifestyle, family habits, and values vary from site to site, from country to country. So, the impact of ecological context also limits this approach. Therefore, the classical psychometric approach does not offer great prospects for developing a possible computer evaluation.

Recent studies proposed computer-aided approach to a cognitive development assessment that relies on social interaction evaluation [2]. Developmental delays possess a common feature: children's lack of interaction ability [4–10]. It is common for children with developmental delays to have difficulty with social and emotional skills, i.e., a shortage of social bonds. So, the social bond evaluation in the mother–child dyads can provide an understanding of cognitive development in pre-verbal children. According to received view in cognitive sciences, the mother-newborn interaction appears due to shared intentionality [11], when children still lack communication skills [e.g., 11–16]. This mother-newborn bond provides cognitive development in young children at initial developmental stage even without sensory cues (meaningful stimuli) [2, 11–16]. Recent theoretical articles contributed to knowledge about the neurophysiological mechanisms of the social interaction at the beginning of cognition [2, 17, 18]. They constituted theoretical grounds for the new assessing cognition method in children by evaluating shared intentionality magnitude. Finally, research empirically proved the computer-aided method of assessing children's cognitive abilities by emulating the mother-newborn communication model. They stimulated ongoing interpersonal dynamics in the mother–child dyads for evaluating shared intentionality magnitude in this bioengineering system [19–22]. However, the proposed high-tech method is limited by the applied mathematical model incorporated in probabilistic terms.

The remainder of this paper is as follows: Sect. 2 gives a concise literature review of the original computer-aided method showing methodological components. The method of the current study is explained in Sect. 3. Section 4 discusses the results. Finally, the paper concludes with Sect. 5.

2 Objectives and Related Literature Review

2.1 Original Method Establishment

Because the translational impact of basic neuroscience research relies on how we define brain functions, definitions of these functions should be carefully considered [23]. About thirty years ago, shared intentionality was defined as collaborative interactions in which participants share psychological states [24, 25]. Empirical evidence of inter-brain neuroscience research [26–28] contributed to the new definition development accounting for neurophysiological mechanisms underlying shared intentionality. They showed increasing coordinated activity of neurons in participants without exchanging sensory cues between them [26–28]. According to recent research [17, 20], shared intentionality is collaborative interactions in which a recipient organism chooses one shared stimulus from many irrelevant stimuli due to intentionality of a contributor organism. These individuals share the essential sensory stimulus for solving the actual cognitive problem. It is pre-perceptual communication via nonlocal neurons coupling. This social bond enables ecological training of the immature organism, starting at the reflexes stage of development, for processing the organization, identification, and interpretation of sensory information in developing perception. In nature, shared intentionality appears in mother-newborn dyads in increasing interpersonal dynamics due to mechanisms of cell coupling that provide the ecological developmental template. This updated definition constitutes the framework for translational research on assessing cognitive development in preverbal children with the following methodological components.

Assessing Objective. The original computer-aided method employs human–computer interaction to provide psychophysiological coherence in the mother–child dyads by stimulating their ongoing interpersonal dynamics. This bioengineering system emulates the "mother-newborn" communication model in the dyads for detecting meaningful interaction within these pairs via shared intentionality since this communication model eliminates other meaningful interactions. Shared intentionality is pre-perceptual communication, i.e., an implicit variable. Therefore, the assessing cognition method needs a manifest variable of shared intentionality that the method can precisely assess. The studies reported that before testing, they asked recipients to guess, and the method registered recipients' intuitive responses to the unintelligible items in the condition of primed contributors [19–22]. During testing, the method stimulated participants' intuition by asking recipients to solve unintelligible tasks in the primed condition of contributors. Recipients did not know the correct responses; they could not solve these problems independently. While the recipients showed greater accuracy in answering unintelligible items, believing they were using their intuition, they answered better due to shared intentionality. From this perspective, the recipients' intuitive responses to the unintelligible items are the manifest variable of shared intentionality. The primary indicator of the original computer-aided method is the superior number of intuitive responses to items over chance.

Object of Influence. The original computer-assisted method assesses the cognitive development of children from 2 years old by shaping the "mother-newborn" communication model in the dyads. Significantly, assessing cognitive abilities means the necessity to assess the social bond between the mother and a child from 2 years old but not a newborn. Therefore, it is expected to develop a bioengineering system to form the "mother-newborn" communication model features in dyads. This bioengineering system should stimulate such specific organisms' properties and put them in such conditions that would emulate the mother-newborn model. Research [2, 17, 19–22] showed that the necessary prerequisites for shaping the mother-newborn communication model are psychophysiological coherence in organisms indwelling in social entrainment, which are motivated to solve a common cognitive task (unintelligible for the newborn). The bioengineering system addressed the challenge of shaping the mother-newborn model by stimulating interpersonal dynamics in the mother and a child during solving unintelligible tasks [2, 17, 19–22]. That is, the object of influence is both the mother and her child. Recent psychophysiological research studies tested this method to assess shared intentionality in mother–child dyads with children of different ages [19–21] and in groups of adults [22]. These studies revealed evidence of statistically significant shared intentionality impact on the collaborative performance of human–human biological systems with large overall effect sizes. The association of children's diagnoses (aged 18 months to 10 years) with the shared intentionality magnitude in the dyads allowed these studies to argue a relationship between the cognitive development trajectories in children and the shared intentionality magnitude in dyads [19, 20].

Stimuli. According to the "mother-newborn" communication model, recipients could solve a common cognitive task (shared with the contributors) in a lack of meaningful sensory interaction due to shared intentionality [2, 17]. During testing, the interaction by sharing intentionality with contributors was the only way for recipients to complete the unintelligible tasks. By asking the unintelligible tasks, the bioengineering system put the participants in conditions where they were forced to muster all their properties. The empirical studies reported that the recipients showed better accuracy in answers on unintelligible items while recipients and primed contributors were solving identical cognitive problems in contrast to the condition of unprimed contributors [22] and the outcome by chance [19–21]. The recipients solved (cooperatively with contributors) unintelligible tasks, receiving clues that were not sensory cues. As mentioned above, the mother-newborn interaction appears due to shared intentionality [11] when children lack communication skills [e.g., 11–16]. That is, the bioengineering system stimulated recipients to choose the correct option via shared intentionality, and the unintelligible tasks encouraged this interaction. So, the first component of stimuli for the bioengineering system is unintelligible test items capable of stimulating shared intentionality in recipients and contributors under specific conditions.

According to Premi et al. [29], low frequency-pulsed electromagnetic fields can modulate cortical excitability in human brains, even after a single-shot application. This neuroscience study reported that electromagnetic impulses of 1 Hz (60 ipm, by generating a magnetic field up to 2 T with an impulse duration of 5 ms and a period of 1000 ms) modulated long-term corticospinal excitability in healthy brains [29]. They observed a persistent increase of more than 60% in corticospinal excitability (as an index of Long-Term Potentiation-Like Cortical Plasticity), recording the motor-evoked potential from the contralateral first dorsal interosseous muscle [29]. This perturbation lasted for at least 30 min after the stimulation protocol, potentially maintaining a significant difference (at least 30%) for an even longer time [29]. That is, a single harmonic oscillator can modulate excitability in neurons of different organisms. Neuroscience inter-brain research [26–28] showed increasing cooperative neuronal activity while participants were solving the shared cognitive problem in contrast to the condition when subjects solved the similar problem alone. Together, these experimental data support the hypothesis of neurobiological processes occurring during social interaction [17], supposing nonlocal coupling of their nervous systems while solving identical cognitive problems. With the reason to form the mother-newborn model in the pairs of participants for strengthening interpersonal dynamics, the research studies reported generating low frequency-pulsed electromagnetic fields of wavelengths of 700 and 400 nm alternately of 1.3 Hz (80 ipm, with impulse duration of 750 ms and a period of 1500 ms each) using their smartphones [19–22]. The wavelengths of 700 and 400 nm were chosen for stimulating emotional arousal (they correspond to red and violet colors). This emotional stimulation contributed to ongoing interpersonal dynamics. Therefore, a single harmonic oscillator–low frequency-pulsed electromagnetic fields–is a second component of stimuli of the bioengineering system for modulating cortical excitability in human brains, providing neuronal coherence in participants.

Data Collection Process. The research studies [19–22] employed the within-subjects forced choice research design. Two groups of participants–unprimed recipients and primed contributors–solved unintelligible cognitive tasks simultaneously. The smartphones simultaneously showed one-by-one ten unintelligible questions for all recipients and contributors. The design of each item was the same for all: recipients and contributors simultaneously saw the same picture with the equal mapping of the task and options of an answer. Furthermore, the design of each task promoted the same geometrical navigation of the correct answer on the screen. In such a manner, all screens showed the same geometrical point of the correct answer. Only contributors in the primed condition could solve the tasks independently. The research design excluded communication between participants. Therefore, the recipients had no sensory clues to solve these unintelligible problems. They could choose the correct option on the same topological point on the screen as the primed contributors due to shared intentionality. The research studies registered the children's answers to the test items [19–22]. While the recipients and contributors saw the quiz-test on their smartphones, only the recipient could independently respond to items by pushing the options on the screen. The software processed the quiz test and collected the recipients' inputs. Just before the quiz test, the recipients were instructed that they needed to guess the correct answer for each task, following the "like/dislike" feeling. After completing the quiz test, the software eliminated results with possible regular sequences of responses that could contain any rational strategy (e.g., when only the first option is correct in each item). The data was recorded for further processing if the recipient answered at least 70% of questions of the quiz test and did not follow any rational strategy. The difference between the number of correct recipients' responses on unfamiliar items and the number of correct answers given by chance showed the magnitude of the shared intentionality effect.

Factors of Shared Intentionality. Many factors contribute to facilitating or depressing shared intentionality in dyads on the day of the test. The most influential ones are presented in Table 1 [20]. Subjects' motivation and concentration of oxytocin hormone in the mother can dramatically impact shared intentionality [20]. Motivation depends on many circumstances, such as mood and the subject's healthy state. For assessing shared intentionality, the bioengineering system can modulate test attractiveness, one of these circumstances. As noted above, the quiz test items are unintelligible tasks without any feedback. Evidence reveals that task difficulty affects performance, e.g., [30–32], and limiting feedback reduces the incidence of trial and error problem-solving strategies, e.g., [33, 34]. Scores in low-stakes tests (purposes of no consequence to the test-taker) correlate with motivation [35]. Because unintelligible tasks reduce motivation, the method is forced to stimulate children's motivation. Research reported [19, 20] to present children with a concise fairy tale dedicated to the test tasks to stimulate motivation. Entertainment in pauses between items was another agency to impact motivation. These studies showed nine video spots of 6 s each between items [19, 20].

A high concentration of oxytocin hormone is associated with pro-social behavior. In humans, the level of oxytocin molecules correlates with the expressions of reciprocity in interaction, social recognition, and social bonding [36, 37]. In addition, it is associated with establishing affective links and affiliative behaviors [37, 38]. Empirical data showed a significant increase in oxytocin concentrations in a woman during the menstrual cycle from the early follicular phase to ovulation [39]; the average length of the follicular phase is 16 days. Conversely, low oxytocin concentration is a

Table 1 The 11 Endogenous and exogenous factors of shared intentionality (SI)

	Endogenous factors	Exogenous factors
Facilitating SI	(1) Social entrainment (2) Normal level of hormones concentration in children (3) An increase of oxytocin concentration in woman	(1) Emotional arousal (2) Interactional synchrony (3) Unintelligible cognitive problem (4) Pleasant social impact (5) Motivation to solve the cognitive problem
Depressing SI	(1) Low oxytocin concentration in woman	(1) Too exciting emotional arousal (2) Depressed psycho-physiological state

factor for depressing social interaction (Table 1). The most common causes of lower-than-normal oxytocin levels in children, for instance, are ASD, depressive symptoms, and panhypopituitarism [40]. Empirical data showed a significant decrease in oxytocin concentrations in women during the menstrual cycle from ovulation to the mid-luteal phase [39]; this depressing period for shared intentionality lasts 6–8 days in a month.

Mathematical Model. The original computer-aided method relies on detecting the difference between recipients' answers in the condition of primed contributors and their answers given by chance. The studies [19–22] reported registering this magnitude for computing the relative value of shared intentionality to provide a universal rating allowing comparison of the different individuals' assessments. The relative value of shared intentionality was denoted by R (Eq. 1). The integrated factor index F evaluated crucial factors influence. The loading of each factor affected the assessing outcome. This F-index minimized the input data deviation from the "true" value of the latent variable in the calculation. It is calculated according to Eq. 2. The R-value compared the current input data with the value calculated from the Bernoulli equation for independent events (Eq. 3). Specifically, observed scores were denoted by Xo, and expected scores were denoted by Xe, i.e., the number of events with the highest probability of occurring calculated by the Bernoulli Eq. (3).

$$R = \frac{F}{n} \sum \frac{\mathrm{Xo} - \mathrm{Xe}}{\mathrm{Xe}} \tag{1}$$

$$F = l(\iota, 1)\, f(\iota, 1) + \cdots + l(\iota, j)\, f(\iota, j) \tag{2}$$

whereby, $l(\iota,j)$ is the loading for the ι-th observation of the j-th factor, $f(\iota,j)$ is the value of the j-th factor of the dyad.

$$P(k) = C^k p^k q^{n-k} \tag{3}$$

The Bernoulli Eq. (3) shows a probability of a number of events (correct responses on items) made in independent trials, where: C–number of combinations n by k; p–the probability in each task; n–independent trials (items), the probability of each is p ($0 < p < 1$); k–events, how many items the child answers correctly; $q = 1 - p$. In the math model of the original method, the baseline is equal to the Bernoulli value Xe $= P(k)$.

Algorithm. The original method began the assessing cognitive development with Step 1 "Registration": the mother and child introduce the child's data through the smartphone interface to register the data and monitor the child's willingness to complete the quiz test. The data entry verified the intention of the child to proceed with testing. Step 2 "Quiz-test": the bioengineering system started the psychophysiological stimulation of shared intentionality. At the same time, the child entered responses to the items (unintelligible intellectual tasks) into the smartphone through the smartphone interface. Step 3 "Outcome": the bioengineering system computed the outcome–the relative value of shared intentionality–by comparing the obtained input value with the random choice value. In parallel, the bioengineering system evaluated the validity of the results based on the assessment components.

2.2 Original Method Limitations

However, the original computer-aided method bears limitations. Assessing cognitive development in children by comparing the shared intentionality outcome with the random choice value had been chosen [19–21]. This mode to assess shared intentionality limits the assessing cognition method since a method based on probabilistic tools consistently bears an extent of uncertainty when applied to assessing the psychological construct of a single individual. Another limitation appears from unintelligible testing content that can affect cognitive performance by producing unexpectable associations. The original method does not consider the environmental and content impact on cognitive performance. Finally, shared intentionality appears in human pairs only in some interpersonal dynamics and not always to the same extent [20]. On the test day, 11 endogenous and exogenous factors contribute to facilitating or depressing shared intentionality [20]. The difference in factors' values on the test day affects the measured manifested variables showing the deviation. Therefore, the assessing system should observe a set of defined factors on the day of the data input and compute their loadings. The integrated factor index F can minimize the input data deviation. However, accounting for each factor loading is a problem. Among 11 factors, six of them require a sophisticated evaluation procedure because four factors represent hormone concentration in the mother and child, and two factors reflect the motivation impact and the current psychophysiological states of subjects.

2.3 The Current Study Objectives

So, the original computer-aided method bears limitations conditioned by applied probabilistic mathematical tools. The study aims to develop an enhanced computer-assisted method of assessing cognitive development by improving a mathematical model and algorithm. These advancements decrease (or even remove) the limitations of the original high-tech method–problems of the probabilistic tools and the impact of the factors mentioned above–also considering the environmental and items' content impact on performance.

3 Approach

This theoretical study analyzes recent findings from neuroscience, medicine, and research on assessment of cognitive development. It constitutes crucial components of the concept design for translational research on bioengineering systems for assessing children's cognitive development by computer-assisted evaluation of shared intentionality for preventing a cognitive delay in children.

4 Results and Findings

The new computer-aided method encompasses the significant components of the original computer-assisted one. The main similarity lies in the fact that the new method duplicates the original version idea, i.e., for assessing cognitive development, it also evaluates shared intentionality in the bioengineering system, which emulates the "mother-newborn" communication model in the mother with a child from 2 years old [2, 19–22]. Again, the new method employs human–computer interaction to provide psychophysiological coherence in the dyads by stimulating their ongoing interpersonal dynamics [19–22]. This bioengineering system shapes the "mother-newborn" model in the dyads for detecting meaningful interaction within these pairs via shared intentionality since this communication model eliminates other meaningful interactions [2, 19–22]. Although the new method uses the main methodological components of the original high-tech method, there are several changes that are described below.

Assessing Objective. The new computer-aided method still addresses assessing cognitive development through evaluating shared intentionality; the manifest variable of shared intentionality is the recipients' intuitive responses to the unintelligible items. However, the primary indicator of the new method is the superior number of intuitive responses to test items of the recipient over her/his responses in unprimed condition of the contributor. This is the main contribution of the new method–it introduces the mean baseline value (MBV) based on recipients' answers in the condition of unprimed contributors in contrast to the baseline equal to the Bernoulli value in the original version. This improvement reduces limitations of the original computer-aided method that appear from the applied tools of probability theory since the new method only partially employs the probabilistic tools (the mean baseline value partially concerns the Bernoulli value).

Further, the new method (hereafter the MBV method) addresses another original method's limitation: the testing content can bear implicit associations (helpful or erroneous) for subjects. The environment can also promote implicit associations by creating the context for the testing content. Because these implicit associations can increase or decrease an obtained baseline magnitude compared to the chance value, it is proposed to account for the mean baseline value instead of data from obtained baseline magnitude (responses in unprimed condition of the contributor). The MBV method proposes two-times assessment cognition ensured by the MBV design. Two times repeated assessment procedure involves similar mental activity in a different environment, and the difference in the outcomes mirrors the environmental effect. The unintelligible testing content can also affect answers. Therefore, the difference between obtained baseline magnitude and the expected by-chance value shows both the influence of content and environmental effect. The MBV method considers this difference in the computation of relative value of shared intentionality.

The next issue, while the MBV method excludes such significant factors as the impact of the probabilistic tools, content, and environment, it can only partially eliminate the impact of other factors. For instance, motivation and concentration of oxytocin hormone in the mother still can significantly impact the outcome of the cognition assessment. The following considerations show how the MBV method accounts for these two factors in computing the relative value of shared intentionality.

Factor of Shared Intentionality: Motivation. The two times repetition of the same task decreases motivation. The method cannot change the testing items in the second phase, and for test validity, it is necessary to repeat exactly the same items to the dyads. The modification of the task can impact the outcome. A change of entertainment in pauses between items can maintain motivation in subjects (18 video spots of 6 s for 2×10 items). So, during two phases of testing, the MBV method presents the child with a brief fairy tale dedicated to the items and 18 video spots in pauses.

Factor of Shared Intentionality: Oxytocin. To reduce this factor, the new method suggests performing assessing cognition during the first two weeks after the day of the mother's menstrual cycle beginning. These days are preferable for testing since research shows high oxytocin hormone concentration in women during that period [20]. That is, the enhanced method consists of the new procedure introducing data collection twice (at breakfast and after launch) in one day for two weeks after the day of the mother's menstrual cycle begins.

Mathematical Model. The mean baseline value (MBV) is calculated by subtracting the influence of the factors from the probability value following the Bernoulli equation for the total independent events (i.e., 10 baseline items from the total 20). Again, the MBV is based on recipients' answers in the condition of unprimed contributors in contrast to the baseline equal to the Bernoulli value in the original version. It is equal to the subtraction: the baseline value (derived from the Bernoulli equation) minus the real mean magnitude deviation (the difference between obtained mean baseline magnitude and the Bernoulli equation value) denoted $X(\Delta)$. This MBV is computed from the two successive measures (totally, 5 MBV items and 5 primed items of the first phase + 5 MBV items and 5 primed items of the second). The MBV is denoted *X(b)* and calculated by Eq. (4).

$$X(b) = P(k) - X(D) \tag{4}$$

The real mean magnitude deviation $X(\Delta)$ of two data collections is calculated by the Eq. 5:

$$X(\Delta) = \frac{b1 - P(k) + b2 - P(k)}{2} \quad (5)$$

whereby, *b1* and *b2* are the mean baseline value of recipients' answers in the condition of unprimed contributors obtained in two data collections. The MBV method also considers the real mean magnitude deviation $X(\Delta)$ for computing the mean recipient answers in the condition of primed contributors denoted $X(o)$, see Eq. 6:

$$X(o) = \frac{(o_1 + o_2)}{2} - X(\Delta) \quad (6)$$

Finally, the relative value of shared intentionality with the mean baseline value (MBV) is calculated by Eq. 7:

$$R^{\mathrm{MBV}} = \frac{X(o) - X(b)}{X(b)} \quad (7)$$

The mathematical model defines the algorithm of the MBV computer-aided method for assessing cognitive development in children (Table 2). In every algorithm, it is essential to maintain the trade-off between the exploration and exploitation processes [41]. The term exploration refers to the limits of the possible solution, whereas exploitation is the practice of selecting the most optimal answer from among several possible ones. The MBV method establishes the balance between the exploration and exploitation by incorporating the mean magnitude deviation $X(\Delta)$ (Eq. 5). The MBV algorithm procedure follows three simple rules: (1) only the day with the high oxytocin concentration in the mother is correct for testing (line 3); (2) the child should show her attention toward the tasks, demonstrating an understanding of the story (line 5); (3) the child's motivation and task comprehension should be expressed by the number of answers to at least 16 questions out of 20 (lines 7, 29, 31).

Validity. According to Devon et al. [42], the soundness of psychometric tests depends on many important factors such as underlying theory, the consequences of weak measures, and the implications of the findings. If it is pointed to become valid, psychometric assessment methods must be designed and researched according to strict scientific principles. Many social science tests are imprecise tools since no one can ever know what is on a person's mind and can only have an observable test score. One of the approaches to estimating the method's validity is confronting its results with estimation from another valid method. The research studies [22, 43] used both methods for assessing shared intentionality for the same subjects: comparing their results with chance and comparing their results in primed and unprimed conditions. The obtained data in both methods were consistent with the hypothesis and relevant to each other. Comparing the outcomes of the two methods reveals the data parity that may mean the methods' validity [22, 43].

Another argument supports the method's validity: at the initial stages, cognition develops due to statistical learning mechanisms and can be described only in

Table 2 The MBV algorithm for assessing cognitive development of children

Algorithm 1. The mean baseline value method algorithm
1 Assessing cognitive development in 2-to-3-year-children
2 Introducing child's date and the mother's menstrual cycle date;
3 Choosing the calendar date. If the date is chosen then go to 4;
4 Beginning. The 1 min fairy tale with the ticker;
5 "Do you like the story?"If yes, then go to 6; if no–go to 4;
6 The rhythmically (80 ipm) changed lightings;
7 Counting balls: While counting go to 8;
8 First Item "baseline" for 30 s
9 Spot for 6 s
10 Second Item "primed" for 30 s
11 Spot for 6 s
12 Third Item "baseline" for 30 s
13 Spot for 6 s
14 Fourth Item "primed" for 30 s
15 Spot for 6 s
16 Fifth Item "baseline" for 30 s
17 Spot for 6 s
18 Sixth Item "primed" for 30 s
19 Spot for 6 s
20 Seventh Item "baseline" for 30 s
21 Spot for 6 s
22 Eight Item "primed" for 30 s
23 Spot for 6 s
24 Ninth Item "baseline" for 30 s
25 Spot for 6 s
26 Tenth Item "primed" for 30 s
27 end for
28 The lights stimulation is finished
29 If less then 10 resp. go to 30; if more, 10–go to 31
30 While confirm, if more then 5 go to 4; if less–go to 2
31 If more than 16 resp. then go to 32. If not–go to 3
32 Computing the results for presenting the assessing score
33 end while. "The test is completed, thanks!"

probabilistic terms [19]. Therefore, the tools of the probabilistic theory are likely appropriate for the phenomenon being measured.

At the same time, even though this new computer-aided method does not use psychometric tools in assessment, this assessment method still assesses a psychological construct. Again, we will never be able to understand what exactly a person keeps in mind. This is another reason for the proposed two-times assessment cognition by the MBV design. If the results differ dramatically, there is still a classical psychometric method for checking the outcome. The MBV method does not employ behavior markers; its indicator is entirely different from the classical psychometric assessment by the behavior marker questionnaire. On the other hand, the computer-aided method assesses the same psychological construct as the psychometric method, which employs behavior markers. The methodological difference and common aim together raise an opportunity to join both approaches in one assessment protocol to improve assessing validity. Suppose the metrological component of the new high-tech cognition assessment method is established (see below translational research perspectives). In that case, one assessment protocol can implement a complex of two approaches (classical psychometric and MBV bioengineering) for assessing cognitive development independently through different indicators. Because the high-tech method can diagnose children online, the negative outcome of assessing can be obtained earlier and through a more straightforward procedure. This easy checking would help parents identify the problem efficiently and react as fast as they receive this asking pediatrists to provide assessment by the psychometric method. Again, the new MBV assessing method is a tool for specialists that parents can use only as a developmental indicator to pay closer attention to the child's behavior in case of a negative result and ask the pediatrists to intervene.

5 Conclusion and Future Scope

This theoretical study developed crucial components of the concept design for translational research on the MBV bioengineering system. It proposed an enhanced computer-aided assessing method by improving a mathematical model and algorithm.

According to metrology, any measurement can be divided into three essential overlapping domains: (i) the definition of units of assessment, (ii) the method of how to apply these units for assessing the biological system in practice, and (iii) the reference to measurement standard that provides a universal correlation of all assessment outcomes in practice. The measurement standard is a material measure, measuring instrument, reference material, or measuring system intended to define, realize, conserve, or reproduce a unit or one or more values of a quantity to serve as a reference. Further translational research can establish the metrological component of the MBV method of assessing cognitive development by defining the measurement standard related to other measurement systems that would be used to define the standard value of shared intentionality.

The MBV method develops great perspectives for diagnosing children in a multicultural environment when parents cannot explain behavior markers relevantly because of social, cultural, and linguistic differences.

References

1. EDJNet., The European data journalism network. https://www.europeandatajournalism.eu/News/Data-news/Mental-illness-costs-Europe-600-billion-euro (assessed on 10 January 2023).
2. Val Danilov I, Mihailova S (2022) A new perspective on assessing cognition in children through estimating shared intentionality. J Intell 10:21. https://doi.org/10.3390/jintelligence10020021
3. Volkmar FR, Langford Booth L, McPartland JS, Wiesner LA (2014). Clinical evaluation in multidisciplinary settings. https://doi.org/10.1002/9781118911389.hautc26
4. Erikson EH (1963) Childhood and society, 2nd edn. W. W. Norton & Company, New York
5. Winnicott DW (1973) The child, the family, and the outside world. Penguin, London. www.goodreads.com/book/show/202755.The_Child_the_Family_and_the_Outside_World
6. Prior M, Ozonoff S (2007) Autism and pervasive developmental disorders. Cambridge University Press, Cambridge, pp 69–128
7. Sternberg RJ, Sternberg K (2012) Cognitive psychology, 6th edn. Wadsworth, Cengage Learning, Belmont
8. Vygotsky L (2016) Myshlenie i rech (Russian). Natsional'noe Obrazovanie, Moscow. ISBN 5445407233
9. Pierce K, Gazestani VH, Bacon E, Carter Barnes C, Cha D, Nalabolu S, Lopez L, Moore A, Pence-Stophaeros S, Courchesne E (2019) Evaluation of the diagnostic stability of the early autism spectrum disorder phenotype in the general population starting at 12 months. JAMA Pediatr 173(6):578–587. https://doi.org/10.1001/jamapediatrics.2019.0624
10. Rojas-Torres LP, Alonso-Esteban Y, Alcantud-Marín F (2020) Early intervention with parents of children with autism spectrum disorders: a review of programs. Children 7(12):294. https://doi.org/10.3390/children7120294
11. Tomasello M (2019) Becoming human: a theory of ontogeny. Belknap Press of Harvard University Press, Harvard
12. Gopnik A (1981) Index of authors. J Child Lang 8:495–499
13. Trevarthen C (1989) Signs before speech. In: Sebeok TA, Umiker-Sebeok J (eds) The semiotic web. De Gruyter Mouton, Berlin and Boston, pp 689–756. https://doi.org/10.1515/9783110874099.689
14. Trevarthen C (1998) The concept and foundations of infant intersubjectivity. Intersubjective Comm Emotion Early Ont 15:46
15. Yingling JM (1990) "Does that mean 'no'?": Negotiating proto-conversation in infant-caregiver pairs. Res Lang Soc Interact 24:71–108. https://doi.org/10.1080/08351819009389333
16. Bråten S, Manstead A, Oatley K (1998) Intersubjective communication and emotion in early ontogeny, No. 3. Cambridge University Press, Cambridge
17. Val Danilov I, Mihailova S (2021) Neuronal coherence agent for shared intentionality: a hypothesis of neurobiological processes occurring during social interaction. OBM Neurobiol 5(4):26 (2021). https://doi.org/10.21926/obm.neurobiol.2104113
18. Val Danilov I (2022) Smartphone in detecting developmental disability in infancy: a theoretical approach to shared intentionality for assessment tool of cognitive decline and e-learning. In: Arai K (ed) Proceedings of the SAI 2022, LNNS 508. Springer Nature Switzerland AG (2022), pp 1–11. https://doi.org/10.1007/978-3-031-10467-1_19
19. Val Danilov I, Mihailova S, Svajyan A (2022) Computerized assessment of cognitive development in neurotypical and neurodivergent children. OBM Neurobiol 6(3):18. https://doi.org/10.21926/obm.neurobiol.2203137
20. Val Danilov I (2022) A bioengineering system for assessing children's cognitive development by computerized evaluation of shared intentionality. In: International Conference on Computational Science and Computational Intelligence (CSCI) 2022, proceedings IEEE Xplore, Conference Location: Las Vegas, NV, USA
21. Val Danilov I, Mihailova S, Reznikoff I (2021) Frontiers in cognition for education: coherent intelligence in e-learning for beginners aged 1 to 3 years. Springer Nature—Book Series:

Transactions on Computational Science & Computational Intelligence. The 20th Int'l Conf on e-Learning, e-Business, Enterprise Information Systems, and e-Government, along with CSCE congress
22. Val Danilov I, Mihailova S, Perepjolkina V (2019) Unconscious social interaction, coherent intelligence in learning. In: Proceedings of the 12th annual conference ICERI Seville (Spain), pp 2217–2222. https://doi.org/10.21125/iceri.2019.0606
23. Rust NC, LeDoux JE (2023) The tricky business of defining brain functions, 46(1):3–4, January 1. https://doi.org/10.1016/j.tins.2022.10.011
24. Gilbert M (1989) On social facts. Routledge, New York
25. Searle JR (1992) The rediscovery of the mind. MITPress, London
26. Fishburn FA, Murty VP, Hlutkowsky CO, MacGillivray CE, Bemis LM, Murphy ME, Huppert TJ, Perlman SB (2018) Putting our heads together: interpersonal neural synchronization as a biological mechanism for shared intentionality. Soc Cog Affect Neurosci 13(8):841–849
27. Valencia AL, Froese T (2020) What binds us? inter-brain neural synchronization and its implications for theories of human consciousness. Neurosci Consci 1:niaa010. https://doi.org/10.1093/nc/niaa010
28. Painter DR, Kim JJ, Renton AI, Mattingley JB (2021) Joint control of visually guided actions involves concordant increases in behavioural and neural coupling. Comm Biol 4:816. https://doi.org/10.1038/s42003-021-02319-3
29. Premi E, Benussi A, La Gatta A, Visconti S, Costa A, Gilberti N, Cantoni V, Padovani A, Borroni B, Magoni M (2018) Modulation of long-term potentiation-like cortical plasticity in the healthy brain with low frequency-pulsed electromagnetic fields. BMC Neurosci 19(1):1–6
30. Huber VL (1985) Effects of task difficulty, goal setting, and strategy on performance of a heuristic task. J Appl Psychol 70(3):492
31. Wood RE, Mento AJ, Locke EA (1987) Task complexity as a moderator of goal effects: a meta-analysis. J Appl Psychol 72(3):416–425. https://doi.org/10.1037/0021-9010.72.3.416
32. Gellatly IR, Meyer JP (1992) The effects of goal difficulty on physiological arousal, cognition, and task performance. J Appl Psychol 77(5):694–704. https://doi.org/10.1037/0021-9010.77.5.694
33. Salmoni AW, Schmidt RA, Walter CB (1984) Knowledge of results and motor learning: a review and critical reappraisal. Psychol Bull 95(3):355–386. https://pubmed.ncbi.nlm.nih.gov/6399752/
34. Cope P, Simmons M (1994) Some effects of limited feedback on performance and problem-solving strategy in a Logo microworld. J Educ Psychol 86(3):368–379. https://doi.org/10.1037/0022-0663.86.3.368
35. Duckworth AL, Quinn PD, Lynam DR, Loeber R, Stouthamer-Loeber M (2011) Role of test motivation in intelligence testing. Proc Nat Acad Sci 108(19):7716–7720
36. Magon N, Kalra S (2011) The orgasmic history of oxytocin: love, lust, and labor. Ind J Endocrinol Metabol 15(Suppl 3):S156–S161
37. Mera JCC, Molano MAC, López CCG, Triana CA, Cotrina JM (2021) Discussions and perspectives regarding oxytocin as a biomarker in human investigations. Heliyon 7(11):e08289. https://doi.org/10.1016/j.heliyon.2021.e08289
38. Zik JB, Roberts DL (2015) The many faces of oxytocin: implications for psychiatry. Psychiatr Res 226(1):31–37. https://doi.org/10.1016/j.psychres.2014.11.048
39. Engel S, Klusmann H, Ditzen B, Knaevelsrud C, Schumacher S (2019) Menstrual cycle-related fluctuations in oxytocin concentrations: a systematic review and meta-analysis. Front Neuroendoc 52:144–155. https://doi.org/10.1016/j.yfrne.2018.11.002
40. Osilla EV, Sharma S (2021) Oxytocin. In: StatPearls [Internet]. StatPearls Publishing. https://pubmed.ncbi.nlm.nih.gov/29939625/

41. Bacanin N, Bezdan T, Tuba E, Strumberger I, Tuba M (2020) Optimizing convolutional neural network hyperparameters by enhanced swarm intelligence metaheuristics. Algorithms 13:67. https://doi.org/10.3390/a13030067
42. Devon HA, Block ME, Moyle-Wright P, Ernst DM, Hayden SJ, Lazzara DJ, Savoy SM, Kostas-Polston E (2007) A psychometric toolbox for testing validity and reliability. J Nurs Scholarsh 39(2):155–164
43. Val Danilov I, Mihailova S (2021) Intentionality vs chaos: brain connectivity through emotions and cooperation levels beyond sensory modalities. In: The Thirteenth International Conference on Advanced Cognitive Technologies, and Applications, COGNITIVE 2021, April 18–22 in Porto, Portugal. https://www.thinkmind.org/index.php?

An Analysis of Real-Time Number Plate-Based Verification System with Insurance Processing Using OCR Techniques

P. Pandiaraja, P. Biranav Kumar, N. Jaisaran, and V. Karthick Ram

Abstract If all aspects of vehicle transportation management are handled manually, it is a time-consuming operation that produces significant errors. To address the issues raised above, it is important to design an automatic license plate recognition system that will automatically identify numbers from an image of the vehicle's front side. Recognition of license plates (LPR), commonly known because ANPR, has emerged as one of the most reliable techniques for vehicle surveillance in recent years. It can be utilized to achieve a range of objectives in numerous public spaces, including traffic safety regulation, automatic toll text collection, parking systems, and automatic car park systems. To automatically recognize license plates, different image processing, and techniques must be employed in a single application. Text localization, extraction, enhancement, segmentation, and identification algorithms are utilized to locate the number plate number in each frame of an image or video. The previous studies only partially covered the complete process of a typical LPR device, from picture acquisition to verification. This project developed an entire real-time, restriction-based license plate identification system. They put this mechanism in place to find lost automobiles and find out how vehicle insurance is doing.

Keywords Transportation system · Character segmentation · Number plate recognition · Optical character recognition · Number plate detection

1 Introduction

Using cameras to read license plates and identify vehicles is called license plate recognition (LPR) technology. Although LPR is frequently connected to police enforcement, it is yet another useful tool for businesses to interact with their customers and customize their experience in-store [1, 2]. Car dealers in particular can use the service

P. Pandiaraja (✉) · P. B. Kumar · N. Jaisaran · V. K. Ram
Department of Computer Science and Engineering, M.Kumarasamy College of Engineering, Thalavapalayam, Karur, Tamil Nadu 639113, India
e-mail: sppandiaraja@gmail.com

G. Rajakumar et al. (eds.), *Intelligent Communication Technologies and Virtual Mobile Networks*, Lecture Notes on Data Engineering and Communications Technologies 171,
https://doi.org/10.1007/978-981-99-1767-9_34

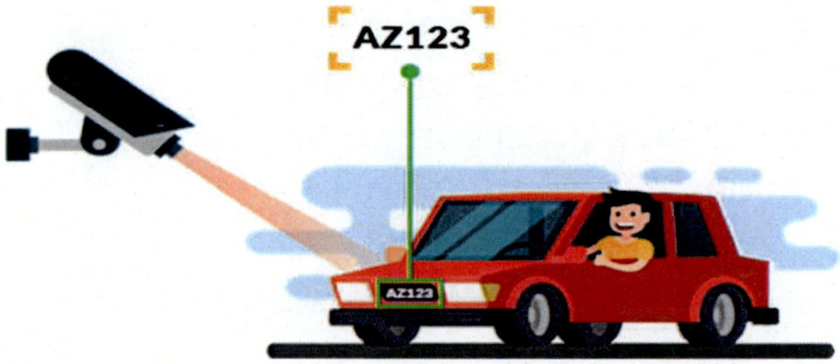

Fig. 1 Number plate detection using OCR Techniques. *Source* www.10pointer.com

drive to uncover extra sales and service possibilities, deploy automated greetings, track client behavior, alert management, and record client activity. Law enforcement, traffic cops, merchants, and car dealers have all expressed a strong interest in license plate recognition in recent years [3, 4]. By utilizing this technology, pertinent data can be quickly accessed with minimum effort or supervision needed. Several advantages also include the following elements:

- An automatic record-keeping based on the vehicle's history
- A subsystem for its management and task scheduling
- Used for immediate identification and real-time tracking of the automobile
- Customer activity heat mapping and their traveling conditions can be analyzed.

To create and test a real-time detection, tracking, and license plate recognition system which will function effectively in the presence of slow-moving objects, objects that temporarily stop moving and return to the foreground, is adaptable to various traffic environmental conditions, robustness against gradual or unexpected illumination changes, and has the shortest possible identification time. The system should be capable of identifying all vehicle types, and all license plates across the nation, and it should also be impervious to image distortions as well as mechanical plate damage that might actually occur [5, 6]. The characteristics of the license plates are crucial to the identification process. In industrialized nations, rigorous regulations are upheld regarding the size, color, and font face—that is, the size, color, and spacing between each character as well as the line count and the height and width of the characters [7]. The plate number detection mechanism is depicted in Fig. 1.

2 Related Work

Shashirangana et al. [8] gradually explore and analyze the existing approaches and techniques that were previously used in ALPR solutions in the recent literature. With a variety of datasets, single neural network learning-based systems have demonstrated strong performances. Large datasets can be used to pre-train inter-object detection sysystem-base deep learning algorithms; however, they have shown to be less accurate and computationally efficient than single-stage methods. The need for data sets in practice has been recognized, and this study has conducted a thorough comparison of the associated studies. Additionally, we have outlined the unresolved issues and provided ideas for future ALPR research directions.

Weihong and Jiaoyang [9] discussed some unavoidable problems of license plate recognition in real situations. For photographs of license plates obtained with such a single camera along the side of the road that may have motion blur, wobbling, or deflection, as well as for weather situations including rain, snow, and fog, as well as use times during the day and at night. Lighting, shading, camera angle, shot distance, pixel elements, background complexity, noise, and other elements all have an impact on how accurately license plates may be read. In order to recover the best image possible, license plate preparation is divided into three sections: tilt correction, image de-noising, and resolution improvement. From 2005 to 2018, research on these three factors was highlighted in pertinent literature. One of them is the study of image denoising, which involves removing noise from the image, such as snow and rain that obscure the license plate, and the other is the study of improving resolution, which involves converting a blurred image brought on by a camera or a moving object into a high-resolution discriminative image.

Zou et al. [10] utilizes a combination of techniques to address the challenges faced during license plate recognition in most common and complex scenarios. It includes three modules: character feature extraction, placement of license plate characters, and feature extraction from the license plate. The number plate module, based on Xception, MobileNetV3, and a spatial attention mechanism, is used to extract the character characteristics of license plates. The license plate letter localization module uses a Bi-LSTM approach along with license plate context location data. The experimental results show that this approach is able to accurately detect and identify characters on license plates, regardless of whether they are standard or irregular, or in simple or complex scenarios.

Henry et al. [11] presented a generalized solution for multinational license plate recognition. Our method includes phases for multinational LP layout identification, unified character segmentation, and LP detection. The text segmentation, as well as character recognition procedures, are combined into the object recognition issue known as LP recognition. Using our suggested transnational LP layout detection technique, our solution is suited to license plates from different nations. As far as we are aware, LPs from the majority of nations can be broadly split into single-line as well as double-line LPs. The suggested layout recognition algorithm is straightforward but is capable of accurately classifying a variety of LP configurations. Our method can successfully extract this same right order of LP numbers from a picture given the proper bounding boxes. A brand-new dataset for Korean license plates (KarPlate) has been made accessible to the public for research.

Zhang et al. [12] presented a robust model for license plate recognition in an unconstrained environment. The suggested model is based on a two-dimensional oriented RNN module with sequence decoding and an Xception CNN component for feature extraction. CycleGAN is designed to produce synthetic LP images with various deformation styles and more balanced area codes, which provides a straightforward yet efficient technique to supplement the existing actual data. This is to address the deficiency or imbalance of real training data. Our methods are superior, especially when dealing with deformed license plates or with little training data, according to extensive testing results. To more fully assess LP recognition algorithms, an LP dataset with photos taken in various ways from diverse locations is gathered.

Pustokhina et al. [13] proposed the OKM-CNN technique which is a three-phase approach for license plate detection and recognition. The first phase uses IBA and CCA models for the localization and detection of license plates. In the second phase, an OKM-based clustering algorithm is used to segment the license plate image, and the characters are then identified using a CNN model. The technique can be used for toll charge collecting, parking management, and traffic surveillance. The model was tested on three datasets and achieved an overall accuracy of 0.981. The model may be effective under certain conditions, such as constant illumination, moderate vehicle speed, designated pathways, and static backgrounds.

Tourani et al. [14] proposed a unified license plate detection (LPD) and character recognition (CR) system for various sets of Iranian vehicle license plates that boasts both real-time performance and high accuracy. The approach uses two successive YOLO v.3 deep networks to gather data under different lighting, noise, and weather conditions from operational driveway security cameras. This training procedure covers a wide variety of photos taken in both difficult and simple realistic settings. The proposed approach was evaluated on different realistic data, and the results showed that it was accurate for the LPD stage with better precision value and a recall value of around 0.979 and 0.991 respectively, and also for the Persian character recognition stage.

Selmi et al. [15] proposed a system for detecting, segmenting, and recognizing license plates with multi-orientation and multi-languages, using Mask Region CNNs (MaskRCNNs). They created a new difficult database containing 610 Tunisian LP photos with various orientations, weather conditions, and complicated backgrounds for performance evaluation. They found that their proposed system performed better in terms of LP identification and character recognition than current methods, and observed an improvement in the results even as the number of characters increased. They attributed this to their data preprocessing technique but noted that the system may still fail to recognize a small number of characters due to the high image clarity or the full failure of the letters in the LPs.

Chene et al. [16] proposed a methodology that combines the use of both convolutional neural networks (CNNs) and k-means clustering to effectively detect and recognize license plates in vehicles. This system is composed of three steps: localization and detection of the license plate using the CCA and IBA model, segmentation of the license plate image using clustering, and identification of the characters using a CNN model. The approach is tested using three different datasets and is found to have a high accuracy compared to other similar models. Additionally, it can recognize license plates in various languages and is useful for applications such as parking control, traffic monitoring, and toll tax collection.

Zhou [17] introduces a powerful false-alarm filtering algorithm for license plate recognition: FAFNet, a deep convolutional neural network-based artificial intelligence system. A compact neural net that can be trained from beginning to end is FAFNet. In the paper, humans first go over the specifics of FAFNet's design. By significantly reducing the number of input constants and calculation speed as well as by deepening the network, the model is able to recognize objects with a high degree of precision. Then, using experiments, we confirm that FAFNet has outstanding false-alarm filtering performance and that it can recognize objects quickly on a variety of devices and recognition rate on various hardware platforms.

3 Character Segmentation Methods

The precision of number plate extraction is necessary for character segmentation and recognition. Therefore, this processing stage is crucial and emphasizes the development of different strategies to ensure accurate number plate detection. The histogram, morphological processing, texture, edge detection, and transformation are the bases for the number plate detection techniques [18, 19].

3.1 Texture-Based Candidate Extraction

The proposed approach for license plate recognition combines techniques such as character feature extraction, placement of the license plate characters, and feature extraction from the license plate. It uses a combination of Xception, MobileNetV3, and a spatial attention mechanism for character feature extraction, a Bi-LSTM approach for character placement, and reference line detection for feature extraction from the license plate [20]. This approach is able to accurately detect and identify characters on license plates, regardless of their standard or irregular shape and in various lighting conditions. Additionally, it uses a combination of projection properties, binarization, Rank filter, Robert's operator, autocorrelation, and projection algorithms for the rejection of false candidates, making it capable of detecting different types of Chinese license plates in real-world scenarios [21].

3.2 Edge Detection-Based Plate Extraction

In order to extract car number plates, a new technique that combines the Line Grouping (LG) and Edge Density (ED) algorithms are presented. Using a set of geometric criteria, the very first algorithm is used to extract the line segments as well as group them. Accurately, it detects a rectangular at the plate boundary. The image is disregarded if no LG rectangle group was created [22]. A second opportunity using the ED approach is given to the rejected image. The second approach locates plate areas where the vertical edges are most densely distributed. The verification procedure is assessed both with and without the double chance framework. The segmentation module is utilized in the verification process [23].

3.3 Morphological-Based Plate Extraction

First, boundaries in the input photos are found. The edge image's vertical projections are then discovered. They can be chosen as potential regions since backgrounds and fences have vertical borders. Authors have suggested the compact factor as a way to avoid this [24]. These structures feature placed parallel in a wider range than a number plate, which has narrow vertical edges. Compact factors employ this functionality. It is used to determine the luminance of pixels in columns and the local maxima in that determine the potential candidate area. The extraction of plates involves morphological processes [25].

3.4 Plate Extraction by Smearing Algorithm

This technique is utilized to determine the plate area since smearing is a technique for extracting text sections from mixed images [26]. According to the specified lowest and highest thresholds, the image is transformed along horizontal and vertical scan lines, and for both types of blurring, white pixels are turned black. The image is then subjected to morphological processing in order to determine the plate location [27].

3.5 Plate Extraction by Image Transformation

The approach developed combines two techniques the Hough transform and the contour algorithm from the processed edge image to identify the approximate boundaries of the individual objects [28]. It describes an approach for identifying and verifying potential license plates using a combination of the Hough transform and contour algorithm, as well as a width-to-height ratio and horizontal crosscut evaluation. The approach aims to accurately identify license plates in complex scenarios, while also taking into account the language and orientation of the license plate [29–31]. The approach is supported by the use of predefined acceptable width-to-height ratio limits and roughly defined criteria for evaluation, ensuring high accuracy in detecting license plates [32].

4 Character Recognition Methods

Isolated characters are necessary for character identification in license plate recognition systems. The process of character segmentation, which involves separating the individual characters from the license plate image, plays a crucial role in the overall accuracy of the system [33]. If the characters are not properly segmented, it can lead to errors in character recognition and negatively impact the overall performance of the system [34]. Various methods such as statistical classifiers, computational intelligence architecture, and pattern matching have been proposed to improve the character segmentation process in license plate recognition systems [35].

4.1 Statistical Classifiers

Statistical classifiers use support vector machines for recognition to overcome the shortcomings of OCR as well as pattern-matching approaches then use SVM to identify numbers. The method begins with a set of examples of numbers taken from license plates [36]. An SVM recognizes each character after being educated by a few known samples beforehand. The trained model is evaluated on each number one by one until the number plate is accurately identified. Getting the highest benefit between SVM outputs is how the identification results are obtained [37].

4.2 ANN Classifiers

Instead of using a cascaded action or feedback mechanism, this method combined recognition and segmentation. It is suggested to combine these two objectives into a single statistical framework using a two-layer Markov network [38]. In terms of probabilities, both low-grade cues and high-level previous information are taken into account. For state estimation, an effective non-iterative belief propagation approach is employed. The recognition algorithm makes use of both pattern recognition and NN approaches (Perceptron). When compared to template matching, NN has offered superior matching for recognition than these two strategies [39].

4.3 Pattern/Template Matching

Shape-invariant, non-rotated characters can be recognized using the pattern-matching method. Additionally, cross-correlation can be used to determine whether normalized characters and templates match. Italian cars traveling through tollgates use this algorithm. For two-letter provinces and character strings, single prototypes are designed as templates [40]. The algorithm tested over 3,000 genuine photographs taken in a variety of lighting and weather conditions and achieved an identification rate close to 91%. Then, using an essential elements-based strategy, an OCR system was developed to recognize the car license plates, which typically have intricate shapes. Based on the fundamental components of characters, this algorithm can identify character patterns [41–43]. With this technique, the bulk of characters is condensed into a handful of crucial strokes. The first approach combines the Hough transform and contour algorithm to identify the approximate boundaries of the license plate and uses criteria such as width-to-height ratio and horizontal crosscuts for verification. The second approach uses a pattern-matching method to recognize shape-invariant, non-rotated characters. The third approach uses an OCR system based on the fundamental components of characters, using Edge Hausdorff Distance for character matching. All of these approaches have been tested on various datasets and have shown promising results in terms of recognition accuracy [44].

5 Proposed System for Number Plate Recognition

All across the world, number plates are used to identify individual automobiles. The photo handling technology is used by the system that recognizes number plates to identify cars based on their license plates. Recognition of number plate systems is used for practical traffic management and security functions like restricting entry to certain areas and tracking down the stolen vehicle [45].

The steps involved in detecting the number plate are:

Step 1. Take a picture of the license plate.

Step 2. Distinguish between character groups.

Step 3. The recognized license plate is displayed on a graphic user interface and is saved with the time and date in a database for further use.

Step 4. Send an SMS to an appropriate individual who has already filed a complaint about the car if a stolen the vehicle is found.

The user may incur additional fines for violating any driving regulations. Based on the number plate recognition, the owner will receive a notification about the applied fine information and insurance obligations. We find it handy because we no longer have to worry about carrying our paperwork with us everywhere. Owners of vehicles can use this program to extract information about their vehicles and pay fines using it [46].

By incorporating CNNs with OCR, the entire process of number plate detection and recognition can be carried out seamlessly, from image preprocessing to character recognition. This approach results in higher accuracy and better performance in comparison to methods that use individual components for each step. The robustness of CNNs to variations in image size, orientation, and lighting conditions makes them ideal for number plate detection, especially when capturing images of moving vehicles in varying conditions. With the ability to learn complex representations of image data, CNNs are effective in character recognition tasks and when combined with OCR, they can accurately detect and identify characters on a number plate, even in the presence of low-quality images or angled shots.

5.1 Condition Random Field

Connected-component labelling is a method used in computer vision to identify connected components in binary digital images. It is often used in image processing for tasks such as object recognition, tracking, and filtering. The process involves examining the labels of neighboring pixels and assigning a label to the current pixel based on certain criteria, such as connectivity. It can also be used in optical recognition systems or human–computer interfaces. It should be noted that segmentation and connected-component labeling are different techniques; segmentation is the process of separating an image into different regions or segments, while connected-component labeling is the process of assigning labels to different connected regions [47].

1. Go through each data point iteratively, first by column and then by column (Raster Scanning)
2. If the component isn't the background,
    (1) Gather the items that are close to the current element.
    (2) If you have no neighbors, identify the current element specifically and go on.
    (3) In the event that is not, assign the neighboring element with the smallest label to the present element.
    (4) Save the equivalent values between nearby labels.

On the second pass

1. Go through each data point iteratively, first by column and then by row.
2. If the component isn't the background,
    (1) Change the element's label to its lowest equivalent.

5.2 Optical Character Recognition or Optical Character Reader (OCR)

For many years, there has been a peek at the interest shown in the field of optical character recognition (OCR). It is described as the method of breaking down a picture of a document into its individual characters. Even after decades of intensive research, creating OCR with human-like abilities is still a difficult task. Researchers from both academic and industrial circles have focused on optical character recognition because of its difficult nature [48]. The number of educational labs and businesses conducting research on character segmentation has drastically expanded during the past several years. This study tries to summarize the research that has already been conducted in the area of OCR [49]. A software program called optical characters recognition (OCR) transforms printed text and graphics into electronic form so that a machine can handle them. Machines really aren't intelligent enough to understand the information contained in images, unlike the human brain, which is able to recognize words and characters from images quite readily. As a result, numerous research projects have been launched in an effort to convert a document image into a machine-understandable format. Due to the wide range of languages, fonts, and writing styles, as well as the intricate linguistic regulations, etc., OCR is a complex task [50].

The Optical Character Recognition (OCR) process usually consists of these steps:

Step 1. Pre-processing: The image undergoes filtering and optimization to get it ready for recognition.

Step 2. Segmentation: The image is divided into segments to separate individual characters.

Step 3. Feature extraction: The character's features, such as shape, size, and pattern, are extracted for recognition purposes.

Step 4. Character recognition: The extracted features are compared to a database of known characters to identify each segmented character.

Step 5. Post-processing: The recognized characters are put together to form words, lines, and paragraphs, resulting in the final text output.

6 Experimental Results and Discussion

In this study, a Python-based framework was implemented in real-time to process number plate images and evaluate its performance in terms of success rate (%). Here the success rate was calculated for three different phases of automatic number plate recognition (ANPR): number plate extraction, character segmentation, and character recognition. The success rate is determined by the ratio of successfully processed plates to the total number of input vehicle images. A higher success rate indicates a better performance in comparison to an ANPR approach with a lower success rate. The proposed method in this study resulted in a higher success rate compared to the existing method of ANPR. The study employed techniques from various branches of computer science, such as image processing, pattern recognition, and natural language processing, to address various issues in the OCR process. These include acquiring, pre-processing, breaking images into segments, extracting features, classification, and post-processing. As a potential next step, combining these methods could lead to the development of an effective OCR system. The system could be used in various real-time applications, such as smart libraries and number plate recognition, and their comparison was mentioned in below Table 1 and represented in Fig. 2.

Figure 2 shows the success rate of existing models that are widely used and the success rate of the proposed model which has a better percentage than the existing models. The proposed model would provide a better success rate in predicting the license plate characters accurately and would prove to be more efficient than the previously implemented models. Out of all researched models, the LPR model, and YOLO model and the R-CNN models have a lower success rate of 50% which shows that the predicted result is not always feasible. The OpenCV and the ALPR methods show a better success rate than the previously mentioned models. The proposed model overall has the capacity to provide a better overall success rate.

Existing systems have limitations, as they are only designed to work with a specific type of license plate format and may not be able to recognize license plates from other countries. Adverse weather conditions like rain, snow, or fog can make it challenging

Table 1 Comparison of proposed work with various methods

Model used	Number of images	Success	Failure	Success (%)
LPR model	10	5	5	50
YOLO model	10	5	5	50
R-CNN model	10	5	5	50
OpenCV	10	6	4	60
ALPR	10	6	4	60
Proposed model	10	8	2	80

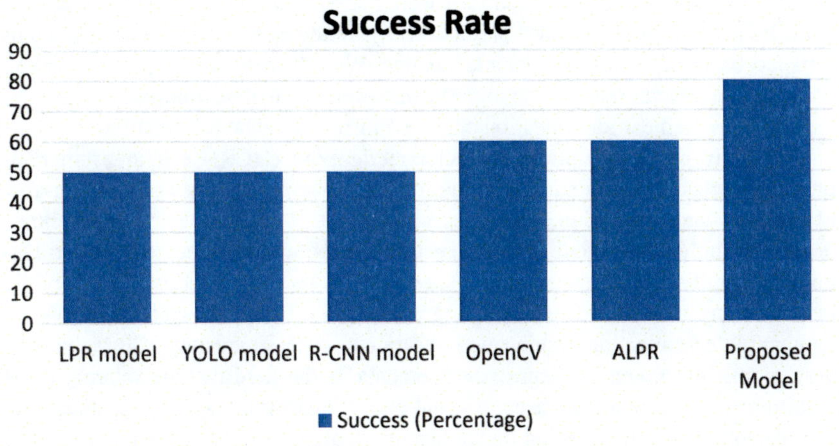

Fig. 2 Success rate

for cameras to capture clear images of license plates, leading to incorrect readings. The speed and movement of vehicles can also impact the accuracy of the systems. Additionally, there may be privacy concerns among individuals regarding the use of certain technologies and the collection of their personal information.

The selection of a suitable model for a particular task in computer vision depends on the specific requirements and objectives of that task. When it comes to text recognition, using OCR in conjunction with a CNN can produce better results than other models such as LRP, YOLO, R-CNN, OpenCV, and ALPR. This is because CNNs are highly effective in recognizing patterns in images, while OCR technology is specifically designed for text recognition. On the other hand, for object detection tasks, models such as YOLO or R-CNN may be a better choice. Toolkits like OpenCV and ALPR can be used to build computer vision applications, but specialized models may provide more accurate results for specific tasks. Layer-wise Relevance Propagation (LRP) is a method to interpret the workings of deep learning models, but it is not a computer vision model by itself. It can be used with other models, including OCR + CNN, to gain a better understanding of their inner workings and make informed decisions.

7 Conclusion

The scholars working on these advances can benefit from the thorough analysis of current trends and potential developments in ANPR provided and mentioned in this paper clearly. The methods used for number plate recognition, character segmentation, and character recognition are covered briefly in this essay. Although there are a number of commercial LPR systems, discussing the strategies employed in them is outside the purview of this work because their functioning is completely private. The LPR algorithms examined were nation-specific. Algorithms described in the literature for number plate prediction are limited by the working parameters of distance, backdrop, light, and vehicle location. With the results of this study, we can use optical character recognition to perform number plate detection and recognition for real-time theft car identification and to confirm insurance information.

The study of using optical character recognition (OCR) techniques in real-time number plate verification systems for insurance processing is a thriving area of research and innovation. The integration of OCR technology with insurance processing holds the potential to enhance the speed and accuracy of the verification process. The trend towards automated systems in various industries, including insurance and automotive, is increasing, making the future prospects of real-time number plate verification systems with insurance processing using OCR techniques very promising. Potential future advancements in the field include refining OCR algorithms to improve their accuracy and speed, linking the system to other relevant databases, and broadening its application in areas such as traffic management, toll collection, and parking management. In summary, the outlook for the future of real-time number plate-based verification systems with insurance processing using

OCR techniques is positive, with numerous opportunities for additional research and development.

The proposed study on real-time license plate-based verification system with insurance processing utilizing OCR techniques is appropriate for both future research and practical implementation due to several reasons. Firstly, its relevance, as the need for an effective and dependable license plate verification system is growing with the increasing usage of vehicles for transportation. Secondly, the use of OCR and CNNs in the proposed study represents the latest advancements in computer vision and machine learning, making it suitable for future research as these technologies continue to advance. Thirdly, the study's real-time implementation capability makes it relevant for practical applications, such as traffic management, border control, and insurance processing. Lastly, the combination of OCR and CNNs has the potential to greatly enhance the accuracy of license plate detection and recognition compared to current systems, making it a valuable area for future research and development.

References

1. Pandiaraja P, Muthumanickam K, Kumar PR (2023) A graph-based model for discovering host-based Hook attacks. Smart Technol Data Sci Comm. Lecture Notes in Netw Syst 558:1–13. Springer, Singapore
2. Lin C-H, Lin Y-S, Liu W-C (2018) An efficient license plate recognition system using convolution neural networks. In: IEEE International Conference on Applied System Invention (ICASI)
3. Roy AC, Hossen MK, Nag D (2016) License plate detection and character recognition system for commercial vehicles based on morphological approach and template matching. IEEE
4. Sumathi K, Pandiaraja P (2023) E-health care patient information retrieval and monitoring system using SVM. In: Smart Technologies in Data Science and Communication. Lecture Notes in Networks and Systems, Springer, pp 558, 1–13
5. Omran SS, Jarallah JA (2017) Iraqi car license plate recognition using OCR. In: Annual conference on new trends in information & communications technology applications (NTICT). IEEE
6. Shankar A, Sumathi K, Pandiaraja P, Stephan T, Cheng X (2022) Wireless multimedia sensor network QoS bottleneck alert mechanism based on fuzzy logic. J Circuits, Syst Comp 31(11)
7. Rana N, Dahiya PK (2017) Localization techniques in ANPR systems: A-state-of-art. Int J Adv Res Comp Sci Soft Eng 7(5)
8. Shashirangana J, Padmasiri H, Meedeniya D, Perera C (2021) Automated license plate recognition: a survey on methods and techniques. IEEE Access 9:11203–11225
9. Weihong W, Jiaoyang T (2020) Research on license plate recognition algorithms based on deep learning in complex environment. IEEE Access, 91661–91675
10. Zou Y, Zhang Y, Yan J (2020) A robust license plate recognition model based on Bi-LSTM. IEEE Access, 211630–211641
11. Henry C, Ahn SY, Lee S-W (2020) Multinational license plate recognition using generalized character sequence detection. IEEE Access, 35185–35199
12. Zhang L, Wang P, Li H, Li Z, Shen C, Zhang Y (2021) A robust attentional framework for license plate recognition in the wild. IEEE Trans Intell Transp Syst, 6967–6976
13. Pustokhina IV, Pustokhin DA, Rodrigues C, Gupta D, Khanna A, Seo C, Joshi GP (2020) Automatic vehicle license plate recognition using optimal K-means with convolutional neural network for intelligent transportation systems. IEEE Access **8**, 92907–92917

14. Tourani A, Shahbahrami A, Soroori S, Khazaee S, Suen CY (2020) A robust deep learning approach for automatic iranian vehicle license plate detection and recognition for surveillance systems. IEEE Access 8:201317–201330
15. Selmi Z, Halima MB, Pal U (2020) DELP-DAR system for license plate detection and recognition. Pattern Recog Lett 129:213–223
16. Chen JIZ, Zong JI (2021) Automatic vehicle license plate detection using K-means clustering algorithm and CNN. J Elect Eng Auto 3(1):15–23
17. Zhou Q (2021) FAFNet: a false-alarm filtering algorithm base on deep learning for license plate recognition. J Phys: Conf Ser 2023
18. Wang J, Bacic B, Yan WQ (2018) An effective method for plate number recognition. Multi Tools Appl, 1679–1692
19. Nafchi HZ, Shahkolaei A, Hedjam R, Cheriet M (2017) CorrC2G: color to gray conversion by correlation. IEEE, 1651–1655
20. Hedjam R, Nafchi HZ, Kalacska M, Cheriet M (2015) Influence of color-to-gray conversion on the performance of document image binarization: toward a novel optimization problem. IEEE, 3637–3651
21. Lin C-H, Li Y (2019) A license plate recognition system for severe tilt angles using mask R-CNN. In: International Conference on Advanced Mechatronic Systems (ICAMechS). IEEE
22. Pandiaraja P, Muthumanickam K (2022) Convolutional neural network-based approach to detect COVID-19 from chest X-ray images. In: Cyber Security, Privacy and Networking. Lecture Notes in Networks and Systems, vol 370, pp 231–245. Springer, Singapore
23. Pustokhina IV et al (2020) Automatic vehicle license plate recognition using optimal K-means with convolutional neural network for intelligent transportation systems. IEEE, 92907–92917
24. Zhang L, et al (2020) A robust attentional framework for license plate recognition in the wild. IEEE, 6967–6976
25. Arafat MY, et al (2019) Systematic review on vehicular licence plate recognition framework in intelligent transport systems. IET Intell Transp Syst, 745–755
26. do Dalarmelina NV, Teixeira MA, Meneguette RI (2019) A real-time automatic plate recognition system based on optical character recognition and wireless sensor networks for ITS. Sensors 20(1):55
27. Pandiaraja P, Boopesh KB, Deepthi T, Laksmi Priya M, Noodhana R (2020) An analysis of document summarization for educational data classification using NLP with machine learning techniques. In: Applied Computational Technologies. ICCET 2022. Smart Innovation, Systems and Technologies, vol 303, pp 127–143. Springer, Singapore
28. Akilandeswari V, Kumar A, Thilagamani S, Subedha V, Kalpana V, Kaur K, Asenso E (2022) Minimum latency-secure key transmission for cloud-based internet of vehicles using reinforcement learning. Comp Intell Neurosci
29. Gu Z, et al (2020) Adversarial attacks on license plate recognition systems. Comp, Mat Contin 65(2):1437–1452
30. Murugesan M, Thilagamani S (2021) Bayesian feed forward neural network-based efficient anomaly detection from surveillance videos. Intell Auto Soft Comp 34(1):389–405
31. Luo X, Ma D, Jin S, Gong Y, Wang D (2019) Queue length estimation for signalized intersections using license plate recognition data. IEEE 11:209–220
32. Lin HY, Dai JM, Wu LT, Chen LQ (2020) A vision-based driver assistance system with forward collision and overtaking detection. Sensors 20:5139
33. Thangallapally SK, Maripeddi R, Banoth VK, Naveen C, Satpute VR (2018) E-security system for vehicle number tracking at parking lot (Application for VNIT Gate Security). In: IEEE International Students' Conference on Electrical, Electronics and Computer Science (SCEECS), pp 1–4
34. Pandiaraja P, Aishwarya S, Indubala SV, Neethiga S, Sanjana K (2022) An analysis of e-commerce identification using sentimental analysis: a survey. Appl Comp Techn. ICCET 2022. Smart Inno Syst, Technol 303:742–754. Springer, Singapore
35. Alamelu V, Thilagamani S (2022) Lion based butterfly optimization with improved YOLO-v4 for heart disease prediction using IoMT. Info Technol Cont 51(4):692–703

36. Shreyas R, Kumar BP, Adithya H, Padmaja B, Sunil M (2017) Dynamic traffic rule violation monitoring system using automatic number plate recognition with SMS feedback. In: 2nd International Conference on Telecommunication and Networks (TEL-NET), pp 1–5
37. Chaithra B, Karthik K, Ramkishore D, Sandeep R (2017) Monitoring traffic signal violations using ANPR and GSM. In: International Conference on Current Trends in Computer, Electrical, Electronics and Communication (CTCEEC), pp 341–346
38. Felix AY, Jesudoss A, Mayan JA (2017) Entry and exit monitoring using license plate recognition. In: IEEE International Conference on Smart Technologies and Management for Computing, Communication, Controls, Energy and Materials (ICSTM), pp 227–231
39. Ibrahim M, Du S, Shehata M, Badawy W (2012) Automatic license plate recognition (ALPR): a state-of-the-art review. IEEE 23:311–325
40. Birgillito G, Rindone C, Vitetta A (2018) Passenger mobility in a discontinuous space: modelling access/egress to maritime barrier in a case study. J Adv Transp, 6518329
41. Saravanan S, Abirami T, Pandiaraja P (2018) Improve efficient keywords searching data retrieval process in cloud server. In: International Conference on Intelligent Computing and Communication for Smart World (I2C2SW), Erode, India, pp 219–223
42. Srividhya SR, et al (2022) A machine learning algorithm to automate vehicle classification and license plate detection. Wireless Comm Mobile Comp
43. Hajare G et al (2022) Automatic number plate recognition system for indian number plates using machine learning techniques. ITM Web Conf 44. EDP Sciences
44. Tom RJ et al (2022) Car license plate detection and recognition using modified U-Net deep learning model. In: 8th International Conference on Smart Structures and Systems (ICSSS). IEEE
45. Tu C, Du S (2022) A hierarchical RCNN for vehicle and vehicle license plate detection and recognition. Int J Elect Comp Eng 12(1):731
46. Padmasiri H, et al (2022) Automated license plate recognition for resource-constrained environment. Sensors 22(4):1434
47. Antar R, et al (2022) Automatic number plate recognition of Saudi license car plates. Eng, Technol Appl Sci Res 12(2):8266–8272
48. Jain P, Arya D, Singh AK (2022) Vehicle license plate detection and their count for traffic management. Int J Prog Res Sci Eng 3(5):170–175
49. Lee Y, Halim AZ, Wahab MNB (2022) License plate detection using convolutional neural network–back to the basic with design of experiments. IEEE Access 10:22577–22585
50. Li W, Pun C-M (2022) A single-target license plate detection with attention. In: International Workshop on Advanced Imaging Technology (IWAIT), Vol 12177. SPIE

Cannotation MeasureUp to Detect Deepfake by Face Recognition via Long Short-Term Memory Networks Algorithm

L. Rahunathan, D. Sivabalaselvamani, A. PriyaDharshini, M. Vignesh, and G. VinithKumar

Abstract In this computerized world, any individual can create deep fake videos, images, and audios, and it's becoming a challenge to differentiate between the real and the fake ones. Such fake videos and images of a person that spread like wildfire on social media ultimately destroys the person's life by leaving a deep mark that is impossible to remove. Hence, it's important to discern the fake and stop the circulation of these types of videos and images. Different technologies with different algorithms have been used to identify fake images. In this research, to distinguish fake and real images, long short-term memory network methods in deep learning technology are used to recognize faces from real-time dataset for better result because of its hidden layers and some inbuilt functions of this algorithm. Haar Cascade is used for detecting the images from real time; even though it's an oldest one, it still has its own place in detecting face by providing more accuracy than others. Backpropagation is then used to feed both forward and backward, so that the values of bias can be reduced according to the predicted output. Sequence transformer makes the images get into cell state with continuous step-up time, so that the image will not collide while capturing. By using all these conditions, the output will be predicted with actual value which gives the desired outcome. Detecting the swapped and manipulated images from the real one is very difficult even though different algorithms are used. Therefore, long short-term memory networks are used in this research, because it's easy to identify fake images due to the continuous networks in the layer.

Keywords Deep fake detection · Haar cascade · Long short-term memory (LSTM) · Sequence transformer · Challenges in face fake detection · Back propagation

L. Rahunathan (✉) · D. Sivabalaselvamani · A. PriyaDharshini · M. Vignesh · G. VinithKumar
Department of Computer Applications, Kongu Engineering College, Perundurai, Erode, Tamil Nadu, India
e-mail: lrahunathan@gmail.com

G. Rajakumar et al. (eds.), *Intelligent Communication Technologies and Virtual Mobile Networks*, Lecture Notes on Data Engineering and Communications Technologies 171,
https://doi.org/10.1007/978-981-99-1767-9_35

1 Introduction

With the advent of new technology, social media usage has increased in people's daily lives to the point where, despite occasionally being helpful, it is becoming a severe problem in this contemporary society. Recent studies show that fake videos and images are spreading fast and until now proper solution to solve this issue is not available in this sophisticated world. The term "Deepfake" has been introduced by artificial intelligence which is susceptible to manipulate anyone's images or videos without their knowledge. Without any new technology, it cannot be identified that easily. Deep learning is one of the technologies that have emerged to solve this complicated problem. The term deep learning has its name from the fact that more layers cannot be added to get accurate output. Deepfake has landed in many fields such as politics, and social media. By changing images or videos of people and providing misleading information, many people have lost their lives by these fallacious activity. Many studies have been conducted to know how deepfake works. Also, deep learning provides an advanced solution by training and testing the dataset with different algorithms and providing good accuracy. In this research, the data is trained and tested by recognizing the images of a person by facial landmarks.

2 Literature Review

Sinaga [1] proposed the technologies of machine learning, to detect deep fake's images by using biometric features of the person's images such as eyes, mouth, and nose, and to identify the fake images from the videos. The dataset from YouTube videos was used by cropping the videos into frames using the convolutional neural network (CNN) algorithm which provides better accuracy. Heo et al. [2] used a vision transformer and effective net technique to detect deep fake images from videos using DFDC dataset which gives better performance, and used a distillation method which gives fake images from videos in the form of frames from the images. Nguyen et al. [3] presented a broad challenging one to detect deep fake creation and detection by using CNN and the recurrent neural networks (RNN) algorithm. The datasets of UADFV, DARPA, and some other dataset from a period of year were used to show how deep fake increases over a period of time and how it was created by using GAN methods.

Guarnera et al. [4] described detecting deepfake images from various datasets such as DFDC, UADFV, Celeb, etc. by combining all these datasets and compressing in the zip format by cropping and detecting the images into frames in different angles, by using GAN reconstruction and transformer to predict the accurate result. Hsu et al. [5] used GAN to create realistic images to make fool of the discriminator for detecting the real images. The extension of GAN was also used to detect the fake images from the real one such as Cycle GAN and some other GAN including CFFN to make detection with better performance. Two-pair algorithm was used for detection of fake images from all generators of GAN dataset.

Bhilare et al. [6] used the deepfake CLI method to classify, localize and interfere with the fake images by using FPGAs, MesoNet dataset to quantize the images from the real one. These methods use both hardware and software domains to detect the dataset from google DFD, Face Forensics++, Celeb DF and DFDC to get more accuracy but it takes a lot of time to train the dataset by using CLI method. By using DFDC dataset, the error in images as well as videos were reduced during cropping the images from the dataset and FPGA produced better and faster accuracy. Nadimapalli et al. [7] discussed fake detection on gender based on datasets such as FaceForsencis++ and Celeb DF which had the labels for those images during training and testing the dataset. The fake images from datasets were identified easily by using GBDF training and by using MTCNN algorithm the datasets were divided and their sizes were changed into 256 × 256-pixel frames to determine the fairness of deepfake datasets from the work and provide an awareness of fairness fake detection in their dataset.

Ko et al. [8] revealed fake image detection by using masks in their dataset. The dataset was trained again by using a mask in face. In masking, two methods such as face patches and face cropping methods were used. In the first method, the landmarks in their nose and mouth were patched with black, and then the dataset was trained. In face cropping, the face under eye was cropped, and then the same images were trained by using different datasets such as Celeb DF, Xception, MobileNet V2, and some others to determine the fake images from real images. The work produced better accuracy and high performance by using masks in the face. The fake images were clearly identified by suppressing face landmarks. Mittal et al. [9] proposed real-time face detection by using the Gotcha method that detected the fake images from the image. Two challenges active and passive provocations using machine learning technology to detect the image were suggested. Active provocation uses face swapping and modifying the images and passive includes describing fake images from digital world. Real-time dataset was used, such that the image can be preprocessed and facial landmarks can be obtained. The encoded images provided color combination of the images and then the real image was detected from the fake images due to digital images trained from the real time videos.

Zhao et al. [10] developed a way to identify deep fakes by exposing the multi-attentional approach and texture facial feature of the photos. Different datasets such as Celeb DF, Face to Face swap, and DFDC were used for the detection. For better classification, BAP was used for feature extraction from real images. A more convolutional layer with dense connection in between the hidden layer was used. Also, Efficient Net methodology and Xception were used for higher performance. Badale et al. [11] used CNN for classifying images from videos which used 900 fake videos such that by training and testing the images in convolutional layers with ReLu activated the function of testing the images. To get better accuracy, a back propagation algorithm was used to adjust the bias values according to the weight in the hidden layer in dense connectivity of CNN networks.

3 Methodology and Implementation of Lstm

System Model

The long short-term memory (LSTM) network is one of the method which stores the data for a long time in sequence order. Therefore, it overcomes the RNN by storing for a longer period of time, and also there are more recurrent units present in this hidden layer. Using LSTM takes less time for identification; if again the same image is used in training the dataset, the data will remain in their neural network which gives better performance for larger dataset and better accuracy while testing the dataset.

Proposed Work

In this research, own real-time dataset is used, and the dataset is trained by using the Haar Cascade algorithm and LSTM networks, so that the image will remain in the static cell state for a longer period of time. The layers in the network can also upgrade itself when a new image enters the cell. These layers present in the networks play a major role for attaining accurate output of images.

Deepfake Creation

As known, the world is becoming more perilous due to the spread of deep false images on social media and generative adversarial networks (GAN) are used to produce fake photos and videos. The GAN algorithm generates the generator and discriminator for creating the image model that do not exist in this world. But by using GAN, less accuracy is attained when compared to LSTM, and hence LSTM algorithm has been used in this work. In GAN, the discriminator classifies the images that are generated by the generators. Also this is an effectual technology that produces fake images, and the extensions of GAN such as Cycle GAN and Style GAN are also being used nowadays for producing fallacious images and videos which makes everyone in the world alarm about their future.

4 Data Preprocessing

The preprocessing plays an important role, because it processes the images by removing the noisy data from the raw data to get a legible image for output. Also, it is the first and most significant step in training the raw data which has been described in Fig. 1. LBPH face classifier is used to capture the image in whichever angle the face images are turned, so that its accuracy is not reduced. If this classifier is not used, then while image capturing, face should not be moved to any side and should only focus the webcam.

When an image is captured by the camera, it will preprocess the images by removing the background using Haar Cascade algorithm. The captured face images are transferred in histogram format, and other inbuilt processing is also done in this step. Finally, these images are stored in the dataset in 30 different angles and in a

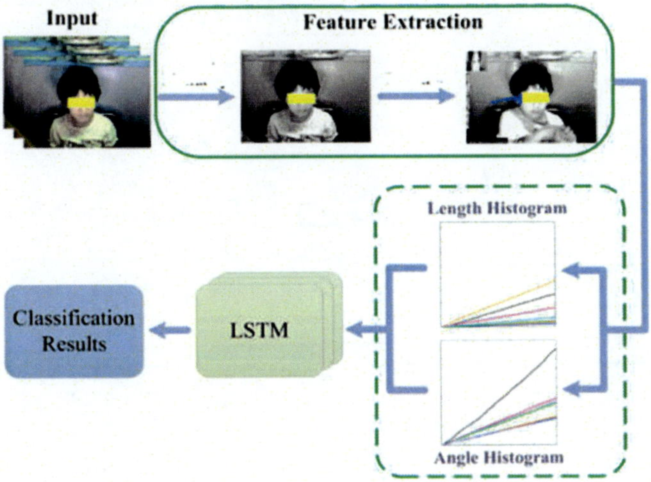

Fig. 1 Data preprocessing

static cell state in the neural network for a period of time. If again an image enters into the network layer, it is updated in the cell state. Then it will give the processed images into the next layer of the cell state.

Long Short Term Memory

Recurrent neural networks have several types, and one of which is the advanced LSTM. Another type of RNN is a sophisticated algorithm called gated recurrent. Even though the application of LSTM algorithm is a bit complex in deep learning, it allows to store information for a period of time. There are three gates in this network layer, they are: the forget gate, input gate, and output gate, which are described in Fig. 2. These gates play an important role in storing the images and is responsible for transferring the images from one gate into another gate. Apart from these gates, there are inbuilt functions such as, multi scale variate which have predefined functions of LSTM algorithm. Even though there are many predefined functions, a particular one is used in this work. So, whenever the functions were called, they automatically use the function because of the inbuilt code behind this function to get the processed output. Cell state is one such part in which the images are given as input, and it upgrades itself. Also, the images from the cell state in these layers can be removed whenever needed.

The solution for the gradient issue is to overcome by using LSTM by adjusting the input values and bias values using activation functions, and these network layers are not present in the recurrent network. Also, there are various LSTM ways that can be used such as one to one, many to one, one to many, and many to many. The algorithm used in this work is one to many, and it achieves higher accuracy. Also, using this algorithm helps capturing the images faster from the live video. Then the histogram helps in identifying the list of the images because these histogram images

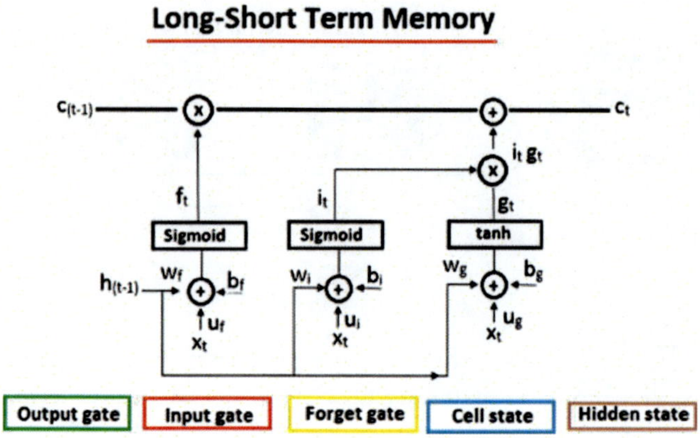

Fig. 2 LSTM

will be different if same image is displayed with some modification or if morphing of image has occurred. By the list of images in this algorithm, the result can be obtained as real or unknown images with its accuracy.

Mathematical Formulas Used By LSTM

Before preprocessing the images, the packages must be imported such as cv2 for video capturing, IL for resizing and other inbuilt features used under this package are, Numpy and OS. Haar Cascade algorithm is used to detect the images in different angles for accurate prediction of the faces or any other objects. When an image is captured by camera, the images are resized by using package PIL, then are divided into pixels by eliminating the background of the images, for a clear classification of result to be obtained. The normalization of the image dataset is done by processing these images, and then the highest value (max) pixels and lowest value (min) pixels are calculated for rescaling the image to enter the cell state using formula (1).

$$Z_i = X_i - \text{low}(x) \div h(x) - \text{low}(x) \tag{1}$$

In the above formula, Z_i represents the captured image. The images are classified into pixels, only then they can enter into the forget layer and check for the image to enter into the next layer of the cell state. The images enter from $h_{t\text{-}1}$ and x_t that are from previous hidden layer, and the input image values are calculated with bias value. These layers then store the image into the forget gate for a period of time to keep this image in memory. This is summed up to the weight (WG) of this layer with bias value (BIA) by using the below formula (2).

$$\text{FOG}_t = \sigma\left(\text{WG}_f \cdot \left[H_{t-1}, X_t\right] + \text{BIA}_f\right) \tag{2}$$

After entering into the forget gate, the images that have mostly occurred are stored, and the image that have not been used for a long time are forgotten, and then the image moves to the next layer. The images in the input layer, sequence the dataset in pre-order format from the forget layer within the time. And this image is transferred to a cell state image. For fast movements of images, sigma function by using formula (3) is calculated. This sigma function makes the output value between 0 and 1 without reducing its efficiency in the images. This process continues for all input images that entered into the gates.

$$\mathrm{INP}_t = \sigma(\mathrm{WG}_i \cdot [H_t - 1, X_t] + \mathrm{BI}_i) \tag{3}$$

After deciding input from cell state, all values are combined with tanθ which then replaces the images with new images and upgrades the systemic flow of state in this second cell used from forget gate, using the formula (4). When $\tan\theta$ is bigger, then the sigmoid function carries the values between -1 and $+1$. If the value is -1, then that image will not be considered, and if the value is $+1$, then the image will be transferred to the next layer in the network. Also, the images that were removed from the forget gate also remains for some period in this network layer for some inbuilt functions.

$$C_t = \tan\theta(\mathrm{WG}_c * [H_t - 1, X_t] + B) \tag{4}$$

Finally, the output layer cleans the unused chunks of the dataset and provides the output in a sequence order because of the sigmoid function and tanh function used in this LSTM network in the cell state, and this function is greatly used for reducing the gradient issue problem. Doing all these processes in this algorithm doesn't consume much time to point out the images from the testing data. Thus, it identifies the fake images by face histogram of the person that was compared at each time in the loop with input layer for final identification of its process by using formula (5).

$$O_t = \sigma(\mathrm{WG} \cdot [H_t - 1, X_t] + B_0) \tag{5}$$

Flowchart of LSTM

The application of the LSTM algorithm is shown in the following flowchart.

In this research, at first, all the packages needed to be used by this program are imported. And then, by using LSTM method, the images that were captured at real time are preprocessed. Then these images are divided into 30 different frames according to the face recognition by using pillow package and some of its features. After that, the image stored in the dataset are trained. Again, the camera captures the image, and the testing process is performed. The recognition of image is then compared to the trained dataset images. If the two images are same, the result as REAL is obtained with its accuracy, or the result FAKE with less accuracy is obtained, by comparing the hist images of the person that were captured during training. Figure 3 shows the systemic flow of LSTM.

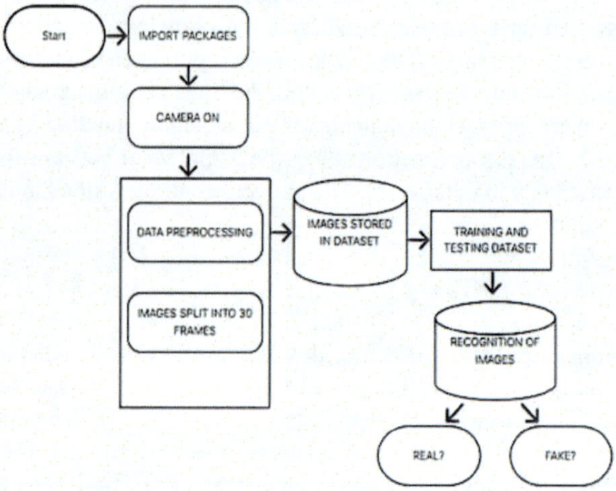

Fig. 3 Systemic flow of LSTM

Experimental Results of LSTM

Pycharm tool is used for implementing and an accuracy of over 95% is obtained by applying this LSTM algorithm to identify fake images. Face swapping and manipulation of real images can be detected easily because of the sequence of layers present in the neural network and Haar Cascade algorithm. A real image is trained, and after that the image will be compared with the tested image each time in the loop, and the fake image is identified. Fake images can be anything such as an unknown person, photos of the person, or the same person with some manipulation in face, which can be recognized by histogram of the images. The dataset in this research used to detect the real-time images for training and testing has been described from Figs. 4, 5 and 6.

By capturing the images from real time, these images are used for training by extracting some features in the image and particularly to recognize the face landmarks in the histogram images, which have been described in the sample code of Fig. 5. Also, the output is predicted based on the recognition of the testing, whether the image is REAL or FAKE, with its accuracy.

Algorithm of LSTM

STEP 1: Import necessary packages from Pycharm tool that has inbuilt function.

STEP 2: Start the webcam and capture the image.

STEP 3: After capturing the image, train and test the images.

STEP 4: For recognition of images, compare the image with the captured image.

```
import cv2
import os
cam = cv2.VideoCapture(0)
cam.set(3, 640) # set video width
cam.set(4, 480) # set video height
#make sure 'haarcascade_frontalface_default.xml' is in the same folder as this code
face_detector = cv2.CascadeClassifier('haarcascade_frontalface_default.xml')
# For each person, enter one numeric face id (must enter number start from 1, this is the lable of person 1)
face_id = input('\n enter user id end press <return> ==>  ')
print("\n [INFO] Initializing face capture. Look the camera and wait ...")
# Initialize individual sampling face count
count = 0
#start detect your face and take 30 pictures
while(True):
    ret, img = cam.read()
    gray = cv2.cvtColor(img, cv2.COLOR_BGR2GRAY)
    faces = face_detector.detectMultiScale(gray, 1.3, 5)
    for (x,y,w,h) in faces:
        cv2.rectangle(img, (x,y), (x+w,y+h), (255,0,0), 2)
        count += 1
        # Save the captured image into the datasets folder
        cv2.imwrite("dataset/User." + str(face_id) + '.' + str(count) + ".jpg", gray[y:y+h,x:x+w])
        cv2.imshow('image', img)
    k = cv2.waitKey(100) & 0xff # Press 'ESC' for exiting video
    if k == 27:
        break
    elif count >= 30: # Take 30 face sample and stop video
         break
# Do a bit of cleanup
print("\n [INFO] Exiting Program and cleanup stuff")
cam.release()
cv2.destroyAllWindows()
```

Fig. 4 Sample screenshot of training

STEP 5: If the captured image is the same as the recognized image, then it will display REAL.

STEP 6: Else it will display FAKE, with low accuracy.

Achievement of LSTM Algorithm

By using this algorithm, capturing and identifying the images from any angle is a quick process, and also better performance which gives the accuracy more than 95% using inbuilt functions in this algorithm is achieved.

Result of LSTM Algorithm

From Fig. 7, the accuracy of over 97% when a real image was detected during testing process can be observed. The result fake is obtained when an image was detected without any training, which is shown in Fig. 8. Figures 9 and 10 give the result as fake with less accuracy when the trained images, by doing some manipulation or morphing of the images in the form of hard and soft objects, were given as input. It also predicts the face manipulated person during testing by comparing it with the trained images which gives more accurate result than using other methods such as CNN, which is because of the inbuilt function used in the LSTM algorithm. Therefore, by choosing this algorithm, faster result and efficiency for the real-time dataset are

```
import cv2
import numpy as np
from PIL import Image #pillow package
import os
# Path for face image database
path = 'dataset'
recognizer = cv2.face.LBPHFaceRecognizer_create()
detector = cv2.CascadeClassifier("haarcascade_frontalface_default.xml");
# function to get the images and label data
def getImagesAndLabels(path):
    imagePaths = [os.path.join(path,f) for f in os.listdir(path)]
    faceSamples=[]
    ids = []
    for imagePath in imagePaths:
        PIL_img = Image.open(imagePath).convert('L') # convert it to grayscale
        img_numpy = np.array(PIL_img,'uint8')
        id = int(os.path.split(imagePath)[-1].split(".")[1])
        faces = detector.detectMultiScale(img_numpy)
        for (x,y,w,h) in faces:
            faceSamples.append(img_numpy[y:y+h,x:x+w])
            ids.append(id)
    return faceSamples,ids
print ("\n [INFO] Training faces. It will take a few seconds. Wait ...")
faces,ids = getImagesAndLabels(path)
recognizer.train(faces, np.array(ids))

# Save the model into trainer/trainer.yml
recognizer.write('trainer/trainer.yml') # recognizer.save() worked on Mac, but not on Pi

# Print the numer of faces trained and end program
print("\n [INFO] {0} faces trained. Exiting Program".format(len(np.unique(ids))))
```

Fig. 5 Sample screenshot of recognition

obtained. Gradient issue is also reduced by using this LSTM because of activation function.

5 Conclusion

Previous research shows that CNN and GAN are preferred by using recurrent neural network and machine learning technology with bigger datasets to predict the fake images and videos. In this paper, fake images are detected from the real images by using LSTM algorithm, which is unique from other methods because of the predefined functions used in this method and achieves an accuracy of above 95%. Although deepfake creation is a key component of artificial intelligence's creation of fake images that are used for dangerous purposes, there are also more technologies available to make everyone aware of these fake images. However, there is a need for more efforts and techniques to identify these fake images to be an expert in this technology.

```
import cv2
import numpy as np
import os

recognizer = cv2.face.LBPHFaceRecognizer_create()
recognizer.read('trainer/trainer.yml')   #load trained model
cascadePath = "haarcascade_frontalface_default.xml"
faceCascade = cv2.CascadeClassifier(cascadePath);

font = cv2.FONT_HERSHEY_SIMPLEX

def LSTM(img_captured):
     if len(img.shape) == 2:
          plt.imshow(img, cmap='gray')
          plt.show()
     im = binary_img
     ret, thresh = cv2.threshold(im, 150, 255, cv2.THRESH_BINARY)
     kernel = np.ones((5, 5), np.uint8)
     opening = cv2.morphologyEx(thresh, cv2.MORPH_OPEN, kernel)
     cleaned = morphology.remove_small_objects(opening, min_size=62, connectivity=2)
     #cv2.imshow("cleaned", cleaned)
     binary_img=cleaned
     got=""
     found=0
     img1=cv2.imread(source)
     # Convert it to HSV
     folder='dataset'
     for filename in os.listdir(folder):
          img2 = cv2.imread(os.path.join(folder,filename))
          #cv2.imshow("Input image",img2)
          #print(filename)
          '''
          if img2 is not None:
```

Fig. 6 Sample screenshot of testing

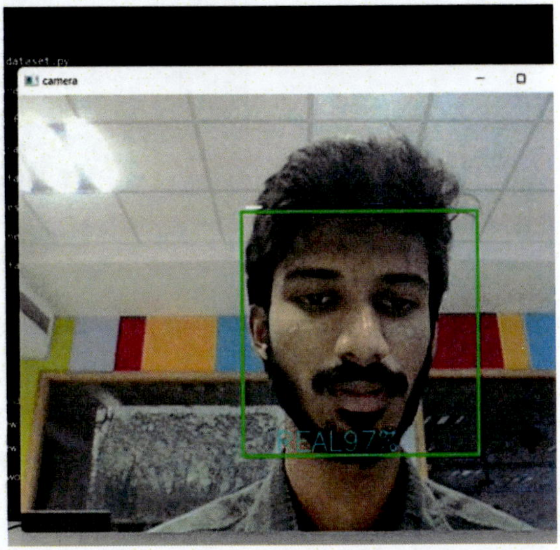

Fig. 7 Accuracy

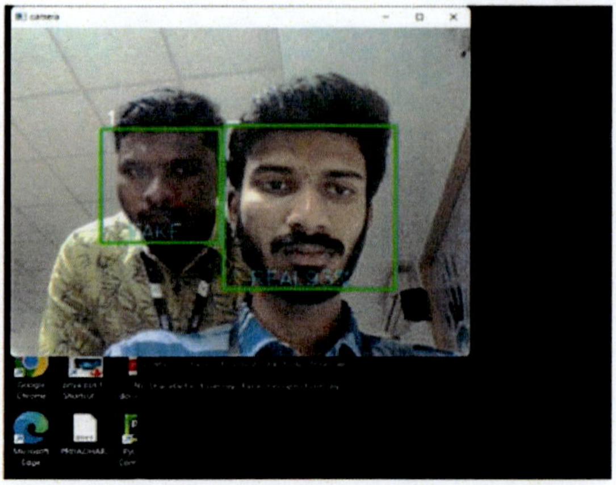

Fig. 8 Comparison with another person

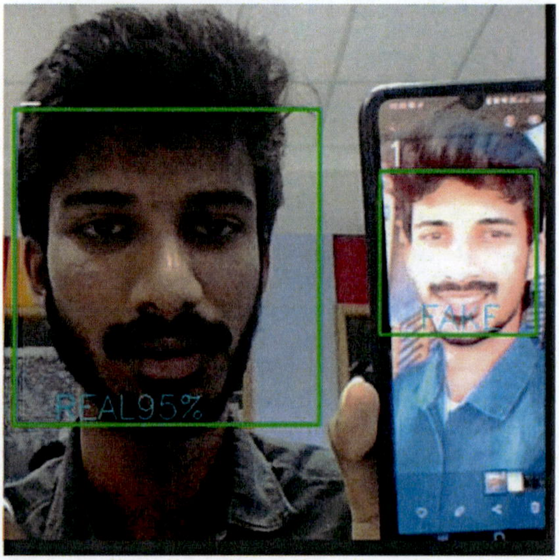

Fig. 9 Comparison of original with soft object

6 Future Enhancement

Though LSTM produces better accuracy, the emerging of fake images continues to grow. Therefore, it's better to use blockchain technology to store the images so that no one can manipulate or swap the real images and it's also hard to create fake image from a real image without the person's knowledge. Because by using this technology, the third person cannot find the original images of the person without their permission. Thus, it will be helpful in a lot of fields to avoid the forgery in this AI-based modern world.

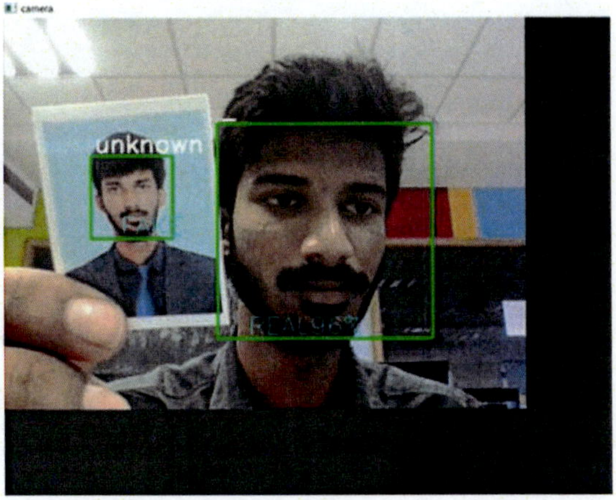

Fig. 10 Comparison of original with hard object

References

1. Sinaga AS (2022) Identification of biometric deepfakes using feature learning and deeplearning
2. Heo Y-J, Yeo W-H, Kim B-G (2022) DeepFake detection algorithm based on improved vision transformer made with generative deep learning technologies. Authors under exclusive licence to Springer Science+Business Media, LLC, part of Springer Nature, Sookmyung Women's University, Seoul, Republic of Korea
3. Nguyen TT, Nguyen CM, Nguyen TD, Nahavandi S (2019) Deep learning for Deepfakes creation and detection: a survey. Published in Computer Vision and Image Understanding
4. Guarnera L, Giudice O, Guarnera F, Ortis A (2022) The face deepfake detection challenge. J Imag 8(10):263
5. Hsu C-C, Zhuang Y-X, Lee C-Y (2020) Deep fake image detection based on pairwise learning. Appl Sci 10(1):370
6. Bhilare O, Singh R, Paranjape V, Chittupalli S, Nguyen ST, Nguyen CM, Nguyen TD, Nahavandi S (2019) Deep learning for deepfakes creation and detection. Published in Computer Vision and Image Understanding
7. Nadimpalli AV, Rattani A (2022) GBDF: gender balanced deepFake dataset towards fair DeepFake detection. In: International Conference on Pattern Recognition
8. Ko D, Lee S, Park J, Shin S, Hong D, Simon (2022) Deepfake detection for facial images with facemasks using deep learning techniques. ACM J Conf Sungkyunkwan University
9. Mittal G, Yenphraphai J, Hegde C, Memon N (2022) A challenge-response system for real-time Deepfake detection using deep learning techniques
10. Zhao H, Zhou W, Chen D, Wei T, Zhang W, Yu N (2021) Multi-attentional Deepfake detection model as a vanilla binary classification problem techniques. University of science and technology in China Microsoft Cloud AI
11. Badale A, Darekar C, Castelino L, Gomes J (2000) (NTASU-2020 Conference Proceedings) published in Inter J Eng Res Technol (IJERT) ISSN: 2278-0181

Control Strategy for Modified Quasi-Admittance Source Inverter

A. Angeline Esther, P. S. Manoharan, J. Nivedita, and P. Deepamangai

Abstract An avant-garde and futuristic Modified Quasi-Admittance Source Inverter (MQYSI) is put forward in this paper. Two administering modes govern how an admittance source inverter works. The redesigned q-YSI has a stronger boost voltage inversion and a lower duty ratio value to enhance output voltage quality compared to the previous YSIs. When compared to inverters of a more traditional design, this topology minimises the stress from voltage athwart the capacitors, solves the issue of a larger early inrush current, and makes decentralised generating systems possible. The inverter working with space vector-based pulse width modulation (SVPWM) is designed and analysed in MATLAB/Simulink environment and further contrasted with sine PWM technique in order to justify the better control technique.

Keywords Admittance source inverter (YSI) · Modified quasi-admittance source inverter (MQYSI) · Space vector pulse width modulation (SVPWM) · Sine pulse width modulation (SPWM) · Boosting ability

1 Introduction

The inverter, a type of DC-to-AC converter, is well known. Switching components are capable of controlling inverters. Typically, transistor and antiparallel diode are used to create inverter switches. There are two popular types of inverters: inverters

A. A. Esther (✉) · P. S. Manoharan
Thiagarajar College of Engineering, Madurai, Tamil Nadu, India
e-mail: angelineantony1999@gmail.com

P. S. Manoharan
e-mail: psmeee@tce.edu

J. Nivedita
Sir Isaac Newton College of Engineering and Technology, Nagapattinam, Tamil Nadu, India

P. Deepamangai
SRM TRP Engineering College, Trichy, Tamil Nadu, India

G. Rajakumar et al. (eds.), *Intelligent Communication Technologies and Virtual Mobile Networks*, Lecture Notes on Data Engineering and Communications Technologies 171,
https://doi.org/10.1007/978-981-99-1767-9_36

with voltage as their source and with an incessant input voltage and inverters with current as their source and also with perpetual input current. These inverters are powered by batteries, a storage cell stack, and voltage and current source [1]. The primary issue with VSI is that it serves as both a boost converter and a buck converter, respectively, for the transmission of energy from alternate current to direct current and from direct current to alternative current. This results in an accessory DC-DC power converter design being needed, which raises the cost of setup, to obtain the voltage of the necessary ac signal. In order to safeguard against the failure of semiconductor devices, the semiconductor switches on one side cannot be activated and turned OFF synchronously. Similar to this, CSI serves as an enhanced converter for converting alternative current to uninterrupted current and a resisting converter for turning the direct current to alternating current in transfer of power. VSI motor regulators and CSI motor regulators are incapable of operating at high voltage as a result. ZSI, a buck-boost inversion model with a broad working voltage level, was developed to address the aforementioned drawbacks. By reducing dead time, the ZSI increases dependability; nevertheless, it has several drawbacks, including inconsistent input current and the need that the capacitor can resist soaring potential and significant inrush current passage during startup.

Enhanced ZSI, quasi ZSI, Switched Inductor-ZSI, and EB-ZSI were developed in the field to address these drawbacks using ZSI as a basis [2]. ZSI and SL-ZSI, however, only have a few drawbacks. Inverters with magnetically coupled transformers were therefore recommended in contemplation to achieve high voltage gain. Snubber circuits are required, and couplings should be sturdy. The problem with the magnetically coupled transformer is weak coupling. These demerits finally led to the achievement of admittance source inverter (YSI). The extended q-Y-source inverter that has been developed in [3] has better boost factor, continuous input current, mitigation of discharge effects, and lessened voltage spikes—achievements that a Z-source inverter could not match. Due to their strong boost capabilities, two novel Y-source inverters with fastened direct current-link voltage are presented in [4]. These inverters are ideal for one stage greater gain reversal. In addition to this, the overall power loss is diminished as a result of the considerable drop in ST current amplitude. All of these factors go into the proposed converters' high efficiency. In order to assess the gain in voltage level, voltage strain over the switches, and converter design across three SPWM approaches, three-phase QYSI was used in [5]. Here, simulation is used to analyse the effectiveness of several Sine PWM control strategies. In [6, 7], a novel type of enhanced q-Y-source DC-to-DC converter model based on the conventional q–admittance design–source structure is introduced. This converter has all that of the favourable circumstances of the prevailing Y–source DC-to-DC converter in addition to having enormous voltage gaining ability, unceasing flow of input current, and a very little starting inrush current. For rooftop utilisations of PV grid with battery storage capacity capability, Xupeng et al. [8] suggests a new upgraded γ type admittance-source inverter. The inverter permits battery charging from a photovoltaic source during times of meagre-peak loads and exports any extra power to the distribution grid. It can sustain high voltage gain even under partial shaded circumstances. Test cases are investigated in various scenarios (stand-alone working condition without the usage of cell stack, private operational mode with the usage of battery). The elimination of dc-coupling voltage spikes without the need for additional switches or lowering voltage gain has been presented in [9] using a high escalated Y-source inverter with a novel absorption circuit design. The suggested inverter design provides a significant boost with a steady dc-associated voltage that is facilitated by appropriate leakage energy recycling. By contrasting the modelling and trial and error findings of an existing Y-source with the suggested inverters, these properties have been confirmed. In [10–16], the admittance source inverters own the demerit of higher component rating and lesser boosting ability but a higher voltage increase when compared to impedance based inverters. In this proposed policy, a modified quasi-admittance-source DC-to-AC converter is developed from the existing Y-source network with a higher voltage boosting capability in a way by designing the transformer windings accordingly. The main intention of this development is to increase or boost the voltage at the output in comparison to the impedance source inverters and extended quasi-impedance source inverters. This brought out quasi-admittance source inverter satisfies the need of higher voltage boost at the output stage, thereby enabling its use in distribution generation systems.

2 Principle of Methodology

2.1 Admittance Source Inverter

A three-phase bridge inverter, a Y-source network, and a filter make up the proposed inverter's input, middle, and output stages. In its most basic form, the Y-source impedance network is composed of a passive diode (D1) as a measurement device, a capacitive device, and—most significantly—a three-winding turns based transformer with turns number ($N1, N2, N3$) for injecting a strong voltage level.

The coupling between the transformer and the bridge in the inverter and $D1$ must be compact to limit leakage from the inductors that can be noticed at the transformer's windings. The Y-source inverter, like a standard impedance-source inverter, features a supplementary zero state in shoot-through and an exact opposite state that are built up of six in-action phases and two level zero modes.

As long as the transformer's winding turns provide the desired K value while still adhering to the design limitations imposed by specific users, they can be chosen with versatility. Figure 1 is the circuit configuration of admittance source inverter.

The pre-existing impedance source inverters suffer from the disability to sustain greater voltage levels and current. The coupling design remains a demerit in magnetically coupled transformer source inverters. To subdue all these weaknesses, the MQYSI has been introduced.

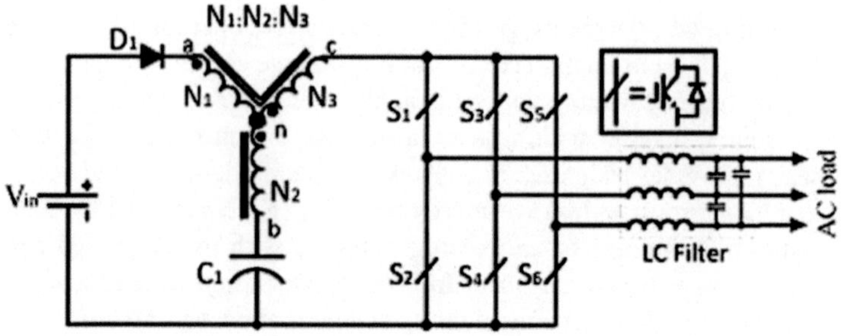

Fig. 1 Circuit configuration of admittance source inverter

2.2 *Modified Quasi-Admittance Source Inverter*

The modified quasi-admittance source inverter is achieved by making slight changes in the circuit configuration of the already established admittance source inverter. Its primary supremacy is that it yields higher voltage level boosting at the output stage of the DC to AC conversion and thereby remaining a key source to instruments or devices that operate in industries under high gain. They are mostly favoured for renewable energy systems involving sun power and as well as wind power. The circuit configuration of the MQYSI is shown in Fig. 2.

The inverter employs shoot-along mode and the non-shoot along mode which are its two operating concepts. The inverter will tend to operate in six functioning circumstances and two original zero-level situations while in shoot-across condition, however, it will short circuit along any one of the three-phase legs or any two-phase leg or any one-phase leg when working under non-shoot-through mode.

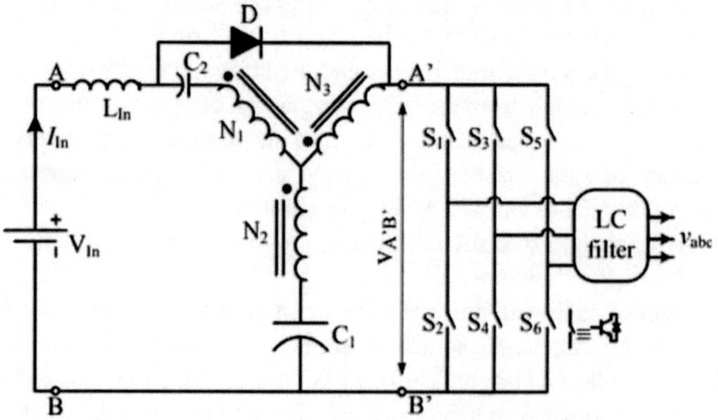

Fig. 2 Circuit configuration of modified quasi-admittance source inverter

The gate signals produced by IGBT switches are being delivered to the switching topology in the DC-to-AC converter which is usually aided by any pulse width modulation strategy using a space vector or sine PWM techniques.

The LC filter, which consists of an inductor and a capacitor does its role as a frequency filter and produces an output voltage wave that is smoother.

In the layout of the modified inverter, the notations $N_{1,2}$ and N_3 represent the three phase transformer winding that are designed in a way so that a higher voltage level is reached by the inverter at the output.

3 Problem Formulation

The input DC voltage source may likely arise from a battery or a fuel stack or from a specific PV module as considered in this problem. The admittance source usually comprises inductors, capacitors, diode, and a three-phase transformer winding with turns number designed as per the requirements. The key circuit is nothing but a six switches comprised bridge inverter which is connected with the admittance design. Here the DC signal is reshaped to an AC signal that can be utilised by electric devices.

4 Results and Discussion

In the problem formulated, the DC source is considered to be generated from a specific PV module instead of a battery or a fuel stack. The whole formulation is analysed according to the selected PV module called SunTech Power STP270- VRM-1.

The specifications of SunTech Power STP270-VRM-1 are given in Table 1, according to which the boosting capability of MQYSI is determined.

In this problem formulation, three modules connected in series are being considered. One array kept at 25 °C and specified irradiation are contemplated, and the Power Versus Voltage estimate and Current Vs Voltage estimate are being analysed as in Fig. 3.

Table 1 Specifications of SunTech power STP270-VRM-1 PV module

Parameters	Value
Maximum power (W)	269.85
Voltage at MPP (V)	35
Current at MPP (A)	7.71
Open circuit voltage Voc (V)	44.5
Short circuit current Isc (A)	8.2
Temperature quantum of Voc (%/deg.C)	−0.313
Temperature quantum of Isc (%/deg.C)	0.054

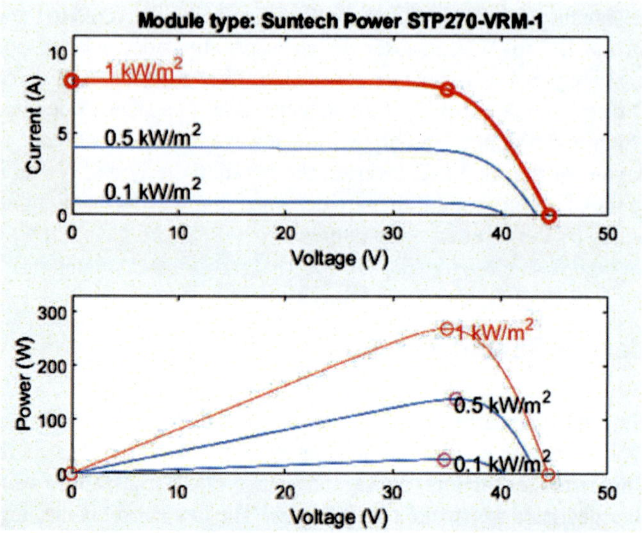

Fig. 3 Corresponding PV and IV curves of SunTech power STP270-VRM-1 PV module

The entire Modified Quasi-Admittance Source Inverter has been expanded in MATLAB/Simulink Environment to justify the voltage boosting ability of the inverter at the output end. The simulation of the developed inverter is depicted in Fig. 4.

The design criteria for the MQYSI are given below in Eqs. (2–7).

$$\text{Modulation Index}M = 0.9 \tag{1}$$

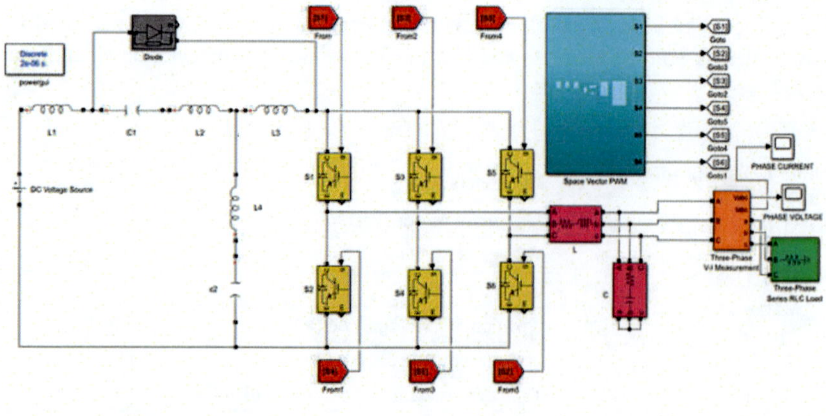

Fig. 4 Simulation of MQYSI

$$V_{ph} = M\frac{\mathrm{BVd_c}}{2} \tag{2}$$

$$N_1 : N_2 : N_3 = 1 : 2 : 3 \tag{3}$$

$$k = \frac{n3 + n1}{n3 - n2} = 4 \tag{4}$$

$$\text{Boost Factor } B = 1.453 \tag{5}$$

$$C_1 = C_2 = 470\ \mu F \tag{6}$$

$$L_1 = 2\ \text{mH} \tag{7}$$

In SVPWM strategy, the reference signal is first converted to dq voltages with the utilisation of Park's Transformation. The angle and switching time of each of the eight vectors are being calculated from which the gate signals are fed to the switching devices that are nothing but IGBTs.

The voltage is being measured across the SunTech Power STP270-VRM-1 PV array so that the voltage boosting level at the output stage can be analysed. The voltage developed across the array is as in the below Fig. 5.

From Fig. 5, it is seen that the voltage across SunTech Power STP270-VRM-1 PV Module for one second time step is nearly about 10 V which is lesser when considered for large distributed generation systems.

The corresponding AC signal at the output side due to the presence of the modified quasi-admittance source inverter is as depicted in Fig. 6.

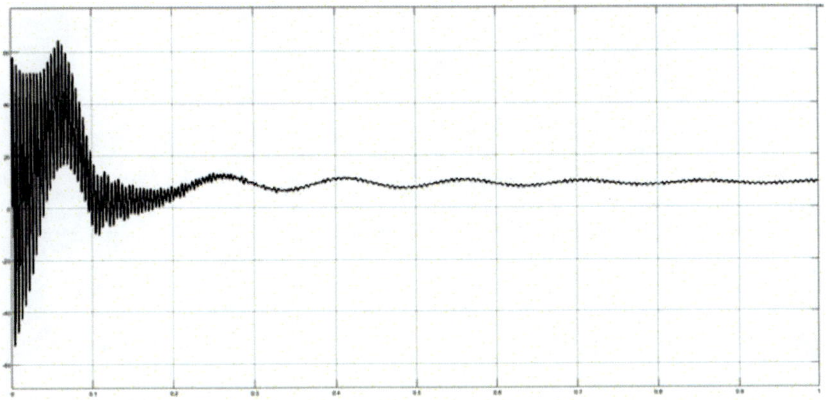

Fig. 5 Voltage seen across the PV array

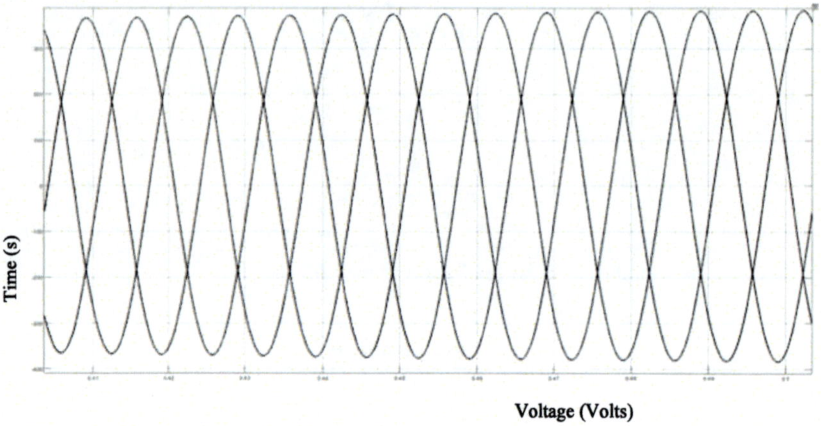

Fig. 6 Output three-phase voltage of MQYSI with SVPWM

From Fig. 6, it is observed that the voltage waveform at the output stage has been boosted drastically due to the presence of the modified admittance inverter.

The input side DC voltage level which was nearly only 10 V has been boosted to nearly 350 V AC at the end side and the waveform is smoother without distortions due to the LC filters playing their major role as frequency filters.

The output has been attained for an *R* load and by the usage of SVPWM control strategy.

The same developed inverter working with sine pulse width modulation control strategy gives an output three-phase voltage which is lesser than that obtained through the former control technique. The simulation of MQYSI with SPWM control strategy is depicted in Fig. 7.

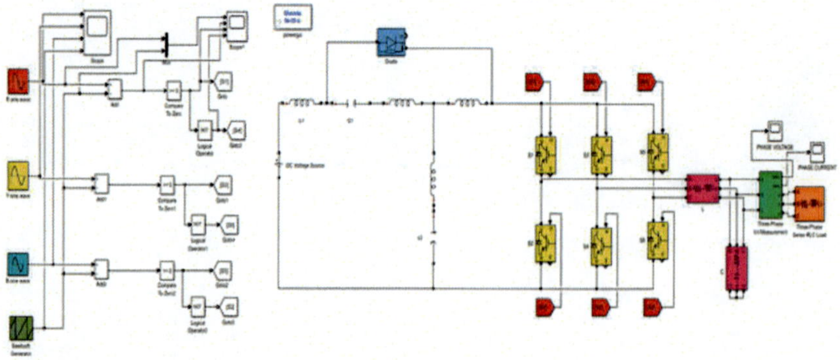

Fig. 7 Simulation of MQYSI with SPWM control strategy

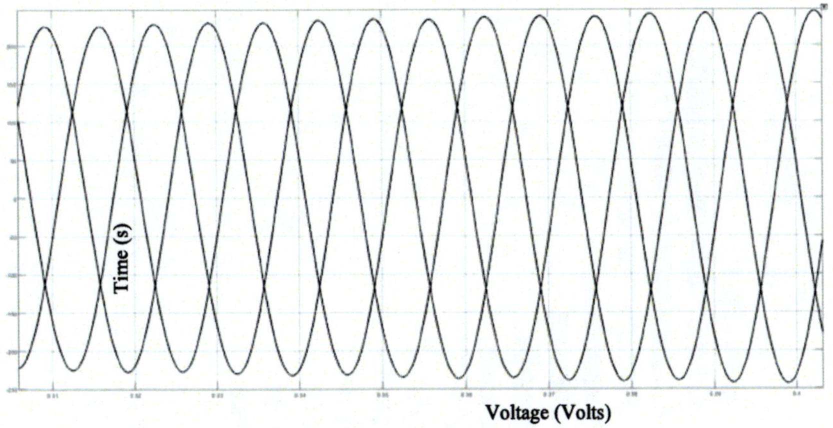

Fig. 8 Output three phase voltage of MQYSI with SVPWM

The corresponding voltage in three phase produced by the MQYSI using the sine pulse width modulation produces a lesser voltage boosting at the output side as given in Fig. 8.

Thus, the term higher boosting ability by the MQYSI holds good and also the results are better than traditional impedance source inverters, magnetically coupled inductors, switched inductor inverters, etc.

This higher boosting of voltage by the designed model thus aids its usage in system that require higher voltage gain and distributed generation systems.

The extended quasi-impedance source inverters existing possess a lesser boosting ability, i.e., the output three-phase voltage was boosted to about 120 Vrms with an input of 10 V. On the other hand, the MQYSI boosts the voltage to about 350 V which is nearly three times more than that of the extended quasi-Z source inverter.

The harmonics for the obtained three-phase output voltage waveform are also analysed, and it is found to be 1.41% as in Table 2 which is much lesser than that of conventional type inverters.

Figure 9 is obtained by the FFT Analysis function in MATLAB that represents the amount to which the voltage at the end has been distorted, thus indicating the quality of the waveform generated.

Table 2 Comparison of MQYSI with existing impedance source inverter

	Quasi-impedance source inverter	Modified quasi-admittance source inverter
Total harmonic distortion (in %)	3.77	1.41

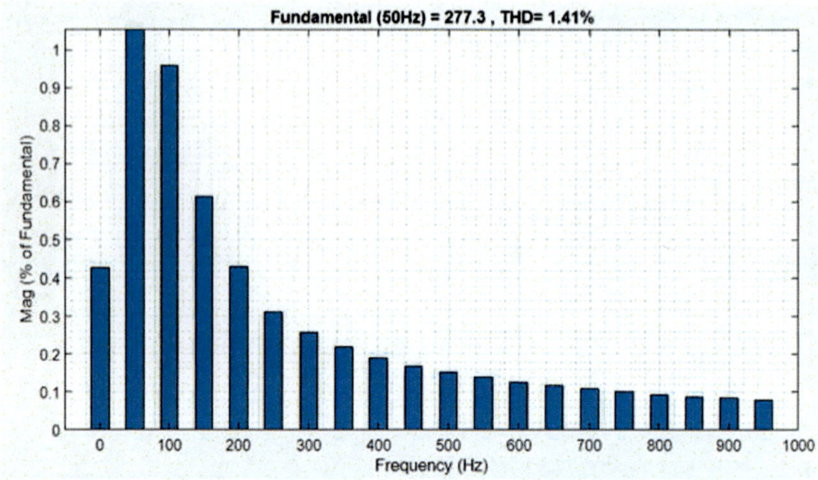

Fig. 9 THD (in %) for the three-phase output voltage attained by MQYSI

5 Conclusion

When compared to earlier produced inverters, for instance the impedance based inverter and the quasi-level impedance source inverters, the modified quasi-admittance source inverter (MQYSI) suggested here offers a greater voltage at the output. Traditional models, such as voltage and current source inverters, simply increase voltage. As a result, a second or an auxiliary DC-to-DC converter model is required to adjust the voltage in accordance with the provide load, making them unsuitable for distributed generating systems. All of these drawbacks are solved by the suggested model, which yields a high boosting capability.

This novel inverter can be administered further with robust reinforcement learning algorithms based maximum power point tracking methods involving an agent, action, space, and reward that tracks the maximum power at an instant of temperature and irradiation condition.

References

1. Lazzarin TB, Bauer GAT, Barbi I (2013) A control strategy for parallel operation of single-phase voltage source inverters: analysis, design and experimental results. IEEE Trans Ind Electron Sol 60(6):2194–2204
2. Jagan V, Kotturu J, Das S (2017) Enhanced-boost quasi-Z-source inverters with two switched impedance network. IEEE Trans Ind Electron 64(9):6885–6897
3. Deepamangai P, Manoharan PS (2019) Robust controller for grid-connected quasi-admittance source inverter for photovoltaic system. Elect Pow Syst Res 175, October

4. Liu H, Li Y, Liu K, Loh PC, Wang W, Xu D, Blaabjerg F (2019) Extended quasi-Y-source inverter with suppressed inrush and leakage effects. IET Power Electronics 12(4):719–728
5. Nikbahar A, Monfared M (2022) Smooth DC-Link Y-source inverters: suppression of shoot-through current and avoiding DC magnetism. IEEE Trans Pow Elect
6. Santos R, Gonçalves FAS (2021) Sinusoidal pwm techniques comparison for the quasi-y-source inverter. J Cont, Auto Elect Syst 32(5):1399–1407
7. Chen M, Loh PC (2020) A single-phase high voltage-gain differential Y-source inverter. IEEE J Emerg Select Topics Pow Elect 9(2):2027–2037
8. Xupeng F, Xiaoli W, Song W (2020) A new type of three-phase switched-quasi-Y-source inverter. In: 2020 Chinese Automation Congress (CAC), pp 2790–2794
9. Reddivari R, Jena D (2022) Improved Gamma type Y-source inverter for rooftop PV based VG applications. Int J Electr Power Energy Syst 142:108261
10. Siwakoti YP, Blaabjerg F, Loh PC (2015) Quasi-Y-source boost DC–DC converter. IEEE Trans Pow Elect 30(12):6514–6519
11. Liu H, Zhou Z, Liu K, Loh PC, Wang W, Xu D, Blaabjerg F (2018) High step-up Y-source inverter with reduced DC-link voltage spikes. IEEE Trans Power Electron 34(6):5487–5499
12. Rajkumar MV, Prakasam P, Manoharan PS (2016) Investigational validation of PV based DCD-MLI using simplified SVM algorithm utilizing FPGA tied with independent sources. Circuits Syst 7(11):3831
13. Fang X, Ding X, Zhong S, Tian Y (2019) Improved quasi-Y-source DC-DC converter for renewable energy. CPSS Trans Pow Elect Appl 4(2):163–170
14. Forouzesh M, Baghramian A, Salavati N (2015) Improved Y-source inverter for distributed power generation. In: 2015 23rd Iranian Conference on Electrical Engineering, pp 1677–1681. IEEE
15. Liu H, Li Y, Zhou Z, Wang W, Xu D (2019) A family of low-spike high-efficiency Y-source inverters. IEEE Trans Industr Electron 66(12):9288–9300
16. Peng FZ, Shen M, Qian Z (2005) Maximum boost control of the Z-source inverter. IEEE Trans Pow Elect 20(4):833–838

The Information and Communication Technology's Impact to Enhance the Tourism Industries in Indonesia

Kendrew Huang, William Gunawan, Rivaldi, Ford Lumban Gaol, and Tokuro Matsuo

Abstract The purpose of this study is to examine the integration of information communication and technology and the development of eTourism as a field in tourism industry, as well as the implications of ICT usage on the Tourism industry in Indonesia. Using various types of literature such as scientific articles, articles, and books, this study emphasizes and examines the importance of ICT and how these acts contribute to the tourism and hospitality sector. Tourism sector businesses should be praised for their use of ICT and use of current technology such as social networks for client engagement. The use of ICT has been carefully monitored and deployed at a low cost, needing little technical knowledge from the workforce. The purpose of this article is to raise awareness among scholars, educators, legislators, leisure entrepreneurs, and public officials on the usefulness of ICT applications in the tourism and hospitality industries in Indonesia. The work is limited to Indonesia, and the conclusion is based on previous research. The report is totally prepared using Indonesian culture. The ICT application is only used as an independent variable in the study. As a result, there may be some other factors influencing the tourist and hospitality industries in Indonesia.

Keywords ICT · Etourism · Tourism and hospitality

K. Huang · W. Gunawan · Rivaldi
School of Information System, Bina Nusantara University, Alam Sutra, Jakarta 13960, Indonesia
e-mail: kendrewhuang@binus.ac.id

W. Gunawan
e-mail: William.gunawan004@binus.ac.id

Rivaldi
e-mail: Rivaldi@binus.ac.id

F. L. Gaol (✉)
Computer Science Department, BINUS Graduate Program—Doctor of Computer Science, Bina Nusantara University, 11480 Jakarta, Indonesia
e-mail: fgaol@binus.edu

T. Matsuo
Advanced Institute of Industrial Technology, Tokyo, Japan
e-mail: matsuo@aiit.ac.jp

G. Rajakumar et al. (eds.), *Intelligent Communication Technologies and Virtual Mobile Networks*, Lecture Notes on Data Engineering and Communications Technologies 171,
https://doi.org/10.1007/978-981-99-1767-9_37

1 Introduction

Tourism encompasses a vital economic impact at a global, domestic, and regional level. This impact is underlined by applied mathematics proof demonstrating the importance of business enterprise in terms of gross domestic product, employment, and economic development [1].

The business enterprise business may be seen collectively in the primary business sectors wherever business functions are nearly completely mistreatment info and communications technologies (ICT) [2]. Information technology (IT) and ICT have a vital role within the development of business enterprise. Computerized reservations systems (CRS) were among the primary applications of IT worldwide [2].

The business is one among the additional triple crown areas of e-commerce as a result of it's for the most part client orienting and since services and therefore the provision of data is at its center [3]. Recommend business enterprise could be a hybrid business since although it's dominated by the availability of data, basically it's a couple of physical products. this needs the 'seamless integration of data and physical service, with versatile configurations of the physical and therefore the informational parts"[4].

ICT facilitates this integration and permits customization of business enterprise merchandise to suit the wants of people thanks to changes in client behavior of the holidaymaker the market is turning into additional divides with every potential client happiness to a variety of market segments at the same time. Holidaymaker operators have to be compelled to bear in mind those changes and be equipped to retort, or higher still, take a proactive approach [5].

The challenge for the business enterprise operator is the provision of correct, localized information, progressively via IT, while maintaining a relationship with the holidaymaker. Instead of being simply group action primarily-based long run relationships have to be compelled to be fostered, and IT will play a task during this relationship [6].

As we know travel destination in around the world, especially in Indonesia is very huge, in Indonesia there are so many travel destination. We all know that tourism sector contributing mostly for Indonesian foreign exchange and country income. In this modern world, enterprise system and also information technology are giving a huge impact into tourism sector all around the world. Because of technology, we can see a hotel booking app, booking an airplane ticket online, and many more [7].

But with rapidly growing of technology from all around the world is impacting the stubborn people that stay in their way to book a plane ticket, book a hotel, etc. The detail of research question as follow: Does the implementation of data Technology have any positive impact in tourism? Base on the above research question, below is the research hypothesis: There are advantages of using information technology in tourism. There are blessings of mistreatment info technology in business enterprise [8].

The Internet modified data exchange and insemination; step by step, the data is developing into a sort direct sales channel path, where needs a business enterprise distributing into reset the model and its performance. This analysis is predicted to be a supply of data once tourists create a plan of action, and it conjointly shows that once tourists rummage around for and steel oneself for the longer term business enterprise expertise [9].

By doing this research, we can providing the data about impact of technology into tourism sector, and by doing this research, we can find a way to increasing the amount of tourist that travel into our country. Also the benefit of this research is we can increase the effectiveness and efficiency of technology into tourism in our country.

2 Literature Review

Information Technology and commercial enterprise aims primarily to contribute to the method of theory building, and thus to the advancement of analysis and scholarship during this growing field. As an associate knowledge domain journal, it supports industry-oriented analysis still as tutorial theory-targeted analysis [10, 19]. ICT can feature each empirical case studies associated technical-theoretical papers staring at tourism, travel, hospitality from associate IT purpose of read and at IT from an applied perspective. The journal contains analysis papers, progressive reviews, analysis notes, and analyzes of business applications. It's conjointly supposed to incorporate papers relevant to the business in numerous national contexts [11, 18].

This is wherever technology because the key tool within the creation of the traveler expertise comes into play [12, 20] claim that technology has notably inspired the shift from the passively receiving shopper to the actively concerned co-creator of his/her own expertise [13, 21] underline that ICTs became a serious component within the co-creation of traveler experiences by permitting corporations to interact with shoppers through websites, mobile devices, transportable town guides, travel guides, virtual life environments, or increased building rooms. In conceptualizing technology-enabled increased traveler experiences, it's imperative to know that traveler experiences can't be staged and delivered to the shoppers, however, should otherwise be co-created at the side of the patron [14, 22]. Technology has to be considered the catalyst that functions as a way to co-create meaning interrelations and experiences between the corporate and therefore the shopper [23]. Taking the argument under consideration that co-creating experiences generates price for the patron, technology can so become the key instrument for the improvement of experiences by facilitating and empowering co-creation and thereby generating other prices for the patron [24]. This can be notably relevant, considering the multi-phase nature of the traveler expertise, i.e., prior/during/post travel. Technology doesn't solely enhance the physical commercial enterprise house on-the-spot [25]. However, conjointly facilitates engagement and knowledge co-creation within the virtual house already before still as once the travel within the tourist's home setting [26]. As a result, by conflating the two areas of traveler expertise and technology, this paper proposes a holistic conceptualization of technology-enabled increased traveler experiences that is crucial for understanding and managing this new variety of expertise in theory and application [15, 27].

The data gap from this analysis is whether or not there's a relationship between the implementation of data technology within the commercial enterprise sector and therefore the results of tourists World Health Organization travel [16]. To shut this gap, knowledge searches square measure disbursed from the web associated with the subject being mentioned, as an example graphs of increase or decrease within the results of victimization technology in sorting out traveler attraction [17].

3 Research Result

The questionnaire is used to collect all of the data used in this study. The questionnaire is distributed to a large number of responders. There are multiple items in the questionnaire, "yes" and "No"; if the responder picks yes, it implies they agree with the statement; if the respondent chooses no, it means they disagree with the statement. After the yes/no choice, there is a scale 1–10 quiz; if respondents choose 1–4, it means they disagree, and if they choose 5–7, it means they agree.

If they select 8–10, it indicates that they are neutral.

a. Gender. range of responses: fourteen responses

Base on Fig. 1, there are 28.6% Female and 71.4% Male.

b. The age distribution

Base on Fig. 2, the most Ages is in 15–19 years.

c. The advantageous of traveling

Based on Fig. 3, the most answers measure affirmative, which means technology would profit for traveling.

d. The facilitate of tourism

Most answers square measure in the facilitates commercial enterprise is shown in Fig. 4.

Fig. 1 Gender distribution

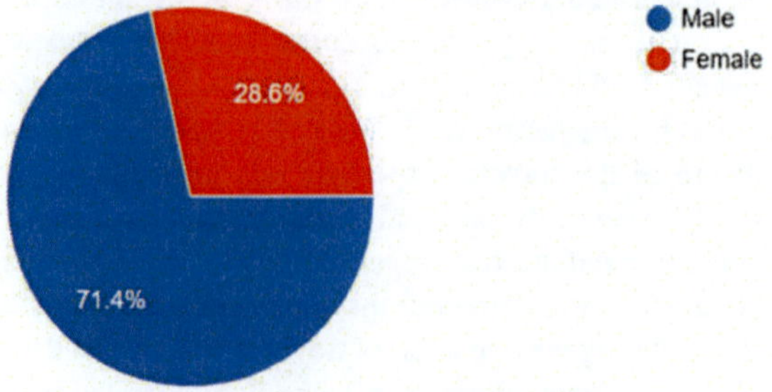

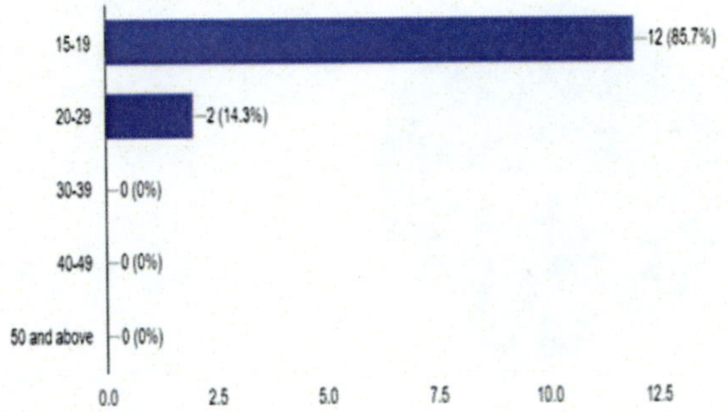

Fig. 2 Distribution of age

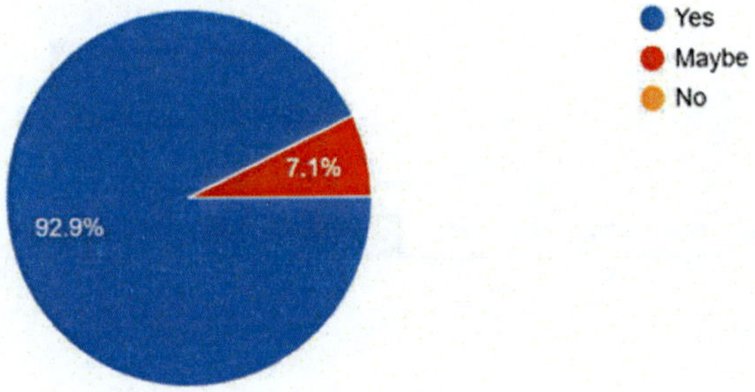

Fig. 3 Benefit of traveling

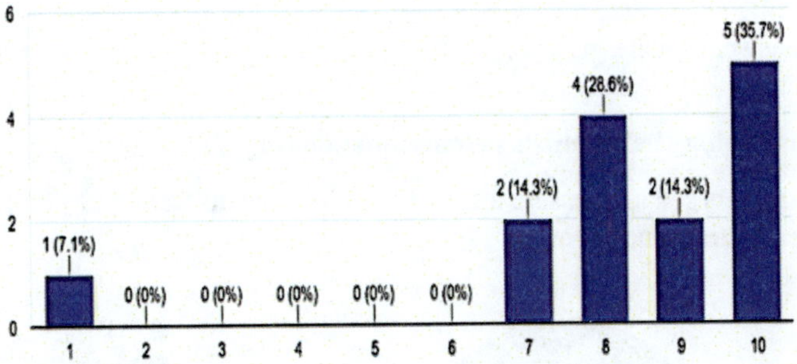

Fig. 4 Facilitate of tourism

e. The use technology after you travel

Base on Fig. 5, all the answers square measure affirmative, which means tourists use technology for traveling.

Fig. 5 The use technology after travel

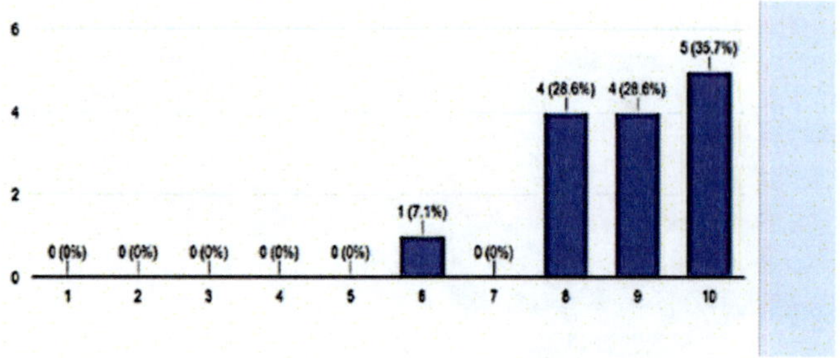

Fig. 6 Rank importance of technology to travel

f. The Rank the importance of technology to travel (Fig. 6)

Most answers square measure in ten, which means technology is very important for traveling.

g. The technology help tourists in transportation (Fig. 7)

Most answers square measure affirmative, which means technology helps tourists a great deal in transportation.

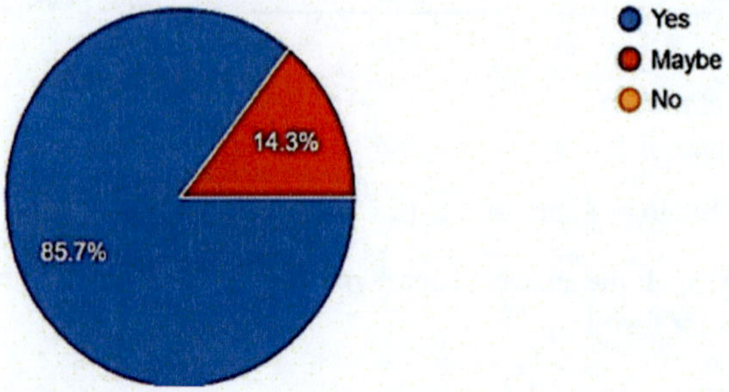

Fig. 7 Use technology after travel

4 Conclusion

By using the technology in tourism people can traveling effortlessly. As we know in the past, buying a flight ticket or a hotel ticket is quite complex, we must open the official web or, etc., the payment is also confusing, by using the technology we can easily book flight ticket, hotel ticket, etc., also the payment method is easy.

References

1. Aldebert B, Dang RJ, Longhi C (2011) Innovation in the tourism industry: the case of Tourism. Tour Manage 32(5):1204–1213
2. Berné C, García-González M, García-Ucedac ME, Múgica JM (2015) The effect of ICT on relationship enhancement and performance in tourism channels. Tour Manage 48(June):188–198
3. Buhalis D (1998) Enhancing the competitiveness of small and medium-sized tourism enterprises. Elect Mark 6(1)
4. Buhalis D, Licata MC (2002) The future eTourism intermediaries. Tour Manage 23(3):207–220
5. Coopers C, Fletchair J (2005) Tourism principles and practices. Pearson, UK
6. Cox C, Burgess S, Sellito C, Buultjens J (2009) The role of user generated content in tourists' travel planning behaviour. J Hosp Market Manag 18(8):743–764
7. Cunill OM, Forteza CM (2010) The franchise contract in hotel chains: a study of hotel chain growth and market concentrations. Tour Econ 16(3):493–515
8. Gomis JM, González-Reverté F (2012) Amadeus: global distribution system's new paradigm. eTourism Case Studies 399
9. Gruescu R, Nanu R, Pirvu G (2009) Information and communications technology and internet adoption tourism. Bull UASVM Horticul 66(2):407–413
10. Hudson S, Gilbert D (2006) The internet and small hospitality businesses: B&B marketing in Canada. J Hosp Leis Mark 14(1):99–116
11. Jennifer XM, Dimitrios B, Haiyan S (2003) Information. J Inf Manag 23(6):451–467
12. Khan MYH, Hakeem SMA, Hossain A (2018) The impact of tourism development on Greenwich Community in the post-Olympic games era. J Tour Manage Res 3(3):348–360
13. Kuttainen C, Lexhagen M (2012) Overcoming barriers to Sme e-commerce adoption using blended learning: a Swedish action research case study. Info Tech Tour 13(1):13–26
14. Main HC (2002) The expansion of technology in small and medium hospitality enterprises with a focus on net technology. Info Techn Tour 4:167–174
15. Michalić T, Praničević DG, Arnerić J (2015) The changing role of ICT competitiveness: the case of the Slovenian hotel sector. Econ Res-Ekonomska Istraživanja 28(1):367–383
16. Mupfiga (2015) Adoption of ICT in the tourism and hospitality sector in Zimbabwe. Inter J Eng Sci (IJES) 4(12):72–78
17. Peppard J (1993) IT strategies for business. Pitman Publishing, London
18. Poon A (1993) Tourism technology and competitive strategies. Oxford: CAB Inter SocioEcon Chal 2(4)
19. Porter ME (2001) Strategy and the internet. Harvard Business Review, March
20. Scheidegger E (2006) Can the sate promote innovation in tourism? Should it? In: OECD (ed.) Innovation and growth in tourism, OECD, Paris, pp 11–16
21. Soteriades M, Aivalis C, Varvaressos S (2004) E-marketing and e-commerce in the tourism industry: a framework to develop and implement business initiatives. Tourism Today 4:1–18
22. Stiakakis E, Georgiadis ChK (2011) Drivers of a tourism e-business strategy: the impact of information and communication technologies. Oper Res Int J 11(2):149–169

23. Suomi R, Li H (2010) Internet adoption in tourism industry in China. IFIP Inter Federation Inform Process 286(1)
24. UNCTAD (2004). UNCTAD's eTourism Initiative, Doc. TD (XI)/BP/6), 26 April. Retrieved October 11, 2018 from: http://www.unctad.org/en/docs/tdxipbd6_en.pdf
25. Werthner H, Klein S (1999) Information technology and tourism—a challenging relationship. Springer, New York
26. WTO (2001) eBusiness for tourism: practical guidelines for destinations and businesses. World Tourism Organization, Madrid
27. Zhou Z (2004) E-commerce and information technology in hospitality and tourism. Delmar Publishing, New York

Effect of PV Penetration on the Steady-State Voltage on Grid-Integrated PV System

Gaurav B. Patil, Santosh S. Raghuwanshi, and L. D. Arya

Abstract In the recent few decades, due to the rapid and extensive growth of renewable sources of generation, the use of solar photovoltaics is at a large extent. Many individuals prefer a grid integrating system instead of a stand-alone system. However, this large usage of energy from solar photovoltaic with grid integration brings some severe challenges along with an increase in PV penetration. Generally, there are two major categories of these challenges: feeder level (distribution) and grid level. In this paper, the feeder-level challenges of solar photovoltaic integrated systems are discussed, and remedies are suggested to overcome these problems. In addition, analysis of PV penetration on several loads after the application of the capacitor and the impact of the on-load tap changer are presented.

Keywords Unity power factor · Feeder · Harmonics · Photovoltaic

1 Introduction

There are mainly two ways in which electricity is produced by solar photovoltaics. One of the ways is the stand-alone system, where the energy is stored in batteries or may be in other forms of storage and then local loads utilize it. The other way of utilizing solar energy electricity is where the electricity produced by solar is directly fed to the grid and then it is utilized by other loads connected on the same feeder, the nearby feeder, or further away from the grid. The second one is the most popular system because it uses the existing grid infrastructure. Therefore, there is no additional cost of extra storage required as in the case of mode one. Although it can use the existing infrastructure, it could potentially create challenges for grid operations, protection, etc. [1]. Therefore, in this paper, inverters that are supposed to control when the system starts the operation has been discussed. With those kinds of preliminary controls, several kinds of challenges are faced by the usual grid distribution

G. B. Patil (✉) · S. S. Raghuwanshi · L. D. Arya
Department of Electrical Engineering, Medi-Caps University, Indore, Madhya Pradesh, India
e-mail: en19ee6e1001@medicaps.ac.in

G. Rajakumar et al. (eds.), *Intelligent Communication Technologies and Virtual Mobile Networks*, Lecture Notes on Data Engineering and Communications Technologies 171,
https://doi.org/10.1007/978-981-99-1767-9_38

system. This paper demonstrates some of the basic principles, issues, and challenges arising from solar integration with the grid.

2 State of Art of PV Penetration in Grid-Integrated Systems

Earlier researchers studied different PV penetration scenarios of a single feeder during steady state and concluded that reverse power flow can result in overvoltages and also obtained solutions to this issue [2–5]. In [6], it was demonstrated that reverse power flow resulting from equally distributed generation exceeded transformer thermal ratings on a sample feeder. Numerous additional publications have concentrated on power reduction or the use of reactive power to address feed-voltage issues brought on by significant PV penetration on a single typical feeder.

When solar PV originated, expectations and obvious prizes of the panel were high. They still did not reach grid parity. So it was supposed that solar can be connected to the grid, but they were not economically viable at that point. It was presumed that the total amount of solar installed would be a minor fraction of the total grid install capacity and predominantly either by coal and gas or nuclear. An initial requirement of the grid-feeding solar inverters was also set accordingly.

Previous reports have looked at various penetrations of PV on a single feeder in steady state, concluded that reverse power flow can cause overvoltages, and examined methods of resolving this problem [4, 5]. In [6], reverse power flow due to evenly distributed generation was shown to exceed transformer thermal ratings on a sample feeder. Many other papers have focused on power curtailment or provision of reactive power to alleviate feeder voltage problems that arise with high PV penetrations on single representative feeder.

Previous reports have looked at various penetrations of PV on a single feeder in steady state, concluded that reverse power flow can cause overvoltages, and examined methods of resolving this problem [4, 5]. In [6], reverse power flow due to evenly distributed generation was shown to exceed transformer thermal ratings on a sample feeder. Many other papers have focused on power curtailment or provision of reactive power to alleviate feeder voltage problems that arise with high PV penetrations on single representative feeder.

Previous reports have looked at various penetrations of PV on a single feeder in steady state, concluded that reverse power flow can cause overvoltages, and examined methods of resolving this problem [4, 5]. In [6], reverse power flow due to evenly distributed generation was shown to exceed transformer thermal ratings on a sample feeder. Many other papers have focused on power curtailment or provision of reactive power to alleviate feeder voltage problems that arise with high PV penetrations on single representative feeder.

Fig. 1 PV panel connected to the grid

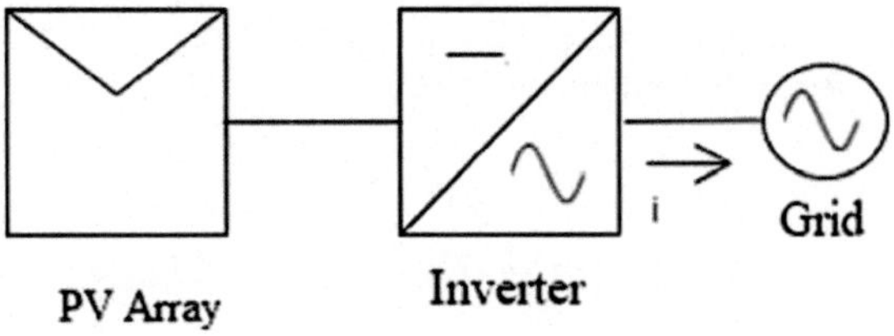

Fig. 2 Harmonics fed by solar inverter SI_1

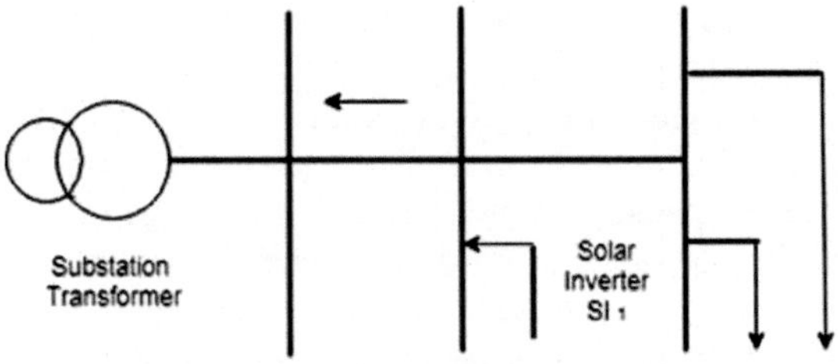

2.1 *Total Harmonic Distortion and the Individual Current Harmonics*

If a panel is connected to the feeding inverter as shown in Fig. 1, it is the feeding power to the grid. It was expected that the current feeding into the grid would have a low value of total harmonic distortion and the individual harmonics, especially lower ones like third, fifth, and seventh would also be restricted to a low value [7].

This is needed because by any chance current fed by the inverter has harmonics, as shown in Fig. 2. The harmonics will interact with the impedance of the feeder and will create harmonics in the voltage available at the load end.

This can then affect other loads and other systems connected to that feeder. So that was the predominant reason for the initial thought that let the solar inverter not pollute the grid. It was ensured good by the mean smaller value of the total harmonic distortion of current.

2.2 *Unity Power Factor*

The reactive power was also not allowed to be supplied or absorbed by the inverter. The initial idea was that these inverters should operate at the unity power factor. If there is a certain load and the inverter starts drawing certain reactive power, then additional capacitor banks must be connected. Therefore, initially, it was suggested that all the inverters must stick to the feeding power at the unity power factor, so that it does not create any additional voltage drop because of the consuming reactive power. Moreover, since the unity power factor was set, deteriorating the voltage profile or improving the voltage profile was not recommended. So it was just expected to feed power at the unity power factor.

2.3 Anti-islanding, Over-Under-Voltage, and Over Frequency

Anti-islanding is that if the grid fails, then the system or the inverter connected to the system also disconnects. It should not produce an island of its own and should disconnect as soon as the grid fails. This is the grid failure part, but usually the transient momentary grid magnitude frequency voltage might go out of the bound. In those cases, the inverter must disconnect as soon as possible.

For example, if there is a disturbance in a grid, then the solar inverter might end up making the situation worse; so, it is best if the solar inverter disconnects and the grids continue to operate with their basic controllers and basic power generation system. Therefore, fast-acting overvoltage, under-voltage, and over-frequency protections which are supposed to be in the inverter were found. For example, if the voltage is 1.1pu, then it certainly disconnects. During under-voltage, if the grid voltage goes below 0.85pu, it disconnects. During over-frequency, if it goes above 50.3 Hz, it disconnects [8].

All of these were intentionally meant so that the inverters were considered as external devices, and it was evaluated that, in case of any disturbance, these inverters get disconnected, which is good, and the main process is handled by the conventional generators.

2.4 Change of PV Penetration

Over time, these things were good if there is a low value of percentage penetration system when compared to total generation capacity. But over time, that has not remained true. The main reasons have been the continuous reduction in the cost of solar PV panels, the concerns for climate change, and so on.

Nowadays, PV is mostly considered on the grid, and there is continuous increasing penetration of PV in the system, which brings many challenges [9–12].

The two major categories of these challenges are feeder level (distribution) and grid Level. In this paper, some feeder-level challenges are discussed.

Unlike conventional generation where all get connected at relatively higher voltage into the system, these are not necessarily close to the load end. Furthermore, they were generally connected using a step-up transformer connected to a relatively higher voltage and the power grid transfers to the load end. But solar, on the other side, gets connected predominantly at the low voltage side of the distribution side [13].

Figure 3 is the graph, which shows solar connected to the system grid in Germany at different voltage levels. It can be seen that a majority of solar gets connected at

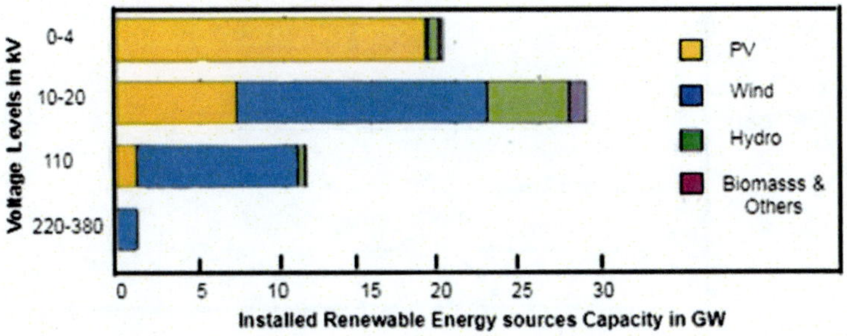

Fig. 3 RES connected to the system grid at different voltage levels

the low voltage which is 400 V level. Some solar systems get connected at levels between 10 and 20 kV, but as moved up in the voltage level, the contribution of solar becomes very small, although things have changed now [14].

Recently, there are many articles and news about connecting and building large solar power plants that should directly connect to the significantly higher voltage. But it is true that it would exist, but at the same time there would be a lot of solar penetration at the end distribution system. When a lot of houses and small industries have solar rooftops, they all may feed power into the grid. This specific feature of solar brings additional challenges [15, 16].

3 Different Modes of PV Integration at Low or Feeder Levels and Their Issues

3.1 Voltage Profile in Feeder

Transformer feeding to some load L_1, L_2, and some generation at another end with solar photovoltaic is considered. Solar photovoltaic feeding several loads is shown in Fig. 4.

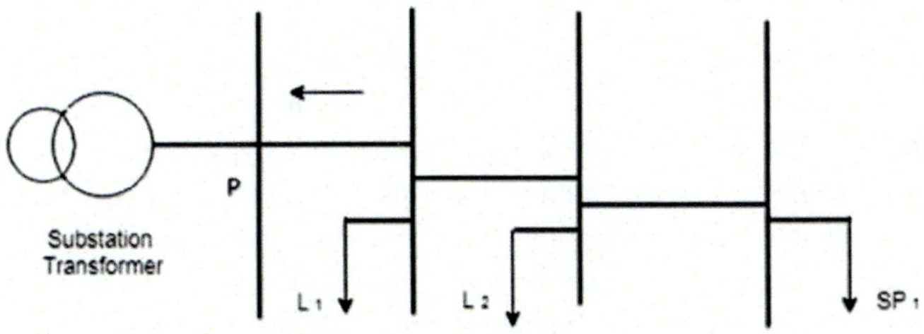

Fig. 4 Solar photovoltaic feeding several loads

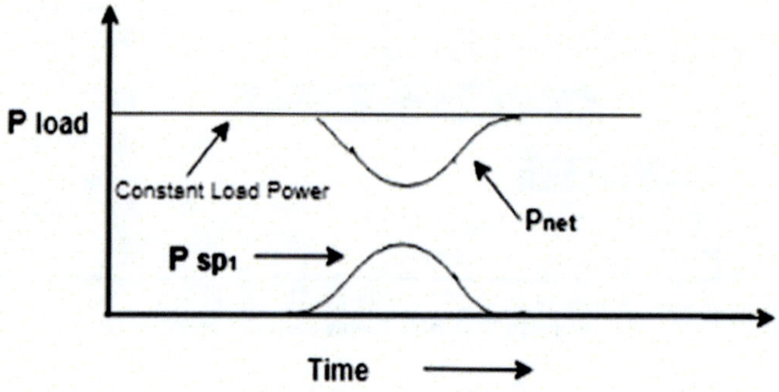

Fig. 5 Ideal power vs time curve

At substation level, where the power is monitored at that position, a power vs time constant graph which is most probably dependent on weather season is obtained. Mostly, it is constant and may have some small variation.

The power which is generated by solar PV Psp1 will rise during the day, and it would become zero in the evening and in the early morning. The net power graph shown in Fig. 5 is the difference between load power and power developed by solar photovoltaic. If the load profile would have been constant, then there would have been a drop in the demand for power during the day, because in day time, the solar generates the most.

The actual graph is a little different from Fig. 6 which is not flat and hence gives an indicative graph. The graph line corresponding to 0% is the load requirement of a particular feeder. The feeder's load requirement varies during the day. In the evening, it goes up, since residential and commercial places' consumption of electricity and power is at peak. The graph keeps changing when solar is added to this particular feeder. Initially, 10% of solar is added and the graph changes. When 20% solar is added, then the curve deepens further, and a green curve as shown in Fig. 6 is obtained. This graph looks like a duck and hence called a duck curve. Eventually, it tells that the majority of power or peak demand occurs in the evening. Unfortunately, solar does not supply power in the evening, but brings down the power requirement during the day. Finally, when more solar is added, the power starts becoming negative during the day. It gives the instant time at which solar produces power in this feeder. The load in this feeder is much less. Therefore, the total power becomes negative. And total power starts feeding back to the grid. At the same time, the peak requirement of the power of the feeder which comes in the evening remains more or less the same. Although solar helps in terms of producing energy, it is not of much help to that particular feeder, since that feeder needs power in the evening. Because of this negative flow of power during the day, e.g., 11 am to 3 pm, if power flows back into the feeder, the voltage profile of the bus may be affected.

To understand the impact, the system as shown in Fig. 7 which has some feeders which are connected to the solar photovoltaic system is considered. Now in the conventional feeder, there is no generation. In that case, power flows from left to

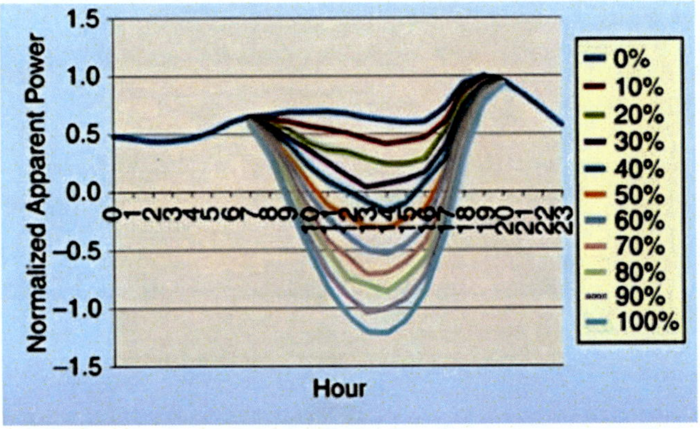

Fig. 6 Duck curve

right. Then as moving further, the voltage reduces, where each load draws some amount of active (*P*) and reactive (*Q*) power. Due to this, V_1, V_2, V_3, V_4, and so on keep reducing as each load drop gets subtracted. Therefore, conventionally capacitors or capacitor banks are added in between the feeder, so that the voltage level is pulled back.

In Fig. 8, the distance from the substation is plotted on the x-axis and voltage at the y-axis. The substation voltage is well maintained which is 1 pu. This is done by an on-load tap changer transformer at the substation. After that, moving along, the blue line can be seen. The voltage continuously reduces when moving further and further away in the feeder. The blue line represents a system without a capacitor. In case of addition of capacitors at specific locations, the voltage profile maintains almost constant. But now if PV is also present and a scenario where net power is negative is considered, that means the feeder exports power. In that case, if the solar photovoltaic is distributed into multiple buses in the feeder, then the total power goes out; therefore, the voltage at the other end of the feeder starts going up, as compared to the substation. This has now become very specific to the generation happening within the distribution system. The voltage profile within the feeder has changed. The overall voltage has also changed; moreover, the profile has changed as well. So,

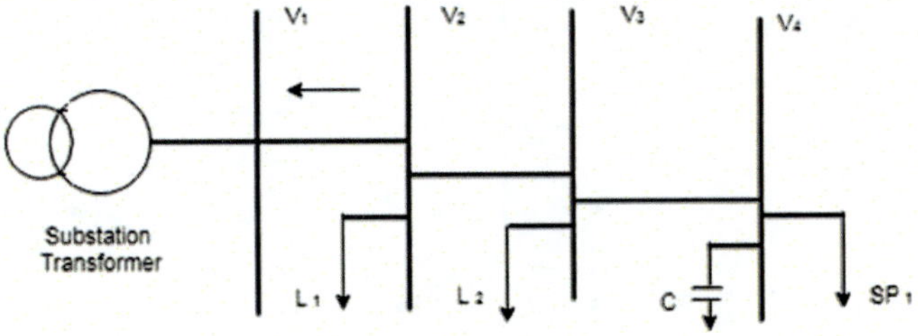

Fig. 7 Solar photovoltaic feeding several loads after application of capacitor

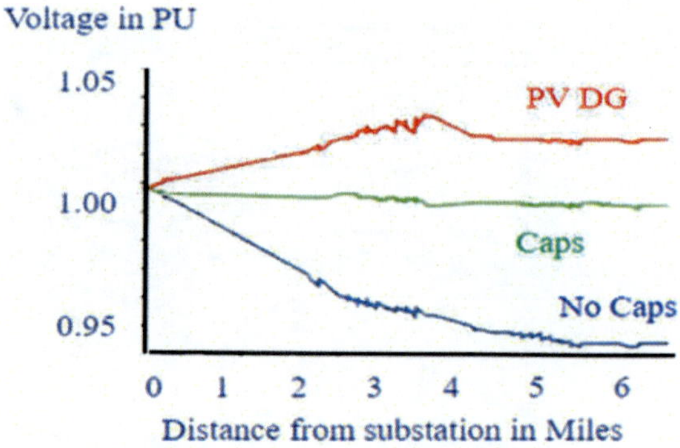

Fig. 8 Feeder voltage distribution

earlier where the capacitor was used to solve the problem, the capacitor bank may not be of help in this kind of situation. That capacitor probably would be deactivated. Still, the problem of voltage, i.e., increasing when moved along the feeder, might remain. So, the solar photovoltaic inverter operates with the basic principles of unity power factor operation.

3.2 *Loading Different Sections of the Feeder*

When the same system where few loads are connected along photovoltaic at the endpoint is considered, then conventionally if there is no solar photovoltaic, then power will flow from left to right. As moving from left to right, the power handled by that particular section is reduced as shown in Fig. 9. This is because, the first section S_1 has to supply power to load L_1, L_2, and L_3, whereas section S_2 has to supply power to load L_2 and L_3, similarly, section S_3 has to supply L_3.

As moving further, the load handling reduces. But on the contrary, if solar photovoltaic is present in the same case, a duck curve is obtained and total power is negative. In that case, solar PV produces significant power, and small part of it gets

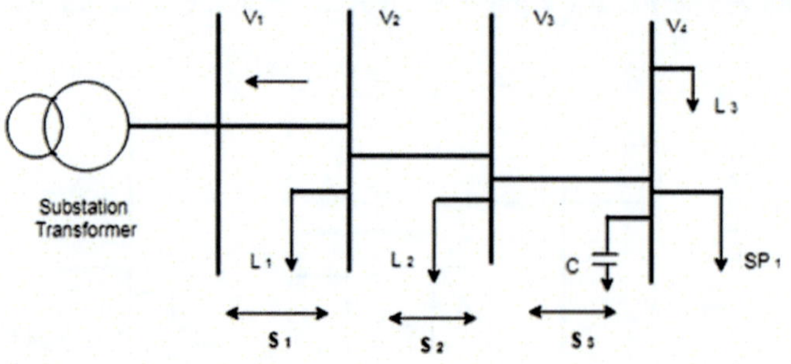

Fig. 9 Interconnected solar PVs along with loads

utilized in load L_3 and L_2, some of it gets utilized by load L_1 and the remaining goes back to the grid. So section S_3 handles the most amount of current, because the power flowing through the section has become maximum. It is in the reverse direction, but the magnitude is maximum. Further, it gets reduced in S_2 and S_1. Initially, the feeder is designed to feed only the loads, in that case, it may not necessarily be suitable if solar photovoltaic is connected at the end of the feeder. So this kind of problem may require reinforcement of the feeder section or reconductoring of the different feeder sections.

3.3 Connection of Solar Photovoltaics to the System

In both the previous problems, the kind of curve where solar power during the day was maximum, but the generation during the day did not necessarily ramp up and down smoothly, was considered. However, a situation where a day is cloudy such as a monsoon season is considered now. Then there is going to be a sudden change in radiation, which would cause a sudden change in power produced by the solar photovoltaic grid-feeding systems.

The graph of power production by one of the solar plants is depicted in Fig. 10, where there are lots of ups and downs throughout the day of solar radiation which is expected. Because of these fluctuations, the power generation and the voltage profile in the distribution system may also fluctuate, which would certainly depend on the value of impedances of the network system and the capacity of solar power. The y-axis in the first graph is kilowatts. To understand the impact of substantial power, this data is used to simulate the system. The second graph shows the simulated system's voltage profile. Here the voltage of the feeder is found to be continuously changing. In this case, a fast-acting on-load tap changer is present. The taps would also change rapidly while they maintain the voltage within the band.

An example of a solar photovoltaic system with a load tap changer as shown in Fig. 11 is considered. The multiple tapings on the primary take feedback on bus voltage. Then the controller action is taken, and the tappings of the transformer are changed accordingly. There is impedance in the upstream as well, and the feeder is connected to solar photovoltaic and Load L_1.

Now, when peak power is fluctuating, the drop also fluctuates, and the upstream also fluctuates. Therefore, voltage at V_1 fluctuates. Now this feeder will try to estimate the voltage in nearby bus V_2. Using the impedance in power flow, V_1 is estimated and accordingly the taps are controlled so that V_1 remains within the profile. If OLTC is allowed to operate very fast, it would not be allowed to. In that case, a lot of switching would be visible in this graph. If it is allowed to operate, then the voltage remains within the profile but again the life of OLTC and transformer reduces significantly. And if it is not allowed to operate fast, then the voltage goes out of the normal operating bands which is also not acceptable. So again this specific problem is because of solar. This also occurs potentially because of load fluctuation.

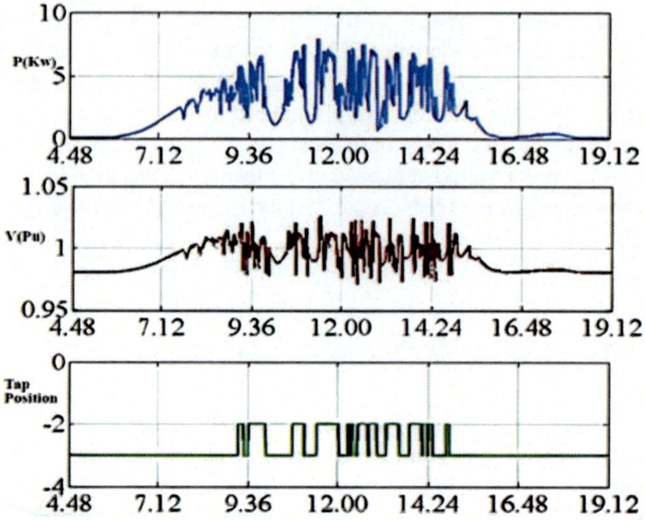

Fig. 10 Intermittent solar energy

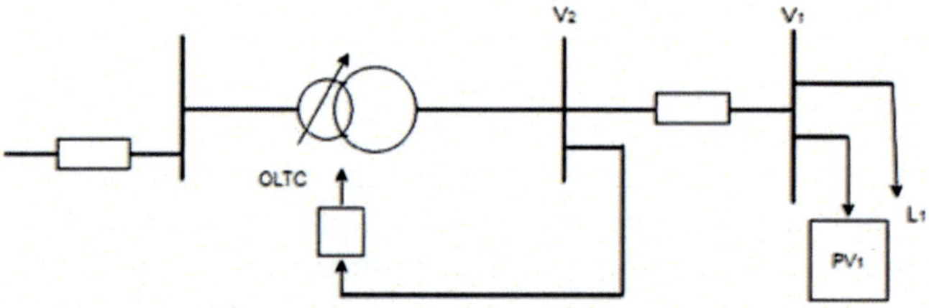

Fig. 11 Solar photovoltaic system with on-load tap changer

The feeder as shown in Fig. 12 where many houses are connected to it is considered. There is some amount of randomization, which is inherently present in terms of loads; therefore, the system would not see the sudden repeated change in power because of the loads. There may be small perturbation occurring with time, but not the major perturbation where all these feeder loads go down, and sometime later again comes over, and again goes down, and so on.

Nowadays, each house has some amount of solar added to the grid. All houses have solar photovoltaics on the rooftop. The bigger cloud while traveling, would likely cover a few houses in one go. Therefore, when cloud comes up and covers about 50% of rooftop solar-installed houses, then suddenly the power being produced by the photovoltaic panel will get reduced. That would create a much bigger perturbation in the power when compared to the perturbation created by the switching of loads, because it is more synchronized. Again when this cloud cover goes away, suddenly all panels start getting solar radiation and start pumping power. Therefore, this power

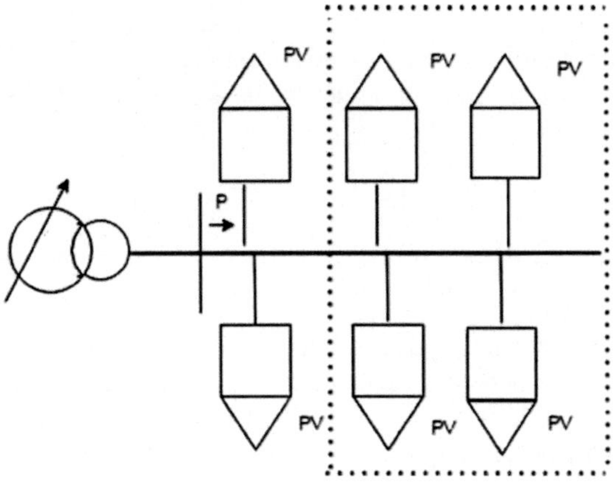

Fig. 12 Feeder supplying a number of houses with rooftop solar PV

produced by the panel within the distribution system has more synchronization when compared to the loads which are distributed in a feeder.

3.4 Unbalance

The fourth problem that might come from solar photovoltaics in the distribution system in the feeder is unbalance. It is very similar to the third problem because there are houses as shown in Fig. 12, and the feeder is distributed among different houses. It is considered that a three-phase feeder gets each of the single phase, then goes to a few houses, and so on. If the cloud covers during transient, some of the houses may get covered, which might cause a bigger unbalance in the system because that particular feeder gets lesser power when compared to other phases. It does depend on a lot of factors like how the houses are distributed on to different feeders and so on, but this is a major problem, especially in a system where solar photovoltaic is connected to a single-phase line and not to a three-phase line where such a problem may not occur.

3.5 Harmonics

The fifth one is harmonics. If there is a power electronic converter connected to the grid, it would create an issue in terms of harmonics. The inverter should inject current and have a very low harmonic distortion value. But this may be found as a smaller problem and more of an advantage, because these inverters, do not just themselves, feed nice sinusoidal current, but if there is a neighboring load that draws harmonics, these inverters can help the neighboring loads in correcting the overall total distortion in the system. As of now, it is not popular yet, since a large utility is not accepting it as the solution. But as moved from UPF to the voltage supports, it is quite possible that these inverters would also be allowed to perform harmonic compensation in the system.

4 Conclusion

From the above discussion, it can be concluded that the grid-connected solar integration at the distribution level has many issues related to the voltage profile in the feeder, loading of a section of the feeder, fluctuations in the feeder, unbalance, and harmonics. It is also observed that when the percentage of solar photovoltaics gets added to the system, the power flows back into the feeder and forms a duck curve, which may consequently lead to voltage collapse of the system. Among all these issues, the first three issues, i.e., voltage profile in feeder, loading different sections of the feeder, and the connection of solar photovoltaics to the system, are more severe concerning issues in the distribution system. The fourth issue, i.e., unbalance, may or may not exist depending upon the situation, architecture, inverter, single phase, three phase, and so on. And the last issue is harmonics, which is not actually a problem but it is more of an opportunity. Also, these problems can be overcome by reinforcement of the feeder section or the implementation of FACTs devices into the grid-integrated system.

References

1. You S, Jiang Z, Li H, Bennett M, Zhang (2022) In: 7th IEEE Workshop on the Electronic Grid, pp. 1–5. Auckland, New Zealand
2. Jithin K, Mayadevi N, Hari Kumar R, Mini PV (2021) The effect of multiple PV and battery penetration on stability of DC microgrid with single bus topology. In: 9th IEEE International Conference on Power Systems (ICPS), pp. 1–5, Kharagpur, India
3. Mamun AA, Ahmed KIU, Chowdhury SA (2022) Detection of power quality disturbances in distribution grid network with photovoltaic penetration. In: Advancement in Electrical and Electronic Engineering (ICAEEE), pp. 1–6. Gazipur, Bangladesh
4. Drishya RK, Jayaprakash P (2021) Impact of solar photovoltaic penetration on voltage stability of power network. In: International Conference on Communication, Control and Information Sciences (ICCISc), pp 1–6. Idukki, India
5. Joshı N, Nema P (2018) Analysis of stability of wind-based distributed energy generation system. Inter J Energy Smart Grid 3:1–16
6. Cipcigan LM, Taylor PC (2007) Investigation of the reverse power flow requirements of high penetrations of small-scale embedded generation. IET Renew Power Gene, 160–166
7. Khandelwal A, Neema P (2019) The state of art for power quality issues in PV grid-connected system. In: International Conference on Nascent Technologies in Engineering (ICNTE), pp. 1–4. Navi Mumbai, India
8. Khandelwal A, Nema P (2021) A 150 kW grid-connected roof top solar energy system—case study. In: Baredar PV, Tangellapalli S, Solanki CS (eds) Advances in clean energy technologies. Springer Proceedings in Energy. Springer, Singapore

9. Naqvi SBQ, Kumar S, Singh B (2021) Weak grid integration of a single-stage solar energy conversion system with power quality improvement features under varied operating conditions. IEEE Trans Indus Appl 57(2):1303–1313
10. Khandelwal A, Nema P (2020) Harmonic analysis of a grid-connected rooftop solar energy system. In: Fourth International Conference on I-SMAC (IoT in Social, Mobile, Analytics, and Cloud) (I-SMAC), pp 1093–1096
11. Xie B, Zhou L, Liu T et al (2020) Harmonic resonance analysis and stability improvement for grid-connected inverters. J Power Electron 20:221–235
12. Zhang Q, Qian J, Zhai Z et al (2022) Control stability of inverters with series-compensated transmission lines: analysis and improvement. J Power Electron 22:1746–1757
13. Gadal R, Elmariami F, Oukennou A, Agouzoul N, Tarraq A (2022) An assessment of line voltage stability indices to select the best combination for voltage stability prediction. In: Bendaoud M, Wolfgang B, El Fathi A (eds) The Proceedings of the International Conference on Electrical Systems & Automation. ICESA 2021. Springer, Singapore
14. Mohammadi F, Neagoe M (2020) Emerging issues and challenges with the integration of solar power plants into power systems. In: Visa I, Duta A (eds) Solar energy conversion in communities. Springer Proceedings in Energy. Springer, Cham
15. Hung DQ, Shah MR, Mithulananthan N (2016) Technical challenges, security and risk in grid integration of renewable energy. In: Jayaweera D (eds) Smart power systems and renewable energy system integration. Studies in Systems, Decision, and Control, vol 57. Springer, Cham
16. Mlilo N, Brown J, Ahfock T (2021) Impact of intermittent renewable energy generation penetration on the power system networks—a review. Technol Econ Smart Grids Sustain Energy 6:25

Efficient Sentiment Classification Model of Tweets Using an Adaptive Megaptera Whale Optimization LSTM Classifier

Priya Vinod and S. Sheeja

Abstract User sentiment analysis from the online social media has become an interest, and importantly, Twitter opens a valuable source for user emotions, opinions, attitudes, confessions, and so on. However, analyzing the massive reviews and posts dumped in the Twitter media is a tedious and computationally ineffective scenario, raising the demand for implementing the automatic sentiment analysis models. Accordingly, Megaptera whale optimization-based adaptive long short-term memory (MWO-based adaptive LSTM) classifier is proposed for sentiment analysis, where the significance of the research relies in the proposed adaptive Megaptera whale hunt optimization that optimizes the classifier parameters toward the effective classification performance. The analysis of the model based on the accuracy reveals that the proposed model acquires the maximal performance with 98.48% of accuracy.

Keywords Megaptera whale optimization · Long short-term memory · Whale hunt mechanism · Tweets · Sentiment analysis

1 Introduction

Social media platform offers an amazing foundation for huge data analytics in many real-world applications. When individuals share their thoughts or opinions while interacting with one another on different social media sites like Twitter, Facebook, Myspace, etc., continuously, a large volume of data is produced [1]. For instance, after watching a movie, individuals will immediately submit their reviews online and begin a series of discussions regarding the acting displayed in the film. This kind of information is the standard by which people assess, judge the worth of any film as well as the quality of other goods, and decide whether it would have been a commercial success [2]. Social media platforms must integrate into daily life

P. Vinod (✉) · S. Sheeja
Computer Science, Karpagam Academy of Higher Education, Srishti, Thottathil House, West Yakkara, Palakkad 678001, Kerala, India
e-mail: piavinod2189@gmail.com

G. Rajakumar et al. (eds.), *Intelligent Communication Technologies and Virtual Mobile Networks*, Lecture Notes on Data Engineering and Communications Technologies 171,
https://doi.org/10.1007/978-981-99-1767-9_39

due to the rapid advancement of web-based technology [3]. Text-based documents have grown rapidly in popularity on social media in recent years. More than 500 million tweets were sent daily by 326 million active Twitter users [4]. Short text messages called tweets can be created, posted, updated, and viewed on Twitter, a rapidly expanding Internet platform. As journalists collect information from Twitter and rewet user input during significant events, the Twitter technology may even indirectly affect the agenda shaping of the traditional media [4, 5].

Sentiment analysis is the computer study of people's opinions, feelings, emotions, and attitudes toward things like products, services, topics, events, subjects, and their features [6]. Sentiment classification is a method of classifying and determining the polarities of texts [7] with the objective of determining whether a specific document has a negative or a positive value in accordance with a specified classification [4, 8]. One of the most effective methods for determining the sentiment of the public is sentiment categorization based on machine learning. It performs admirably and has good precision [9, 10]. However, the sentiment classification technique has certain challenges due to the noisy data and the feature space's complexity. To solve the issues, a feature selection is therefore required [3, 10]. The feature selection process can choose crucial features, eliminate superfluous features, reduce feature dimensions, and speed up calculation [11–13]. Four separate domains can be used to categorize the classification task that goes into sentiment analysis: subjectivity categorization, opinion extraction, and document and word sentiment classification [1]. Twitter sentiment classification follows up on machine learning methods that use polarity labels of manually annotated sentiments to create classifiers from tweets [14]. By classifying Twitter sentiment, it is possible to determine if tweets are mostly negative or positive [3].

The research's objective is to recognize and categorize the sentiment using an adaptive LSTM classifier. The Twitter sentiment analysis database is where the data is gathered. This work focuses on the requirement for choosing the features and creating a useful model, and as a result, a proposed Megaptera whale optimization-based deep LSTM is developed, adding value to this research. These collected data are preprocessed by using the steps such as tokenization, word to vector embedding, stop word removal, stemming, and lemmatization. After that the essential information is exposed to feature extraction using the TF-IDF. The extracted features are forwarded LSTM classifier for the classification of sentiment polarity.

The research's primary contribution is as follows:

- Megaptera whale optimization: The proposed adaptive Megaptera whale hunt optimization ensures the adaptive parameter selection in optimization with intelligent whale hunt optimization, which boosts the global optimal convergence of the algorithm toward selecting the classifier hyper parameters.
- MWO-based adaptive LSTM classifier: The suggested adaptive LSTM classifier maximizes training effectiveness and performance by memorizing the relevant information that is crucial and identifying patterns.

The research is categorized as follows: The necessity of the sentiment analysis is discussed in Sect. 2. The suggested approach for sentiment categorization using

LSTM based on Megaptera whale optimization is shown in Sect. 3. Results are presented and discussed in Sect. 4, and the conclusion part is discussed in Sect. 5.

2 Motivation

Numerous machine learning algorithms are used for sentiment analysis. The pros and cons of the current models are evaluated in this part to highlight the necessity of developing a new one. The following topics are covered in relation to the study's challenges and the literature review.

2.1 *Literature Review*

The existing works undertook in the classification of sentiment analysis are enumerated as follows.

A technique for performing hurricane-specific sentiment analysis on tweets was developed by Yao and Wang [15]. DSSA-H extracted tweets about storms and used domain-adversarial neural network for determining their sentiment with the aid of the trained supervised learning classifier random forest (RF) (DANN). In a certain data-limited domain, this technique helped in reducing on sentiment classification time and labor requirements. The disadvantage of the method produced overlapping while gathering data from tweets. Chugh et al. [3] classified a sentiment using deep recurrent neural network trained by spider monkey crow optimization algorithm (SMCA). This method used stemming and telecom evaluation that removed stop words and inappropriate content which helped in shorten user search times. Although the strategy dramatically increased recall and precision, it was only applicable to a small number of attributes. Tam et al. [16] devised a sentiment categorization model by combining the convolutional neural network structure with Bi-LSTM models. This model examined both local and global relationships in contextual phrase usage, which effectively helped the model in text sentiment classification. The complexity and calculation time are decreased by this model. This network was impacted by overfitting problems, which can occur when models perform exceptionally well on training data but fail to generalize to new data. A multi-layer convolutional neural network was initiated by Zhang et al. [17] to represent the context several times while performing parallel context analysis. Through the application of an attention approach, this network was able to directly learn the interactive model of emotion. This method avoided the overfitting problem. The drawback of the method was computational complexity. Akhtar et al. [18] developed a technique for multitask learning that identified and categorized the terms in a single model. For both languages, the multitask approach produces competitive performance with less complexity. Through the use of shared representations, this technique increased data efficiency and decreased overfitting. The method's issue was its lack of accuracy.

2.2 Challenges

This research faced certain challenges, and it is described as follows.

- Social media offers data sources for emergencies and disasters. It is still difficult to use these resources to get relevant data for making decisions [15].
- One of the largest obstacles is end-user authentication, where there is a chance that noise will be incorporated into the data collected. Social media data consistency is a significant barrier as well [1].
- Sentiment analysis at the sentence level using classifiers is performed using a feature selection approach and the voting model. The labeled feature sets cannot transform across the vectors, although this strategy enhanced accuracy with high entropies [3].

3 Sentiment Classification Using the Proposed Adaptive Megaptera Whale Optimization Using LSTM Classifier

The objective of the research is to use an adaptive LSTM classifier to recognize and categorize the feelings. Data is initially gathered from the Twitter sentiment analysis database, and it is then preprocessed to make it acceptable for classification of emotions using the Megaptera whale hunt optimization-based LSTM classifier. The preprocessing of data is carried using several steps such as tokenization, word to vector embedding, stop word removal, stemming, and lemmatization. Tokenization is the process of converting the raw data into small chunks, which consist of the words or sentences. Tokenization also aids in understanding the text's meaning. From this text, collected words with similar meaning are grouped under a vector, and this process takes place through word to vector embedding. After embedding, the stop words removal is performed where the low-level information is eliminated in order to focus on the high-level information along with that the repute words also get eliminated. After this, the stem of the word is derived by stemming, and context of the word is extracted using lemmatization. After preprocessing, the key data is extracted using the term frequency-inverse document frequency (TF-IDF), and the categorization of the sentiment as positive, negative, or neutral is done using a LSTM classifier. The systematic representation of sentiment classification model is illustrated in Fig. 1.

3.1 Input Data for Sentiment Analysis

The input data is gathered from the sentiment analysis of Twitter [19], which contains 140 sentiment datasets with 1.6 million tweets. The input is mathematically represented by

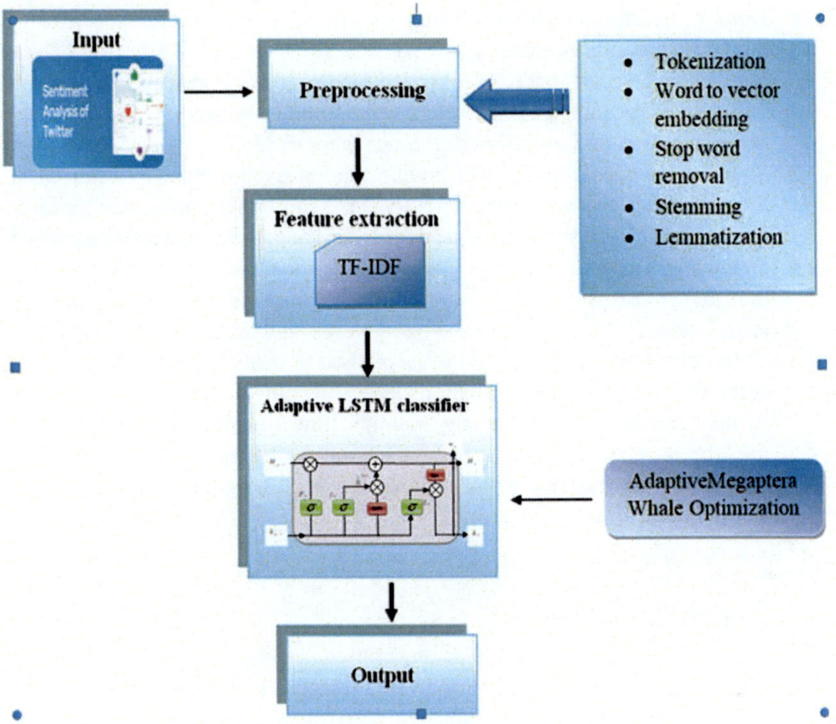

Fig. 1 Systematic representation of sentiment classification using LSTM

$$G = \{G_1, G_2, \ldots G_n\}, \tag{1}$$

where G denotes the dataset and the $\{1, 2, 3, \ldots n\}$ denotes the number tweets present in the data.

3.2 *Preprocessing*

Many opinions may be expressed in a tweet in various ways by various users. The tweets were improperly organized, written in quasi-language, and the data is prone to duplication and inconsistency. The main purpose of preprocessing is to increase the data quality and remove noise from the data. There are several steps involved in preprocessing the data such as tokenization, word to vector embedding, stop word removal, stemming, and lemmatization.

3.2.1 *Tokenization*: It is defined as the process of dividing a stream of textual data into tokens, which can be words, terms, sentences, symbols, or other significant components.

3.2.2 *Word to vector embedding*: Words are mapped into vector space using language model's process. Each word is represented as a vector of real numbers in a vector space. Additionally, it allows the representation of words with related meanings. As a result, training time is reduced, word vectors are better encoded, and overall performance is increased.

3.2.3 *Stop word removal*: A word with no information is represented by a stop word. Stop words are removed using extensive texts through the process of removal. To save up storage space and speed up processing, the stop word is removed.

3.2.4 *Stemming*: The main task of the stemming process is to break down words into their most basic forms using morphology stemming. For instance, the term "advice" is formed from the basic words "advising" and "advised".

3.2.5 *Lemmatization*: Lemmatization is the method of collecting words that share a common root, or lemma, but have different meaning derivatives or inflections so they can be studied as a single entity. Inflectional suffixes and prefixes must be dropped in order to expose the word's dictionary form. Reducing morphological variation or inflectional forms is the goal of both stemming and lemmatization.

3.3 Feature Extraction

The method used for feature extraction in this research is TF-IDF, which is more accurate. Tweets are effectively interpreted by deep learning when texts are converted into vectors using TF-IDF. A word's usage frequency in a tweet is determined by its term frequency, or TF. IDF measures word significance in the meantime. Multiple occurrence words are not accepted. The mathematical formula of TF-IDF is expressed as

$$\mathrm{TF} - \mathrm{IDF} = \mathrm{he}_h \times \mathrm{ise}_h, \tag{2}$$

where the term frequency is denoted as he_h and the inverse document frequency is denoted as ise_h.

3.4 Proposed Adaptive LSTM Classifier for Sentiment Classification

An LSTM is a special type of recurrent neural networks (RNNs). The LSTM mechanism essentially employs a variety of gates to control the flow of sequential data. Like a regular RNN, an LSTM network controls the propagation of activations throughout the network, allowing it to learn when to eliminate a current input, recall previous hidden state, or emit a nonzero input. Because these networks are effective in storing information for both long and short periods of time, they are known as LSTM. The LSTM design is shown in Fig. 2 mathematically.

The LSTM has three gates such as forget, input, and the output gate. The output gate determines how much the current node impacts the external network, while the input gate decides how much relevant details from the current step should be added. The forget gate also controls how much of the previous state can pass through. The mathematical relationship between the LSTM's input, forget, and output gates is determined by the following formulae.

$$j^{(s)} = \sigma_a\left(X^{(j)}y_s + V^{(j)}k_{s-1} + o^{(j)}\right) \tag{3}$$

$$p^{(s)} = \sigma_a\left(X^{(p)}y_s + V^{(p)}k_{s-1} + o^{(p)}\right) \tag{4}$$

$$g^{(s)} = \sigma_a\left(X^{(g)}y_s + V^{(g)}k_{s-1} + o^{(g)}\right) \tag{5}$$

The values of the input, forget, and output gates are determined using Eqs. (3), (4), and (5), respectively.

$$\hat{h}^{(s)} = \tanh\left(X^{\left(\hat{h}\right)}\right)y_s + V^{\left(\hat{h}\right)}k_{s-1} + o^{\left(\hat{h}\right)} \tag{6}$$

$$h^{(s)} = p^{(s)} * h^{(s-1)} + j^{(s)} * \hat{h}^{(s)} \tag{7}$$

$$k^{(s)} = g^{(s)} * \tanh\left(h^{(s)}\right) \tag{8}$$

Equation (6) is used to determine the proposed state value $\hat{h}^{(s)}$, which is replaced to determine the present state $h^{(s)}$ in (7). The hidden state $k^{(s)}$ is then calculated

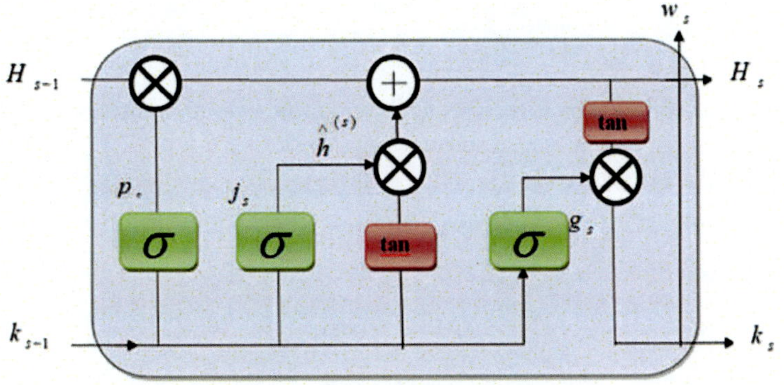

Fig. 2 Architecture of Megaptera whale optimization-based LSTM

using (8). The symbol g denotes an element-wise product, where each of the three gates and the suggested state have a unique bias vector o and weights matrices X and V that are learned through training. Normally, the Adam optimizer is employed for tuning the performance of the classifier, and for improving the convergence issues, the adaptive Megaptera whale optimization is proposed to tune the parameters of weight and bias, which is detailed in below section.

3.4.1 Adaptive Megaptera Whale Optimization

A metaheuristic optimization approach is Megaptera whale optimization. The unusual hunting methods of Megaptera whales are their most alluring feature. The term bubble net foraging refers to this method of foraging. Megaptera whales like to hunt krill or small fish when they are near the ocean. It is significant that Megaptera whales are the only species known to prey on bubble nets. The three stages of the whale optimization algorithm are encircling predation, bubble attacking, and prey seeking.

Mathematical model for the proposed Megaptera whale hunt optimization: In MWO algorithm, M stands for whale population size, r represents the search space dimension, and the position (Z) of the jth whale in the dimensional space is denoted as

$$Z_j = \left(Z_j^1, Z_j^2, ...Z_j^R\right), j = 1, 2, ...M. \tag{9}$$

i. *Encircling predation*

At the beginning of the algorithm, Megaptera whales can detect and encircle their prey. The ideal option is considered to be where the prey is located in the current population. Since there is no precedence in the global optimal position, the ideal individual is surrounded by other Megaptera whales in the community. Mathematically, the location of the Megaptera whale is represented as,

$$Z(u+1) = Z_q(u) - B \times \left|Q \times Z_q(u) - Z(u)\right| \tag{10}$$

where the current iteration is represented by u, the local optimal solution is $Z_q(u) = \left(Z_p^1, Z_p^2, \ldots Z_p^R\right)$, and the step of surroundings is given by $B \times \left|Q \times Z_q(u) - Z(u)\right|$. The B and Q are expressed as

$$B = 2b \times \text{rand}_1 - b \tag{11}$$

$$C = 2 \times \text{rand}_2 \tag{12}$$

where the random number is represented by rand_1 and rand_2 lies in the range between 0 and 1. As the number of iterations increases, b represents the convergence factor, which decreases linearly from 2 to 0. The expression is

$$b = 2 - 2u/u_{\max}, \tag{13}$$

where the highest number of iterations is represented by $u_{\max}$.

Bubble attacking

Bubbles are used in the process of Megaptera whale predation to attack in order to accomplish the goal of local Megaptera whale optimization. Spiral renewal position and shrinking encirclement resemble the behavior of whale predation bubbles.

(a) *Mechanism of contraction encircling*

According to Eq. (10), whale population is decreased to and surround. When $|B| < 1$, the individual Megaptera whale reaches close to the individual Megaptera whale present in the optimal position, the greater the value of $|B|$, or smaller the value, the slower the whale individual swims.

(b) *Position update in spiral*

Individual whales measure their distance from their current prey before beginning a spiral search for it. The mathematical model for spiral walk mode is interpreted as follows:

$$Z(u+1) = R' \times e^{cd} \times \cos(2\pi c) + Z_q(u), \tag{14}$$

where the distance between the j whale and its prey is represented by $R' = |Z_q(u) - Z(u)|$, to define the shape of the logarithmic spiral is represented by d which is a constant, and the random number -1 and 1 is designated by c. In the optimization process, there is an equal chance of choosing the spiral position update and the shrinkage encirclement mechanism, which is 0.5.

Hunting stage

Megaptera whales can randomly search for food. Individual Megaptera whales conduct random searches based on where they are in relation to one another. The random search made by the Megaptera whale is represented as

$$Z(u+1) = Z_{\text{rand}}(u) - B|Q \times Z_{\text{rand}}(u) - Z(u)|, \tag{15}$$

where $Z_{\text{rand}}(u)$ represents the position of a randomly chosen whale in the current population. The pseudocode of the proposed optimization is given in Table 1.

Table 1 Pseudocode of the proposed MWO optimization

S. no.	Pseudocode for the proposed MWO optimization
1	Input: Z_j
2	Output: $Z(u+1)$
3	Initialize: Z_j
4	**Encircling predation**
5	Determine the location: $Z(u+1)$
6	**Bubble attack**
7	Determine the contraction encircle:
8	If $\|B\| < 1$
9	Close to the whale individual
10	else
11	Slower the whale individual
12	end
13	Determine the position update in spiral
14	**Hunting stage**
15	Randomly search the food

4 Results and Discussion

The sections below provide an interpretation of the sentiment classification findings obtained using the MWO-based adaptive LSTM.

4.1 Experimental Setup

The experiment is conducted utilizing the Twitter sentiment analysis database and the Python programming language in a Windows 10 computer with 8 GB of RAM [19].

4.1.1 Experimental Results

Illustration of sentiment analysis results is shown in Fig. 3.

STRING :waiting in line at safeway.

PREPROCESSED STRING :wait line safeway

TOKENIZED STRING :[13, 4, 12, 1, 22, 5, 9, 5, 3, 13, 10, 2, 5, 13]

SENTIMENT :positive

STRING :saw night at the museum 2 last night.. pretty crazy movie.. but the cast was awesome so it was well worth it. Robin Williams forever!

PREPROCESSED STRING :saw night museum last night pretticrazimovi cast awesom well worth robin williamforev

TOKENIZED STRING :[16, 1, 3, 2, 14, 3, 21, 5, 3, 5, 23, 12, 3, 5, 7, 1, 21, 1, 1, 8, 8, 5, 18, 1, 16, 2]

SENTIMENT :Positive

Fig. 3 Illustration of sentiment analysis results

4.2 Dataset Description

The 140 datasets from the Twitter sentiment analysis database were used in this study's dataset. This data was obtained using the Twitter API, which has 1,600,000 tweets and has enabled researchers to learn more about sentiment analysis.

4.3 Parameter Metrics

The parameters used for proving the supremacy of the research are accuracy, sensitivity, and specificity.

Accuracy: The percentage of samples successfully detected by the proposed MWO-based adaptive LSTM in identifying the emotion polarity is known as accuracy and is represented as

$$J_{\text{acc}} = \frac{J_{\text{TP}} + J_{\text{TN}}}{J_{\text{TP}} + J_{\text{FP}} + J_{\text{TN}} + J_{\text{FN}}}. \tag{16}$$

Sensitivity: According to the proposed MWO-based adaptive LSTM, sensitivity is the probability that a test result is actually positive and is denoted by

$$J_{\text{Sen}} = \frac{J_{\text{TP}}}{J_{\text{TP}} + J_{\text{FN}}} \tag{17}$$

Specificity: According to the proposed MWO-based adaptive LSTM, sensitivity is the probability that a test result is actually negative and is denoted by

$$J_{\text{Sen}} = \frac{J_{\text{TN}}}{J_{\text{TN}} + J_{\text{FP}}}. \tag{18}$$

4.4 Performance Analysis—Training Percentage

Figure 4a–c illustrate the observation and performance analysis in terms of accuracy, sensitivity, and specificity. The proposed MWO-based adaptive LSTM classifier first measures the accuracy rate for the different epochs 5, 10, 15, 20, and 25, and during the training period of 90, it achieves values of 92.33%, 95.84%, 97.01%, 97.79%, and 99.74%. Similar to this, for 90% of the training data, the proposed MWO-based adaptive LSTM classifier's sensitivity rate is measured, and the values are interpreted as 92.84%, 93.32%, 98.64%, 99.73%, and 99.74%. The specificity values for the various populations are then calculated using 90% training data, and the results are given as 91.84%, 95.50%, 96.02%, 98.63%, and 99.74%. The observation shows that the proposed MWO approach works effectively in a variety of epochs.

4.5 Comparative Analysis for the Proposed MWO-Based Adaptive LSTM

The comparative analysis is done to demonstrate the value of the suggested method with different portions of previous literature.

4.6 Comparative Method

The proposed method of the MWO-based adaptive LSTM is compared with artificial neural network (ANN), deep convolutional neural network (CNN), simple recurrent neural network (RNN), and LSTM.

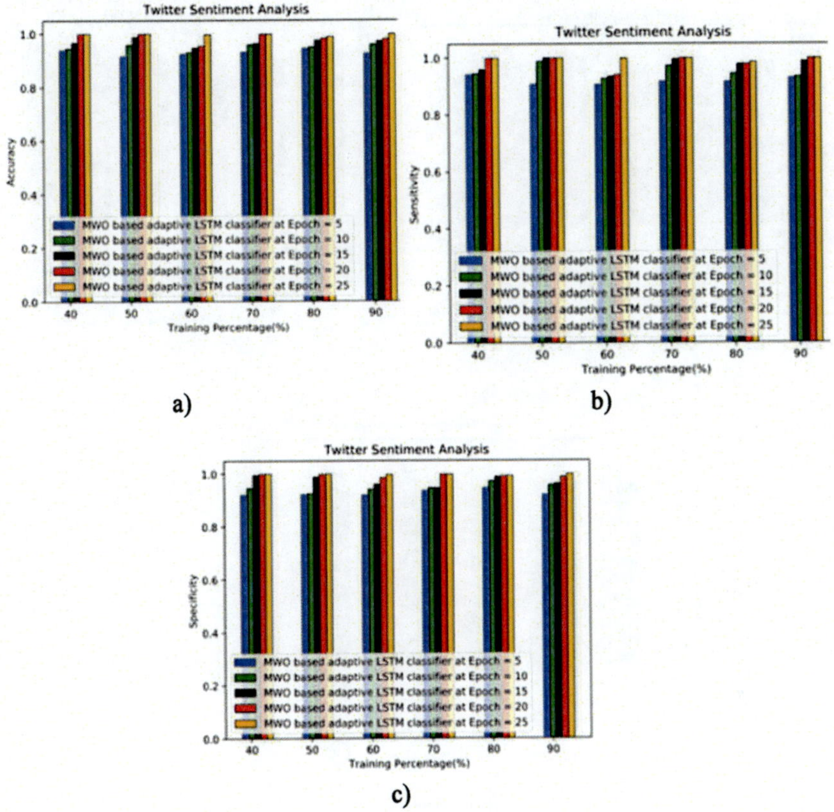

Fig. 4 Performance analysis—training percentage in terms of **a** accuracy, **b** sensitivity, and **c** specificity

4.6.1 Comparative Analysis—Training Percentage

The proposed MWO-based adaptive LSTM classifier is compared against other classifiers using ANN, deep CNN, simple RNN, and LSTM to determine how much improvement was made. In the simplified approach, the accuracy, sensitivity, and specificity during training percentage 80 are used to interpret the improvement rate as illustrated in Fig. 5a–c. The proposed technique achieved an improvement rate of 0.30% as compared to the LSTM classifier when the accuracy improvement was first measured. The proposed technique achieved an improvement rate of 1.07% when compared to the LSTM classifier when the improvement in sensitivity was measured similarly. Finally, the suggested MWO-based LSTM classifier's specificity is evaluated. When compared to LSTM classifier, it registered a 1.40% improvement. Table 2 shows the comparative analysis.

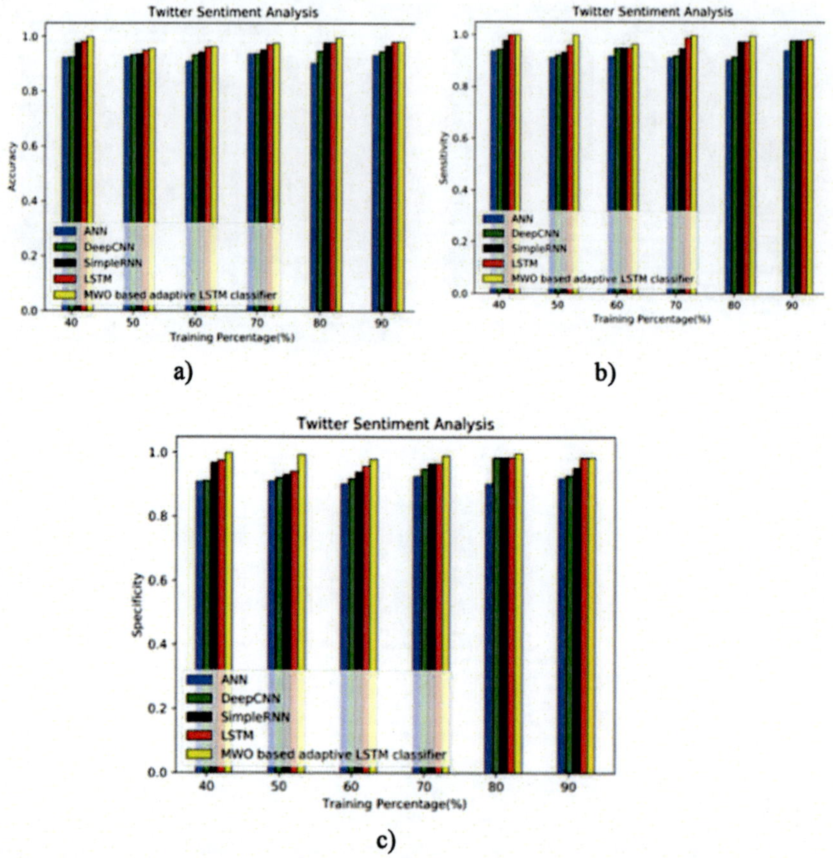

Fig. 5 Comparative analysis—training percentage in terms of **a** accuracy, **b** sensitivity, and **c** specificity

Table 2 Comparative discussion

Methods	Training percentage		
	Accuracy (%)	Sensitivity (%)	Specificity (%)
ANN	93.37	93.90	91.99
DeepCNN	94.67	97.76	92.86
SimpleRNN	96.75	97.91	95.25
LSTM	98.18	97.91	98.45
Proposed	98.48	98.97	99.85

5 Conclusion

This paper conducts research on the most effective machine learning classifier for categorizing sentiment polarity using Twitter sentiment analysis. The suggested Megaptera whale optimization is used to train the deep LSTM classifier optimally. Data is first collected and preprocessed to remove unneeded noise in the data and enhance model accuracy. The suggested Megaptera whale optimization-based adaptive LSTM classifier is used to do the classification once the features are selected using TF-IDF. The MWO, in which the classifier is precisely tuned for achieving the desired findings, determines the significance of the research. The adaptive LSTM built on MWO delivered effective results while lowering computational complexity. The research is superior because of its metrics for measuring values, and its accuracy, sensitivity, and specificity parameters achieved values of 98.48%, 98.97%, and 99.85%, which are more effective than state-of-the-art techniques.

References

1. Behera RK, Jena M, Rath SK, Misra S (2021) Co-LSTM: convolutional LSTM model for sentiment analysis in social big data. Inf Process Manage 58(1):102435
2. Roy A, Ojha M (2020) Twitter sentiment analysis using deep learning models. In: 2020 IEEE 17th India Council International Conference (INDICON), pp 1–6
3. Chugh A, Sharma VK, Kumar S, Nayyar A, Qureshi B, Bhatia MK, Jain C (2021) Spider monkey crow optimization algorithm with deep learning for sentiment classification and information retrieval. IEEE Access 9:24249–24262
4. Ruz GA, Henríquez PA, Mascareño A (2020) Sentiment analysis of Twitter data during critical events through Bayesian networks classifiers. Futur Gener Comput Syst 106:92–104
5. Valenzuela S, Puente S, Flores PM (2017) Comparing disaster news on twitter and television: an intermedia agenda setting perspective. J Broadcast Electron Media 61(4):615–637
6. Ligthart A, Catal C, Tekinerdogan B (2021) Systematic reviews in sentiment analysis: a tertiary study. Artif Intell Rev 54(7):4997–5053
7. Coletta LFS, da Silva NFF, Hruschka ER, Hruschka ER (2014) Combining classification and clustering for tweet sentiment analysis. In: 2014 Brazilian Conference on Intelligent Systems. IEEE
8. Tellez ES, Miranda-Jimenez S, Graff M, Moctezuma D, Siordia OS, Villasenor EA (2017) A case study of Spanish text transformations for twitter sentiment analysis. Expert Syst Appl 81:457–471
9. Shahid R, Javed ST, Zafar K (2017) Feature selection based classification of sentiment analysis using biogeography optimization algorithm, In: 2017 Int Conf Innov Electr Eng Comput Technol, IEEE, pp 1–5
10. Parlar T, Özel SA, Song F (2018) QER: a new feature selection method for sentiment analysis. Human-Centric Comput Inf Sci 8:81–19
11. Tian W, Li J, Li H (2018) A method of feature selection based on Word2Vec in text categorization. In: 2018 37th Chinese Control Conf., Technical Committee on Control Theory, Chinese Association of Automation, 9452–9455
12. Bidi N, Elberrichi Z (2016) Feature selection for text classification using genetic algorithms. In: 2016 8th International Conference Modelling, Identification and Control. University of MEDEA, Algeria, 806–810

13. Prastyo PH, Hidayat R, Ardiyanto I (2021) Enhancing sentiment classification performance using hybrid query expansion ranking and binary particle swarm optimization with adaptive inertia weights. ICT Express 8(2):189–197
14. Jianqiang Z, Xiaolin G, Xuejun Z (2018) ‘Deep convolution neural networks for Twitter sentiment analysis. IEEE Access 6:23253–23260
15. Yao F, Wang Y (2020) Domain-specific sentiment analysis for tweets during hurricanes (DSSA-H): a domain-adversarial neural-network-based approach. Comp, Environ Urban Syst 83:101522
16. Tam S, Said RB, Tanriöver ÖÖ (2021) A ConvBiLSTM deep learning model-based approach for Twitter sentiment classification. IEEE Access 9:41283–41293
17. Zhang S, Xu X, Pang Y, Han J (2020) Multi-layer attention based CNN for target-dependent sentiment classification. Neural Process Lett 51(3):2089–2103
18. Akhtar S, Garg T, Ekbal A (2020) Multi-task learning for aspect term extraction and aspect sentiment classification. Neurocomputing 398:247–256
19. Twitter sentiment analysis is from https://www.kaggle.com/datasets/kazanova/sentiment140
20. Diamantini C, Mircoli A, Potena D, Storti E (2019) Social information discovery enhanced by sentiment analysis techniques. Futur Gener Comput Syst 95:816–828
21. Zeng D, Peng J, Fong S, Qiu Y, Wong R (2018) Medical data mining in sentiment analysis based on optimized swarm search feature selection. Australas Phys Eng Sci Med 41:1087–1100

Electric Vehicle Charge Scheduling Based on Circle-Inspired Optimization Algorithm

Durga Mahato, Vikas Kumar Aharwal, and Apurba Sinha

Abstract Electric vehicle (EV) is seen as a feasible approach to lowering pollution. Because the number of EVs is increasing, it is critical to include charging stations (CS). Conversely, many unscheduled EVs survive owing to a lack of charging options or adequate energy. This paper proposes an optimization-aware method for EVs charge scheduling. In this work, the EVs simulation is the initial step. The revelation of changing requests from EVs and available charging stations is carried out. Afterward, the charge scheduling algorithm is called for scheduling the EV. The charge scheduler algorithm termed as adaptive circle-inspired optimization algorithm (Adaptive CIOA) is proposed here. In this case, a new multi-objective fitness function is created with parameters such as distance parameter, remaining power, user preference, and charging cost. The EVs are assigned to the CS based on the scheme. Moreover, the characteristics of the EV charging scheduling systems are then revised to reveal the method's efficiency. The proposed adaptive CIOA tends to outperform with the lowest charging cost of 16.22, lowest fitness of 0.0091, maximum user convenience of 0.826, and maximum power of 10.06 J.

Keywords Electric vehicle · Charge scheduling · Circle-inspired optimization algorithm · Deep maxout network · Energy management

D. Mahato (✉)
Research Scholar, Department of Electrical Engineering, Dr. A.P.J. Abdul Kalam University, Indore, India
e-mail: durgamahato379@gmail.com

V. K. Aharwal
Assistant Professor, Department of Electrical Engineering, Dr. A.P.J. Abdul Kalam University, Indore, India

A. Sinha
Assistant Professor, Department of Computer Science Engineering, GGSESTC, Bokaro Technical University, Jharkhand, India

G. Rajakumar et al. (eds.), *Intelligent Communication Technologies and Virtual Mobile Networks*, Lecture Notes on Data Engineering and Communications Technologies 171,
https://doi.org/10.1007/978-981-99-1767-9_40

1 Introduction

With the increase in population density in various countries, as well as the projected growth of EV saturation, this is critical to recognize the necessity of acceptable charging network and management. The CS can be categorized based on their required to charge levels or placement. Residence, worksite, and public charging systems are the three types of CS [1]. EVs have received a great deal of attention as a result of growing social awareness of environmental concerns and a desire to reduce the use of fossil fuels [2]. EVs charging network is a normal cyber-physical system that comprises a power grid as well as a huge number of EVs and aggregators that gather data and monitor the charging operation [3]. The "smart grid" theory has become increasingly popular as a power grid with two-way network technologies that provide end-users with acceptable signals, that is kWh price incentives, to stimulate them to influence their load profile and effectively respond to grid necessities [4]. Micro-grids were initiated as key components of smart grids. A micro-grid is a collection of interlinked loads as well as distributed energy resources (DERs) with well-defined electrical bounds that operate as a single monitoring entity in relation to the grid [5]. In power distribution networks, EVs are considered as a newly running load from grid-to-vehicle (G2V) including storage capabilities from vehicle-to-grid (V2G). Even if large-scale EV requirement is not accepted adequately, these moving demands have a significant impact on power networks [6]. When grouping of EVs for possible grid congestion management and planning the sequential access of EVs, as well as their location information, is a significant criterion to suggest [7].

EVs also received a lot of attention recently as an ecologically aware and value substitute to usual vehicles powered by internal combustion engines (ICEs). People have cheaper costs than ICE vehicles and can also be powered by locally produced renewable energy sources (RES) [7]. The EV with the on unit (OBU), which is a Global Positioning System (GPS)-based navigator. Using GPS and map installed OBU, EV is instructed to charge the next CS. This CS-selection choice is based on the EV's location (as determined by GPS), the remaining electricity (as determined by the EV meter), and the CSs (as determined by a pre-installed map) [8]. Efficient charging control in EVs is essential for both electric grid operators and drivers. Recently, numerous analyses for EV charging and battery charging methods have been created. The majority of the strategies concentrated on EV charging methods both are controlled and uncontrolled. The uncontrolled EV charging can exacerbate the grid concern, which already involves substation overloads, energy loss, and voltage fluctuations. Synchronized charging methods are intended to resolve the shortcomings of existing methods like time of use (TOU), distributed control, and centralized control [9]. An excellent particle swarm optimization technique is utilized in [10], to optimize EV charging. However, this slow charge optimizer for EVs at night is considered, and it may have an impact on performance. In [11], an optimizer with the lowest charging cost is introduced, and this model is addressed by the non-dominated sorting genetic algorithm II (NSGA-II), which fails to change to load variation. To achieve a near-optimal solution, utilize the model predictive control (MPC) [12]. The charging behavior of EV batteries is optimized with the goal of minimizing charging costs, achieving satisfied state-of-energy levels, and achieving optimum power controlling [13].

The goal of this study is to locate a suitable CS for EVs. At first, an EV simulation is performed, and the charging requests from EVs and available CS are detected here. The load is then calculated using the deep maxout network (DMN) [14]. Following this, the charge scheduling algorithm is used to allocate the EV that employs the adaptive CIOA. In this case, a multi-objective fitness function with some factors such as distance parameters, user preference, charging cost, and remaining power is developed. Following that, the EVs were allotted to the CS according to its priority scheduling approach. In this case, the priority schedule is based on adaptive CIOA, and the factors of the EV charging scheduling model are adjusted to reveal the method's efficiency.

The main contribution of the paper is illustrated below:

- **Developed adaptive CIOA for EV charge scheduling with multi-objective function:** The charge scheduling algorithm is employed for scheduling the EV using adaptive CIOA. Moreover, the adaptive CIOA is the combination of CIOA [15] with an adaptive concept. In addition, the multiple objective functions are newly modeled by taking into account certain parameters such as distance parameter, user preference, charging cost, and remaining power. The EVs are allotted to the CS by the adaptive CIOA.

The rest of the paper's structure is illustrated as follows: Sect. 2 covers a literature review of conventional techniques of charge scheduling. Section 3 contains the system model of charge scheduling among EV. Section 4 describes the adaptive CIOA for charging scheduling. Section 5 computes the efficiency of the proposed model by comparing with classical techniques. The conclusion part is explained in Sect. 6.

2 Motivation

Regardless of the fact that the solution is provided with a conjunction of information models, monitoring is provided to reduce the levels of charging vehicles with the grid load. Several survey results are limited to the EV charging path and its interface with the charging resources, as well as the evaluation of EV planning paths that include the information of road conditions. Therefore, the problems encountered by traditional methods are employed to develop a novel EV charging scheduling scheme.

2.1 Literature Review

Chung et al. [16] developed EV fee scheduling concern on a micro-grid scale for vehicular ad hoc network (VANET) charging scheduling. A series of CS is controlled by a central aggregator. It was the major issue in this case. The CS operator is the primary stakeholder, and the person is motivated to decrease charging station costs. For jointly optimizing two factors that interact with these stakeholders, a bi-objective optimization was also modeled. An online centralized scheduling method was created to decrease both data transmission cost and system computation cost. While the centralized scheduling technology offered excellent performance, it carries a high cost of storage. Lee et al. [17] devised an adaptive charging network (ACN) for the EV charging process. The ACN used an adaptive scheduling algorithm (ASA) based on model-predictive control and convex optimization for allowing an essential over-subscription of the power network. Furthermore, the ASA was used as a flexible model-predictive control-based algorithm with pre- and post-processing that could be simply designed to gather various objectives. Cao et al. [18] devised a technique for adapting the globally optimal strategy for charging EVs using greedy-choice property. In this case, the problem was solved using a heuristic technique that included a multi-commodity network flow framework for charge scheduling. However, the method failed to account for a number of constraints under different settings. Gupta et al. [1] developed a decentralized charging (DES) control system for EV scheduling in a multi-aggregator circumstance. The EV arrival at the CS was modeled using mobility-aware scheduling. This method was created based on the consumer's willingness to evaluate scheduling factors such as charging duration, vehicle arrival/scheduling time, and charging cost. These factors were used to efficiently schedule the EVs. Nonetheless, this method does not take into account the driver's preferences; it only takes into account the performance of the scheduling factors.

2.2 Challenges

The concerns faced in the classical EV charging scheduling approaches are described below.

- As even more EVs are auctioned, more charging stations are being constructed to resolve the EV charging issue. Nevertheless, the EV's features are challenging to anticipate precisely, which has hampered EV charging optimization.
- In [16], a low-scope distributed procedure was created to decrease both the rate of data transmission and the computation cost while charging an EV. In online, charge schedulers, on the other hand, survived from load issues and charging price limitations.
- To decrease the burden, a strategy based on the greedy-choice property was devised. The introduction of EV charging has the potential to raise peak-time

demand and overload the grid, putting stress on generating units and resulting in unusual voltage sag and reduced power quality [18].

- The decentralized strategies in [1] needed continuous access to a two-way communication network between both the EVs' communication systems and their synchronization, which increased the cost of the system.

3 System Model

Due to the large quantity of EVs on the road, effectual EV battery charging has become a major problem. Cooperated charging is regularly favored over uncoordinated charging, which has a negative impact on the power grid through enhancing peak load and the overall cost. A coordinated charging task [16] is regarded here, in which a large number of parking stations are controlled by a central aggregator (CA). In the CS, a sub-aggregator (SA) is established to exchange the data in CA. It is accountable for EV charge scheduling via controlling the charging rates by evaluating the starting and ending periods of charging. Moreover, scheduling is done by assembling the necessary information, such as charging requests, arrival time, and deadlines set by SA. Figure 1 depicts the EV charging scheduling model.

Consider H is the charging stations quantity, G indicates the number of EVs in a micro-grids, and F signifies a number of SAs for H CSs and a CA for controlling SA. The time slots are represented as $s = \{1, 2, \ldots, S\}$. Here, the state of charge of EV_b^c that is bth EV under cth CS is given as SOC_b^c, , which relies in [0, 1] wherein 1 represents full battery and 0 is the empty battery. The arrival time e_b^c refers arrival time of EV_b^c, the deadline of EV_b^c is given as f_i^j, battery capacity of EV_b^c is given as $V_{b,c}^{\mathrm{cpt}}$, the arrival time's initial battery energy level is given as $V_{b,c}^{\mathrm{int}}$, and the battery energy level at finish time is expressed as $V_{b,c}^{\mathrm{fns}}$. The arrival time's initial battery energy level is formulated as

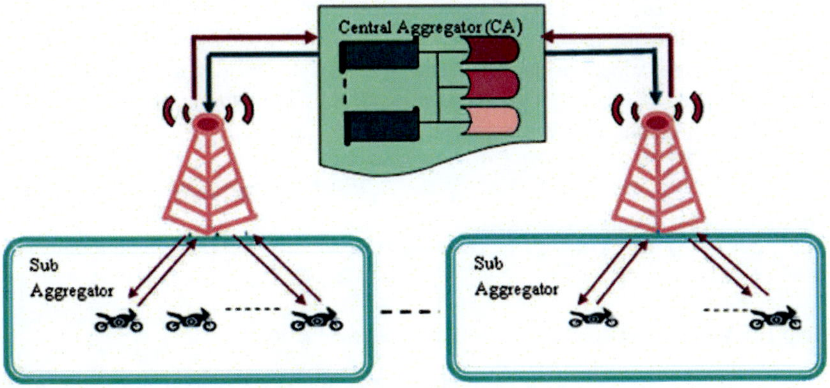

Fig. 1 System model of EV charging scheduling

$$V_{b,c}^{i\,\mathrm{int}_{b,s_{b,c}}^{c^{\mathrm{cpt}}}}. \tag{1}$$

The value of fit factor $g_{b,s}^{c}$ is given as

$$g_{b,s}^{c} = \begin{cases} 1; \text{ If } \mathrm{EV}_b \; is \text{ incth station at times} \\ 0; \qquad\qquad \text{Otherwise} \end{cases}. \tag{2}$$

The power available in EV_b^c is formulated as

$$R_{b,s}^{c} = \sum_{c=1}^{H} \sum_{b=1}^{F} \left(\mathrm{SOC}_{b,c}^{\mathrm{fns}} - \mathrm{SOC}_{b,s}^{c}\right) V_{b,c}^{\mathrm{cpt}}, \tag{3}$$

where $\mathrm{SOC}_{b,c}^{\mathrm{fns}}$ represents SOC finish time of EV_b. The power value relies on $R_{b,\min}^{c_{b,s_{b,\max}}^{c^c}}$. The state of charge of EV_b^c is given as

$$\mathrm{SOC}_{b,s+1}^{c} = \mathrm{SOC}_{b,s}^{c} + \frac{R_{b,s}^{c}}{V_{b,c}^{\mathrm{cpt}}} \tag{4}$$

4 Proposed CIO for EV Charging Scheduling

EVs take a lot of time queuing at public CSs, especially at peak times. As a result, an effective way for reducing total charging time for EVs is required. The EV charging scheduling (EVCS) system is designed with a multi-objective function that utilizes adaptive CIOA. Initially, an EV simulation is run, from which charging requests from EVs and accessible charging stations are detected. The load is then calculated utilizing DMN [14]. Following this, call charge scheduling is carried out using adaptive CIOA. Following the charge scheduling phase, EVs are scheduled to CS based on the charging station's priority scheduling approach. Priority scheduling is based on the proposed adaptive CIOA in this case. The block diagram of adaptive CIOA for charge scheduling of EV is displayed in Fig. 2.

4.1 Steps of EV Charging Scheduling

Step 1: EV-based network simulation calculates S the load.

Step 2: Detecting EV charging requests and finding the available scheduling of charges.

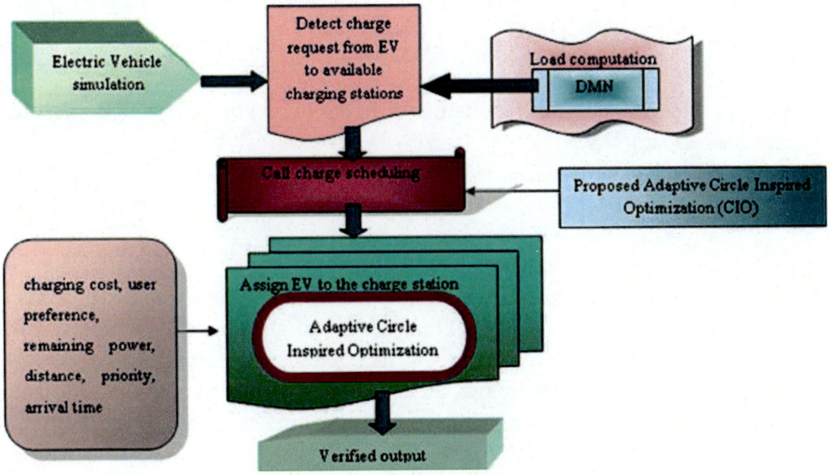

Fig. 2 Block diagram for EVCS with multi-objective function based on proposed adaptive CIOA

Step 3: Calling the CS algorithm using adaptive CIOA.

Step 4: Assigning EVs to CSs in accordance with the proposed scheduling method.

Step 5: Updating the parameters $\left(SOC^{c}_{b,s+1} \text{ and Energy}\right)$.

A network is operated as a connected graph $M = (Q, B, T)$, in which $G = \{1, \rightleftarrows \ldots, Y\}$ symbolizes nonzero set of Y nodes that reveals the intersection of the road and the capable location of EV charging station, B symbolizes set of edges to indicate the length of road segment, and T implies the adjacency matrix that implies if nodes are associated. For set of edges, term $k^{ij} > 0$ signifies the transition from node i to node j with a positive weight. Moreover, the graph indicates a highway wherein the traffic is assumed the capable to flow from all ways exclusive of losing generality.

4.2 *Load Computation Using DMN*

The load in the CS is indefinite; the forecasting approach can be employed to enhance the scheduling routine, which is not above the load generated at every instance. Thereby, load forecasting is required to improve the efficiency and accuracy of the scheduling. The preliminary values of state of charge (SOC) are generated randomly and constantly in the range [0, 1], and the SOC target is tuned to 1. Therefore, base load information is derived from load forecasting results. Thus, the base load unit is changed as

$$Z^s_{\text{base}} = \frac{Z^s_{\text{fore}} \times Z_{\text{peak}}}{\max_s\left(Z^s_{\text{fore}}\right)}, \tag{5}$$

where Z_{peak} refers to the peak load under different number of EV and Z_{fore} is the forecasted load using DMN.

4.2.1 Architecture of DMN

DMNs are made up of multiple layers that use the maxout function to create hidden activations. A DMN can be built by connecting various maxout layers in a row. Figure 3 depicts the DMN architecture; it facilitates the optimization process by limiting the hidden units. Furthermore, the maxout unit represents a trainable activation function although it is not an arbitrary function estimator. Thus, the input offered to DMN is $Z^s_{b,c}$, is the load of bth EV in cth CS.

The features are assumed as input $J \in D^u$, where J implies the raw input vector, the activation of the hidden unit is expressed as

$$u^1_{b,c} = \max_{c\in[1,q_1]} J^S \psi_{b,c} + w_{bc} \tag{6}$$

$$u^2_{2,c} = \max_{c\in[1,q_2]} u^1_{b,c}{}^S \psi_{bc} + w_{bc} \tag{7}$$

$$u^j_{b,c} = \max_{c\in[1,q_j]} u^{j-1^S}_{b,c} \psi_{bc} + w_{bc} \tag{8}$$

$$u^x_{b,c} = \max_{c\in[1,q_x]} u^{x-1^S}_{b,c} \psi_{bc} + w_{bc} \tag{9}$$

$$v_b = \max_{c\in[1,q_x]} u^x_{b,c} \tag{10}$$

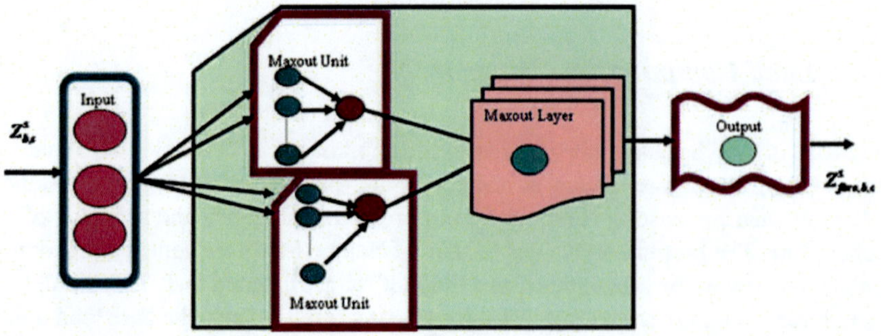

Fig. 3 Structure of DMN

where x symbolizes total layers,e indicates bias, and $\psi_{b,c}$ refers the weights of DMN. The DMN output is utilized to forecast load. The total load calculation is computed on the basis of a certain condition.

If the condition, $L^{s}_{\text{base}} + \sum_{b} Rc^{c}_{b,s} g^{c}_{b,s} \leq \delta^{b,c}_{s}$, then the total load of EV^{c}_{b} can be computed as

$$\delta^{b,c}_{s} = Z^{s}_{base} + \sum_{b} R^{c}_{b,s} g^{c}_{b,s} \tag{11}$$

4.3 Multi-Objective Fitness Computation

The proposed CIOA with charge scheduling is utilized for modeling a descriptive explanation of fitness function. CIOA fitness function was designed from the ground up to choose the best charging system. Fitness function in this case includes user convenience, distance, charging cost, and available power, which is formulated as

$$\text{Fitness} = \sum_{c=1}^{H} \mu^{c}_{b,s} + \left(1 - \partial^{c}_{b,s}\right) + A^{c}_{b,s} + \left(1 - R^{c}_{b,s}\right), \tag{12}$$

where $\mu^{c}_{b,s}$ signifies charging cost of EV^{c}_{b}, $\partial^{c}_{b,s}$ refers to the user convenience, $A^{c}_{b,s}$ is distance, and $V^{c}_{b,s}$ symbolizes available power.

The charging cost of EV^{j}_{i} [16] is given as

$$\mu^{c}_{b,q} = \sum_{s=1}^{S} \left[q_0 \left(\delta^{b,c}_{s} - Z^{s}_{\text{base}} \right) + q_1 \left(\delta^{b,c^2}_{s} - Z^{s^2}_{\text{base}} \right) \right], \tag{13}$$

where q_0 and q_1 indicate the constants. It is a minimization function.

The user convenience is given as

$$\partial^{c}_{b,s} = \frac{1}{\rho^{c*}_{b,s} \rho^{c}_{b,s}}, \tag{14}$$

where $\rho^{c*}_{b,s} = \frac{(\text{SOC}^{\text{fns}}_{b,s} - \text{SOC}^{c}_{b,s}) V^{\text{cpt}}_{b,c}}{R^{c}_{b,\max}}$ and $\rho^{c}_{b,s} = f^{c}_{b} - s$. It is a maximization function.

The distance is expressed as

$$A^{c}_{b,s} = \frac{1}{G \times o} \sum_{b=1}^{G} \left|\left| v^{s}_{b,c} - v_p \right|\right|, \tag{15}$$

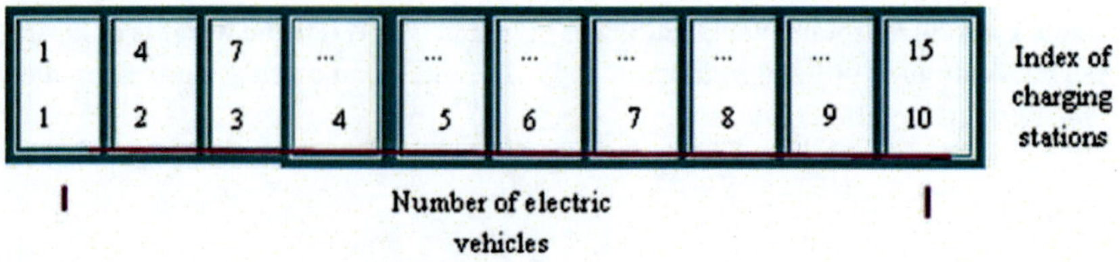

Fig. 4 Solution encoding of using adaptive CIOA

where o is the normalizing factor, v_p expresses the location of CS, and $v^s_{b,c}$ represents the location of EV^c_b. It is a minimization function.

4.4 *Charge Scheduling Strategy Using Proposed Adaptive CIOA Algorithm*

The objective is to use the adaptive CIOA algorithm to find more capable CS for EVs. Generally, this method focuses on the EVs charging point scheduling. The user convenience, distance, charging cost, and available power are used as key attributes to reveal the technique's efficiency. The parked EV waits till the CS charging schedule. The solution encoding and algorithmic steps of the adaptive CIOA algorithm are discussed below.

4.4.1 Solution Encoding

Figure 5 depicts a representation of a solution with adaptive CIOA. The solution is a charging station index for charging the EV. Adaptive CIOA is used in this case to choose the finest CS. In this case, the CS is chosen arbitrarily while adaptive CIOA searches for the best CS. Figure 4 depicts the CIOA algorithm's solution encoding scheme. Consider the following: There are ten electric vehicles and fifteen charging stations. The solution encoding is in the size of $1 \times N$.

4.4.2 Proposed Adaptive CIOA for Charging Scheduling

Adaptive CIOA is a metaheuristic optimization algorithm. It is ideal for optimization problems that are either constrained or unconstrained. CIOA can be used to resolve complex optimization issues such as real-time optimization issues, benchmark function optimization, structural truss optimization, and particularly in engineering problems. Furthermore, the CIOA provides the merits of quick convergence and a lowered amount of variables that the user defined. As a result, the adaptive

CIOA improves the model performance. The steps for creating an adaptive CIOA are explained below.

Step 1: Initialization

Each search agent in the adaptive CIOA runs across the arcs controlled by two main criteria: The radius angle specified by the user is estimated by this algorithm, and the rate of which depends upon the objective function analysis, the superior estimation of the objective function is done via search agent, and the lesser rate in this agent for the coming iteration.

The adaptive CIOA is initialized by creating a vector, $\vec{a}$ where the value a_m in every factor is deliberated using Eq. (16).

$$a_m = \frac{E_a \cdot m^2}{W_{\text{ag}}}, 1 \leq m \leq W_{\text{ag}} \tag{16}$$

where W_{ag} is the number of search agents and E_a is a constant that is calculated by Eq. (19). The element in the radius vector is ordered in ascending manner, where the first component is given as

$$a_1 = E_a / W_{\text{ag}}, \tag{17}$$

where the last element is given by

$$a_{W_{ag}} = E_a \cdot W_{\text{ag}} \tag{18}$$

$$E_a = \frac{\sqrt{U_d - L_d}}{W_{\text{ag}}} \tag{19}$$

Here, U_d is the upper bound, and L_d is the lower bound of every variable. After, every search agent allocates the arbitrary values to design factors in the first solution. Next to this, the objective function is evaluated, and every search agent is labeled in a ranking based on the solution's quality.

Step 2: Determination of fitness

The optimum solution is determined using the fitness function and is deemed as minimization concern; hence, the solution producing the least fitness is finding as the optimum solution. The fitness function is already described in Sect. 4.3.

Step 3: Evaluation of update equation

It is noted that, in best-classified agents, the agents take up the horrible classifications with long scheduling. The search agent is ordered as mth finest solution in iteration n have the updated coordinates in $n + 1$ iteration is calculated by Eq. (20).

$$I(n+1) = I(n) - t_1 \cdot a_m \cdot \sin(n.\theta) + t_2 \cdot a_m \cdot \sin(En+1).\theta) \tag{20}$$

In the above equation, t_1 is made as adaptive is given by

$$t_1 = 3(0.5 - a) * \lambda, \tag{21}$$

where $\lambda = 20 * \exp\left(-12 \cdot a\left(\frac{\text{itr}}{\max\_\text{itr}}\right)\right)$.

Therefore, Eq. (20) updates $\vec{I} = [I_2; I_4; I_6, \ldots,]$ variables are random numbers with a uniform distribution between zero and 1; the angle θ is a parameter provided by the user; the variable a_m corresponds to the mth element of the vector $\vec{a}$.

Step 4: Local search phase

The central loop of adaptive CIOA supports the local and global searches concurrently. This is due to the agents, which generate the superior solutions illustrated in tiny movements corresponding to local search, while the agents produce the worst solutions illustrated in big movements corresponding to global search. A parameter Glob^{It} implies the iterations proportion before the insertion of exclusively local search in the algorithm, and its rate varies freely in (0, 1) interval. Still, the use of $0.75 \le \text{Glob}^{\text{It}} \le 0.95$ is suggested.

The exclusively local search starts on iteration "*k*", while the ratio of the total number of iterations and "*k*" is greater than Glob^{It}. In this instant, every search agent is restarted assuming the coordinates that produced the best solution so far. In addition, a change is made to the upper and lower bounds of each variable so that the new limits of *i*th dimensions are given by U_{d1_r}, and L_{d1_r}, calculated using Eqs. (22) and (23).

$$U_{d1_r} = L_{r\text{best}} + \frac{U_d - L_d}{10000} \tag{22}$$

$$L_{d1_r} = L_{r\text{best}} - \frac{U_d - L_d}{10000} \tag{23}$$

where $L_{r\text{best}}$ is the variable in r dimension that generated the optimum solution so far. Following the initialization process, the local search step is managed by similar equations as the main loop expressed and replacing U_d with U_{d1_r} and L_d with L_{d1_r}. Therefore, the search space of the agents is limited, forcing the rates to the design variables. The design variable is close to the values that created the optimum solution in main loop. In specific cases,$L_{d1_r}\langle L_d or U_{d1_r}\rangle U_d$ where the value of the design variable lies in $L_{d1_r}\langle I_r \langle L_d or U_{d1_r}\rangle I_r \rangle U_d$, its value is updated automatically, and the value of $I_r = L_d or I_r = U_d$.

Step 5 Re-evaluation of fitness function

The fitness of the new solution is estimated, and the solution showing minimum fitness value is taken as the optimal solution.

Table 1 Pseudocode of developed adaptive CIOA

Input: Population N_{ag}
Output: Best solution $I(n+1)$
Begin
Define $Glob^{It} and \theta$
Initialize a radius of vector a *using Eq.* (16)
Allot random values to design variables
Estimate the objective function for each search agent
While 1 $\left(n \leq Gob^{It} \times \max imumnumberofiterations\right)$
Sort the search agents while the quality of the solutions attained
Update the position of agents by Eq. (18)
Validate the design variables of search agent exceeds the limits imposed
If n is a multiple of value rounded down from $360/\theta$
Update $\vec{a}$
End If
End While1
Allot to all agents the position created the best solution so far
Update the variable's range using Eqs. (22) and (23)
While 2 ($n \leq \max i\ mum\ number\ of\ iterations$)
Repeat the process expressed in While 1, using the new range of variables
End While 2
Visualization of result
End

Step 6: Termination

The optimal weights are generated until the utmost iteration is attained. Table 1 defines pseudocode of the adaptive CIOA algorithm.

5 Results and Discussion

The efficiency of adaptive CIOA with other conventional approaches is described. The efficiency is evaluated by estimating the methods through charging cost, fitness, available power, and user convenience with 100 and 150 vehicles.

Table 2 Simulation parameters

Parameter	Value
Number of iterations	500
Number of lanes	2
Lane width	29
Lane length	290
Number of roads	4
Number of intersections	2

5.1 Experimental Setup

The adaptive CIOA is implemented in MATLAB. Table 2 shows the parameters of the proposed method.

5.2 Simulation Results

The simulation results using adaptive CIOA are examined. Figure 5 depicts the outcome of the simulation by the proposed adaptive CIOA based on vehicle count and simulation time. Figure 5 shows red nodes in the network indicating vehicles moving west–east. The EV simulated model is shown at various simulation times. CS denotes each EV with a battery deficiency recharges its battery via the nearest CS. Figure 5a depicts the VANET model with 100 vehicles at a time of 10 s. The CS is initially idle, and after a few seconds, the EV is filled to the respective CS to charge its EV. Figure 5b depicts the VANET model with 150 vehicles after 15 s. Here, the CS is in an idle position at first, and then it is filled to the respective CS to charge its EV.

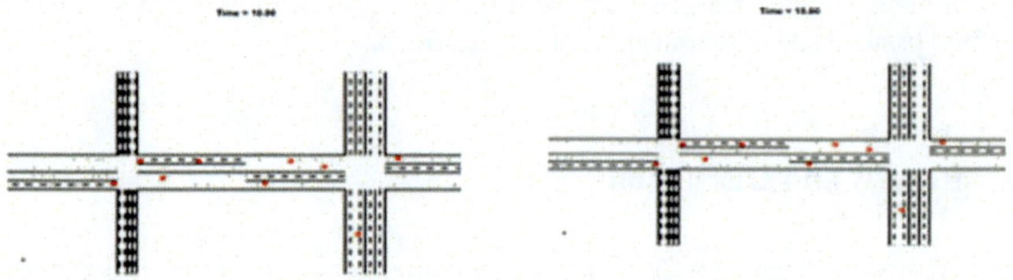

(a) (b)

Fig. 5 Simulation results of proposed adaptive CIOA using **a** 100 vehicles and **b** 150 vehicles

5.3 *Performance Metrics*

The estimation of the proposed adaptive CIOA-based charge scheduling design is made with four metrics, namely charging cost, fitness, available power, and user convenience, which are already described in Sect. 4.

5.4 *Competing Methods*

The existing centralized scheduling [16], multi-aggregator collaborative scheduling [17], prediction-based charging [18], model-predictive control optimization [1], Jaya-based multi-verse optimizer (JMVO) [19], and proposed adaptive CIOA [5] are taken for analysis.

5.5 *Comparative Analysis*

The analysis of performances based on charging cost, fitness, available power, and user convenience is described using 100 and150 vehicles.

5.5.1 Analysis with 100 Vehicles

Figure 6 displays the evaluation of techniques by 100 vehicles that consider the charging cost, fitness, available power, and user convenience. The analysis of techniques using charging cost is shown in Fig. 6a. For 101 iterations, the charging cost estimated by centralized scheduling, multi-aggregator collaborative scheduling, prediction-based charging, model-predictive control optimization, JMVO, and adaptive CIOA is 27.01, 28.25, 26.43, 28.05, 22.77, and 21.86. The analysis of methods with fitness is displayed in Fig. 6b. For 51 iterations, the fitness evaluated by centralized scheduling, multi-aggregator collaborative scheduling, prediction-based charging, model-predictive control optimization, JMVO, and adaptive CIOA is 0.509, 0.201, 0.195, 0.363, 0.159, and 0.152, respectively. The assessment of techniques with power are shown in Fig. 6c. For 151 iterations, the power evaluated by centralized scheduling, multi-aggregator collaborative scheduling, prediction-based charging, model-predictive control optimization; JMVO, and adaptive CIOA is 2.719 J, 2.932 J, 3.415 J, 3.933 J, 5.193, and 5.297 J. The analysis of techniques with user convenience is shown in Fig. 6d. For 201 iterations, the user convenience evaluated by centralized scheduling, multi-aggregator collaborative scheduling, prediction-based charging, model-predictive control optimization, JMVO, and proposed adaptive CIOA are 0.609, 0.664, 0.676, 0.723, 0.761, and 0.776, respectively.

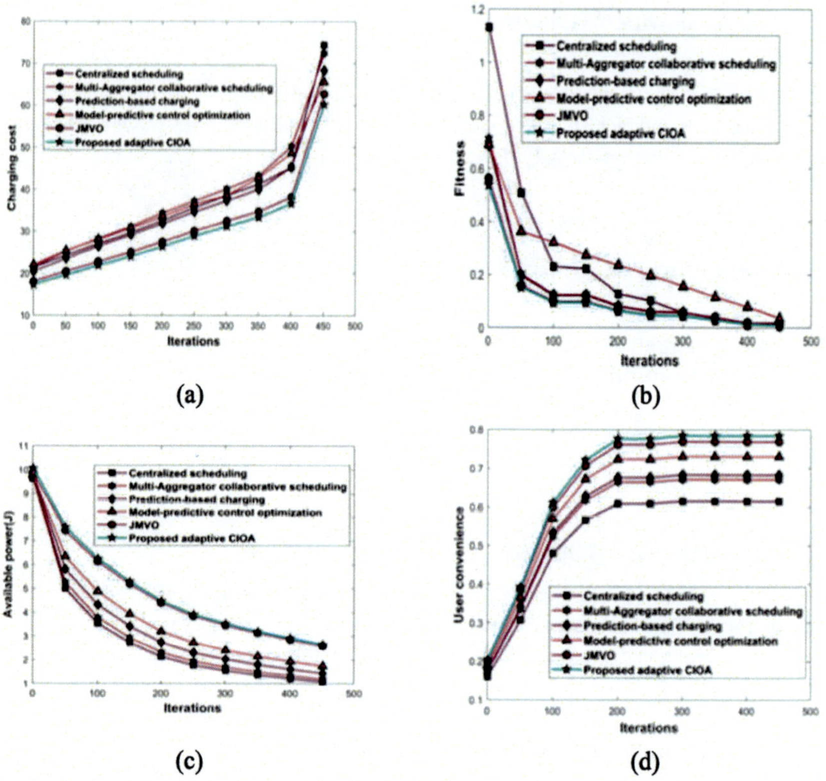

Fig. 6 Assessment of techniques using 100 vehicles **a** charging cost, **b** fitness, **c** power, and **d** user convenience

5.5.2 Analysis with 150 Vehicles

Figure 7 deliberates the analysis of techniques using 100 vehicles via charging cost, fitness, available power, and user convenience. The evaluation with charging cost is shown in Fig. 7a For 151 iterations, the charging cost evaluated by centralized scheduling, multi-aggregator collaborative scheduling, prediction-based charging, model-predictive control optimization, JMVO, and proposed adaptive CIOA is 30.81, 33.04, 29.73, 35.24, 25.98, and 24.94. The assessment of methods with fitness is portrayed in Fig. 7b. For 101 iterations, the fitness obtained by centralized scheduling, multi-aggregator collaborative scheduling, prediction-based charging, model-predictive control optimization, JMVO, and proposed adaptive CIOA is 0.399, 0.761, 0.551, 0.397, 0.275, and 0.264, respectively. The assessment of power is shown in Fig. 7c. For 301 iterations, the power attained by centralized scheduling, multi-aggregator collaborative scheduling, prediction-based charging, model-predictive control optimization, JMVO, and proposed adaptive CIOA is 1.526 J, 1.468 J, 1.422 J, 1.546 J, 1.722 J, and 1.757 J, respectively. The assessment of methods with

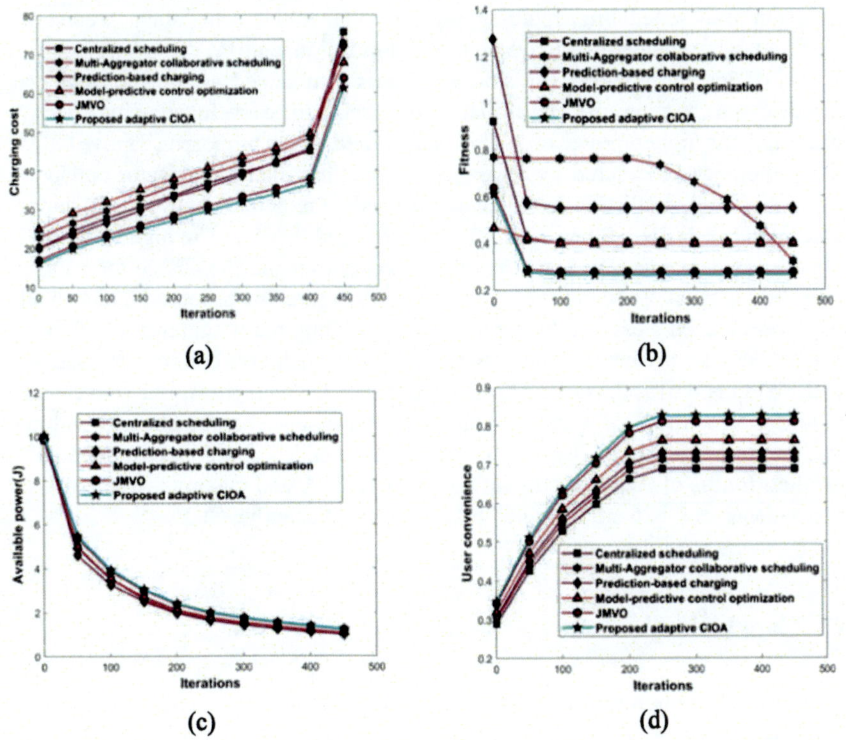

Fig. 7 Assessment of techniques using 150 vehicles **a** charging cost, **b** fitness, **c** power, and **d** user convenience

user convenience is deliberated in Fig. 7d. For 251 iterations, user convenience by centralized scheduling, multi-aggregator collaborative scheduling, prediction-based charging, model-predictive control optimization, JMVO, and proposed adaptive CIOA is 0.688, 0.713, 0.729, 0.761, 0.810, and 0.826, respectively.

5.6 Comparative Discussion

Table 3 explains the comparative analysis for using 100 and 150 vehicles considering user convenience, charging cost, fitness, and power. Using 100 vehicles, the minimum charging cost of 17.30% is measured by the proposed adaptive CIOA, while the charging cost evaluated by centralized scheduling, multi-aggregator collaborative scheduling, prediction-based charging, model-predictive control optimization, and JMVO is 21.38, 21.49, 20.39, 22.00, and 18.02. When there are plenty of CS, the proposed adaptive CIOA selects nearest CS and thus the charging cost tends to be less. The smallest fitness of 0.0091 is measured by CIOA, while the fitness evaluated by centralized scheduling, multi-aggregator collaborative scheduling, prediction-based charging, model-predictive control optimization, and JMVO is 0.017, 0.012, 0.011, 0.036, and 0.0095. The smallest fitness is obtained as user convenience, distance, and charging cost, and available power tends to be low for attaining effective charging. The highest power of 10.06 J is measured by the proposed adaptive CIOA, while the power measured by centralized scheduling, multi-aggregator collaborative scheduling, prediction-based charging, model-predictive control optimization, and JMVO is 9.620 J, 9.656 J, 9.723 J, 9.777 J, and 9.865 J. The highest power is obtained by adaptive CIOA as it poses the ability to retain the charge for a longer time. The highest user convenience of 0.784 is measured by proposed CIOA, while user convenience measured by centralized scheduling, multi-aggregator collaborative scheduling, prediction-based charging, model-predictive control optimization, and JMVO is 0.615, 0.671,0.683, 0.730, and 0.769. The highest user convenience is identified as the EVs with maximum user convenience can be charged with the maximum charging rate. Using 150 vehicles, the smallest charging cost of 16.22%, smallest fitness of 0.263, maximum power of 10.04 J, and maximum value of user convenience of 0.826 are measured by the proposed adaptive CIOA.

6 Conclusion

In this paper, an optimization-driven technique is developed for charge scheduling in EV. The preliminary step is the simulation of EV, and the detection of the changing request from EVs and available charging stations is carried out. Thereafter, the calling of charge scheduling technique is performed to schedule the EV in which the charge scheduling algorithm is adaptive CIOA. Here, a multi-objective fitness function is newly modeled using certain attributes that involve charging cost, user preference, remaining power, and distance parameter, which is considered as a minimization function. As per the scheduling technique, the EVs are allocated to the charging station. Then, the parameters of EV charging scheduling model are updated to reveal the effectiveness of the method. The adaptive CIOA technique can concurrently generate a minimal charging cost and minimize the time taken for charging. The developed adaptive CIOA provided superior performance with the smallest charging cost of 16.22, minimum value of fitness of 0.0091, highest power of 10.06 J, and maximum value of user convenience of 0.826. In the future, other advanced optimization technique can be considered for scrutinizing the flexibility of the proposed model.

Table 3 Comparative analysis

Vehicles	Metrics	Centralized scheduling	Multi-aggregator collaborative scheduling	Prediction-based charging	Model-predictive control optimization	JMVO	Proposed adaptive CIOA
100 vehicles	Charging cost (%)	21.38	21.49	20.39	220.00	18.02	17.30
	fitness	0.017	0.012	0.011	0.036	0.0095	0.0091
	Power (J)	9.620	9.656	9.723	9.777	9.865	10.06
	user convenience	0.615	0.671	0.683	0.730	0.769	0.784
150 vehicles	Charging cost (%)	20.04	22.90	20.09	25.05	16.90	16.22
	Fitness	0.397	0.320	0.549	0.397	0.274	0.263
	Power (J)	9.812	9.918	9.973	10.00	9.850	10.04
	user convenience	0.688	0.713	0.729	0.761	0.810	0.826

References

1. Gupta V, Kumar R, Panigrahi BK (2020) User-willingness-based decentralized EV charging management in multiaggregator scheduling. IEEE Trans Ind Appl 56(5):5704–5715
2. Yao W, Zhao J, Wen F, Dong Z, Xue Y, Xu Y, Meng K (2014) A multi-objective collaborative planning strategy for integrated power distribution and electric vehicle charging systems. IEEE Trans Power Syst 29(4):1811–1821
3. Jin C, Tang J, Ghosh P (2013) Optimizing electric vehicle charging: a customer's perspective. IEEE Trans Veh Technol 62(7):2919–2927
4. Soares A, Antunes CH, Oliveira C, Gomes A (2014) A multi-objective genetic approach to domestic load scheduling in an energy management system. Energy 77:144–152
5. Farzin H, Fotuhi-Firuzabad M, Moeini-Aghtaie M (2016) A stochastic multi-objective framework for optimal scheduling of energy storage systems in microgrids. IEEE Trans Smart Grid 8(1):117–127
6. Alonso M, Amaris H, Germain JG, Galan JM (2014) Optimal charging scheduling of electric vehicles in smart grids by heuristic algorithms. Energies 7(4):2449–2475
7. Mukherjee JC, Gupta A (2014) A review of charge scheduling of electric vehicles in smart grid. IEEE Syst J 9(4):1541–1553
8. Yang SN, Cheng WS, Hsu YC, Gan CH, Lin YB (2013) Charge scheduling of electric vehicles in highways. Math Comput Model 57:2873–2882
9. Yagcitekin B, Uzunoglu M (2016) A double-layer smart charging strategy of electric vehicles taking routing and charge scheduling into account. Appl Energy 167:407–419
10. Clement-Nyns K, Haesen E, Driesen J (2010) The impact of charging plug-in hybrid electric vehicles on a residential distribution grid. IEEE Trans Power Syst 25:371–380
11. Richardson P, Flynn D, Keane A (2012) Optimal charging of electric vehicles in low-voltage distribution systems. IEEE Trans Smart Grid 27:268–279
12. Tang W, Zhang YJ (2016) A model predictive control approach for low-complexity electric vehicle charging scheduling: optimality and scalability. IEEE Trans Power Syst 32(2):1050–1063
13. Zhang T, Chen W, Han Z, Cao Z (2013) Charging scheduling of electric vehicles with local renewable energy under uncertain electric vehicle arrival and grid power price. IEEE Trans Veh Technol 63(6):2600–2612
14. Battula TK, Eppili J (2022) Visual features and Deep Maxout Network-based moving target detection using radar signals under sea clutter. Int J Intell Syst 37(1):424–449
15. de Souza OAP, Miguel LFF (2022) CIOA: Circle-Inspired Optimization Algorithm, an algorithm for engineering optimization. SoftwareX 19:101192
16. Chung HM, Li WT, Yuen C, Wen CK, Crespi N (2018) Electric vehicle charge scheduling mechanism to maximize cost efficiency and user convenience. IEEE Trans Smart Grid 10(3):3020–3030
17. Lee ZJ, Lee G, Lee T, Jin C, Lee R, Low Z, Chang D, Ortega C, Low SH (2021) Adaptive charging networks: a framework for smart electric vehicle charging. IEEE Trans Smart Grid 12(5):4339–4350
18. Cao Y, Li D, Zhang Y, Chen X (2020) Joint optimization of delay-tolerant autonomous electric vehicles charge scheduling and station battery degradation. IEEE Internet Things J 7(9):8590–8599
19. Mahato D, Aharwal VK, Sinha A (2023) Multi-objective optimisation model and hybrid optimization algorithm for Electric Vehicle Charge Scheduling. J Exp Theoret Artificial Intell

Fuzzy Logic and ANN in an Artificial Intelligent Cloud: A Comparative Study

Pooja Chopra and Munish Gupta

Abstract Artificial intelligence has a remarkable effect in every field. Various tools of artificial intelligence have proved themselves in different sectors. Whether it is the banking sector, health industry, or any other sector, it is. One such sector is cloud computing. Cloud computing is Internet-enabled technology and has proved to be a boon in information technology. The aim of this research is to study the two tools of AI in the field of cloud computing. In this paper, we are targeting fuzzy logic and ANN as a tool. A comparative investigation has been made by applying these tools in cloud computing-based resource scheduling. Fuzzy logic works very well in cloud computing because it handles uncertainties efficiently. ANN, on the other hand, provides a trained expert machine for predicting future behavior. After comparing the results in the case of both tools, we get that fuzzy logic outperforms artificial neural networks in the field of resource scheduling. This research is very helpful for researchers researching scheduling cloud-based jobs by setting up their priorities based on chosen attributes.

Keywords Cloud computing · Artificial intelligence · Artificial neural network · Fuzzy logic · Quality of service

1 Cloud Computıng-Introduction

Cloud computing is a developing technology based on the Internet where anyone can access information from any place from any device at any time. Cloud computing is a computation method comprising many hardware and software resources that can be provided to consumers as per their demand [1]. Customers can use these resources as

P. Chopra
Lovely Professional University, Punjab, India
e-mail: pooja.27304@lpu.co.in

M. Gupta (✉)
Chandigarh University, Punjab, India
e-mail: gupta.munish2005@gmail.com

G. Rajakumar et al. (eds.), *Intelligent Communication Technologies and Virtual Mobile Networks*, Lecture Notes on Data Engineering and Communications Technologies 171,
https://doi.org/10.1007/978-981-99-1767-9_41

long as needed, and based on usage, they are charged as per resources utilized. The lower investment cost for infrastructure and resources is the main benefit of cloud computing. The customers who require these costly hardware and software resources need not to spent while procuring and maintaining them. Rather, they can arrange these resources from the cloud service provider. The cloud service provider allocates resources when required by the customers and de-allocates them when they are no longer required [2]. Various types of clouds, i.e., private cloud, community cloud, public cloud, and hybrid cloud, provide unlimited service. Broadly, its services are divided into three types.

SAAS: As the name implies, software as a service (SAAS) is another cloud-based service through which software resources can be procured on request by customers on a rental basis. They need not worry about software procuring costs and other related costs like licensing costs, maintenance costs, etc. This will remain a headache for the service provider. A well-known example of SAAS is Gmail.

IAAS: Infrastructure as a service (IAAS) providers provide the customers with demanded hardware resources. Customers are charged periodically for this.

PAAS: The application development's operating environment is provided on demand to consumers in platform as a service (PAAS).

2 Artificial Intelligence Approaches

The first approach in our review is the fuzzy logic that Zadeh proposed in 1965. It proposes humble means to reach a firm inference based on imprecise, unclear, unknown, or mislaid input information. It is a kind of logic that identifies more than simple "yes" and "No" values [3]. Fuzzy logic is widely used in the domains where human expertise is embedded into real-world applications like ACs, automatic washing machines, microprocessors, microcontrollers, and many others. Fuzzifier controls the input and output variables, and linguistic variables are allocated to them with the help of membership functions [4]. The fuzzified values are matched against rules stored in rule base. On the base of that, it attains the fuzzy output. Defuzzification converts this fuzzy output into a single crispy value. ANN is also an artificial intelligence tool that emulates the learning and decision-making abilities of the human brain. When trained, it works as an expert while answering queries. As these AI-based tools have brilliant responses in various applications and their competence to handle uncertainty, fuzzy logic and ANN are also used in cloud computing [5].

3 Applications of Fuzzy Logic and Artificial Neural Network in Cloud Computing

Based on existing literature, we have noticed various claims of fuzzy logic and ANN in different fields of cloud computing. Load balancing is a practice of sharing the data load among the servers, thereby minimizing response time [6] while eliminating a condition in which some nodes are overburdened. In contrast, some others are under burdened [7]. A cloud partitioning concept has been used to generate a load balancing model to improve performance and maintain stability [8]. For load balancing, a fuzzy logic-based firefly algorithm in cloud computing also has been offered where the cloud is segregated into substantial and less burdened nodes. For this, the best suitable fragment for the cloud has opted, and the firefly algorithm makes it possible for the load to get attracted toward that fragment. Fuzzy logic is used in this approach to deal with uncertainties. The planned approach is equated against the genetic algorithm, and the results show that the proposed approach does not have as much execution time while incurring low execution costs [9]. A different fuzzy logic-based approach for effective load balancing also has been offered, where the researcher has fuzzified the load and speed of the processor. Rules have been laid down in the rule base, and after defuzzification, the output comes in the form of a balanced load. The proposed algorithm is evaluated against already existing approaches, and the outcomes have proved that the projected technique minimizes the load, response time, and processing time while enhancing resource utilization [10]. Scheduling is a process by which scarce computer resources are arranged for processes [11]. In this context, fuzzy logic and ANN-based algorithm has been offered where fuzzy logic is employed to fuzzify the categorized inputs. The genetic algorithm in ANN is used to map the computer resources with the processes. Crisp output is obtained with the help of defuzzification process. The offered functionality is deployed on CloudSim and is evaluated against the existing techniques and minimizes bandwidth utilization and completion time while enhancing performance [12]. Another approach also has been proposed, which is based on the allocation of resources and the length of the process. For this, genetic algorithm and fuzzy logic have been used. The proposed approach is evaluated against existing approaches, and the results have shown that the execution cost and execution time are less as compared to existing approaches [13]. Another multiple queue job scheduling approach based on fuzzy logic has been offered in the field of cloud computing, thereby reducing waiting time and response time compared to conventional techniques [14]. In this context, an algorithm has been offered by using fuzzy logic for the arrangement of resources while improving the consistency of cloud computing. For this, input parameters, i.e., cost, job dimension, and trust values, are fuzzified. The resulting outcome of the proposed method is the precedence based on which resources will be assigned to the processes. The offered system gets compared with other existing techniques, and the outcomes have revealed minimization in waiting time, turnaround time, correctness, and dependability [15]. Another abstract model has been designed to evaluate and prioritize the hazard that may happen if the company wants to opt for cloud computing services. For this, fuzzy logic is employed to handle uncertainties and evaluate the risk rate with the possibility and effect of risk. The results have been debated by several beings and proved that decision taken using fuzzy logic is more advantageous for assessing risk and managing processes [16]. For this concern, a fuzzy logic-based technique has been proposed for predicting the behavior, load, and performance based on imprecise information. The fuzzification and the defuzzification processes have been carried out for the scale control module that increases and decreases the use of VMs. This approach has been evaluated against the conventional approach and results in an improved scaling process of VMs while reducing SLA violation and refining QOS parameters [17].

Another expert has offered an estimation approach for gauging the quality of cloud service (QOCS) where the cloud service providers are evaluated following their services provided per user demands. With this approach, the uncertainties in the services provided can also be calculated based on QOCS data. Comparisons have been made with existing techniques, i.e., "QOSC" and "SLC". Results have proved that the projected approach performs better than the conventional approach concerning correctness, time, and price [18]. For trust assessment, another model has been offered depending upon the cloud-based services. The trust value is obtained with the help of fuzzy logic. Various inputs are considered, and at every request, these are measured and stored in the database at different time slots. Then their normalized values are fuzzified and defuzzied, resulting in the trust value for the time slot. Then, the global trust value is calculated using induced weighted aggregating operator (IOWA) to calculate the dependability of the service. Fuzzy logic is used in dynamic cloud-based services to deal with uncertainties such as traits [19]. Fuzzy logic has also been used in cloud computing to design and implement power monitoring and control systems to automatically adjust the running time of electrical appliances we utilize in our daily lives. The fuzzy logic part helps to calculate the running time of appliances. With cloud computing, several computers and other devices can be connected and utilized to get evidence about the power that electrical appliances consume. The proposed system helps in efficient energy utilization while saving power [20]. The neural network has applications for future workload prediction and adequate resource provisioning. Research shows that the neural network model does almost nine times less over-allocation than the static model [21–23]. An error and recovery monitoring system based on neural networks are provided for cloud computing while reducing costs compared to existing models [24]. Artificial neural networks and fuzzy logic are used to process huge amounts of data, i.e., big data. For this, a multilayer neural network based on MapReduce to process big data is used. For training back, the propagation algorithm is used. The proposed model results in improved speed and convergence rate compared to existing algorithm [25].

4 Comparison of Fuzzy Logic and ANN in Cloud-Based Resource Scheduling

For a comparison of fuzzy logic and artificial neural networks, we have used the resource scheduling field to predict the sequence in which CPU-intensive resources will be allocated to the jobs. The results of both cases are compared. While comparing, we have checked which one is more accurate for predicting the priority in cloud-based resource scheduling. Based on input parameters, i.e., job's importance, waiting time, and burst time, and output parameters, i.e., urgency, it is calculated with the help of both models. For this, linguistic terms have been allocated to all input and output parameters in terms of extreme, high, moderate, less, and very less. Implementing membership functions in the case of input and output parameters has been done in MATLAB. Mathematically, we are defining them as follows.

4.1 *Importance (Input Parameter)*

The membership function (denoted by μ) for low importance has been defined as follows in Eq. (1). Here, we have taken a triangular membership function where the low importance range is between 0.02684 and 0.5261. It is zero if it goes beyond this range; otherwise, its value will be calculated as given in the following equation.

$$\mu_L(x) = \begin{cases} 0 & 0 \text{ if } X \leq 0.02684 \\ \frac{X-0.2684}{0.2765-0.02684} & \text{if } X \in (0.02684,\ 0.2765) \\ \frac{0.5261-X}{0.5261-0.2765} & \text{if } X \in (0.02765,\ 0.5261) \\ 0 & 0 \text{ if } X \geq 0.5261 \end{cases} \tag{1}$$

4.2 *Waiting Time (Input Parameter)*

The Gaussian membership function has defined the membership functions for low waiting and high waiting times. They are defined as follows in Eqs. (2 and 3).

$$\mu_1(x) = \left[\exp -(x - 0.25)^2/2 * 0.1062^2\right] \tag{2}$$

$$\mu_H(x) = \left[\exp -(x - 0.75)^2/2 * 0.1062^2\right] \tag{3}$$

4.3 Burst Time (Input Parameter)

The membership function for medium burst time has been defined as follows in Eq. (4). Here, we have taken a triangular membership function where the range for medium importance is between 0.02499 and 0.7501. It is zero in case it goes beyond this range; otherwise, its value will be calculated as given in the following equation.

$$\mu_M(x) = \begin{cases} 0 & \text{if } x \leq 0.2499 \\ \frac{x-0.2499}{0.5-0.2499} & \text{if } x \in (0.02499,\ 0.5) \\ \frac{0.7501-x}{0.7501-0.5} & \text{if } x \in (0.5,\ 0.7501) \\ 0 & \text{if } x \geq\ 0.7501 \end{cases} \quad (4)$$

Rule base has been defined to get the result. The above approach was with fuzzy logic.

In the ANN approach, Levenberg–Marquardt optimization has been used to train the model. This is one of the fastest back propagation algorithms that adjust weights and bias values until it gets an optimal solution. Both cases, i.e., fuzzy logic and artificial neural networks have been implemented in MATLAB [26]. Both cases' results and expert opinions are presented in Tables 1 and 2.

The average consistency score in the case of artificial neural networks is 12.05714. The graph for the above comparison based on consistency scores is as follows in Fig. 1.

Figure 1 shows that concerning maintaining consistency with human experts, fuzzy logic-based system outperforms artificial neural network-based system.

5 Results and Findings

The results of both approaches have been compared with industrial experts' opinions. Few expert opinions are shown in Tables 1 and 2. For this, we have given consistency scores to both approaches, as shown in both tables. If the system output is the same as the expert output, it will get the maximum score, i.e., 5. If the system output is one less than the experts' output, it will get one score less than the maximum, i.e., four, and so on. Then these scores are added and averaged. The same approach is followed for both artificial intelligence approaches. After calculating averages in consistency tables and looking at pictorial representation, we conclude that out of the two intelligence techniques, fuzzy logic performs well in cloud-based resource scheduling. Also, the system is designed with fuzzy logic responses following the expert's intelligence. We are trying to make our system mimic human beings. Human nature is uncertain. Fuzzy logic is best for dealing with such types of uncertainties. However, artificial neural network, part of deep learning, is best suited for predictions.

Table 1 Fuzzy logic approach

S. no.	Inputs			FL-output	Expert opinion			Total consistency score
	Importance	Waiting time	Burst time		1	2	3	
1	EX	EX	MOD	HI	HI (5)	EX (4)	HI (5)	14
2	EX	EX	LE	HI	EX (4)	EX (4)	HI (5)	13
3	EX	HI	LE	HI	EX (4)	EX (4)	HI (5)	13
4	EX	MOD	VL	EX	HI (4)	EX (5)	EX (5)	14
5	EX	LE	HI	MOD	HI (4)	HI (4)	HI (4)	12
6	EX	LE	LE	HI	MOD (4)	HI (5)	HI (5)	14
7	HI	HI	EX	HI	MOD (4)	MOD (4)	HI (5)	13
8	HI	LE	EX	MOD	MOD (5)	MOD (5)	MOD (5)	15
9	HI	LE	MOD	HI	MOD (4)	MOD (4)	MOD (4)	12
10	MOD	HI	EX	LE	LE(5)	LE(5)	LE(5)	15
11	MOD	HI	MOD	MOD	MOD (5)	MOD (5)	MOD (5)	15
12	MOD	HI	VL	HI	HI (5)	HI (5)	HI (5)	15
13	MOD	MOD	LE	MOD	MOD (5)	MOD (5)	MOD (5)	15
14	MOD	LE	MOD	LE	LE(5)	LE(5)	LE(5)	15
15	LE	HI	MOD	LE	LE(5)	LE(5)	LE(5)	15
16	LE	MOD	LE	MOD	LE(4)	LE(4)	MOD (5)	13
17	LE	MOD	VL	MOD	MOD (5)	MOD (5)	MOD (5)	15
18	LE	LE	VL	MOD	VL(3)	VL(3)	MOD (5)	11
19	LE	VL	HI	VL	VL(5)	VL(5)	VL(5)	15
20	VL	HI	MOD	LE	LE(5)	VL(4)	LE(5)	14
21	VL	MOD	HI	VL	VL(5)	VL(5)	VL(5)	15
22	VL	MOD	VL	MOD	LE(4)	LE(4)	MOD (5)	13
23	VL	LE	EX	LE	VL(4)	VL(4)	LE(5)	13
24	VL	VL	EX	LE	VL(4)	VL(4)	LE(5)	13
25	EX	HI	MOD	HI	EX (4)	EX (4)	HI (5)	13
26	EX	LE	VL	HI	EX (4)	EX (4)	HI (5)	13
27	HI	EX	HI	HI	EX (4)	EX (4)	HI (5)	13

(continued)

Table 1 (continued)

	Inputs			FL-output	Expert opinion			Total consistency score
S. no.	Importance	Waiting time	Burst time		1	2	3	
28	MOD	LE	VL	MOD	MOD (5)	MOD (5)	MOD (5)	15
29	MOD	VL	VL	MOD	MOD (5)	MOD (5)	MOD (5)	15
30	LE	EX	LE	MOD	MOD (5)	MOD (5)	MOD (5)	15
31	LE	HI	LE	MOD	LE(4)	LE(4)	LE(4)	12
32	LE	HI	VL	MOD	MOD (5)	MOD (5)	MOD (5)	15
33	LE	LE	MOD	LE	LE(5)	LE(5)	LE(5)	15
34	VL	HI	HI	MOD	VL(3)	VL(3)	MOD (5)	11
35	VL	MOD	LE	MOD	VL(3)	VL(3)	MOD (5)	11
Average consistent score								13.71

Source 1 Author's Computation
Note 1 Extreme = EX, Moderate = MOD, High = HI, Less = LE, Very Less = VL
The average consistency score in the case of fuzzy logic is 13.71.

6 Conclusion and Future Work

As a future work, the results calculated with fuzzy logic will be implemented in CloudSim, where highly prioritized jobs will be scheduled in hybrid cloud environments. For that, we will be deploying results from the fuzzy logic approach for setting priorities for the jobs and routing them to the appropriate cloud in case of a hybrid cloud-based environment. We are planning to take public and private cloud models where the proposed scheduling scheme is used to route process to the appropriate cloud that will eventually serve the jobs better by reducing the response time.

Table 2 Artificial neural network approach

	Inputs			ANN-output	Expert opinion			Total consistency score
S. no.	Importance	Waiting time	Burst time		1	2	3	
1	EX	EX	MOD	MOD	HI (4)	EX (3)	HI (4)	11
2	EX	EX	LE	HI	EX (4)	EX (4)	HI (5)	13
3	EX	HI	LE	HI	EX (4)	EX (4)	HI (5)	13
4	EX	MOD	VL	HI	HI (5)	EX (4)	EX (4)	13
5	EX	LE	HI	HI	HI (5)	HI (5)	HI (5)	15
6	EX	LE	LE	EX	MOD (3)	HI (4)	HI (4)	11
7	HI	HI	EX	HI	MOD (4)	MOD (4)	HI (5)	13
8	HI	LE	EX	MOD	MOD (5)	MOD (5)	MOD (5)	15
9	HI	LE	MOD	HI	MOD (4)	MOD (4)	MOD (4)	12
10	MOD	HI	EX	HI	LE (3)	LE (3)	LE (3)	9
11	MOD	HI	MOD	MOD	MOD (5)	MOD (5)	MOD (5)	15
12	MOD	HI	VL	HI	HI (5)	HI (5)	HI (5)	15
13	MOD	MOD	LE	HI	MOD (4)	MOD (4)	M(4)	12
14	MOD	LE	MOD	MOD	LE (4)	LE (4)	LE (4)	12
15	LE	HI	MOD	MOD	LE (4)	LE (4)	LE (4)	12
16	LE	MOD	LE	HI	LE (3)	LE (3)	MOD (4)	10
17	LE	MOD	VL	HI	MOD (4)	MOD (4)	MOD (4)	12
18	LE	LE	VL	MOD	VL (3)	VL (3)	MOD (5)	11
19	LE	VL	HI	MOD	VL (3)	VL (3)	VL (3)	9
20	VL	HI	MOD	HI	LE (3)	VL (2)	LE (3)	8
21	VL	MOD	HI	MOD	VL (3)	VL (3)	VL (3)	9
22	VL	MOD	VL	MOD	LE (4)	LE (4)	MOD (5)	13
23	VL	LE	EX	MOD	VL (3)	VL (3)	LE(4)	10
25	EX	HI	MOD	HI	EX (4)	EX (4)	HI (5)	13
26	EX	LE	VL	HI	EX (4)	EX (4)	HI (5)	13
27	HI	EX	HI	HI	EX (4)	EX (4)	HI (5)	13

(continued)

Table 2 (continued)

	Inputs			ANN-output	Expert opinion			Total consistency score
S. no.	Importance	Waiting time	Burst time		1	2	3	
28	MOD	LE	VL	HI	MOD (4)	MOD (4)	MOD (4)	12
29	MOD	VL	VL	MOD	MOD (5)	MOD (5)	MOD (5)	15
30	LE	EX	LE	MOD	MOD (5)	MOD (5)	MOD (5)	15
31	LE	HI	LE	HI	LE (3)	LE (3)	LE (3)	9
32	LE	HI	VL	HI	MOD (4)	MOD (4)	MOD (4)	12
33	LE	LE	MOD	LE	LE (5)	LE (5)	LE (5)	15
34	VL	HI	HI	MOD	VL (3)	VL (3)	MOD (5)	11
35	VL	MOD	LE	MOD	VL (3)	VL (3)	MOD (5)	11
Average Consistent score								12.057

Source 2 Author's Computation
Note 2: Extreme = EX, Moderate = MOD, High = HI, Less = LE, Very Less = VL

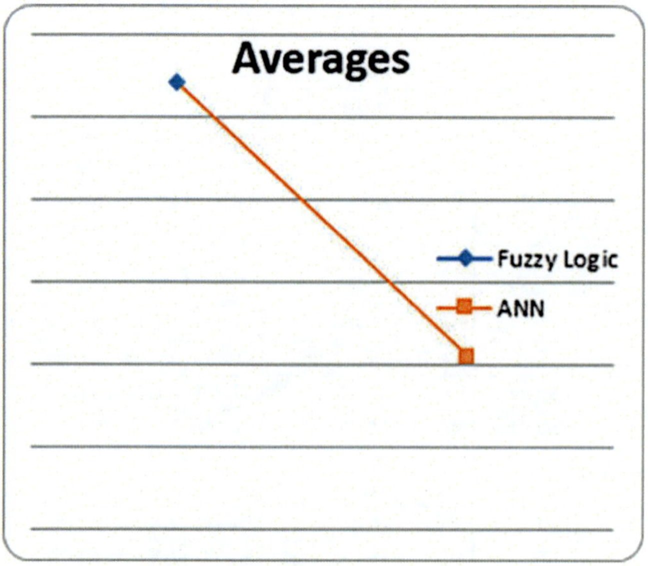

Fig. 1 Consistency scores. *Source* 3: Author's generation

References

1. Shawish, Salama M (2014) Cloud computing: paradigms and technologies. Studies Comp Intell 495:39–67. https://doi.org/10.1007/978-3-642-35016-0_2/COVER
2. Ali M (2009) Green cloud on the horizon. Lecture Notes in Computer Science (including subseries Lecture Notes in Artificial Intelligence and Lecture Notes in Bioinformatics), vol 5931. LNCS, pp. 451–459. https://doi.org/10.1007/978-3-642-10665-1_41/COVER
3. Gelfond M, Lifschitz V (1993) Representing action and change by logic programs. J Logic Program 17(2–4):301–321. https://doi.org/10.1016/0743-1066(93)90035-F
4. Thaker S, Nagori V (2018) Analysis of fuzzification process in fuzzy expert system. Procedia Comput Sci 132:1308–1316. https://doi.org/10.1016/J.PROCS.2018.05.047
5. Mellit A, Kalogirou SA (2008) Artificial intelligence techniques for photovoltaic applications: a review. Prog Energy Combust Sci 34(5):574–632. https://doi.org/10.1016/J.PECS.2008.01.001
6. Xu M, Tian W, Buyya R (2017) A survey on load balancing algorithms for virtual machines placement in cloud computing. Concurr Comput 29(12):e4123. https://doi.org/10.1002/CPE.4123
7. Rao A, Lakshminarayanan K, Surana S, Karp R, Stoica I (2003) Load balancing in structured P2P systems. Lecture Notes in Computer Science (including subseries Lecture Notes in Artificial Intelligence and Lecture Notes in Bioinformatics) 2735:68–79. https://doi.org/10.1007/978-3-540-45172-3_6/COVER
8. Remesh Babu KR, Samuel P (2016) Enhanced bee colony algorithm for efficient load balancing and scheduling in cloud. Adv Intell Syst Comp 424:67–78. https://doi.org/10.1007/978-3-319-28031-8_6/COVER
9. Wang J et al (2017) Planning virtual infrastructures for time critical applications with multiple deadline constraints. Futur Gener Comput Syst 75:365–375. https://doi.org/10.1016/J.FUTURE.2017.02.001
10. Gawali MB, Shinde SK (2018) Task scheduling and resource allocation in cloud computing using a heuristic approach. J Cloud Comp 7(1):1–16. https://doi.org/10.1186/S13677-018-0105-8/FIGURES/15
11. Pradhan P, Behera PK, Ray BNB (2016) Modified Round Robin Algorithm for resource allocation in cloud computing. Procedia Comput Sci 85:878–890. https://doi.org/10.1016/J.PROCS.2016.05.278
12. Rekha PM, Dakshayini M (2019) Efficient task allocation approach using genetic algorithm for cloud environment. Cluster Comput 22(4):1241–1251. https://doi.org/10.1007/S10586-019-02909-1/METRICS
13. Albelwi S, Mahmood A (2017) A framework for designing the architectures of deep convolutional neural networks. Entropy 19(6):242. https://doi.org/10.3390/E19060242
14. ben Alla H, ben Alla S, Touhafi A, Ezzati A (2018) A novel task scheduling approach based on dynamic queues and hybrid meta-heuristic algorithms for cloud computing environment. Cluster Comput 21(4):1797–1820. https://doi.org/10.1007/S10586-018-2811-X/METRICS
15. Devaraj AFS, Elhoseny M, Dhanasekaran S, Lydia EL, Shankar K (2020) Hybridization of firefly and improved multi-objective particle swarm optimization algorithm for energy efficient load balancing in cloud computing environments. J Parallel Distrib Comput 142:36–45. https://doi.org/10.1016/J.JPDC.2020.03.022
16. Abreu A, Martins JDM, Calado JMF (2018) Fuzzy logic model to support risk assessment in innovation ecosystems. In: 13th APCA International Conference on Control and Soft Computing, CONTROLO 2018 - Proceedings, pp. 104–109, Oct. https://doi.org/10.1109/CONTROLO.2018.8514281
17. Mubeen A, Ibrahim M, Bibi N, Baz M, Hamam H, Cheikhrouhou O (2021) Alts: an adaptive load balanced task scheduling approach for cloud computing. Processes 9:1514. https://doi.org/10.3390/PR9091514
18. McNally S, Roche J, Caton S (2018) Predicting the price of bitcoin using machine learning. In: Proceedings—26th Euromicro International Conference on Parallel, Distributed, and Network-Based Processing, PDP 2018, pp 339–343. https://doi.org/10.1109/PDP2018.2018.00060

19. Moura BMP, Schneider GB, Yamin AC, Santos H, Reiser RHS, Bedregal B (2022) Interval-valued Fuzzy Logic approach for overloaded hosts in consolidation of virtual machines in cloud computing. Fuzzy Sets Syst 446:144–166. https://doi.org/10.1016/J.FSS.2021.03.001
20. Ahdan S, Susanto ER, Syambas NR (2019) Proposed design and modeling of smart energy dashboard system by implementing IoT (internet of things) based on mobile devices. In: TSSA 2019—13th International Conference on Telecommunication Systems, Services, and Applications, Proceedings, pp 194–199. https://doi.org/10.1109/TSSA48701.2019.8985492
21. Yang R, Ouyang X, Chen Y, Townend P, Xu J (2018) Intelligent resource scheduling at scale: a machine learning perspective. In: Proceedings—12th IEEE International Symposium on Service-Oriented System Engineering, SOSE 2018 and 9th International Workshop on Joint Cloud Computing. JCC 2018:132–141. https://doi.org/10.1109/SOSE.2018.00025
22. Ali Kumar DSNK, Newaz SHS, Rahman FH, Lee GM, Karmakar G, Au TW (2022) Green demand aware fog computing: a prediction-based dynamic resource provisioning approach. Electronics 11:608. https://doi.org/10.3390/ELECTRONICS11040608
23. Mostofi VME, Krishnamurthy D, Arlitt M (2021) Fast and efficient performance tuning of microservices. In: IEEE International Conference on Cloud Computing, CLOUD, September, pp 515–520. https://doi.org/10.1109/CLOUD53861.2021.00067
24. Ragmani A, Elomri A, Abghour N, Moussaid K, Rida M, Badidi E (2020) Adaptive fault-tolerant model for improving cloud computing performance using artificial neural network. Procedia Comput Sci 170:929–934. https://doi.org/10.1016/J.PROCS.2020.03.106
25. Yao J, Qi J, Zhang J, Shao H, Yang J, Li X (2021) A real-time detection algorithm for Kiwifruit defects based on YOLOv5. Electronics 10(14):1711. https://doi.org/10.3390/ELECTRONICS10141711
26. Fathi M, Parian JA (2021) Intelligent MPPT for photovoltaic panels using a novel fuzzy logic and artificial neural networks based on evolutionary algorithms. Energy Rep 7:1338–1348. https://doi.org/10.1016/J.EGYR.2021.02.051

Harnessing the Power of AI to Create Intelligent Tutoring Systems for Enhanced Classroom Experience and Improved Learning Outcomes

Ashraf Alam

Abstract Artificial intelligence (AI) is used to personalise learning experiences for students, adapt to their individual needs and abilities, and provide real-time feedback on their progress, whereas virtual and augmented reality (VR/AR) are used to create immersive learning experiences that allow students to explore and interact with virtual environments and simulations. Online learning platforms provide students with access to educational resources and courses from anywhere in the world and allow for greater flexibility in terms of when and where they learn. Adaptive learning is a form of technology-enabled learning that adjusts to the student's learning style, pace, and progress. This is done using algorithms that analyse student data, such as their performance on assessments, and adjust the content or pedagogy accordingly. There is also a wide-ranging emphasis on gamification for teaching–learning. Incorporating game-like elements into the learning process makes it more engaging and interactive for students. AI has completely revolutionised the formal education space. Thus, in this article, the researcher investigates how the most advanced technologies are currently being developed and used to enhance the way we educate students, and how it is integrated into the curriculum and classroom.

Keywords Classroom · Curriculum · Teacher · University · Pedagogy · Artificial intelligence · Educational technology · ICT in education · Learner

1 Introduction

Integration of AI into the Curriculum, Pedagogy, and Classroom Learning

Artificial intelligence today is implemented and effectively integrated into the curriculum, pedagogy, and classroom learning in several ways [1]. AI is used to personalise the learning experience for each student by analysing data on their

A. Alam (✉)
Rekhi Centre of Excellence for the Science of Happiness, Indian Institute of Technology Kharagpur, Kharagpur, West Bengal, India
e-mail: ashraf_alam@kgpian.iitkgp.ac.in

G. Rajakumar et al. (eds.), *Intelligent Communication Technologies and Virtual Mobile Networks*, Lecture Notes on Data Engineering and Communications Technologies 171,
https://doi.org/10.1007/978-981-99-1767-9_42

strengths and weaknesses and adapting the curriculum and teaching methods accordingly. AI-based intelligent tutoring systems provide students with immediate feedback and guidance, as well as help teachers identify areas where students need extra support [2]. AI is used to create adaptive assessments that adjust the difficulty level and content of questions based on the student's performance. AI helps automate administrative tasks such as grading, attendance, and scheduling, freeing up more time for teachers to focus on instruction [3]. AI is used to generate personalised learning materials, such as quizzes, practice problems, and summaries that are tailored to each student's needs. AI-powered virtual teaching assistants assist teachers in monitoring student progress, providing feedback, and answering student questions in real time.

2 Personalised Learning Experience

There are several ways in which AI is used to personalise the learning experience for each student by analysing data on their strengths and weaknesses and adapting the curriculum and teaching methods [4]. AI-powered adaptive learning systems analyse student data such as performance on assessments, progress, and learning style and use this information to adjust the curriculum, teaching methods, and learning materials accordingly [5]. AI-based student modelling is used to create a representation of each student's knowledge, skills, and preferences, which is used to personalise the learning experience. AI-based learning analytics is used to analyse data on student behaviour, such as how long they spend on a particular task or how often they access certain resources [6]. This information is used to identify areas where students need extra support and adjust the curriculum and teaching methods accordingly. AI-based predictive modelling is used to predict student performance and identify areas where students are at risk of falling behind. This information is used to provide targeted interventions and support. AI-powered natural language processing is used to analyse student writing and speech, providing teachers with insights into student understanding and language proficiency [7]. By using these AI-based techniques, teachers gain a more detailed understanding of each student's strengths and weaknesses and use this information to personalise the learning experience.

3 Intelligent Tutoring Systems for Immediate Feedback and Guidance to Students

AI-based intelligent tutoring systems (ITS) provide students with immediate feedback and guidance by using AI techniques [8]. ITS use natural language processing to understand students' responses and provide feedback that is tailored to their level of understanding. ITS use rule-based systems to provide students with step-by-step guidance on solving a problem and give feedback on their progress. ITS use case-based reasoning to provide students with examples of similar problems and solutions and give feedback on how well they are applying the concepts [9]. ITS use machine learning algorithms to adapt to each student's learning style and provide feedback that is tailored to their individual needs. These AI techniques help ITS provide students with immediate feedback and guidance as they work through problems, helping them to identify and correct mistakes, and providing them with a deeper understanding of the material [10]. In addition to providing students with immediate feedback and guidance, ITS also help teachers identify areas where students need extra support by collecting data on student performance, such as which problems are being answered correctly and which are not [11]. Teachers then use this data to identify areas where students are struggling and provide targeted interventions, such as additional practice problems or one-on-one tutoring [12].

4 Adaptive Assessments

AI is used to create adaptive assessments that adjust the difficulty level and content of questions based on the student's performance [13]. Item response theory (IRT) is a statistical model that is used to create adaptive assessments. It is used to determine the difficulty level of each question and adjust the difficulty level of subsequent questions based on the student's performance. Machine learning algorithms are used to analyse student performance data and adjust the difficulty level and content of questions based on the student's level of understanding. Bayesian knowledge tracing (BKT) is a machine learning algorithm that is used to track student knowledge and understanding of specific concepts over time. This information is used to adjust the difficulty level and content of questions based on the student's current level of understanding. Predictive modelling is used to predict student performance based on the previous performance and adjust the difficulty level and content of questions accordingly. Natural language processing (NLP) is used to understand student responses and adjust the difficulty level and content of questions based on the student's level of understanding. By using these AI-based techniques, adaptive assessments are created that adjust the difficulty level and content of questions based on the student's performance, providing a more personalised and efficient learning experience.

5 Automating Administrative Tasks: Grading, Attendance, and Scheduling

AI helps teachers automate administrative tasks such as grading, attendance, and scheduling [14]. AI-powered systems are used to grade written assignments, such as essays and short-answer questions, by using techniques such as natural language processing and machine learning. These systems automatically grade the assignments and provide feedback to students, allowing teachers to focus on more complex tasks. AI-powered systems are used to track student attendance by using techniques such as facial recognition and machine learning. These systems automatically mark students as present or absent and generate reports for teachers, freeing up time for other tasks. AI-powered systems are also used to schedule classes, meetings, and other events by using techniques such as natural language processing, machine learning, and optimisation algorithms. These systems automatically schedule events based on the availability of teachers, students, and resources, reducing the need for manual scheduling. AI-powered systems analyse student data and provide teachers with insights into student progress, identify areas where students need extra support, and generate reports. By automating these administrative tasks, AI helps free up time for teachers to focus on instruction, allowing them to spend more time interacting with students and providing personalised support.

6 Personalised Learning Materials

AI is used to generate personalised learning materials, such as quizzes, practice problems, and summaries that are tailored to each student's needs [15]. Adaptive learning systems use machine learning algorithms to analyse student performance data and adjust the content and difficulty level of quizzes, practice problems, and summaries based on the student's level of understanding. NLP is used to understand student responses and generate personalised feedback and summaries based on their understanding. Predictive modelling is used to predict student performance based on the previous performance and generate personalised quizzes and practice problems that are tailored to the student's current level of understanding. AI-powered knowledge maps are used to generate personalised learning materials by mapping out the relationships between different concepts and identifying areas where the student needs extra support. AI-powered chatbots are used to generate personalised quizzes and practice problems by engaging students in natural language conversations and adjusting the content and difficulty level based on their responses. By using these AI-based techniques, personalised learning materials are generated that are tailored to each student's needs, providing a more efficient and effective learning experience.

7 Virtual Teaching Assistants

AI-powered virtual teaching assistants assist teachers in monitoring student progress, providing feedback, and answering student questions in real time [16]. AI-powered virtual teaching assistants use machine learning algorithms to analyse student performance data, such as test scores, homework assignments, and engagement metrics, to provide teachers with insights into student progress. This helps teachers identify areas where students need extra support and adjust their instruction accordingly. AI-powered virtual teaching assistants use natural language processing to understand student responses and generate personalised feedback in real-time [17]. This helps students receive immediate feedback on their understanding and improve their performance. AI-powered virtual teaching assistants use natural language processing and machine learning to understand student questions and provide accurate, relevant answers in real time. This helps students receive immediate support and reduce the burden on teachers [18]. AI-powered virtual teaching assistants use machine learning algorithms to identify students who are having difficulties and flag them to the teacher. AI-powered virtual teaching assistants also provide students with additional resources, such as videos, articles, and interactive simulations to help them understand difficult concepts. By using these AI-based techniques, virtual teaching assistants assist teachers in monitoring student progress, providing feedback, and answering student questions in real time.

8 Adaptive Learning Systems for Analysing Student Data to Adjust the Curriculum, Teaching Methods, and Learning Materials

AI-powered adaptive learning systems use machine learning algorithms to analyse student data such as performance on assessments, progress, and learning style [19]. This data is then used to adjust the curriculum, teaching methods, and learning materials to better suit the needs of individual students. When a student interacts with the adaptive learning system, the system collects data on the student's performance, such as their answers to questions, their response time, and the difficulty level of the questions they are answering. This data is then analysed by the system's machine learning algorithms to identify patterns and trends in the student's performance. Based on this analysis, the system adjusts the curriculum, teaching methods, and learning materials to better meet the needs of the student [20]. For example, if the system identifies that a student is struggling with a particular concept, it provides additional resources and explanations to help the student understand the concept better. If the system identifies that a student is excelling in a particular subject, it provides more challenging material to help the student continue to progress. The system also considers the student's learning style when adjusting the curriculum, teaching methods, and learning materials. For example, if a student is an auditory learner, the system provides more audio resources to supplement the material. Overall, the goal of an AI-powered adaptive learning system is to personalise the learning experience for each student, by providing them with the resources and support they need to succeed.

9 Student Modelling for the Representation of Each Student's Knowledge, Skills, and Preferences

AI-based student modelling is a process used to create a representation of each student's knowledge, skills, and preferences [21]. This representation is used to personalise the learning experience for the student, by providing them with resources and support that are tailored to their specific needs. The process of AI-based student modelling begins with the collection of data on the student's performance, such as their answers to questions, their response time, and the difficulty level of the questions they are answering. This data is then analysed by machine learning algorithms to identify patterns and trends in the student's performance. Based on this analysis, the system creates a representation of the student's knowledge, skills, and preferences, which is known as a student model. The student model includes information such as the student's strengths and weaknesses, their learning style, and their interests. Once the student model is created, it is used to personalise the learning experience for the student. For example, if the student model indicates that the student is struggling with a particular concept, the system provides additional resources and explanations to help the student understand the concept better. If the student model indicates that the student is excelling in a particular subject, the system provides more challenging material to help the student continue to progress [22]. Overall, AI-based student modelling is a powerful tool for personalising the learning experience for each student, by providing them with the resources and support they need to succeed.

10 Learning Analytics

AI-based learning analytics is a process used to analyse data on student behaviour, such as how long they spend on a particular task or how often they access certain resources [23]. This information is used to understand how students are interacting with the learning materials and to identify patterns and trends in their behaviour. The process of AI-based learning analytics begins by collecting data on student behaviour, such as how long they spend on a particular task, how often they access certain resources, and how they interact with the learning materials. This data is then analysed by machine learning algorithms to identify patterns and trends in the student's behaviour. Based on this analysis, the system generates insights into the student's learning process and the effectiveness of the learning materials. For example, if the data shows that a student is spending a lot of time on a particular task, it could indicate that the task is challenging for them. On the other hand, if the data shows that a student is quickly completing a task, it could indicate that the task is too easy for them. The system also uses this data to identify students who may be at risk of falling behind or who may need additional support. For example, if the data shows that a student is not accessing certain resources or is not spending enough time on a particular task, it could indicate that the student needs additional support to understand the material. Additionally, the system uses learning analytics data to identify patterns across a group of students, such as common misconceptions or areas where most students are struggling. This help educators adjust the curriculum and teaching methods to address these issues more effectively. Overall, AI-based learning analytics is a powerful tool for understanding how students interact with learning materials and for identifying areas where students may need additional support. It helps educators to make data-driven decisions on how to improve the learning experience for their students.

11 NLP to Analyse Student Writing and Speech

AI-powered natural language processing (NLP) is a technology that uses machine learning algorithms to analyse text and speech data, providing teachers with insights into student understanding and language proficiency [24]. NLP is used to analyse a wide range of student data, such as written essays, speech recordings, and dialogue transcripts. The process of using NLP to analyse student writing and speech begins with the collection of the student data, such as written essays, speech recordings, and dialogue transcripts. This data is then processed by NLP algorithms, which extract various features such as grammar, vocabulary, and sentence structure. Once the data is processed, the NLP algorithms analyse the student's language proficiency by comparing their writing and speech to a reference dataset of text written by native speakers. This provides teachers with information on the student's mastery of grammar, vocabulary, and sentence structure, as well as their fluency in the language. Additionally, NLP is used to analyse the content of the student's writing and speech, providing teachers with insights into their understanding of the subject matter. For example, NLP is used to identify key concepts and themes in the student's writing, as well as their ability to use specific vocabulary and terminology related to the subject. Furthermore, NLP is used to analyse the sentiment and tone of the student's writing and speech, which provide teachers with information on the student's attitude and engagement with the subject matter. Overall, AI-powered NLP is a valuable tool for teachers, providing them with insights into student understanding and language proficiency.

One example of an NLP algorithm that is used to analyse student writing and speech is sentiment analysis. Sentiment analysis is a process of determining the emotional tone behind a piece of text, such as a sentence or a paragraph. The algorithm uses natural language processing techniques to identify and extract subjective information from the text, such as opinions, evaluations, appraisals, and emotions. The output of the algorithm is a score or a label that indicates the overall sentiment of the text, such as positive, negative, or neutral. Sentiment analysis is used to evaluate student writing and speech, such as essays, speeches, and presentations, to provide feedback on the emotional tone and impact of the communication.

NLP is used in the following ways: (1) Automated Essay Grading: by understanding the student's writing style and level of understanding, the algorithm adjusts the difficulty level and content of the next writing prompt. (2) Dialogue-Based Tutoring Systems: NLP is used to understand student responses in dialogue-based tutoring systems, such as chatbots. The algorithm analyses the student's responses and provides feedback that is tailored to their level of understanding. For example, if a student is struggling with a particular concept, the algorithm may provide additional resources and explanations to help them understand. (3) Speech Recognition for Oral Assessments: NLP is used to analyse speech and understand student responses in oral assessments. The algorithm recognises the student's voice, transcribe their speech, and analyse the content of their response. Based on the student's level of understanding, the algorithm adjusts the difficulty level and content of the next question.

12 Sentiment Analysis Algorithms

Sentiment analysis algorithms determine the emotional tone behind a piece of text by analysing the words and phrases used in the text. There are several techniques that are commonly used in sentiment analysis, including:

Lexicon-based methods: This technique uses a predefined lexicon, or dictionary, of words and their associated sentiment scores to analyse the text. The algorithm counts the number of positive, negative, and neutral words in the text and calculates an overall sentiment score based on the counts.

Machine Learning-Based Methods: This technique uses machine learning algorithms to learn the sentiment of a text from a training dataset. The algorithm is trained on a dataset of labelled texts (i.e., texts that have been labelled as positive, negative, or neutral) and learns to classify new texts based on the patterns, it has learned from the training dataset.

Neural network-based methods: This technique uses neural networks, a type of machine learning algorithm, to analyse the text. Neural networks are trained on large amounts of labelled text data and learn to understand the meaning and sentiment of the text.

The most common approach is to use a combination of these techniques and to use pre-trained models that have been trained on large datasets of labelled text. Once the algorithm has determined the sentiment of the text, it will output a score or a label indicating the overall sentiment of the text, such as positive, negative, or neutral.

13 Intelligent Tutoring Systems (ITS) to Understand Students' Responses and Provide Feedback That Is Tailored to Their Level of Understanding

When a student interacts with an ITS, the system uses NLP algorithms to analyse the student's responses. The system extracts various features such as grammar, vocabulary, and sentence structure, allowing it to understand the student's level of language proficiency. Additionally, NLP is used to identify key concepts and themes in the student's responses, providing insight into their understanding of the subject matter. Once the ITS have analysed the student's response, it uses this information to provide feedback that is tailored to the student's level of understanding. For example, if the student's response indicates that they have a strong understanding of the subject matter, the ITS may provide more advanced or challenging feedback. Furthermore, ITS also use NLP to understand the sentiment and tone of the student's responses, which provide insight into the student's attitude and engagement with the subject matter. For example, if the student's responses indicate a negative attitude or low engagement, the ITS may provide feedback or resources that are designed to motivate or re-engage the student. Additionally, ITS use NLP to generate feedback in a natural, human-like way, making it more easily understandable and relatable for the students.

14 Rule-Based Systems on Providing Students with Step-By-Step Guidance on Solving a Problem

Intelligent tutoring systems (ITS) use rule-based systems to provide students with step-by-step guidance on solving a problem and give feedback on their progress [25]. Rule-based systems are a type of expert system that uses a set of predefined rules to make decisions and provide guidance. When a student interacts with an ITS, the system uses a set of predefined rules to guide the student through the problem-solving process. For example, if the student is working on a math problem, the ITS may use rules to guide the student through the process of solving the problem step-by-step, providing guidance on concepts and procedures that the student may not be familiar with. The ITS also use rules to evaluate the student's progress and provide feedback. For example, the system may use rules to determine if the student has correctly applied a concept or procedure and provide feedback on how to improve. The ITS also use rules to provide hints and suggestions to help the student to get over a difficulty they are facing. For example, if the student is stuck on a particular step, the ITS may use rules to provide a hint or suggestion that will help the student to continue solving the problem. Additionally, the ITS use rules to monitor the student's progress over time, providing feedback on areas where the student is excelling or struggling. Furthermore, ITS also use rule-based systems to provide feedback that is tailored to the student's individual learning style. For example, if the student is a visual learner, the ITS may use rules to provide feedback that includes diagrams and illustrations.

15 Case-Based Reasoning on Giving Feedback and Applying the Concepts

Intelligent tutoring systems (ITS) use case-based reasoning (CBR) to provide students with examples of similar problems and solutions and give feedback on how well they apply the concepts. CBR is a type of problem-solving method that uses past experiences or cases to solve new problems. When a student interacts with an ITS, the system uses CBR to provide the student with examples of similar problems that have been solved in the past [26]. For example, if the student is working on a math problem, the ITS may use CBR to retrieve examples of similar problems and their solutions from its database. These examples help the student to understand the problem-solving process and apply the concepts in a practical way. The ITS also use CBR to evaluate the student's progress and provide feedback. For example, the system may use CBR to compare the student's solution to the retrieved examples and provide feedback on how well the student has applied the concepts. Additionally, ITS also use CBR to adapt the examples and the feedback to the student's level of understanding. For example, if the student is struggling with a concept, the ITS may retrieve examples that are more basic or that use simpler language. Furthermore, ITS use CBR to improve its own performance over time. As the ITS are used by more students and more examples are added to its database, it continuously learns and improves its ability to retrieve relevant examples and provide effective feedback.

16 Machine Learning Algorithms for Students' Diverse Learning Styles

ML algorithms are mathematical models that learn from data and improve their performance over time [27]. When a student interacts with an ITS, the system uses machine learning algorithms to analyse data on the student's performance, such as their responses to questions and their progress on tasks. The ITS use this data to create a representation of the student's learning style, which includes factors such as the student's prior knowledge, their preferred learning methods, and their strengths and weaknesses. Once the ITS have a representation of the student's learning style, it uses this information to adapt the curriculum, teaching methods, and learning materials to suit the student's individual needs. For example, if the student is a visual learner, the ITS may use this information to provide more diagrams and illustrations in the learning materials. The ITS also use machine learning algorithms to provide feedback that is tailored to the student's learning style. For example, if the student is struggling with a concept, the ITS may use this information to provide feedback that is more visual or that uses simpler language. In addition, ITS use machine learning algorithms to monitor the student's progress over time and make predictions about their future performance [10]. For example, the ITS use this data to identify areas where the student is at risk of falling behind and provide targeted interventions to help the student to catch up. Moreover, ITS use ML algorithms to adapt its own performance over time. As the ITS are used by more students and more data is collected, it continuously learns and improves its ability to adapt to different learning styles and provide effective feedback.

ML algorithms are used to analyse student performance data and adjust the difficulty level and content of questions based on the student's level of understanding [28]. This approach is known as adaptive learning. One example of how this is done is using a reinforcement learning algorithm. This type of algorithm uses trial-and-error to learn from the student's responses to questions. It starts by presenting the student with easy questions, and as the student answers correctly, the algorithm gradually increases the difficulty level of the questions. If the student answers incorrectly, the algorithm will present easier questions [29]. Over time, the algorithm learns the student's proficiency level and adjusts the difficulty of the questions accordingly. Another example is using a supervised learning algorithm, such as a decision tree or a neural network, to predict a student's proficiency level based on their performance on a set of questions. The algorithm is trained on a dataset of student performance data, and it learns to identify patterns in the data that are associated with different proficiency levels. Once the algorithm is trained, it uses this information to predict a student's proficiency level based on their performance on a new set of questions. A common use case is using a supervised learning algorithm to predict a student's proficiency level in a particular subject. The algorithm is trained on a dataset of student performance data, such as scores on assessments or answers to multiple-choice questions [30]. The training dataset also includes information about the student's background, such as their prior knowledge of the subject or their demographic information. Once the algorithm is trained, it uses this information to predict a student's proficiency level based on their performance on a new set of questions. For example, if the algorithm predicts that a student is not proficient in a particular subject, the student will be presented with questions that are designed to help them build their understanding of the subject.

17 Item Response Theory (IRT) for Creating Adaptive Assessments for Students

Item response theory (IRT) is a statistical method that is used to create adaptive assessments for students [31]. It is based on the idea that the difficulty of an assessment item (such as a multiple-choice question) and the ability of the student are separate factors that interact to determine the student's performance on the item.

IRT models are used to estimate the difficulty of each item on an assessment and the ability of each student. The estimated item difficulty and student ability are then used to create an adaptive assessment that is tailored to the student's level of ability. An example of how IRT is used in practice is an adaptive test that is given to students to measure their proficiency in a particular subject. The test is made up of a set of multiple-choice questions, each of which has been calibrated using IRT. At the beginning of the test, the student is presented with a set of easy questions that are designed to quickly determine their proficiency level. As the student answers these questions, the test uses IRT to estimate their ability level. Based on this estimated ability level, the test then presents the student with a set of questions that are at the appropriate difficulty level. For example, if the student is performing well, they will be presented with more difficult questions to further assess their proficiency. On the other hand, if the student is struggling, they will be presented with easier questions in order to help them to build their understanding of the subject. This way, the adaptive assessment adjusts to the student's proficiency level and provides a more accurate measure of their abilities. Additionally, it also provides a more challenging experience for students who have a higher proficiency level and a more supportive experience for students who have a lower proficiency level.

In summary, item response theory (IRT) is a statistical model used to create adaptive assessments for students. The IRT model estimates the probability of a student answering a question correctly based on their ability level. Here is an example of how IRT is implemented using Python:

```
from irt import Irt

# Initialise the IRT model

irt = Irt()

# Define the question bank.

question_bank = [
    {'question': 'What is the capital of France?', 'difficulty': -1, 'discrimination': 1},
    {'question': 'What is the square root of 9?', 'difficulty': 0, 'discrimination': 1},
    {'question': 'What is the derivative of x^2?', 'difficulty': 1, 'discrimination': 1},
    {'question': 'What is the chemical formula for water?', 'difficulty': 1, 'discrimi-
nation': 1},
    ]

# Define the student abilities

student_abilities = [−0.5, 0.5, 1.5, 2.5]

# Generate adaptive assessment

assessment = irt.generate_assessment(question_bank, student_abilities, 3).
print(assessment)
```

Output: ['What is the capital of France?', 'What is the square root of 9?', 'What is the derivative of x^2?']

In this example, we first import the IRT package. Then, we define a question bank, which contains a list of questions with their difficulty and discrimination parameters. Next, we define a list of student abilities. Finally, we use the generate_assessment method to generate an adaptive assessment for each student based on their ability level. The output is a list of questions that are most appropriate for the student's ability level. This is a simple example to demonstrate the basic idea of how IRT is implemented using Python, and there are much more advanced IRT packages available for use in teaching and learning applications.

18 Bayesian Knowledge Tracing (BKT) for Tracking Student Knowledge and Understanding of Specific Concepts Over Time

Bayesian knowledge tracing (BKT) is a machine learning algorithm that is used to track student knowledge and understanding of specific concepts over time. It is based on Bayesian probability theory and is commonly used in educational technology to personalise learning and adapt to the student's needs. The BKT algorithm models student learning as a series of transitions between different states of knowledge. There are two main states: 'knowing' and 'not knowing' a concept. The algorithm uses a set of parameters to represent the probability of transitioning between these states based on student performance on assessments or other activities. For example, let us say a student is learning about the concept of photosynthesis in a biology class. Initially, the student may not know anything about the concept. As they learn more, they may start to understand some aspects of it, but they may not yet have a complete understanding. Over time, as they learn more and practice applying the concept, they may develop a solid understanding of it. The BKT algorithm uses student performance data, such as answers to multiple-choice questions or scores on assessments, to update the student's probability of 'knowing' the concept. As the student answers more questions correctly, the algorithm updates the probability of the student 'knowing' the concept, and as the student answers more questions incorrectly, the algorithm updates the probability of the student 'not knowing' the concept. Another example is, when a student is taking an online course, the algorithm tracks their progress over time by analysing their interactions with the course materials, such as how much time they spend on each module or how often they access certain resources. As the student progresses through the course, the algorithm uses this information to update the student's probability of 'knowing' the concepts covered in the course.

Here is an example of a simple implementation of the Bayesian knowledge tracing algorithm using Python:

import necessary libraries

```
import numpy as np
# define the initial knowledge state for each student
# 0 represents not knowing the concept, 1 represents knowing the concept
student_knowledge = np.array([0, 0, 0, 0, 0])
# define the transition probabilities
# probability of transitioning from not knowing to knowing the concept
p_transition = 0.2
# probability of transitioning from knowing to not knowing the concept
p_forgetting = 0.1
# define the prior probabilities
# probability of a student answering correctly on a question they do not know
p_guess = 0.1
# probability of a student answering correctly on a question they know
p_slip = 0.9
# define the student's answers for each question
answers = np.array([1, 0, 1, 1, 0])
# Iterate through each question
  for i in range(len(answers)):
    # calculate the likelihood of the student's answer
        likelihood = p_guess + (student_knowledge * p_slip) + ((1–
student_knowledge) * (1–p_guess))
    # update the student's knowledge state
      student_knowledge = student_knowledge + (p_transition * (answers[i]–
student_knowledge))–(p_forgetting * student_knowledge * (1–answers[i]))
    student_knowledge = np.clip(student_knowledge, 0, 1)
    print(student_knowledge)
```

This is a basic example that is used to track the student's performance data during their learning process of photosynthesis concept in a biology class. The student's initial knowledge state is defined as not knowing the concept, and the transition probabilities are defined as the probability of transitioning from 'not knowing' to 'knowing' the concept is 0.2, and the probability of transitioning from knowing to not knowing the concept is 0.1. Then, we define the prior probabilities, which are the probability of a student answering correctly on a question they do not know and probability of a student answering correctly on a question they know. Then, we have an array of student's answers for each question, and using these answers, we calculate the likelihood of the student's answer and update the student's knowledge state. This is a simple example, and in practice, the algorithm would be more complex, considering multiple factors such as the difficulty level of the question, the student's performance on previous questions, and so on.

19 Predictive Modelling on the Previous Performance and Adjusting the Difficulty Level and Content of Questions

Predictive modelling is a type of machine learning that is used to predict future outcomes based on past data. In the context of education, predictive modelling is used to predict student performance based on the previous performance and adjust the difficulty level and content of questions accordingly [32]. Here are three examples of how predictive modelling is used in education:

Adaptive Testing: Predictive modelling is used in adaptive testing to adjust the difficulty level of questions based on the student's previous performance. For example, if a student answers a series of questions correctly, the algorithm may increase the difficulty level of the next question to challenge the student. Conversely, if a student answers a question incorrectly, the algorithm may decrease the difficulty level of the next question to provide additional support.

Student Retention: Predictive modelling is used to predict which students are at risk of dropping out of a course or programme based on the previous performance and other factors such as attendance and engagement. By identifying students at risk, educators take proactive steps to provide additional support and resources to keep them on track.

Personalised Learning: Predictive modelling is used to personalise learning for each student by predicting which concepts and materials they are likely to struggle with and providing additional resources and support [33]. For example, an algorithm may analyse a student's performance data and predict that they will have difficulty understanding a particular concept. The algorithm then recommends activities, videos, or other resources that are specifically designed to help students understand that concept.

AI-powered systems are used to grade written assignments, such as essays and short-answer questions, by using NLP and ML [34]. These systems are used to automate the grading process, which saves time and resources for teachers and educators. Here are some examples of how AI-powered systems are used to grade written assignments:

Automated Essay Grading: These systems use NLP techniques to analyse the grammar, vocabulary, coherence, and cohesiveness of the essay and provide feedback to the student. This is used to improve student writing skills and help them understand their strengths and weaknesses.

Short-answer question grading: AI-powered systems are used to grade short-answer questions. These systems use NLP techniques to understand the student's response and determine whether it is correct or not. They also provide feedback on the student's answers and suggest alternative answers. These systems are used to grade questions in subjects such as math and science, where the answers are typically specific and objective.

ML-based systems: ML-based systems are trained to grade written assignments. These systems use supervised learning techniques to train on a dataset of annotated essays or short-answer questions. Once trained, the system automatically grades new essays or short-answer questions, providing a score and feedback. This is used to improve student writing skills and help them understand their strengths and weaknesses.

20 Tracking Student Attendance Using Facial Recognition

AI-powered systems are used to track student attendance using techniques such as facial recognition and machine learning (ML). These systems automate the process of tracking attendance, which saves time and resources for teachers [35]. One example of an AI-powered system used for tracking student attendance is a facial recognition system. These systems use computer vision techniques to capture images of students as they enter the classroom. The system then uses facial recognition algorithms to match the captured image to a pre-existing database of student images. This is used to automatically record attendance for each student as they enter the classroom. Another example of an AI-powered system used for tracking student attendance is an ML-based system. These systems use supervised learning techniques to train on a dataset of student images and corresponding attendance records [36].

Once trained, the system automatically detects and recognises students based on their images and records their attendance. AI-powered systems are also integrated with mobile-based attendance tracking applications. These applications use the camera of a mobile device to capture images of students and use facial recognition algorithms to match the image to a pre-existing database of student images. This is used to automatically record attendance for each student as they enter the classroom. Facial recognition is becoming more popular in schools, universities, and other educational institutions, due to its ease of use, accuracy, and ability to integrate with other systems.

21 Optimization Algorithms to Schedule Classes, Meetings, and Other Events in Universities

AI-powered systems are used to schedule classes, meetings, and other events in universities by using techniques such as optimization algorithms. These systems automate the process of scheduling, which saves time and resources for administrators and educators. Here are some examples of how AI-powered systems are used to schedule events in universities:

Room Scheduling: One example of an AI-powered system used for scheduling classes and meetings is a room scheduling system. These systems use optimization algorithms such as linear programming or mixed-integer programming to schedule classes and meetings in the most efficient way possible. These systems take into account factors such as room capacity, availability, and the schedules of both instructors and students to find the best possible schedule.

Class Scheduling: Another example of an AI-powered system used for scheduling classes is a class scheduling system. These systems use optimization algorithms to generate a schedule that meets the needs of both students and instructors. For example, it takes into account the availability of instructors, the number of students enrolled in each class, and the preferred class times of students.

Meeting Scheduling: AI-powered systems are also used to schedule meetings. These systems use optimization algorithms to find the best time and location for a meeting based on the availability of attendees. For example, it takes into account the time zones of remote attendees and finds a time that works for everyone.

22 Predictive Modelling to Foresee and Forecast Student Performance: Linear Regression, Decision Trees, and Neural Networks

Predictive modelling is a powerful tool that is used to predict student performance based on the previous performance data. There are several different techniques that are used for this purpose, including linear regression, decision trees, and neural networks. One example of how predictive modelling is used to predict student performance which is using linear regression. In this case, a linear regression model is trained on a dataset of past student performance data, including factors such as the previous test scores, grades, and attendance records. The model is used to predict a student's future performance based on their past performance data. Another example is using the decision tree algorithm, where it is used to predict student performance by analysing factors such as the previous test scores, grades, and attendance records. The decision tree algorithm constructs a tree-like model based on this data, with the branches of the tree representing different combinations of factors that are thought to influence student performance. By analysing the tree, the algorithm identifies patterns in the data that are associated with high or low student performance and uses these patterns to make predictions about future performance. A third example is using neural network, where it is trained on a dataset of past student performance data, including factors such as the previous test scores, grades, and attendance records. The neural network would then be used to make predictions about future student performance based on the patterns, it has identified in the data.

23 Knowledge Maps for Generating Personalised Learning Materials

The 'concept mapping' approach involves mapping out the relationships between different concepts in a subject area and identifying areas where a student may need extra support. The AI system analyses the student's performance on assessments, their progress, and their learning style to determine which concepts they have difficulty with and create a personalised learning plan to address those areas. Another example is using a technique called 'knowledge tracing'. This approach involves tracking a student's understanding of specific concepts over time, by analysing their performance on assessments and activities. The AI system then uses this information to identify areas where the student needs extra support and generate personalised learning materials to address those areas. A third example is using a technique called 'adaptive learning'. This approach involves the use of ML algorithms to analyse student performance data and adjust the difficulty level and content of questions based on the student's level of understanding. The AI system then uses this information to generate personalised learning materials that are tailored to the student's individual needs.

24 AI-Powered Chatbots for Generating Personalised Quizzes and Practice Problems

Chatbots are used to generate personalised quizzes and practice problems by engaging students in natural language conversations. One example of an AI-powered chatbot used for this purpose is ALEKS, which stands for assessment and learning in knowledge spaces. ALEKS uses NLP to understand student responses and generate personalised quizzes and practice problems that are tailored to the student's level of understanding. The chatbot uses a combination of rule-based systems and machine learning algorithms to generate questions that are appropriate for the student's current level of knowledge. Another example is Querium's 'The Virtual Writing Tutor' which is an AI-powered chatbot that helps students improve their writing skills. It uses natural language processing to understand student responses and provide feedback on grammar, punctuation, and style. The chatbot also generates personalised quizzes and practice problems to help students improve their writing skills. A third example is the AI-powered chatbot called 'Jouko' developed by the University of Jyväskylä, Finland. Jouko provides students with personalised feedback on their understanding of mathematical concepts and problems. The chatbot uses NLP to understand student responses and generates personalised quizzes and practice problems that are tailored to the student's level of understanding.

25 Conclusion

AI has completely revolutionised the formal education space in several ways. AI is used to analyse student performance data and adapt the curriculum and teaching methods to the individual needs of each student. This has led to a more efficient and effective learning experience. AI-powered intelligent tutoring systems provide students with immediate feedback and guidance and help teachers identify areas where students need extra support. AI is used to create adaptive assessments that adjust the difficulty level and content of questions based on the student's performance. AI helps automate administrative tasks such as grading, attendance, and scheduling, freeing up more time for teachers to focus on instruction. AI-powered virtual teaching assistants assist teachers in monitoring student progress, providing feedback, and answering student questions in real time. AI is used to generate personalised learning materials, such as quizzes, practice problems, and summaries that are tailored to each student's needs. With the help of AI, the learning process is continuous and lifelong, as AI tracks students' progress and adjusts the learning path accordingly, which helps in keeping the students motivated and engaged. AI is used to predict student performance based on the previous performance and generate personalised quizzes and practice problems that are tailored to the student's current level of understanding. AI-based on-demand learning platforms provide students with customised learning experiences based on their interests, goals, and learning styles. AI is used to enhance the remote learning experience by providing interactive, engaging, and personalised content, as well as real-time support and feedback. It is important to note that whilst AI has revolutionised the formal education space, its implementation should be done with care, taking into consideration the ethical, privacy, and equality implications.

References

1. Alam A (2022) A digital game based learning approach for effective curriculum transaction for teaching-learning of artificial intelligence and machine learning. In: 2022 International Conference on Sustainable Computing and Data Communication Systems (ICSCDS), pp 69–74. IEEE
2. Bhutoria A (2022) Personalized education and artificial intelligence in United States, China, and India: a systematic review using a human-in-the-loop model. Comput Educ: Artificial

Intelligence, 100068
3. Alam A (2022) Investigating sustainable education and positive psychology interventions in schools towards achievement of sustainable happiness and wellbeing for 21st century pedagogy and curriculum. ECS Trans 107(1):19481
4. da Silva LM, Dias LP, Barbosa JL, Rigo SJ, dos Anjos J, Geyer CF, Leithardt VR (2022) Learning analytics and collaborative groups of learners in distance education: a systematic mapping study. Inform Educ 21(1):113–146
5. Alam A (2022) Social robots in education for long-term human-robot interaction: socially supportive behaviour of robotic tutor for creating robo-tangible learning environment in a guided discovery learning interaction. ECS Trans 107(1):12389
6. Kaddoura S, Popescu DE, Hemanth JD (2022) A systematic review on machine learning models for online learning and examination systems. PeerJ Comp Sci 8:e986
7. Alam A, Mohanty A (2022) Metaverse and posthuman animated avatars for teaching-learning process: interperception in virtual universe for educational transformation. In: International Conference on Innovations in Intelligent Computing and Communications, pp 47–61. Springer, Cham
8. Dai CP, Ke F (2022) Educational applications of artificial intelligence in simulation-based learning: a systematic mapping review. Comp Educ: Artificial Intelligence, 100087
9. Alam A (2020) Pedagogy of calculus in India: an empirical investigation. Periódico Tchê Química 17(34):164–180
10. Xu W, Ouyang F (2022) The application of AI technologies in STEM education: a systematic review from 2011 to 2021. Inter J STEM Educ 9(1):1–20
11. Alam A (2022) Positive psychology goes to school: conceptualizing students' happiness in 21st century schools while 'Minding the Mind!' are we there yet? evidence-backed, school-based positive psychology interventions. ECS Trans 107(1):11199
12. Kshirsagar PR, Jagannadham DBV, Alqahtani H, Naveed NQ, Islam S, Thangamani M, Dejene M (2022) Human intelligence analysis through perception of AI in teaching and learning. Comp Intell Neurosci
13. Alam A (2022) Mapping a sustainable future through conceptualization of transformative learning framework, education for sustainable development, critical reflection, and responsible citizenship: an exploration of pedagogies for twenty-first century learning. ECS Trans 107(1):9827
14. Tan SC, Lee AVY, Lee M (2022) A systematic review of artificial intelligence techniques for collaborative learning over the past two decades. Comp Educ: Artificial Intelligence, 100097
15. Alam A (2022) Employing adaptive learning and intelligent tutoring robots for virtual classrooms and smart campuses: reforming education in the age of artificial intelligence. In: Shaw RN, Das S, Piuri V, Bianchini M (eds) Advanced computing and intelligent technologies. Lecture Notes in Electrical Engineering, vol 914. Springer, Singapore
16. Umutlu D, Gursoy ME (2022) Leveraging artificial intelligence techniques for effective scaffolding of personalized learning in workplaces. Artificial Intelligence Education in the Context of Work, 59–76
17. Alam A (2022) Platform utilising blockchain technology for eLearning and online education for open sharing of academic proficiency and progress records. In: Asokan R, Ruiz DP, Baig ZA, Piramuthu S (eds) Smart data intelligence. Algorithms for Intelligent Systems. Springer, Singapore
18. Holmes W, Tuomi I (2022) State of the art and practice in AI in education. Eur J Educ 57(4):542–570
19. Alam A (2023) Cloud-based e-learning: scaffolding the environment for adaptive e-learning ecosystem based on cloud computing infrastructure. In: Satapathy SC, Lin JCW, Wee LK, Bhateja V, Rajesh TM (eds) Computer communication, networking and IoT. Lecture Notes in Networks and Systems, vol 459. Springer, Singapore
20. Namatherdhala B, Mazher N, Sriram GK (2022) A comprehensive overview of artificial intelligence tends in education. Inter Res J Modern Eng Techn Sci 4(7)

21. Alam A (2022) Cloud-based e-learning: development of conceptual model for adaptive e-learning ecosystem based on cloud computing infrastructure. In: Kumar A, Fister I, Gupta PK, Debayle J, Zhang ZJ, Usman M (eds) Artificial intelligence and data science. ICAIDS 2021. Communications in Computer and Information Science, vol 1673. Springer, Cham
22. Limna P, Jakwatanatham S, Siripipattanakul S, Kaewpuang P, Sriboonruang P (2022) A review of artificial intelligence (AI) in education during the digital era. Adv Knowl Execut 1(1):1–9
23. Alam A (2021) Possibilities and apprehensions in the landscape of artificial intelligence in education. In: 2021 International Conference on Computational Intelligence and Computing Applications (ICCICA), pp 1–8. IEEE
24. Xia Q, Chiu TK, Zhou X, Chai CS, Cheng M (2022) Systematic literature review on opportunities, challenges, and future research recommendations of artificial intelligence in education. Comp Educ: Artif Intell, 100118
25. Alam A (2022) Impact of university's human resources practices on professors' occupational performance: empirical evidence from India's higher education sector. In: Inclusive Businesses in Developing Economies. Palgrave Macmillan, Cham, pp 107–131
26. Zafari M, Bazargani JS, Sadeghi-Niaraki A, Choi SM (2022) Artificial intelligence applications in K-12 education: a systematic literature review. IEEE Access
27. Alam A (2022) Educational robotics and computer programming in early childhood education: a conceptual framework for assessing elementary school students' computational thinking for designing powerful educational scenarios. In: 2022 International Conference on Smart Technologies and Systems for Next Generation Computing (ICSTSN), pp 1–7. IEEE
28. Alam A (2020) Challenges and possibilities in teaching and learning of calculus: a case study of India. J Educ Gifted Young Scient 8(1):407–433
29. Ouyang F, Zheng L, Jiao P (2022) Artificial intelligence in online higher education: A systematic review of empirical research from 2011 to 2020. Educ Inf Technol 27(6):7893–7925
30. Alam A (2020) Possibilities and challenges of compounding artificial intelligence in India's educational landscape. Inter J Adv Sci Tech 29(5):5077–5094
31. Alam A, Mohanty A (2022) Business models, business strategies, and innovations in EdTech companies: integration of learning analytics and artificial intelligence in higher education. In: 2022 IEEE 6th Conference on Information and Communication Technology (CICT), pp 1–6. IEEE
32. Akinwalere SN, Ivanov V (2022) Artificial intelligence in higher education: challenges and opportunities. Border Cross 12(1):1–15
33. Alam A, Mohanty A (2022) Foundation for the future of higher education or 'misplaced optimism'? being human in the age of artificial intelligence. In: International Conference on Innovations in Intelligent Computing and Communications. Springer, Cham, pp 17–29
34. Mijwil MM, Aggarwal K, Mutar DS, Mansour N, Singh RS (2022) The position of artificial intelligence in the future of education: an overview. J Appl Sci 10(2)
35. Alam A (2020) Test of knowledge of elementary vectors concepts (TKEVC) among first-semester Bachelor of Engineering and Technology students. Periódico Tchê Química 17(35):477–494
36. Alam A (2021). Should robots replace teachers? mobilisation of AI and learning analytics in education. In: 2021 International Conference on Advances in Computing, Communication, and Control (ICAC3), pp 1–12. IEEE

Human Activity Recognition Using CNN-Attention-Based LSTM Neural Network

P. Pravanya, K. Lakshmi Priya, S. K. Khamarjaha, K. Buela Likhitha, P. M. Ashok Kumar, and R. Shankar

Abstract In the past years, understanding human behavior and classification of human actions and intentions by researching human activity recognition (HAR) using traditional pattern recognition has made great progress. In this paper, a novel approach is proposed based on convolutional neural network (CNN) and attention-based long short-term memory (attention LSTM) architecture. Human activity recognition (HAR) or recognizing human behavior is one of the challenging tasks due to human tendencies as the activities are not only complex but also multitasking. This deep learning-based long short-term memory network using convolutional neural networks (CNN-attention LSTM) architecture predicts the activities performed by humans and improves the accuracy by reducing the complexity of raw data and also by removing unnecessarily complex data. The convolutional layers act as a feature extractor, where they learn hierarchical representations of the image by applying multiple filters to the input image and passing the resulting feature maps through multiple activation functions. These learned features are then used as input to another classifier, such as a attention LSTM networks, to make the final prediction. On the internal UCF50 dataset, the proposed model achieves an 84.43% accuracy. The outcomes demonstrate that the suggested model is more robust and capable of activity detection than some of the results that have been reported.

Keywords Human activity recognition · Deep learning · Convolutional neural network · Activity recognition

1 Introduction

Human activity recognition (HAR) is a field of study that involves the identification of specific movements or actions of a people. HAR is crucial to daily life since it

P. Pravanya · K. L. Priya · S. K. Khamarjaha (✉) · K. B. Likhitha · P. M. A. Kumar · R. Shankar
Department of Computer Science and Engineering, Koneru Lakshmaiah Education Foundation, Vaddeswaram, India
e-mail: khamarjahashaik04@gmail.com

G. Rajakumar et al. (eds.), *Intelligent Communication Technologies and Virtual Mobile Networks*, Lecture Notes on Data Engineering and Communications Technologies 171,
https://doi.org/10.1007/978-981-99-1767-9_43

can gather information about human behavior from datasets [1]. Research on human activity recognition has become popular as it is one of the human and computer interaction applications. The main purpose of human activity recognition is to recognize a person through different technologies and devices like cameras, sensors, location sensors, inertial sensors, etc. [2]. People can classify human motion just by obtaining the information that is used for feature extraction of the activities that are performed daily [3]. Human activity recognition has numerous applications in real-time, and some of them involves the detection of human activities through surveillance camera (CCTV), home behavior analysis, medical assistance, etc. [4]. Figure 1 is the proposed CNN-attention LSTM framework.

Due to the rapid increase in technologies and advancement of computer vision, field researchers have been studying the various architectures in activity recognition

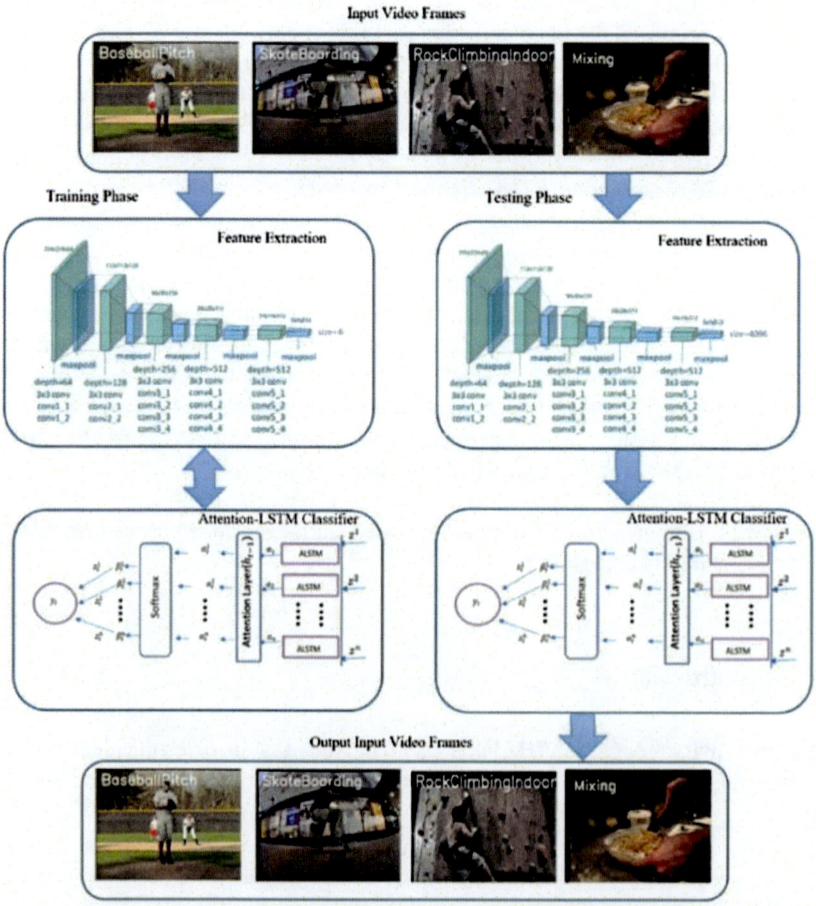

Fig. 1 Illustration of the whole proposed CNN-attention LSTM framework

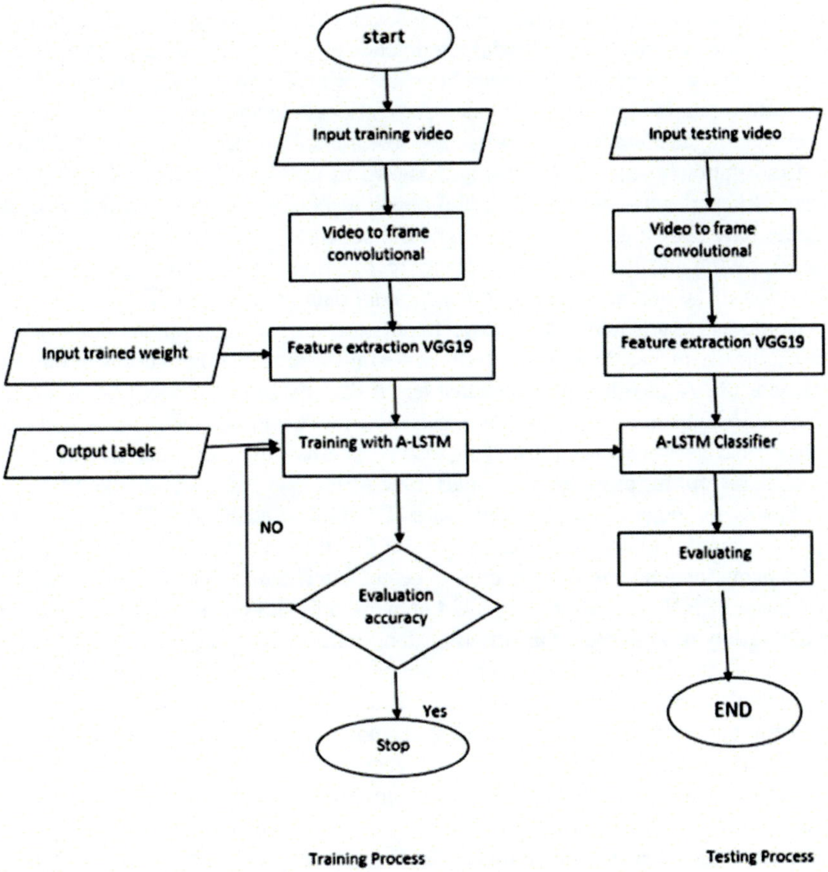

Fig. 2 Flowchart of the whole proposed CNN-attention LSTM network

to improve the accuracy of the recognition. The advancements in AI have provided a great opportunity for the evolution of HAR in various application domains [5]. The basic neural network strategy will not provide you with the level of accuracy that you will receive from a deep neural network (DNN) methodology, machine learning techniques for evaluating the act detection tasks in the primary trust heuristic manual feature extraction [6]. Figure 2 is the proposed CNN-attention LSTM network.

2 Literature Survey

Recently, there has been a significant amount of research which is carried in this area of study in the computer vision field. Authors have developed a variety of methods like deep learning methods that use in CNN and attention-based methods and classical machine learning and artificial intelligence to identify human actions in video streams. In the last ten years, researchers have primarily developed effective HAR via features engineering systems using conventional machine learning approaches. Deep learning techniques are now used by researchers to extract the sequential data.

Convolutional neural networks, also known as CNNs, are a form of frequently used AI neural network for object and object detection and classification. Recent studies have shown that convolutional neural network (CNN) has a great scope for the improving accuracy of human activity recognition. CNN is greatly used for the detection of human activities by taking spatial data as input [7]. The capacity of convolutional neural networks to extract both complex and basic human movements hierarchically makes them particularly powerful for recognizing patterns in human behavior [8]. A deep learning spec that learns directly from knowledge could be a convolutional neural network CNN which is terribly useful for recognizing objects, classes, and classes in photos by trying to find patterns within the pictures. They will be quite helpful for categorizing signals, time series, and audio knowledge [9].

Among the most effective architectures for HAR is attentıon LSTM. Attentıon LSTM is similar to conventional methods in that it does not merge time steps like the convolutional neural network does to detect relationships in data in the temporal dimension (CNN). The attentıon LSTM network provides outstanding performance and a variety of possible applications. A long-term memory network is sequential system that facilitates information persistence. It can handle the vanishing gradient issue that the recurrent neural network experience [10]. To avoid the vanishing gradient, attentıon LSTMs were developed, and as a result, an attentıon LSTM cell may retain context for extended input sequences.

The attentıon LSTM architecture can be utilized to reliably solve tasks involving sequential data, such as time series prediction, speech recognition, and language translation [11]. Because attentıon LSTM has a specific structure for remembering information for a longer period of time, gradients can progress smoothly over time. By comparing the information already stored in the internal memory with the recently entered information, input gateways and forget gated regulate how to overwrite the knowledge [12].

Attentıon LSTM learns the patterns every 12 periods and not only incorporates the past predictions but also preserves a longer-term context which will resolve the long-term reliance issue that is faced by other models several continual neural networks can learn long-run dependencies, notably in tasks involving sequence prediction. Apart from singular knowledge points like photos, and attentıon LSTM feedback connections, creating it capable of processing the entire sequence of information. This has uses in artificial intelligence and speech recognition, among others. A singular version of RNN known as attentıon LSTM exhibits exceptional performance on a large variety of problems.

The CNN attentıon LSTM design combines attentıon LSTMs, a dense layer that allows sequence prediction, with convolutional neural network (CNN) layers at the front end to aid with extracting features from training dataset. The combination of CNN and attentıon LSTM is a deep learning neural network that could quickly separate and categorize active features while emulating the feature extraction process within the approach by only using a few fundamental parameters the CNN-attention LSTM architecture is the best approach for processing input data in which the image passes through several filters of CNN that extract several features. Video labeling, image labeling, and activity recognition are the three main uses of the CNN-attention LSTM.

3 Methodology

Several neural network techniques are available at the moment. The aim of this paper is to ensure the CNN-attention LSTM model's ability to detect different actions. A training dataset will be used by CNN's supervised algorithm to capacity to efficiently deep knowledge. The following are the strategies used for this proposed approach: 1. feature extraction using CNN, 2. convolutional attention LSTM network.

3.1 Feature Extraction Using CNN

The procedure of feature extraction original cast the content of the original data gathering while translating made from raw information into numerical features that may be processed. Instead of doing it manually, CNN applies a feature extractor in the training phase. The proposed work uses VGG19, which is a deep convolutional neural network (CNN) architecture that was proposed by the visual geometry group (VGG) at the University of Oxford. It is a well-known and widely used architecture for image classification tasks due to its high accuracy and robust features.

When using VGG19 (as shown in Fig. 3) for feature extraction, the basic idea of the present work is to remove the fully connected layers of the network and keep only the convolutional layers. The convolutional layers act as a feature extractor, where they learn hierarchical representations of the image by applying multiple filters to the input image and passing the resulting feature maps through multiple activation functions. These learned features are then used as input to another classifier, such as a attention LSTM networks, to make the final prediction.

It is a 19-layer deep network, consisting of 16 convolutional layers and 3 fully connected layers. VGG19 consists of multiple blocks of convolutional layers, each followed by a max pooling layer, which reduces the spatial resolution of the feature maps. This results in a reduction of computational complexity and helps the network to focus on the most important features of the image.

The summary of the VGG 19 architecture:

- Input layer: The input layer takes a $224 \times 224 \times 3$ color image as input, where the three channels represent the red, green, and blue color channels.

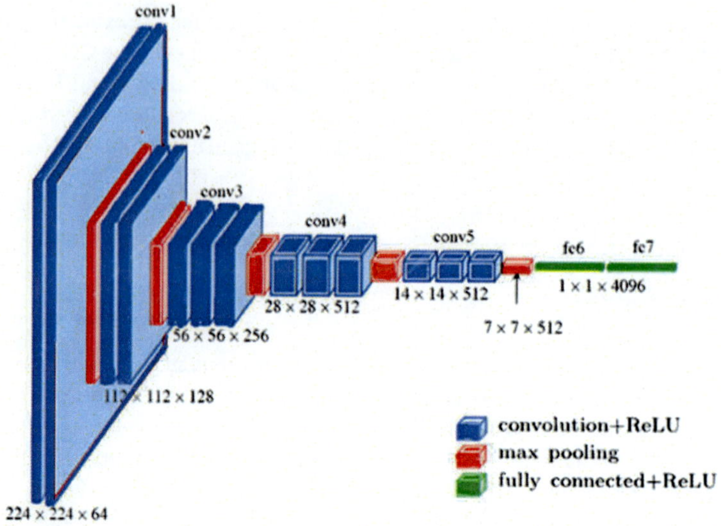

Fig. 3 VGG19 convolutional

- Convolutional layers: The convolutional layers apply multiple filters to the input image and learn hierarchical representations of the image. In VGG19, there are 16 convolutional layers in total, where each layer consists of multiple filters of different sizes and shapes.
- Max pooling layers: The max pooling layers reduce the spatial resolution of the feature maps, resulting in a reduction of computational complexity. There are 8 max pooling layers in VGG19, each followed by a convolutional layer.
- Fully connected layers: The fully connected layers connect every neuron in the previous layer to every neuron in the next layer. In VGG19, there are 3 fully connected layers, each followed by a ReLU activation function.

First, a pre-trained VGG19 model is loaded. A pre-trained model is one that has already been trained on a large dataset, and its parameters have been optimized. In this case, the model has been trained on the ImageNet dataset, which contains over a million images. The input image is then passed through the pre-trained VGG19 model. The input image is first resized to the required size (224 × 224 pixels) and normalized to have values between 0 and 1. The activations of the convolutional layers are extracted and used as features. The activations represent the feature maps of the image, and these feature maps capture the hierarchical representations of the image at different levels of abstraction.

The extracted features are then used as input to a classifier, such as an attention LSTM networks. This classifier is then trained on a smaller dataset to classify the images based on the learned features. Finally, the trained classifier is used to make predictions on new, unseen images. The image is passed through the pre-trained

VGG19 model to extract the features, and the classifier is then used to make the final prediction based on these features.

3.2 *Attention LSTM Network*

Attention LSTM networks are a type of recurrent neural network (RNN) that incorporate the concept of attention mechanisms. Attention mechanisms allow the network to dynamically focus on the most important parts of the input sequence, rather than using a fixed-length representation.

In the case of an attention LSTM network, the attention mechanism is applied to the hidden state of an LSTM network, allowing the network to dynamically attend to different parts of the input sequence at each time step. This helps the network to better capture the dependencies between the input elements and the output, leading to improved performance on tasks such as language translation and text classification.

The attention LSTM network (as shown in Fig. 4) consists of an encoder and a decoder, where the encoder processes the input sequence, and the decoder produces the final output. In this paper, we decoder as two nodes, with each node representing two classes. At each time step, the attention mechanism computes a weight for each element in the input sequence, indicating its importance for the current task. These weights are used to compute a weighted sum of the hidden states, which serves as the input to the next time step of the decoder.

The input for the attention LSTM cell can be a set of time series data or a 3D tensor with shape (samples, time steps, or features). A 4D tensor with shape is provided as an input to the convolution layer from a collection of images (samples, channels, rows, cols). The samples, time steps, channels, rows, and cols fields of a 5D tensor that reflects a set of images that has been gathered over time.

Attention-LSTM provides a summary of the procedure for creating attention weights and additional data and input data features based on attention LSTM.

$$a_t = f_1(a_{t-1}, z_t) \tag{1}$$

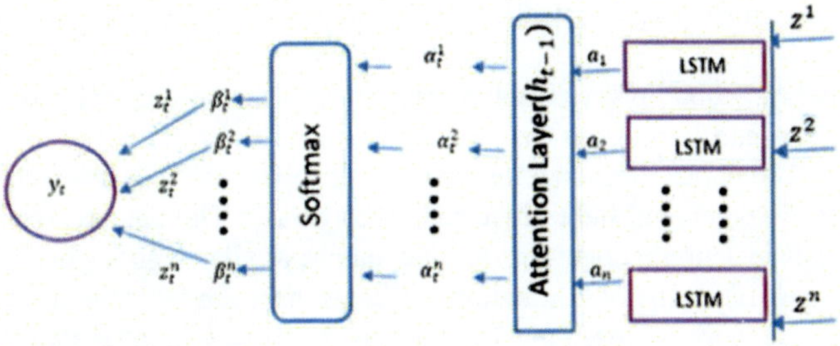

Fig. 4 Attention-LSTM

In attention LSTM and a, f_1 are non-linear activation functions. It is the size of the concealed state at time t and is the hidden state at time t, and attention LSTM is used as f_1 to overcome the attention LSTM dependence issue that shows in time series prediction.

$$\alpha_t^p = v^T \tanh(K_1.[a_{t-1}, C_{t-1}] + K_2 z^p) \tag{2}$$

$$\beta_t^p = \text{softmazx}(\alpha_t^p) = \frac{\exp(\alpha)}{\sum_{i=1}^{n} \exp\left(\alpha \frac{p}{t}\right)} \tag{3}$$

In this model, we include an attention mechanism for a given features sequence using a deterministic attention model. The vector v and the metrics are the model's learnable parameters. Comprising of parts and elements, the i-th item measuring the weight of the p-th input feature sequence at time t. The vector v and the matrices K_1, K_2 make the model's learnable parameters.

An attention weight, or p, indicates how much emphasis should be paid to the kth feature sequences. The weighted input feature sequence y_t, which represents the output of the attention model at time t, can also be obtained as follows:

$$y_t = \left(\beta_t^1 y_t^1, \beta_t^2 y_t^2, \ldots, \beta_t^n y_t^n\right)^T \tag{4}$$

the attention-LSTM model is then improved by substituting the new computed y_t for z_t in equations through model.

4 Experimental Result and Discussions

In order to evaluate our model, we collected the multiple different human activities images with a camera. There are different people with different activity as mentioned in the image. Firstly, we will describe the use model with the combination of CNN and attention LSTM methods. Next, we created the model and implemented it. Finally, the experimental results are presented from this study's investigation of human activity recognition.

We install a Python library pafy to retrieve metadata. And imported the libraries of TensorFlow and keras with the sequential, to_categorical, EarlyStopping, and plot_model. The parameters used for the proposed work are shown in Table 1.

Datasets: We download and extract the UCF50 dataset. This dataset is traditionally used for realistic action recognition. Get the names of all categories in UCF50 data sets to and generate a list of 20 random values as there are 50 action categories. In this prospective work, we used the 16 photos of humans with different activity. We used sample images of different activity. Iterating through all the generated random values and retrieve a class name with random values. Converting the frames from

Table 1 Experimental parameters

Parameter	Value
Size of input vector	4096
Size of batch	32
Epochs	20
Learning rate	0.001
Regularization rate	0.025
Probability of dropout	0.2
Activation function	ReLU
Optimization	Adam
Output layer	Softmax

BGR into RGB format to display the frames. Specifying the frames of the video as that will fed to model as the sequence. Projected a list to store the video frames. The video capture object is used to read the video file stored in disk. Gathering the total number of frames in the video. If the frame is not read successfully, then normalize the resize frame. As normalize the resize frame by dividing it with the 255 pixel so that each pixel values between 0 and 1. Proclaim an empty list and fill with the values for the features, labels, and video file paths. Here, keras's to_categorical method is used to convert the labels into one-hot-encoded vector.

i. Normalized confusion matrix

A confusion matrix which is presented as a tabular used to demonstrate how the model works. The number of inputs that are labeled correctly are represented in diagonal of the matrix, and the number of inputs that are labeled incorrectly are represented in the non-diagonal of the matrix. In the normalized confusion matrix, the sum of each row in a balanced is 1.00.

The neural networks are proven that are very effective in the recognition of activities. To built the model of the human activity recognition, we use the Keras ConvLSTM. The CNN along with LSTM the TensorFlow is used and keras are the more effective of implementing of neural network by these approaches the code is implemented in functional. Some of the methodology are used in this research which is ConvLSTM. By capturing the image of the activity storing the image, in the disk to extract the frame of the image. The image classifier objective is to take the input as an. The constants such as IMAGE_HEIGHT, SEQUENCE_LENGTH, and IMAGE_WIDTH are increased for the better results. After the constants, create a function for dataset creation.

Normalize the resize the frame by dividing it with 255 pixel. Labels were converted into one-hit-encoded vectors using the categorical method of a keras package. Split our data to create testing sets and training. To avoid bias, the dataset is rearranged before splitting. In this research, we implement the first approach by LSTM cells. The attention LSTM with the convolution included within this architecture, as it is capable of recognize spatial features of the data. The Con ALSTM is of taking in

3-dimensional input. By the ConLSTM for the first layer, the convolution structure is applied, sequentially for the second layer, the attention LSTM layer is applied. The attention LSTM is used to overcome the gradient problems, and the attention LSTM cell is capable of remembering the context of the data for long input sequences.

The time distributed is a temporary storage of the slices of the input as the input is in 3-dimensional. MaxPooling3D accomplishes downsampling of cuboidal pooling regions by dividing 3-dimensional input, then computing the maximum of each region of the input. Converting the input of multidimensional into flatten, where it is easy to handled models trained consistent. Commonly used to get the full connected layer. You can add as many dense layer for the process. That every neuron in the dense layer can take the all type of input from all other neurons of the previous layer as each neural network contains the all the neurons which are connected deeply within themselves. As the dense layer classify the image of the input based on the output of the convolutional layers, compute the weighted average of its input by the activation function. A plot_model() method is used to find the structure of the research model that was created. To prevent the overfitting, add an early stopping callback in instance. During model compilation, the loss function, the optimizer, and the values will be specified. Save the model, here the plot metric() function is used to view visualize training and validation loss metrices.

By val_loss methodology, the value of the cost function and loss of the value of cost function are validation data and your training data. To avoid the overfitting of the data add some noise by the drop out and use more complete training data. In order to reduce the validation loss, decrease network size or to increase dropout.

In this paper, we use the CNN method along with attention LSTM in human activity recognition (HAR). The CNN is the best for image data, as to extract spatial features in the input, and attention LSTM is the very efficient when working with the data of sequence as it consider of all the previous input of the data while bring out an output. As CNN can able to recognize the activity of the image where input is in 3-dimension. As the CNN architecture is capable of accepting the 3-dimensional input, the attention LSTM is not capable of taking the 3-dinmensional input. Some of the methodologies convert the 3-dimensional input into flatten or 1-dimensional. The image is divided into small pixels and arranged in the order to detect the activity. As the image is passed, then each-every pixel is observed in a sequence way that the CNN-attention LSTM architecture will recognize what is action which is happening in the image. The combination of both CNN and attention LSTM approaches in the TensorFlow to work very in efficient way in the recognition of the moment in the image. The model_evaluation_history function is applied to determine the loss and accuracy of the model. The graph represents the loss of the model with the two components loss and val_loss. Here, the loss is constant at some place (as shown in Figs. 5 and 6).

Firstly, we used UFC50 to find accuracy in activity. The deep CNN-attention LSTM with self-attention model was designed to: function as anticipated, we carried out a number of tests to thoroughly evaluate the effectiveness of the model discussed above. We used one dataset to implement various human activities and the models to make some of the classifications on the dataset was initially divided into many

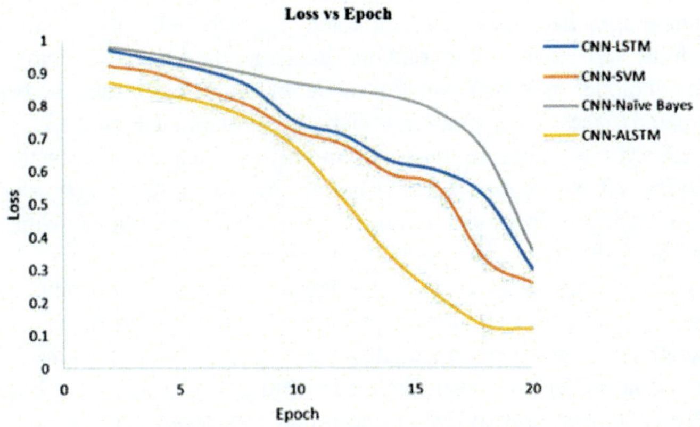

Fig. 5 Loss curve for CNN-LSTM, CNN-SVM, CNN-Naive Bayes, CNN-attention LSTM

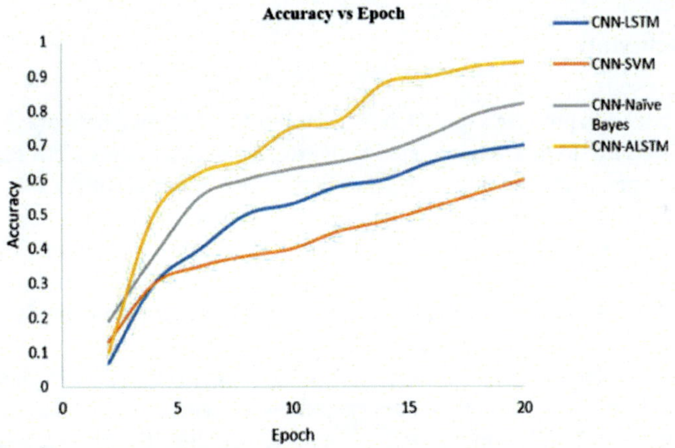

Fig. 6 Accuracy for CNN-LSTM, CNN-SVM, CNN-Naive Bayes, CNN-attention LSTM

activities like walking, jogging, sleeping, eating, running, sitting, downstairs, etc. We used some methods called convolution neural network (CNN) and attention-long short-term memory (attention LSTM) to design and are implemented in our paper. Before output parameters, a parallel dense network was used. This parallel processing enables the hidden layer to select the best path. Additionally, issues with vanishing gradients are avoided via a traces of the previous layer's connection. We show how, given a deep framework, convolution methods are reliable sufficient to be immediately applied to raw sensor data in order to extract features that outperform prior findings. By omitting hand-made or heuristic features, CNNs have the advantage of minimizing engineering bias. This is particularly important when utilizing activity identification techniques in fields with more complex actions or open-ended scenarios, where classifiers must adjust to a fluctuating number of classes. To solve research challenges like these, we employed the innovative deep CNN-attention LSTM with the self-attention model. Our chosen method has been compared with other different models that we have. Our model of choice performed the best testing when compared to these three models, which were sequential attention LSTM layers (attention LSTM-CNN) before the convolution layer, sequential convolution, and dropout layers (CNN-attention LSTM) before the convolution layer, and parallel attention LSTM layers (parallel attention LSTM-CNN) with the convolution layer results, with the UFC50 dataset and 84.44% overall.

Hence all of the mentioned to match our dataset, all of the aforementioned models were therefore constructed with the exact parameters specified in the relevant literature. Although all models performed adequately on the CNN-attention LSTM with the self-attention model beat other models in testing data. Thus, it has been proven that the deep CNN-attention LSTM with self-attention model is useful for the human activity recognition technique.

5 Conclusion

The main goal is to propose a novel CNN-attention-LSTM neural network to improve the HAR's ability to detect associated activities. Since attention LSTM can handle long-term dependencies in time series data and CNN can successfully capture spatial information, the proposed model can detect various types of complex human activities. In the proposed work, CNN is used to extract features, and attention LSTM is used to classify the human activities. In training phase, we used pre-trained weights for VGG19, and Adam optimizer is used as optimization technique for learning weights in ALSTM. We got accuracy of 84.4% within 20 epochs. In testing stage, unknown test video samples are given to VGG19, followed by ALSTM classifier for prediction purposes. We compared the existing work with CNN-SVM, CNN-Naive Bayes, and CNN-LSTM models and is found to be better than all state-of-the-art techniques.

References

1. Zabihi S, Rahimian E, Asif A, Mohammadi A (2022) Light-weighted CNN-attention based architecture for hand gesture recognition via ElectroMyography, pp 1–5
2. Khodabandelou G, Kheriji W, Selem FH (2021) Link traffic speed forecasting using convolutional attention-based gated recurrent unit. Appl Intell 51:2331–2352
3. Kamyab M, Liu G, Rasool A, Adjeisah M (2022) ACR-SA: attention-based deep model through two-channel CNN and Bi-RNN for sentiment analysis. Peer J Comput Sci 8:e877
4. Wang Y, Nie X, Shi Y, Zhou X, Yin Y (2021) Attention-based video hashing for large-scale video retrieval. IEEE Trans Cogn Develop Syst 13:491–502
5. Wall C, Zhang L, Yu Y, Kumar A, Gao R (2022) A deep ensemble neural network with attention mechanisms for lung abnormality classification using audio inputs. Sensors (Basel) 22(15):5566

6. Zhang Y, Yin Z, Nie L, Huang S (2020) Attention based multi-layer fusion of multispectral images for pedestrian detection. IEEE Access 8:165071–165084
7. Zhang P, Li F, Du L, Zhao R, Chen X, Yang T, Fang Z (2021) Psychological stress detection according to ECG using a deep learning model with attention mechanism. Appl Sci 11:2848
8. Mahamoud IS, Voerman J, Coustaty M, Joseph A, d'Andecy VP, Ogier JM (2021) Multi-modal attention-based learning for imbalanced corporate documents classification. In: Lladós J, Lopresti D, Uchida S (eds) Document analysis and recognition—ICDAR 2021. ICDAR 2021. Lecture Notes in Computer Science(), Vol 12823. Springer, Cham
9. Hamdan M, Chaudhary H, Bali A et al (2022) Refocus attention span networks for handwriting line recognition. IJDAR
10. Altun K, Barshan B (2010) Human activity recognition using inertial/magnetic sensor units. In: Salah AA, Gevers T, Sebe N, Vinciarelli A (eds) Human behavior understanding. HBU 2010. Lecture Notes in Computer Science, vol 6219. Springer, Berlin, Heidelberg
11. Győrbiro N, Fábián Á, Hományi G (2009) An activity recognition system for mobile phones. Mob Netw Appl 14:82–91
12. Arifoglu D, Bouchachia A (2017) Activity recognition and abnormal behaviour detection with recurrent neural networks. Proc Comp Sci 110:86–93

Impact of Online Meeting Applications in Indonesia During the Pandemic

Fannisa Hanin Nabilah, Nabilah Putri Intaka, Ford Lumban Gaol, and Tokuro Matsuo

Abstract The influence of globalization and the rapid development of technology is certainly very helpful for our daily lives. Every day people use technology for their daily needs, such as seeking knowledge. Developed technology makes it easier for us to live our daily lives, especially during a pandemic. These technological advances also help the community in dealing with problems that occur during a pandemic. During this pandemic, we are required to stay at home. Therefore, the government implemented large-scale social restrictions (PSBB). Stay at home if you do not have an important need, so that the virus does not spread. The government must find a way so that daily activities can be carried out even at home. One of the easiest ways to do this is by using an online meeting application. This online meeting application is an application that can be used to communicate with one or many people indirectly with the help of the Internet. This study aims to determine the impact of using online meeting applications during the pandemic and to see the impacts that occur in society. The use of easy-to-use online meeting applications is expected to increase consumer comfort in daily activities such as studying or working online during a pandemic.

Keywords COVID-19 · Online learning · Online meeting · Collage · Technology

F. H. Nabilah · N. P. Intaka
School of Information System, Bina Nusantara University, Alam Sutra 13960, Indonesia

F. L. Gaol (✉)
Computer Science Department, BINUS Graduate Program—Doctor of Computer Science, Bina Nusantara University, Jakarta 11480, Indonesia
e-mail: fgaol@binus.edu

T. Matsuo
Advanced Institute of Industrial Technology, Tokyo, Japan
e-mail: matsuo@aiit.ac.jp

G. Rajakumar et al. (eds.), *Intelligent Communication Technologies and Virtual Mobile Networks*, Lecture Notes on Data Engineering and Communications Technologies 171,
https://doi.org/10.1007/978-981-99-1767-9_44

1 Introduction

The impact of the COVID-19 pandemic that hit the world not only has an impact on human health but also has an impact on other aspects of life. One aspect that is affected is the social aspect of education. Schools and campuses were closed simultaneously to prevent the spread of the virus. During this COVID-19 pandemic, the Indonesian government implements large-scale social restrictions (PSBB) and has a very large impact on our daily activities, and we are unable to go to school, work, and other activities because of physical distancing [1]. Because there is physical distancing, everything goes online such as online learning, work from home, and others. And all using online meeting applications [2].

Based on the research background that has been described, here are the issues that will be raised in the research:

1. What is the impact of using online meeting applications during this pandemic? Do they have a good or bad impact on community activities?

In some situations, rather than stating research objectives, researchers will prefer to use research questions. In the example below, the objectives stated in the previous example are reframed as research questions:

1. What is online meeting applications?
2. What is pandemic COVID-19?
3. What the impact of online meeting applications during pandemic in Indonesia?
4. How do meeting applications adapt during a pandemic?
5. What benefits of online meeting applications during a pandemic in Indonesia?

A research aim will usually be followed by a series of statements describing a project's research objectives. Research objectives indicate in more detail the specific research topics or issues the project plans to investigate, building on the main theme stated in the research aim [3]. Normally, at least two or three research objectives will be stated. It is a good practice to put these in a numbered list so they can be clearly identified later in a proposal or report. Here is an example of a set of research objectives:

- To explore online meeting applications even more during a pandemic.
- To understand the status and trends of online meeting applications during a pandemic.
- Explore the impact of online meeting applications today whether it is having a positive or negative impact.
- Find out the impact of online meeting applications for daily activities in Indonesia
- The impact of using online meeting applications in education.

2 Research Methodology

This research uses a quantitative approach research method. As mentioned before that the quantitative data is data in the form of numbers, or extrapolated quantitative data (scoring) [4], so quantitative data is data that has trends that can be analyzed by means or statistical techniques. The data can be in the form of numbers or scores and is usually obtained using a data collection tool whose answers are ranges of scores or weighted questions [5].

This study uses primary data and secondary data. Primary data was obtained using a questionnaire filled in by 110 respondents online, while secondary data was the research method used to collect some of the data obtained by means of the literature study method, the method of collecting data obtained by secondary data that are collected from published study, research paper, books, journals, Internet, official statistical documents and theses, that are related to the impact of online meeting application in Indonesia during pandemic [6].

3 Literature Review

The education sector is experiencing changes due to the pandemic COVID-19. The government through Letter of the Minister of Education and Culture No. 36962/MPK.A/HK/2020 about online learning and working from home to prevent the spread of COVID-19 as an effort to prevent the spread of COVID-19. The concept of e-learning is an ICT-based learning technology (Information and Communication Technology) that has been accepted in education. Several studies state that the quality of information technology implementation such as e-learning will always be related to voluntary user acceptance [7]. With the rapid development of technology so that the government is easy to solve the problem, online meeting applications are one way that learning can be done even though it is online. Online meetings applications are discussion and collaboration activities through the website using audio, video, and screen sharing [8]. This activity is organized by the host and attended by participants invited by the host. Online meetings are intended for meetings with more interactive sessions like offline meetings [9].

The participants in the meeting can interact with each other or make voice interruptions. In this case, the online meeting organizer must be able to control the traffic sound and images that will appear. There are several video conferencing applications that you can use, such as

- Zoom
- Microsoft Team
- Webex Meeting
- Google Hangout
- Skype
- GoToMeeting

- Blackboard Collaborate
- Adobe Connect.

Performance expectations are defined as the extent to which individuals believe that the system used can help them achieve benefits in work [10]. Around the world, millions of people have left offices following government instructions so that employees work from home (work from home). Many people have taken advantage of developed technology, especially during a pandemic like now, the existing system in online meeting applications is very much needed, for example, using Google Meet, which is a feature from Google that can be used to work from home while doing social distancing to prevent the spread of the COVID-19 virus. Google Meet has added various features so that users can take advantage of the system provided by Google Meet. After more and more people started using video conferencing applications to work from home during the COVID-19 pandemic, many technology companies are updating the features of their teleconferencing applications, including Google [11].

In the application of e-learning, performance expectations are a construct that can improve individuals to use technology [12]. Today's technology is increasingly developing, even Internet security companies that offer anti-virus and firewall products, this is proof that technology is getting advanced. Here are some easy tips you can use to keep online safe, including

1. Make sure the online meeting application is always updated
2. OS update
3. Use the registration facility for the online meeting
4. Please use the waiting room if available
5. Use a password for event access
6. Lock the event if all participants are present so that it is not infiltrated by outsiders
7. Use Internet security products that are reliable and updated and do not burden computer performance.

Advantages of using application meeting online are as follows:

- Large space capacity

A virtual meeting application must be able to accommodate many participants in one conference session. Like meetings in general, virtual meetings usually have a large number of participants.

- Support presentation

It is for those, who want to explain or understand a material more deeply, in the online meeting application that can present material files to other participants. Both participants and presenters can scribble on the presentation so that it will make it easier to deepen the material.

- Video and sound quality

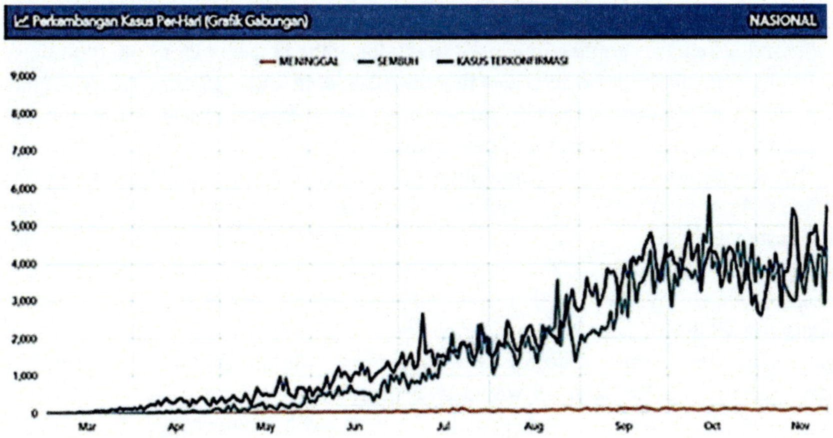

Fig. 1 COVID-19 pandemic situation [11]

The video and sound quality in the online meeting application must be good quality, because these two things are very important in an online meeting application so that users can use it comfortably.

According to [13], coronavirus is a virus with pandemic potential that can cause severe acute respiratory distress and considerable human fatalities, leading to public health emergencies. It first declared when it was found at Wuhan, China, on December 2019 that later spread quickly to several countries since the deployment of the virus is easy only by using fluids from humans or even just a simple physical contact with other people that have been contained with those fluids (such as, saliva, and other bodily fluids and excretions), and it was confirmed by WHO that COVID-19 has become a global pandemic on March 2020. As of November 25, 2020, the virus was still out there with more than 59.7 million confirmed cases, with more than 1.4 million deaths attributed to COVID-19 as shown in Fig. 1.

4 Experimental Result

The data above is official data that has been collected by the Indonesian government regarding the pandemic case that occurred in 2020 nationally in Indonesia. In the data that has been collected in the graph, there is a significant increase in active cases of corona infection every month, although this is accompanied by an increase in healing cases which also continues to increase every month, but still the Indonesian people themselves still have a high risk of having this virus due to a lot of active cases going on.

In November 2020, the total of active cases that has been occurred is 65,804 cases. With the progress of recovered cases, this might sound like a lot but a relief. But considering that the number of active cases is still quite high and even penetrates thousands of cases, people still cannot freely leave the house or carry out their daily activities, especially those activities that are associated with gathering many people. Therefore, most people prefer to carry out activities online or from home to avoid being infected with the unwanted coronavirus.

Based on data collected from a questionnaire that has been made regarding the impact of online meeting applications in Indonesia during the pandemic, there are 103 respondents who respond to the questionnaire with different gender and ages, but based on the data collected there were more male respondents than female respondents. There are 59.2% of male respondents, namely 61 people among 103 respondents and 40.8% of female respondents, namely 42 respondents. Respondents who gave responses to the questionnaire were also of different ages. There are 61.2% respondents aged 19–22 years with a total of 63 out of 103 respondents who gave responses. While the rest were aged 13–15 years with a percentage of 11.7% (12 respondents), 16–18 years with a percentage of 17.5% (18 respondents), and also aged more than 22 years with percentage of 9.7% (ten respondents). Figures 2 and 3 provide the gender distribution and age distribution, respectively.

A. **Gender**

See Fig. 2.

B. **Age**

See Fig. 3.

Based on the questionnaires distributed, it can be seen from the data above that of the 103 respondents, on average the respondents are used to using online meeting applications. Only 6.8% (seven respondents) are not used to do it, the remaining 93.2 (96 respondents). So it can be concluded that online meeting applications are very often used during this pandemic.

C. **Are you accustomed to using online meeting applications?**

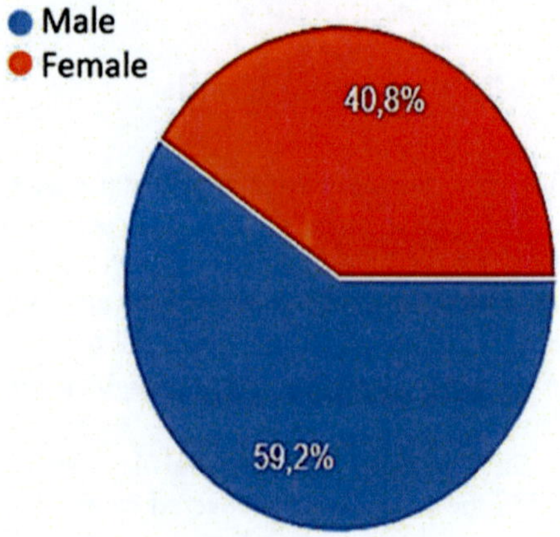

Fig. 2 Gender distribution

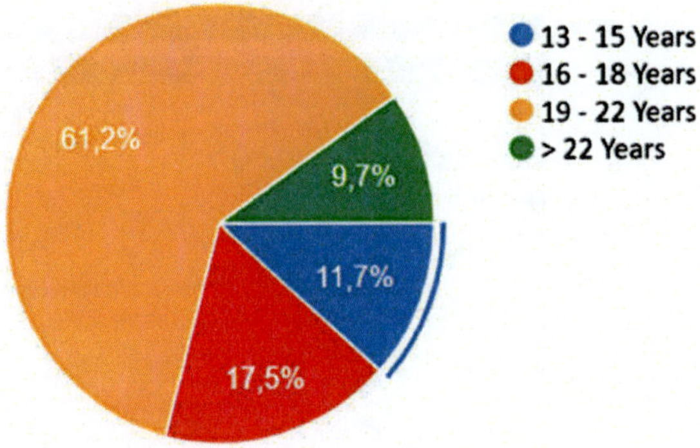

Fig. 3 Age distribution

See Fig. 4.

Based on the questioners that we have been collected before, online meeting applications are often used, we can see that from 103 respondents, 87.4% (90 respondents) are using Zoom, then at number two there is Google Meet with a percentage of 58.3% (60 respondents), and number three is Microsoft Teams with a percentage of 20.4% (21 respondents). Figure 4 shows the accustomed online meeting applications.

D. **Online meeting application that you use**

See Fig. 5.

Online meeting applications are often used for the above activities by the respondents like the diagram above, but what respondents most often do is for school (education) with a percentage of 73.8% (76 respondents) that is because during a pandemic like this, from kindergarten to university, they hold online learning, then in number two is for seminars with a percentage of 50.5% (52 respondents) even though we hold social distancing, we can still be productive by attending online seminars, and number three is for taking part in online events with a percentage of 30.1% (31 respondents) although we cannot attend events in person we can see online events,

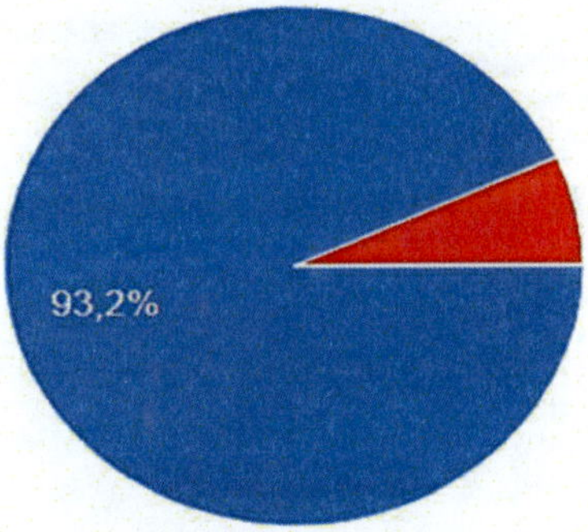

Fig. 4 Accustomed online meeting applications

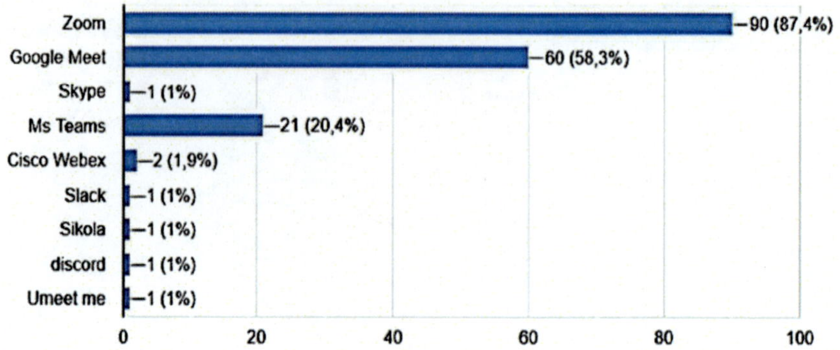

Fig. 5 Online meeting application that you use

such as online music events. Then in other activities that were attended by respondents, such as organizational meetings, thesis guidance, work from home, and even to online reunions, the online meeting application that you use is shown in Fig. 5.

E. **What activities do you do using the online meeting application?**

See Fig. 6.

Based on data collection that has been done before, from 103 respondents there are 45.6% (47 respondents) who answered that the existence of online meeting applications during the pandemic really helped them in meeting the needs of daily activities. While 35% (36 respondents) answered that the existence of online meeting applications during the pandemic helped, 17.5% (18 respondents) answered that it was neutral, and 1.9% (two respondents) thought it was not helpful. This shows that the online meeting application helps people fulfill their daily activities during a pandemic because many people cannot or there is a ban from the government to go out of the house. In addition, only two respondents chose "unhelpful" and none chose the "very unhelpful" option which means that online meeting applications are

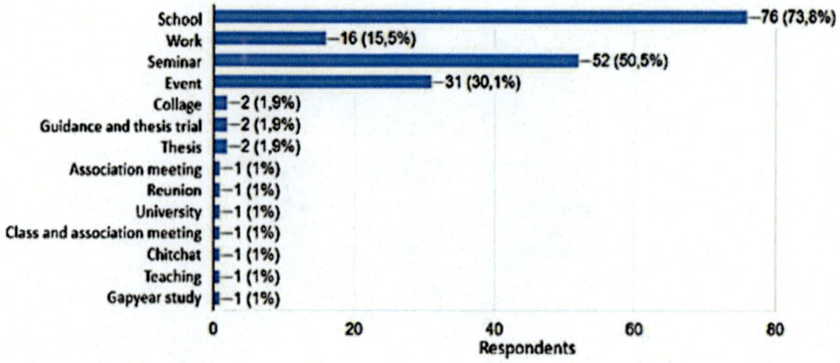

Fig. 6 Activities using the online meeting application

useful and needed by people. Figure 6 shows the activities using the online meeting application.

The use of online meeting applications during the pandemic

Based on data collection that has been done before, we asked respondents what concerns were when using online meeting applications, and the highest concerns chosen by respondents were like the diagram above. The highest concern that the respondents are worried about is the unstable network problem with a percentage of 78.6% (81 respondents), then the concern about buying Internet quota with a percentage of 39.8% (41 respondents), then the third is the concern about personal data security with a percentage of 27.2% (28 respondents), and the last is supporting facilities such as a computer or laptop with a percentage of 17.5% (18 respondents). So from the data above, many respondents are worried about networks, costs and devices as well as the problem of spreading their personal data, because there was a data leak from one of the online meeting applications. Figure 7 is the use of online meeting applications during the pandemic.

The risks using online meeting applications during a pandemic

Based on data collection that has been done before, we asked what were the advantages of using online meeting applications during this pandemic, many of their 103 respondents answered that the use of online meeting applications was more flexible and efficient, because in this pandemic it could help work remotely by communicating

Fig. 7 Use of online meeting applications during the pandemic

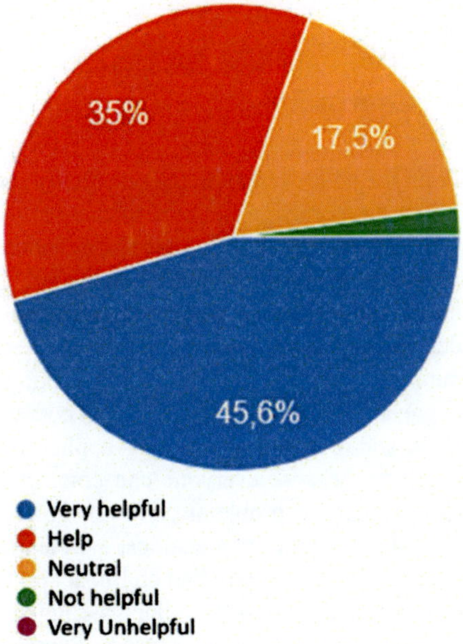

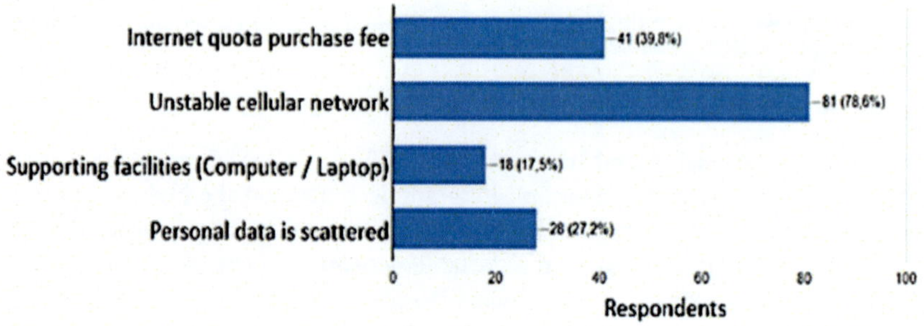

Fig. 8 Risks using online meeting applications

indirectly using the application. Then because of using the online meeting application, the respondents also find it easy to attend online events or seminars because they can be used anywhere and of course can also save time and costs for traveling, then we can also keep social distancing but can still run or attend events, schools, nor work from home. Figure 8 shows the risks using online meeting applications.

Based on data collection that has been done before, we asked what the disadvantages of using online meeting applications were, and most of the 103 respondents mentioned disadvantages such as unstable networks, wasteful quotas, and security of personal data, and there are also those who say it is like when the event is running, the vibes are not felt, when online learning is difficult to understand the material, it is difficult to ask about it, or there is a miscommunication in class.

5 Conclusion

From the above research, the researcher can conclude that development technology is getting more sophisticated so that if this pandemic can be used properly so that activities that are usually carried out continue to be supported using online meeting applications. Even companies always use online meeting applications to keep business running. Online meeting application is a necessity that is experiencing high demand as the only safest and proven alternative to staying in the office amidst the outbreak. Participants in this online meeting only need an optional video conferencing application installed and an Internet connection.

In addition, online meeting application is also able to overcome geographic problems because everyone can connect without having to think about distance. Communication problems can be resolved, and jobs can also be sent via email.

Online meeting applications, as a learning tool, are widely used by companies or individuals that have had an impact on the smooth running of meeting activities. In a pandemic situation, online meeting applications help Indonesian citizens to do learning, lectures, and of course business.

References

1. Enriquez MAS (2014) Students' perceptions on the effectiveness of the use of edmodo as a supplementary tool for learning. In DLSU Research Congress, pp 1–6
2. Iftakhar S (2016) Google classroom: what works and how. J Educ Soc Sci 3(1):12–18
3. Kassymova GK, Duisenbayeva Sh. S., Adilbayeva UB, Khalenova AR, Kosherbayeva AN, Triyono MB, Sangilbayev OS (2019) Cognitive competence based on the e-learning. Int J Adv Sci Technol 28(18):167–177
4. Kassymova G, Arpentieva M, Kosherbayeva A, Triyono M, Sangilbayev O (2019) Science, education and cognitive competence based on e-learning. Bull Nat Acad Sci Republic of Kazakhstan 1(377):269–278
5. Kassymova G, Bekalaeva A, Yershimanova D, Flindt N, Gadirova T, Duisenbayeva SH (2020) E-learning environments and their connection to the human brain. Int J Adv Sci Technol 29(9s):947–954
6. Kassymova G (2018) Competence and its implications. Chall Sci. https://doi.org/10.31643/2018.063
7. Moszkowicz D, Duboc H, Dubertret C, Roux D, Bretagnol F (2020) Daily medical education for confined students during COVID-19 pandemic: a simple videoconference solution. Clin Anat 33(5)
8. Naciri A, Baba MA, Achbani A, Kharbach A (2020) Mobile learning in higher education: unavoidable alternative during COVID-19. Aquademia 4(1):ep20016
9. Oranburg S (2020) Distance education in the time of coronavirus: quick and easy strategies for professors. Legal Studies Research Paper Series. Duquesne University School of Law Research Paper, 2020-02
10. Sharma A (2020) Is zoom safe? Yet another country doesn't seem to think so. Accessed https://www.androidauthority.com/Zoom-india-unsafe-advisory-1108248/
11. Sicat AS (2015) Enhancing college students' proficiency in business writing via schoology. Int J Educ Res 3(1):159–178
12. Triyono BM, Mohib N, Kassymova GK, Pratama GNIP, Adinda D, Arpentieva MR (2020) The profile improvement of vocational school teachers' competencies. Vysshee obrazovanie v Rossii=Higher Education in Russia 29(2):151–158
13. United Nations of Educational (2020) COVID-19 and higher education: today and tomorrow. Accessed http://www.iesalc.unesco.org/en/wpcontent/uploads/2020/04/COVID-19-EN-090420-2.pdf

Implementation Artificial Neural Network on Identification System of Neurological Disorder

Rismayani, Suci Rahma Dani Rachman, Sri Wahyuni, Asmanurhidayani, Joe Y. Mambu, and Martina Pineng

Abstract Applying an artificial neurological network algorithm to a Web-based system, this study hopes to discover neural diseases. Critical to this study is identifying neurological states based on symptom data and analyzing them with an artificial neural network algorithm. The artificial brain network method is implemented; the artificial neural network algorithm is utilized to conduct an analysis that begins with the gathering of data in the form of symptoms and kinds of neural illnesses. The results of this study are based on an analysis of the identification system for neural disorders using an artificial neural network algorithm with an accuracy of 92%.

Keywords Artificial neural network (ANN) · Identification · Neurological disorder

1 Introduction

System neurology disorders are disorders or damage that occurs in the human neural system that can affect its function. The human neural system is a complex link in humans [1]. One of the diseases caused by disorders of the nervous system is stroke. Bimo Adi Laksono et al. explain that stroke attacks the brain and then explain the

Rismayani (✉) · S. R. D. Rachman · S. Wahyuni
Dipa Makassar University, Makassar, Indonesia
e-mail: rismayani@undipa.ac.id

Asmanurhidayani
Puangrimaggalatung University, Sengkang, Indonesia
e-mail: pelangiasma@gmail.com

J. Y. Mambu
Univeristas Klabat, Airmadidi, Indonesia
e-mail: joeyuan.mambu@unklab.ac.id

M. Pineng
Kristen Indonesia Toraja University, Makale, Indonesia
e-mail: mpineng@gmail.com

G. Rajakumar et al. (eds.), *Intelligent Communication Technologies and Virtual Mobile Networks*, Lecture Notes on Data Engineering and Communications Technologies 171,
https://doi.org/10.1007/978-981-99-1767-9_45

cognitive function in post-ischaemic stroke patients [2]. The research problem is how to perform an analysis using an algorithm for artificial neural networks in the identification system of neurological system disorders. This study aims to apply a neural network algorithm to identify neural disorders and determine their accuracy. As for some of the references used as follows, the education system is a strategy used in the learning process to develop the potential that exists in students [3]. The education system also impacts the community by providing information or educating the public about specific problems, such as disasters [4, 5] and disease in humans [6]. So health education is an effort to increase one's health knowledge about diseases and human behaviour to prevent disease and recover from an illness [7]. Then, genetic conditions, autoimmune diseases, trauma, mechanical injury, infection, and malignant growths such as tumours, cell degeneration, stroke, and structural defects can cause nervous system disorders [8]. Although anyone can experience neurological disorders, it does not mean everyone understands this condition's symptoms. So that many symptoms are experienced but are not recognized as neurological disorders; however, if left for too long and not treated immediately, this condition can become more severe and fatal. Common symptoms of neurological diseases are headache, pain, tingling and numbness, weakness, and more. Next, a massively distributed parallel processor is an artificial neural network (ANN) comprised of basic processing units that may store information in the form of experience that is employed in other processes [9]. The fundamentals of an artificial neural network (ANN) are governed by three key components of a neural model [10], namely:

- The weight or strength of a group of synapses, or linkages, is used to categorize them.
- A summarizing adder for the input signals measured each neuron's synaptic strength.
- The purpose of activation is to limit how much the neuron can send out. This function is meant to limit the output signal's amplitude to a certain number.

The ANN layer has three main layers, namely the input layer, the hidden layer, and also the output layer. The input layer is data that is processed in the network, then the hidden layer is a bridge from the input and output layers, which can change the input into something that is used by the output unit. Furthermore, the output layer is a series of transformations in the hidden layer to produce output.

2 Literature Review

As for other authors, the suggested automated diagnostic approach has the capacity to distinguish Parkinson's disease (PD) patients from healthy ones, as demonstrated by a number of similar investigations done by the researchers before [11–13]. Furthermore, this study shows that with (5-12-3) MLP ANN implemented on the FPGA, epilepsy can be diagnosed with a high degree of accuracy (95.14%); the findings illustrate the steps towards correct implementation of ANN on the FPGA, which can subsequently be utilized as the foundation for application-specific integrated circuit (ASIC) design [14]. The Connectome mapping for Alzheimer's illness, autism spectrum disorder, attention deficit hyperactivity disorder, and my electroencephalogram signals and granular genetic fingerprints [15]. This research conducted experiments with residual neural networks in performing AD classification tasks and compared them to the previous studies in the domain with better results with fine-tuned models and for all stages of AD with an accuracy of 100%, 96.85%, 97.38%, 97.40%, and 98.01% for CN, SMC, EMCI, LMCI, and MCI, respectively [16]. In addition, eight neuropsychiatric and neurological illnesses, including stroke, Alzheimer's, Parkinson's, epilepsy, autism, migraine, cerebral palsy, and multiple sclerosis, are predicted by a comprehensive assessment of deep learning methodologies [17–19]. Discussions provide technical knowledge about popular methods, algorithms, and applications in medical diagnosis that have emerged in recent years [20]. The use of preprocessing methods before the CNN model improves the accuracy of dementia classification [21]. Conduct a neuro-fuzzy systems (NFS) approach to the diagnosis of neurologic illnesses in order to identify the most effective NFS technique for clinical practice [22]. The discussion used different CNN-based transfer learning approaches for classifying Alzheimer's disease with additional parameters and excellent accuracy on the ADNI benchmark dataset [23]. The study uses a convolutional network architecture employing frozen features taken from the ImageNet source data set for binary and ternary classification in an experiment utilizing the National Alzheimer's Disease Initiative (ADNI) dataset of MRI images [24]. Research that explores the challenges of examining paediatric brain development in the context of neurological illnesses and providing insight into the possibilities of deep learning approaches on tasks including variables and uncertainties [25]. Further, research provides A logical introduction to this jargon-filled topic and methods for neuroscientists to comprehend, evaluate, and utilize discoveries from this burgeoning discipline [26]. Research deep neural network-based portable diagnostic device (DNN) for improved diagnosis and evaluation of patients with autism spectrum condition (ASD) [27].

3 Method

3.1 Data Collection and Analysis

Data were obtained based on interviews with competent experts. In addition, the authors conducted a literature study from books, the Internet, and journals related to artificial neural networks and nervous system disorders. Furthermore, disease symptom data as test data variables were processed as many as 20 variables with a target of 5 nervous system disorders in the form of Peripheral Neuropathy, Cell Phone Elbow Claw (Paresthesia), Carpal Tunnel Syndrome (Text Claw), iPhosure or Neck Text (Neck Pain), and Primary Lateral Sclerosis (PLS).

Information is transmitted through neuronal synapses. The output is determined by the activation function using the weights of the connections. The model of the artificial neural network is backpropagation, which minimizes error rate by altering the weight of each neuron in response to deviations between production and the desired outcome.

3.2 Data Processing

The system architecture specifies data processing with two users, namely the administrative user and the user. The administrator inputs the type of nervous system disorder and its symptoms, which constitute the data processing parameter to be analyzed by the artificial neural network, as well as the treatment. This solution provides findings if there is a neurological system disorder based on ANN; data processing is visible from the user's system contact with the nervous system utilizing an artificial neural network.

3.3 Data Acquisition

In addition, data acquisition in the form of hardware consisting of 1 Laptop Unit with AMD Ryzen 7 processor specifications, HD 1 TB, 8 GB DDR4 RAM, and software consisting of Windows Operating System 10, PHP programming language, and MySQL database used in the construction of a nervous system disorder recognition system employing an artificial neural network. The research method is supported by software and hardware that are capable of and compatible with the constructed nervous system disease education system.

3.4 Data Testing

The testing data used in this study is data on symptoms of nervous system disorders, which are used as variables and given a value which can be seen in Table 1 as follows:

Table 2 is a grouping table the targets of nervous system disorders as parameters.

Table 1 Variables testing

Variables	Value	Symptoms
NV1	1 = Yes 0 = No	Pain in the legs, thighs, back, hands, or face
NV2	1 = Yes 0 = No	Difficulty walking, coordination problems, cramps, muscle weakness, or reduced muscle mass
NV3	1 = Yes 0 = No	Numbness
NV4	1 = Yes 0 = No	Stiff or tense muscles
NV5	1 = Yes 0 = No	The affected limb will feel weak
NV6	1 = Yes 0 = No	The limbs that describe will feel weak
NV7	1 = Yes 0 = No	The more you feel when you move a standing body
NV8	1 = Yes 0 = No	Feel the cold sensation in the painful part
NV9	1 = Yes 0 = No	A stabbing feeling in the limbs of the branch
NV10	1 = Yes 0 = No	Symptoms of pain in the arm
NV11	1 = Yes 0 = No	Reduced strength in the affected hand or finger
NV12	1 = Yes 0 = No	There is a tingling feeling
NV13	1 = Yes 0 = No	Pain in three fingers (middle finger, index finger, and thumb)
NV14	1 = Yes 0 = No	Dizziness or headache
NV15	1 = Yes 0 = No	If the position of the head is in one place for a long time, it will feel excruciating
NV16	1 = Yes 0 = No	The head is difficult to be moved
NV17	1 = Yes 0 = No	Stiffness when walking due to weakened legs
NV18	1 = Yes 0 = No	Weakness or stiffness in one body part that extends to other body parts, such as the arms, hands, tongue, and mouth
NV19	1 = Yes 0 = No	Changes in speech ability (hoarse, slow, or slurred) and saliva often comes out of the mouth due to weakened facial muscles
NV20	1 = Yes 0 = No	Swallowing or breathing in the advanced stages of the disease

Table 2 Disorder target

Variable	Value	Information
ND1	1	Neuropati Peripheral,
ND2	2	Cell Phone Elbow Claw (Paresthesia)
ND3	3	Carpal Tunnel Syndrome (Text Claw)
ND4	4	iPhosture or Text Neck (Neck Pain)
ND5	5	Primary Lateral Sclerosis (PLS)

4 Result and Discussion

4.1 Experiment

Furthermore, the symptom variable of nervous system disorders was made for the experimental analysis. A variable NV was allocated to the goal or category of nervous system disorders and the parameter limitations for the disorder's symptoms. The objective or kind was varied ND; the experiment was carried out using determinants based on variables results which is shown in Table 3.

4.2 Test Accuracy

Moreover, the accuracy of the tests is dependent on variables and goals evaluated by the system's artificial neural network for neurological system disorders, follow:

$$\text{Accuracy} = \frac{\text{Data Valid}}{\text{Testing Total}} \times 100\%$$

$$\text{Accuracy} = \frac{23}{25} \times 100\% = 92\%$$

Table 4 is a test for the accuracy of the education system in recognizing neurological diseases utilizing ANN, which results in an accuracy of 92%. Twenty-five tests were conducted, resulting in 23 valid values and two invalid tests. Five tests were conducted for every category/target of nervous system problems.

Logic testing uses a white box where the neural network algorithm is used in the implementation of the education system to identify neurological system disorders which is described in the form of a flowchart with the results of the testing process on the main page, and 3 regions are obtained, namely R1, R2, and R3, meaning that there are 3 areas that bounded by edges and nodes. Then, we get 3 cyclomatic complexity (CC), meaning that there are 3 conditions in the programme that connect the initial node to the final node.

Table 3 Experiment

Variables	Target				
	ND1	ND2	ND3	ND4	ND5
NV1	1	0	0	0	0
NV2	1	0	0	1	1
NV3	0	1	1	1	1
NV4	0	1	0	1	1
VN5	0	1	0	1	0
NV6	0	1	0	1	0
NV7	0	1	0	0	0
NV8	0	1	0	0	0
NV9	0	1	0	0	0
NV10	0	0	1	0	0
NV11	0	0	1	0	0
NV12	0	0	1	0	0
NV13	0	0	1	0	0
NV14	0	0	0	1	0
NV15	0	0	0	1	0
NV16	0	0	0	1	0
NV17	0	0	0	0	1
NV18	0	0	0	0	1
NV19	0	0	0	0	1
NV20	0	0	0	0	1

Establishment of training variables
Neuronal density in the hidden layer = 20
Learning rate (a) = 0.1
Time limit = 1000
Target error = 0.01

5 Conclusion

Using the artificial neural network method, this research creates the public which can utilize a Web-based technology to identification neurological system disorders. Based on the white box test, it was determined that there were no logical mistakes in the test findings. In addition, valid findings were achieved from the functional testing that was conducted. The implementation of an artificial neural network (ANN) method in the educational system may evaluate symptom data to diagnose nervous system disorder so that it may be adopted as a preventative or therapeutic intervention. Based on the accuracy test, 25 attempts with each target yielded a score of 92%. Five times with various symptoms, these categories were examined. As for future system work,

Table 4 Accuracy testing

Number	Symptom	System	Target	Accuracy
1	NV1,NV2	Neuropati Peripheral (ND1)	Neuropati Peripheral (ND1)	Valid
2	NV1, NV2, NV3	Neuropati Peripheral (ND1)	Neuropati Peripheral (ND1)	Valid
3	NV1, NV2, NV3, NV4	Neuropati Peripheral (ND1)	Neuropati Peripheral (ND1)	Valid
4	NV1, NV2, NV3, NV4, NV5	Cell Phone Elbow Claw (Paresthesia) (ND2)	Neuropati Peripheral (ND1)	Invalid
5	NV1, NV2, NV3, NV4, NV5, NV5	Cell Phone Elbow Claw (Paresthesia) (ND2)	Neuropati Peripheral (ND1)	Invalid
6	NV3, NV4, NV5, NV6, NV7, NV8, NV9, NV10	Cell Phone Elbow Claw (Paresthesia) (ND2)	Cell Phone Elbow Claw (Paresthesia) (ND2)	Valid
7	NV3, NV4, NV5, NV6, NV7, NV8, NV9, NV10, NV11	Cell Phone Elbow Claw (Paresthesia) (ND2)	Cell Phone Elbow Claw (Paresthesia) (ND2)	Valid
8	NV3, NV4, NV5, NV6, NV7, NV8, vV9, NV10, NV11, NV12	Cell Phone Elbow Claw (Paresthesia) (ND2)	Cell Phone Elbow Claw (Paresthesia) (ND2)	Valid
9	NV3, NV4, NV5, NV6, NV7, NV8, NV9, NV10, NV11, NV12, NV13	Cell Phone Elbow Claw (Paresthesia) (ND2)	Cell Phone Elbow Claw (Paresthesia) (ND2)	Valid
10	NV3, NV4, NV5, NV6, NV7, NV8, NV9, NV10, NV11, NV12, NV13, NV14	Cell Phone Elbow Claw (Paresthesia) (ND2)	Cell Phone Elbow Claw (Paresthesia) (ND2)	Valid
11	NV3, NV11, NV12, NV13, NV14, NV15	Carpal Tunnel Syndrome (Text Claw) (ND3)	Carpal Tunnel Syndrome (Text Claw) (ND3)	Valid
12	NV3, NV4, NV11, NV12, NV13, NV14, NV15,	Carpal Tunnel Syndrome (Text Claw) (ND3)	Carpal Tunnel Syndrome (Text Claw) (ND3)	Valid
13	NV3, NV4, NV5, NV11, NV12, NV13, NV14, NV15	Carpal Tunnel Syndrome (Text Claw) (ND3)	Carpal Tunnel Syndrome (Text Claw) (ND3)	Valid
14	NV3, NV4, NV5, NV6, NV11, NV12, NV13, NV14, NV15	Carpal Tunnel Syndrome (Text Claw) (ND3)	Carpal Tunnel Syndrome (Text Claw) (ND3)	Valid
15	NV3, NV4, NV5, NV6, NV7, NV11, NV12, NV13, NV14, NV15	Carpal Tunnel Syndrome (Text Claw) (ND3)	Carpal Tunnel Syndrome (Text Claw) (ND3)	Valid

(continued)

Table 4 (continued)

Number	Symptom	System	Target	Accuracy
16	NV2, NV3, NV4, NV5, NV6, NV16, NV17, NV18	iPhosture or Text Neck (Neck Pain) (ND4)	iPhosture or Text Neck (Neck Pain) (ND4)	Valid
17	NV2, NV3, NV4, NV5, NV6, NV7, NV16, NV17, NV18	iPhosture or Text Neck (Neck Pain) (ND4)	iPhosture or Text Neck (Neck Pain) (ND4)	Valid
18	NV2, NV3, NV4, NV5, NV6, NV7, NV8, NV16, NV17, NV18	iPhosture or Text Neck (Neck Pain) (ND4)	iPhosture or Text Neck (Neck Pain) (ND4)	Valid
19	NV2, NV3, NV4, NV5, NV6, NV7, NV8, NV9, NV16, NV17, NV18	iPhosture or Text Neck (Neck Pain) (ND4)	iPhosture or Text Neck (Neck Pain) (ND4)	Valid
20	NV2, NV3, NV4, NV5, NV6, NV7, NV8, NV9, NV10 NV16, NV17, NV18	iPhosture or Text Neck (Neck Pain) (ND4)	iPhosture or Text Neck (Neck Pain) (ND4)	Valid
21	NV2, NV3, NV4, NV18, NV19, NV20	Primary Lateral Sclerosis (PLS) (ND5)	Primary Lateral Sclerosis (PLS) (ND5)	Valid
22	NV2, NV3, NV4, NV5, NV18, NV19, NV20,	Primary Lateral Sclerosis (PLS) (ND5)	Primary Lateral Sclerosis (PLS) (ND5)	Valid
23	NV2, NV3, NV4, NV5, NV6, NV18, NV19, NV20,	Primary Lateral Sclerosis (PLS) (ND5)	Primary Lateral Sclerosis (PLS) (ND5)	Valid
24	NV2, NV3, NV4, NV5, NV6, NV7, NV18, NV19, NV20,	Primary Lateral Sclerosis (PLS) (ND5)	Primary Lateral Sclerosis (PLS) (ND5)	Valid
25	NV2, NV3, NV4, NV5, NV6, NV7, NV8, NV18, N19, NV20,	Primary Lateral Sclerosis (PLS) (ND5)	Primary Lateral Sclerosis (PLS) (ND5)	Valid

development may be undertaken using Android-compatible views and algorithms that are more exact and relevant, employing a variety of ANN models.

Acknowledgements The authors would like to thank DRPM SIMLITABMAS KEMDIKBUD RISTEK and LLDIKTI Wilayah IX 2021 at Indonesia for Research Fund Support to the author and the University of Dipa Makassar, who has given the author with the necessary knowledge and assistance to finish this study properly.

References

1. Akil Musi M, Nurjannah (2021) Neurosains: Menjiwai Sistem Saraf dan Otak. Prenada Media

2. Laksono BA, Widyastuti K, Trisnawati SY (2019) Profil gangguan fungsi kognitif pada pasien pasca stroke iskemik di RSUP Sanglah Denpasar Bali, Indonesia periode 2019. Intisari Sains Medis, p 4
3. Miningou EW, Tapsoba MSJ (2017) Education systems and foreign direct investment: does external efficiency matter? International Monetary Fund
4. Rismayani, Paliling A (2021) Buku Sistem Game Edukasi Penanggulangan Bencana Banjir. Buku Deepublish. https://penerbitbukudeepublish.com/shop/sistem-game-edukasi/. Accessed 8 Dec 2021
5. Rismayani AP, Nurhidayani A, Pineng M (2022) Fundamental design of flood management educational games using virtual reality technology. Int J Online Biomed Eng (iJOE) 18(03). https://doi.org/10.3991/ijoe.v18i03.27787
6. Rismayani MP, Herlinda (2022) Using artificial neural network for system education eye disease recognition web-based. J Biomimet Biomater Biomed Eng 55:262–274. https://doi.org/10.4028/p-7z9xpt
7. "bpjs Kesehatan." Accessed 8 Dec 2021. [Online]. Available: https://bpjs-kesehatan.go.id/bpjs/dmdocuments/098efcaf5574de037e9e29e0bfb0f0d5.pdf
8. Davis LE, Kennedy PGE (2000) Infectious diseases of the nervous system. Butterworth-Heinemann
9. HSMPS et al (2022) Artificial intelligence. Media Sains Indonesia
10. Ashil Hiral P, Senthil KAV (2020) Applications of artificial neural networks for nonlinear data. IGI Global
11. Ali L, Zhu C, Zhang Z, Liu Y (2019) Automated detection of parkinson's disease based on multiple types of sustained phonations using linear discriminant analysis and genetically optimized neural network. IEEE J Translation Eng Health Med 7:1–10. https://doi.org/10.1109/JTEHM.2019.2940900
12. Miao Y., Lou X., Wu H. (2021) The diagnosis of Parkinson's disease based on Gait, speech analysis and machine learning techniques. In: Proceedings of the 2021 international conference on bioinformatics and intelligent computing, New York, NY, USA, Jan 2021, pp 358–371. https://doi.org/10.1145/3448748.3448804
13. Vyas T, Yadav R, Solanki C, Darji R, Desai S, Tanwar S (2022) Deep learning-based scheme to diagnose Parkinson's disease. Expert Syst 39(3):e12739. https://doi.org/10.1111/exsy.12739
14. Šarić R, Jokić D, Beganović N, Pokvić LG, Badnjević A (2020) FPGA-based real-time epileptic seizure classification using artificial neural network. Biomed Signal Process Control 62:102106. https://doi.org/10.1016/j.bspc.2020.102106
15. Valliani AA-A, Ranti D, Oermann EK (2019) Deep learning and neurology: a systematic review. Neurol Ther 8(2):351–365. https://doi.org/10.1007/s40120-019-00153-8
16. Ramzan F et al (2019) A deep learning approach for automated diagnosis and multi-class classification of Alzheimer's disease stages using resting-state fMRI and residual neural networks. J Med Syst 44(2):37. https://doi.org/10.1007/s10916-019-1475-2
17. Gautam R, Sharma M (2020) Prevalence and diagnosis of neurological disorders using different deep learning techniques: a meta-analysis. J Med Syst 44(2):49. https://doi.org/10.1007/s10916-019-1519-7
18. Chen X, Li L, Sharma A, Dhiman G, Vimal S (2022) The application of convolutional neural network model in diagnosis and nursing of MR imaging in Alzheimer's disease. Interdiscip Sci Comput Life Sci 14(1):34–44. https://doi.org/10.1007/s12539-021-00450-7
19. Balaji C, Suresh DS (2022) Multiclass recognition of Alzheimer's and Parkinson's disease using various machine learning techniques: a study. Int J Model Simul Sci Comput 13(01):2250008. https://doi.org/10.1142/S1793962322500088
20. Kamdar JH, Jeba Praba J, Georrge JJ (2020) Artificial intelligence in medical diagnosis: methods, algorithms and applications. In: Jain V, Chatterjee JM (Eds) Machine learning with health care perspective: machine learning and healthcare, Springer International Publishing, Cham, 2020, pp 27–37. https://doi.org/10.1007/978-3-030-40850-3_2
21. Janghel RR, Rathore YK (2021) Deep convolution neural network based system for early diagnosis of Alzheimer's disease. IRBM 42(4):258–267. https://doi.org/10.1016/j.irbm.2020.06.006

22. Bali B, Garba EJ (2021) Neuro-fuzzy approach for prediction of neurological disorders: a systematic review. SN Comput. Sci. 2(4):307. https://doi.org/10.1007/s42979-021-00710-9
23. Ashraf A, Naz S, Shirazi SH, Razzak I, Parsad M (2021) Deep transfer learning for alzheimer neurological disorder detection. Multimed Tools Appl 80(20):30117–30142. https://doi.org/10.1007/s11042-020-10331-8
24. Naz S, Ashraf A, Zaib A (2022) Transfer learning using freeze features for Alzheimer neurological disorder detection using ADNI dataset. Multimedia Syst 28(1):85–94. https://doi.org/10.1007/s00530-021-00797-3
25. Sargolzaei S (2021) Can deep learning hit a moving target? A scoping review of its role to study neurological disorders in children. Front Comput Neurosci 15, 2021, Accessed 30 Apr 2022 [Online]. Available https://www.frontiersin.org/article/; https://doi.org/10.3389/fncom.2021.670489
26. Auger SD, Jacobs BM, Dobson R, Marshall CR, Noyce AJ (2021) Big data, machine learning and artificial intelligence: a neurologist's guide. Pract Neurol 21(1):4–11. https://doi.org/10.1136/practneurol-2020-002688
27. Khullar V, Singh HP, Bala M (2021) Deep neural network-based handheld diagnosis system for Autism spectrum disorder. Neurol India 69(1):66. https://doi.org/10.4103/0028-3886.310069

Implementation of HBEA for Tumor Cell Prediction Using Gene Expression and Dose Response

P. Selvi Rajendran and K. R. Kartheeswari

Abstract An important aspect of sustainable drug development is drug-target interaction. In cancer cell lines, the drug response target ratio is critical. It is important to estimate the drug reaction in a cancer cell line. In prior research, we employed ensemble algorithms with voting methods to predict medication response and achieved 97.5% accuracy. A hybrid ensemble algorithm for the revised drug response (HBEA) method is developed to improve drug-target strategy in cell lines. Rather than generating several homogeneous weak learners to generate a single model in the ensemble, this enhanced algorithm uses a diverse collection of weak learners such as random forest, Naive Bayes, and decision tree to create a strong meta-classifier. Cross-validation of hard and soft data would be used to accomplish this. The concentrations of various drugs are used as inputs, and the cell line predicts the relevant drug response. The goal of this enhanced ensemble algorithm is to suggest a new medicine based on a single licensed drug or a combination of drugs. This approach increased the drug responsiveness from 97.5 to 100%, according to our findings. The proposed method is applied in an open-source and freely available at https://decrease.fimm.fi.

Keywords Drug response prediction · Ensemble learning · Cancer prediction · Machine learning · Drug response

1 Introduction

People across the globe are dying from cancer at a record rate. Anti-cancer treatments are an essential part of cancer treatment, and their proper regulation can help prolong the patient's life. Many clinical studies have shown that cancers with different

P. S. Rajendran (✉) · K. R. Kartheeswari
Computer Science and Engineering, School of Computing, Hindustan Institute of Technology and Science, Chennai, India
e-mail: selvir@hindustanuniv.ac.in

G. Rajakumar et al. (eds.), *Intelligent Communication Technologies and Virtual Mobile Networks*, Lecture Notes on Data Engineering and Communications Technologies 171,
https://doi.org/10.1007/978-981-99-1767-9_46

genetic characteristics respond differently to the same treatment or drug [1–3]. Precision medicine tries to precisely pick cancer treatments based on each patient's genetic information [4]. It is difficult to predict the response to anti-cancer treatments for individual patients in precision medicine [5–7]. Several cell line-based datasets are publicly available, including the National Cancer Institute 60, the Cancer Therapeutics Response Portal (CTRP), the Genomics of Drug Sensitivity in Cancer, the Cancer Cell line Encyclopedia, and Genentech Cell Line Screening Initiative (gCSI). This dataset can help to get better outcomes. Medicine's physiochemical properties strongly influence their therapeutic index. In aqueous physiological conditions at pH 7.4, the majority of anti-cancer drugs are poorly soluble. Paclitaxel and docetaxel are axene-based medicines, and 9-nitrocamptothecin is a camptothecin derivative. To develop delivery techniques as well as lung clearance, fundamental features of anticancer drugs, such as log P and pKa values, are critical [8]. Medicinal drug's solubility determined both the permeability and potency of their cancer therapeutic effect, according to Lipinski's rule [9, 10]. Early cancer detection and prediction of survivorship depth can assist patients and healthcare providers in better controlling expenditures, treatment intensity, and time spent in the medical care setting. When such a condition is detected early on, the chances of a positive outcome increase. Although considerable progress has been made in the early detection of cancer [11, 12], much more research is needed to discover strategies for estimating survivorship that is both common and feasible in medical practice [13]. The introduction of targeted anti-cancer therapies based on gene-specific effects may prove a useful tool in the fight against cancer. Many clinical studies are required to develop particular targeted therapy for cancer patients in clinical treatment. However, there are other challenges, including sample limitations, difficult procedures, severe environmental standards, and expensive costs, which prevent the supply from matching demand [14]. To limit the risk of drug-target interactions, computational methods were used for these predictions. Therefore, the focus of research on the drug target in cell lines might be more effective [15, 16]. In general, drugs interact with their target molecules in three ways: (1) Machine learning-based prediction, (2) deep learning prediction, and (3) network-based prediction [16–23]. Recently, multiple classifiers have become popular in research. It is well established that by integrating multiple classifiers, classification performance can be improved over single classifiers [24–26]. Over the last few years, hybrid and ensemble machine learning algorithms have attracted the scientific community's interest [27]. There is strong evidence that multiple, ensemble models perform better than single weak learners in both theory and practice, especially when dealing with multidimensional, difficult regression, and classification problems [28]. The paper is structured as follows: The literature study on similar recent techniques is described in Sect. 2. The proposed algorithm is analyzed in Sect. 3. Section 4 consists of architecture diagram which includes various components involved in drug response prediction. Materials and methodology are discussed in Sect. 5. The result has been analyzed in Sect. 6. Finally, the conclusion is presented in Sect. 7.

2 Literature Survey

Ensemble Learning

Creating and combining multiple inducers to solve a specific machine learning problem is called ensemble learning. The natural reason for the ensemble process comes from human nature and our proclivity to collect and weigh multiple viewpoints to make a complex decision. The primary premise is that evaluating and aggregating multiple individual viewpoints is preferable to choosing one individual's opinion. Matching a medical treatment to sickness is an example of such a decision [29]. The limitation is homogeneous ensemble learning. The absence of a strong classifier to predict drug response [29].

Agarwal et al. used ensemble voting to predict lung cancer survival after 6-months, 9-months, 1-year, 2-years, and 5-years of diagnosis. Five decision tree algorithms were created using data from the SEER program. By taking the average of the probabilities generated by each classifier, they integrated. The researchers used tenfold cross-validation for training and testing and compared the performance of ensemble voting with that of individual classifiers to demonstrate how ensemble data mining can help weak classifiers perform better. The limitation of this research is only based on the clinical outcome [30, 31]. In a study published in science [32], Matlock et al. examined a variety of ensembles that were trained to predict drug sensitivity, including a deep learning method for gene expression data. There were only a limited number of cell lines and compounds used in this study.

Ensemble learning is a machine learning method in which multiple models, referred to as "weak learners," are taught to tackle the same issue and then integrated to improve results, according to the survey on ensemble learning. The core concept is that we can generate more accurate and/or resilient models by correctly integrating weak models.

Machine Learning

Zhang [33] stated that the machine learning approach saves time and effort by eliminating the need for specialists to establish rules and threshold values because the algorithm performs these activities internally. Choosing features simplifies the model and makes it easier to understand and implement. It enhances accuracy while also cutting down on training time. It helps to support generalization by reducing overfitting. It lowers the chances of data mistakes during installation. Algorithms for machine learning are self-improving, which means they learn from examples and experiences. Even if a machine-based method is beneficial, it is necessary to focus on key features for medication analysis. Only cell line data is concentrated in feature selection in this case. It is necessary to keep an eye on the omics analysis. As a result, the reaction is still not clinically meaningful. Complementary ML models, according to Costello et al. [32], improve the predictability of drug response prediction models. It improves the model's robustness. The work's limitation is the tiny number of cell lines. In this research [34], the author Guosheng Lianga et al. examined that artificial intelligence plays a significant part in the discovery of new materials and intensely accelerates anti-cancer drug development. The author stated that machine learning analysis was used to predict the sensitivity of the drug. The drawback of this method is that it is tough to formulate the best treatment.

There are three types of machine learning-based methods:

Feature vector-based approach

Feature vectors are n-dimensional vectors that represent the characteristics of an object. In the field of machine learning, numerical representations of objects are commonly used because they make statistical analysis and processing easier. As an example, feature values can represent pixels of an image, and term occurrences in texts. Feature vectors are the same as explanatory variable vectors, which are utilized in statical techniques like linear regression. By pairing feature vectors with weights, a linear predictor function is constructed, which is used to calculate a prediction score. This is known as the feature space or vector space. A variety of dimensionality reduction approaches can be used to lower the dimensionality of feature space. Feature creation has long been thought to be a useful approach for improving structure accuracy and comprehension, especially in high-dimensional issues [35]. Analyzing which input features contributed most to a given prediction could help identify potential biomarkers of drug response or characteristics of drugs that would induce better drug response. Identifying new biomarkers for precision medicine is particularly important [36].

Similarity-based approach

Treating given similarities as inner products in some Hilbert space or treating dissimilarities as distances in some Euclidean space is a prominent technique for similarity-based categorization. There are a number of common techniques for similarity-based categorization, such as treating given similarity as an inner product in some Hilbert space or treating dissimilarity as distance in some Euclidean space. To begin, the samples are explicitly embedded in a Euclidean space based on the differences (dis)similarities using multidimensional scaling [37]. A second option is to embed the samples implicitly in a Euclidean space based on their similarity or dissimilarity. Santini and Jain [38] found similar functions might fail to satisfy the other mathematical requirements for metrics or inner products—especially when they are asymmetric. Similarity-based classification can be useful in computer vision, bioinformatics, information retrieval, natural language processing, and more. Some simple examples of similarity functions (edit distance) are travel time from one location to another, the compressibility of the random process given a code corresponding to another, and the minimum number of steps needed to transform one sequence into another. As part of bioinformatics, the Smith–Waterman algorithm [39], the FASTA algorithm [40], and the BLAST algorithm [41] are commonly used to determine amino acid sequence similarity in protein classification. Term frequency and inverse document frequency vector (tf-IDF) are commonly utilized in information retrieval and text mining for document categorization. There is a limitation to similarity networks in that they are only based on genome-wide gene expression profiles, but do not take into consideration somatic mutants and copy number variations within cell lines.

Network-based Prediction method

Stanfield et al. [42] propose a network-based prediction method that combines cell line genetic mutation and drug responses with protein–protein interaction (PPI) network and uses random walk with restart (RWR) to calculate feature vectors for drugs and cell lines in order to predict missing response values. Other researchers, on the other hand, are focusing on using similarities across cell lines or medications to make predictions [42, 43]. Zhang et al. [44], e.g., project multiple side information from a cell line and a drug into similarity networks to create an integrated dual-layer network that fills in the gaps [44].

While the methods described above have been proved to outperform previous methods, they do have certain drawbacks. Stanfield's method, e.g., solely takes into account gene mutation information and ignores similarity information, whereas Zhang et al. method ignores genetic variants and does not take gene correlations into account.

Cross-Validation

According to Pedregosa et al. [45], datasets were separated into training and testing sets after dimensionality reduction. To avoid overfitting and account for variance in each classifier, tenfold cross-validation was utilized, in which the dataset was randomly partitioned into training and test sets ten times. Hastie et al. and Duda et al. [46, 47] stated that cross-validation is a data resampling technique. It is used for evaluating prediction model generalization and avoiding overfitting. In [48], Efron et al. stated that cross-validation is like a bootstrap that belongs to the Monte Carlo method family. This research introduces cross-validation and resampling procedures that go with it. Because every observation is used for both training and testing, this technique uses data more "efficiently" than other machine learning algorithms.

Heterogeneous-Based Ensemble Algorithm (HBEA)

In [45], Zhang et al. stated that for solving real-world problems, ensemble machine learning algorithms are extremely powerful and adaptable. Ensemble learning improved the predictability of decision-making systems by increasing accuracy by minimizing variation. The author explained the feature selection, missing features, and data imbalance. Classifiers are integrated into a variety of ways to improve performance measurements [27], including bagging, boosting, and stacking. According to [49–52], Bhardwaj et al. developed a double ensemble machine learning algorithm to predict the pathological response after neoadjuvant chemotherapy. This double ensemble was used to predict multi-criteria decision-making. Based on the above reference to develop a strong learner from a weak learner by generating a more accurate model to predict the drug response, the HBEA technique is introduced. When compared to the homogeneous ensemble approach, the accuracy improvement was significantly higher. In addition, to improve the precision of the analysis, the drug responses in cell lines were absorbed. A specific pair index graph is plotted for a certain drug concentration ratio.

In the following section, pseudocode of the proposed methodology is discussed.

3 Pseudocode

1. Extract drug data set # contains similarity matrix of each pair of drugs and targets
2. Input $x =$ conc1 && conc2 # drug concentration
3. For $i = 1$ to n # n-number of concentration count in drug dataset
4. Applied random Ei forest

$$f_{rf}^{n}(x) = 1/n \sum_{i=1}^{n} T_i(x) \tag{1}$$

$f_{rf}{}^{n}(x)$—Drug Response

5. Applied Gaussian NB

$$P(x_i/y) = 1/\text{sqrt}\left(2 * 3.14 * \sigma_y^2\right) \tag{2}$$

$P(x_i/y)$—probability of drug response

6. Applied Logistic Regression

$$(x) = 1/(1 + \exp(-f(x))) \tag{3}$$

$f(x) = b_0 + b_1 x$

7. Applied SVM $y = f(x)$ # $x \in R^{D}$, R^{D} here is a vector space with D dimension.
8. Applied Decision Irie Classifier

$$y = -\sum_{i=1}^{n} x_i \log_2(x_i) \tag{4}$$

$y =$ Drug Response for x-drug concentration
Enhanced HBEA algorithm

9. Chosen the best result and algorithm among RF, GNB, LR, SVM & DT #Random Forest, Gaussian NB, Logistic Regression, Support Vector Machine, Decision Tree
10. Extract higher accuracy

$$\text{Extract higher accuracy} \tag{5}$$

Hard Voting to obtain drug response

11. Calculate the average

$$(RF, GNB, LR, SVM \,\&\, DT)/5 \tag{6}$$

  # Soft voting to obtain drug response
  # Cross validation
12. Split data set # k number of subsets
13. K-fold cross-validation set # k-1 training parts
14. Cross-validation of all models $E = 1/K \sum_{k=1}^{k} Ei$ # E_i—HBEA of random forest && Gaussian NB && Decision Tree
15. Made more evaluation # training more evaluation on all the subsets
16. Used model stacking
17. Work with dependent/Grouped Data
18. Parameters fine-tuning
19. Cross validation of (5) && (6)
20. Retrieve Drug Response # target mapping

4 Architecture Diagram

The process flow diagram of the proposed approach is presented in Fig. 1. This model comprises two modules as genomic module (Phase #1) and the computational module (Phase #2).

Genomic Module

Here, bio genomic and biomechanical features are the input dataset of the model. This dataset is freely available on the web for research purposes [1] and carries drug and cell line features. A model can predict a particular drug response using

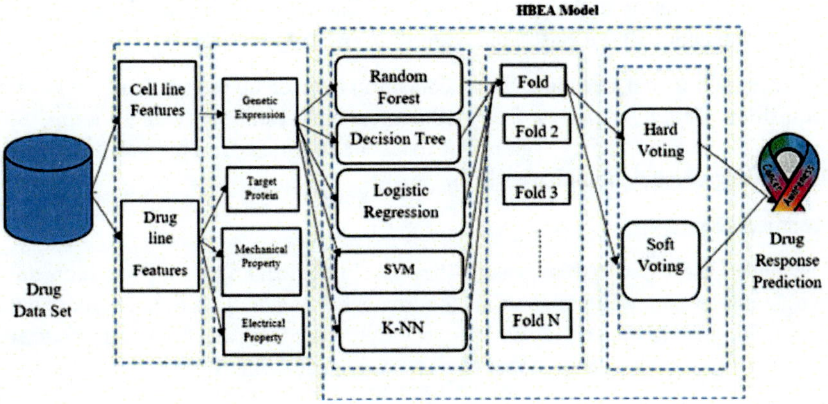

Fig. 1 Architecture diagram of proposed HBEA model

multiple molecular features and their combinations due to the considerable molecular heterogeneity observed across tumors [53]. Hence, the genetic expression has been considered an important feature in cell lines. As well as target protein, mechanical, and electrical properties are also considered drug line features.

Detailed process is explained in the materials and methods section.

Computational Models

A computational model incorporates several variables that describe the system under investigation. Adjusting each of the variables individually or in combination will monitoring the effect of changes in the outcomes. Five different types of weak learners in machine learning models are combined to fabricate an HBEA model which are logistic regression, decision tree, support vector machine, *K*-nearest neighbor, and Naïve Bayes. Cross-validation helps to estimate the performance of the model where *N*-fold cross-validation is used in this analysis. *N*-fold cross-validation is performed by partitioning the original training dataset into *N* equal subsets [54]. Each subset is called a fold which is denoted as fold 1, fold 2, …, fold *n*. Generally, the value of k is taken to be 10, whereas k can be any value. The voting classifier estimator is assembled by combining different classification models which turn out to be a strong meta-classifier that stabilizes the individual classifier's weakness on a particular dataset. The hard voting classifier takes the majority voting based on weights, and the soft voting classifier takes the probabilities of all the predictions made by different classifiers. The target attribute is a binary variable indicating whether the drug is responding in a cell line or not [55, 56].

5 Materials and Methodology

A comprehensive set of genetic, molecular, and electro-mechanical features for cancer cell lines are collected from CCLE, GDSC, and DECREASE [1] dataset and it has been tested in the proposed machine learning framework.

Heterogeneous-Based Ensemble Algorithm (HBEA) Model:

Data acquisition and selection: Data acquisition is made up of two words: data and acquisition. Data refers to raw facts and numbers that can be structured or unstructured, and acquisition refers to gathering data for a specific goal. This web link carries 23,595 drug combination metrics [57].

Preprocessing

The models discussed in this review utilize past biological knowledge like route data to filter out less relevant variables and optimize the models. This drug response data is multidimensional and highly noisy, so some preprocessing and filtering is required, particularly for omics datasets that characterize the cell lines [58].

Applying algorithms

HBEA enhanced ensemble learning model is created using five different types of machine learning algorithms. The most common of these are logistic regression, decision tree, support vector machine, K-nearest neighbor, and Naive Bayes. In other ensemble models, a homogeneous collection is used. However, in HBEA, a heterogeneous collection is used. In other ensemble models, a homogeneous collection is used. However, in HBEA, a heterogeneous collection is used.

Analysis

Drug response prediction from a cancer patient is used to predict the response of a patient's cell line to a drug. This dataset comprises the cancer patient's details including the drug names and their concentration level and also the response to the concentration level. It comprises the record of 1152 cancer patient details with 3 different attributes in three different cell lines. The target attribute is a way of predicting drug response in a specific cell line. It indicates whether the patient has cancer or not.

In Hex, Hela, and Hep cell lines, two different drug concentrations are compared with a specific pair index. In comparison to Hex cell lines, the medication appears to respond quickly in Hela and Hep cell lines.

Confusion Matrix

The confusion matrix displays the number of patients with various actual and predicted drug responses in the cell line. The classification findings can be interpreted in a variety of ways. For example, as demonstrated in the confusion matrices (Figs. 2 and 3), "misclassified" samples for a given medicine could be an indicator of its potential for novel usage, or repurposing, in these "incorrectly" assigned conditions. As a result, misclassification may lead to surprising new findings. This method covers the way for the use of machine learning in the field of drug response.

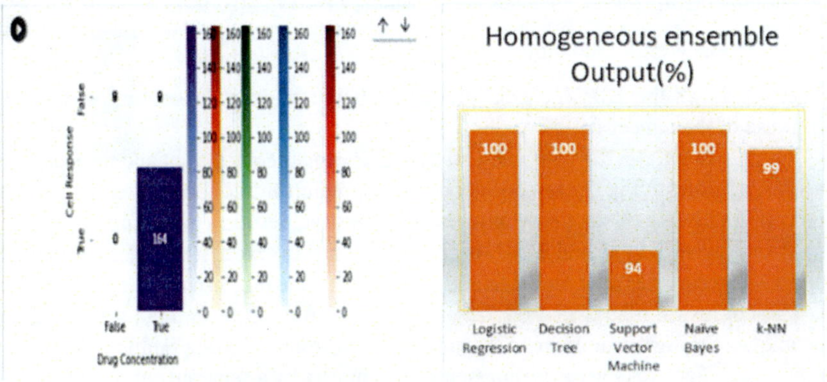

Fig. 2 **a** Homogeneous confusion matrix output. **b** Accuracy measure of homogeneous

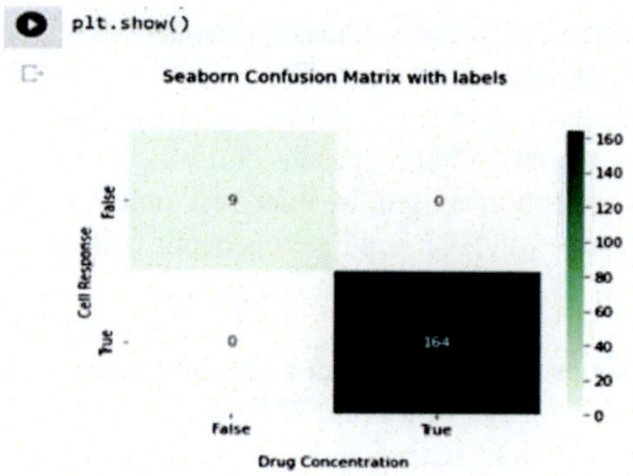

Fig. 3 Heterogeneous confusion matrix

6 Experimentation

The heterogeneous and homogeneous outputs are compared in this study. When compared to the homogeneous model, the heterogeneous ensemble model produces more accurate findings. The next sections detail the experimentation and analysis of homogeneous and heterogeneous output.

Homogeneous Output

The accuracy prediction analysis utilizing the homogeneous ensemble method is shown in Fig. 2. In this study, weak learners, logistic regression, decision trees, support vector machines, K-NN, and the Naive Bayes model were employed.

The confusion matrix output of homogeneous ensemble weak learners and their accuracy level are depicted in Fig. 2a and b. The drug concentration in cell response is predicted using the confusion matrix in Fig. 2a. Figure 2b shows the accuracy of logistic regression, decision trees, support vector machines, naive Bayes, and K-NN. In a homogeneous mode, each learner's accuracy is evaluated individually.

HBEA Model Result Outcome

Figure 3 shows the HBEA model prediction between drug concentration and the response of the cancer cell line.

Figure 3 illustrates the confusion matrix of heterogeneous output and its accuracy level. According to Fig. 3, the confusion matrix predicts the concentration of drugs in the cell. In this Fig. 3, accuracy of logistic regression, decision trees, support vector machines, naive Bayes, and K-NN is combined. In this graph, it can be seen that all algorithm accuracies are combined and evaluated as one plot in strong learner mode.

The confusion matrix of Figs. 2a and 3 gave information on data visualization techniques through heat maps. A heatmap is a two-dimensional graphic representation of data. Each data value is represented in a matrix by a separate color. Confusion matrix describes the performance of a classification model (or "classifier") on a set of test data which already contains the true values. The output of the heatmap is

Table 1 Confusion matrix output

173	Predicted: NO	Predicted: YES
Actual: NO	9	0
Actual: YES	0	164

explained in Table 1. It has been shown that 1152 patient's medication concentrations and cell line responses were taken as an input. 15% of the testing data is used in the classification after training. As a result, 173 data points were provided for testing. The following findings were discovered throughout the testing. In this cell line, 164 patients had a good response to the drugs. Nine patients showed a poor response to the drugs. We calculated the categorization model's performance based on that observation. The findings below prove that the HBEA algorithm is 100% accurate. The explanation is given in Table 1.

Table 1 has given that out of 173 cancer patients, 164 are diagnosed with cancer. The remaining nine patients are cancer free. True and false positives and negatives are used to assess this, where

True Positives (TP) indicate that these are the cell lines in which we predicted yes (), and they are not leaving the network. The value is 164.

True Negatives (TN) indicate that no cancer prediction in cell lines, and they are not leaving the network. The value is 9.

False Positives (FP): We predicted yes, but they are not leaving the network (a drug not responding to the cell line). It is also known as a **"Type 1 error."** Here no one meets this criterion.

False Negatives FN: Although we predicted no, they left the network (drug responding to the cell line). Type 2 errors occur when this happens. None of the records in this dataset meet this criterion.

Therefore, accuracy obtained in this validation is given below in Eq. (7).

$$\text{Accuracy} = (\text{True positive} + \text{True negative})/\text{Total} \tag{7}$$

$$\text{Accuracy} = 164 + 9/173 = 100\%$$

From the above validation, it is clear that the heterogeneous ensemble is obtaining good accuracy in the cancer prediction.

Inputs and Outputs

The input data for this proposed approach is given in Table 2. Two medications are used as inputs, and their concentrations are taken into account using a specific pair index. Finally, the drug response in a specific cell line is anticipated. The cell lines HEK 293, HeLa, and HepG2 are showing rapid pharmacological responsiveness.

Table 2 Input data

Pair index	Drug1	Drug2	Conc1	Conc2	Response	Cell
1	BMS-754807	LY3009120	0	0	0	HEK293
1	BMS-754807	LY3009120	0	10	0	HEK293
1	BMS-754807	LY3009120	0	30	0	HEK293
1	BMS-754807	LY3009120	0	100	0	HEK293
1	BMS-754807	LY3009120	0	300	0	HEK293
1	BMS-754807	LY3009120	0	1000	11.45047	HEK293
1	BMS-754807	LY3009120	0	3000	25.40777	HEK293

Table 3 Output: drug response in different cell lines for particular concentration level for a particular pair index

Pair index	Response	Cell line	Pair index	Response	Cell line
5	11.450471	HEK293	5	26.325592	HeLa
6	25.407765	HEK293	6	37.922136	HeLa
7	39.124137	HEK293	7	9.532144	HeLa
8	20.587264	HEK293	8	8.474513	HeLa
9	7.484891	HEK293	9	10.639141	HeLa
10	13.882070	HEK293	10	10.335465	HeLa
11	20.333125	HEK293	11	16.718651	HeLa
12	22.442476	HEK293	12	32.197165	HeLa
13	36.298115	HEK293	13	30.485807	HeLa
14	40.358234	HEK293	14	48.800628	HeLa

Table 3 gives the drug response in different cell lines for particular concentration level for a particular pair index.

The graph between drugs (BMS-754807, BGB324, Cisplatin, and NVP-LCL161) and drug concentration is shown in Fig. 4a. A graph is plotted between drugs (LY3009120, Trametinib, Everolimus, Ipatasertib) and drug concentration in Fig. 4b. Pair index and their numbers are displayed in Fig. 4c. Figure 4d depicts the drug response in HEK 293, HeLa cell, and HepG2 cell lines. The maximal drug response in the HEK cell line is 91.78%, as shown in Fig. 4d. In Hela and Hep G2, the maximum medication response was 100%. In the HEK293 cell line, the pair index below 7 indicates that there is no pharmacological response. If the pair index was 7 or above, the drug response in HeLa cells was zero. If the pair index was greater than 13, the medication response was nil in Hep G2.

Considering the context of the present research, ensemble models are considered to be useful tools to enhance anti-cancer drug response. In contrast to previous studies of the use of homogeneous ensembles for such types of problems, the present research concentrates on the use of heterogeneous ensembles.

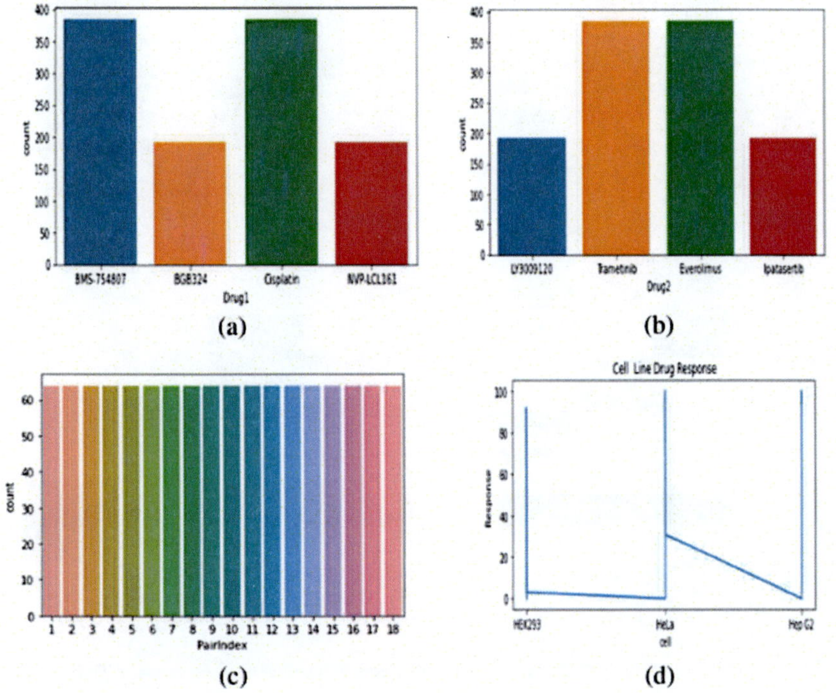

Fig. 4 **a** and **b** drug versus drug concentration count in nm(nanomole). **c** Pair index versus and pair index count in cell line. **d** Cell line versus drug response

The results obtained suggest that the application of HBEA modeling for anti-cancer drug prediction is effective. For cancer diagnostics, heterogeneous ensembles provide reliable predictions.

It is obvious from the output that inhomogeneous mode, not all of the classifiers perform well in the drug response prediction analysis. As a result, it has been demonstrated that HBEA (heterogeneous) classifiers are effective in predicting anti-cancer medication response in cell lines.

7 Result and Discussion

HBEA was developed with the goal of predicting drug response in cancer cell lines, assuming that cancer pathways would properly represent drug therapeutic effects (Fig. 5). We estimated drug responses for 1152 drugs across 3 cell lines collected from the DREAMZ dataset using available drug concentration and several pair indices. In our prediction model, we need two drug-based data types and three cell line-based data types to predict drug response values.

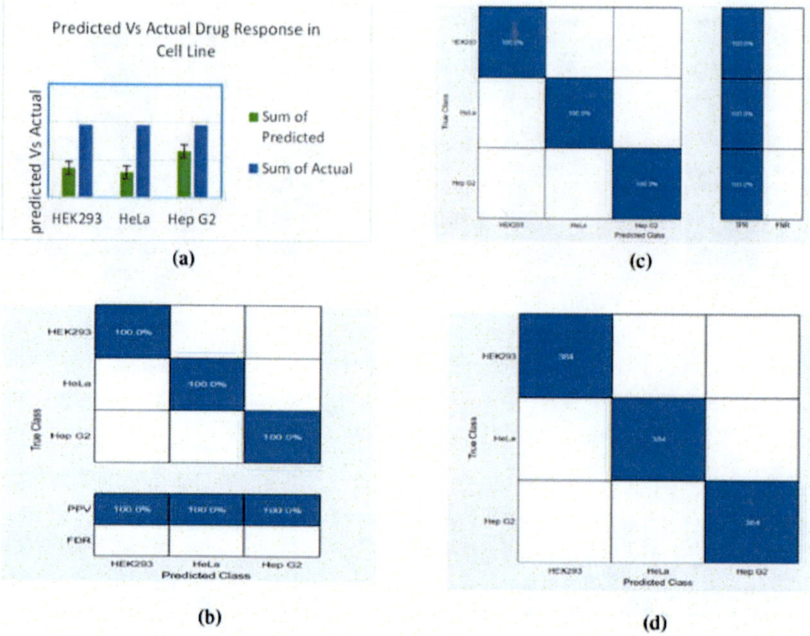

Fig. 5 **a** Predicted versus actual drug therapeutic effects. **b** Number of observation. **c** True positive rates (TPR) and false negative rates (FNR). **d** Positive predicted values (PPV) and false discovery rates (FDR)

The actual drug response is compared to the predicted drug response in the cell line in Fig. 5. The number of observations in the cell line, TPR with FNR, and PPV with FDR can all be seen in the same fig. The model was dubbed HBEA after a study of five machine learning algorithms: logistic regression, decision tree, support vector machine (SVM), Naïve Bayes, and K-NN. We compared the results of subsets of the data categories as well. Drug concentration1 and 2 are drug-based qualities for a specific pair index. The cell line-based features include pathway enrichment scores in Hep G2, Hela, and HEK 293 cells. Figure 6 shows that it is working.

Figure 6 shows the drug reactions in cell lines, including the pair index, drug1 concentration level, and drug2 concentration level. As a result, in Fig. 6 for the particular pair index, the use of a specific drug and a specific concentration to improve drug response is clearly detailed.

The main metrics in the performance test results relating to the application's stability are standard deviation, range, L2 norm, zero mean, and unit variance are explained in Fig. 7.

As shown in Fig. 7, standard deviation is a key performance test result analysis metric that is related to application stability. There is a possibility of making a mistake when calculating standard deviation when a large number of data points are available.

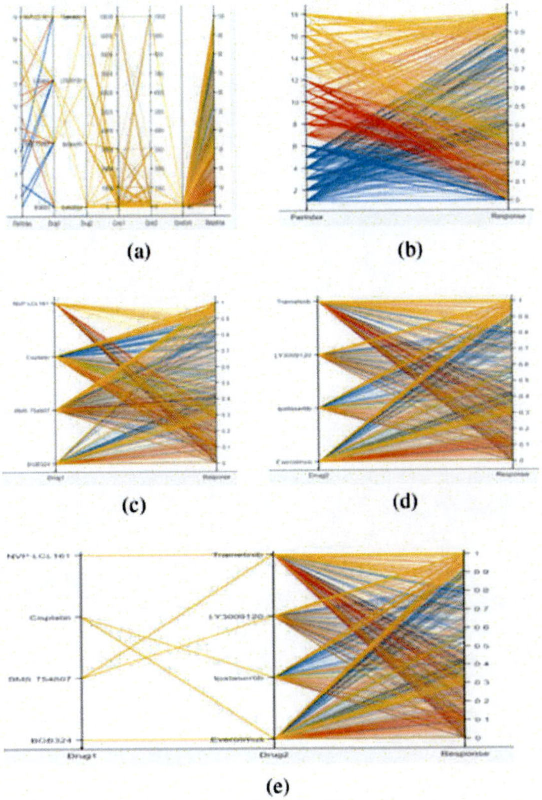

Fig. 6 **a** Dataset. **b** Pair index versus response. **c** Cell line response of Drug 1. **d** Cell line response of Drug 2. **e** Drug 1 and Drug 2 responses in cell line

The standard deviation is a metric for determining the range of values in a sample. The standard deviation of a sample is calculated using the following formula:

$$\sqrt{(x_I - x_b)(x_I - x_b)(x_I - x_b)\left(x_I - x_{(n-1)}\right)} \tag{8}$$

where: x_I is the I-th value in the sample; x_b is the sample mean; n is the sample size x_I The I-th value in the sample; x_b. From the calculation, it is observed that the higher the standard deviation, the more uniformly dispersed the data in the sample. Figure 7b shows the L2 norm. It is known as least squares. It is basically minimizing the sum of the square of the differences (D) between the target value (X_i) and the estimated values ($f(Y_i)$

$$D = \sum_{i=1}^{n} (X_i - f(Y_i))^2 \tag{9}$$

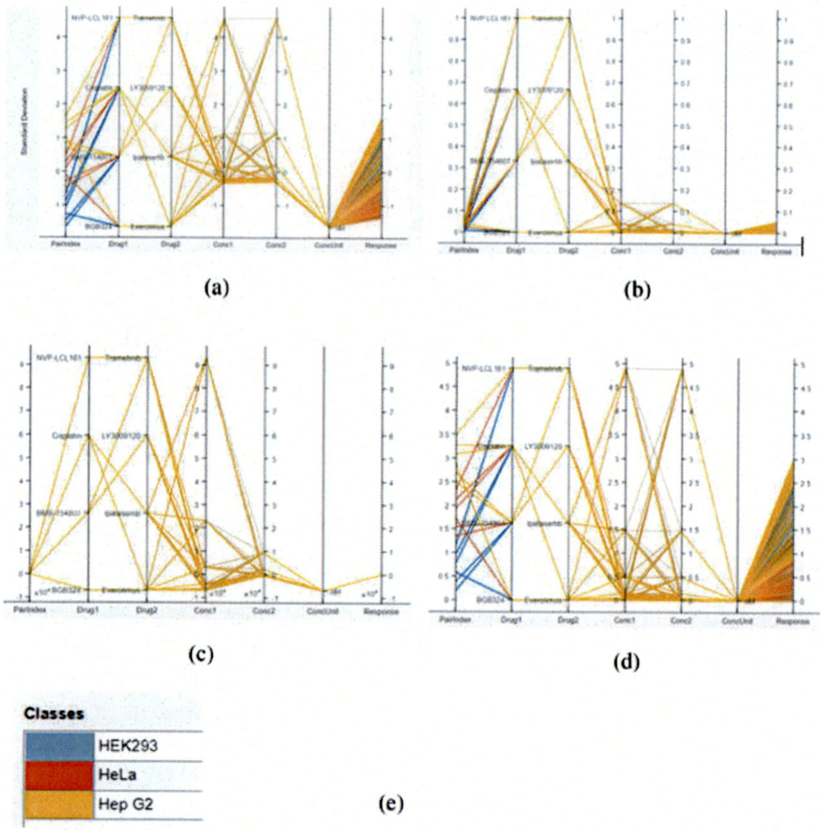

Fig. 7 **a** Standard deviation analysis. **b** L2 norm evaluation. **c** Zero mean calculation. **d** Unit variance. **e** Different colour indication of cell line

Figure 7c shows the zero mean calculation. Figure 7d explains the unit variance. Variance is expressed in significantly bigger units (e.g., meters squared). Because the variance units are much greater than the units of a normal dataset value, it is more difficult to grasp the variance number intuitively. As a result, the standard deviation is frequently used as a primary measure of variability. Figure 7e explained the varied color indications in the graph for each cell line.

8 Conclusion

Customized medicine seeks to determine the most effective way to treat each patient while minimizing side effects. Several machine learning-based methods have been proposed to solve the problem for currently available data, but the problem remains challenging in terms of predictability and interpretability. There is a need for a better classification of anti-cancer medication response prediction using complex networks. HBEA is proposed for this purpose and is based on not only the neighboring drugs and cell lines, but also all other drugs and cell lines. Therefore, it can be used to examine the similarities between drugs, cell lines, and known drug responses around the world. Using the HBEA technique, the suggested method was tested on the DREAMZ dataset, which contains 1152 cancer patient details with 5 different features in three different cell lines. The proposed method's findings were compared to ensemble approaches. When applied to the DREAMZ dataset with the heterogeneous-based ensemble algorithm, the accuracy results of the comparison with other methods revealed that the proposed technique was more aggressive than any other methods.

In this challenge, various types of machine learning algorithms were put together to solve a categorization problem. From the output, it is observed that the HBEA ensemble algorithm improves the drug response prediction in HeLa and HepG2 Cell lines. In this research, heterogeneous ensemble models were more appropriate than single homogenous ensembles. Source code and output graph are available at https://github.com/KartheeswariRamasamy/drug_response.git.

Future Scope

In forecasting anti-cancer treatment response, the heterogeneous ensemble is critical. Using the methodology, HeLa and Hep G2 cell lines respond favorably to the medication when compared to HEK 293 cell lines in this research. Plan to compare medication responses in several cancer cell lines using the same heterogeneous ensemble approach in the future.

Acknowledgements This research is funded by the Indian Council of Medical Research (ICMR). (Sanction no: ISRM/12(125)/2020 ID NO.2020-5128 dated 10/01/21).

References

1. Ashley EA (2015) The precision medicine initiative: a new national effort. JAMA 313:2119–2120
2. Collins FS, Varmus H (2015) A new initiative on precision medicine. N Engl J Med 372:793–795
3. Adam G, Rampášek L, Safikhani Z, Smirnov P, HaibeKains B, Goldenberg A (2020) Machine learning approaches to drug response prediction: challenges and recent progress, NPJ Precis Oncol
4. Wang L, Li X, Zhang L, Gao Q (2017) Improved anticancer drug response prediction in cell lines using matrix factorization with similarity regularization. BMC Cancer 17:513
5. Lu X, Gu H, Wang Y, Wang J, Qin P (2019) Autoencoder based feature selection method for classification of anticancer drug response. Front Genet 10:233
6. Azuaje F (2017) Computational models for predicting drug responses in cancer research. Brief Bioinf 18:820–829
7. Costello JC et al (2014) A community effort to assess and improve drug sensitivity prediction algorithms. Nat Biotechnol 32:1202

8. Lee W-H, Loo C-Y, Daniela Traini P, Young M (2015) Inhalation of nanoparticlebased drug for lung cancer treatment: advantages and challenges. Asian J Pharma Sci
9. Patton JS, Fishburn CS, Weers JG (2004) The lungs as a portal of entry for systemic drug delivery. Proc Am Thorac Soc 1:338–344
10. Schanker LS, Less MJ (1977) Lung pH and pulmonary absorption of nonvolatile drugs in the rat. Drug Metabol Dispos 5:174–178
11. Lai Y-H, Chen W-N, Hsu T-C, Lin C, Tsao Y, Semon W (2020) Overall survival prediction of non-small cell lung cancer by integrating microarray and clinical data with deep learning. Sci Rep 10(1):1–11
12. Doppalapudi S, Qiu RG, Badr Y, Lung cancer survival period prediction and understanding: deep learning
13. SEER Program (2019) National Cancer Institute (NCI), SEER Incidence Data, 1975–2017. Available https://seer.cancer.gov/data/
14. Fangfang Xia "A cross-study analysis of drug response prediction in cancer cell lines", 23(1), 2022, 1–12.
15. Yıldırım MA et al (2007) Drug—target network. Nat Biotechnol 25:1119
16. Heba E-B, Attia A-F, Nawal E-F, Torkey H (2021) Efficient machine learning model for predicting drug-target interactions with case study for Covid-19. Comput Biol Chem
17. Ding H et al (2014) Similarity-based machine learning methods for predicting drug-target interactions: a brief review. Brief Bioinform 15(5):734–747
18. Nath A, Kumari P, Chaube R (2018) Prediction of human drug targets and their interactions using machine learning methods: current and future perspectives. Methods Mol Biol 1762:21–30
19. Ezzat A et al (2018) Computational prediction of drug-target interactions using chemogenomic approaches: an empirical survey. Brief Bioinform 20(4):1337–1357
20. Sachdev K, Gupta MK (2019) A comprehensive review of feature-based methods for drug-target interaction prediction. J Biomed Inform 93:103159
21. Zhou L et al (2019) Revealing drug-target interactions with computational models and algorithms. Molecules 24(9):1714
22. Zhang W et al (2019) Recent advances in machine learning-based drug-target interaction prediction. Curr Drug Metab 20(3):194–202
23. Thafar M, Raies AB, Albaradei S, Essack M, Bajic VB (2019) Comparison study of computational prediction tools for drug–target binding affinities. Front Chem 7:782
24. Kuncheva LI (2004) Combining pattern classifiers: methods and algorithms. Wiley, Hoboken, NJ
25. Windeatt T, Roli F (eds) (2003) Multiple classifier systems, in Lecture Notes in Computer Science, vol 2709, Springer
26. Roli F, Kittler J, Windeatt T (eds) (2004) Multiple classifier systems, in Lecture Notes in Computer Science, vol 3077, Springer
27. Zhou Z-H (2012) Ensemble methods: foundations and algorithms (Chapman & Hall/CRC data mining and knowledge discovery series). Chapman & Hall/CRC, Boca Raton, FL
28. Brazil P, Giraud-Carrier C, Soares C (2009) Meta-learning: applications to data mining. Springer, Berlin Heidelberg
29. Polikar R (2006) Ensemble-based systems in decision making. IEEE Circ Syst Mag 6(3):21–45
30. Agrawal SM, Narayanan R, Polepeddi L, Choudhary A (2012) Lung cancer survival prediction using ensemble data mining on SEER data. Sci Progr 20(1):29–42
31. Safiyari A, Javidan R (2017) Predicting lung cancer survivability using ensemble learning methods. Intell Syst Conf (IntelliSys)
32. Costello JC, Heiser LM, Georgii E et al (2014) A community effort to assess and improve drug sensitivity prediction algorithms. Nat Biotechnol 32(12):1202–1212
33. Zhang C, Ma Y (2012) Ensemble machine learning: methods and applications, Springer
34. Lianga G, Fanb W, Luo H, Zhua X (2020) The emerging roles of artificial intelligence in cancer drug development and precision therapy, vol 128, p 110255
35. http://www.definitions.net

36. Iorio F, Knijnenburg TA, Vis DJ et al (2016) A landscape of pharmacogenomic interactions in cancer. Cell 166(3):740–754
37. Borg I, Groenen PJF (2005) Modern multidimensional scaling: theory and applications, 2nd edn. Springer, New York
38. Santini S, Jain R (1999) Similarity measures. IEEE Trans Patt Anal Mach Intel 21(9):871–883
39. Smith F, Waterman MS (1981) Identification of common molecular subsequences. J Mole Biol 147(1):195–197
40. Lipman J, Pearson WR (1985) Rapid and sensitive protein similarity searches. Science 227(4693):1435–1441
41. Gish AW, Miller W, Myers EW, Lipman DJ (1990) Basic local alignment search tool. J Mole Biol 215(3):403–410
42. Stanfield Z et al (2017) Drug response prediction as a link prediction problem. Sci Rep 7:40321
43. Yang J, Li A, Li Y, Guo X, Wang M (2019) A novel approach for drug response prediction in cancer cell lines via network representation learning. Bioinformatics
44. Zhang N et al (2015) Predicting anticancer drug responses using a dual-layer integrated cell line-drug network model. PLoS Comput Biol 11:e1004498
45. Pedregosa F, Varoquaux G, Gramfort A, Michel V, Thirion B, Grisel O, Blondel M, Prettenhofer P, Weiss R, Durbourg V, Vanderplas J, Passos A, Cournapeau D, Brucher M, Perrot M, Duchesnay E (2011) Scikit-learn: machine learning in python. J Mach Learn Res 12:2825–2830
46. Hastie T, Tibshirani R, Friedman J (2008) The elements of statistical learning, 2nd edn. Springer, New York/Berlin/Heidelberg
47. Duda R, Hart P, Stork D (2001) Pattern classification. John Wiley & Sons
48. Efron B, Tibshirani R (1993) An Introduction to the Bootstrap. Chapman & Hall
49. Bhardwaj R, Hooda N (2009) Prediction of pathological complete response after Neoadjuvant Chemotherapy for breast cancer using ensemble machine learning, pp 2352–9148
50. Chen JIZ, Hengjinda P (2021) Early prediction of coronary artery disease (CAD) by machine learning method-a comparative study. J Artif Intell 3(01):17–33
51. Manoharan S (2019) Study on Hermitian graph wavelets in feature detection. J Soft Comput Paradigm (JSCP) 1(1):24–32
52. Kening Li, Yuxin Du, Lu Li, Dong-Qing Wei. "Bioinformatics Approaches for Anti-cancer Drug Discovery" , Current Drug Targets, 2019.
53. Kuenzi BM, Park J, Fong SH, Sanchez KS, Lee J, Kreisberg JF, Ma J, Ideker T (2020) Predicting drug response and synergy using a deep learning model of human cancer cells. Cancer Cell
54. Anithaashri TP, Rajendran PS, Ravichandran G (2022) Novel intelligent system for medical diagnostic applications using artificial neural network, Lecture Notes on Data Engineering and Communications Technologiesthis link is disabled, vol 101, pp 93–101
55. Niroomand N, Bach C, Elser M (2021) Segment-based CO emission evaluations from passenger cars based on deep learning techniques. IEEE Access
56. Prajwal K, Tharun K, Navaneeth P, Anand Kumar M (2022) Cardiovascular disease prediction using machine learning. In: International conference on innovative trends in information technology (ICITIIT)
57. Ianevski A, Giri AK, Gautam P, Kononov A, Potdar S, Saarela J, Wennerberf K, Aittokallio T (2019) Prediction of drug combination effects with a minimal set of experiments, vol 1, pp. 568–577
58. Baptista D, Ferreira PG, Rocha M (2020) Deep learning for drug response prediction in cancer. Brief Bioinform
59. Rajendran PS, Kartheeswari KR (2022) Anti-cancer drug response prediction system using stacked ensemble approach, Lecture Notes in Networks, vol 436, pp 205–218

Implementing an Integrated Network Load Balancer for Minimizing Weighted Response

Apoorv Kumar Sinha, Sanskriti Sanjay Kumar Singh, Shreyas Sai, and M. Sivagami

Abstract The network load balancer acts as a reverse proxy and reroutes the traffic across a cluster of instances of the application service. The proposed integrated algorithm with a novelty of load sharing approach disperses the oncoming requests such that an optimal server ends up fulfilling the request. Along with that, the load balancer has a number of other utility functions that enable it to work similarly in a dynamic Web application. This proposed algorithm is evaluated with the horizontally-scaled architecture on the basis of latency and throughput as recommended in the RFC 2544 for evaluation on a LAN network. Through this research work, it was found that the ability of this integrated load balancing algorithm was successful in reducing the load for a set of four particular use-cases based upon reducing the response time while simultaneously maintaining all the characteristics of a distributed system.

Keywords Network load balancer · Latency · Throughput · Reduced responsive time · Integrated load balancing · Horizontal-scaled algorithm

1 Introduction

A network load balancer (NLB) automatically distributes incoming traffic across multiple targets such as instances of a server according to their availability or an induced priority. Doing so, it also monitors the health of said targets and routes traffic accordingly by scaling as the incoming traffic changes over time, even as the workload increases exponentially. As such, it functions at the fourth layer of the OSI model and follows an explicitly programmed and rule-based programming algorithm. Similarly, it can also function with both TCP and UDP requests using the same algorithm irrespective of the type of traffic or assignment.

This is done by adjusting the load such that it can disperse the assignments to the servers efficiently to provide the services at a faster rate. Along with this, it keeps

A. K. Sinha · S. S. K. Singh · S. Sai · M. Sivagami (✉)
Vellore Institute of Technology, Chennai, India
e-mail: msivagami@vit.ac.in

G. Rajakumar et al. (eds.), *Intelligent Communication Technologies and Virtual Mobile Networks*, Lecture Notes on Data Engineering and Communications Technologies 171,
https://doi.org/10.1007/978-981-99-1767-9_47

a track of the status of all the resources while distributing requests. This enables it to achieve the same goals of achieving the extremely low level of latencies and thereby, maximizing the throughput of any Web service. Moreover, this application's distributed nature will have a greater number of advantages over a singular server other than mere traffic congestion due to the data transfer and security with authentication across the servers and client and better logging. In this paper, the NLB will be centralized such that data is gathered by a single hub, and the framework would divide it into the other servers and then be returned to the server.

The decided strategy to perform this task is by hosting the servers of our own Website through three different servers. With this setup, the HTTP requests from these servers are distributed to three different machines. All these requests would be profiled and logged onto a terminal. This enables real-time visualization of how the requests are handled on three different, horizontally-scaled machines, and how quickly they are dealt with in terms of profiling. Similarly, the SSL termination proxy is decrypted and encrypted in the network traffic by generating self-signed SSL certificates for the local hosts.

2 Load Balancer as a Distributed System

For the network load balancing algorithm to function as it is, there is a need to establish that the proposed load balancer architecture is a distributed system. For this purpose, several common parameters and characteristics of a distributed system have been discussed to showcase how the proposed model mimics the original load balancer and how some aspects of the same are incrementally improved by changes made to the novelty algorithm presented by the authors.

Starting with openness and heterogeneity, the proposed model is an algorithm that can have several modules removed from it or appended to it. For instance, if the desired framework does not support one of the modules that create the novel algorithm, the user has the choice of removing the same without fear of losing functionality. At the same time, the off-loading function of the load balancer defends the system against common attacks such as distributed denial-of-service (DDoS) among others.

Where load balancers primarily shine in such parameters are scalability, fault tolerance, and concurrency. Load balancers in general are built to handle organic and planned network growth by distributing incoming workloads to multiple end nodes based on the algorithm chosen. For the novel algorithm, this happens to be instances of the same server which can be further adapted to scale-out architecture which adds more parallel server. At the same time, the proposed model detects and masks the failure when it comes to fault tolerance. In Module I discussed later in the paper, the algorithm presented by the authors states that it will not attach a weight to the response time if there is no transmission of packets from that instance. Concurrency is largely self-explanatory as the resource of the server is used by the desired application, and the servers are present similarly to how scaled-out architecture presents itself.

As for other aspects such as quality of service (QoS) and transparency, the novel model's resource-based aspects that are based on the adaptability of the local machines make it better than usual algorithms, while at the same time, the system can be expanded in terms of its scalable transparency with negligible change to the system structure or the application algorithms.

3 Literature Review

To understand the work done by the previous researchers on the topic of load balancing and expand upon the domain space, the works of the following researchers are considered and applied the existing research works as a new way to perceive the topic while also identifying existing research gaps and room for improvement.

Mishra et al. [1] discussed the various performance parameters for load balancing in the literature and their effects. Following this, they then developed a taxonomy for load balancing algorithms suited for cloud environment and also provided simulation results of the few heuristic-based load balancing algorithms.

Neghabi et al. [2] provided a systematic literature review (SLR) of the studies relevant to load balancing mechanisms in the software defined network (SDN). They selected the articles based on their three stage article selection process and then categorized the selected load balancing algorithms using deterministic and non-deterministic criteria. They then describe the algorithm used, advantages, and disadvantages of the load balancing techniques in each of the groups.

Pit-Claudel et al. [3] proposed SHELL, a stateless application-aware load balancer which made used power-of-choices scheme for dispatching new flows to the suitable application instance and a covert channel for recording or reporting the assigned channel. The load balancing scheme's performance evaluation showed that its performance was comparable to other stateless load balancing schemes, while enabling application instance-load-aware dispatching and significantly increasing per-connection consistency resiliency.

Rathore [4] and Rathore et al. [5] proposed a hybrid load balancing solution for grid network, the design of which is a mixture of client/server and peer-to-peer models. It used least connection algorithm, round robin algorithm, least load algorithm, and fastest first algorithm for server selection and used round robin algorithm for client selection.

Zhong et al. [6] proposed a load balancing scheme under SDN architecture, based on server response time. This method claims to have effectively overcome the drawbacks faced by traditional load balancing algorithms.

Afzal and Kavitha [7] focused on exploring the various problems that lead to load balancing in cloud computing. They then reviewed the existing load balancing approaches and then provided an in-depth classification of the same.

Al-Rahayfet et al. [8] proposed a novel approach for load balancing in cloud computing, which made use of dominant sequence clustering (DSC) for task scheduling and then used weighed least connection (WLC) algorithm for load balancing.

Rajagopalan [9] provides a detailed overview of how load balancers and application delivery controllers can help small to medium-sized businesses (SMB) achieve scalability, reliability, performance, and Website management that is on par with their competitors, i.e., small to medium-sized enterprises (SME) and large businesses.

Nurwasito and Rahmawati [10] proposed a weighted response time algorithm for Web server load balancing in SDN. The proposed model showed promising results since it was found that the networks which use the proposed algorithm needs significantly lesser response time and produces better throughput than a network using any response time-based algorithms.

Guo et al. [11] proposed Libra, a stateful layer-4 (L4) load balancer that makes its decisions based on their proposed weighted M-least-connection first (WMLCF) mechanism in order to enable load balancing fairness.

Praveenchandar and Tamilarasi [12] proposed a load balancing approach for dynamic resource allocation which made use of resource tables and task tables to optimize waiting time and worked on resource scheduling, taking in account the task size and the bidding value coded by each customer.

Shafiq et al. [13] provided a comprehensive review of the various load balancing techniques suitable for cloud environment. They performed an analytical review of the load balancing algorithms and discovered the strengths and weaknesses of each of them, based on the various performance metrics.

Kumar and Kumar [14] provided a review of current load balancing techniques in cloud computing and the associated challenges, to aid in developing more effective algorithms. They also reported various methods for enhancing performance and resource utilization based on load balancing, task scheduling, resource management, quality of service, and workload management.

Priya et al. [15] focused on improving the efficiency of cloud service provisioning by introducing an integrated resource scheduling and load balancing algorithm. The algorithm used a fuzzy-based multidimensional resource scheduling model and a multidimensional queuing load optimization algorithm to increase virtual machine utilization and reduce latency. Simulation analysis of the proposed method leads to a 7% increase in resource scheduling efficiency and a 35.5% decrease in response time as compared to the previous works.

The rise of Amazon Web Services and cloud computing with its applications have brought a larger need to decongest servers so as to reduce latency and maximize the throughput. A number of applications are latency sensitive, that is, their functionalities become useless, and thus, the network load balancer preserves the client-side source IP and allows the backend to see the IP address of the client. The crux of this research paper sees that the latency and throughput have been improved by utilizing various algorithms which optimize execution time, along with space and bandwidth utilization. This paper also presents an overview of various load balancing algorithms, and how they can be integrated together to create an algorithm with the novelty of load sharing that functions optimally.

4 Proposed Methodology

As mentioned previously, the novelty in the work is to balance the loads in load balancers. It lies in incorporating ideas found in existing algorithms and creating a new one that finds the optimal servers consistently to handle requests without significantly increasing computation costs. However, at the same time, there is a need to have a basic understanding of how the existing methodologies work and can be further implemented in the proposed algorithm. As a result, the proposed algorithm is preceded by a brief overview of the methodologies adopted and combined in creating the algorithm.

4.1 Modules of the Algorithm

Weighted Response. The response of the application servers determines which application server receives the next request. In the meantime, the application server response time to a health check is used to calculate the application server weights. Figure 1 depicts the algorithm of the integrated load balancer.

As a result, the application server which responds the fastest receives the next request. This is the most appropriate for scenarios where throughput and latency are most important parameters, and the application response time is a paramount concern.

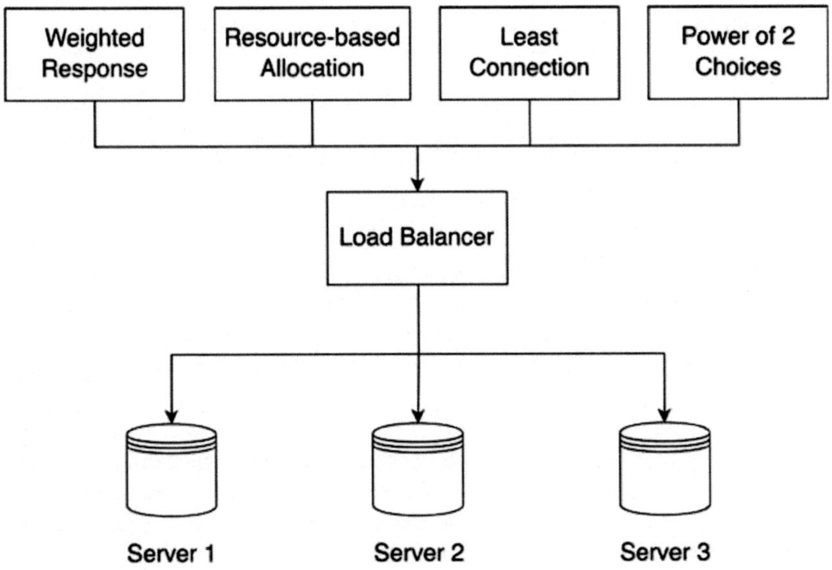

Fig. 1 Algorithm of the integrated load balancer

Resource-Based Allocation. Adaptive load balancing makes decisions based on status indicators retrieved from the backend servers. Every node has an API endpoint named /mem and /CPU which regularly checks for this status information and sets the dynamic weight of the server.

(a) Memory Usage: Based on the memory storage
(b) CPU Usage: Based on the CPU load.

Given these two metrics, the weights to be set have a greater affinity for low values considering these parameters. Therefore, this is almost like a 'health check' which is why this is useful for situations like applications where the workload is varied and detailed application performance.

Least Connection. Dynamic load balancing algorithm where client requests are distributed to the least number of active connections at the time client request is received. This is preferred where application servers have similar specifications, this algorithm takes the active connection load into consideration. The servers are relatively similar in terms of processing power which is used.

By adding weights to the application server based on the relative processing power, this algorithm can be altered at the same time.

Power of 2 Choices. Most load balancing algorithms have one guide, that is, load balancer, with a view of queues and response time. However, distributed servers have multiple guides but they may see the same queue with the least number of processes so multiple guides queue the same process.

Therefore, this algorithm does not make the best choice using incomplete data but picks two server queues at random and choses the better option of the two. Because of this, it is efficient since you don't need to see all the queues and avoids the undesired herd behavior by the approaching the worst queue and distributing traffic with large entropy.

4.2 Specificity of Devised Algorithm

The algorithm based on weighted response decides the weight depending on the latency of the nodes when the load balancer pings them. As per empirical evaluations, that latency measure is quite less information for the update and by giving more information, primarily devised from the resources in the local machine, the weight updates provide better results since they consider more parameters.

Similarly, the use of the resource-based algorithm locally has become primitive in recent times especially as more practices shift to a cloud infrastructure where adaptive resource-based allocation is used to check multiple instances of the server. By combining the local variant of this algorithm in the novel algorithm created, the primitive aspect of the load balancer is handled.

As for the least connections module, it is more unreliable than others due to the way the framework is built and how it interacts with other frameworks such as

NodeJS while the module of power of 2 choices is based on gaining more information through entropy so the new updated weights increase the information gain for the framework.

Thus, the devised algorithm tackled these issues of the four standard algorithms for load balancing and decided ways of how the merits of one algorithm can solve the drawbacks of the other. A thorough explanation of how these four modules interact with each other to create the novel algorithm is presented in the next section.

5 Algorithm

Onward to the novel design proposed by the authors, the integrated algorithm comprising the aforementioned modules. The end goal was to minimize the latency and maximize the throughput of the server and attend to incoming requests such that they are distributed to many servers. This makes sure that there is no pooling and that a single server is not overworked. This is the crux of this paper, and the novelty designed by the authors to ensure that the optimal algorithm is chosen.

The proposed algorithm developed in this work paper follows the given logic:

- Upon receiving the requests, the weights for the servers are updated based on the response time, following the logic discussed in the section talking about weighted response.
- At the same time, the API calls the /mem and /cpu parameters from the local machine of the server, getting the information of the hosts of the servers to understand the parameters.
- This allows information of memory storage and average CPU load to be available to the backend, which it utilizes further in its calculation of the integrated algorithm.
- The updated weights are stored in the array as a function that was created by the authors of this paper. The equation designed is as follows:

  Equation 1. Weights for each node

$$W(x,\ y) = \text{CPU\_usage} + \text{Memory}/x + \text{Response\_time} * y \tag{1}$$

where x and y are any natural numbers with 10^n.

- Every 5 s, weights are updated and checks whether the target servers are down, thereby satisfying the fault tolerance characteristic of a distributed system.
- After another request, algorithms select two random operating servers and choose the best among them depending on the newly calculated weight (based on the function $W(x, y)$).

Furthermore, the function's basic logic is arithmetic, reducing the computational cost of constant calculations, while at the same time, constantly maintaining the status of the application server's weight as prescribed in the least connections algorithm.

6 Pseudocode

```
SET response_weights = []
SET load_avg = []
SET mem = []
SET available_servers = []
SET weights = []
FUNCTION updateWeights():
    FOR i in servers:
        available = serverAvailable(servers[i])
        IF available:
            load_avg[i] = getLoad (servers[i])
            mem[i] = getFreeMem(servers[i])
            response_weights[i] = ping(servers[i])
            available_servers[i] = isAvailable(servers[i])
            //calculating the weights using the formula
            weight[i] = response_weights[i]*10^n + mem[i]/10^n + load_avg[i]
FUNCTION chooseServer():
    //using power of two
    SET [a, b] = [random(servers), random(servers)]
    //check if the servers are available
    IF isAvailable(a) == true && isAvailable(b) == false:
        return a
    IF isAvailable(a) == false && isAvailable(b) == true:
        return b
    //return server with the least weight
    IF weights[servers.indexOf(a)] > weights[servers.indexOf(b)]:
        return b
    ELSE:
        return a
WHILE true:
    //wait for 5 seconds
    delay(5000)
    updateWeights()
```

7 Results and Discussion

This section outlines the results achieved from analysis. For the environment to achieve the results, a NodeJs framework was used, implemented via express. The network load balancer was applied into three domains, and the evaluation metrics were the two out of the four network performance criteria described in RFC 2544. They are throughput and latency, respectively.

To begin with the network load balancer was set up on multiple networks connected to a LAN. By pinging the address of the machines, they can be incorporated into the array with the server details, and since this network is a LAN, the evaluation metrics are evaluated and compared to a single machine.

Upon the selection of servers and weights, a random server is chosen out of two to handle the HTTP request, and after every five seconds, the weights of the servers are updated, and it continues to perform the requested task. Furthermore, to demonstrate the liveness, the load balancer calls up the endpoints to each of the servers so as to check if it is responsive to the balancer's directives.

This ensures that for a responsive Website, different servers are called to process a requirement from the client side. This reduces the load for a particular instance of the server. Latency was minimal and throughput maximal consistently upon multiple comparisons which will be elucidated upon next. For the purposes of this study, the authors picked four tasks for evaluation. The results are given below for more inferential viewing where the response time in milliseconds has been tabulated for the prescribed tasks in Table 1.

In the use-case termed as Task 1, the Website was a resource-intensive dynamic Web app that uses the NextJS framework and has API calls within the server. The proposed model rerouted each request to a different server using the proposed algorithm and kept on doing that for a large number of HTTP requests. Here, the proposed model outperformed the base model by a significant measure. According to the workflow designed, the multiple instances of the same application worked in such a way that no server was called consecutively and barring a breakage in the endpoints of the servers, the load was managed expeditiously by the novel algorithm. Chart

Table 1 Table showcasing the performance of various load balancing algorithms for the prescribed tasks

Task	Without load balancer	Proposed load balancer	Power of two choices	Weighted response	Resource-based allocation	Least connection
1	36	14	26	23	21	18
2	22	16	19	17	20	18
3	12	10	11	9	14	12
4	17	15	15	16	18	19
Average	21.75	13.75	17.75	16.25	18.25	16.75

describing the performance of the server for the chosen tasks with different load balancing approaches is presented in Fig. 2.

Task 2 was chosen to be using API calls to retrieve system information, tested on both physical and virtual computer systems using the REST API. Since the newer GET call implemented by IBM returns the details of existing computers as well as historical data about computers that were removed, the multiple servers aided in more comprehensive extraction of data especially when virtualized machines were hosted on a physical machine which the base algorithm had problems with.

To compare performance of computationally inexpensive and standalone procedures, Task 3 was chosen to be a routine from within an SQL database. Usually, there was next to no difference in the execution by both the improvised algorithm and the original architecture, but upon adding schema and routine clauses, the proposed model showed better performance incrementally.

Task 4 was imitating a serializable isolated transaction. For reference, these transactions occur concurrently but give the impression of running in serial order by acquiring locks for read and write operations which makes them particularly useful for complex queries and updates. As a result, whenever bulk loads are created in the eventuality of using multiple instances, the environment automatically starts blocking queries which somewhat equalizes the time gained by using multiple servers.

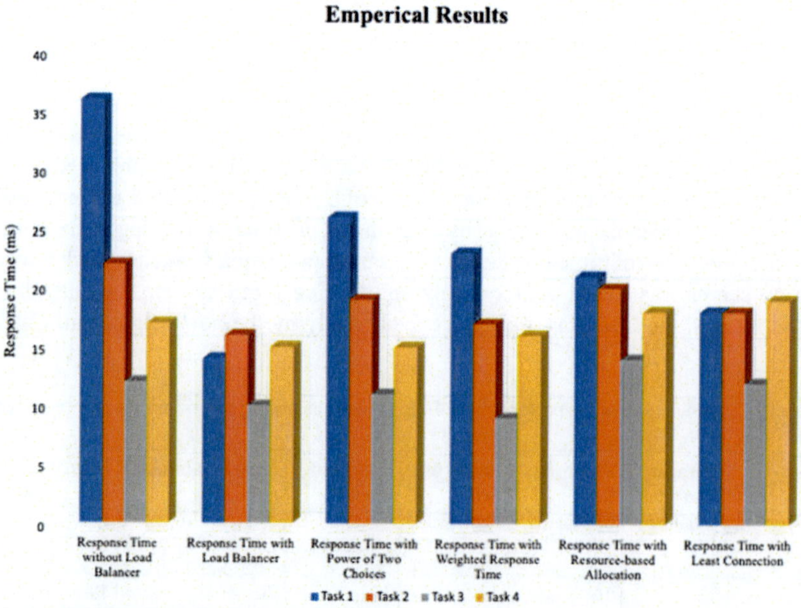

Fig. 2 Chart describing the performance of the server for the chosen tasks with different load balancing approaches

8 Conclusion and Future Enhancements

From the given findings, it is evident that the load balancer outperforms the servers that do not use the load balancer. As such, it accomplishes the goals of minimizing latency and maximizes the throughput set out at the beginning. A small caveat also shows which particular use-cases require the use of load balancing more so than others. As is evident by the description of the tasks, whenever the number of tasks increases or when there are multiple calls, the advantage of using the load balancer becomes more significant which is largely due to the novel algorithm.

The integrated algorithm ends up superseding the existing algorithms in its application and works in accordance with all the applications it is supposed to do. Along with this, its capabilities to distribute the data into multiple servers is implemented by updating weights adequately. As a result, the premise of the project is satisfied.

For further work, the algorithm can be optimized further by implementing a stronger weightage of least connections method by implementing a framework other than NodeJS. Along with that, it is also possible to make a better equation that optimizes the algorithm depending on the data received by incorporating machine learning into the said algorithm, which furthermore can be improved upon by an application built on the concepts of federated learning.

References

1. Mishra SK, Sahoo B, Parida PP (2020) Load balancing in cloud computing: a big picture. J King Saud Univ Comput Inf Sci 32(2):149–158. https://doi.org/10.1016/j.jksuci.2018.01.003
2. Neghabi NJN, Hosseinzadeh M, Rezaee A (2018) Load balancing mechanisms in the software defined networks: a systematic and comprehensive review of the literature. IEEE Access 6:14159–14178. https://doi.org/10.1109/ACCESS.2018.2805842
3. Pit-Claudel YD, Pfister P, Townsley M, Clausen T (2018) Stateless load-aware load balancing in P4. In: 2018 IEEE 26th international conference on network protocols (ICNP), 2018, pp 418–423. https://doi.org/10.1109/ICNP.2018.00058
4. Rathore N (2018) Performance of hybrid load balancing algorithm in distributed web server system. Wirel Pers Commun 101(3):1233–1246. https://doi.org/10.1007/s11277-018-5758-6
5. Rathore NK, Rawat U, Kulhari SC (2020) Efficient hybrid load balancing algorithm. Natl Acad Sci Lett 43(2):177–185. https://doi.org/10.1007/s40009-019-00834-w
6. Zhong H, Fang Y, Cui J (2018) Reprint of 'LBBSRT: an efficient SDN load balancing scheme based on server response time.' Future Gener Comput Syst 80:409–416. https://doi.org/10.1016/j.future.2017.11.012
7. Afzal S, Kavitha G (2019) Load balancing in cloud computing—a hierarchical taxonomical classification. J Cloud Comput 8(1):22. https://doi.org/10.1186/s13677-019-0146-7
8. Al-Rahayfeh SA, Abuhussein A, Almiani M (2019) Novel approach to task scheduling and load balancing using the dominant sequence clustering and mean shift clustering algorithms. Future Internet 11(5). https://doi.org/10.3390/fi11050109
9. Rajagopalan S (2020) Server load balancing for small to medium-sized enterprises. In: 2020 4th international conference on electronics, communication and aerospace technology (ICECA), 2020, pp 739–743. https://doi.org/10.1109/ICECA49313.2020.9297381
10. Nurwasito H, Rahmawati R (2021) Weighted response time algorithm for web server load balancing in software defined network. In: 2021 3rd international conference on electronics

representation and algorithm (ICERA), Jul 2021, pp 143–148. https://doi.org/10.1109/ICERA53111.2021.9538792
11. Guo X, Zhu L, Zhang D, Wu C (2022) Libra: a stateful Layer-4 load balancer with fair load distribution. In: 2022 IEEE international performance, computing, and communications conference (IPCCC), 2022, pp 246–253. https://doi.org/10.1109/IPCCC55026.2022.9894309
12. Praveenchandar J, Tamilarasi A (2022) An enhanced load balancing approach for dynamic resource allocation in cloud environments. Wirel Pers Commun 122(4):3757–3776. https://doi.org/10.1007/s11277-021-09110-x
13. Shafiq DA, Jhanjhi NZ, Abdullah A (2022) Load balancing techniques in cloud computing environment: a review. J King Saud Univ Comput Inf Sci 34(7):3910–3933. https://doi.org/10.1016/j.jksuci.2021.02.007
14. Kumar P, Kumar R (2019) Issues and challenges of load balancing techniques in cloud computing: a survey. ACM Comput Surv 51(6). https://doi.org/10.1145/3281010
15. Priya V, Kumar CS, Kannan R (2019) Resource scheduling algorithm with load balancing for cloud service provisioning. Appl Soft Comput 76:416–424. https://doi.org/10.1016/j.asoc.2018.12.021

Insights of Deep Learning-Based Video Anomaly Detection Approaches

Dipak Ramoliya and Amit Ganatra

Abstract Deep learning is a powerful computing strategy that has changed the landscape of computer vision. It has been used to tackle complicated cognitive tasks such as detecting abnormalities in videos. Anomalies in the video are events or objects in the footage that don't fit the typical, learned patterns. Using deep learning, it is possible to automatically and in real-time identify unusual actions and objects like fights, riots, traffic rule violations, abrupt rushes, and the presence of weapons in restricted areas or abandoned luggage. Despite the challenges posed by video anomaly detection, this review offers a comprehensive assessment of published deep learning algorithms for the task. Future research can build on this work by understanding the existing methods to create more effective solutions. First, the challenges of video anomaly identification are discussed as the benefits of deep learning in anomaly detection. Furthermore, several types of abnormalities were explored, followed by diverse methodologies for anomaly identification. Furthermore, significant aspects of anomaly detection using deep learning, including learning approaches, were presented. Finally, numerous datasets used in anomaly detection were examined, followed by a discussion of deep learning-based algorithms for spotting video anomalies.

Keywords Anomaly detection · Outlier detection · Deep learning · Computer vision

D. Ramoliya (✉)
Department of Computer Science and Engineering, Devang Patel Institute of Advance Technology and Research (DEPSTAR), Faculty of Technology and Engineering (FTE), Charotar University of Science and Technology (CHARUSAT), Changa 388421, India
e-mail: dipakramoliya.ce@charusat.ac.in

A. Ganatra
Department of Computer Engineering, Devang Patel Institute of Advance Technology and Research (DEPSTAR), Faculty of Technology and Engineering, Charotar University of Science and Technology (CHARUSAT), Changa, Anand 388421, India

Computer Science and Engineering Department, Faculty of Engineering and Technology, Parul University (PU), Waghodia, Vadodara 391760, India

G. Rajakumar et al. (eds.), *Intelligent Communication Technologies and Virtual Mobile Networks*, Lecture Notes on Data Engineering and Communications Technologies 171,
https://doi.org/10.1007/978-981-99-1767-9_48

1 Introduction

The usage of surveillance cameras has grown significantly in recent years to improve public safety. Unfortunately, the number and rate of production of surveillance data have outpaced the capacity of security forces to monitor this footage. As a result, there is a noticeable gap in the application of cameras used for surveillance, and the ratio of cameras to human monitors which is unacceptable. However, it is conceivable to create systems that automate the process of detecting anomalies using the computational power and resources available today. Such systems can handle issues in other relevant areas of computer vision and machine learning, such as activity recognition, and video annotation in addition to reducing (or perhaps eliminating) the requirement for user intervention in anomaly detection [1–3]. Figure 1 provides the general flow of anomaly detection.

The term "anomaly identification" refers to the process of uncovering an object or occurrence of any event that diverges from the norm, either inside a dataset or in relation to other datasets. Anomaly detection is becoming more and more important, as recognized in data mining, artificial intelligence, machine learning, computer vision, and deep learning communities. This is due to the rising demand in extensive domains, such as financial surveillance, health, medical risk, agriculture, safety, and security. Identifying abnormal events, such as traffic accidents, crimes, or unlawful activity, is a crucial duty in video surveillance. Compared to regular life, anomalous occurrences are extremely rare. Therefore, developing efficient computer vision algorithms for automatic video anomaly identification is crucial for saving time and manpower. The goal of any useful anomaly detection system is to provide an early warning of any activity that breaks with typical patterns, as well as pinpoint when this deviation first occurred [4, 5].

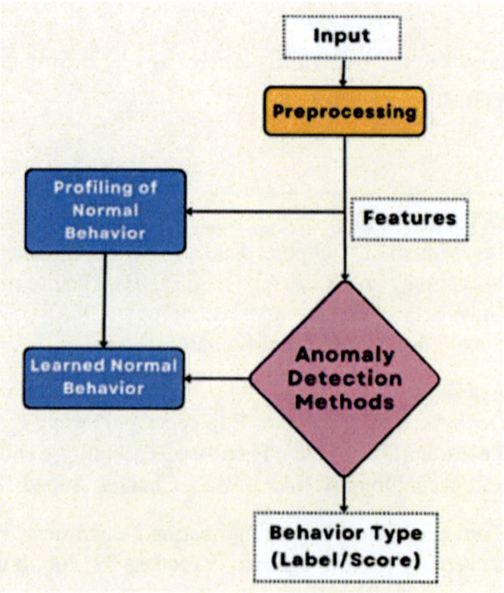

Fig. 1 General flow of anomaly detection

2 Background Theory and Related Work

2.1 Different Types of Video Anomalies

(1) Anomalies both locally and globally: Local anomalies are activities that differ greatly from what is happening in the area around them. For example, a vehicle driving in the wrong direction. Individuals who act strangely here are different from their neighbors. Global anomalies are activities that happen worldwide and are abnormal or unexpected. Global anomalies are caused by a group or collective anomalies. If a group of data points shows normal behavior when examined individually, but anomalous behavior when looked at as a group, that set is called a collective or group anomaly [1, 6].

(2) Anomalies in point and interaction: Some data points are very different from the rest, and these are called point anomalies. To put it another way, a point anomaly is a form of random abnormality that has the potential to be misunderstood as deviant conduct, such as loitering. Interaction anomalies, such as a fight between two people, are examples of abnormal interactions between separate entities [7].

(3) Anomalies that are contextual or conditional: These are data items that don't fit in with the rest of the dataset, creating anomalies. Contextual data, such as time and place, is used to determine the context [8]. Behavioral characteristics, such as normal activities, are used to determine the anomaly. This category includes most video oddities such as violence, riots, stampedes, and so on. Spatial and temporal abnormalities can also be considered contextual anomalies. Contextual abnormalities are generally best represented by spatio-temporal anomalies [1, 7].

3 Video Anomalies Detection Using Deep Learning-Based Approaches

3.1 Trajectory-Based Methods

To understand the trajectory of the item or topic of interest, the researchers followed it between frames. They then used this information to deduce any abnormal behavior the object may have been carrying out [9]. The accuracy of trajectory-based approaches is highly dependent on object recognition and tracking. Tracking effectiveness can be negatively impacted by factors such as high crowd density, poor video quality, rapid camera movement, and occlusion. As a result, the trajectory approaches were discovered to provide better accuracy in sparsely packed environments, rather than in moderately or highly congested environments. Furthermore, while determining the anomaly, these approaches lack contextual information [10].

3.2 Global Pattern-Based Methods

There are a growing number of studies on convenience stores and grocery stores that use video footage to understand how these stores work. These studies use low- or medium-level features to examine the footage as a whole [11]. They rely on global pattern-based approaches, which are successful in both moderately and densely populated areas. Space–time gradients, motion energy, and the direction of light flux are all utilized in these methods. Using a crowd behavior-based social attribute-aware force model, we can identify unusual occurrences [8]. Statistical hypothesis tests and Gaussian Noise Model approximation are applied to discover abnormal events [12]. A model for anomalous crowd behavior is given using kinetic energy derived using the assistance of optical flow and crowd dispersion index [13]. Detecting anomalous behavior in crowds can be difficult. Several models have been created to try and make this process easier, including ones that use motion maps or hierarchical feature representation [14]. However, locating abnormalities can be time-consuming using these methods.

3.3 Grid Pattern-Based Methods

Grid pattern techniques are frequently employed because they can shorten the amount of time required for analysis. This is accomplished by restricting the number of characteristics that are gathered from a particular spatio-temporal region. This means that anomalies in each cell, or grid, are studied separately without respect to their neighbors [15]. Furthermore, similarities are retrieved from the frames' split blocks rather than interpreting the frames as a singular entity. These methods do not need the use of STIPs or saliency detection. It is possible to improve video anomaly detection by examining numerous attributes or grid patterns. Spatial anomaly detection using keypoint detection [16], temporal anomaly detection using background removal using the Gaussian Mixture Model (GMM) [17], and sparse combination learning [17] is just a few of the methods that have been used. Detecting spatial and temporal abnormalities in busy settings can be tricky. Some researchers have explored using maps that show both spatial and temporal anomalies [3]. Another approach is to look at object speed, texture, and size in order to identify changes in a particular area. Grid pattern-based approaches are similar to local region-based methods in that they use a probabilistic model to identify anomalies. Markov Chains, Gaussian mixture models, and Bag-of-words [9] are used to compile a set of compact characteristics for use in creating a grid with cells of varying sizes. Recently, grid-based video anomaly detection algorithms have begun to incorporate deep learning techniques, such as deep Gaussian mixture models and spatio-temporal convolution neural networks.

3.4 Representation Learning Models

Represent learning is the process of learning effective representations of raw video data while counting the fact critical previous information about the situation. It reduces the impact of the dimensionality curse by translating very high-dimensional video data into manageable d-dimensional vectors, making computational operations easier [15]. Classifiers and predictors built on representation learning models could be applied in contexts as diverse as action recognition, anomaly detection, and object recognition. Additionally, this method aids in the extraction of a generic, compact, efficient, and simple video descriptor or representation [11].

(1) Deep neural networks inspired by sparse coding: One of the most extensively utilized strategies for anomaly identification is normal distribution learning. The training set of video data serves as a dictionary for the sparse coding-based video anomaly detection algorithm. It then discovers the aberrant actions that the learned dictionary is unable to rebuild [8]. Adaptive sparse representations combine a combined sparse model for multiple-item anomaly detection with nonlinearity for class separation. These approaches seldom took into account the encoding of both spatial and temporal links between different blocks and frames. The accuracy of anomaly detection is improved when an intra-frame classification approach is used [11].

(2) Models based on reconstruction: Video anomaly detection reconstruction models can only learn what is typical behavior or activity from the training video dataset. Anomalous or aberrant behaviors deviate from the acquired model, resulting in poor reconstruction. Auto-encoders and Principal Component Analysis (PCA) are among the techniques employed. Because of its hierarchical feature extraction technique, Deep auto-encoder (Deep-AE) may provide superior feature representations. A deep CAE was trained with an end-to-end framework to maintain spatio-temporal properties independent of encoding dynamics [9]. It is noise-proof, can detect visual anomalies, and can even pick them out by itself. A sparse AE-based system sufficient for dynamically identifying and localizing video anomalies is described, which reduces the memory required at the time of execution and the false positive rate [12]. The concept, which is founded on the basis of subspace clustering, structured AE, has the ability to make use of nonlinear transformations in order to seamlessly translate the input to the output while still maintaining the local and global subspace structures [13].

(3) Slow feature analysis: The slowness concept underpins slow feature analysis (SFA) [9]. Human activity recognition uses SFA-based video feature representations at first. For video anomaly detection, D-IncSFA integrates the process of extraction of features and detection of anomaly into a single step [17].

3.5 Discriminative Models

The techniques which attempt for finding out the distinguishing traits between classes of data are discriminative modeling. After these models have been trained using supervised learning techniques, they are successful with balanced datasets. However, these techniques are rarely applied for video anomaly identification because of the scarcity of well-balanced video anomaly datasets and the ambiguity surrounding what exactly constitutes an anomalous action or entity. There is a need for a framework to identify video abnormalities using conventional density estimation techniques [16]. However, this has not been extensively explored and a commonly acknowledged definition for the anomaly does not exist. Discriminative models could be more effective in this case, but they are not widely used due to the lack of a consensus on what an anomaly is.

3.6 Predictive Models

A video can be thought of as a spatio-temporal signal, with each frame creating its own unique pattern. Because of their ability to incorporate both geographical and time-based information, predictive models (also known as spatio-temporal models) are frequently employed for video anomaly identification (or motion data) [18].

3.7 Generative Models

Joint probability $P(X, Y)$ can be found via generative modeling, and from there the conditional posterior probability $P(X/Y)$ can be calculated. In general, generative models model each class's actual distribution, while discriminative models model the line of separation between classes [18]. There is no restriction on how the likelihood should be represented for deep generative models; they are still able to learn using the maximum likelihood principle. In order to overcome difficulties caused by limited data and imbalanced data, deep generative models are gaining popularity for application in video anomaly detection. However, more research is needed before these models can be adopted for video anomaly detection.

3.8 Deep Hybrid Models

Auto-encoders are used to learn hidden representations, which are then used to find outliers in deep hybrid models for anomaly detection. The transfer learning models are tremendously successful in extracting the features from the pre-trained models which inspire the use of these models as in hybrid models. Outlier detection using deep learning algorithms is a growing field. However, current hybrid models that use coupled feature extractor training do not perform well. This is due to the absence of a trainable goal specifically for anomaly detection [19]. Researchers have proposed employing singular anomaly detection goals such deep one-class classification or single-class neural networks to counteract this shortcoming [20].

3.9 One-Class Neural Networks

There is increasing enthusiasm in one-class classification methods for anomaly detection, and Chalapathy et al. [21] present an approach that is based on kernel-based deep networks. This technique blends the deep networks' ability to take out a richer data representation with a single-class objective of creating a secure perimeter around normal data. The OC-NN objective dictates the data format in the hidden layer, which is optimized for anomaly detection, making this technique unique. Deep neural networks (DNNs) are a sort of neural network that is more advanced than ordinary neural networks. They are trained to extract common patterns of variation in data. This is done by mapping common factors of variation using Singular Value Decomposition (SVDD). This technique was applied to the MNIST and CIFAR-10 datasets, and it was found that data instances moved to the center of the sphere improved performance [21, 22].

3.9.1 Statistical Techniques

There is a statistical signal processing method known as the Hilbert transform which can be used to obtain a real-valued signal's analytic representation [14]. This attribute is highly promising for detecting abnormalities in health-related time series datasets in real time. The method discovers real-time anomalies by sequentially integrating wavelet analysis, neural networks, and the Hilbert transform. To completely comprehend the potential and usefulness of statistical approaches and deep learning methodologies for anomaly detection, more research is required [19].

4 Performance Evaluatıon Methodologıes

4.1 Error Matrix

Video anomaly detection, along with other areas of computer vision, shares the same aim of classifying data points as normal or abnormal, although using quite different approaches to accomplish this. Video anomaly detection is therefore a binary classification problem that can be addressed. Additionally, any of the fundamental components of performance analysis, such as True Positive (TP), False Positive (FP), True Negative (TN), and False Negative (FN), can be used to describe each binary classifier choice.

4.2 *Receiver Operating Characteristic Curve*

To evaluate the efficacy of detection at varying false positive rates (FPRs), a receiver operating characteristic (ROC) plot is constructed with the TPR (sensitivity) on the *Y* axis and the FPR (probability of false alarm) on the *X* axis (or thresholds). The Area Under the Curve (AUC) of the Receiver Operating Characteristic curve is used to evaluate the effectiveness of an anomaly detector across a variety of evaluation metrics (AU-ROC). The AU-ROC value ought to be maximized as closely to one as possible (between zero and one) [23].

4.3 *Precision-Recall Curve*

A plot of precision and recall is known as a precision-recall (PR) plot. With regards to solving the anomaly detection problem, the Area Under the Precision-Recall curve (AU-PR) is superior to the Area Under the Receiver Operating Characteristics (AU-ROC). This is a result of anomaly detection's data imbalance issue, where TNs are larger than TPs. A further feature of the PR curve is its weighting of optimistic and out-of-the-ordinary forecasts. The AU-PR should have the highest possible value between 0 and 1 [23].

4.4 *Equal Error Rate*

The percentage of incorrectly labeled frames is denoted by the acronym "EER" (for "equal error rate") where both the TPR and FNR are equivalent. When looking for video anomalies, it performs admirably [21].

4.5 *Detection Rate*

The ratio of detected anomalies to all abnormalities included in the data, expressed as a percentage, is known as the detection rate (DR). The detection rate, or accuracy rate, as determined by the Equal Error Rate, is useful for pinpointing video anomalies.

$$\mathrm{DR} = \frac{\mathrm{TP}}{\mathrm{TP} + \mathrm{FP}}$$

4.6 Reconstruction Error

The reconstruction error error_reconst(t) for a given frame at the instant in time t is calculated using the equation.

$$\text{error\_reconst(t)} = \sum_{(x,y)} e(x, y, t)$$

where $e(x, y, t)$ is the pixel level reconstruction error for intensity level I at a position (x, y) in the specific frame at time instant t.

Higher reconstruction error scores indicate a higher possibility of spotting a video anomaly [21].

4.7 Anomaly Score

An abnormality indexes, the scoreano(t), is a numeric value between 0 and 1 that describes the degree of an anomaly.

$$\text{Scoreano(t)} = \frac{\text{error\_reconst(t)} - \text{error\_reconst}_{\text{min}}\text{min(t)}}{\text{error\_reconst}_{\text{max}}\text{min(t)}}$$

Higher anomaly score values reflect a higher level of anomaly.

4.8 Regularity Score

Regularity score Scorereg(t), which is determined using the opposite of the anomaly score is calculated using the following equation

$$\text{Scorereg}(t) = 1 - \text{Scoreano}(t)$$

The lower the level of anomaly, the greater the regularity score.

Table 1 Datasets

Dataset name	Total video clips
UCSD dataset	98
Avenue dataset	37
ShanghaiTech Campus dataset	437
UCF-Crime dataset	1900
LV dataset	30

5 Datasets

5.1 UCSD Dataset

This dataset (in Table 1) was captured in 2010 at UC San Diego. The data was captured by a high, stationary camera that looked down on the walkways below. Two separate cases, Peds1 and Peds2, are represented in the dataset. In every scene, there is a distinct train test, with normal samples only in the train and abnormal ones in the test. The Peds1 dataset includes 34 training video clips and 36 testing video clips, while the Peds2 dataset only includes 16 training video clips and 12 testing video clips [11].

5.2 Avenue Dataset

Another extensively used dataset is The Avenue dataset, which was published in 2013 and was photographed in CUHK campus avenue. It is comprised of brief video clips that were taken outdoors. The camera faces a sidewalk that runs alongside a building. The majority of anomalies include actions, such as running away, stopping briefly, or tossing something into the air [8].

5.3 ShanghaiTech Campus Dataset

The dataset was made public in 2017. In order to build one of the largest datasets for video abnormalities, CCTV cameras were installed all around the ShanghaiTech campus. There are a broad variety of settings and camera perspectives included in the 13 included scenarios. The 330 normal event movies have a resolution of 480 × 856 pixels, while the 107 mixed resolution recordings have a size of 720 × 576 pixels [12].

5.4 *UCF-Crime Dataset*

The UCF-Crime dataset was published in 2018. It is a huge collection of online videos shot with a large number of different cameras, producing a variety of situations. It consists of 1900 uncut surveillance videos, 950 of which are of ordinary happenings, and the remaining 950 are from 13 preset categories of criminal activity. Arrest, Assault, Accident, Abuse, Burglary, Explosion, Fighting, Arson, Robbery, Shooting, Stealing, Shoplifting, and Vandalism are a few examples of the events that are covered [6, 7, 24].

5.5 *ARENA Dataset*

ARENA stands for "Architecture for the REcognition of threats to mobile assets using Networks of multiple Affordable sensors." This dataset was built as a part of the PETS2014 challenge. The dataset was captured by deploying four cameras (frame rate: 30 FPS, resolution: 1280 $\times$ 960 pixels) at the University of Reading's crossing path and parking lot without any overlap [9].

5.6 *LV Dataset*

The live videos (LV) dataset is a comprehensive collection of real-world videos that were recorded by surveillance cameras under challenging environmental conditions such as varying lighting and camera movements. It was made available to the public in 2017 and comprised of thirty video snippets that were culled from various online places in order to illustrate thirty distinct scenarios. The entire time span covered by this dataset is 3.93 h [13]. Figure 2 is the sample of anomalies.

6 Limitations and Open Challenges

6.1 *The Necessity of Improved Datasets*

Given that this research area is very young, there are significantly fewer publicly accessible datasets regarding anomaly detection from videos. Additionally, the adoption of supported learning-based models is hindered by the data imbalance between anomalous occurrences (Positive samples) and normal events (Negative samples). Good benchmarks are needed to assess the effectiveness of the algorithms used to locate and recognize video anomalies [1, 25].

Fig. 2 Sample of anomalies from different datasets

6.2 Reduction in Computational Complexity

The feature representation process used in video anomaly detection is typically computationally expensive and takes a long time, which is a significant impediment to its use in real-world applications. The currently used algorithms for video anomaly detection have a high cost of space and time complexities. Consequently, these techniques are not appropriate for real-world applications [25].

6.3 Methodology Incompleteness

The technique used by the current methods for anomaly detection in the video is insufficient, indicating that no technique is capable of finding every kind of anomaly. As a result, there is a requirement for an effective anomaly detection platform for the video that can solve the issue of incompleteness by utilizing hybrid models [25].

6.4 Accuracy and Processing Time Trade-Off

It is challenging to find the best balance between accuracy in detection and execution time in video anomaly detection systems by using a limited number of highly descriptive features to achieve competitive online performance. Furthermore, the high computational complexity and lengthy processing times are required by deep learning-based algorithms to accomplish the high accuracy of identification and localization of video anomalies. Therefore, there is always an adjustment between the required detection exactness and the processing time along with computing complexity [6, 7].

6.5 *Environmental Factors*

The changes in object viewpoint and the distance between the object of interest and the monitoring camera are two sensitive areas of video anomaly detection methods. The mentioned issue has been discussed using grid patterns, although there remains room for development. There is a pressing need for efficient approaches to deal with environmental changes such as shifting illumination and backdrops, occlusion problems, noisy input, and the working state of the surveillance camera [4, 5].

7 Conclusion

In this article, the approaches for the detection of video anomaly detection using deep learning have been discussed and it shows the great promise in detecting abnormal events in surveillance footage. Various architectures such as auto-encoder, RNN, and 3D CNN have been proposed and evaluated for their performance on different datasets. The results have shown that these methods can achieve high accuracy in detecting anomalies, but there is still room for improvement in terms of computational efficiency and generalizability to different types of videos. In the future, other architectures such as transformer-based models, and to explore the use of unsupervised learning techniques for anomaly detection can be explored. Overall, the insights gained from this paper demonstrate the potential of deep learning for video anomaly detection and pave the way for further research in this area.

References

1. Nayak R, Pati UC, Das SK (2021) A comprehensive review on deep learning-based methods for video anomaly detection. Image Vis Comput 106:104078
2. Wang L et al (2022) Unsupervised anomaly video detection via a double-flow ConvLSTM variational autoencoder. IEEE Access 10:44278–44289
3. Luo W et al (2019) Video anomaly detection with sparse coding inspired deep neural networks. IEEE Trans Patt Anal Mach Intell 43(3):1070–1084
4. Wang H-G, Li X, Zhang T (2018) Generative adversarial network based novelty detection usingminimized reconstruction error. Front Inform Technol Electr Eng 19(1):116–125
5. Franklin RJ, Dabbagol V (2021) Anomaly detection in videos for video surveillance applications using neural networks. In: 2020 fourth international conference on inventive systems and control (ICISC). IEEE

6. Li N et al (2015) Anomaly detection in video surveillance via gaussian process. Int J Patt Recogn Artif Intell 29(06):1555011
7. Sultani W, Chen C, Shah M (2018) Real-world anomaly detection in surveillance videos. In: Proceedings of the IEEE conference on computer vision and pattern recognition
8. Lu C, Shi J, Jia J (2013) Abnormal event detection at 150 fps in Matlab. In: Proceedings of the IEEE international conference on computer vision 2013, pp 2720–2727
9. Xu K, Jiang X, Sun T (2018) Anomaly detection based on stacked sparse coding with intraframe classification strategy. IEEE Trans Multimedia 20(5):1062–1074
10. Munir M et al (2019) A comparative analysis of traditional and deep learning-based anomaly detection methods for streaming data. In: 2019 18th IEEE international conference on machine learning and applications (ICMLA). IEEE
11. Mahadevan V, Li W, Bhalodia V, Vasconcelos N (2010) Anomaly detection in crowded scenes. In: Proceedings of the IEEE computer society conference on computer vision and pattern recognition, IEEE, 2010, pp 1975–1981
12. Kumari P, Bedi AK, Saini M (2021) Multimedia datasets for anomaly detection: a survey. arXiv preprint: arXiv:2112.05410
13. Piciarelli C, Micheloni C, Foresti GL (2008) Trajectory-based anomalous event detection. IEEE Trans Circ Syst Video Technol 18(11):1544–1554
14. Kiran BR, Thomas DM, Parakkal R (2018) An overview of deep learning based methods for unsupervised and semi-supervised anomaly detection in videos. J Imag 4(2):36
15. Zhao Y et al (2017) Spatio-temporal autoencoder for video anomaly detection. In: Proceedings of the 25th ACM international conference on Multimedia
16. Wu P, Liu J, Shen F (2019) A deep one-class neural network for anomalous event detection in complex scenes. IEEE Trans Neural Netw Learn Syst 31(7):2609–2622
17. Leyva R, Sanchez V, Li C-T (2017) Video anomaly detection with compact feature sets for online performance. IEEE Trans Image Process 26(7):3463–3478
18. Liu W, Luo W, Lian D, Gao S (2018) Future frame prediction for anomaly detection—a new baseline. In: Proceedings of the IEEE conference on computer vision and pattern recognition 2018, pp 6536–6545
19. Leyva R, Sanchez V, Li C-T (2017) The lv dataset: a realistic surveillance video dataset for abnormal event detection. In: Proceedings of the 5th international workshop on biometrics and forensics (IWBF), IEEE 2017, pp 1–6
20. Del Giorno A, Andrew Bagnell J, Hebert M (2016) A discriminative framework for anomaly detection in large videos. In: European conference on computer vision. Springer, Cham
21. Sabokrou M et al (2017) Deep-cascade: cascading 3d deep neural networks for fast anomaly detection and localization in crowded scenes. IEEE Trans Image Process 26(4):1992–2004
22. Zhou JT et al (2019) Anomalynet: an anomaly detection network for video surveillance. IEEE Trans Inform Forensics Secur 14(10):2537–2550
23. Chalapathy R, Chawla S (2019) Deep learning for anomaly detection: a survey. arXiv preprint: arXiv:1901.03407
24. Ramachandra B, Jones M (2020) Street scene: a new dataset and evaluation protocol for video anomaly detection. In: Proceedings of the IEEE/CVF winter conference on applications of computer vision, 2020
25. Berroukham A et al (2023) Deep learning-based methods for anomaly detection in video surveillance: a review. Bull Electr Eng Inform 12(1):314–327
26. Ullah W et al (2021) CNN features with bi-directional LSTM for real-time anomaly detection in surveillance networks. Multimedia Tools Appl 80(11):16979–16995

Intelligent Indoor Positioning Systems: The Case of Imbalanced Data

Firuz Kamalov, Sherif Moussa, and Jorge Avante Reyes

Abstract The ubiquity of Wi-Fi over the last decade has led to increased popularity of intelligent indoor positioning systems (IPS). In particular, machine learning has been recently utilized to develop intelligent IPS. Most of the existing research focus on developing intelligent IPS using balanced data. In this paper, we investigate a hitherto unexamined issue of imbalanced data in the context of machine learning-based IPS. We consider several traditional machine learning algorithms to determine the optimal method for training IPS on imbalanced data. We also analyze the effect of imbalance ratio on the performance of the IPS. The results show that the k-nearest neighbors algorithm provides the best approach to developing intelligent IPS for imbalanced data.

Keywords Indoor positioning systems · Imbalanced data · Intelligent systems · Machine learning

1 Introduction

Indoor positioning systems (IPS) based on Wi-Fi signals have recently gained popularity. There exist several traditional approaches to indoor positioning including received signal strength indicator (RSSI), channel state information (CSI), time-reversal location fingerprints, and others. Machine learning has also been applied to derive positioning systems based on Wi-Fi signals. The key advantage of machine learning is that intelligent systems trained on Wi-Fi signal data are able to

F. Kamalov (✉) · S. Moussa
Department of Electrical Engineering, Canadian University Dubai, Dubai, UAE
e-mail: firuz@cud.ac.ae

S. Moussa
e-mail: smoussa@cud.ac.ae

J. A. Reyes
Faculty of Engineering, Canadian University Dubai, Dubai, UAE
e-mail: jorge_avante@outlook.com

G. Rajakumar et al. (eds.), *Intelligent Communication Technologies and Virtual Mobile Networks*, Lecture Notes on Data Engineering and Communications Technologies 171,
https://doi.org/10.1007/978-981-99-1767-9_49

autonomously learn rules for identifying the location of the wireless receiver devices. As a result, AI-based IPS has attracted increased interest from the researchers.

One of the major obstacles in applying machine learning algorithms is imbalanced data which refers to skewed distribution of target labels. Imbalanced data can distort the optimization process of the learning algorithm. It affects a wide range of applications. In particular, imbalance data can occur in networking applications such as IPS. In this paper, we investigate the ability of machine learning algorithms to cope with imbalanced data in the context of IPS based on Wi-Fi signals. We consider the performance of machine learning algorithms to learn the location of the receiving device based on the Wi-Fi signal strength, where the model training data exhibits a skewed label distribution.

Machine learning algorithms have been applied in many fields to create intelligent systems that learn and operate based on given data. In particular, AI has been used to develop IPS based on Wi-Fi signals. While researchers have demonstrated that AI can effectively identify the location of a receiving device based on the Wi-Fi signal, most of the studies utilized balanced training data. However, in a real-life scenario, the available data is not always perfectly balanced. Imbalanced data can cause bias in classification and other adverse consequences. Machine learning algorithms have shown varying degrees of success in dealing with imbalanced. There are two key questions that arise in dealing with imbalanced data: (i) what is the optimal machine learning algorithm?, and (ii) what is the relationship between the imbalance ratio and classification accuracy? We investigate these two questions in the context of Wi-Fi signal data. Our goal is to identify the optimal machine learning algorithm that can be used to analyze Wi-Fi signal data for indoor positioning. We also study the relationship between the imbalance ratio and performance of machine learning classifiers.

To study the effects of imbalanced data on IPS, we consider the classic wireless indoor localization dataset [28]. We sample the original dataset to obtain several subsets with various degrees of imbalance. The ratio of minority to majority labels in the sampled datasets ranges from 0.05 to 0.3. We train and test several machine learning algorithms to identify the best method to handle imbalanced Wi-Fi data. The algorithms considered in our study include popular learning models such as logistic regression, linear discriminant analysis, quadratic discriminant analysis, k-nearest neighbors, Naive Bayes, and quadratic logistic regression [14, 30]. The results show that the k-nearest neighbors (KNNs) algorithm achieves the optimal performance among all tested methods. We also find that while low imbalance ratio (m/M) has a negative effect on the classification accuracy, the magnitude of the impact differs between the models.

Our paper is composed as follows. In Sect. 2, we provide a brief overview of the related literature. In Sect. 3, we describe the methodology including the machine learning models and the dataset used in our study. Section 3 presents the results of our numerical experiments. We conclude with final remarks in Sect. 4.

2 Literature

Indoor positioning systems (IPS) is a dynamic field that is continuously evolving [5]. The IPS can be grouped into two categories: conventional and intelligent systems. The conventional approaches use deterministic, human-engineered protocols to identity the position of an object, while intelligent systems rely on machine learning-based algorithms. Both conventional and intelligent systems employ various types of signals to locate a mobile device. The major approaches used in IPS are based on the inertial measurement unit [21], Bluetooth [25], Wi-Fi [32], ultra-wide band [24], and light [22]. In addition, chanel state provides useful information regarding the Wi-Fi signal that can also be used effectively for indoor positioning [20]. More sophisticated approaches combine several techniques (fusion) to build a multi-step algorithm to build the positioning system [12].

Recently, machine learning algorithms have been applied to design IPS [10, 29]. In [36], the authors combine KNN and Bayesian estimation to design a localization system using Bluetooth low energy. The KNN algorithm was also employed in [25] to design a Bluetooth-based positioning system. Several machine learning algorithms including SVM, DT, RF, and KNN are explored in [35] to design a light-based IPS. Deep learning was used in [2] for indoor localization to overcome the limitations of RF-based approaches. In [31], the authors employ convolutional neural network (CNN) to process magnetic field fingerprints to identify the position. Similarly, a CNN-based localization method was proposed in [27] A deep neural network (DNN) is utilized in [19] for indoor positioning based on magnetic field data. It was shown in [33] that random forest has better accuracy than the minimum Euclidean distance approach for indoor localization. A semi-supervised approach based on variational autoencoder was proposed in [26] to deal with unlabeled data. Similarly, a DNN and generative adversarial network-based approaches were proposed in [23] to deal with limited data. A location prediction system was proposed in [4] based on neural network embeddings. The previous location coordinates were used as inputs in recurrent neural network to compute more accurate current location coordinates in [3]. A simple DNN combined with the Kriging algorithm was proposed in [18] that achieved better accuracy than other DNN-based localization approaches.

While the issue of imbalanced data has been investigated in several applications including medicine, fraud detection, network intrusion detection, and others, it has received little attention in the context of IPS [1]. In general, there exist two approaches for dealing with imbalanced data: sampling [34] and cost-sensitive learning [7, 15]. Sampling refers to balancing the ratio of class labels by either increasing the number of minority instances or decreasing the number of the majority instances with former approach considered as more effective [11]. Cost sensitive learning involves adjusting the loss function to increase the penalty for misclassification of the minority instances. The most popular existing sampling algorithm is synthetic minority oversampling technique (SMOTE) which produces new points through random linear interpolation between existing neighboring points [6, 9]. The SMOTE algorithm has been extended by several authors in attempt to improve its performance [13,

16]. Recently, generative approaches based on a trained neural network capable of synthesizing new samples from the latent space have been utilized [37]. Similarly, a generative approach based on variational autoencoder was proposed in [17]. DeepSMOTE, a deep learning model based on SMOTE, is proposed in [8].

The review of the literature shows that there has been a considerable amount of research dedicated to the topic of IPS using both the conventional and machine learning-based approaches. The proposed IPS designs utilize a variety of signal sources including Wi-Fi, Bluetooth, and light. However, the topic of imbalanced data in the context of intelligent IPS has received little interest in the literature.

3 Methodology

In this section, we provide the details about the machine learning algorithms and the dataset considered in our study.

3.1 *Machine Learning Algorithms*

In our study, we consider several traditional machine learning algorithms [14, 30]:

1. Logistic regression (LR)
2. Linear discriminant analysis (LDA)
3. Quadratic discriminant analysis (QDA)
4. K-nearest neighbors (KNNs)
5. Naive Bayes (NB)
6. Quadratic logistic regression (QLR).

In logistic regression (LR), each sample is assigned a likelihood using the logistic function. The goal of LR is to find the parameters of the logistic function that maximize the overall likelihood of the data. It results in a linear decision boundary between two classes. LR is a high bias and low variance classifier. The parameters of LR can be calculated efficiently with little computational resource. On the other hand, LR is an older model that often performs poorer than the more recent methods. Nevertheless, it is a good benchmark model for our study.

Linear discriminant analysis (LDA) uses the Bayesian approach to calculating $P(Y|X)$. In particular, LDA estimates $P(X|Y = y)$ as a multivariate Gaussian distribution where the mean and standard deviation are calculated from the given data. LDA assumes constant standard deviation (covariance matrix) of X for all values of Y which results in a linear classifier. Similar to LR, LDA is efficiently calculated and is convenient for large datasets. On the other hand, it is not as accurate as modern algorithms.

Quadratic discriminant analysis (QDA) employs the same approach as LDA. However, in QDA, the standard deviation of the Gaussian representing $P(X|Y = y)$ is not assumed to be constant for all values of Y. As a result, the boundary between the classes becomes nonlinear (quadratic). A quadratic decision boundary provides more flexibility than a linear boundary to separate different class labels. However, the additional flexibility of QDA may result in overfitting, if the data is sparse. On the other hand, given sufficient amount of data, QDA often yields more accurate classification than LDA, albeit at a higher computational expense.

The k-nearest neighbors algorithm is a popular nonparametric classification method. Given a new observation x, its label is assigned based on the majority vote of its k-nearest neighboring points. It is easy to implement, albeit it requires nontrivial computational resources to identify the neighboring points for the entire data. The choice of the value of k controls the trade-off between bias and variance. A low value of k produces a low bias and high variance classifier and vice versa [14].

The Naive Bayes (NB) classifier aims to calculate $P(X|Y = y)$ in a simplified manner by assuming conditional independence of the individual attributes. In particular, the NB algorithm makes a simplification assumption that

$$P(x_1, x_2, ..., x_k|Y, x_{k+1}, x_{k+2}, ..., x_{k+m}) = P(x_1, x_2, ..., x_k|Y), \quad (1)$$

for all k and m. The NB algorithm aims to solve the problem of the sparcity of data in estimating conditional probabilities. Despite the strong simplification assumptions, the NB algorithm achieves a relatively good performance, especially in case of limited or high dimensional data.

The quadratic logistic regression applies the logistic regression to the second order combinations of the original attributes. Given a set of predictors $\{x_1, x_2, ..., x_k\}$, the QLR applies the logistic regression on the transformed set of attributes $\{x_1^2, x_1x_2, ..., x_kx_{k-1}, x_k^2\}$. Thus, QLR is a nonlinear classifier. It provides higher accuracy than the traditional LR, but with higher variance.

3.2 Dataset

In our study, we employ the classic wireless indoor localization dataset [28]. The dataset consists of 7 predictors, where each predictor represents a Wi-Fi signal strength observed on smartphone. The original target label is a categorical variable that can take one of 4 possible values representing the location (room) of the smartphone. The dataset contains 2000 observations with 500 observations for each location. We convert the original target variable into a binary variable—Room 3 or not. A sample of the dataset is presented in Table 1. Thus, the classification task at hand is to determine whether or not the smartphone is in Room 3.

Table 1 Sample of the dataset

WS1	WS2	WS3	WS4	WS5	WS6	WS7	Class (old)	Class (new)
−64	−56	−61	−66	−71	−82	−81	1	0
−68	−57	−61	−65	−71	−85	−85	1	0
−12	−60	−68	−15	−74	−73	−70	2	0
−12	−64	−63	−18	−69	−72	−74	2	0
−51	−55	−54	−49	−60	−79	−84	3	1
−51	−58	−51	−51	−62	−79	−86	3	1
−58	−55	−53	−59	−52	−81	−84	4	0
−59	−59	−52	−60	−51	−87	−86	4	0

Original class labels are replaced with a binary label (Room 3 or not)

3.3 Experimental Setup

To obtain an imbalanced dataset, we sample the original data according to the desired ratio. Concretely, for each class ratio (m/M) from 0.05 to 0.3, we sample the data based on the given ratio and train the classifiers on the sampled data. Then, we use the remaining (unsampled) data to test the performance of the classifiers. This procedure is carried out 50 times for each class ratio. In the end, we take the average accuracy, sensitivity, and specificity over the 50 bootstrap trials. The experimental procedure provides a definitive analysis of the classifier performance at each class ratio.

4 Results and Analysis

In this section, we discuss the performance of machine learning models in identifying the location of a smartphone based on Wi-Fi signals. The machine learning models are trained on imbalanced Wi-Fi signal data to study their accuracy under imbalanced class distribution. We analyze the performance of the models over a range of class ratios obtained from the Wi-Fi dataset.

4.1 Results

As shown in Fig. 1, the KNN algorithm achieves the highest accuracy among all the tested models. The KNN algorithm at $k = 3$ is slightly more accurate than at $k = 1$. Remarkably, the KNN algorithm achieves high accuracy even at low class ratio. At class ratio 0.05, the accuracy of KNN is over 92%. It indicates that KNN is robust against imbalanced class distribution in Wi-Fi data. We note that the accuracy of all the models improves as the class ratio increases. The biggest improvement in

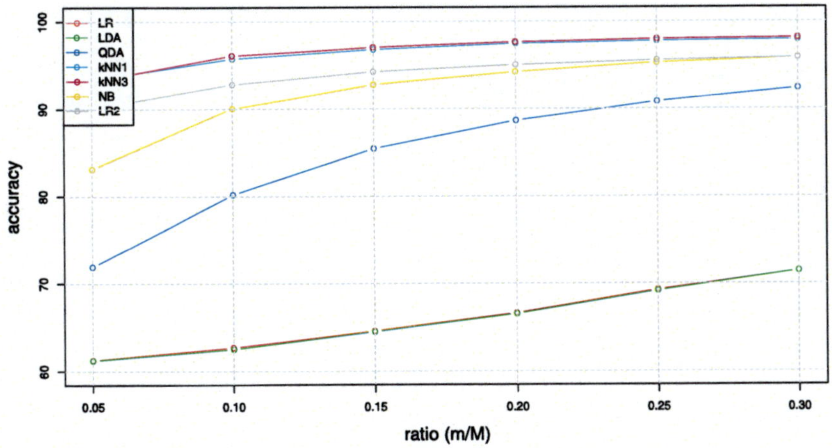

Fig. 1 Classification accuracy of IPS based on different machine learning algorithms for different imbalance ratios

accuracy is achieved by the QDA and NB. As the class ratio increases from 0.05 to 0.1, the accuracy of these models improves by approximately 8%. Unsurprisingly, the linear classifiers LR and LDA do not perform well as data is rarely linearly separable in real life. In addition, the considerable size of the dataset helps to avoid issues with overfitting more flexible models.

We use sensitivity to evaluate the performance of the classifiers with respect to the minority class. Sensitivity is given by the equation TP/T, where T is the total number of positive labels and TP is the number of correctly labeled positive samples (true positive). As shown in Fig. 2, the sensitivity of the classifiers is in line with the accuracy results. In particular, KNN achieves the highest sensitivity among all the tested models with $k = 3$ slightly outperforming $k = 1$. It means that KNN can correctly identify most of the minority observations. The KNN algorithm performs very well even at low class ratios. The sensitivity of all the models improves as the class ratio increases.

4.2 Discussion

In our numerical experiments, we studied the effectiveness of machine learning models with regard to identifying the location of a smartphone based on Wi-Fi signals. The results lead us to three main conclusions:

1. The top 3 models—KNN, LR2, and NB—achieve high accuracy and sensitivity rates despite data imbalance.

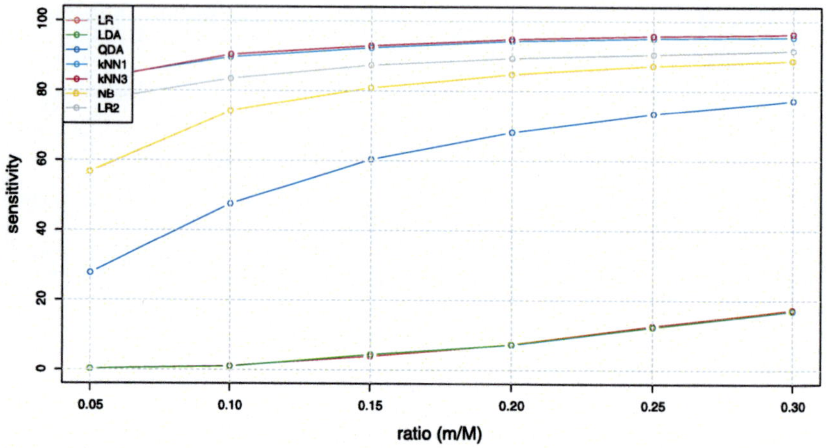

Fig. 2 Sensitivity rates of IPS algorithms for different imbalance ratios

2. The KNN algorithm provides the best performance among all the tested models. Remarkably, it achieves high overall accuracy and sensitivity even at low class ratios.
3. The performance of all models improves as the class ratio increases.

The performance of KNN leads us to conclude that the Wi-Fi data does not have a parametrizable distribution. Most of the machine learning algorithms, except KNN, considered in our study assume conditional normal distribution of the data. If the data does not meet the requirements, it leads to poor results in this case. The second observation above indicates that the minority class data is clustered, and the minority points are in close proximity with each other. Finally, it is not surprising to observe improvement in accuracy as the class ratio increases. Balanced data reduces classification bias and improved overall accuracy of the predictions.

5 Conclusion

In this paper, we studied the performance of machine learning algorithms in identifying the location of a smartphone based on Wi-Fi signals. In particular, we focused on the scenario where the models are trained on imbalanced data. Our numerical experiments reveal that KNN achieves the best performance. It is highly accurate even at low class ratios. Other nonlinear models such as NB and QLR also perform well achieving high accuracy and sensitivity rates despite class imbalance data. The performance of the models improves as the class ratio increases, which shows that imbalanced data is indeed detrimental to the classification outcome.

We find that machine learning models can be used effectively to identify the location of a mobile device based on Wi-Fi signals even when trained on imbalanced data. Since machine learning models are capable of learning the necessary rules autonomously with little supervision from an expert user, they provide a lucrative alternative to the traditional IPS.

References

1. Alhomayani F, Mahoor MH (2021) Oversampling highly imbalanced indoor positioning data using deep generative models. IEEE Sens 1–4
2. Ayyalasomayajula R, Arun A, Wu C, Sharma S, Sethi AR, Vasisht D, Bharadia D (2020) Deep learning based wireless localization for indoor navigation. In: Proceedings of the 26th annual international conference on mobile computing and networking, pp 1–14
3. Bai J, Sun Y, Meng W, Li C (2021) Wi-Fi fingerprint-based indoor mobile user localization using deep learning. Wire Commun Mob Comput
4. Bilbao-Jayo A, Almeida A, Sergi I, Montanaro T, Fasano L, Emaldi M, Patrono L (2021) Behavior modeling for a beacon-based indoor location system. Sensors 21(14):4839
5. Brena RF, García-Vázquez JP, Galván-Tejada CE, Muñoz-Rodriguez D, Vargas-Rosales C, Fangmeyer J (2017) Evolution of indoor positioning technologies: a survey. J Sens
6. Chawla NV, Bowyer KW, Hall LO, Kegelmeyer WP (2002) SMOTE: synthetic minority over-sampling technique. J Art Intell Res 16:321–357
7. Chen X, Gong C, Yang J (2021) Cost-sensitive positive and unlabeled learning. Inf Sci 558:229–245
8. Dablain D, Krawczyk B, Chawla NV (2022) DeepSMOTE: fusing deep learning and SMOTE for imbalanced data. IEEE Trans Neur Netw Learn Syst
9. Elreedy D, Atiya AF, Kamalov F (2023) A theoretical distribution analysis of synthetic minority oversampling technique (SMOTE) for imbalanced learning. Mach Learn 1–21
10. Feng X, Nguyen KA, Luo Z (2022) A survey of deep learning approaches for WiFi-based indoor positioning. J Inf Telecommun 6(2):163–216
11. Fu Y, Du Y, Cao Z, Li Q, Xiang W (2022) A deep learning model for network intrusion detection with imbalanced data. Electronics 11(6):898
12. Guo X, Ansari N, Hu F, Shao Y, Elikplim NR, Li L (2019) A survey on fusion-based indoor positioning. IEEE Commun Surv Tutor 22(1):566–594
13. Han H, Wang WY, Mao BH (2005) Borderline-SMOTE: a new over-sampling method in imbalanced data sets learning. In: International conference on intelligent computing. Springer, Berlin, Heidelberg, pp 878–887
14. Hastie T, Tibshirani R, Friedman JH, Friedman JH (2009) The elements of statistical learning: data mining, inference, and prediction, vol 2. Springer, New York, pp 1–758
15. He S, Li B, Peng H, Xin J, Zhang E (2021) An effective cost-sensitive XGBoost method for malicious URLs detection in imbalanced dataset. IEEE Access 9:93089–93096
16. Kamalov F, Denisov D (2020) Gamma distribution-based sampling for imbalanced data. Knowl Based Syst 207:106368
17. Kamalov F, Ali-Gombe A, Moussa S (2022) Conditional variational autoencoder-based sampling. In: ICT analysis and applications: proceedings of ICT4SD 2022. Springer Nature Singapore, Singapore, pp 661-669
18. Labinghisa BA, Lee DM (2021) Neural network-based indoor localization system with enhanced virtual access points. J Supercomput 77(1):638–651
19. Lee N, Ahn S, Han D (2018) AMID: accurate magnetic indoor localization using deep learning. Sensors 18(5):1598

20. Liu W, Cheng Q, Deng Z, Chen H, Fu X, Zheng X, Wang S et al (2019) Survey on CSI-based indoor positioning systems and recent advances. In: 2019 International conference on indoor positioning and indoor navigation (IPIN). IEEE, pp 1–8
21. Lu C, Uchiyama H, Thomas D, Shimada A, Taniguchi RI (2019) Indoor positioning system based on chest-mounted IMU. Sensors 19(2):420
22. Maheepala M, Kouzani AZ, Joordens MA (2020) Light-based indoor positioning systems: a review. IEEE Sens J 20(8):3971–3995
23. Njima W, Bazzi A, Chafii M (2022) DNN-based indoor localization under limited dataset using GANs and semi-supervised learning. IEEE Access 10:69896–69909
24. Poulose A, Eyobu OS, Kim M, Han DS (2019) Localization error analysis of indoor positioning system based on UWB measurements. In: 2019 eleventh international conference on ubiquitous and future networks (ICUFN). IEEE, pp 84–88
25. Pu YC, You PC (2018) Indoor positioning system based on BLE location fingerprinting with classification approach. Appl Math Model 62:654–663
26. Qian W, Lauri F, Gechter F (2021) Supervised and semi-supervised deep probabilistic models for indoor positioning problems. Neurocomputing 435:228–238
27. Qin F, Zuo T, Wang X (2021) Ccpos: Wifi fingerprint indoor positioning system based on cdae-cnn. Sensors 21(4):1114
28. Rohra JG, Perumal B, Narayanan SJ, Thakur P, Bhatt RB (2017) User localization in an indoor environment using fuzzy hybrid of particle swarm optimization & gravitational search algorithm with neural networks. In: Proceedings of sixth international conference on soft computing for problem solving. Springer, Singapore, pp 286–295
29. Roy P, Chowdhury C (2021) A survey of machine learning techniques for indoor localization and navigation systems. J Intell Robot Syst 101(3):1–34
30. Sarker IH (2021) Machine learning: algorithms, real-world applications and research directions. SN Comp Sci 2(3):1–21
31. Shao W, Luo H, Zhao F, Ma Y, Zhao Z, Crivello A (2018) Indoor positioning based on fingerprint-image and deep learning. Ieee Access 6:74699–74712
32. Subedi S, Pyun JY (2020) A survey of smartphone-based indoor positioning system using RF-based wireless technologies. Sensors 20(24):7230
33. Suroso DJ, Rudianto AS, Arifin M, Hawibowo S (2021) Random forest and interpolation techniques for fingerprint-based indoor positioning system in un-ideal Environment. Int J Comput Digit Syst
34. Thabtah F, Hammoud S, Kamalov F, Gonsalves A (2020) Data imbalance in classification: experimental evaluation. Inf Sci 513:429–441
35. Tran HQ, Ha C (2019) Improved visible light-based indoor positioning system using machine learning classification and regression. Appl Sci 9(6):1048
36. Yadav RK, Bhattarai B, Gang HS, Pyun JY (2019) Trusted k nearest bayesian estimation for indoor positioning system. IEEE Access 7:51484–51498
37. Zhai J, Qi J, Shen C (2022) Binary imbalanced data classification based on diversity oversampling by generative models. Inf Sci 585:313–343

Mechanism for Digital Transformation of Intelligent Transport Systems

Nozima Akhmedova

Abstract The mechanism for digital transformation of transportation systems is described in this article. In particular, cloud technologies, distributed ledger technology, blockchain technologies, big data technology, the Internet of Things, augmented and virtual reality technologies, technologies of artificial intelligence, and additive technologies were considered for their use in an intelligent transport system.

Keywords Intelligent transport systems · Cloud technologies · Distributed ledger technology · Blockchain technologies · Big data · Internet of Things · Augmented and virtual reality technology · Artificial intelligence · Additive technology

1 Introduction

The intensive development of information and communication technologies determines the variety of digital transformation tools that provide data analytics and provide an opportunity to comprehensively address issues related to the effective development of the digital ecosystem.

One of the priorities for the digital transformation of transport systems as a formative element of the economy and an infrastructural basis is to give impetus to the creation and development of a single digital transport system that unites all types of transport and all its participants. It provides coordination and synchronization of interaction between participants in the transport market, integration of information resources in the transport sector, increases the efficiency of transport infrastructure management, and optimizes budget expenditures at all levels of its development. It is the only digital space that provides openness and reduces transportation costs, allowing to increase the profitability of manufacturers and consumers of transport

N. Akhmedova (✉)
Tashkent University of Information Technologies Named After Muhammad Al-Khwarizmi, Tashkent, Uzbekistan
e-mail: nozimaaxmedova@tuit.uz

G. Rajakumar et al. (eds.), *Intelligent Communication Technologies and Virtual Mobile Networks*, Lecture Notes on Data Engineering and Communications Technologies 171,
https://doi.org/10.1007/978-981-99-1767-9_50

services. Big data technology, the emergence of which is associated with the need to ensure the processing, analysis, and storage of a constantly growing and updated large amount of data, the volume of which exceeds the capabilities of traditional databases and analytical tools and systems. Currently, the main value of information is determined by the possibility of participation in improving their competitiveness and efficiency, creating new products, ensuring the validity, and quality of management decisions.

2 Main Part

2.1 Cloud Technologies

Cloud technologies (Cloud technologies) provide the ability to provide the user with convenient and prompt access to a wide range of computer resources and capacities via the Internet, including various types of servers, storage systems, services, and applications, implying the ability to release them from the load without the provider's participation and significant operating costs. Thus, cloud technologies enable the transfer of functions from personal computers to Internet servers for processing and storing data, allowing for the organization of instant remote access to information resources at any level from anywhere in the world and on any device, radically changing existing business models and increasing efficiency and business efficiency.

The development of cloud technologies is currently going at a very fast pace. According to the forecasts of the consulting company McKinsey, the total economic effect from the introduction of cloud technologies by 2025 will be estimated at 1.7–6.2,trillion USD annually.

Currently, cloud market products are most in demand in the segments of Internet services, retail, banking, and telecommunications. Cloud computing has a high growth potential in the transportation industry because it allows for the optimization and management of traffic flows, as well as the creation of automated systems for planning and modelling transport routes, including urban public transportation in large metropolitan areas.

At the same time, it should be noted that there is the possibility of numerous problems when using cloud computing associated with threats to confidentiality and reliability of information storage, as well as the development of destructive processes as a result of the spread of malicious software or software containing significant errors, which requires the implementation of appropriate protective measures aimed at reducing these risks [1].

2.2 Distributed Ledger Technology

Distributed ledger technology (DLT) allows organizing, storing, and exchanging data shared by all participants in the trusted environment, and each of which has a full copy of the data registry and access to the entire transaction history. At the same time, the reliability and security of information are ensured by cryptographic protection using special keys and an electronic signature, which allows you to control the actions of all participants. Distributed registry technologies provide synchronization of data registry copies based on the agreement (consensus) of participants to add new information as well as reliable data protection from possible changes, partial deletion, or complete deletion.

Depending on the access modes and options for agreements between participants during transaction validation, several types of distributed registries are distinguished:

- open: access to participation is not limited, there is no full identification, and the possibility of validating transactions is provided to all participants (e.g. in cryptocurrency networks);
- closed: access to participation is limited in accordance with established membership criteria; identification of participants is provided; and persons responsible for validating transactions are established (e.g. private distributed registries);
- mixed (hybrid): combines the properties of open and closed types of distributed registries.

One of the most common distributed ledger technologies is the blockchain technology, which is a sequential chain of data blocks containing information, built according to certain rules, where each new block is built into a chain of chronologically and cryptographically linked blocks. Thus, blockchain technology provides reliable storage of the entire data chain in the absence of centralized control.

The founder of the Institute for Blockchain Studies, M. Swan, identified three main areas that can fundamentally change under the influence of this technology:

- Blockchain 1.0—currency (performing financial transactions using cryptocurrency);
- Blockchain 2.0—contracts (various applications that work with financial instruments such as stocks, bonds, futures, as well as the so-called smart assets and contracts);
- Blockchain 3.0—applications for non-financial markets and transactions, primarily in the field of public services, healthcare, education and science [2].

Thus, blockchain technology, due to its reliability, transparency, and efficiency of use, has a high development potential and is an important component of the emerging digital economy. According to a survey conducted in 2015 as part of the International Economic Forum, by 2027, 10% of global GDP can be stored in blockchain technologies [3].

In the field of transport systems, the main areas for using distributed ledger technologies are:

- storage of information about the operation and repairs of vehicles, replaceable spare parts;
- supply chain management, ensuring their transparency, and improving the efficiency of operations;
- ensuring the availability of smart contracts ("smart" contracts), which provide for the implementation of a mechanism for the programmed and consistent fulfilment of the clauses and conditions of contracts;
- support of loyalty programmes by providing monitoring and analysis of actions performed within the framework of relevant programmes.

However, despite the high importance of these technologies in the development of the digital economy, their large-scale application is hampered by the lack of common standards and standard design solutions, as well as technical limitations associated with the need to ensure sufficient network bandwidth and the ability to integrate with existing systems built on the basis of other technologies [3].

2.3 Big Data

The term "big data" refers to the need to ensure the processing, analysis, and storage of large volumes of constantly growing and updated data, and the size of which exceeds the capabilities of traditional databases and analytical tools and systems. Currently, the main value of data is determined by the possibility of their participation in improving competitiveness and performance, creating new products, and ensuring the validity and quality of managerial decisions, both on the basis of identified cause-and-effect relationships and on the basis of established correlations. In order to understand the essence of big data technology, the META group analyst identified three of its main characteristics (the so-called three V concept):

- large volume of data (terabytes and petabytes);
- high speed (velocity) of updating and changing data, as well as generating new data;
- variety of data types: structured (coming from established sources, well-ordered and uniform) and unstructured (coming from different sources and in different formats, unordered).

Later, four additional essential characteristics were added to this concept: variability (variability)—dependence of data interpretation on context—reliability (veracity); visualization, which provides accessibility of perception—and value (value).

One of the most promising software solutions in the field of big data is Apache Hadoop software, which is a complete platform for processing, storing, and managing a variety of large data sets and includes the Hadoop Distributed File System (HDFS), the Apache Hadoop YARN cluster resource scheduling and resource management platform, and the platform for parallel processing large amounts of data, Apache Hadoop MapReduce [4].

The technologies of big data are universal and can be used in various industries and fields of activity that are characterized by large amounts of data on customers, operations, and processes, in particular, the financial sector, e-commerce, retail, medicine, education, systems in public administration, and solving social issues, etc.

The use of big data technology in the transport industry allows for the optimization, planning, and management of traffic flows, greater mobility of citywide transport, as well as the efficient use of transport infrastructure and other resources, by collecting, processing, and analysing data on road congestion in order to optimize travel routes and timetables, about problem areas that limit throughput and prevent continuous uniform movement, about the technical condition of vehicles, units, mechanism, and related equipment in order to predict malfunctions in a timely manner and provide preventive maintenance, etc.

The use of big data technology is most productive in interaction with the concept of the "Internet of Things," as a result of which most of the generated data is transmitted, which will be discussed in more detail below.

Speaking about the possibility of wide distribution of big data, it should be noted the significant contradictions between this technology and the basic principles of personal data legislation, in particular, the principle of limiting the processing of personal data only to a predetermined purpose, the principle of informed, specific, and conscious consent of the subject to the processing of his personal data, as well as the possibility of ensuring anonymity through the depersonalization of personal data, which requires a corresponding improvement in the norms of legislation aimed at eliminating existing contradictions [5].

2.4 Internet of Things

The concept of "Internet of Things" (IoT) provides for the interaction of physical and virtual objects and systems with each other and the external environment based on built-in information and communication technologies and standards through communication channels, and the decisive role in the emergence of this concept is played by the development of cloud technologies and big data technologies, widespread computerization, and the reduction in the cost of computing power and data transmission.

The Internet of Things, which actually represents a set of interconnected devices, devices, sensors, and sensors, allows you to collect various kinds of information about the connected control object and transfer it for further processing and storage via the Internet. The architecture of the Internet of things is represented by four levels: the level of sensors and sensor networks, which provides the collection and processing of information in real time; the level of gateways and networks, which is a single network platform; a service layer designed to automate technological and business operations; and the level of applications for various sectors and fields of activity.

The introduction of the concept of the Internet of Things into business processes and activities of companies, as well as entire industries, makes it possible to increase the efficiency of resource use, ensure security and mobility, and make optimal management decisions based on the data obtained to solve economic and social problems.

With regard to the transportation industry, the main directions for implementing the concept of the "Internet of Things" are:

Real-time remote control and monitoring systems, including traffic congestion, the location and movement of vehicles, the state of transport infrastructure facilities, transported goods, weather and sea conditions, etc.;

Systems of safety sensors installed in vehicles (cars, boats, and aircraft, as well as railway rolling stock) and allowing diagnostics, monitoring, and control of the operation of all important systems and devices with the possibility of informing and alerting (e.g. in case of emergency, failure or malfunction, or if a planned replacement or repair of parts, mechanisms, etc.);

Systems of sensory sensors (sensors of circular view, on-board sensors), which allow for the possibility of autonomous control of the vehicle.

In addition, the Internet of Things acts as a technological platform for projects in the transport sector implemented within the framework of the ideology of the sharing economy (sharing economy), which provides for the implementation of the model of joint consumption and the concept of development of the modern urban environment of large megacities ("Smart City").

Despite the growing popularity and importance of the concept of the Internet of Things, the possibility of its widespread implementation is currently limited by a number of technical, organizational, legal, and psychological factors, which include:

Incompatibility of various modern technologies and data exchange devices, which makes it challenging to integrate them;

The need to ensure the protection and confidentiality of transmitted data, including those related to both commercial secrets and individual information and data on the state and behaviour of a person;

The need to ensure the timely transmission of data based on the algorithm for distributing and fixing responsibility for their reliability, as well as the rationality and adequacy of decisions made on their basis.

The solution to these problems lies in the development of various levels of technical standards, state regulation of relations in the field of the Internet of things, cybersecurity issues, and the use of personal data, ensuring the digital literacy of the population [6].

2.5 Augmented and Virtual Reality

Augmented and virtual reality technologies open up new possibilities for perceiving the surrounding world, enriching it and making it more valuable and informative. Augmented reality (AR) technology allows you to supplement the real world with virtual elements distributed in space in real time by overlaying special content (text, graphics, and audio). Web cameras of tablets and personal computers, smartphones, as well as AR glasses, in particular, Google Glass from Google, HoloLens from Microsoft, Project Morpheus from Sony, and others, are used as devices that allow the implementation of augmented reality technologies.

The scope of augmented reality is constantly expanding and now covers not only the gaming and media industries but also such areas as navigation, medicine, architecture, art, cinema, education, design, and advertising. Augmented reality technologies are actively used in the transportation sector in aviation to control air traffic based on information received by air traffic controllers using AR applications integrated with the radar complex, to control aircraft, including through the installation and use of technical vision systems, to improve aircraft maintenance efficiency by presenting to the user tied to real virtual instruction objects, and to train and test employees.

The phenomenon of augmented reality is also widely used in the process of interaction of a vehicle (cars, sea and river vessels, railway trains) with the person who controls it through the visualization in his field of vision of information about traffic and the environment around the moving vehicle, about the objects of the road transport environment and infrastructure, adverse weather conditions, etc., allowing you to improve the efficiency of vehicle management and traffic safety.

In order to ensure the effective implementation and use of augmented reality technologies, it is necessary to solve a number of conceptual problems, which include ensuring the accuracy of the geolocation tools used and the necessary computing power of mobile devices, improving computer vision technologies, and ensuring the realism of virtual objects.

Unlike augmented reality, virtual reality (VR) is completely created using modern technical means and provides for the complete replacement of the real world with a set of virtual models and objects perceived by a person through the senses. Virtual reality is characterized by interactivity, which provides an imitation of the impact of virtual objects on the user and his response to this impact.

The use of virtual reality technologies in the transportation industry is due to the possibility of using various kinds of simulators that allow supplementing the training process of industry specialists with the formation and development of the necessary practical skills as a result of full immersion in virtual reality that reflects real situations and the environment, including by modelling critical and extreme conditions and various emergency situations.

In addition, thanks to the ability to visualize and evaluate ongoing processes in real time, augmented and virtual reality technologies are becoming one of the defining trends in the construction of intelligent transport systems [7].

2.6 Artificial Intelligence

Technologies of artificial intelligence (artificial intelligence, or AI) based on the use of various algorithms that simulate the processes of human thinking to support decision making. Artificial intelligence, being a cognitive tool, allows, based on the analysis of the capabilities of the human mind and modelling the internal structure of the system, to make decisions depending on the problem and context in real time, which allows you to automate a significant part of production processes and ensure the digitalization of economic and social processes.

In the field of artificial intelligence, which is extremely heterogeneous, various directions of research are developing depending on the research objectives, the tools used, and the models of thinking.

One of the areas of development of artificial intelligence is machine learning (ML), the main principle of which is to identify dependencies and patterns, including those hidden from humans, based on initial data and many other external factors. With this approach, the solution of problems is provided not only according to the established algorithm, but also on the basis of experience gained as a result of solving similar problems.

Training based on artificial neural networks has recently become widely used within the framework of machine learning, allowing for the closest possible simulation of the processes of functioning of the human nervous system, the work of brain neurons, a feature of which is the ability to self-learn based on previous experience, reducing the number of mistakes made.

Within the framework of machine learning, training based on artificial neural networks has recently been widely used, which makes it possible to provide the closest possible simulation of the processes of functioning of the human nervous system, the work of brain neurons, a feature of which is the ability to self-learn based on previous experience, which reduces the number of mistakes made. Artificial intelligence technologies are actively used in many industries and areas of activity, in particular for object recognition in images, in voice interfaces, in video analytics and monitoring systems, in automated control systems for devices and production processes, as "smart" assistants, in robotic and self-learning systems.

In the transport industry, due to the possibility of combining various modern digital technologies (cloud computing technologies, big data, artificial intelligence, the Internet of things, etc.) and engineering solutions, intelligent transport systems are widely used, which allow, based on the collection and analysis of data from transport infrastructure facilities, to provide management, forecasting and supporting the adoption of both operational and strategic decisions that allow solving the problem of reducing the capacity of the transport network, ensuring the efficiency and safety of the transportation process, optimizing and saving financial and material resources, and minimizing the negative consequences of environmental impact [8–10].

2.7 *Additive Technologies*

Additive technologies (Additive Manufacturing, AM; Additive Fabrication, AF) use computer three-dimensional technologies to create objects by layer-by-layer adding (building) material based on the data of their digital model. Thus, additive technologies, or so-called 3D printing, are based on the construction of an object by successively applying layers that form its contours and appearance, which is an alternative to traditional methods of production and processing products.

Additive technologies differ depending on the following features: methods of formation and fixation of the layer, building materials used, key technologies (with and without the use of a laser), energy supply methods for fixing the layer of the construction, and others.

The unique capabilities of additive technologies make it possible to reproduce complex structures, parts, shapes, and objects. At the same time, the obvious advantages of using these technologies include waste-free production, an increase in the speed of manufacturing products and the possibility of making adjustments at any stage of their creation, cost reduction in mass and mass production, as well as an increase in the "environmental friendliness" of the process.

The scope of application of additive technologies is extremely wide and tends to expand further. Today, 3D printing is used in such industries and areas of activity as architecture, industrial design, construction, medicine, bioengineering, education, furniture, clothing and footwear, jewellery, mechanical engineering, aircraft building, the military-industrial complex, etc.

With regard to the transport industry, the use of additive technologies ensures the rapid production of the necessary products and parts for vehicles and transport infrastructure facilities, both for the timely replacement of worn-out elements, and for the creation of new modern products and mechanism that meet the requirements of functionality, quality, and safety [8, 11, 12].

The above tools for the digital transformation of transport systems allow not only changing individual business processes, but also ensuring the restructuring of the entire industry as a whole and setting the trajectory for its future development. At the same time, the greatest effect will be achieved with the synergy of technologies, the possibility of their application in the aggregate.

3 Conclusion

Thus, the study allows us to formulate the following main conclusions:

Based on the analysis of economic literature, the features of the change in the technological order under the influence of innovative, breakthrough technologies are revealed, and the decisive role of infrastructure in the process of transition to each new technological order is emphasized, since such a transition is carried out on the basis of the existing infrastructure by expanding and transforming it, as a result of which the existing infrastructure barriers are overcome. Restrictions and a fundamentally new infrastructure are being created that takes into account scientific and technical achievements and meets the goals and needs of the effective functioning of the economy.

Taking into account the identified features of the current stage of development, the definition of the transport system is formulated as a transport and logistics system that has innovative features and is an infrastructural element of the high-speed economy, while ensuring the generation of added value at all stages of the transport and logistics chain through the integration of products and services provided within the framework of a new technological structure.

As a result of the analysis of global trends in the economy of the future, which form new challenges and opportunities, the main promising directions for the development of transport systems are identified, related to the formation of new concepts and technologies for the movement of goods and passengers, the development of unmanned vehicles, electric transport technologies, intelligent transport, market "uberization" passenger and freight transportation, the development of multimodal integration of transportation by various modes of transport in regional, interregional and international traffic, and the creation of large multimodal centres.

The necessity of creating and developing an appropriate infrastructure, the foundation of which should be high-tech facilities and specialized complexes, taking into account scientific knowledge and achievements of advanced equipment and technologies, was emphasized.

Based on the analysis of existing approaches to understanding the digital economy, the main characteristics inherent in it, as a complex system of socio-economic relations, are identified, which include:

- the emergence as a result of the emergence of a new technological order and the impact of breakthrough technologies;
- use as a basis of digital (information) technologies and digital infrastructure that ensures their functioning;
- transformation of global and national economies, industries and services, the public sector, enterprises and human society as a whole in accordance with changing conditions;
- focus on achieving economic and social benefits, including ensuring competitiveness, increasing the efficiency of economic processes, social well-being, accelerating economic growth, increasing labour productivity, creating added value, etc.

Taking into account the specifics of the functioning of the transport industry, the main stages and directions of the digital transformation of transport systems, the consistent implementation of which will ensure a large-scale transformation of technological and organizational processes aimed at improving the efficiency of traffic organization and management of the transportation process. The main tools for the digital transformation of transport systems, which include cloud technologies, distributed registry technologies, big data, the concept of the Internet of Things, augmented and virtual reality technologies, artificial intelligence, and additive technologies, are analysed, along with their features and characteristics, existing limitations of a technological and legal nature, that prevent their wide distribution at the present stage of development. Based on the analysis of digital transformation tools, promising areas of their application in transport systems, as well as the main effects from their implementation, generated by the synergy of interaction, are identified and described.

References

1. The NIST Definition of Cloud Computing. NIST Special Publication 800-145. September, 2011. https://bigdatawg.nist.gov/_uploadfiles/M0006_v1_3333767255.pdf
2. Hancock M, Vaizey E (2016) Distributed ledger technology: beyond block chain. London, p 88. https://assets.publishing.service.gov.uk/government/uploads/system/uploads/attachment_data/file/492972/gs-16-1-distributed-ledger-technology.pdf
3. McNulty E (2014) Understanding big data: the seven V's. 22 May 2014. http://dataconomy.com/seven-vs-big-data/
4. Savelyev AI (2015) Problems of application of legislation on personal data in the era of "Big Data" (Big Data) // Pravo. J High Sch Econom 1:43–66
5. Roslyakov AV, Vanyashin SV, Yu A (2015) Internet of Things: textbook [text]. Grebeshkov. Samara: PGUTI, p 200
6. Gorbunov AL, Nechaev EE, Terentsi G (2012) Augmented reality in aviation. Appl Inform 4:67–80
7. Burakov MV (2013) Neural networks and neurocontrollers: textbook. Manual. St. Petersburg: GUAP, p 284
8. Zlenko MA, Nagaytsev MV, Dovbysh VM (2015) Additive technologies in mechanical engineering. Manual Eng. M.: SSC RF FSUE "NAMI", p 220
9. Komil T, Saydulla K, Nozima A (2021)Big data technologies in transport planning. In: 2021 international conference on information science and communications technologies (ICISCT), Tashkent, Uzbekistan, 2021, pp 1–3. https://doi.org/10.1109/ICISCT52966.2021.9670328
10. Akhmedova N (2020) A study of security problems in big data and their solutions. Chem Technol Contr Manage 2020(4):81–85
11. Tashev K, Agzamova M, Axmedova N (2021) Application of the Aho-Corasick algorithm to create a network intrusion detection system. Bull TUIT Manage Commun Technol 4(4):6
12. Zaynutdinova M, Sayfullaev S (2021) Analysis and modeling of information security systems in Industry 4.0. In: Aliev RA, Yusupbekov NR, Kacprzyk J, Pedrycz W, Sadikoglu FM (Eds) 11th world conference "intelligent system for industrial automation" (WCIS-2020). WCIS 2020. Advances in intelligent systems and computing, vol 1323. Springer, Cham. https://doi.org/10.1007/978-3-030-68004-6_2

Metaverse 3C: Concept, Components, and Challenges in Travel and Tourism Sector

S. Poonkuzhali, J. Sangeetha Archana, and T. P. Prem Anand

Abstract The metaverse is a global and immersive virtual shared environment formed by the fusion of virtual and physical reality, made possible through the use of augmented reality and virtual reality headsets. The vision of an immersive Internet as a massive, determined, united, and communal realm is vital to metaverse technology. In the context of digital reality, the metaverse was conceptualized as Web 3.0 or 3D Internet. A new open and decentralized virtual reality will emerge as a result of Web 3.0. The metaverse, which has entirely opened up in numerous fields, is made up of three key technical advancements: artificial intelligence, augmented reality, and virtual reality. The traditional tourism industry is having a serious impact because of the COVID-19 pandemic, which has a higher influence on lowering the country's GDP. As a result, it is critical to address this issue through digital technology with enhanced user experiences. This paper focuses on the fundamental concepts, components, and key challenges of metaverse technology in the travel and tourism sectors. This technology is still being developed and will require extensive research before it can be realized to its full potential.

Keywords Augmented reality · Metaverse · Travel · Tourism · User experience · Virtual reality

S. Poonkuzhali (✉) · J. S. Archana · T. P. P. Anand
Rajalakshmi Engineering College, Chennai, India
e-mail: poonkuzhali.s@rajalakshmi.edu.in

J. S. Archana
e-mail: archanajanakar@gmail.com

T. P. P. Anand
e-mail: premanand.tp@rajalakshmi.edu.in

G. Rajakumar et al. (eds.), *Intelligent Communication Technologies and Virtual Mobile Networks*, Lecture Notes on Data Engineering and Communications Technologies 171,
https://doi.org/10.1007/978-981-99-1767-9_51

1 Introduction

The "Second Life" game offers players an online social platform where they can engage themselves with other players and carry out in-game activities that are exactly like those that occur in real life, such as interacting physically, having conversations, trading, and building, in order to build a "real world on the Internet" [1]. The term "metaverse" first appeared in a piece of speculative fiction called Snow Crash, published by Neal Stephenson in 1992. The term "metaverse" is a composite word made up of "meta" (transcendence, fictitious, and abstract) and "universe" (the real world, meaning a three-dimensional virtual world) [2]. The metaverse technology is an invention that gave rise to the recent phenomenon of new tourism experiences derived from a smart tourism destination's virtual reality.

When it comes to prospective employment opportunities for millions of people, tourism is frequently seen as the most significant sector in a major rising market like India. Combining several emerging technologies is crucial for this industry to grow in a way that allows it to meet tourist demand at the right time [3]. The COVID-19 problem in China caused a severe underestimation of the pandemic's effects on the travel and tourism sectors. Even now, the scenarios and repercussions of the crisis, which will have an unprecedented impact on the tourism industry, are not fully understood by policy creators and tourism professionals [4].

Globally, the generation of jobs, socioeconomic progress, and cultural advancement is all significantly influenced by travel and tourism. Travel and tourism are important contributors to the GDP of many regions, cities, and nations [5]. The metaverse will be the next pivotal point for the travel sector. The metaverse is conceived of as a continuum that includes a diverse set of digitally augmented realities, business methods, and universes. From consumers to employees, from 2 to 3D, from reality to virtuality and back, from cloud and AI to extended reality, digital twins, blockchain technologies, and beyond, it applies across the entire organization. It also applies to the entire company, from the consumer to the worker. The travel industry has the opportunity to rethink how people travel. The abilities of the metaverse in travel and tourism include virtual tourism, enhanced engagement of tourists, personalized virtual experiences, improved accessibility to visit virtual representations of destinations and attractions from anywhere in the world and new marketing opportunities.

People can interact socially in the metaverse by having conversations, playing games, working together on projects, and solving or learning from difficulties. In the metaverse, one's companions or friends may be real people or imaginary characters. Additionally, just like in the actual world, there can be many different types of activities or events in the metaverse, including political events, natural disasters, and economic activity. The only restriction in such a digital space is an individual's imagination. Additionally, the lifelogging feature allows for complete documentation of metaverse life [6]. Figure 1 depicts various forms of tourism. Tourism is classified into three forms: leisure, corporate, and specialist tourism.

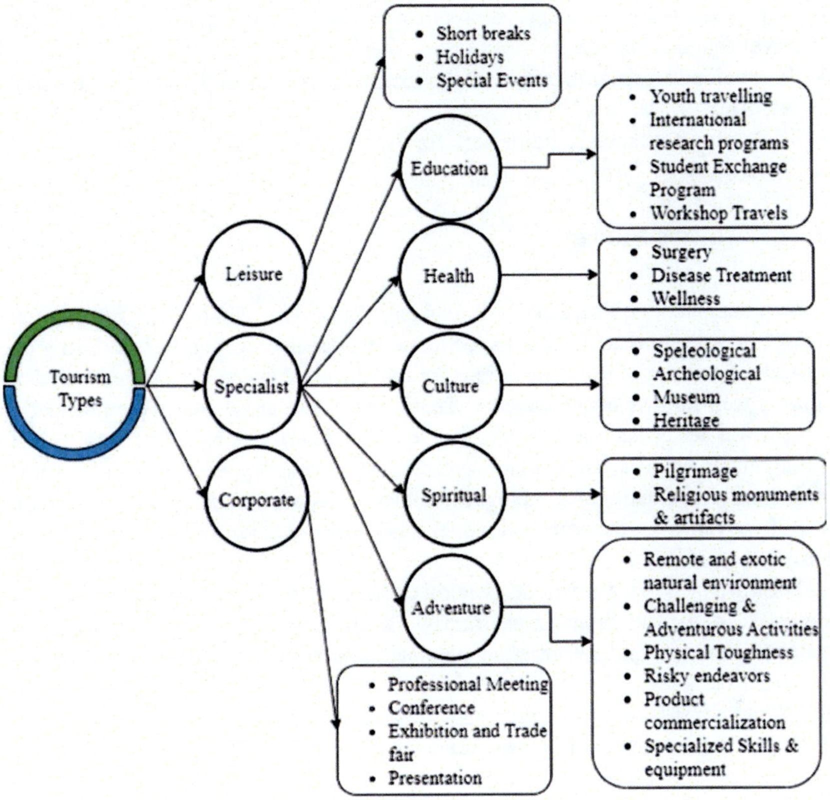

Fig. 1 Types of tourism

Travel that is made for relaxation, pleasure, specific interests, or enjoyment is known as leisure tourism. Short breaks, holidays, and special events are examples of leisure tourism. Any travel related to a person's work is referred to as "corporate tourism," also known as "business tourism." Corporate teams may or may not be required to spend the night away from home. Professional meetings, presentations, conferences, exhibitions, and trade shows are all examples of corporate travel.

Numerous diverse types of specialized tourism fall under the umbrella term "specialist tourism." It frequently relates to a personal interest, pastime, or sport. The most popular forms include health tourism, adventure tourism, cultural tourism, educational tourism, and spiritual tourism.

Infrastructure such as transportation systems, limited accommodation options, and insufficient tourist facilities can make it difficult for the travel and tourism sector to grow and attract tourists. Also, the technological changes, such as the rise of online travel platforms, can disrupt the traditional business model of the travel and tourism sector, requiring businesses to adapt to remain competitive. The overall contribution of this work is summarized as follows

- To provide various concepts, components, and challenges of the metaverse in the travel and tourism sectors.
- To provide various utilizations of the metaverse related to the travel and tourism sectors.
- To develop a metaverse framework for the travel and tourism sectors.

2 Literature Survey

A deep autoencoder was used by Ratnakanth and Poonkuzhali [7] to forecast user ratings in order to recommend a specific tourist destination to travelers. Using two datasets consisting of user evaluations for two different eccentric locations obtained from a well-known travel website, TripAdvisor, the suggested recommendation system was tested and confirmed. The second dataset was based on various sites throughout Europe that are available in the Kaggle repository, whereas the first dataset includes user evaluations and ratings for numerous locations throughout Jaipur, India. The effectiveness of the proposed model was evaluated in terms of root mean square error.

For a well-known tourism destination in India, Ratnakanth and Poonkuzhali [8] suggested a deep learning-based framework that assessed the various sentiments stated in reviews and information obtained from renowned travel websites like TripAdvisor. The Taj Mahal and Periyar Tiger Reserve were two ethnic locations taken into consideration for sentiment research. LSTM and GRU algorithms were deployed in the framework to categorize these retrieved reviews. Based on the text analysis and classification, the major tourist attractions were understood, and those were promoted toward increasing tourism. Dalila Martins et al. [9] focused on the case study of Amiais, a trivial Portuguese community that was recreated in the Second Life metaverse so that tourists could experience it electronically. They recreated and shared Amiais' cultural legacy in the Second Life metaverse.

Dongying Wei's [10] blockchain-based web platform Gemiverse provides public services for professional certification and tourism. Tourists can travel to national parks and sites by paying the entrance charge online and receiving unique NFT access tickets that can be exchanged in the market by paying the admission fee to the builders. By creating a simulation of the urban environment using a Roblox video game engine, Cecile Meier et al. [11] detailed the user experience to combine the production of virtual pathways about the sculptural legacy of a city in the classroom. Pannee Suanpang et al. [12] used cutting-edge information expertise to serve the tourism and hospitality industries and adopted the concept of a "mart travel destination approach." This "extensible metaverse" was an open metaverse platform that allowed new designers to individually expand the functionality of the metaverse technology. Tae Hwan Yoon et al. [13] developed hyper-immersive interactive content based on the metaverse, including the virtualization of a surfing community, by recognizing and evaluating the components through three-dimensional examination. For the development of integrated informational materials related to regional tourism and vacation organization using AI-powered smart tourism services. Myung and Michael [14] developed a theoretically complete model based on customers' hedonistic behaviors during sustained use, with the moderating visitor or non-visitor role at the VR tourism site.

3 Metaverse Concepts

3.1 Metaverse

The term "metaverse" denotes a network of 3D immersive virtual worlds made possible by the usage of augmented reality and virtual reality headsets, in which avatars socially connect. The factors that contribute to the ease of customization and flexibility in the metaverse include user-friendly interfaces, customizable avatars, collaborative experiences, dynamic environments to create and manipulate digital environments, providing a high degree of flexibility and control over their virtual experiences.

3.2 Avatar

An avatar is a metaverse manifestation of a user that performs specific roles. They are a person's identity in 2D or 3D form, allowing them to participate in various adventures and experiences in the metaverse.

3.3 Digital Twin

A "digital twin" is a digital form that is designed to precisely imitate a physical entity. These digital forms function in a way that is consistent with the physical world. The virtual world updates to match changes in the actual physical world.

3.4 Virtual Reality

Users can explore and engage with virtual settings in a 3D simulation known as "virtual reality". VR transports users to a new artificial environment that differs from realism, providing them with experiences that make them feel as if they are in that particular scene.

3.5 Augmented Reality

The superimposition of virtual things onto the physical world is referred to as "augmented reality" (AR). Augmented reality (AR) allows people to view virtual objects by overlaying digital information on the real world using cameras and displays, such as smartphones or AR glasses. AR technology tracks the user's environment and integrates 3D graphics or animations into the live view. This creates an illusion of virtual objects existing in the physical space, providing an interactive and immersive experience [15].

3.6 Mixed Reality

Mixed reality (MR) is the combination of AR and VR. It is a theory that enables the creation of digital objects that allow people to engage with the three-dimensional world. The metaverse is a virtual world that incorporates elements of both virtual reality and the physical world, creating a blended or "mixed" reality experience for users. Some of the different types of mixed reality in the metaverse are AR, which adds virtual elements to the real world, providing a blended experience that enhances the physical environment and VR, which creates metaverse a fully immersive digital environment that replaces the physical world [16]. The tourism metaverse combines physical reality and MR to meet the demands and interests of stakeholders in a shared, 3D virtual space. It also converts physical spaces into MR spaces, effectively turning the internet into a parallel virtual universe.

3.7 Extended Reality

XR is a newly coined umbrella term for all immersive technologies. The unification of all currently existing realities is referred to as "extended reality" (XR). By fusing the virtual and "real" worlds together or by generating a completely immersive experience, all immersive technologies expand the reality we experience.

4 Components for Creating Metaverse Application

4.1 Physical Devices and Sensors

Head-mounted displays superimpose an image on the user's view of the real world and play sound through the speaker. The gloves with sensors are wrapped around the user's hand for providing input using hand-based methods.

4.2 Recognition and Rendering

Recognizing sounds and processing speech aid in understanding the environment and communicating with other avatars in the metaverse. The conversation is a direct way of training non-player characters in the metaverse as well as connecting with other avatars. A technique that can distinguish between one's own voice without noise and the background noise is needed since the metaverse connection is created in varied contexts. In addition, depending on distance, the sound's volume varies. Voice recognition technology is used for providing realistic atmosphere in metaverse environment by considering the surrounding area into account and adjusts the volume based on distance.

The structural features of the human body help more accurately in understanding the reason for the action by superimposing the human body in 3D motion capture and pose estimation. High accuracy in the metaverse for tourism can be achieved through various technologies and techniques such as high-quality 3D modeling of real-world tourist destinations and landmarks, real-time tracking of user's movements in virtual environment, high-resolution displays, network connectivity, and integration with real-world data. Although a single-color camera can separate the structure of a physical body, such as shaking hands and capture real-time 3D motion in difficult scenes, it has limitations when it comes to capturing close encounters, such as hugs.

There are several places where the generation of entities and background scenes is applied. The name of the game character was used as an object for scene and object identification. The 3D modeling technology is being used by tourism companies to give users a better view of tourist attractions without having to leave their current location. Users of the 3D modeling platforms can upload 3D designs or scans to create 3D visuals that provide a 360-degree view of the locations, luring customers with such immersive experiences. Implementing 3D development in the metaverse involves determining VR or AR platform, creating 3D assets using Blender or Maya, assemble the scene and adding details such as lighting, textures, and animations, interactivity implementation in virtual environment by adding scripting and programming elements, such as logic, triggers, and events. The technology stack listed below was used to implement the 3D development.

Unreal Engine. Unreal Engine is a cutting-edge open-source real-time 3D development tool used to create custom VR projects, interactive visualizations, and projects for the next generation of 3D technology using C++ and Blueprints. The following features are available in Unreal Engine 5's most recent version:

- Toolset for an open environment;
- Real-time rendering of the future;
- Built-in characters and powerful animation tools;
- Best UI editor.

Unity. Unity is a full real-time 3D, AR, and VR solution with tools, a community, and support to help experts create immersive 3D projects. The benefits of Unity are:

- Sophisticated 3D editor;

- Unity Gaming Services offers tools to enhance gaming projects;
- In-house artist tools and well curated documentation.

CRYENGINE. CRYENGINE is a popular 3D development tool for game developers using C++ and Lua programming language. This tool also functions as a powerful VR development tool, supporting a wide range of platforms and providing the following features:

- Creation of stunning visuals using high-quality 3D models, adding particles and special effects, advanced lighting techniques, post-processing effects such as bloom and depth of field.
- Assist in the creation of simple and feature-rich VR apps;
- Provides built-in audio solutions for creating highly immersive digital experiences.

4.3 Scenario and Story

Content creation is the fundamental component of the metaverse for multimodal representation. Content is created using well-ordered stories and events to provide an immersive experience in the metaverse. Story realism, environment design, and conceptual comprehensiveness are all important in creating content. Color and illumination scenes, video, audio, sampling and aliasing, real-life content, and environmental navigation are all required for designing the virtual environment. Multimodal content depiction of tourist locations is represented using surveillance, personal photographs, audio and video recording, and avatars as depicted in Fig. 2. The user's motions, characters, and avatar personas all have an impact on behavioral modeling. Virtual non-player representative characters are depicted as an imitation curator who assists with introductions and scenarios, as well as an avatar of a user who advances the story of a tourist attraction. The scenario is filled with visual representations of real-world tourist destinations.

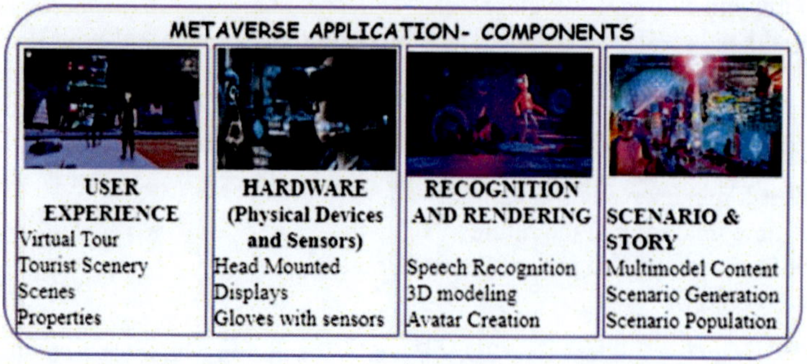

Fig. 2 Components for creating metaverse application

4.4 User Interaction

The user can see a 360-degree view of tourist attractions and communicate with the non-player characters for tour guidance in the metaverse so that travelers can gain first-hand experience of their desired destination before visiting it in person. Furthermore, AR and VR technologies combined with the metaverse can allow travelers to test out potential hotels before booking.

5 Challenges in Metaverse

5.1 Fusion Between Dissimilar Technologies

The metaverse need seamless fusion, coordination, and collaboration between different technologies on various techniques to support application. It is necessary to manage the superposition, compatibility, and integration between multiple technologies, to keep the technical system stable and reliable.

5.2 Energy and Resource Management

The developer needs to solve the contradiction between technological development and energy consumption and resource imbalance.

5.3 Theoretical and Practical Research Gap

Most of metaverse research stays in the primary stage of theoretical explorations. It is essential to consider the development of infrastructures, the imbalance of technical resources, the lack of the recognized industry standard, the difficulties in large-scale production, etc. In addition, it also needs to explore the most appropriate forms of business models and interactive ways used in the metaverse.

5.4 Security

The personal information of avatars is at risk of leakage and invasion in the digitalized world. It is important to establish security and confidentiality protection mechanism in the metaverse by adopting resolutions to strengthen identity authentication and security management, etc.

5.5 Balance Between Reality and Virtuality

The metaverse is a digitalized computer-generated world with strong interaction between virtuality and reality. Users switch freely between virtual and real world to achieve a balance. Also, users have to deal with the conflicts and coordination between virtuality and reality.

6 Metaverse Usage in Travel and Tourism Sectors

6.1 Flight Experience

Passengers go on board the imitation of an airplane seat in the metaverse, which provides a demo of safety flight measures and serving food details. This idea is to provide them with a realistic travel experience on a flight to a destination.

Passengers are then given a virtual tour of the destination via Head-Mounted Displays upon "arrival." While this type of experience is still in its early stages, it has the potential to give people access to travel opportunities that they find difficult to afford.

6.2 Virtual Landmark Destinations

Popular tourist landmarks frequently have negative environmental consequences as a result of excessive visitors. The number of users can be controlled by creating metaverse experiences of these landmarks, reducing the environmental impact.

6.3 Booking Interface

Another recent development in the tourism metaverse is a virtual booking interface as depicted in Fig. 3. Users can book a vacation while wearing a head-mounted display. The entire booking process for avatars are made through the metaverse. The user uses the head-mounted displays to do everything from choosing a hotel to making travel arrangements. Although this only has a few uses, we can see how businesses and travel agencies might use it to boost conversions. It seems to be the logical next step from offering one-time travel experiences to managing the entire booking process. Virtual booking interfaces are still relatively new, but they are becoming more widespread in the travel sector.

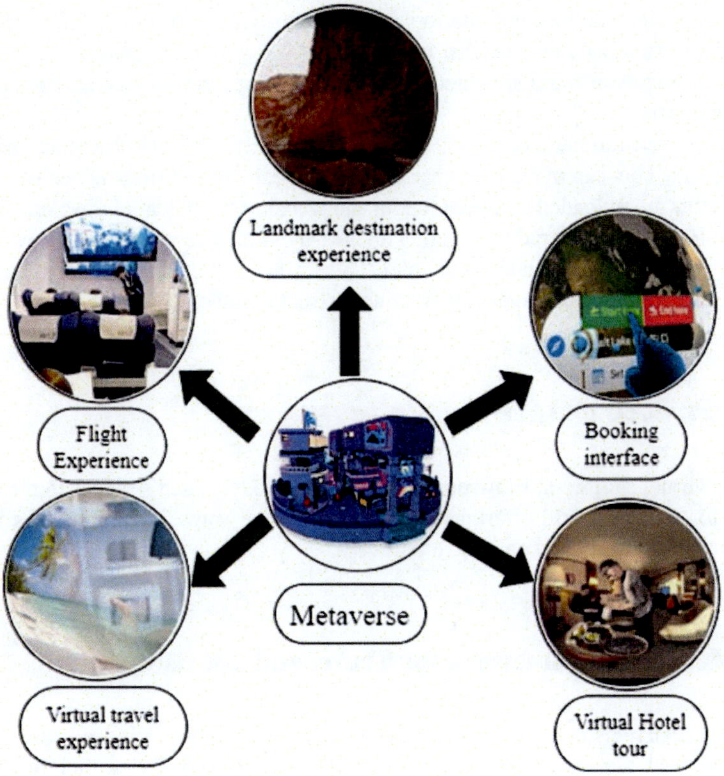

Fig. 3 Metaverse usage in travel and tourism sectors

6.4 Metaverse Travel Experiences

Virtual travel experiences are commonly used to describe metaverse tourism videos created for head-mounted displays. These virtual travel experiences strive to produce an experience that is as similar to really being in the actual destination as possible. Metaverse travel opportunities, at the cutting edge of 360 recognition, give the user something truly unique and memorable. This technology promises a bright future for the number of travel agencies and companies.

6.5 Virtual Hotel Tour

Users can now experience a hotel and its grounds in a way that is considerably more immersive than ever before using virtual hotel tours. Virtual tours are transforming the hotel industry in the same way that they are transforming the real estate industry. Hotel exteriors and interiors can be captured in exquisite detail using high-resolution cameras and specialist equipment. After choosing the room to explore, the photographs of room are stitched together to form a 360-degree interactive tour for the viewer.

Using head-mounted displays, the avatars can virtually visit the hotel and take hotel tours. This can make the experience more realistic and immersive. The metaverse tours are uploaded to digital media and websites for potential travelers to view at their leisure. Unlike traditional hotel images, these metaverse tours allow avatars to visualize themselves in the virtual hotel. This immersive experience leaves a lasting impression on future travelers by creating distinct brand engagement.

6.6 Products and Services

Create virtual shops and showrooms for existing products and services in the metaverse to sell real-world offerings in more immersive ways. Creating NFTs in the metaverse for utility and collective purposes.

7 Metaverse Framework for Travel and Tourism Sector

Travelers can register in the metaverse application for authentication purposes. Once authenticated, the authorized users create their own avatars and view their metaverse profiles. The metaverse environment contains 3D characters, 3D scenes, and 3D objects for tourism as depicted in Fig. 4. The traveler will be able to view the digital twinning of destinations, locations, hotels, and resorts for travel that was stored in the metaverse engine. A digital twin metaverse is a virtual replica of the physical world that can be used to simulate performance and maintenance. The performance simulation used for testing and simulation of various scenarios, such as crowd management, guest flow, and resource utilization. The maintenance planning is used for prioritizing tasks, and allocates resources more effectively. Virtual simulations can be used to train employees on a wide range of skills, from customer service to emergency response. The metaverse prompts the metaverse virtualization engine to provide virtual tour services using 2D and 3D rendering and context awareness.

Avatars view the virtual destinations in the form of three-dimensional details such as objects, scenes, or virtual services. By viewing and choosing the destination landmark, the user can view the nearby digitally twinning hotels and resorts and make their choices. After choosing the destination and hotels, they view the packages so that they can plan for the trip.

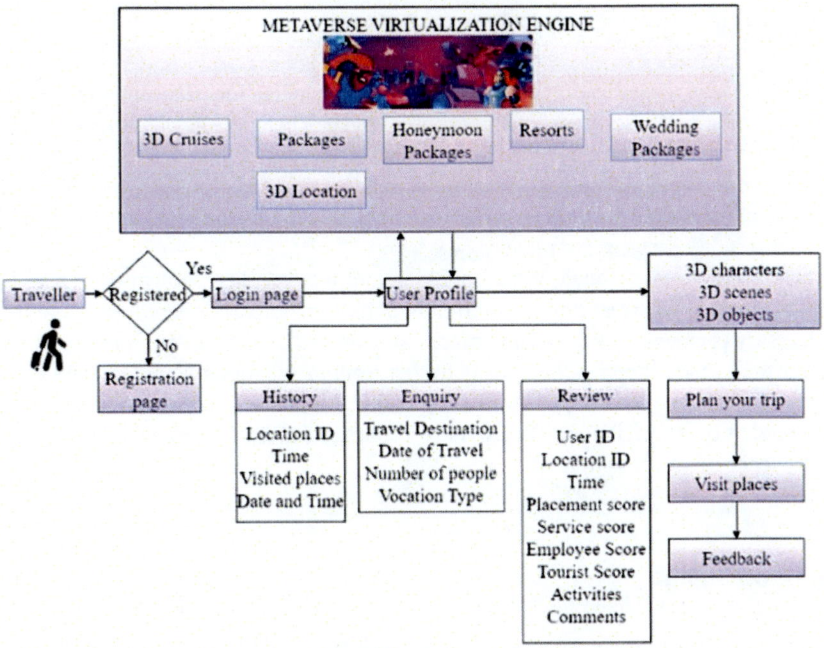

Fig. 4 Metaverse framework for tourism industry

7.1 *Registration and Login*

Users can register themselves as avatars in metaverse tourism by providing details such as their name, sex, age, address, email ID, phone number, and password. Once registered, the avatars can login to the system via email and their password.

7.2 *User Profile*

The user profile contains the following details about the avatars:

History. Location ID, time, user ID, and visited places log.

Review. Location ID, date and time, user ID, placement score, service score, employee score, tourist score, activities, and feedback. The score details have the range of 0–5 (0-low, 5-high).

Enquiry. Travel destination, date of travel, number of people, and vocation type.

7.3 Metaverse Virtualization Engine

The metaverse virtualization engine contains 3D visualization of destinations, locations, hotels, and resorts for travel purposes. Avatars can decide which locations to explore in the metaverse and visualize them as a 360-degree interactive tour formed by stitching together the high-resolution images. After exploration, the user can choose the preferred destination, hotel or resorts for staying, the necessary packages, and hospitality services for planning the trip.

Once they have planned, virtual booking interfaces will be shown to the user in the metaverse, where avatars can virtually book the destination landmarks and the necessary packages. After visiting places, the avatars have to fill out a review form about their travel experiences and any further recommendations. These reviews were stored in the database, and retrieved using SQL commands. Those recommendations are used by the travel industry to provide prospective travelers with a more immersive experience of various sites.

8 Conclusion

During COVID-19 pandemic, the travel and tourism sectors faced severe challenges for promoting their destination that affected country's GDP growth. So, implementing metaverse in travel and tourism sector will overcome the challenges in the development of travel and tourism sector. This paper analyzed various concepts, components and challenges for creating metaverse application in travel and tourism sectors. Then, for tourism sectors, a travel destination metaverse framework is created, which allows avatars to visualize unrealistic trips designed in the metaverse framework. This virtual journey allows the avatars to listen to their speech and visualize text and 3D objects around the digital world. This framework naturally attracts tourists by virtually visiting ancient sites, historical monuments, museums, etc., in preparation for visiting these locations. The metaverse generates vast amounts of data that can be analyzed and used to gain insights into travel and tourism behavior, preferences, and patterns. This information can be used to improve travel and tourism offerings, as well as to inform marketing and promotional efforts in our future work.

References

1. Lv Z, Qiao L, Li Y, Yuan Y, Wang FY (2022) BlockNet: beyond reliable spatial digital twins to parallel metaverse. Patterns 3
2. Lee LH, Braud T, Zhou P, Wang L, Xu D, Lin Z, Kumar A, Bermejo C, Hui P (2021) All one needs to know about metaverse: a complete survey on technological singularity, virtual ecosystem, and research agenda. arXiv preprint arXiv:2110.05352
3. Nazmfar H, Eshghei A, Alavi S, Pourmoradian S (2019) Analysis of travel and tourism competitiveness index in middle-east countries. Asia Pacific J Tour Res 3(6):501–513

4. Skare M, Soriano DR, Porada-Rochon M (2021) Impact of COVID-19 on the travel and tourism industry. Technol Forecast Soc Change 163
5. Abbas J, Mubeen R, Iorember PT, Raza S, Mamirkulova G (2021) Exploring the impact of COVID-19 on tourism: transformational potential and implications for a sustainable recovery of the travel and leisure industry. Curr Res Behav Sci 2
6. Allam Z, Sharifi A, Bibri SE, Jones DS, Krogstie J (2022) The metaverse as a virtual form of smart cities: opportunities and challenges for environmental, economic, and social sustainability in urban futures. Smart Cities 5(3):771–801
7. Ratnakanth G, Poonkuzhali S (2023) Indian tourist recommendation system using collaborative filtering and deep autoencoder. In: Information and communication technology for competitive strategies. Springer. Singapore, pp 341–356
8. Ratnakanth G, Poonkuzhali S (2021) Sentiment classification of Indian tourist reviews for sustainable tourism development. J Environ Prot Ecol 22(6):2535–2547
9. Martins D, Oliveira L, Amaro AC (2022) From co-design to the construction of a metaverse for the promotion of cultural heritage and tourism: the case of Amiais. Proc Comp Sci 204:261–266
10. Wei D (2022) Gemiverse: the blockchain-based professional certification and tourism platform with its own ecosystem in the metaverse. Int J Geoherit Parks 10(2):322–336
11. Meier C, Saorin J, de Leon AB, Cobos AG (2020) Using the roblox video game engine for creating virtual tours and learning about the sculptural heritage. Int J Emerg Technol Learn 15(20):268–280
12. Suanpang P, Niamsorn C, Pothipassa P, Chunhapataragul T, Netwong T, Jermsittiparsert K (2022) Extensible metaverse implication for a smart tourism city. Sustainability 14
13. Yoon TH, Do JK, Jeong SC (2022) Developing Songjeong metaverse surfing village: development of metaverse-based platform specialized for marine tourism. In: IEEE/ACIS 7th international conference on big data, cloud computing, and data science, pp 276–278
14. Kim MJ, Hall CM (2019) A hedonic motivation model in virtual reality tourism: comparing visitors and non-visitors. Int J Inf Manage 46:236–249
15. Arena F, Collotta M, Pau G, Termine F (2022) An overview of augmented reality. Computers 11(2)
16. Buhalis D, Karatay N (2022) Mixed Reality (MR) for generation Z in cultural heritage tourism towards metaverse. In: Information and communication technologies in tourism. Proceedings of the ENTER 2022 eTourism conference, pp 16–27

The Impact of Online Technology on Increase of Learning Motivation of Information System Student During Covid-19 Pandemic

Jerald Glad, Ardi Satyo Ramadhan, Ikrar Gama Raditya, Ford Lumban Gaol, and Tokuro Matsuo

Abstract Student motivation to learn becomes the encouragement that comes from himself. This encouragement makes students able to perform the learning activities to increase their knowledge. However, during this pandemic, students are studied at home using video conferencing facilities which could affect students' learning motivation. This is the motivation of this paper to analyze the learning motivation of junior high school students in Information Systems during Covid-19. This study aims to determine the student's learning outcomes, one of which is the learning motivation. The method used in this research is a survey method, wherein the data is collected from a sample or respondents of a population to represent the entire population using a questionnaire or interview. After the data needed is collected, in the form of a questionnaire, the analysis is discussed to determine the learning motivation. Likert scale is used to assist in analyzing the data. Based on the analysis of the data that has been collected, it is observed that junior high school students majoring in Information Systems during Covid-19 have relatively strong learning motivation. They have the willingness to succeed, have a driving force and need to learn, have hopes and dreams for the future, have an appreciation for learning, have

J. Glad · A. S. Ramadhan · I. G. Raditya
Information Systems Department, School of Information Systems, Bina Nusantara University, Jakarta, Indonesia 11480
e-mail: jerald.glad1@binus.ac.id

A. S. Ramadhan
e-mail: Ardi.ramadhan1@binus.ac.id

I. G. Raditya
e-mail: ikrar.raditya1@binus.ac.id

F. L. Gaol (✉)
Computer Science Department, BINUS Graduate Program—Doctor of Computer Science, Bina Nusantara University, Jakarta, Indonesia 11480
e-mail: fgaol@binus.edu

T. Matsuo
Advanced Institute of Industrial Technology, Tokyo, Japan
e-mail: matsuo@aiit.ac.jp

G. Rajakumar et al. (eds.), *Intelligent Communication Technologies and Virtual Mobile Networks*, Lecture Notes on Data Engineering and Communications Technologies 171,
https://doi.org/10.1007/978-981-99-1767-9_52

learning activities that are interesting, and have a conductive learning environment, which allow the students to perform well.

Keywords Knowledge · Learning outcomes · Learning motivation · Studied at home · Student

1 Introduction

Humans are learning creatures, because from birth, people learn a lot. Especially as a student, learning is an activity that must be done; hence, they are referred to as students. In addition, learning can also be defined as a process of behavior change that is expressed in the form of control, use, and assessment of attitudes and knowledge values contained in various fields of study, or more broadly in various aspects of life [1]. Gagné also argues that learning is influenced by external factors and internal factors, which interact with each other. However, teaching and learning activities also require basic encouragement that moves a person to learn, which is meant as motivation [2]. Student learning motivation will affect the final outcome of learning, so there needs to be motivation/encouragement for someone to learn. This happens due to various factors both from inside and outside.

With the current condition being unable to meet or have direct contact, students are required to study at home, using the facilities provided by the campus; students learn through video conferencing such as Zoom and Microsoft Teams. In the video conference, students can carry out learning activities led by lecturers or do group work, but with students studying independently in their homes, this can affect the level of motivation to learn, especially the juniors who are new to the learning environment.

Authors have determined that student learning motivation will affect the final outcome of learning; hence, learning motivation is needed by students to learn. This encouragement comes from a variety of factors. The first is family; students will have high motivation if they get support or encouragement from family members. Second is the environment; students will be highly motivated if the environment in which the student lives is suitable for the student's learning method. Third is the association. Interaction in school will have a very big effect in the learning process because if the students have bad associations, they will be less motivated in the learning process. Therefore, student learning motivation can be seen more clearly and every student is obliged to do his best on campus, because if the GPA < 2, then the student must take a short semester which means that after graduation, they have to study again. Therefore, motivation to learn is very important on campus so that the score is equal or above the minimum completeness criteria value in order to graduate.

Whether there is an impact on the motivation of the nature of Information Systems is also analyzed in this paper.

The problem identified in this research is as follows. Knowing all the different factors affecting the learning motivation of junior students of Binus University Alam

Sutera majoring Information System in a pandemic situation, the main driving factor of the motivation can be determined.

The problem formulation has been constructed as follows. Which factors influence the students of Information Systems the most, in the level of motivation to learn?

The research objectives are knowing how much level of learning motivates students at the Binus University Alam Sutera majoring Information System, while the second objective is analyzing what factors affect the most in the motivation to learn in students of Information Systems.

The benefits of research are determined. At first, this scientific work can broaden the readers' insight about the importance of learning motivation. This scientific work can also analyze the driving factors for students in the learning process. Lastly, this scientific work can add to the readers' insight as a guide for the learning process.

2 Literature Review

According to [2], learning motivation is internal and external encouragements for students who are learning to make behavioral changes, which are marked by indicators and supporting elements. According to Iskandar [3], learning motivation is the driving force within the individual to carry out learning activities to increase knowledge. According to [4], learning motivation is the overall driving force within students that causes learning activities, where this motivation ensures continuity and direction in learning activities so that the goals desired by the learning subject can be achieved.

From several definitions according to the experts above, it can be concluded that the student's learning motivation becomes encouragement that comes from himself. This encouragement makes students able to carry out learning activities to increase their knowledge.

According to Bahri, the benefits of learning motivation are that it serves as a spur for action. At first, students do not have a desire to learn, but because there is something they are looking for, there is an interest in learning. This is in line with his curiosity which ultimately encourages students to learn. This attitude ultimately underlies and leads to a number of actions in learning. So, motivation which serves as a driving force affects what attitudes students should take in order to learn. Motivation is a driving force for action. The psychological drive that gives birth to attitudes toward students is an unstoppable force. Students will do activities with all their soul and body. Intellect and mind proceed with physical attitudes that tend to submit to the will of learning.

From the theory above, it can be concluded that there are four benefits of motivation. The first motivation to learn is to boost students' enthusiasm for learning so that the students can be better. The second motivation to learn is to become a driving force to achieve goals more quickly. The third is to boost students' enthusiasm so that students can learn better and faster. The learning motivations are seen as a machine for the brain to run faster.

2.1 Factors Affecting Learning Motivation

According to [5], there are several factors that influence learning motivation, including:

(a) External factors—factors from outside the individual, which are divided into two: social factors include other human factors either directly or indirectly present and non-social factors include air conditions, air temperature, weather, time, place of study, and others.
(b) Internal factors—factors from within an individual, which are divided into two: physiological factors including physical condition and state of physiological functions and psychological factors including interest, intelligence, and perception.

According to Sardiman [4], there are several forms and ways to foster motivation in learning activities at school, including:

(a) Giving numbers is a symbol of learning activities; many students learn only to get good numbers/grades.
(b) Usually, the students chase test scores or grades in report cards.
(c) Gifts can also be used as motivation, but this is not always the case. It is because a reward for a job may not appeal to someone who is unhappy and unworthy of the job.
(d) Rivalry/competition can also be used as motivation; either individual competition or group competition can improve student achievement.
(e) Self-involvement fosters awareness in students to feel the importance of the task and accept it as a challenge and do the task with hard work.
(f) Giving tests: students will study actively to know the results of these tests. By knowing the test results, students will be motivated to learn, and it encourages the progress of students.
(g) Praise is a positive gift and at the same time provides good motivation. Punishment is a negative reward but if given properly and wisely can be a motivational tool.
(h) The desire to learn means that there is an element of deliberation, where there is an intention to study a particular interest. Apart from that, motivation also arises because of the need to control interest, so it is appropriate that if interest is the main motivation, the learning process will run smoothly.
(i) Goals that are recognized and accepted by students are very important motivational tools. By understanding the goals that must be achieved, they feel very useful and profitable, and there will be a passion to continue learning.

According to Farozin [6], the factors that influence learning motivation are psychological factors, which are as follows: curiosity and want to investigate the wider world (environment); creative nature and the desire to always go forward; the desire to get sympathy from parents, teachers, and friends; the desire to fix failures with new efforts; the desire to get a sense of security when mastering the lesson and when there is a reward or punishment as the end of the learning process.

Social factors are human factors (teachers, counselors, and parents), whether present directly or indirectly (photo or voice). The learning process will go well, if the teacher teaches in a fun way, such as being friendly, paying attention to all students, and always helping students who have learning difficulties. When at home, students still get attention from parents, both material attention by providing learning facilities and infrastructure to help and make it easier for students to study at home.

From the theory above, it can be concluded that there are many factors that influence learning motivation. The first is the environment in which students learn. This environment can support and facilitate student learning or inhibit it. The second factor is the psychological condition of the student that if the student does not have a sense of curiosity, the student will not progress because he is not motivated in this regard. Third, the social element of a student will develop with the group where he is; if the group supports him, he will be more advanced. The higher the group support, the higher the student's motivation to learn and vice versa.

2.2 Related Works

Motivation is also used to make students achieve maximum performance. There are various factors that influence learning motivation, ranging from psychological conditions to the social environment. In addition, there are various benefits of learning motivation, namely moving students' enthusiasm for learning and accelerating students' time to achieve their goals. Overall learning motivation is a mental boost that students have to construct their experiences to increase knowledge that will be used to advance.

Fauiziah proposed the relationship between learning motivation and learning interest of class Iv Sdn Poris Gaga 05 Kota Tangerang, wherein the importance of learning motivation to achieve certain goals was discussed. In the paper, Interest and Motivation in Improving Student Learning Outcomes, there is a debate whether teachers/lecturers can increase student motivation. Following these debates, basically all students have the motivation to learn, but also, it is the responsibility for professional teachers to help increase the enthusiasm of students to be motivated while learning. Lectures must be able to explore what can rise the motivation of students, so that the effective teaching and learning process is created in the classroom and students can achieve a goal as a result of their learning.

Authors have determined that student learning motivation will affect the outcome of learning. Thus, student learning motivation can be seen more clearly, and every student is obliged to try their best. That is why the research is needed to know the driving force of how students are motivated to learn. It is interesting to know how students improve better by knowing how to motivate themselves.

Aritonang [7] proposed *Minat dan Motivasi dalam Meningkatkan Hasil Belajar Siswa*, which explained that the factors that affect students' interest in learning are the teacher's way of teaching, the character of the teacher, and others.

The gap in this research is that the student learning motivation will affect the outcome of learning; hence, learning motivation is needed by students to carry out the learning processes. Other gaps are the differences in the target population. This research is conducted among the junior grade IV students in SDN Poris Gaga, majoring in Information System.

3 Research Methods and Data Sources

Based on benefits:

Basic research: Basic research was conducted to investigate the relevant issues with a view to confirm or disconfirm the positions theoretical and empirical. The main objective of basic research is to obtain general information about a phenomenon, with less emphasis, placing applications on real-world examples of the phenomenon [8].

Descriptive research: According to [9], descriptive research is the research that aims to explain or describe a situation, event, object, people, or anything related to variables, that can be explained either with numbers or words.

Based on the method:

Survey research: Survey research is the research in which data is collected from a sample of a population to represent the entire population using a questionnaire or interview [8].

Based on data collection:

Quantitative research: Quantitative research methods can be interpreted as a research method based on the philosophy of positivism used to research on certain populations or samples, where sampling techniques are generally carried out randomly. Data collection uses quantitative/statistical data analysis research instruments with the aim of testing predetermined hypotheses [10].

3.1 Variable

According to Sutrisno [11], variables are symptoms that vary. The variable in this study is the learning motivation in junior students, Department of Information Systems, during the Covid-19 pandemic.

According to [2], there are a few variables that affect the quality of motivation in studying. Those effects are:

1. There is a willingness to succeed.
2. There is a driving force and need for studying.
3. Having hopes and dream for the future.
4. Achieving an award for learning.
5. There are interesting activities in learning.

3.2 Population

According to Sugiyono [12], population is a generalization area consisting of subjects or objects that have certain characteristics and qualities determined by a researcher to be studied which are then drawn to a conclusion. The population in this study are junior students in the Information System department which had 294 students.

3.3 Samples

According to [13], the sample is a part or representative of the population to be studied. If the research is conducted by part of the population, it can be said that the research is a sample study. With this understanding and calculations using the Slovin formula (1), it was found that the sample required is 74.6 rounded off to 75 respondents.

$$n = \frac{N}{1 + N^2} \tag{1}$$

where n is the sample, N is the population, and e is the margin of error (1% was used).

3.4 Types of Data

Types of data in this study are primary data and ordinal data types. According to Kriyantono [14], primary data is the data obtained from first-hand or first-hand data sources in the field. Ordinal data is the data that not only categorizes variables to show differences among various categories, but also sorts them in several ways [15].

3.5 Data Collection Techniques

Steps needed when collecting data and sources from respondents are:

1. Looking for a theory of learning motivation to serve as a guide for indicators to be made.
2. Make and test indicators to get valid statements.
3. Distribute validity test questionnaires to 75 random respondents who are the junior students in the Department of Information System during the Covid-19 pandemic.

4. Enter data from the final questionnaire from the students. A valid questionnaire, which has valid 23 statements, is distributed to the students.
5. Draw conclusions and give suggestions.

The aims and objective of this research is to find out how the junior students from the Information Systems major about their motivation. And, it can have a real impact on the learning desires related to their knowledge or can add and multiply knowledge in an effective way and also provide insight into the learning process, such as factors that influence learning motivation, the benefits of having motivation to learn, and knowing the true meaning of learning motivation.

The implication and contribution to knowledge in this research paper would help the study and learning industries further by telling the way students want to learn and motivate them even more. This contribution can help in pushing and knowing the boundaries of motivation in study.

3.6 *Practicalities and Potential Obstacles*

The practical implementation of this research will be reduced because learning is carried out online; therefore, questionnaires will be carried out via Google Docs or online questionnaires. This increases the potential obstacle of dishonesty from the respondent and the nature of the respondent who is not willing to answer the questionnaires that is given. This makes the sample collection and obtaining an accurate result difficult.

3.7 *Research Design and Methods*

The author uses descriptive analysis method. According to Sugiyono [16], descriptive analysis is a statistic used to analyze data by describing the collected data as it is, without making any generalized conclusions. The data is obtained through a questionnaire in this study. The quantitative descriptive analysis was used to analyze the results of the questionnaire obtained, which is described in the statistical form.

The data obtained is qualitative data which will be converted into quantitative using a Likert scale and includes the type of ordinal data, which is then analyzed again qualitatively.

Examples of answers on a Likert scale as shown in Table 1 are: strongly agree (5); agree (4); doubt (3); disagree (2); strongly disagree (1). By using a Likert scale, the variables to be measured are translated into dimensions, the dimensions are translated into subvariables, and then, the subvariables are translated into measurable indicators. Finally, these measurable indicators can be used as a starting point for making instrument items in the form of statements that need to be answered by

Table 1 Statements

Description	Abbreviation	Result
Strongly agree	SS	5
Agree	S	4
Neutral	RG	3
Disagree	TS	2
Strongly disagree	STS	1

respondents. Each answer is associated with a statement or attitude support which is expressed in the following words.

Practical Implication

The practical implication of this research on the students is that students can find out how to get motivation to learn and later change the way they learn when pursuing knowledge in accordance with the research.

Theoretical Implication

The theoretical implication of this research is related to the learning motivation of junior students majoring in Information Systems. Many of the existing theories have put forward the theory that the motivation to encourage students comes from oneself. This encouragement led students to carry out learning activities to increase their knowledge.

4 Results and Discussion

After the research gets the data needed in the form of a questionnaire, the analysis is discussed to see the motivation to learn. Researchers use a Likert scale to assist in data analysis. The questionnaire was distributed to 75 respondents. Then, the range used is 75 which is obtained by dividing the ideal number of scores by 5 because the interval from the Likert scale is 5, which are STS, TS, RG, S, and SS. So, the results are 75, starting with the number 0, then 75, and finally 75 is added 5 times to get 375 results.

To get the percentage result from the Likert scale, an interval of 20 is used so that the ideal percentage result is 100, and 100 divided by five intervals on the Likert scale is 20. Five Likert scale intervals determine the percentage results such as very weak, weak, sufficient, strong, and very strong.

Numbers 0%–20% = Very weak.
Numbers 21%–40% = Weak.
Numbers 41%–60% = Sufficient.
Numbers 61%–80% = Strong.
Numbers 81%–100% = Very strong.

Table 2 Results of 'willingness to succeed'

Description	Count	Result
SS (5)	164	285
S (4)	88	340
RR (3)	43	351
TS (2)	3	60
STS (1)	2	11
Total	300	1309

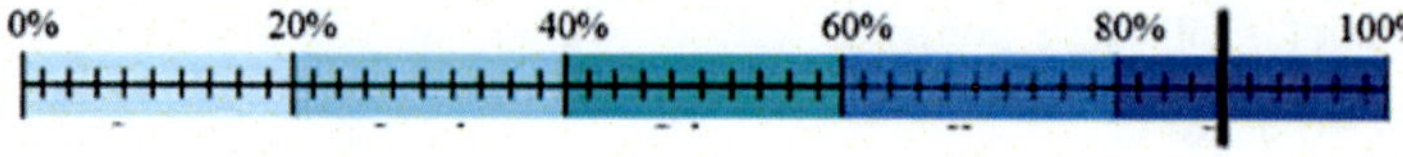

Fig. 1 Willingness to succeed

Overall Learning Motivation

Item no. 1 Uno theory [2]: There is a willingness to succeed.

The ideal score in this indicator is 1500, which is obtained from $75 \times 5 = 375$ in the presence of four statements $375 \times 4 = 1500$. The percentage is calculated as, $1309/1500 \times 100 = 87.27\%$. The details of result are shown in Table 2 and Fig. 1.

Based on Uno's theory [2], the first factor of learning motivation is the willingness to succeed. From the results above, it can be concluded that junior students of the Information System major during Covid-19 pandemic students have a very strong willingness to succeed in their driving force of studying with the percentage of 87.27%. This proves that students have very strong feelings toward wanting to graduate with a good GPA, striving to reach the goals, and studying to get good. The percentage of encouraged to study to get good grades are shown in Fig. 2.

Item no. 2 Uno theory [2]: There are a driving force and need for studying.

The ideal score in this indicator is 1500, which is obtained from $75 \times 5 = 375$ in the presence of four statements $375 \times 4 = 1500$. The percentage is calculated as, $1131/1500 \times 100 = 75.40\%$

Based on Uno's theory [2], the second factor of learning motivation is the presence of driving force and need for studying. From the results above (Table 3), it can be concluded that junior students of the Information System major during Covid-19 pandemic students have a strong agreement to this indicator with the percentage of 75.40%. This proves that the junior students have a strong driving force and need for studying in the shape of paying attention to the lecturer's explanation in class. Having the mindset of studying is a necessity, and encouraged to study to get good

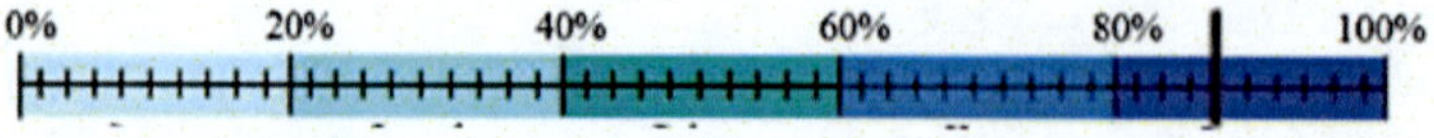

Fig. 2 Encouraged to study to get good grades

grades and encouraged to learn from the people around were observed. The results observed are shown in Tables 4, 5 and 6, and Fig. 3 depicts the learn in order to get rewards.

Item no. 3 Uno theory [2]: Learn in order to get rewards.

Table 3 Results of 'encouraged to study to get good grades'

Description	Count	Result
SS (5)	66	265
S (4)	118	384
RR (3)	99	213
TS (2)	15	8
STS (1)	2	1
Total	300	1131

Table 4 Results of 'learn in order to get rewards'

Description	Count	Result
SS (5)	27	135
S (4)	71	284
RR (3)	127	381
TS (2)	52	104
STS (1)	23	23

Table 5 Results of 'interesting activities in learning'

Description	Count	Result
SS (5)	57	285
S (4)	85	340
RR (3)	117	351
TS (2)	30	60
STS (1)	11	11
Total	300	1047

Table 6 Results of 'conducive learning environment to learn well'

Description	Count	Result
SS (5)	74	370
S (4)	126	504
RR (3)	129	387
TS (2)	37	74
STS (1)	9	9
Total	375	1035

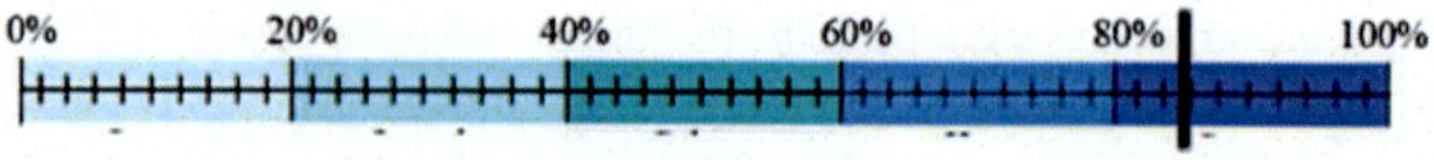

Fig. 3 Learn in order to get rewards

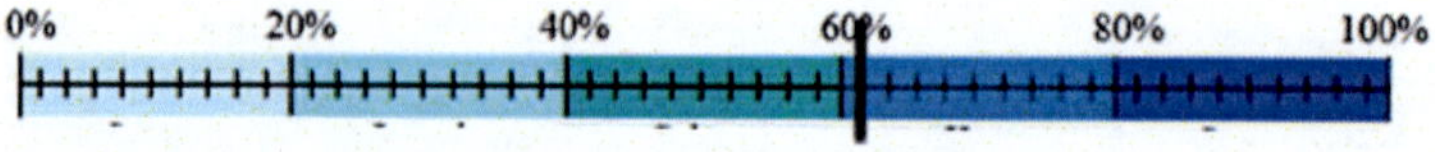

Fig. 4 Interesting activities in learning

Based on Uno's theory [2], the third factor of learning motivation is to learn in order to get rewards. From the results above, it can be concluded that junior students of the Information System major during Covid-19 pandemic have a strong agreement to this indicator with the percentage of 61.8%. This proves that the junior students have a motivation to learn in order to get rewards. The interesting activities in learning are shown in Fig. 4.

Item no. 4 Uno Theory [2]: There are interesting activities in learning.

The ideal score in this indicator is 1500. It is obtained from $75 \times 5 = 375$, in the presence of four statements $375 \times 4 = 1500$. The percentage is calculated as, $1047/1500 \times 100 = 69.80\%$

Based on Uno's theory [2], the fifth factor of learning motivation is the existence of interesting activities in learning. From the results above, it can be concluded that junior students of the Information System major during Covid-19 students have interesting activities in learning, which is indicated by a percentage of 69.80%. This proves that the students agreed strongly in motivation from interesting activities in learning in the shape of agreeing to statements such as easier to understand the lesson if it is presented in an interesting way, have an understanding of learning is an interesting activity, and many other interesting activities when learning.

Item no. 5 Uno theory [2]: The existence of a conductive learning environment, allowing a student to learn well. The results on conducive learning environment to learn are provided in Fig. 5.

The ideal score in this indicator is 1875. It is obtained from $75 \times 5 = 375$ in the presence of four statements $375 \times 5 = 1875$. The percentage is calculated as, $1035/1875 \times 100 = 71.68\%$

Based on Uno's theory [2], the fifth factor of learning motivation is the existence of interesting activities in learning. From the results above, it can be concluded that junior students of the Information System major during Covid-19 students have

Fig. 5 Conducive learning environment to learn

conductive learning environment which is indicated by a percentage of 71.68%. This proves that the students have the existence of a conductive learning environment, allowing a student to learn well. This supports the statement that they agree that classmates support students in studying well, healthy campus life supports studying, home environment helps in studying, and friends and family support students in studying.

5 Conclusion

Based on the analysis of data that has been collected, it can be observed that junior students of the Information System major during Covid-19 have a relatively strong learning motivation. They have the willingness to succeed, driving force and need for studying, hopes and dreams for the future, award for learning, interesting activities in learning, and a conductive learning environment, allowing a student to perform well. From all these factors, the least powerful indicator is the one having an award for learning. So, it can be said that an award does not really affect the motivation of learning. But, it is more important to be motivated from the inside or more specifically with the willingness to succeed.

References

1. Gagné MA (1998) The role of dependency and colonialism in generating trauma in First Nations citizens: Inter Handb Multigenerational Legacies of Trauma. pp 355–372
2. Uno H (2016) Teori Motivasi dan Pengukurannya. Bumi Aksara, Jakarta
3. Iskandar (2009) Metodologi Penelitian Kualitatif. Gaung Persada Press, Jakarta
4. Sardiman (2011) Interaksi dan Motivasi Belajar-Mengajar. Rajawali Pers, Jakarta
5. Suryabrata S (2004) Psikologi Pendidikan. Rajagrafindo Persada, Jakarta
6. Farozin M (2011) Model bimbingan klasikal untuk meningkatkan motivasi belajar sekolah menengah pertama. Disertasi Tidak Dipublikasikan. Yogyakarta: PPs UNY
7. Aritonang KT (2008) Minat dan Motivasi dalam Meningkatkan Hasil Belajar Siswa. Penabur. SMPK 1 Penabur, Jakarta
8. Silalahi U (2009) Metode Penelitian Sosial. Refika Agitama, Bandung
9. Setyosari P (2010) Metode penelitian pendidikan dan pengembangan. Kencana, Jakarta
10. Sugiyono (2015) Metode Penelitian. Kuantitatif, Kualitatif, dan R&D. Penerbit Alfabeta, Bandung
11. Hadi S (2015) Metodologi Riset. Pustaka Pelajar, Jakarta
12. Sugiyono (2008) Metode Penelitian Kuantitatif, Kualitatif, dan R&D. Bandung, Alfabeta
13. Arikunto S (2006) Qualitative research methods. Bumi Aksara, Jakarta
14. Kriyantono R (2010) Teknik praktis riset komunikasi. Prenada media
15. Uma S (2006) Metodologi penelitian bisnis. Jakarta
16. Sugiyono (2004) Metode Penelitian Bisnis. Cetakan Keenam. Bandung, Alfabeta
17. Asrori M (2009) Psikologi Pembelajaran. Wacana Prima, Bandung
18. Danang S (2012) Validitas dan Reliabilita. Nuha Medika, Jakarta
19. Dimyati DM (2009) Belajar dan Pembelajaran. Rineka Cipta, Jakarta

20. Fauziah RA (2017) Hubungan Antara Motivasi Belajar Dengan Minat Belajar Siswa Kelas Iv Sdn Poris Gaga 05 Kota Tangerang. Pendidikan Guru Sekolah Dasar. Universitas Muhammadiyah Tangerang, Tangerang
21. Iskandar (2009) Psikologi Pendidikan. Gaung Persada Press, Jakarta
22. Kriyantono R (2012) Teknik praktis riset komunikasi. Kencana, Jakarta
23. Mulyasa (2003) Kurikulum Berbasis Kompetensi. Remaja Rosda Karya, Bandung
24. Nurocim (2013) Perencanaan Pembelajaran IlmuIlmu Sosial. Rajawali Pers, Jakarta
25. Suharsimi (2010) Dasar Dasar Evaluasi Pendidikan. Bumi Angkasa, Jakarta
26. Husein U (2003) Metodologi Penelitian Untuk Skripsi dan Tesis Bisnis. Gramedia Pustaka, Jakarta
27. Ramadhan GR (2020) Questionnaire regarding student learning motivation, https://forms.gle/aiadJgdLZahQ3LJT9

Modelling Sentiment Analysis on Indonesian Product Reviews Using Machine Learning

Andry Chowanda and Lasmy

Abstract Insight from product reviews from customers is important messages to the business. It provides feedback to the business and enhances the customers' experiences and leading to sustainable competitive advantage for the business. Sentiment analysis can be performed from the customers' product reviews regarding to particular product or service. Capturing sentiment analysis can be done automatically using machine learning techniques. This research aims to explore several machine learning algorithms to model automatic sentiment analysis on Indonesian product reviews. The product reviews data was gathered from an e-commerce platform in Indonesia. All of the product reviews are in the local (i.e. Indonesian) language. The research contributes to a state of the art of automatic sentiment analysis model on Indonesian product reviews. Eight algorithms and 4507 settings are explored to find the best setting and algorithm to model sentiment analysis in Indonesian product reviews. The results demonstrate that the best model was achieved by the one trained with five layers of artificial neural network with 99.1% for testing model accuracy, precision, recall, and F1-score. The best AUC of the model was 99.6%. The model requires 34.68 minutes to be trained.

Keywords Product reviews · Sentiment analysis · Machine learning · Artificial neural network

A. Chowanda (✉)
Department of Computer Science and School of Computer Science, Bina Nusantara University, Jakarta 11480, Indonesia
e-mail: achowanda@binus.edu

Lasmy
Department of Management and BINUS Business School Undergraduate Program, Bina Nusantara University, Jakarta 11480, Indonesia
e-mail: lasmy@binus.edu

G. Rajakumar et al. (eds.), *Intelligent Communication Technologies and Virtual Mobile Networks*, Lecture Notes on Data Engineering and Communications Technologies 171,
https://doi.org/10.1007/978-981-99-1767-9_53

1 Introduction

The fourth revolution industry has changed the way industries do their business models. Most industries have implemented technology and the Internet to enhance customers' experiences, increasing revenue. Nowadays, both customers and industries can easily communicate between them. The customers can provide feedback on the products or services, and the industries can easily gather customer market surveys. One of the platforms for the customer to provide feedback is through reviews, be it product reviews or service reviews. Customer reviews are essential to industries for them to improve their products and services, hence increasing the customers' experiences. Customer reviews can be automatically captured and analysed by using machine learning. An automatic sentiment analysis system can be implemented to capture the customers' sentiments. There are several machine learning algorithms that can be applied to model an automatic sentiment analysis system. Automatic sentiment analysis model can be trained using several machine learning algorithms such as support vector machine [8, 16], decision tree [7], random forest [7], Naïve Bayes [7], generalised linear model, artificial neural network [8], convolutional neural network [13], and long short-term memory [2–4]. The algorithms can be tailored based on the data and use case [5].

This research aims to explore several machine learning algorithms to model automatic sentiment analysis on Indonesian product reviews. The product reviews data was gathered from an e-commerce platform in Indonesia. All of the product reviews are in the local (i.e. Indonesian) language. The research contributes to a state of the art of automatic sentiment analysis model on Indonesian product reviews. Eight algorithms and 4507 settings are explored to find the best setting and algorithm to model sentiment analysis in Indonesian product reviews. The algorithms are Naïve Bayes, fast large margin, generalised linear model, artificial neural network, decision tree, gradient boosted trees, random forest, and support vector machine. The results demonstrate that the best model was achieved by the one trained with five layers of artificial neural network with 99.1% for testing model accuracy, precision, recall, and F1-score. The best AUC of the model was 99.6%. The model requires 34.68 minutes to be trained. SVM provides the fastest training time (12.55 minutes); however, the performance is still inferior compared to the one trained with five layers of ANN architecture. The rest of the manuscript is organised as follows: the state of the art on sentiment analysis modelling in product reviews is demonstrated in the next section. The research methods, architectures, and experiment settings are illustrated in the methodology section. Moreover, the results are comprehensively discussed in the results and discussion section. Finally, the conclusion and future research direction are demonstrated in the last section, the conclusion, and future work section.

2 Recent Work

Several text processing techniques can be applied to extract information from text. Some vectorisation techniques, such as term frequency-inverse document frequency, bag of words, Word2Vec, GloVe, and transformer-based embedding, can be applied to represent information from text [4]. There are also several conventional machine learning algorithms as well as deep learning algorithms that can be implemented to model sentiment analysis from the text. Some research implemented several machine learning algorithms to model online-review sentiment analysis. The best performance was achieved by the model trained by the support vector machine algorithm, which achieved an accuracy score of 98.6%. The machine learning algorithms are also evidently superior that the one modelled by using lexicon-based, with an accuracy score of 95% [11]. Another research on sentiment analysis was performed by Koukaras et al. [9], where they implemented support vector machine and valence aware dictionary and sentiment reasoner. They achieved the best F1-score of 76.3% and the AUC score of 67% [9]. Another sentiment analysis model trained with the support vector machine algorithm was modelled using movie recommendation data with the best accuracy score of 98.63% [12]. The research also shows that the support vector machine algorithm is superior to Naïve Bayes algorithm with the best accuracy score of 97.33%. More research done by Kalarani and Selva Brunda [8] with part of speech, joint sentiment topic, and trained with the support vector machine algorithm. The model achieved the best F1-score of 86%. The research also shows that the support vector machine algorithm is superior to artificial neural network architecture, with the best F1-score of 86.5%.

Moreover, some researchers [1] proposed contextual-based approaches to model sentiment analysis from the text. The best model achieved by support vector machine with the best accuracy score of 75%. Another researcher proposed several machine learning and deep learning algorithms to model sentiment analysis from Twitter [14]. The best performance achieved by the conventional machine learning algorithm was the support vector machine classifier, with an accuracy score of 70.56%. Moreover, the best performance achieved by the deep learning algorithm was the long short-term memory architecture with an accuracy score of 73.81%. Another researchers [6] performed an exploration of several machine learning algorithms such as decision tree (J48), random forest, Naïve Bayes, and support vector machine to model sentiment analysis on student courses. The best model was achieved by the one trained with the support vector machine classifier with the best accuracy of 63.79%. Another researcher [10] proposed a sentiment analysis model from Twitter to capture sentiment analysis on tourism topics in the Thai language. The best model was achieved by the one trained with the random forest algorithm with the best accuracy of 95.4%. The second best model was achieved by the model trained with the random forest algorithm with the best accuracy of 95.4%. Some deep learning algorithms are also performed by researchers [13] to model sentiment analysis from Hindi movie reviews. The best architecture to model the sentiment analysis was the convolutional neural network, with the best accuracy of 95%.

3 Methodology

This research aims to explore several algorithms to predict sentiment from product reviews in the Indonesian marketplace automatically. Figure 1 illustrates the proposed methods in this research. There are three groups of methods: pre-processing, feature engineering and modelling, and evaluation. In the pre-processing, the data was loaded and prepared. The data then splits into training and testing sets with the percentage of 65% and 35%, respectively. The dataset used in this research is the PRDECT-ID [15]. The dataset consists of 5400 annotated product reviews. The dataset was annotated using basic emotions and sentiment analysis. This research focuses on the latter one. The next step is the basic feature engineering process, which handles unknown values, replaces all missing values, and encodes the data. The next group is feature engineering and modelling. In this group, the text features are extracted from the text. Next automatic feature engineering process is applied to the data before the model is trained with several machine learning algorithms. The model is trained with several algorithms, such as Naïve Bayes, fast large margin, generalised linear model, artifi-

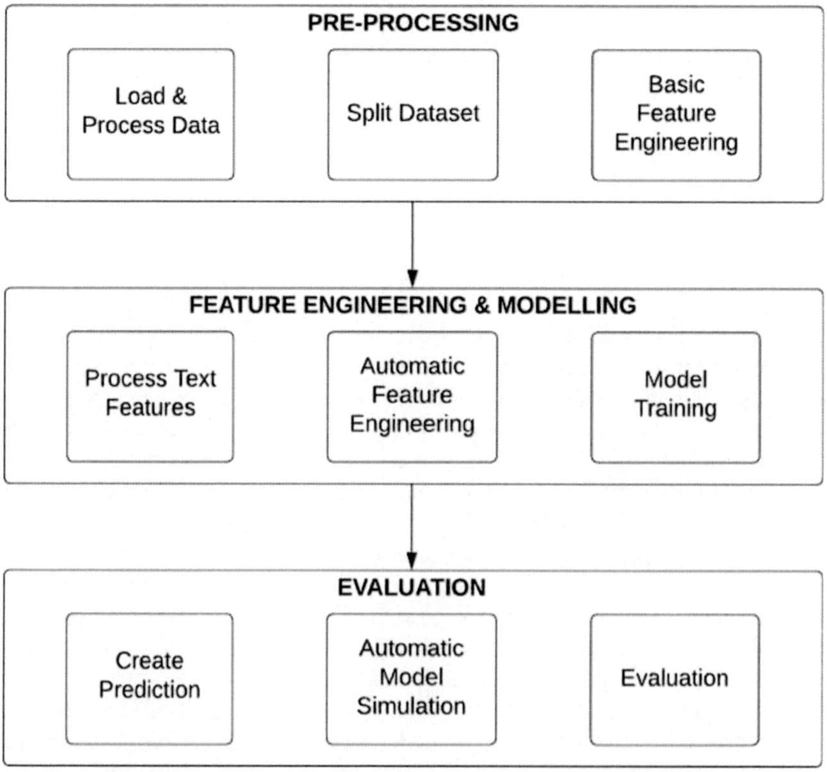

Fig. 1 Proposed methodology

cial neural network, decision tree, gradient boosted trees, random forest, and support vector machine. The algorithms implemented an automatic feature engineering process. Overall, each algorithm automatically performed 118–986 features exploration and 5–62 features generation. Moreover, artificial neural network implemented five layers of the vanilla neural network. The first layer consists of n number of nodes depending on the number of features. The second, third, and fourth layer consist of 50 nodes with ReLU activation function. The rectified linear unit (ReLU) activation function provides non-negative output value with $f(x) = max(0, x)$. The final layer (output) consists of two nodes (positive and negative class) of the Softmax activation function. The $L1$ value is set to 0.000010. Finally, the models were evaluated using several performance metrics, such as accuracy, AUC, precision, recall, and F1-score.

4 Results and Discussion

Hundreds of automatic features explorations and generation from Naïve Bayes, fast large margin, generalised linear model, artificial neural network, decision tree, gradient boosted trees, random forest, and support vector machine learning algorithms. The models automatically perform and evaluate the models trained by the algorithms. Table 1 demonstrates the overview of the results of the best models from each algorithm. Naïve Bayes model resulted in a quite overfitted, where the accuracy score was 63.8, but the F1-score was 40.3. The precision-recall score was 95.4 and 25.6, respectively. The AUC score achieved by Naïve Bayes model was 70.7. The model trained by fast large margin achieved relatively high performance with 87.0 and 87.8 scores in accuracy and F1-score, respectively. The precision-recall score was quite imbalanced; the score was 80.0 and 97.3, respectively. The AUC score achieved by the generalised linear model was 89.8. The model trained by generalised linear model achieved relatively high performance with 98.4 and 98.3 scores in accuracy and F1-score, respectively. The precision-recall score was 99.4 and 97.1, respectively. The AUC score achieved by the generalised linear model was 99.6. Next, the model trained by the artificial neural network model achieved the best performance, where the accuracy, F1-score, and precision-recall score achieved a score of 99.1. The AUC score achieved by the artificial neural network model was 99.6.

Table 1 Overview results

Performance	NB	GML	FLM	ANN	DT	RF	GBT	SVM
Accuracy	63.8	98.4	87.0	99.1	98.4	96.4	98.4	98.4
AUC	70.7	99.6	89.8	99.6	98.4	97.6	97.6	97.6
Precision	95.4	99.4	80.0	99.1	99.3	97.0	97.0	97.0
Recall	25.6	97.1	97.3	99.1	97.4	95.5	95.5	95.5
F1-score	40.3	98.3	87.8	99.1	98.4	96.2	96.2	96.2

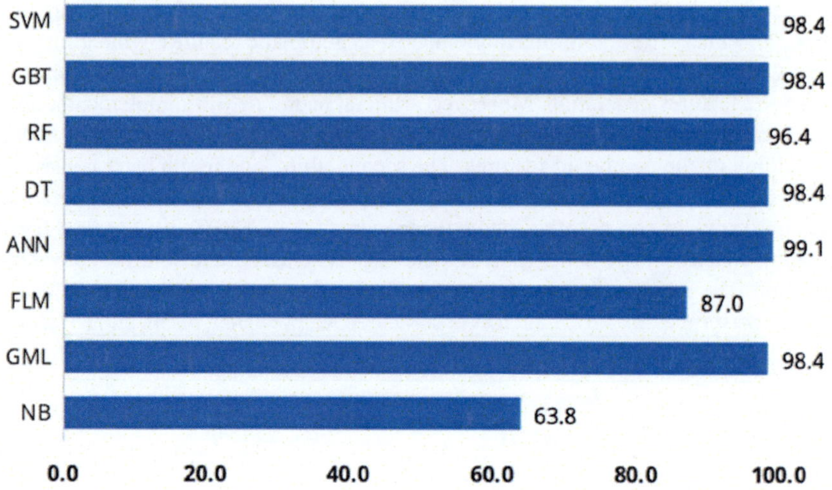

Fig. 2 Best accuracy all models

Moreover, the model trained by decision tree achieved relatively high performance with a 98.4 score in accuracy and F1-score, respectively. The precision-recall score was 99.3 and 97.4, respectively. The AUC score achieved by the decision tree was 98.4. The model trained by random forest achieved 96.4 and 96.2 scores in accuracy and F1-score, respectively. The precision-recall score was 97.0 and 95.5, respectively. The AUC score achieved by the random forest was 97.6. Another tree-based algorithm, the gradient boosted tree algorithm, also provides relatively high performance. The model trained by gradient boosted tree achieved 96.4 and 96.2 scores in accuracy and F1-score, respectively. The precision-recall score was 97.0 and 95.5, respectively. The AUC score achieved by the gradient boosted tree was 97.6. These results were relatively similar to the model trained with random forest. Furthermore, the model trained by support vector machine achieved relatively high performance with a 98.4 and 96.2 score in accuracy and F1-score, respectively. The precision-recall score was 97.0 and 95.5, respectively. The AUC score achieved by the decision tree was 97.6. Figures 2 and 3 demonstrate the overview of the best accuracy and F1-score for all the models. The model trained with artificial neural network achieved the best and the most balanced performance, with a score of 99.1 for both accuracy and F1-score. The model trained with Naïve Bayes achieved the worst performance as well as the most imbalanced performance amongst all the models, with a score of 63.8 and 40.3 for accuracy and F1-score. Figure 4 shows the best training time for all models in minutes. The fastest training time was the support vector machine algorithm, with 12.55 minutes of training time for all training data, whilst the Naïve Bayes algorithms provided the longest training time, with 36.77 minutes of training time for all training data. Moreover, whilst artificial neural network provided the best performance in-term of accuracy and F1-score, the algorithm required quite a large

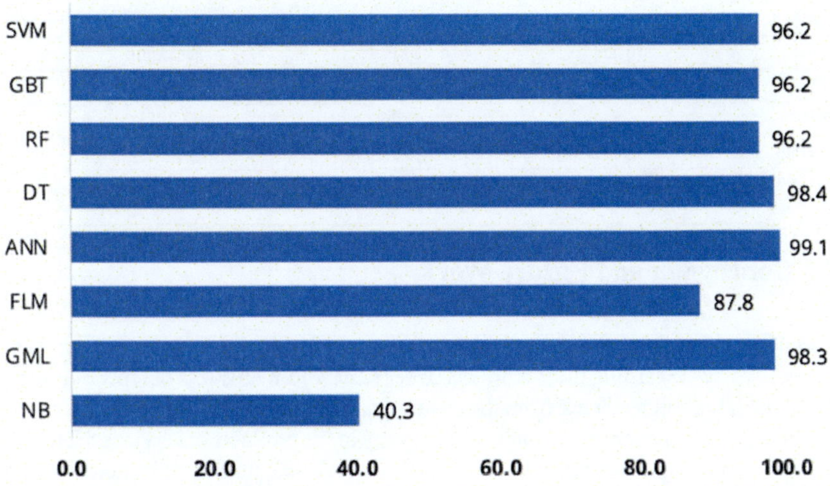

Fig. 3 Best F1-score all models

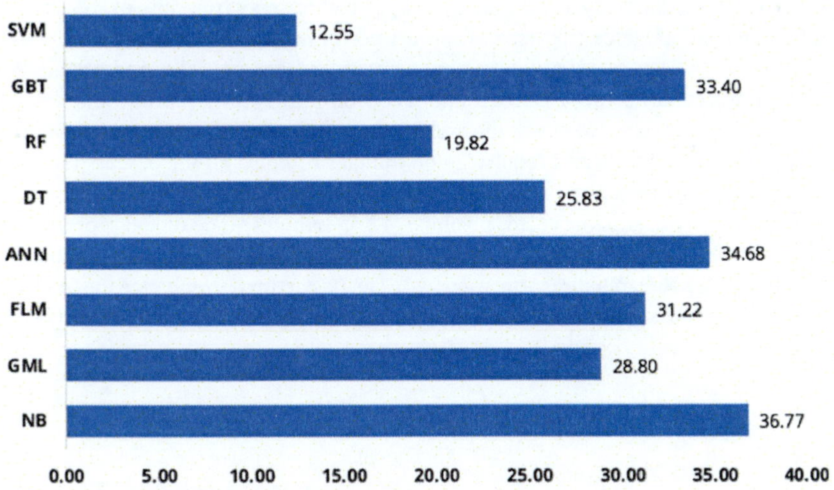

Fig. 4 Best training time all models in minutes

number of resources in the training time, with 34.68 minutes of training time for all training data. Finally, Table 2 illustrates the confusion matrix of the best model. The model was trained with five layers of the artificial neural network architecture. The precision and recall for each class were quite balanced for the model.

Table 2 ANN result

	TN	TP	Precision
PN	798	9	98.84%
PP	5	726	99.32%
Recall	99.38%	98.78%	

5 Conclusion and Future Work

Product reviews from the customers are important for the business to capture customers' voices regarding their products or services. From business perspective, capturing insights and feedback from the customers will improve the customers' experiences leading to the improvement to their loyalty to the products or services. Hence, it is important to capture the customers' voices in a real-time manner to be able to make a real-time decision. Hence, this research proposes an automatic sentiment analysis from product review. Several machine learning algorithms have been explored to model sentiment analysis on Indonesian product reviews from one of the largest Indonesian e-commerce platform. The best performances acheived by the model trained with five layers of artificial neural network architecture. The model achieved 99.1% for accuracy score, F1-score, and precision-recall score; moreover, it scored 99.6% for AUC score. However, the architecture requires a relatively massive number of resources to train and require 34.68 minutes to finish all the process (second slowest training time). The fastest architecture for training time was achieved by the model trained with support vector machine (12.55 minutes). The model proposed in this research has disadvantage, and the model was trained with limited number of data. For future research direction, more data can be captured to enhance the training results. More variation in the review topics and products can be added to the model.

References

1. Aziz AA, Starkey A (2019) Predicting supervise machine learning performances for sentiment analysis using contextual-based approaches. IEEE Access 8:17722–17733
2. Chowanda A, Chowanda AD (2017) Recurrent neural network to deep learn conversation in indonesian. Proc Comp Sci 116:579–586
3. Chowanda A, Chowanda AD (2018) Generative indonesian conversation model using recurrent neural network with attention mechanism. Proc Comp Sci 135:433–440
4. Chowanda A, Muliono Y (2022) Indonesian sentiment analysis model from social media by stacking bert and bi-lstm. In: 2022 3rd international conference on artificial intelligence and data sciences (AiDAS). IEEE, pp 278–282
5. Chowanda A, Sutoyo R, Tanachutiwat S et al (2021) Exploring text-based emotions recognition machine learning techniques on social media conversation. Proc Comp Sci 179:821–828
6. Dake DK, Gyimah E (2022) Using sentiment analysis to evaluate qualitative students' responses. Educ Inf Technol 1–19

7. Fitri VA, Andreswari R, Hasibuan MA (2019) Sentiment analysis of social media twitter with case of anti-lgbt campaign in indonesia using naïve bayes, decision tree, and random forest algorithm. Proc Comp Sci 161:765–772
8. Kalarani P, Selva Brunda S (2019) Sentiment analysis by pos and joint sentiment topic features using svm and ann. Soft Comput 23(16):7067–7079
9. Koukaras P, Nousi C, Tjortjis C (2022) Stock market prediction using microblogging sentiment analysis and machine learning. In: Telecommunication, vol 3. MDPI, pp 358–378
10. Leelawat N, Jariyapongpaiboon S, Promjun A, Boonyarak S, Saengtabtim K, Laosunthara A, Yudha AK, Tang J (2022) Twitter data sentiment analysis of tourism in Thailand during the Covid-19 pandemic using machine learning. Heliyon 8(10):e10894
11. Machova K, Mach M, Vasilko M (2022) Comparison of machine learning and sentiment analysis in detection of suspicious online reviewers on different type of data. Sensors 22(1):155
12. Pavitha N, Pungliya V, Raut A, Bhonsle R, Purohit A, Patel A, Shashidhar R (2022) Movie recommendation and sentiment analysis using machine learning. Glob Trans Proc 3(1):279–284
13. Rani S, Kumar P (2019) Deep learning based sentiment analysis using convolution neural network. Arabian J Sci Eng 44(4):3305–3314
14. Rodrigues AP, Fernandes R, Shetty A, Lakshmanna K, Shafi RM et al (2022) Real-time twitter spam detection and sentiment analysis using machine learning and deep learning techniques. Comput Intell Neurosci
15. Sutoyo R, Achmad S, Chowanda A, Andangsari EW, Isa SM (2022) Prdect-id: Indonesian product reviews dataset for emotions classification tasks. Data in Brief 44:108554
16. Xia H, Yang Y, Pan X, Zhang Z, An W (2020) Sentiment analysis for online reviews using conditional random fields and support vector machines. Electron Comm Res 20(2):343–360

Moving Pictures Feature Extraction Using Scheme Pixel Values Method

D. Saravanan and Dennis Joseph

Abstract Technology allows users to store a variety of datasets. Among these datasets, retrieving the correct source from the vast respiratory system is one of the most critical tasks. This extraction process is used in various applications in the engineering, medicine, and science domains. Extracting the image attributes from the stored respiratory is one of the more complex processes. Due to the intricacy of the input set, many scholars find these techniques among the most difficult undertakings. The process is the consolidation of various other attribute sets. From this combination, extracting the specific domain or specific patterns is more difficult for the investigators. The work is made considerably more complicated inputs like video sets. The proposed work finds the best solution for extracting image datasets using image attributes as an input. Investigational results prove that the suggested system will work with various types of inputs.

Keywords Object mining · Clustering · Image categorizations · Data mining · Image extraction · Knowledge extraction · Image segmentation · Video shots

1 Introduction

Digital video has rapidly become a significant source for engineering, medicine, and science domains. From this huge repository, extracting the specific knowledge, examining the content, grouping the information based on the image attribute or image properties, labeling the extracted image set, and creating an indexation are not an easy job for the researcher. It requires extraordinary attention and requires domain knowledge. This facilitates faster information extraction for the user [1, 2]. Today, with the increasing demand for image data analysis, extracting attribute-related information is one of the more complex operations. Lots of investigators

D. Saravanan (✉) · D. Joseph
Faculty of Operations and IT, The ICFAI Foundation for Higher Education (IFHE) (Deemed to be University U/S 3 of the UGC Act 1956), ICFAI Business School (IBS), Hyderabad, India
e-mail: Sa_roin@yahoo.com

G. Rajakumar et al. (eds.), *Intelligent Communication Technologies and Virtual Mobile Networks*, Lecture Notes on Data Engineering and Communications Technologies 171,
https://doi.org/10.1007/978-981-99-1767-9_54

also highlight object-based extraction, and a few of them deal with object extraction directly. Extracting image-based content is the most challenging area for many researchers. Existing procedures may work with limited sets of input or limited types of datasets. It motivates the researchers to develop a new methodology for extracting the needed content effectively. Extracting image content from image respiratory has been one of the fastest-growing research domains for the past couple of years. While there are many effective techniques in this domain, none work with all image files. Creating the tags for segmented images is one of the significant tasks in object mining [3, 4]. This task is too complex when your input datasets are not arranged properly. In image extraction, creating a tag for video images is one of the important tasks. These tags are created based on the picture properties like text, pixels, time intervals between the frames, and more. In particular, this task is complex for unstructured video datasets. This task is too complex in its global surroundings. This type of environment consists of a huge number of cameras. From each camera, various types of videos are captured. It will increase the time. Due to this, creating tags in this type of environment is difficult to compare with a single camera and has less time duration [5]. In ubiquitous surroundings, finding a particular event or shot is difficult for researchers [6].

2 Pre-request of a Video Extraction

Any video data extraction requires pre-requesting steps they are:

1. This is processed without any domain knowledge.
2. Researchers do not have any clue about the input dataset.
3. It should discover stimulating measures.

The above pre-requesting steps step 2 and step 3 are highly correlated. In step 2, whenever the user wants to extract any exciting knowledge from the video dataset, he or she does not have prior knowledge about the domain or the process [7, 9]. It brings new knowledge for every iteration. Based on the extraction, users use it to uncover knowledge or remove data from the process.

3 Associated Efforts in Video Data Mining

Many of the existing video retrieval investigations are done based on a particular image attribute or a particular event. In this process, additional supporting details such as frame position, pixel values, and other visual features such as text is used for extracting the particular video [9]. For example, if a user wants to extract a specific event or specific shot, respective logo, or any other supporting visual feature, it aids the user in getting rid of necessary content.

3.1 Method of Pretreating for Video Data

Converting the moving file format into static image datasets is the most essential task in video data mining. For converting the dynamic file into a static image, various algorithms are currently available. Most existing techniques work well with certain types of video files only [10, 11]. Based on the use of currents, procedures will be executed. Initially, the process starts with raw video sets. A motion picture is converted into shots, and further, these fragmented shots are converted into frames. Later, applying the data mining preprocessing steps, the moving files are converted into fixed pictures, and duplicate or unwanted shots are removed. It will help the researchers to get the information faster and more precisely.

3.2 Video Segmentation and Assigning Labels

To extract the required information from the archived image database, researchers must produce fixed picture datasets. Most of the existing video data mining methods are based on frame-based or content-based extractions. Similar content is extracted from a stored set using image attributes such as frame pixel, frame average pixel values, the time delay between frames, text positioning in the frame, and other attributes are used. Based on the above qualities, the image threshold used in our existing technique Based on these values, users compare every frame [12]. If the average frame value is more significant than the compared frames means those frames are retained. If the value is less than the average frame value, frames are identified as duplicate frames, and they are removed from the image dataset [13]. After eliminating the duplicate frames, the rest are combined and create a new frame set. This set helps to retrieve the needed content more quickly and accurately. This process is shown in the proposed architecture. Initially, the dynamic video is divided into different schemes based on the users' needs. After segmentation using scheme pixel average value, duplicate or noisy schemes are eliminated. After eliminating the duplicate or noisy frames from the existing frame sets, the reaming frames are collected and stored separately for further processing [14]. Using this set, an input frame or keyframes were selected to retrieve the needed content from the reserved group. This process is shown in Fig. 1. Cartoon frame segmentation is shown in Fig. 2. Cartoon video segmentation process is shown in Fig. 3.

Advantages of the proposed technique:

1. Grouping of information is done more quickly compared to the existing techniques.
2. The system works well in multiple types of video files.
3. The number of outlier elimination is more.
4. The time taken to form the clustering is reduced.
5. More accurate results are obtained.

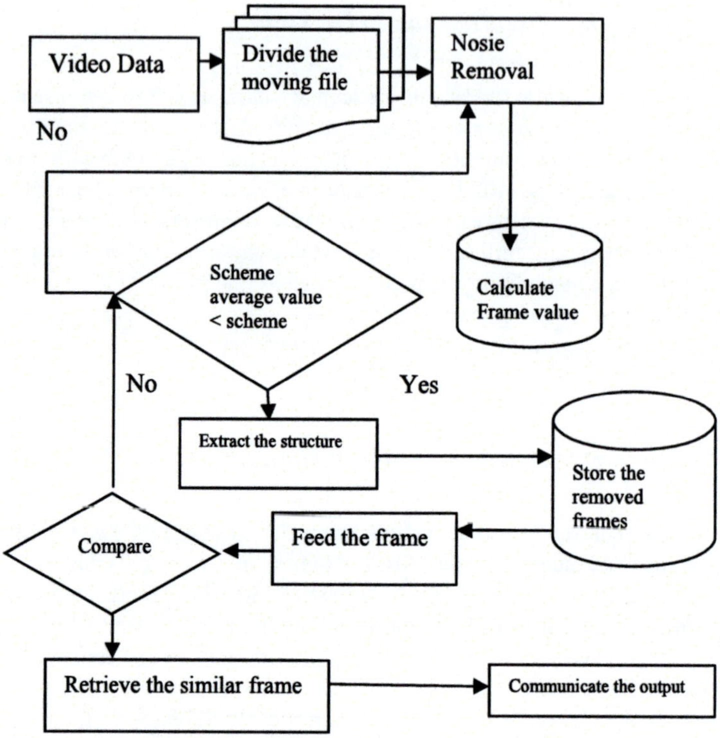

Fig. 1 Proposed feature extraction method

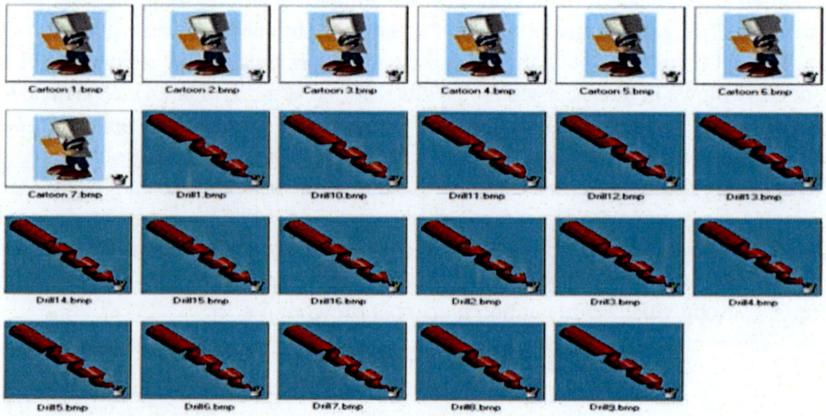

Fig. 2 Cartoon frame segmentation

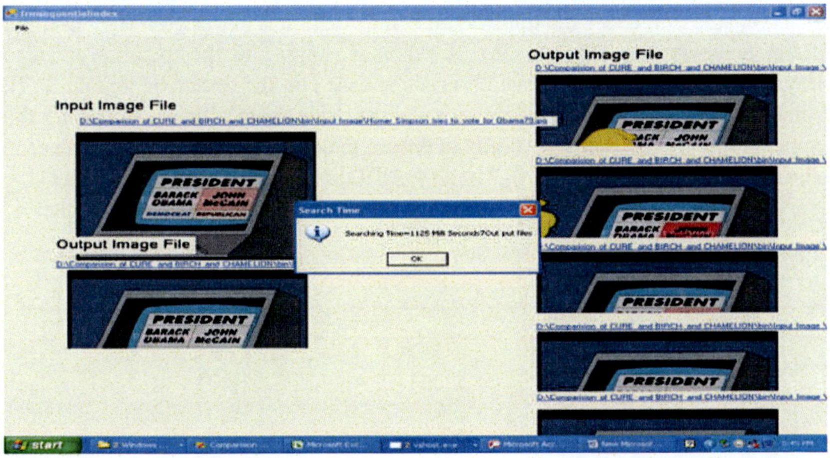

Fig. 3 Cartoon video segmentation process

3.3 Examine the Related Videos

Based on the user's input, the primary function of video data mining is to retrieve relevant or comparable content from the gathered image data collection. This input may be text-based or image-based. Based on the query, the user is expecting a similar type of video set or similar type of image set. User needs to train the picture dataset for this purpose. This process is done before and after the actual data or image extraction starts. The search is based on image properties such as scheme pixel average, frame average, content position, the interval between the frames, and more [15]. Most existing techniques use any of the above image attributes or combinations of those attributes.

1. Using image features, image shots are categorized. This is used as an input frame for extracting relevant video from the image database.
2. The user must specify the most pertinent image from the output, i.e., user needs to create a rank among the shots. Further, the search was refined with users' input.

3.4 Image Decomposition

Segmentation is the process of decomposing dynamic video into static images with an equal number of shots [16]. Dynamic videos are converted into static videos. Each shot is assigned a label or creates an object identifier. Image pixel values are calculated using the image threshold technique to picture each shot or each object these values are calculated and kept individually. This procedure was repeated on the server side too. Using the trained value, information is extracted from the stored dataset. Segmentation supports the researchers in simplifying the operation and also brings the most relevant information or shots based on the researcher's input [17]. It also helps the user to perform the image analysis more efficiently. During the segmentation process, the user needs to ensure the quality of the video input needs to be maintained. In Figs. 4, 5, 6, 7 and 8, and Table 1, this procedure is depicted.

Multiple video files performance is shown in Fig. 9. Image count versus interval of sport is shown in Fig. 10.

id	frame	numclus	time
1	0	35204	3:17:55 PM
2	1	35175	3:18:01 PM
3	2	29162	3:18:07 PM
4	3	29762	3:18:14 PM
5	4	30475	3:18:20 PM
6	5	30975	3:18:26 PM
7	6	31351	3:18:33 PM
8	7	31867	3:18:39 PM
9	8	32339	3:18:45 PM
10	9	32427	3:18:52 PM
11	10	32482	3:18:58 PM
12	11	32440	3:19:04 PM
13	12	32842	3:19:11 PM
14	13	32964	3:19:17 PM
(AutoNumber)	0		

Fig Cluster formation vs time is ms sec

fname	value
D:\Video1\Video Frame Comparision Using CURE Algorithm\bin\Frames \145.bmp	17320874
D:\Video1\Video Frame Comparision Using CURE Algorithm\bin\Frames \146.bmp	17964267
D:\Video1\Video Frame Comparision Using CURE Algorithm\bin\Frames \147.bmp	18608012
D:\Video1\Video Frame Comparision Using CURE Algorithm\bin\Frames \148.bmp	19007717
D:\Video1\Video Frame Comparision Using CURE Algorithm\bin\Frames \149.bmp	19422327
D:\Video1\Video Frame Comparision Using CURE Algorithm\bin\Frames \150.bmp	19898512
D:\Video1\Video Frame Comparision Using CURE Algorithm\bin\Frames \151.bmp	20300463
D:\Video1\Video Frame Comparision Using CURE Algorithm\bin\Frames \152.bmp	20552418
D:\Video1\Video Frame Comparision Using CURE Algorithm\bin\Frames \153.bmp	21017728
D:\Video1\Video Frame Comparision Using CURE Algorithm\bin\Frames \154.bmp	21313259
D:\Video1\Video Frame Comparision Using CURE Algorithm\bin\Frames \155.bmp	21753461
D:\Video1\Video Frame Comparision Using CURE Algorithm\bin\Frames \156.bmp	22026854
D:\Video1\Video Frame Comparision Using CURE Algorithm\bin\Frames \157.bmp	22240212
D:\Video1\Video Frame Comparision Using CURE Algorithm\bin\Frames \158.bmp	22205344
D:\Video1\Video Frame Comparision Using CURE Algorithm\bin\Frames \159.bmp	22329771
	0

Fig. 4 Frame pixel calculation process

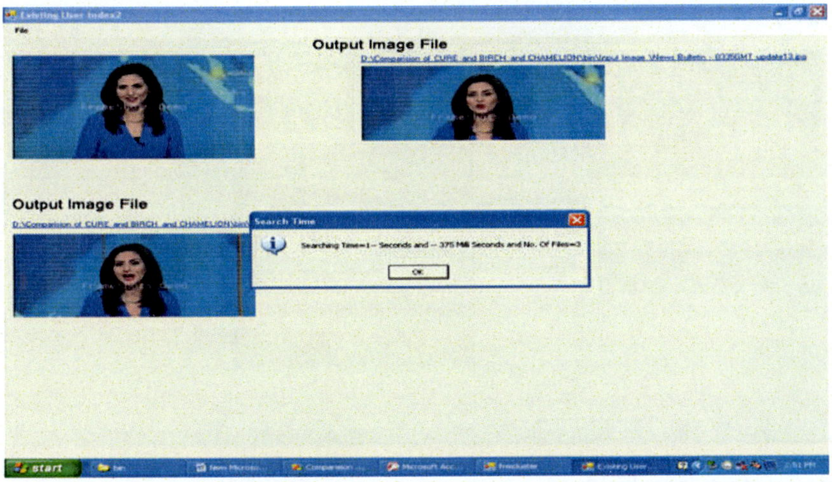

Fig. 5 News video segmentation process

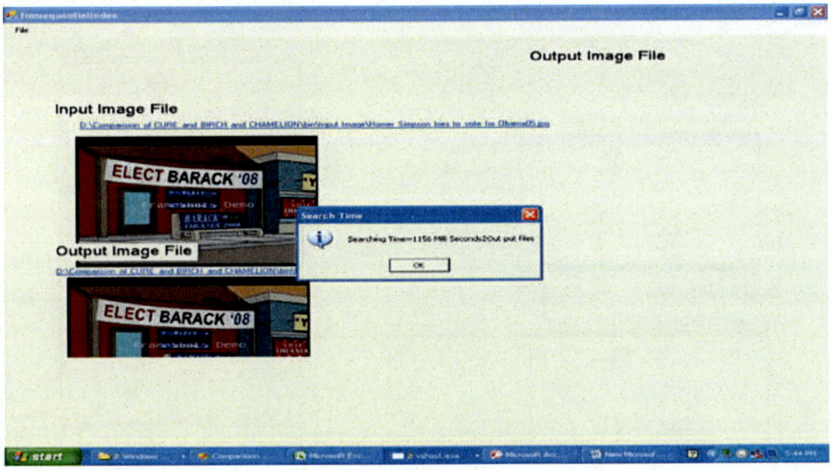

Fig. 6 Comic strip segmentation process

4 Conclusions

The proposed work gives a complete understanding of the video segmentation process using image attributes. This paper also presents the effective segmentation of the dynamic image into static images. There are various existing techniques for video segmentation, each of which works well for particular types of videos. The output of each method varies. None of them produces effective results. The proposed technique

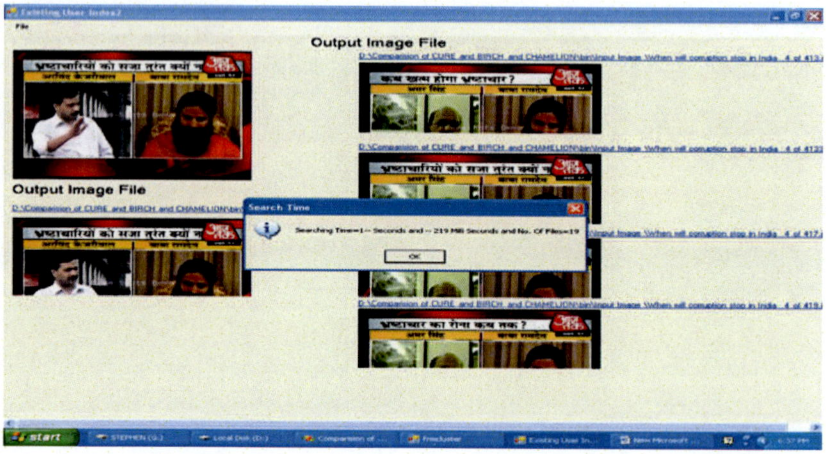

Fig. 7 Convocation segmentation process

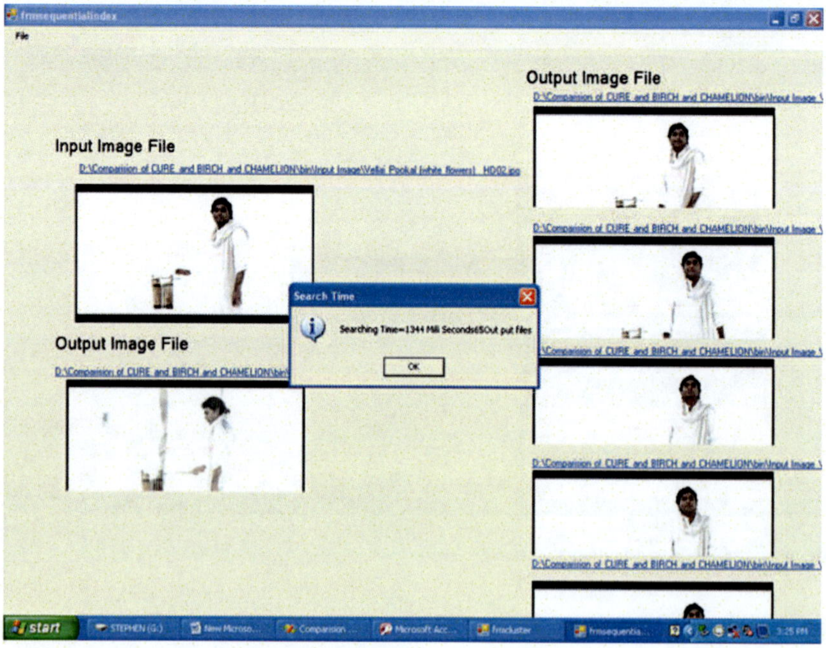

Fig. 8 Dynamic picture segmentation process

Table 1 Frame counts versus duplicate elimination

Video type	Segmented response shots	Segmented productivity shots	Eliminated shots
Comic strip	8	7	1
Visuals	11	11	2
Convocation	10	9	5
Regular site	16	12	3
Dynamic picture	13	10	2

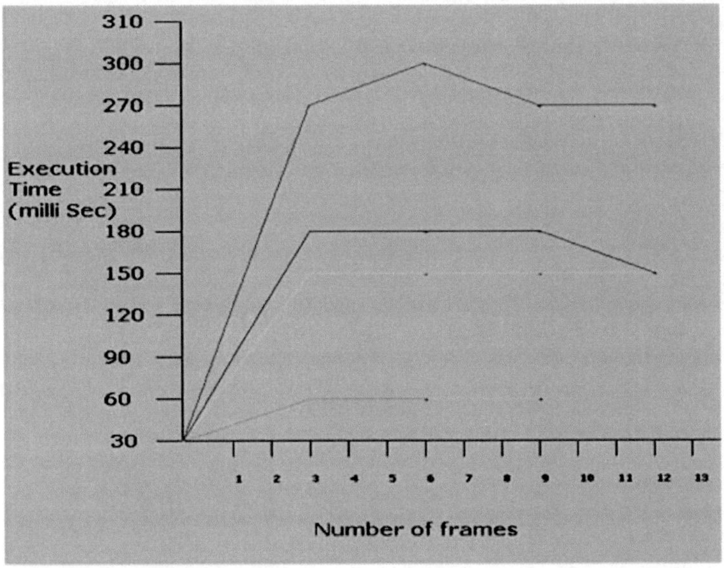

Fig. 9 Multiple video files performance

Fig. 10 Image count versus interval of sport

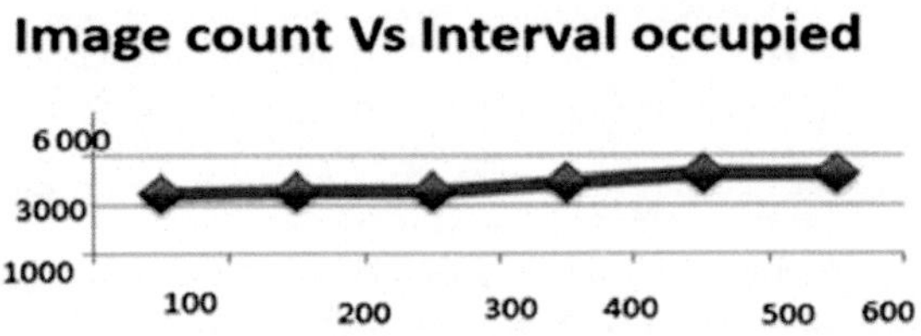

suits all types of image files and produces more effective outcomes than compared to the existing methods. An investigation result verifies this.

References

1. Arijon D (1976) Grammar of the Film Language. Hasting House Publishers, NY. Benitez AB, Rising H, Jrgensen C, Leonardi R, Bugatti A, Hasida K, Mehrotra R, Tekalp AM, Ekin A, Walker T (2002) Semantics of multimedia in MPEG-7. In: IEEE international conference on image processing.
2. Boreczky JS, Wilcox LD (1997) A hidden Markov model framework for video segmentation using audio and image features. In: IEEE international conference on acoustics, speech and signal processing
3. Chang SF, Chen W, Horace H, Sundaram H, Zhong D (1998) A fully automated content based video search engine supporting spatio-temporal queries. IEEE Trans Circ Syst Video Technol 602–615
4. Saravanan D, Vaithyasubaramanian S (2021) Image extraction using image attribute concept technique. Adv Autom Sign Process Instrum Contr Lect Notes Electr Eng vol 700:935–943. ISBN: 978-981-15-8220-2. https://doi.org/10.1007/978-981-15-8221-9_87
5. DeMenthon D, Latecki LJ, Rosenfeld A, Vuilleumier-Stuckelberg M (2000) Relevance ranking of video data using hidden Markov model distances and polygon simplification. In: Advances in visual information systems, VISUAL 2000, pp 49–61
6. Saravanan D, Dennis J (2018) Image data extraction using image similarities. In: Lecture Notes in Electrical Engineering, vol 521, pp 409–420. ISBN: 978-981-13-1905-1
7. Dimitrova N, Agnihotri L, Wei G (2000) Video classification based on HMM using text and faces. In: European conference on signal processing
8. Haering N (1999) A framework for the design of event detections, (Ph.D. thesis). School of Computer Science, University of Central Florida. Haering NC, Qian R, Sezan M (1999) A semantic event detection approach and its application to detecting hunts in wildlife video. IEEE Trans Circ Syst Video Technol
9. Saravanan D (2019) Information retrieval using image attribute possessions. In: Soft computing and signal processing, advances in intelligence systems and computing, vol 898, Springer. pp 759–767. https://doi.org/10.1007/978-981-13-3393-4_77
10. Hampapur A, Gupta A, Horowitz B, Shu CF, Fuller C, Bach J, Gorkani M, Jain R (1997) Virage video engine. In SPIE, storage and retrieval for image and video databases, vol 3022, pp 188–198
11. Saravanan D (2019) Multimedia data retrieval data mining image pixel comparison technique. Lecture notes on Data Engineering and communications Technology 31, Aug 2019, pp 483–489. ISBN 978-3-030-24642-6, Chapter 57
12. Antonie M-L (2001) Application of data minin1g techniques for medical image classification. ACM SIGKDD, San Francisco. USA
13. Saravanan D, Ramesh Kumar A (2013) Content based image retrieval using color histogram. Int J Comp Sci Inform Technol (IJCSIT) 4(2):242–245. ISSN: 0975-9646
14. Saravanan D, Immanuel Y (2021) Video information recantation using picture characteristic method. In: Journal of physics: conference series, vol 1770, 012024, pp 01–07. IOP Publishing, ISSN: 1742-6588. https://doi.org/10.1088/1742-6596/1770/1/012024
15. Chen S-C, Shvu M-L, Zhang C et al (2001) Multimedia data mining for traffic video sequences. ACM SIGKDD, San Francisco, USA
16. Saravanan D, Somasundaram V (2014) Matrix based sequential indexing technique for video data mining. J Theor Appl Inform Technol 67(3)
17. Saravanan D, Vaithyasubramaniyan, Dennis J (2019) Effective utilization of image information using data mining technique. Intell Syst Ref Libr 172(172):207–215. ISBN 978-3-030-32643-2

Multi-model Essay Evaluation with Optical Character Recognition and Plagiarism Detection

Aditya Vijay Kavatage, Shishir Menon, Shreesh Ravindra Devi, Siddhi Patil, and S. Shilpa

Abstract Essay writing is a common component of educational assessments used to evaluate student learning. Automated Text Scoring (ATS) is a collection of statistical and natural language processing techniques used to grade a text automatically on a scoring scale. Grading essays manually is time-consuming, expensive, and prone to inconsistencies. With the current student-to-teacher ratio disparity in the education system, there is an urgent need to develop a grading system. The goal of this project is to automate the essay evaluation process while providing feedback on the quality of work and detecting plagiarism. Students can use it to evaluate themselves. It can be used to evaluate or grade the essay-style responses to a specific question, for which sample answers are provided. The OCR detection and plagiarism checks can also be utilized independently.

Keywords Natural language processing · Automated essay grading · Deep learning · Plagiarism detection · OCR · Machine learning · Neural networks · LSTM · Tokenization

1 Introduction

For institutions with a small number of raters that possess the same domain expertise, grading essays is a lengthy and arduous procedure. A teacher's ability to spend time preparing for class or connecting with students could be compromised if too much time is spent grading essays. Different raters also have diverse personal perspectives and objective biases. Even with consistent grading metrics, the foregoing factors may affect the accuracy of a human-graded product. In addition, several educational institutions offer the service of grading student writings in order to determine their

A. V. Kavatage · S. Menon · S. R. Devi · S. Patil (✉) · S. Shilpa
Department of Computer Science, PES University, Bangalore, India
e-mail: siddhipatil506@gmail.com

S. Shilpa
e-mail: shilpas@pes.edu

G. Rajakumar et al. (eds.), *Intelligent Communication Technologies and Virtual Mobile Networks*, Lecture Notes on Data Engineering and Communications Technologies 171,
https://doi.org/10.1007/978-981-99-1767-9_55

level of writing proficiency. In situations where there is a high volume of essays, human grading becomes tedious and often impossible. If an automated grading system already exists, it might significantly lessen bias and cut down on administrative expenses. It is useful for academic institutions, non-profits, and educators. It is possible for schools to standardize their grading system. As more and more essays are submitted to this grading tool from multiple sources, its accuracy may improve.

The term "automated essay grading system" is used to describe software that assigns a numerical grade to an essay based on one or more machine learning or deep learning models that have been pre-trained. The system's numerical rating for the identical essay should represent the same level of material and quality. Conventional machine learning models rely heavily on the designed characteristics, and it is challenging to identify the most relevant features. To achieve this goal, we suggest switching to a model built on deep learning. As the field of deep learning has advanced, neural networks have become capable of making more precise predictions across a broader range of datasets and pattern types. To be more precise, recurrent neural networks, which can capture the order of the words, produce a very respectable performance on text input data (essays). In the following sections of this article, we will dive into the specifics of the actual implementation.

While initial approaches involved checking for grammatical errors by extracting features from answers such as grammar and vocabulary, in an early model proposed by Yigal Attali and Jill Burstein [1], more modern approaches involved the use of NLP techniques such as the creation of word embeddings [2]. A method that attempts to find similarities between words by converting them into vectors and projecting them over a vector space. Distance between these vectors are an indicator of similarity between the two vectors/words.

The overarching goal of this project is to provide a user-friendly interface for the automated essay grading system, which makes use of the most recent technologies. Each section of this paper is presented in the following order: descriptions of the dataset, its preprocessing, neural network models, user interface, and a few experiments.

2 Related Works

Automated Essay Scoring systems perform much worse when the models used for score prediction are trained on data that is non-objective, i.e., has large inconsistencies due to human error. The authors of [3] proposed a new method to tackle this issue by training AES models using the concept of item response theory-based scores. IRT is a mathematical model that tries to define every recorded score as a function of latent characteristics of the human graders (response) and the features of every record (item). The original dataset does not capture any human rater characteristics, and hence, the study hired human evaluators externally to grade essays from the ASAP dataset. The biases in raters are characterized by: Consistency (how differently were similar essays graded), Severity (how extreme were the assigned scores), and Range restriction (what ranges did raters constrict themselves to). The research made use of the Generalized Many-Facet Rasch Model (GMFRM) which incorporates all three characteristics as opposed to the traditional Rasch model that takes only one into consideration. The results show that this method significantly improved the accuracy of score predictions by traditional AES models. The process is simple and easy to apply to more conventional AES models and can hence be used to significantly enhance performance of existing systems. It effectively improves AES performance and tempers the impact of human biases on the AES models. The proposed methodology is highly acclimatized to educational program applications and low-medium stakes tests where it is difficult to acquire high-quality data. However, this method requires extensive datasets that include rater biases, making acquisition of training data a challenging task. The effectiveness of the proposed methodology also needs to be further tested on more generalized datasets (datasets that cover a wider range of topics). The model also needs to be tested on longer essays that have higher degrees of freedom making accurate score prediction difficult.

In [4], the paper aimed at essay evaluation done with as minimal of a training dataset as possible. It makes use of cognitive information—gaze behavior to grade essays. In broad terms, gaze behavior refers to areas of interest (IA). It finds words and phrases that hold more value over others. The model is based on convolutional neural networks and makes use of LSTM layers as well. As for the dataset, they make use of ASAP's AEG dataset which can be found on Kaggle. The model trains data based on prompts—source prompts for training and target prompts for testing. Cohen's Kappa with Quadratic Weights (QWK) is used as an evaluation metric to get accuracy. While this approach requires a rather small and unlabeled dataset, it has a very low accuracy and does not perform as well for longer essays.

The goal of [5] was to create a system (AutoSAS) that evaluates short answers. The dataset (ASAP-SAS) was augmented by including jumbled content among the normal responses. While training, one low-rated response was included among many high-rated responses to ensure that irrelevant answers were penalized. Conversely, one high-rated response was included among several low-rated responses to make sure that badly written answers were not highly rated. The paper also trains a random forest model on all the extracted features. This model was chosen as it performs well with multiple features and secondly, so that the order of importance of each feature is known (since it is not possible to do so with neural networks). The implemented system utilizes NLP techniques, and the model used supports features such as lexical diversity, Word2Vec, Doc2Vec, prompts, weighted keywords, and content overlap. The system, AutoSAS, was tested on the ASAP-SAS dataset which resulted in a mean QWK score of 0.79. Implements regression analysis and the proposed model is a supervised regression model. The dataset used is a publicly available dataset called ASAP-SAS.

In [6] ontologies were introduced. It can be noted that an ontology is nothing but a formal and explicit specification of a shared conceptualization. Ontologies are usually domain specific, because information extraction is primarily concerned with collecting information for a certain domain, using an ontology to formally and explicitly explain the domain's ideas might be beneficial to the process. In this paper [6], domain knowledge is leveraged through ontologies. The procedure employed for scoring essays relies on similarity matching, i.e., comparing the responses from students against the teachers marking scheme directly. It employs the Wu and Palmer algorithm for semantic similarity, which is based on shortest distance between synets along with their hypernyms. It further extracts information from the related ontology and compares it with the user's response and assigns scores based on degree of matching. The system is divided into three layers, namely—Presentation layer (user input and interaction), Business Logic Layer (System Functionality), and Data Layer (Database). The model has an accuracy of about 80% while also providing feedback to the users. However, the accuracy scales inversely with the length of the essay leading to longer essays having a weaker score.

This particular work [7] incorporates TSLF—two-stage learning framework—aiming to combine the benefits of feature engineering as well as end-to-end models. It assigns scores based on Semantics, Coherence, and prompt-relevance. Two stages are employed—in the initial stage, a score based on semantic data is given to the essay. The following stage of the model emphasizes hand-crafted features like essay length and grammar correction. Herein, the three scores along with hand-crafted features are fed to a XGBoost classifier for further training. While adversarial examples further boost grading which helps in preventing cheating that may be underway through individuals that are familiar with the AES systems, this method worked only for five prompts out of eight and the hand-crafted features used here tend to be difficult to scale.

3 Dataset

The Hewlett Foundations' Automated Student Assessment Prize dataset [8] was used to train our models. There are a total of eight groups of essays, each of which was written in response to a different prompt. Middle school students from grades 7–10 wrote each essay in English. In total, there are approximately 17,450 essays (with an average of over 300 words each response) and nearly 3000 essays (for each prompt). These essays have been graded by three different people to exclude any possibility of bias, and the resulting scores fall anywhere from 0 (very low) to 60 (very high). The details for the scoring mechanism per prompt are given in Table 1.

4 Proposed Methodology

Our approach is split into three parts: OCR, plagiarism, and AES Model. System architecture is shown in Fig. 1.

Table 1 Dataset description

Prompt	Number of essays	Scores
1	1783	2–12
2	1800	1–6
3	1726	0–3
4	1772	0–3
5	1805	0–4
6	1800	0–4
7	1569	0–30
8	723	0–60

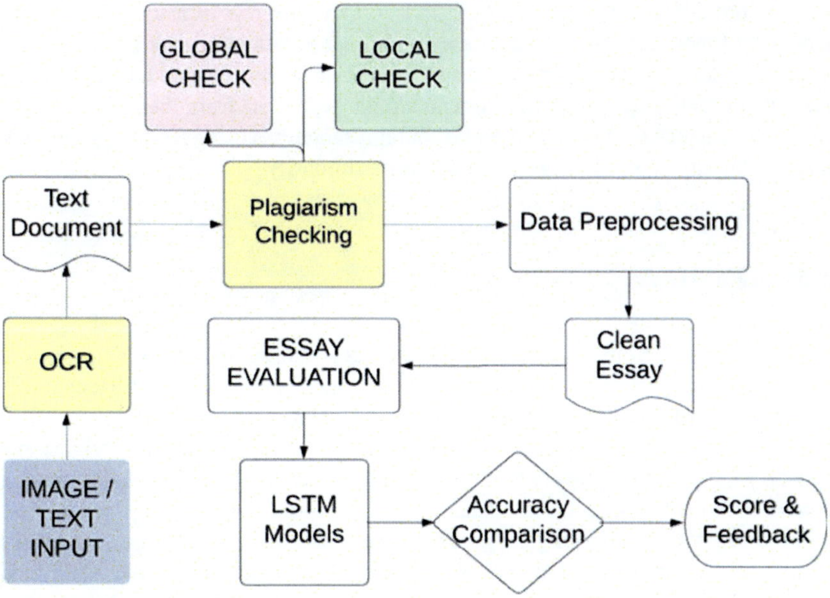

Fig. 1 System architecture

4.1 *Optical Character Recognition*

We have integrated the feature for ease of use. It allows the user to extract text from an image. This enables the user (student) to upload a screenshot/image of their essay instead. To implement this feature, we have built upon the open-source engine Tesseract developed by Google. The image provided by the user is preprocessed before passing to Pytesseract [9]. The preprocessing is done using OpenCV library [10] and includes steps such as de-skewing, converting to grayscale, and noise removal. De-skewing—rotating the image to make it vertical. Noise removal/reduction—the process of removing unwanted/useless information present in the image. This involves converting it to grayscale and other image denoising methods implemented by OpenCV.

4.2 Plagiarism Detection

Detection of plagiarism is an important aspect of evaluation to assess the students' learning capability. There exists two layers of detection—local and global [11]. The data is first preprocessed before it is fed into our model. Preprocessing involves word tokenization, stop word removal, lemmatization, and stemming. The local plagiarism checker compares a given essay with all the previous submissions made for that specific prompt. It converts all submissions to its vector representation. The cosine similarity function is then used to assign an appropriate similarity score. For the global check, the essay submitted is split apart into sentences, and a Google query is made to find articles that may contain similar text. Text from these links is then extracted and then converted to their vector representations using HashingVectorizer, and a similarity score is assigned using cosine similarity.

4.3 AES Model

Preprocessing—before the data is fed to the model for training, the text data needs to be converted into a format that the computer can understand. We used NLP techniques such as stop word removal, lowercasing, special character removal, lemmatization, and tokenization of the essay. This was accomplished using NLTK libraries and functions. The next step involves removing the tagged labels and then finally forming a word list by tokenizing the sentence.

Feature generation—using the generated word list, we create feature vectors. We make use of the pre-trained model, Word2Vec [12] to load vector embeddings and convert the cleaned text into numeric vector representations. Word2Vec helps create vector representations for words such that words that are similar point toward the same direction. The vectors (sentences) are fed to the LSTM layer, wherein each word is treated as a separate step. The LSTM model learns the parameters that best define the relation between the training essays and their respective human scores. We make use of the ADAptive Moment estimation (Adam) optimizer to fine tune the model and find the optimal values of hyperparameters to minimize the loss function in predicted scores and increase accuracy. We used the Mean Squared Error (MSE) metric as our loss function given in Eq. (1)

$$MSE = \frac{1}{n}\sum_{i=1}^{n}(\gamma_i - \widehat{\gamma_i})^2, \tag{1}$$

where

$n =$ the number of observations.

y_i is the ith observed value.

$\widehat{y_i}$ is the corresponding predicted value.

Deep learning approach—Since simply employing statistical methods for essay evaluation can fail when the relationship between the input and output is complex and nonlinear, it is better to utilize deep learning models, such as LSTMs, as they are better equipped to handle the complex relationships and dependencies between words in an essay. By capturing and retaining context information over a longer sequence, LSTMs are able to understand the meaning of an essay and evaluate its quality in a more effective manner. Recurrent Neural Networks (RNNs) [13] are a type of neural network that uses outputs generated in previous steps to process the current inputs. They are capable of processing variable length sequential data and hence suitable for the task of understanding linguistic compositionality. Conventional RNNs face the problem of vanishing gradients and hence are unable to retain information over a longer range. The solution to this is the LSTM network [14] which is a specialized RNN architecture that augments RNN to have "long term memory". It also provides a greater degree of control over context that is to be carried forward, with the help of gates. An LSTM unit consists of three gates: forget (to decide what information to consider from the previous step), input (considers current input), and output (to decide what information has to be sent to the next step). Neural network—below Fig. 2 shows one of the neural network architectures:

- The first layer is an LSTM layer with 300 units and a dropout rate of 0.4 for both the input and recurrent connections. The input to this layer will have the shape of (batch_size, 1300).

```
Model: "sequential_4"
_________________________________________________________________
 Layer (type)                Output Shape              Param #
=================================================================
 lstm_8 (LSTM)               (None, 1, 300)            721200

 lstm_9 (LSTM)               (None, 64)                93440

 dropout_4 (Dropout)         (None, 64)                0

 dense_4 (Dense)             (None, 1)                 65

=================================================================
Total params: 814,705
Trainable params: 814,705
Non-trainable params: 0
```

Fig. 2 Neural network architecture

- The second layer is another LSTM layer with 64 units and a recurrent dropout rate of 0.4.
- The third layer is a dropout layer with a dropout rate of 0.5, which helps to prevent overfitting by randomly "dropping out" (ignoring) some of the units during training.
- The last layer is a dense layer with 1 unit and a Rectified Linear Unit (ReLU) activation function. This is the output layer of the model.
- The model is then compiled with a loss function of "mean_squared_error", an optimizer of "adam", and metrics of "mae" (Mean Absolute Error).

The first layer is the input layer that ingests the cleaned essay text data. The second layer is the embedding layer that processes the textual data with the pre-trained embedding model. The third layer is the long short-term memory layer with all weighted parameters. The fourth layer is adding other input parameters like grade level and essay type. The fifth layer connects all information together and trains data as a whole. The dropout layer and activation layer were added for further improvement and to remove overfitting of data.

This architecture is typically used for sequential data, like a time series or text data, where the order of the input data is important and the model needs to take into account the context of the data.

Other models—to further improve our accuracy, we introduced a Convolutional Neural Network (CNN) layer [15]. CNNs generally are useful in NLP tasks as they can identify localized spatial relations between words. We make use of this layer to identify the features in an essay that contribute the most to accurate score prediction. Its architecture includes a layer that extracts features from input sequence, followed by an LSTM layer and an output layer to make final predictions. As a variation of the LSTM model, we made use of bidirectional LSTMs. LSTMs only capture context from left to right, whereas bidirectional LSTMs consider it in both directions. This model constitutes an input layer to take in the input sequence, two LSTM layers, one for each direction, and a concatenation layer to combine outputs from both layers. The model is then able to consider the relevance of the words preceding the current word as well as the ones that follow it. This greatly improves the performance of the model but comes at the cost of requiring increased computational power.

Performance—after training all the models (LSTM, bidirectional LSTM, and CNN-LSTM), we selected the one with the highest QWK score. The QWK score is calculated by first creating a weighted matrix W using Eqs. (2) and (3).

$$\omega_{i,j} = \frac{(i-j)^2}{(N-1)^2}, \tag{2}$$

where

$\omega_{i,j}$ is the weight determined by the actual and predicted values.
N is the dimension of the matrix.
i is the actual and j is the predicted.

$$\kappa = 1 - \frac{\Sigma_{i,j} W_{i,j} O_{i,j}}{\Sigma_{i,j} W_{i,j} E_{i,j}}, \tag{3}$$

where

κ is the QWK score measuring agreement of outcomes.

$W_{i,j}$ is the weight determined by the actual and predicted values.

$O_{i,j}$ is observed proportion of actual value i by predicted value j.

$E_{i,j}$ is expected proportion of actual value i by predicted value j.

Overfitting—in order to ensure that our models do not over fit the training dataset, we make use of a dropout layer. Dropout layers work by ignoring nodes in the network randomly in order to temporarily create a new network architecture. This method improves internal workings of each individual node, thus making them more independent while decreasing bias of the dominant nodes. As our model consists of recurrent layers, we have made use of recurrent dropout which works between timesteps as compared to just between neurons. As we trained the model, we monitored it for its performance and to find the optimal number of epochs after which the model stabilizes.

5 Results and Discussion

The main goal was to create an essay grading model that could analyze a particular piece of text and based on its content and structuring detect its possible score. To increase ease of grading, we incorporated optical character recognition to extract textual data from student uploaded files. And because essays are generally subjected to plagiarism, in order to avoid grading copied content, we provided feedback in terms of level of plagiarism within your dataset of uploaded essays as well as across the internet. To develop the grading model, since statistical measures such as Term Frequency-Inverse Document Frequency (TF-IDF), Bag-of-Words, and several machine learning approaches proved to be insufficient with respect to deriving semantic content from the essay, we turned to deep learning methods.

To do so, we employed long short-term memory models also known as LSTM's of three kinds, the feedforward LSTM, bidirectional LSTM, as well as convolutional LSTM, and compared their accuracies. Although we have used three different neural network architectures, our accuracy has mostly remained consistent throughout the various models. The performance has been calculated using Quadratic Weighted Kappa (QWK) which measures rater agreement. From the below figures, we can conclude that the bidirectional LSTM performed the best out of the three, since it understands context in both directions. In the case of CNN-LSTM, we got similar results as compared to the basic LSTM model, but the training was less resource intensive and took less time due to reduction in dimensionality of the inputs by the CNN layer. Every model was trained over a specific number of epochs, in our case 50, after which the model performance stops improving on the validation data. A larger dataset will lead to better performance on unseen essays as a greater knowledge base

Table 2 Model performances

	LSTM	BLSTM	CNN_LSTM
QWK	0.95854	0.96328	0.95894

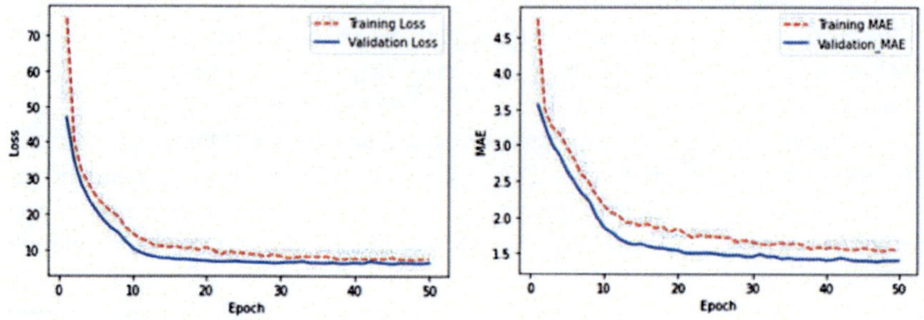

Fig. 3 Training and validation loss comparisons for LSTM

would greatly improve the training model, however since the current dataset only provides 1600 essays per prompt which is not sufficient to extract and understand all features. Unfortunately, getting a bigger dataset is difficult and time intensive as such a dataset would require highly qualified human evaluators to grade and the manual grading is also dependent on individual biases in rating. Model performances are shown in Table 2.

The Fig. 3 represents the relationship between the training loss, a metric used to assess how the model fits the training data over the epochs, a hyperparameter that defines the number of times that the model will work through the entire training data, which in our case is 50. The decline in loss indicates that the amount of errors is reducing exponentially over the course of 50 epochs on our training set.

6 Conclusion

Automatic essay evaluation is a rapidly growing field that has the potential to revolutionize the way writing is taught and assessed. By using natural language processing and machine learning techniques, automatic essay evaluation systems can provide detailed and objective feedback on written work. They can also help to alleviate the burden of grading on instructors and allow for more efficient and consistent assessment of student writing. However, it is important to note that these systems are not perfect and should be used in conjunction with human evaluation to ensure the most accurate and fair assessment of student writing. Latest approaches to the problem use deep learning techniques with models such as LSTMs, and these models traverse through the essay and remember the previous context and future expected words/features. We have discussed a method for developing an AES system that included the usage of LSTMs with plagiarism scoring as an additional feature to indicate authenticity which would aid teachers while grading. Variations are to be expected as we increase the number of training prompts. However, the data reveals that the QWK and MAE metrics have consistent values. The Kappa value remains in the range of 0.95–0.96, thus proving that the LSTM models give a satisfactory performance.

Additionally, more research is needed to fully understand the limitations and potential biases of these systems and to continue to improve their performance. Overall, automatic essay evaluation has the potential to greatly benefit the field of education and improve the writing skills of students.

References

1. Attali Y, Burstein J (2006) Automated essay scoring with e-rater® V.2. J Tech Learn Asses 4(3)
2. Young T, Hazarika D, Poria S, Cambria E (2018) Recent trends in deep learning based natural language processing [review article]. IEEE Comput Intell Mag 13(3):55–75. https://doi.org/10.1109/MCI.2018.2840738
3. Ajetunmobi SA, Daramola O (2017) Ontology-based information extraction for subject-focussed automatic essay evaluation. In: 2017 International Conference on Computing Networking and Informatics (ICCNI), pp 1–6. https://doi.org/10.1109/ICCNI.2017.8123781
4. Liu J, Xu Y, Zhao L (2019) Automated essay scoring based on two-stage learning. ArXiv, abs/1901.07744
5. Mathias S, Murthy R, Kanojia D, Bhattacharyya P (2021) Cognitively aided zero-shot automatic essay grading
6. Uto M, Okano M (2022) Automated essay scoring models trained using item response theory-based scores to decrease effects of rater biases. Electronic ISSN: 1939-1382 CD: 2372-0050
7. Kumar Y, Aggarwal S, Mahata D, Shah RR, Kumaraguru P, Zimmermann R (2019) Get IT scored using AutoSAS—an automated system for scoring short answer system
8. The hewlett foundation: short answer scoring. https://www.kaggle.com/c/asap-sas/data
9. Saoji S, Singh R, Eqbal A (2021) Text recognition and detection from images using Pytesseract. J Interdiscip Cycle Res XIII:1674–1679
10. Jessica SS, et al (2020) Handwritten text detection using OpenCV and CNN. Inter J Eng Res Tech (IJERT)
11. Butakov S, Scherbinin V (2009) The toolbox for local and global plagiarism detection. Comp Educ 52(4):781–788
12. Rong X (2014) word2vec parameter learning explained. arXiv preprint arXiv:1411.2738
13. Sherstinsky A (2020) Fundamentals of recurrent neural network (RNN) and long short-term memory (LSTM) network. Physica D 404:132306
14. Staudemeyer RC, Morris ER (2019) Understanding LSTM—a tutorial into long short-term memory recurrent neural networks. arXiv preprint arXiv:1909.09586
15. Alzubaidi L, Zhang J, Humaidi AJ et al (2021) Review of deep learning: concepts, CNN architectures, challenges, applications, future directions. J Big Data 8:53

OPSUM: An Opinion Summary Generator Model for Customer Feedback on Restaurants

C. Sindhu, Kalluri Shanmukha Sai, Akula Triyan Subramanyan, Lingamaneni Sri Nidesh, and C. Kavitha

Abstract Recently, intense research and development have been conducted in the research area of Artificial Intelligence, to find various techniques to implement this and create more advanced versions of the technology. Emotion Artificial Intelligence or opinion mining happens to be one such technique that provides advancement to Artificial Intelligence. By definition, opinion mining refers to the collection of different opinions or reviews of people with respect to a particular subject. Classification and categorization of these reviews in a consolidated manner of broad categories are known as summarization of these opinions. The representation of this information thus obtained in the forms of models, charts, and graphs is known as visualization of this summary. In the past couple of years, online platforms have increased rapidly. These platforms advanced themselves to not just providing items that can be purchased but also other activities that can be done with the help of the internet such as reservations, financial transactions. The services also include purchasing/booking various kinds of tickets, for traveling or for entertainment, or ordering food using these online portals. In the following work, the various restaurants which deliver food via online portals and their services have been discussed with respect to the feedbacks or opinions received by customers. In this work, the various reviews left by the customers have been summarized and presented and categorized manner classifying them as positive, negative. These have also been visualized for a better understanding and help in comparison to two or more restaurants. The main goal of this work is to ease the difficulty faced by people to decide which place to go and will also increase the efficiency as the time consumed will be reduced as compared to visiting each website and reading the reviews.

C. Sindhu (✉) · K. S. Sai · A. T. Subramanyan · L. S. Nidesh
Department of Computing Technologies, School of Computing, SRM Institute of Science and Technology, Kattankulathur 603203, India
e-mail: sindhucmaa@gmail.com

C. Kavitha
Department of Computer Science and Engineering, Mahendra Engineering College for Women, Tiruchengode, Namakkal, India

G. Rajakumar et al. (eds.), *Intelligent Communication Technologies and Virtual Mobile Networks*, Lecture Notes on Data Engineering and Communications Technologies 171,
https://doi.org/10.1007/978-981-99-1767-9_56

Keywords Sentiment analysis · Sentiment classification · Summary generation · Opinion visualization · Aspect-based summarization · Opinion summarization (OPSUM)

1 Introduction

The study of natural language and speech using computational methods is known as Natural Language Processing (NLP). NLP is sometimes referred to as a computer program's capacity to comprehend spoken human language. Artificial Intelligence (AI) includes the discipline of Natural Language Processing. NLP is a technique which is used for summarization of data. It identifies the words and helps to shorten the complete text. When a business is monitoring online interactions, sentiment analysis, also known as contextual mining of text, finds and extracts subjective information from the source material to assist them understand the social sentiment of their brand, product, or service [1]. Sentiment analysis is frequently referred to as emotion mining or opinion mining. NLP is a notion that is utilized in AI, and it relates to the use of various ways to determine the emotion of a sentence. The challenges encountered while using NLP technique for summarizing documents which have been elaborated in [2, 3] are irony and sarcasm, negations, word ambiguity, and multi-polarity.

Irony and Sarcasm: Irony and sarcasm are another very important challenge involving sentiment analysis. The sentence may be directed toward the negative scale but will be deemed positive by the classifier due to the choice of words. During summarization, it is difficult to identify irony used in the sentences of the datasets used. **Negations:** Phrases such as "not good", "not bad" contain words that contradict each other. It is difficult for the classifier to identify the orientation of the sentence. This can be one of the causes of the lower precision of the algorithm. Stop words are removed before the summarization technique is implemented, and sometimes, the negation words are also removed. **Word Ambiguity**: When words or phrases used in a sentence have many meanings, it becomes confusing for the parser to identify the correct meaning of the sentence, with respect to the context. All of these have been studied deeply in [1–20]. Due to the eradication of stop words, the meaning to the opinion sentences may be misunderstood. **Multi-polarity:** When it comes to multi-polarity, the sentence contains both positive reviews and negative. This confuses the classifier of what the sentence means and when does its orientation lie. Summarization is used to reduce the number of sentences to be read by the user. In all the above challenges, we have found that summarization was the most complex one, so we have worked on summarization elaborately in this work.

Summarization gives the gist of what the whole input paragraph or dataset means. It is a very important technique used in sentiment analysis. The orientation of the summary should be equivalent to the overall result of the dataset. Then, the resultant summary is considered valid. Different types of opinion summarization algorithms are used for processing unstructured review documents and extracting the important features of services or products to analyze the subjectivity or objectivity of the sentences in the reviews. For opinion summary generation, traditional text summarization algorithms could be used. Researchers mostly use the machine learning algorithms (supervised and unsupervised algorithms) for opinion and polarity detection. Some other methods of opinion summarization algorithms aim in handling sparsity constraints and non-negativity constraint. Mainly, the opinion summarization method focuses on strong sentiment text extraction that is either positive or negative. There are basically two tasks in this summarization method, which are (i) sentiment classification and (ii) summary generation. The summary which is generated from this method consists of sentences with either positive or negative polarity by using sentiment analyzers. Poor accuracy, which derives from language comprehension concerns, and scalability problems, which call for far more advanced and complex NLP technology to be handled, are some of the hurdles that are being encountered in opinion summarization.

Visualization is the graphical representation of data and information. Using some of the visual elements like charts, graphs, maps, and data visualization tools provides an accessible way to view and understand trends, outliers, and patterns in data. Some of the data visualization techniques include scatter plots, bar charts, histograms, line graphs, and many more are there. Visualization is effectively used to do a comparative study and to choose the best available option easily. Visualization is a common technique used to represent a large amount of data in a small concise form which can be easily interpreted by the reviewers. The type of visualization used is dependent on the size of data, the end viewers, and the purpose of the visual. Libraries such as matplotlib are used in Python to implement the representation of data. There are various functions included in this library for the different types of charts, plots, and graphs that are used by the coder as and when necessary.

2 Related Work

Opinion mining and summarization comprise areas of Natural Language Processing techniques, feature extraction, data mining, and computational linguistics. There have been a lot of studies on related topics. POS tagging which is an NLP technique using Stanford Parser proposed by [2, 7] is used for preprocessing feature extraction. Multi-Aspect Sentiment (MAS) is a model proposed by [7] to discover contents in customer reviews and extract segments of text that correspond to the relatable aspect to support exponential ratings. The various opinions are required to be summarized and classified into positive and negative polarities. The proposed method finds aspect score (positive, negative, and objective score) of each opinion with the help of SentiWordNet by [9, 12]. The AFINN library used in [7] is considered as the simplest and the most useful lexicons and is used mostly for sentimental analysis.

In [4], NLP and DM techniques are used for review mining and summarization process of Arabic reviews. The proposed study by [8, 10, 11], and [12] uses various machine algorithm techniques which include various classifiers like NB Classifier, Support Vector Machine (SVM), or Maximum Entropy algorithm which are used to map the input data to a specific category. In our work, the Naive Bayes classifier is used. Approaches applied on summarization and visualization are shown in Table 1.

3 Overlooking the Summarization and Visualization System

This paper is to summarize various customer reviews of the restaurant dataset into a simpler and consolidated manner. It then uses SentiWordNet to find the polarity of a word and then the sentence. A sentiment score is generated and clustering is done based on which summarization and visualization of the various reviews are represented. The architecture of this work is given in Fig. 1.

3.1 Input Data

Input data is collected from all possible sources which are appropriate for this research. This module will help in the collection of various customer reviews of a restaurant from the web. The reviews which are extracted are required to be stored in the database, which will then be used as an input to opinion mining techniques. The restaurant reviews' datasets are already available in some of the sites.

Numerous datasets as shown in Table 2 have been considered and used for opinion summarization and visualization in various works. We are using www.kaggle.com/yelp-dataset/yelp-dataset which consists of various customer reviews comprised different aspects about the restaurant.

3.2 Data Preprocessing

Data preprocessing is a significant data mining technique that is used in transforming the raw data into a useful and efficient format [1]. It is required because the real-world data consists of a lot of noisy, incomplete, and inconsistent data. Tasks involved in data preprocessing are data cleaning, data integration, data transformation, data reduction, and data discretization. Data preprocessing helps in getting much better results through the classification algorithms. Besides some of the basic preprocessing techniques, other important techniques that can be used are tokenization, lemmatization, stemming, etc. Among the various preprocessing techniques, in our work, we are using the below-mentioned techniques.

Table 1 Approaches applied on summarization and visualization

Paper Id	Concept	Algorithm	Enhancements	Disadvantages
[4]	Extract features from reviews and summarizes Arabic reviews	Information extraction with sentiment lexicons	Highly relevant measures inaccuracy of both positive and negative summaries are achieved	Previously not many work has been done in extracting feature-based opinion summaries from Arabic
[5]	Multiple opinions of customers are summarized to help users	(1) Neural networks (2) Reinforcement learning	Summarization in languages other than English	Multiple language reviews cannot be summarized
[6]	Understanding the importance of news, a key aspect of tweet, and the interest of people	(1) MDS approximation (2) Summarization methods (3) Approximation algorithms	Identification of relevant tweets, identifying the diversity of opinions of people, the understanding objective of the tweet	Unable to understand the relationship between different objectives
[2]	Recent opinion summarization challenges in social media are considered	(1) POS tagging (2) Machine learning algorithms	More attention is given to TensorFlow, Microsoft Cognitive Toolkit for implementation	Blogs and social media posts lack any established guidelines, are utterly chaotic, boisterous, and have a colloquial dialect
[7]	An overall restaurant star rating is calculated based on the individual aspect rating after text mining is used to analyze customer reviews	(1) POS tagging (2) MAS model	The Stanford CoreNLP API is used to extract noun-adjective pairs from texts and perform POS tagging. The polarity of words is determined using the AFINN library	The application's ability to evaluate just text reviews submitted in English is one of its limitations
[8]	Different methods for sentiment analysis and opinion mining are discussed	(1) NB (2) Maximum entropy (3) SVM	Opinion mining and sentiment analysis for various user's opinions, views, or feedback	Summarizing user reviews is an important problem

(continued)

Table 1 (continued)

Paper Id	Concept	Algorithm	Enhancements	Disadvantages
[9]	Aspect-based opinion mining of customer reviews is considered, and a sentiment profile of its important features is automatically generated	(1) SWN (2) Opinion orientation detection	The use of an aspect-based opinion miner to automatically identify key characteristics or aspects and their opinions	Only explicit aspects are considered, and word sense disambiguation is ignored
[10]	Utilizing the R-tool to automatically anticipate the text based on the values of the data collection	(1) NB classifier	Gives a classification of sentiment classification methodologies in relation to features, benefits, and drawbacks, as well as tools for sentiment analysis in relation to the various sentiment analysis techniques	Multiple language reviews are used
[11]	Various methods visual representation	(1) 3D representation (2) Python libraries	Different types of representation techniques have been explored	Uses complex techniques such as brushing and queuing
[12]	Data is represented dynamically. The representation changes as the data does	(1) R programming	Animation is used to represent the changes in the data	Not appropriate for long-term use. Progress overlaps and becomes slow with passage of time

3.2.1 Tokenization

Tokenization involves separating a stream of texts into tokens, which can be words, phrases, symbols, or other meaningful items. This step is also referred to as segmentation or lexical analysis, and the list of tokens becomes input for further processing. word_tokenize and sent_tokenize in the NLTK Library make it simple to separate a stream of text into a list of words or a list of sentences, respectively.

3.2.2 Stemming and Lemmatization

Both lemmatization and stemming are techniques used to reduce each word's inflectional forms to a basic or root form. Stemming and lemmatization are slightly related.

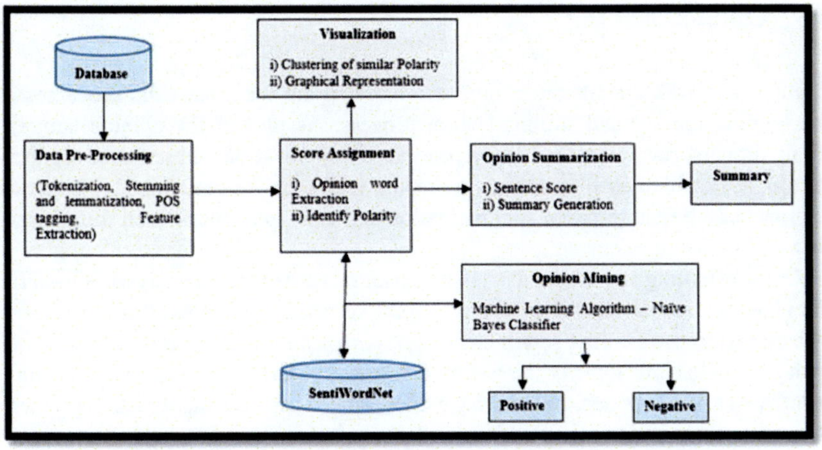

Fig. 1 Summarization and visualization work flow

Table 2 Datasets commonly used for summarization and visualization

Paper Id	Topics	Links
[5]	Travel	Booking.com, TripAdvisor
[9]	Restaurant	www.zomato.com
[12]	Hotel	www.tripadvisor.com
[18]	Restaurant	Yelp dataset challenge

Lemmatization relates words with comparable meaning to one another because it contextualizes the words, in contrast to stemming, which works on a single word without understanding its context. Since lemmatization performs the morphological study of words, it is typically favored over stemming.

3.2.3 Part-Of-Speech Tagging

The process of classifying the words into their part of speech and tagging them accordingly is Part-of-Speech (POS) tagging. A POS tagger goes through sequence of words and assigns a POS tag to each word like a noun, adverb, adjective, verb, etc. With the help of POS tagging, it will become easier to classify the location of words in the sentences which will make the task of feature and opinion extraction easier.

3.3 Opinion Mining

Opinion retrieval, classification, and summarizing are the three main processes in the opinion mining and summarization process. As part of the opinion summarizing process, review text is first preprocessed (tokenization and sentence segmentation), after which each sentence's score and relevance are determined. These two approaches are feature-based summarization and term frequency-based summarization.

Opinion mining as explained in [14], also known as sentiment analysis, is used for the identification of various customer's opinions about a particular topic write-up. The main purpose of using sentiment analysis is to determine the attitude of a writer with respect to some topic or the overall contextual polarity, i.e., negative or positive polarity of a sentence. Opinion mining can be defined as a text classification problem where customer review's text is classified into positive or negative opinion reviews. Machine learning algorithms like Naive Bayes classifier algorithm can be used for the opinion classification of review text, and the SentiWordNet algorithm can be used as a resource-based approach.

3.4 Opinion Summarization Techniques

The main aim of opinion summarization is to summarize long statements written as feedback by customers into short and useful statements. A good summary includes the necessary information, tells us about the entire text in an organized manner. Some of the text summarization techniques could be useful for extracting important features of product or services from the reviews. Opinion summarization emphasizes the sentiment polarities of the texts. The opinion summarization method is generally based on strong sentiment sentence extraction, either positive or negative. Various summarization techniques are discussed in [16, 18].

3.4.1 Traditional-Based Opinion Summarization

Traditional-based opinion summarization methods are useful for extracting the most important sentences in the review document, but it is not concerned with the aspects and its opinions. It generally includes extractive or abstractive based approach for producing a summary. The main problem with this method is that it does not concentrate on the objects or aspects and its opinions. In opinion summarization, usually the polarities of customer opinions are important. However, text summarization methods still can be helpful in opinion summarization in finding the most representative text or sentences from each category, after separating the review data and its polarities.

(i) **Extractive text Summarization**: The extractive text summarization explained in [19] is about finding the key phrases from the original document and combining them to generate a summary. The process of extraction is done according to the proper defined metric without making any changes to the original texts. There are three independent tasks involved in extractive summarization system, which are: (1) Intermediate Representation which reflects the important aspects of the texts, (2) Sentence Score, and (3) Summary Sentence Selection. Extractive text summarization uses various methods to summarize the texts. Some of the extractive summarization methods are cluster-based method, Term Frequency-Inverse Document Frequency (TF-IDF) method, machine learning approach which are discussed below.

(ii) **Abstractive text summarization**: The abstractive text summarization mentioned in [19] approach is somewhat different from extraction technique; it basically involves paraphrasing and compressing parts of the original document. The abstractive text summarization algorithms generally generate new phrases and sentences which contain the most essential information from the original review document. Though abstraction is better than extraction in performance, summarization algorithms which are required to perform abstraction are more difficult to develop. Abstractive text summarization is divided into two categories: semantic-based and structured-based methods. Rule-based method, ontology method, tree-based method, graph-based method, etc. are methods which come under structured-based methods. Methods like multimodal semantic model, semantic text representation model, information-item based model, etc. come under semantic-based approach.

3.4.2 Aspect-Based Opinion Summarization

Reviewers write about different aspects of a particular product or service in a paragraph or sentence. For example, a review about a restaurant might include opinions about food, ambience, location, hygiene, and many more. Opinions in the aspect-based summarization approach are categorized based on objects, aspects, and its opinions, with a separate level like aspect level, document level, or sentence level. Aspect-based summarization involves three important basic stages: (i) aspect extraction, (ii) sentiment classification, and (iii) opinion summaries' generation as shown in Fig. 2. Aspect extraction and sentiment classification use various machine learning algorithms and Natural Language Processing techniques. In this work, we are mainly considering summarization based on different aspects of restaurants like food, ambience, service, location, etc.

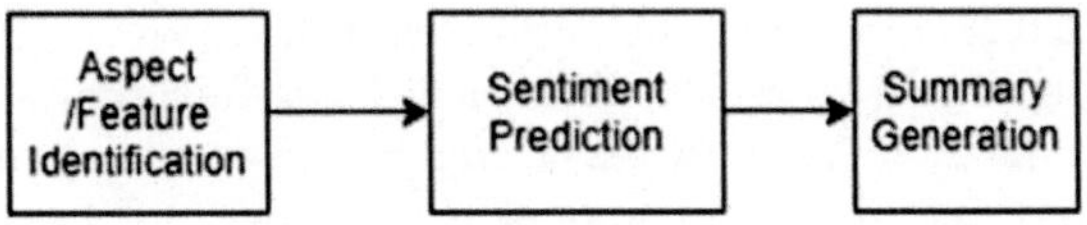

Fig. 2 Aspect-based summary generation

Aspect Extraction

A procedure related to dimensionality reduction called aspect extraction divides a large initial collection of raw data into smaller groupings that may be processed more easily. These huge datasets share the trait of having many variables that demand a lot of computational power to process. The feature extraction phase, which reduces the complexity of the classification task by narrowing the feature space, is a vital step in opinion summarization. The process of selecting and/or combining variables into features is known as feature extraction. This strategy significantly reduces the amount of data that needs to be processed while still accurately and fully characterizing the initial dataset. A lot of data analysis software packages provide for feature extraction and dimensionality reduction. Some common numerical programming environment like Numpy, R language, MATLAB, Scilab deals with some of the simpler feature extraction techniques via built-in functions. Aspect extraction is a technique which can be implemented using machine learning. It involves recognizing the required aspect as required by the user from the sentence. These aspects can include numerals, adjectives, and noun/noun phrases. The aspects extracted help to describe the features of the text and categorize them, respectively.

Opinion Extraction

Opinion extraction is an important step of aspect-based summarization. In opinion extraction, all the opinion words are identified from each sentence. To identify opinion words, extracted features are used to detect the closest opinion words which can be of positive or negative sentiment words with adjectives or adverbs or adjectives along with adverbs, etc., for example: in a review sentence of a restaurant like "Food is good but ambience is bad". Here, the extracted features will be "Food" and "Ambience" and opinion words are "good" and "bad". After extracting opinion words, the next step is assigning opinion word to the nearest extracted feature, which means forming a feature–opinion pair.

Polarity Assignment Based on SentiWordNet Method

Polarity assignment can be done with the help of SentiWordNet. A lexical resource for opinion mining is SentiWordNet [8]. SentiWordNet offers three sentiment scores—positivity, pessimism, and objectivity—to each synset of WordNet. The range of Senti scores is 0–1. The resource was created using vectorial term representations for semi-supervised synset classification and a quantitative analysis of the glosses related to synsets. SentiWordNet is utilized in opinion mining applications to improve the test representation by incorporating knowledge of the sentiment-related characteristics of the various terms in the text. The three scores described above are obtained by combining the outcomes produced by a committee of eight ternary classifiers, and all are characterized by the same level of accuracy but different classification behaviors. In our work, we are using SentiWordNet method for sentiment classification.

Sentiment Classification Based on Machine Learning Methods

Data mining involves the two most frequent modeling goals that are classification and prediction. The classification model classifies unordered discrete values or data. The classification is the initial process of data mining and uses algorithms like decision trees, Bayesian classifiers. The goal of classification is to accurately predict the target class for each case in the data.

(i) **Naive Bayes Classifier**

It is based on the Bayesian theorem and is best applicable when the dimensionality of the inputs is high. There are other classifier techniques like Support Vector Machine and Maximum Entropy technique, but Naive Bayes is simple and easy to implement. Despite its simplicity, Naive Bayes can often outperform the most sophisticated classification algorithm used by [8, 10, 11], and [12].

Bayes theorem can be applied using the following formula:

$$P(y|X) = (P(X|y) * P(y))/(P(X)),$$

where y is class variable and X is a dependent feature vector (of size n), where:

$$X = (x_1, x_2, x_3 \ldots x_n).$$

Here, the class variable (y) has only two outcomes, i.e., Yes or No. But, there are times when classification could be multivariate. To create a classifier model, we find the probability of a given set of inputs for all possible values of the class variable y and pick up the output with maximum probability. Mathematically, it can be expressed as:

$$y = \text{argmax}_y P(y) \prod_{i=1}^{n} P(x_i|y).$$

(ii) **Support Vector Machine**

SVM is a supervised machine learning algorithm that is used for analyzing data required for classification and regression analysis. An SVM classifier model [3, 8] builds a model to predict new examples for a dataset consisting of features and labels' set. It assigns new examples/data points to one of the classes. If there are only two classes, then it can be called a Binary SVM Classifier. It is a discriminative classifier generally defined by a separating hyperplane. In two-dimensional space, this hyperplane is a line dividing a plane into two parts wherein each class lays on either side.

Fig. 3 Text-based opinion summarization example

Lake-View Restaurant

Aspect : Food
  Positive : 12 sentences
  Negative : 2 sentences
Aspect : Service
  Positive : 26 sentences
  Negative :10 sentences
Aspect : Hygiene
  Positive : 8 sentences
  Negative : 3 sentences
....

Summary Generation Methods

After all the aspects and its opinion extraction and sentiment classification, the final step is aspect-based review summary generation. Aspect-based summarization consists of various ways of generating summary which are discussed below:

(i) **Text-Based Summarization**

In text-based summarization, various aspect-based opinion summarization methods generate a text-based summary along with statistical data for each object or aspect. In addition, based on positive and negative opinions, each pair of aspects and their opinions are grouped as explained in [15] which is shown in Fig. 3. With such aspect-based summaries, the customers can understand more accurately how the existing customers feel about the particular restaurant.

Generally, generation of text-based summary is implemented with the help of ranking stage on the basis of weight of the opinion sentences. The highest weighted sentences will be given as the output of a summary. Text-based summarization process uses various different methods for finding the weights of each sentence in the given document. By the linear combination of the derived weights, the sentence score can be calculated. After the calculation of sentence score, ranking of the sentences is done in descending order of their scores. The sentence with the highest score is selected for the generation of summary. Summary generation consists of few important tasks: (1) for the most important features, related opinions are put into negative and positive categories. Then, a count is computed which represents what number of reviews gives positive or negative opinions to the feature. (2) Then, the summary sentences which are extracted are ranked by using the following methods as used in [4] to find out the importance of a sentence.

(a) **Strength**: The first method is to find out the strength of sentiment word by identifying intensity of opinion words as Negative, Mildly Negative, or Strongly Negative. Through summing up the sentiment scores of words of each sentence, the strength of a sentence can be calculated.

(b) **Term Frequency**: Inverse Document Frequency (TF-IDF): TF-IDF is used as a weighting algorithm, and it is a very important method used for text mining to represent the importance of a specific word or phrase in a given document. Term Frequency and Inverse Document Frequency are two statistical methods used in TF-IDF. In TF-IDF, Term Frequency (TF) refers to the number of times a term has appeared in the given document and Inverse Document Frequency refers to the number of times a word has appeared in the given text corpus with the help of which it measures the importance of a word within the document. The output of TF-IDF is a term-score ranked list, and the higher the term-score, the more the term is related to the sentence.

$$\mathrm{tfidf}_s = \frac{\sum_{w \in S} \mathrm{tfidsff}^w}{|S|}.$$

(c) **Cosine Similarity**: Similarity measure is where a number of customers use the sentences or relevant to it. So, if the sentence reviewed is used by more than one customer reviewer, it reflects the importance of that sentence. Cosine similarity is calculated by taking two document vectors A and B, where these vectors represent each distinct word with an index, and the value which is calculated at that index is a measure of how important is the word to the document, as given in below equation.

$$\mathrm{Similarity}\,(A, B) = \frac{\sum_{i=1}^{n} A_i B_i}{\sqrt{\sum_{i=n}^{n} A_i^2}\sqrt{\sum_{i=n}^{n} B_i^2}}$$

Using the above three methods, weight of a sentence can be computed using a linear equation using the below equation. According to the weight value, the summaries are generated.

$$\mathrm{Weight}(S) = \mathrm{Strength}(S) * \mathrm{tf\,idf}\,(S) * \mathrm{Similarity}(S).$$

(ii) **Visual-Based Summarization**

In addition to text-based summarization, review summaries can be presented in a graphical form. Ratings, bars, stars, values, or other information visualization are several previous studies which generate visual-based summaries. There are some of the well-known commercial sites' reviews present a summarization of product with different visualizations such as thumbs up or down, star ratings, positive or negative signs, smileys such as in IMDb, Amazon, TripAdvisor, Yelp. Graphical representation is more understandable and easier to read than representation in text.

(iii) Summary Generation by Scoring Aspects

The review summarization [17] is done by displaying the score of each extracted feature. Methods such as feature occurrences, feature co-occurrence information, rate of reviews, and sentimental polarity of opinion words are required in order to find out the score of each feature. The total score of the opinion-based survey is reduced to a certain scale. For generating a paragraph summary along with the rating, the overall score of the reviews is taken, and based on the polarity ratio, the sentences are selected. Polarity ratio as mentioned in [17] means the ratio of the number of picked positive or negative polarity sentences to the overall positive and negative scores, respectively.

As mentioned in [20], rate of review is derived by combining several opinions regarding features. For example, if the rate of review is 5, then it is mostly derived from several positive and negative opinions about the features of the place. Positive opinion adds to the rate of review, while negative opinion brings down the rate of review. In order to show this process, the below equation is used.

If SP(f_i,R_j) is positive,

$$\text{ownscore}(f_i, R_j) = \text{rate}(R_j) * (1 + P_{\text{neg}}(R_j)) * \text{weight}(f_i)$$

If SP(f_i,R_j) is negative,

$$\text{ownscore}(f_i, R_j) = \text{rate}(R_j) * (1 + P_{\text{pos}}(R_j)) * \text{weight}(f_i)$$

($1 \leq i \leq$ Number of features, $1 \leq j \leq$ Number of reviews).

In the above equation, f_i means the feature at *i*th index and R_j means the review at *j*th index, SP (f_i, R_j) represents the sentimental polarity of the *i*th feature in the *j*th review. P_{neg} (R_j) and P_{pos} (R_j) are the percentages of positive or negative opinions in the *j*th review.

Score of the features used by [13] are calculated by counting the number of occurrences of that feature. A reviewer mentions those features in the comments which he/she felt important than other features. So, the more number of times a feature occurs, the more is its weightage when the score is calculated. The weight of the feature can be calculated by using the below equation.

$$\text{weight(feature)} = 2 - 2^{1-\text{occurence(feature)}}.$$

The result of this equation is used as the opinion strength in the equation used for calculating score.

Representative score of the particular feature after the computation of the score of that feature in all reviews is computed by using the below scoring equation.

$$\text{Score}(f_i) = \frac{\sum_{k=1}^{\text{NR}} \text{ownscore}(f_i R_k)}{\text{NR}}$$

In the above scoring equation, NR represents the number of reviews where the ith feature was utilized and R means the review consisting of the ith feature in the comment.

(iv) **Text Selection**

Reading actual text summaries gives a clearer picture than going through the statistical summaries of overall idea of customer's opinion. There are large volume of opinion on a certain aspect, so displaying a complete list of sentences is not that useful. Using short text as summaries will be a more helpful solution for the users. Topic modeling approach as discussed in [18] is one of the ways for generating a word-, phrase-, and sentence-level summaries. A word-level summary is generally provided for each topic (aspect) since the list of words and their probabilities are common output of topic modeling methods. One way is to rank opinion terms and to select the strongest opinionated term for each aspect. For phrase-level summary, clustering approaches can be used, and for sentence-level summary generation, TF-IDF method can be used for scoring the sentences and selecting the most similar and discriminative sentence as summary.

Topic modeling methods: Topic modeling approaches are useful in improving the quality of opinion summarization. Topic model is referred as a probabilistic topic model which finds topics of texts by using vocabulary distribution. The main aim of topic modeling is to recognize a set of topics from a huge collection of documents. Various representative topic modeling methods are Latent Dirichlet Analysis (LDA) and Probabilistic Latent Semantic Analysis (PLSA). For example, in a restaurant reviews' document collection, some of the topics may be food, cost, ambience, etc. It is obvious that many of these topics represent aspects and each aspect has several opinions which need to be summarized. If opinion words used in positive documents are quite different from opinion words used in negative documents, topic modeling techniques may identify positive and negative topics. Thus, topic modeling methods can be highly useful in automatic extraction of aspects and sentiment classification for opinion summarization.

3.5 Visualization

Visualization is a way of expressing the information or data in order to help the reader understand better and more thoroughly. There are five main factors that affect the type of visualization technique used to represent the data collected. These factors help to narrow down the various visualization techniques and use them more effectively. A visualization technique is used to compare the relative ranks and similarity scores of the data presented. Visualization is most effective in observing the differences between two or more data. Each algorithm uses a different representation technique, and the figures used for the same may vary.

(i) **Audience**—the data visualized is highly dependent on the type of audience viewing it. Such as, the end customers would require a simple interpretation, whereas researchers would require more specifications.
(ii) **Content**—it determines the type of data interpreted. The content of the plots should be relatable to the topic and the context.
(iii) **Context**—it shows how the given data is interpreted and its emphasis. For example, colors are used to differentiate between the various aspects.
(iv) **Dynamics**—the correctness of the plot in the given context is the dynamics of visualization.
(v) **Purpose**—visualization is done to achieve a goal. Each plot or graph used should have its own meaning and value.

Data visualization in general represents the things that we see. It represents the information in a visual manner so that it can be understood easily and quickly. Some common examples of visualization are graphs, charts, maps, images, and movies. The different types of data visualization methods are:

3.5.1 Real-Time Data Visualization

Real-time data visualization is used to describe the time of occurrence of a particular event or action. A similar study related to this was done in [1]. Charts formed for this technique are highly sensitive about time, they require more storage as the granularity gets smaller, and it stands valid only for a certain period of time after that the data value drops. Some subjects that need to be observed are counter, timer, and gauge. This helps identify the frequency of the events and the duration for which they last. A gauge is generally required to identify if the node is live or not, the number of ongoing process, and the usage of memory at that time. Computer games are the most common example for real-time data visualization.

3.5.2 Interactive Data Visualization

It requires input from the user and accordingly updates the images. It is an iterative interactive process between the user and the visualization. The related diagrams are made as the data values are changed. In short, it is a repetitive, bidirectional, and cyclic method of visualization. It allows direct changes to be made on the plots. Some of the commonly used applications of this is paint, identification, scaling.

3.5.3 Geographical Visualization

Geographical visualization is also known as geovisualization which is used to depict the geographical data. It is most commonly done in the form of maps, which can represent the whole earth in a single map, or a particular region at a time, depending on the requirement of the study. This is used to represent the climate, travel routes, wildlife, flora and fauna, soil types, etc. The feature to zoom in or out of the map is also available. A live example of this available in Google Maps, as they can be used to locate a destination, check for traffic, etc.

3.5.4 Multidimensional Data Visualization

Many dimensions are used to represent the data. The data is represented in two-dimensional or three-dimensional method. Sometimes, a four-dimensional representation is also used to visualize the given data. A two-dimensional representation could be a graph or a chart. These objects can also be represented in three-dimensional way. A movie can be considered as a four-dimensional representation object or as a series of three-dimensional objects.

There are various tools that are used to represent the data. Different colors and patterns are used when two or more types of data of the same category are compared. The use of colors and line patterns or shading methods help differentiate the text. A key is provided to help understand the coding and relate to the original text instead of cluttering the visual and to help it remain neat. The most effective tool is then implemented based on the task and data at hand.

(i) **Charts/Graphs**: Charts are the most common way to represent multiple data in a single diagram. It consists of text and is represented most commonly in a tabular form. The different types of charts include different graphs, line charts, bar charts, pie charts.
(ii) **Plots**: Plots are used to compare data of the same category. A plot can be 2D or 3D as per the requirement and the number of dimensions it needs to represent. Plots help to understand trends, correlations, and patterns, e.g., histograms, scatter plots, box, and whisker plots.
(iii) **Maps**: They help determine the position of the objects and their layouts. Maps can be heat map, blueprints of a building, the layout of a complex, online navigation
(iv) **Diagrams**: Diagrams are complex structures which give insight about the data and represent the details of the object. Multiple types of data can be visualized in just a single diagram. The particular diagram can be multidimensional or has a tree structure, etc.

The implementation of many of these models and techniques has been implemented, and the data has been expressed using various methods. It uses datasets from multiple sources and time series has been visualized. A dataset is visualized in a digital platform using codes with the help of libraries that are present in the IDE. Libraries such as matplot, matplotlib, NumPy, SciPy, Seaborn, are used in the Python language.

When the data is summarized, the visualization becomes clearer. In Fig. 4, the same data has been represented in the form of a bar chart. The following graph depicts the three restaurants which have a cumulative value picturized for various

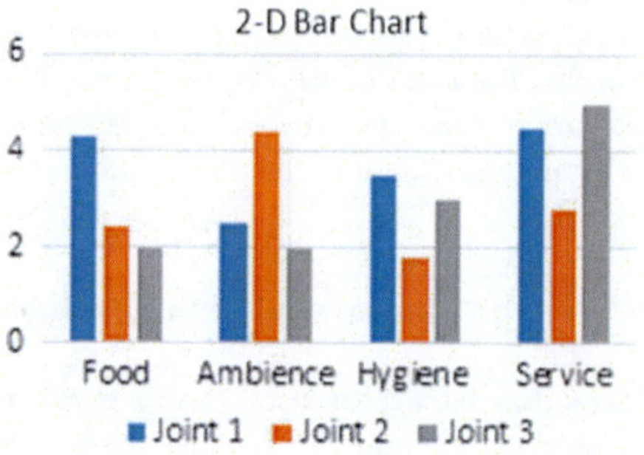

Fig. 4 Bar chart

aspects. The library used in this case was matplotlib, which is a very common library of Python.

4 Performance Evaluation

The performance of this work needs to be evaluated to infer the effectiveness of the system. The performance parameters required are Accuracy, Precision, Recall, and F-measure. To perform the testing, firstly a restaurant dataset with customer reviews is required. Then, each sentence should be annotated with positive or negative polarities which can be done manually. Then, this annotated dataset can be further used for evaluation purposes. Opinion text classified as positive or negative by different opinion classification methods can be evaluated by finding the effectiveness of the method using precision or recall values. These values are used to confirm the accuracy of the algorithm used based on the data corpus used.

$$\text{Recall} = \frac{\text{Total no. of correctly classified positive examples}}{\text{Total no. of positive examples}} = \frac{\text{TP}}{\text{TP} + \text{FN}}$$

$$\text{Precision} = \frac{\text{Total no. of correctly classified positive examples}}{\text{Total no. of predicted positive examples}} = \frac{\text{TP}}{\text{TP} + \text{FP}}$$

where TP = True Positive, TN = True Negative, FP = False Positive, and FN = False Negative.

$$F - \text{Measure} = \frac{2 * \text{Recall} * \text{Precision}}{\text{Recall} + \text{Precision}}$$

The F-measure will always be nearer to the smaller value of Precision or Recall. The accuracy is given by the relation mentioned below.

$$\text{Accuracy} = \frac{\text{TP} + \text{TN}}{\text{TP} + \text{TN} + \text{FP} + \text{FN}}$$

Accuracy Score : 0.73
Report :

	precision	recall	f1-score	support
0	0.82	0.57	0.67	97
1	0.68	0.88	0.77	103
accuracy			0.73	200
macro avg	0.75	0.73	0.72	200
weighted avg	0.75	0.73	0.72	200

Fig. 5 Screenshot of result obtained

5 Result

In this work, the precision, recall value of the dataset, the F1-score, and the accuracy for the same have been calculated using the Naive Bayes algorithm. The performance evaluation is measured between 0 and 1. The accuracy score of this work is 0.73 units, which is 73% and is given in Fig. 5.

6 Conclusion

The main aim of this paper is to find out various efficient and suitable techniques and algorithms which are required for opinion mining, summarization, and visualization. The classification of reviews of various restaurants as positive or negative is preferred using the Naïve Bayes classifier as it is the most easily understandable classifier. The SentiWordNet is trained for the opinion words, and the objective of the sentence is determined for the taken dataset. SentiWordNet contains the various values of a particular keyword, their meanings, synonyms, antonyms, etc. This survey provides various methods required to summarize the multiple customer reviews on the basis of sentiment scores and provides a visualized representation to understand the ratings and various aspects related to restaurants on which the ratings were provided. This survey also looks for techniques to visualize the data that is collected from the summarized opinions. The various papers referred here were used to find the visualization and summarization methods suitable for the data collected. The basis of this survey was to implement opinion mining for various reviews given by the customers by first obtaining a summary and then generating graphs for the same. The summary that is generated is based on sentiment score calculation. The multiple varied reviews are combined into a single paragraph, talking broadly about the different aspects taken into consideration.

References

1. Liu C-Y, Chen M-S, Tseng C-Y (2015) Incrests: towards real-time incremental short text summarization on comment streams from social network services. IEEE Trans Knowl Data Eng 27(11):2986–3000
2. Moussa ME, Mohamed EH, Haggag MH (2018) A survey on opinion summarization techniques for social media. Future Comp Inform J 3:82–109
3. Sindhu C, Vadivu G (2021) Fine grained sentiment polarity classification using augmented knowledge sequence-attention mechanism. J Microprocess Microsyst 81. https://doi.org/10.1016/j.micpro.2020.103365
4. El-Halees AM, Salah D (2018) Feature-based opinion summarization for Arabic reviews. In: Proceedings of International Arab Conference on Information Technology (ACIT)
5. Pecar S (2018) Towards opinion summarization of customer reviews. In: Proceedings of Student Research Workshop, Association for Computational Linguistics
6. Chakraborty R, Bhavsar M, Dandapat S, Chandra J (2019) Tweet summarization of news articles: an objective ordering-based perspective. IEEE Trans Comp Soc Syst 6(4):761–777
7. Cuizon JC, Lopez J, Jones DR (2018) Text mining customer reviews for aspect based restaurant rating. Int J Comp Sci Inform Tech (IJCSIT) 10(6)
8. Sindhu C, Chandrakala S (2012) Opinion mining and sentiment classification: a survey. ICTACT J Soft Comp 3(1)
9. Chinsha TC, Joseph S (2014) Aspect based opinion mining from restaurant reviews. Inter J Comp Appl
10. Spoorthi C, Kumar PR, Adarsh MJ (2018) Sentiment analysis of customer feedback on restaurants. Inter J Eng Res Tech (IJERT), special issue
11. Raut VB, Londhe DD (2014) Survey on opinion mining and summarization of user reviews on web. Inter J Comp Sci Inform Tech 5(2):1024–1030
12. Raut VB, Londhe DD (2014) Opinion mining and summarization of hotel reviews. In: Proceedings of Sixth International Conference on Computational Intelligence and Communication Networks
13. Yang J-Y, Myung J, Lee S-G (2009) A holistic approach to product review summarization. Soft Technol Future Depend Distri Syst, 150–154
14. Momin HG, Shaikh RE, Kondhawale CS, Gawande KG (2017) Feature-based evaluation of hotel reviews and ratings for differently-abled people. In: Proceedings of International Conference on Energy, Communication, Data Analytics and Soft Computing
15. Pabbi K, Sindhu C, Reddy IS, Abhijit BNS (2021) The use of transformer model in opinion summarization. Webology 18:1084–1095
16. Zha Z-J, Yu J, Tang J, Wang M, Chua T-S (2014) Product aspect ranking and its applications. IEEE Trans Knowl Data Eng 26(5):1211–1224
17. Akkamahadevi RH, Patil MM, Patil PM (2016) Summarization of customer reviews for a product on a website using natural language processing. In: Proceedings of International Conference on Advances in Computing, Communications and Informatics (ICACCI)
18. Kim HD, Ganesan K, Sondhi P, Zhai C, Comprehensive review of opinion summarization
19. Andhale N, Bewoor LA (2015) An overview of text summarization techniques. IEEE Trans
20. Sindhu C, Vadivu G (2019) Sentiment analysis and opinion summarization of product feedback. Inter J Recent Tech Eng (IJRTE) 8:59–64

Performance Comparison of Object Detection-Based Deep Learning Techniques: A Review

B. Sudha and Kathiravan Srinivasan

Abstract This paper provides a comprehensive review of the performance comparison of object detection-based deep learning techniques. Object detection is a crucial task in computer vision that involves identifying and localizing objects within an image, signal, or video. Deep learning has emerged as a promising approach for object detection due to its ability to learn and extract features automatically. In this paper, we review the recent literature on object detection-based deep learning techniques, including region-based, edge-based, texture-based, etc. This paper also analyzes the performance metrics used to evaluate these techniques. Through this review, we aim to provide a better view of the strengths and weaknesses of various object detection techniques and their applicability to different applications. Our findings suggest that while some techniques perform better in specific scenarios, there is no single best-performing technique for all scenarios. Therefore, the choice of object detection technique should be based on the specific application and performance requirements.

Keywords Dataset · Deep learning · Detection · Object

1 Introduction

The object detection (OD) framework finds things in the real world by using object models that have already been created [1]. When compared to humans, who can do OD quickly and efficiently, this task is almost impossible for machines to do [2]. This work is a review of the different techniques and methods used to elaborate things in

B. Sudha · S. Kathiravan (✉)
School of Information Technology and Engineering, VIT, Vellore, India
e-mail: kathiravan.srinivasan@vit.ac.in

B. Sudha
e-mail: b.sudha2018@vitstudent.ac.in

S. Kathiravan
School of Computer Science and Engineering, VIT, Vellore, India

G. Rajakumar et al. (eds.), *Intelligent Communication Technologies and Virtual Mobile Networks*, Lecture Notes on Data Engineering and Communications Technologies 171,
https://doi.org/10.1007/978-981-99-1767-9_57

images and video capturing. Object identification is a fundamental topic in machine vision and deep learning and it lays the groundwork for the in-depth development of a variety of research problems such as instance segmentation, object tracking and optimization, trajectory prediction, and image reconstruction. Pedestrian detection is a specific application of the object detection issue, and in recent years, it has become one of the most active study areas. It has significant utility in the areas of intelligent driving and security surveillance. Due to the uniqueness of individuals and the highest safety standards, it is more significant than other types of object detection in the field of intelligent driving. In intelligent driving, a camera, LIDAR, and wireless sensor network collectively sense the surroundings, with vehicle-mounted computers and cloud computing used for decision-making and control.

1.1 Object Detection in Image Handling

Image handling is a technique for converting a image into electronic form and performing a few tasks on it, such as creating a enhanced image or extracting some important information from it. It is a good setup that has a picture, such as a video edge or photo yield, on input, possibly a picture or assigns associated with that image [3]. In general, the image processing structure joins viewing models as two-dimensional signs while employing noteworthy planning techniques. The three stages of preparation are as follows:

- Using an optical scanner, the photographs are foreign or of exceptional quality.
- Examining and interpreting data and diagnostics that are invisible to the naked eye, such as those contained in control data and satellite images.
- The outcome is the image or message that resulted from the film experiments that altered the outcome of the preceding stage.

Figure 1 shows both the calculation and how a camera extracts an image from an image. By utilizing the hypothetical focuses, the form decides the highlights, and the photo can be used to determine the object to be cemented.

1.2 Object Detection Architecture

Support vector machine (SVM) is a type of deep learning algorithm that is used as supervised learning for classification or regression of data groups. SRAM is an abbreviation for static random-access memory; HOG is an abbreviation for histogram of positioned slopes.

Figure 2 depicts the general structure for detection. The twelve scales' pixels are distributed to three parallel finders. The objective of the distribution is for the three sensors to have adjusted the remaining tasks at hand. The initial scale is transmitted to detector.

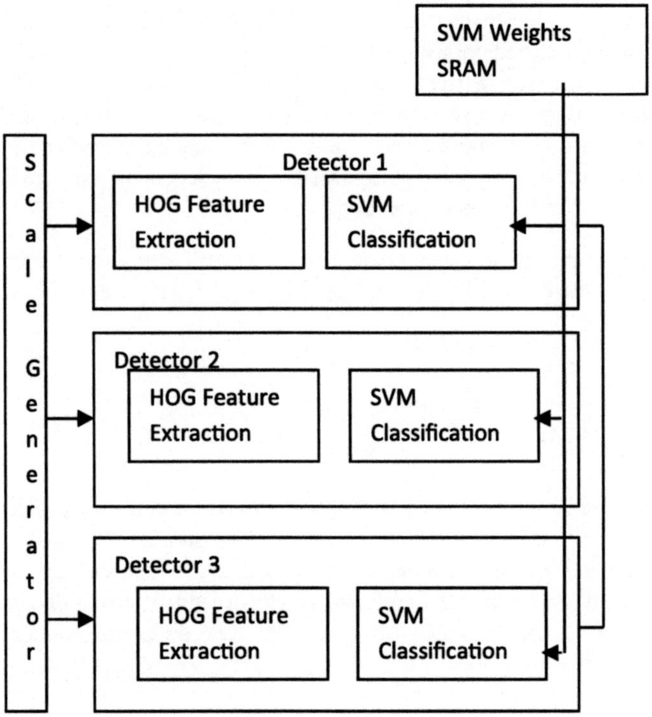

Fig. 1 Block diagram for object detection

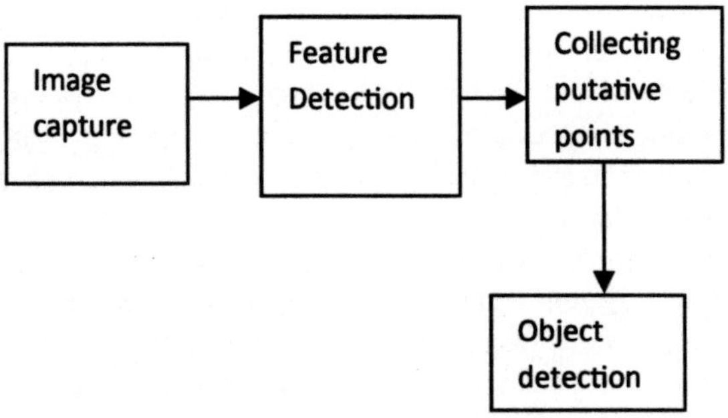

Fig. 2 Object detection system architecture

(1) The two scales to detector are as follows.
(2) The remaining nine detector scales.
(3) The range of the histogram and the aggregator supports are different given the number and size of the scales created by each sensor, but the three locators cannot be distinguished from one another.

(4) Putative points locate the object in the image that calculates the transformation relating the matched points, while eliminating outliers.
(5) HOG features were extracted using encrypted blocks in image. The images are differentiated as vertical and horizontal forms. The gradient direction and strength maps are computed over the small grids called as HOG cells [4].
(6) The HOG blocks are concatenated based on the selection of adjacent HOG cells that defines the features to be extracted.

Every sensor has an adequate amount of memory. Three parallel pointers with voltage scaling save 3.4 times as much energy as a single locator [5] in terms of frames.

1.3 *Justification and Scope of the Study*

Object detection is used in Personal Computer (PC) visual tasks, such as confront detection and acknowledgment, and video object co-division. It is also utilized in subsequent objects [6]. For example, following a ball in a football coordinate, tracking the development of a cricket bat, or following a guy in a video item co-division alludes to the problem of simultaneously dividing many recordings that contain common objects [7, 8].

The primary objective of the object detection system is to take shape, pose, illumination, and occlusion into account. The approach must cluster the several instances of the same item in a manner that improves object detection precision.

Throughout the training phase, it preserves the multi-view samples of the same object. This object can increase detection precision and decrease the error rate.

Adapting intelligent artificial neural networks for object detection, which would replace missing features and classes, is the primary objective. To propose an object identification method that recognizes items using a region-growing method, the number of missing objects should be identified in group object images and to enhance real-time video surveillance.

Face detection's variable lighting, occlusion, and backdrop characteristics used for facial recognition system and their accuracy is poor since they do not properly group the faces of the same object under the aforementioned conditions. The face object clustering was not conducted in an efficient manner. This increases the ratio of incorrect classifications. The majority of algorithms extract facial objects and measure their resemblance to various object groups. Nevertheless, the algorithms miss certain traits and values, which reduce the accuracy of face detection.

The previous methods do not consider and store objects acquired from diverse perspectives, and the absence of a multi-view feature for the same item hinders face detection and tracking performance.

Several techniques have been utilized to implement the posture and occlusion characteristics in object detection. However, they introduce a higher false ratio since they do not preserve the property of distinct situations.

1.4 Challenges

Even after much research in this area, object detection is still an open research question. Even today, an excellent test is one that uses a dynamic, precise, and elite technique [9]. Depending on how you characterize the entity that needs to be located and followed, this problem might range in difficulty.

It is not too difficult to identify all the pixels with the same color from the intended object if only a few visual features, such color, are used to represent the object. In every scenario, another object or foundation with comparable shading data is always an option. Additionally, the different lighting in the scene does not guarantee that the shading for the same object will be the same in each instance [10]. Incorrect division results from this when only visual highlights are used (e.g., shading). In video objects and, for the most part, in things, this kind of fluctuation shift is quite noticeable [11]. It begins with one edge and progresses to the next, completing the camera's field of view [12]. Three common causes of this fluctuation—change in target posture or distortions, variation in illumination, and partial or complete obstruction of the objective—are to blame [13]. The following list contains the typical issues with foundation subtraction that relate to video surveillance.

1.5 Illumination Changes

The foundation display's ability to adapt to the earth's ongoing changes is appealing. For instance, the bright force varies frequently throughout the day in outdoor areas. Likewise, the scene may experience abrupt changes in lighting. This type of advancement can occur, for example, when a light is abruptly turned on or off inside. Similar things could happen in external stages (quick pace from shady to splendid daylight). This explicitly affects the foundation's presence and results in false-positive detections. With the use of deep learning, numerous methods for object detection have been developed. However, the majority of algorithms are divided into the following four classes.

The advantages and disadvantages of each method for object detection are listed in Table 1.

(1) Dynamic Condition

A few portions of the landscape may contain development (a spring, mist events, the swaying of tree branches, the rushing of water, etc.), but they must be regarded as a foundation due to their significance. This development may be routine or intermittent. Dealing with such fundamental elements is a challenging task.

(2) Occlusion

Impediment (fractional/complete) may affect the process of determining the foundation outline. However, situations and difficulties can arise when a subject goes behind an object in relation to a camera.

Table 1 Classification of object detection techniques

Techniques	Brief description	Advantages	Disadvantages
Knowledge based	Encode human Generally, the formation of a common face is the relationship between knowledge and facial features Mainly, spreading face is used	(i) Simple rules are easy to establish a face and a child's relationship with their relationships (ii) In an input film, the face is first extracted, and the faces are on the face Has been identified (iii) Facial work spreads well on uncontrolled backgrounds	(i) Difficult to translate human knowledge into the rules: detailed rules cannot find the faces and public rules can be found in many falsifications (ii) This approach to detect faces All the possibilities are unlikely in the account: it is difficult to differentiate
Feature invariant	To locate facial features which are invariant under different conditions and using these to locate face pose, intention or even when the lighting variables are intended to find the structural elements of a face	(i) Features are invariant to pose changes	(i) Difficult to locate facial features in the presence of various illumination changes and complex background
Template matching	Several standard patterns of face are stored and correlation computed with a test image for detection. Those patterns are used to describe an entire or shown individual facial features	(i) Easy to determine the face location (nose, eyes) based on the correlation values (ii) Simple to implement	(i) Templates have to be opened closer to the images of face (ii) The templates that take into account different difficulties
Appearance based	Models are learnt from face images by training and learnt models are used to perform detection	(i) Use powerful machine learning methods (ii) Models have proven good experience results (iii) Find faces in different pro and extended bass	(i) Usually, you have to search over size and space (ii) It has had lot of true and false examples (iii) Limited view approach

(3) Clutter

The proximity of foundation mess complicates the task of division. It is challenging to provide a foundation that reliably provides the messy base and separates the moving closer view items from that.

(4) Concealment

Whether intentionally or unintentionally, a few objectives may differ ineffectively from the existence of experience, making precise classification difficult. This is especially important for applications involving inquiry. Disguise is largely a problem for ephemeral differencing systems.

(5) Presence of Shadow

Shadows cast by lead objects often make it hard to handle things after the foundation has been subtracted. Covering the shadows of frontal area districts, for example, makes it hard to tell them apart and put them in groups. Analysts have come up with different ways to find shadows.

(6) The Motion of the Camera

The video may be captured by unstable cameras. The level of jitter greatness varies from one video to the next.

(7) Bootstrapping

If the statement statistics that can be gathered from closer-look objects are not available, a bootstrapping method must be used to set up the foundation shows.

(8) Video Noise

Disturbance is often overlaid on video flags. Video reconnaissance foundation subtraction methods must be flexible enough to accommodate corrupted signals influenced by noise sources such as sensor commotion and pressure antiquities. The rate at which influencing and repeating objects will be taken as parameter. The speed at which an object is traveling plays a crucial role in establishing its presence. Because of the object's slow motion, the transient differencing technique will fail to identify the parts of the object that are uniformly preserving their space. However, in the observed frontal area covered, a fast-moving object leaves a trail of apparition locale behind it. When objects move in a way that is not completely regular, we see "ghosting" artifacts in the apparent motion; this happens when things move, then halt for a while, and then start moving again. It is possible for a film to feature both items that were previously at rest but which suddenly begin moving, such as a parked car speeding off, and objects that were previously in motion but which are now abandoned.

(9) Challenging Weather

When recordings are captured in testing climate settings (winter climate conditions, e.g., snowstorm, snow on the ground, fog), air disturbance, etc., detecting moving objects becomes a testing task.

(10) Object Detection in Object Following

Object detection in recordings entails confirming an object's proximity in an arrangement of visual contours. Object following is the process of putting things where they can be seen. It is closely related to video production. There are numerous applications for object recognition and tracking in Personal Computer (PC) vision, including video surveillance, vision-based control, video stress, human–PC interactions, and mechanical autonomy. Additionally, it contributes to an increase in vision-related projects, such as 3D recreation and depiction. It is also a key part of video databases, where it is used for things like content-based ordering and retrieval [14].

1.6 Artificial Neural System Under the Deep Learning

Inspired by the composition and organization of real neural networks, a synthetic neuron arranges data in a computer programmer. It is also possible to refer to the artificial neural network, or ANN, as a neural system.

(1) Recurrent Neural System

RNNs are a type of state-of-the-art ANN that makes use of memory cycles that are coordinated with one another. Broken neural networks are able to build on well-known models of architectures with information vectors and output vectors that are all the same size.

(2) Convolutional Neural System

A convolutional neural network (CNN) is a type of artificial neural network that uses a specialized form of supervised learning to analyze data through the use of visual knowledge representation. CNNs can be used for many different kinds of mental errands, including image preparation, normal dialect handling, and so on.

(3) Deep Learning Difficulties

Deep learning is a technique that, rather than actually making use of it, models human conceptual reasoning. Nonetheless, despite its many benefits, this technology also has a number of significant drawbacks. The method involves feeding the input feature into a neural network and then using that network's output to make a classification.

Pseudo Code of Object detection in neural network:

Input: Video v, Neural network Nn

Output: Detected object Ob

Step 1: Start

Step 2: Read input video v

Step 3: for each object Ob
    Obi = perform segmentation (img)

Neural network = Feature Extraction (Obi)
End

Step 4: Boolean b = perform Deep learning (Dl)

Step 5: stop

The aforementioned pseudo code detects the presence of an object and then returns an appropriate result.

(4) Continuous Information Data Administration

In deep learning, the development of a product requires the examination of massive amounts of data. Even though information is constantly flowing and changing, there is still time to ensure an effective preparation process. To solve this problem, data scientists need to change how they calculate deep learning so that neural systems can handle large amounts of database information reliably.

(5) Ensuring End Transparency

Programming for deeper learning is another complex barrier that lacks the evidence to support its conclusion. In the case of archaic machine learning variants, one cannot continue to be computed in order to determine why your framework was a photographer, a bird, and a cat. To remedy errors in computations involving deep learning, you must rethink the entire calculation.

(6) Deep Learning Future

As advancements in deep learning progress, there is a risk that this technology will become incomprehensible to regular engineers, making it impossible to comprehend without elevated consideration. Currently, experts in deep learning have an abundance of tool options, the majority of which are open source. TensorFlow, BigDL, Open Deep, Caffe, Theano, Torch, and MXNet are the most well-known.

(7) Deep Learning Will Increase Local Help Inside Spark

The Spark people group will boost the stage's local profound learning capacities over the next coming years. The network is obviously leaning toward better support for TensorFlow after hearing about it at the Spark Summit, which is currently going on, with funding also going to BigDL, Caffe, and Torch.

For speedy coding, improved programming systems will be combined on rooted learning devices.

The application engineer network will need application programming interfaces (APIs) and other programming reflections to construct the basic algorithmic capabilities more quickly and with less lines of code. The usage of integrated, open, cloud-based development environments by deep learning engineers will give them access to a variety of commercially accessible and pluggable computation libraries in the future. These will make it possible to create microservices that may be combined with APIs to enhance deep learning applications. The technologies will automate all of the more intricate learning improvement pipeline functions and offer a blank canvas for group collaboration and a shared point of view. When this pattern gets stronger, we will see more features, like "Generative Adversarial Nets" in 50 Lines of Code "(PyTorch)."

(8) Comparative analysis are shown in Table 2.

2 Discussion

Research into object detection has been conducted in many different industries, such as health care, transportation, welding, and robotics.

2.1 Object Detection in Medical Field

Object detection in deep learning is providing promising solutions to medical imaging analysis challenges, and it is widely regarded as an important future-proofing technique. The calculations that deep coordinate learning as a central part has been connected to therapeutic image detection, division and enrollment, and Personal Computer-assisted examination, using a wide variety of application domains, are all clearly understood here, providing a solid foundation for further research. Scholars and practitioners in the field of medical imaging research will find deep learning for medical image analysis to be an invaluable resource, as will graduate students enrolled in programs that focus on machine learning and deep learning in the contexts of computer vision and medical image analysis.

- Addresses the common challenges encountered in therapeutic image analysis research.
- Features an explanation of deep learning approaches and related theories for restoring images through analysis.
- Explains the connections between mathematics and a wide variety of fields of application, such as X-rays of the chest, breast CAD, microscopy and pathology, computed tomography (CT), and more. In this article, we will take a look at how object detection is used in CT scans for diagnostic purposes.

(1) Object Detection in CT scan

A computerized tomography (CT) scan is a compilation of multiple X-ray images taken at different depths. This beam has the ability to zero in on specific targets. The paper will discuss the application of CT data to machine learning and other fields of study. It will also discuss the challenges faced by analysts and other research methods, as well as how CT tests are being used to improve education at all levels.

Images from a CT scan are pieced together to form a three-dimensional representation of the body. The way medical care is delivered is now being drastically altered thanks to this technological advancement.

Table 2 Comparative analysis

Author	Year	Report	Method	Implementation
Xiang Wang, Huimin Ma	2018	Firstly, region-based CNN methods do not have enough environment to find out precisely the main object because they can handle each zone individually	Contextual Convolutional Neural Network (CoConv)	This framework clearly identifies the boundary and multiple-level environmental stability at the same time
Xiaozhi Chen, Kaustav Kundu	2018	The material size priors the problem that the objects are likely to encode is an estimation of the energy function	KITTI Object Detection Dataset	This framework clearly states depth information features
Francois Belletti, Daniel Haziza	2017	To perform other tasks that are currently at the core of engineering cyber physical systems	Mutual Weight Regularization (MWR)	To specialize its action policy so as to tailor it to the local parameters of the part
Sidike, Asari, Alam	2016	The Joint Conversion Correct (JTC) is based on meaning and non-object spectrum signatures	The class-sub-spectrum margin is adjusted for the correction (CSFJTC)	The CSFJTC output is a matched target and is negligible or has no connection with no compromise and gives a couple of sharp contact peaks for the peaks
Rodriguez-Serrano, Larlus	2016	Our main novelty is designed or apparently uses general represent and uses unobtrusive similarities. Unlike previous work you need to know some of the transformation tuning the image similarities	Data-driven detection (DDD)	Despite the simplicity of its conception and the timing of performance—tests show these two contributions in some cases resulting in comparable or better results than standard sliding window detectors

(continued)

Table 2 (continued)

Author	Year	Report	Method	Implementation
Matteoli, Corsini	2015	To this aim, laser-induced fluorescence (LIF) spectroscopy is exploited	Light detection and ranging (LIDAR)	Implementation of results from a laboratory testbed shows that the proposed processing chain is effective at automatically recognizing objects
Paglieroni, Pechard, Beer	2015	The GPR data from the trajectories of the previous path was found to be stored in a galaxy of network surfaces buried objects	Ground-penetrating radar (GPR)	Detection-False Warning Rate Performance Our transformation point is significantly better than our detection point
Pang, Zhang, Yuan	2014	Two important goals are to find the location of video objects as less computationally complex and high generalization in vision and learning	Block-based histograms of oriented gradients (BHOG)	The diagnosis of the video hand, face, and pedestrian results shows the superiority of the proposed method
Gang Wang, Forsyth	2013	Due to the inherent long tail distribution of material in the real world	Regularized kernel machine algorithm	It can make significant improvement on object categorization
Zhao, Ngo	2013	Large-scale video copy detection	Flip-invariant scale-invariant feature transform (F-SIFT)	Consistent improvement across different kinds of key point detectors is observed
Del-Blanco, Jaureguizar, & Garcia	2012	Automatic visual object counting and video surveillance	Bayesian tracking model	On future consumer electronics, proving its superior performance
Subudhi, Nanda, Ghosh	2011	These programs are used to identify and track the moving objects together	Change detection mask (CDM)	The proposed external spatial approach is compared to the future JSEG system

(continued)

Table 2 (continued)

Author	Year	Report	Method	Implementation
Youngmin Park, Lepetit	2010	One-time objects that can be observed in different 3D simultaneous objects	Method combines object detection and tracking	Implementation done with real time experimental data with sensors.
R. Kasturi, D. Goldgof, Soundararajan	2009	A variety of application domains	Object detection and tracking algorithm	In the coming years, the future of computer science has researched many real time system.
Leibe, Schindler, Cornelis	2008	A moving vehicle passing through crowded city areas	Subsequent trajectory estimation module analyzes moving frames	To guide object detection in future frame sequence.
Akcay, Aksoy	2008	The object-oriented inspection provides valuable spatial depiction of distant images	Segmentation algorithm	The results show that the future methods are able to detect automatically.
Wei Zhang, Xiang Zhong Fang	2007	A novel method for moving cast shadows detection	Moving cast shadows' detection	The work significantly outperforms state-of-the-art methods
Manjunath Bhagavathy	2006	The model is learned in a two-layered framework	Texture segmentation	The work performs comparitive study of four texture segmentation methods.
V. Kaftandjian, Zhu	2005	This framework, in the study of X-ray and UVs	Fuzzy logic Image Processing Used.	Image segmented to different region by edge detection.
O. Carmichael, Hebert	2004	The cascade is to discard clutter edge pixels and group	Shape-based recognition of wiry objects	Efficient at runtime

(2) Data from CT Filters

The majority of CT scans are made available in Digital Imaging and Communications in Medicine (DICOM) files, which contain 2D images with pixel forces and are therefore measured in Hounsfield units. However, these are not in the regular 0–255 range, but rather in Hounsfield units or CT numbers.

Example pseudo code:

Step 1: Input Ds data initialization

Step 2: Prefect logs for each Class $Cl \rightarrow$ Ts

Identify search term attribute for frequent query Fvi

Attribute For each $Cl \rightarrow A_i$ of Fvi

Semantic compute data

$$Cl = \int_{i=1}^{N} \sum (\text{Ai(Fvi)} - \text{Ai(Fv)})^2$$

End

$\text{Ds}(i) = \sum \text{Ds}i + Cl$

End

End

Step 3: Identify each class Cl of data request set Ts

Ai $\rightarrow$ for each case attribute

Compute the semantic similarity count.

$$\text{SC} = \int_{i=1}^{N} \sum \text{Ds}i(\text{Ai}) \geq \text{STh}$$

End

Measure relative similarity case

$$\text{Dm} = \frac{\text{SC}}{\text{size}(cli)} \times 100$$

End

Step 4: Read end

Dynamic causal mapping (DCM) records are digital images stored in the Digital Imaging and Communications in Medicine (DICOM) image format. A medical image, like a CT scan or an ultrasound, is saved. Patient information may also be saved. Different media have different radio-force values, which are measured in Hounsfield units (HUs). This is because different media have different constriction coefficients.

(3) Deep Learning for CT Checks

Different models are possible depending on the data analysis technique used:

Basic 2D convolutional systems like DenseNet and ResNet were used to complete the single-image processing. These can be integrated into a start-to-finish pipeline to generate the output directly, or they are typically used to process and concentrate highlights before passing them on to a different model. When given sufficient data, end-to-end models have proven to be more effective than using multiple models for these tasks. Preparing a series or multiple images, these can be stacked to make a "3D image" (or 3D point cloud). This data could then be processed by 3D convolutional systems. The alternative method of dealing with this information is to examine a 2D Kernel.

2.2 *Object Detection Application CNN Engineering in Medical*

A Convolutional Neural Network (CNN) records highlights from privately associated voxels (i.e., neighborhood voxels); as a result, it is commonly used as the standard display for medical image analysis. In general, an image undergoes multiple convolution layers, and the highlights are extracted from multiple segments. The eliminated characteristics are also used for other projects. Table 3 shows that the CNN engineering is used in CADx/CADe applications. CADx and CADe are abbreviations for Computer-Aided Analysis and Computer-Aided Detection, respectively. MRI stands for magnetic resonance imaging, and CT stands for computed tomography.

Detection: The detection task involves locating the location associated with the disease. In some instances, the detection task is assigned to the control work, such as locating the vocals of disease detection. The arrangement of occupational therapy is required. Expectation tasks include estimating imaging investigation results for object evaluation (usually not plain-eye apparent). Prediction is frequently based on benchmark imaging, where benchmark imaging findings are used to determine that the object's properties have not been predetermined. Several studies have been done to show the likelihood of medical factors, such as a drug or alternative reaction, the patient's life, and the imaging test results, in this flawed evaluation. Data on Imaging the number of class-based job arrangements. Conditions and positions (e.g., sound and sickness) or multiple courses can be paired (e.g., subdivisions within the given unfortunate position). CNN is associated with a vast array of options, effectively employing CT imaging of various organs (i.e., cerebral cysts, lungs, and liver).

When the framework joins, the convolution arranges parameters prepared by the RPN as a contribution to the Fast-(R-CNN) Regions with the Convolution Neural Network for preparation. Finally, the parameters of each system layer are the contributions to the model for Vehicle-Type Target Recognition. The best way to get them ready for a similar system architecture is to make a display that can do more than one thing and uses the same convolution. Table is comparing the Faster R-CNN and R-CNN vehicle-based strategies. Quicker R-CNN means Faster Region with the Convolutional Neural Network, while R-CNN means Region with the traditional neural network. The case for a few vehicles examines the distinction between faster R-CNN and R-CNN and determines the velocity level. Vehicles are autos, cycles, prepared for transport.

Table 3 CADx/CADe applications utilizing convolutional neural system

Task	Modalities	Object	Clinical goal
Detection	MRI	Brain Heart	Lesion detection disease
Prediction	MRI	Brain Lung	Disease prediction Survival prediction
Classification	CT	Brain Liver	Disease classification

Table 4 Examination table for faster R-CNN and R-CNN techniques utilizing the vehicle

Objector vehicle	Faster-CNN (%)	R-CNN (%)
Car	78.7	60.8
Cycle	48.6	34.89
Train	82.6	72.40
Bus	80.4	67.44

Table 2 displays the estimated speeds of all vehicles. All vehicles are identified by deep learning techniques. The vehicles have increased precision and velocity, while R-CNN vehicles are minimally moderate and have some effectiveness.

Because group movement behaviors are more common in pedestrians, which is also a significant barrier restricting the application of pedestrian identification, occlusion is more likely to occur in pedestrian detection compared to general object detection. The amount of pedestrian occlusion in the suggested image collection has a significant impact on the precision of pedestrian positioning, which is more sensitive to the threshold, making it simple to suppress the candidate frames of comparable pedestrians. The amount of model calculations has dramatically risen, and the detection results have improved to some level [15]. Examination table for Faster R-CNN and R-CNN techniques utilizing the vehicle is shown in Table 4.

2.3 *Mechanical Object Detection in Welding Process*

Welding is a production process that joins at least two materials, typically metals and non-metals. It is also the tried-and-true, efficient, and conservative method for permanently joining comparable and non-comparable metals and plastic. Welding is utilized extensively in all facets of the manufacturing industry. Recently, a variety of welding techniques have developed. The way in which heat and weight (when used) are connected, as well as the type of equipment employed, varies greatly between these processes. Numerous welding techniques are accomplished by weight without the use of extreme heat, others by a combination of temperature and weight, and still others by heat alone without the use of weight.

Welding is utilized in numerous fields, including bridges, weight vessels, building structures, airship, automobile, railway track, railway mentor plants, nuclear establishments, security organizations, pipelines, electrical and electronic, etc.

Types of welding joints are listed as follows:

- Deep welding joint.
- Lading welding joint.
- Edge welding joint.
- Garner Welding Compound.
- Butt welding joint.
- Square welding joint in square.

- Tea welding joint.

Two individuals cross 90 degrees to meet the edges of a plate or section to form the edges of the sheet welding. The tea joints can be viewed as a form of shear ignition, and all the objects are additionally welded into the base plate (such as pipe or tube). Additional consideration must be taken to ensure adequate ignition top entrance.

Welding Styles Used to Build a D-Joint

- Plug ignition, pilot burner, radius ignition, slot ignition, flare angle crack ignition, J-groove weld, Melt through the weld. Considering the mechanical oriented side, to employ an optimized orthogonal wavelet for the wavelet feature extraction. The below Fig. 3 shows different types of welding joints.

The concept of an orthogonal transform originates from wavelet theory, where it is defined as a transform that has a wavelet transform that is also orthogonal.

$$\frac{1}{2}s_f\left(\frac{Z}{2}\right) = \sum_{m\in x} w_f s_f\left(x_{a,b} - c\right)$$

where (w_f) stands for wavelet transform, the (s_f) scaling factor, the pixel value (c), and the image data ($x_{a,b}$), Z is orthogonal Wavelet. After retrieving an image, it is then subjected to a recognition procedure.

Pseudo Code Extraction:

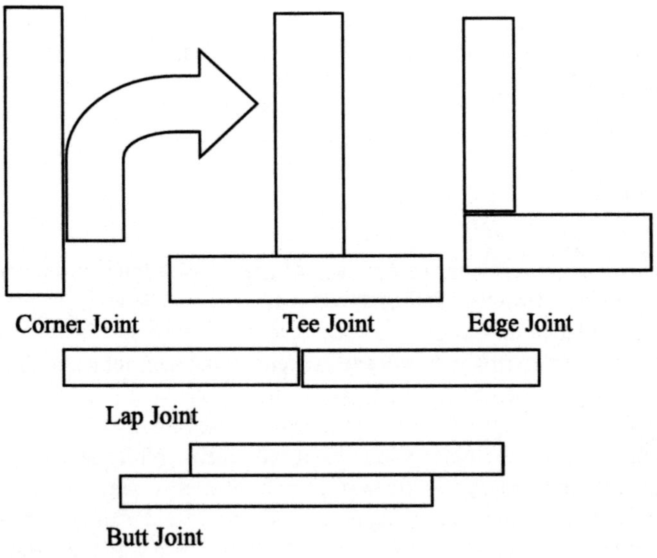

Fig. 3 Block diagram types of welding joint

Input: Image img

Output: Feature set Fs

Step 1: Start

Step 2: Read input image Img

Step 3: Obtain edges Es = Apply canny edge detector on img

Step 4: Extract local tetra pattern LTrP

Step 5: Extract wavelet features W_f

Step 6: Add to feature set Fs

Step 7: Stop

Using the aforementioned algorithm, we can see how the image's various features were retrieved.

2.4 *Identification of Welding*

Welding rod identified in E6011 electrode indicates arc welding. Each electrode designed for various applications. The type of electrode used depends on particular properties of weld used.

Gas metal arc welding used in Automobile industry, Robotics etc. Stick Shielded Metal arc Welding used for thin metals. Gas Tungsten Welding used in Industries for metals. Flux coated arc welding used for thick metals. Gas welding is portable. Different types of weld defects can be identified using various machine learning and deep learning techniques.

3 Conclusion

In conclusion, the performance comparison of object detection based deep learning techniques has been reviewed and analyzed in this paper. The various deep learning architectures and their application in object detection have been discussed and evaluated based on their accuracy, speed, and computational requirements. Through this review, it has become apparent that deep learning-based object detection techniques have shown remarkable progress in recent years, achieving high levels of accuracy and efficiency in detecting objects in images and videos. While each deep learning architecture has its advantages and disadvantages, it is clear that the field is rapidly advancing, with new techniques being developed and refined regularly. As such, it is essential for researchers and practitioners to stay up-to-date with the latest advancements in object detection to choose the most appropriate technique for their specific application. Overall, this paper has provided valuable insights into the current state of object detection using deep learning techniques and has highlighted the importance of ongoing research in this field to address the challenges that remain. The knowledge gained from this review will undoubtedly contribute to the continued advancement of object detection in a wide range of applications, including autonomous vehicles, surveillance systems, and robotics.

References

1. Wang MC (2018) Edge preserving and multi-scale contextual neural network for salient object detection. IEEE Trans Image Process 27(1):121–134
2. Chen K, Zhu M, Fidler, Urtasun (2018) 3D object proposals using stereo imagery for accurate object class detection. IEEE Trans Pattern Anal Mach Intell 40(5):1259–1272
3. Li L, Li W, Xu FY (2018) Multistage object detection with group recursive learning. IEEE Trans Multi 20(7):1645–1655
4. Huston HH, Clarke CP (1992) Reliability defect detection and screening during processing-theory and implementation. In: 30th Annual Proceedings Reliability Physics, pp 268–275
5. Le SA (2018) Video salient object detection using spatiotemporal deep features. IEEE Trans Image Process 27(10):5002–5015
6. Sultani W, Mokhtari S, Yun H-B (2018) Automatic pavement object detection using super pixel segmentation combined with conditional random field. IEEE Trans Intell Transp Syst 19(7):2076–2085
7. Millard, Liu (2018) Simulation of near-surface detection of objects in Layered Media by the BCGS–FFT method. IEEE Trans Geosci Remote Sens 42(2):327–334
8. Omid-Zahoor Y, Ta, Merman (2018) Toward always-on mobile object detection: energy versus performance tradeoffs for embedded HOG feature extraction. IEEE Trans Circuits Syst Video Tech 28(5):1102–1115
9. Vergel A-C, Santelli, Moreno-Nogoer (2018) Boosted random ferns for object detection. IEEEE Trans Patten Anal Mach Intell 40(2):272–288
10. Mahmud KH, Vasarely (2018) Applications of deep learning and reinforcement learning to biological data. IEEE Trans Neural Netw Learn Syst 29(6):2063–2079
11. Pan W, Cheng W, Yu (2018) Multisource transfer double DQN based on actor learning. IEEE Trans Neu Netw Learn Syst 29(6):2227–2238
12. Bellotti H, Gomes, Bayne (2018) Expert level control of ramp metering based on multi-task deep reinforcement learning. IEEE Trans Intell Transp Syst 19(4):1198–1207
13. Yang L, Chen (2018) Supervised learning of semantics-preserving hash via deep convolutional neural networks. IEEE Trans Pattern Anal Mach Intell 40(2):437–451
14. Kitayama M, Kiya H (2019) HOG feature extraction from encrypted images for privacy-preserving machine learning. In: 2019 IEEE International Conference on Consumer Electronics—Asia (ICCE-Asia), Bangkok, Thailand, pp 80–82
15. Tian D, Han Y, Wang B, Guan T, Wei W (2021) Neural network-based machine learning in data mining for big data systems. Comp Intell Neurosci, Article ID 5410049

Regularized Information Loss for Improved Model Selection

Firuz Kamalov, Sherif Moussa, and Jorge Avante Reyes

Abstract Information criteria are used in many applications including statistical model selection and intelligent systems. The traditional information criteria such as the Akaike information criterion (AIC) do not always provide an adequate penalty on the number of model covariates. To address this issue, we propose a novel method for evaluating statistical models based on information criterion. The proposed method, called regularized information criterion (RIL), modifies the penalty term in AIC to reduce model overfitting. The results of numerical experiments show that RIL provides a better reflection of model predictive error than AIC. Thus, RIL can be a useful tool in model selection.

Keywords Information criteria · Model selection · Akaike information criterion · Computational statistics · Intelligent systems

1 Introduction

Statistical models are rarely exact when estimating the underlying processes based on random sample data. Therefore, identifying the best statistical model to fit the data is a one of the primary tasks in data science [11]. A good statistical model should simultaneously fit the data at hand and generalize to unseen data. There exist several approaches to select the right candidate model including F-statistic, residual errors, information criteria, and others. In particular, information criteria are a popular approach for model evaluation. It is based on estimating the amount

F. Kamalov (✉) · S. Moussa
Department of Electrical Engineering, Canadian University Dubai, Dubai, UAE
e-mail: firuz@cud.ac.ae

S. Moussa
e-mail: smoussa@cud.ac.ae

J. A. Reyes
Faculty of Engineering, Canadian University Dubai, Dubai, UAE
e-mail: jorge_avante@outlook.com

G. Rajakumar et al. (eds.), *Intelligent Communication Technologies and Virtual Mobile Networks*, Lecture Notes on Data Engineering and Communications Technologies 171,
https://doi.org/10.1007/978-981-99-1767-9_58

of information lost by the model from the data. However, the existing information criteria do not always accurately estimate the generalization ability of the statistical models. Therefore, we propose a new information criterion with the aim of improving the measurement of model performance on unseen data. Numerical experiments demonstrate the effectiveness of the proposed method over existing approaches.

Information criteria are a widely used method for evaluating the goodness of a statistical model. In particular, the Akaike information criterion (AIC) and the Bayesian information criterion (BIC) are two of the most popular tools for model selection in statistics and machine learning [3, 5, 36]. The AIC and the BIC estimate the amount of information loss for a given dataset and thereby the quality of statistical models. Information criteria reward the likelihood of explaining the data by the model, while penalizing for additional model parameters. The AIC and BIC have the same reward for the likelihood of the model, but they assign different penalties for additional model parameters. Given the ease of calculating the information criteria together with the solid theoretical basis, the AIC and BIC are frequently used in model evaluation. Indeed, most of the modern statistical packages and software provide the ability to quickly and conveniently compute the AIC and BIC values.

Although the AIC and BIC are frequently used in model selection, they do not always correctly identify the best statistical model. The difference in the penalty for the additional model parameters can lead to diverging outcomes when using the AIC and BIC. Striking the right balance between the the reward and penalty is an intricate issue that does not have an exact analytical solution. Increased penalty for model parameters leads to increased bias, while decreasing the penalty leads to higher variance. In this paper, we propose a new model evaluation criterion dubbed regularized information loss (RIL). The proposed criterion is based on the AIC with a modified penalty for additional parameters. Concretely, we suggest to increase the penalty for the number of parameters in the model. The increased penalty reduces model variance and leads to better out-of-sample performance. We compare the proposed model against the AIC and BIC using numerical experiments. The results show that the RIL significantly outperforms the benchmark criteria.

Our paper is structured as follows. In Sect. 2, we provide a brief overview of the literature related to model evaluation. Section 3 provides the description of the proposed information criterion as well as the necessary background. Section 4 contains the details of our numerical experiments to compare the proposed approach to the existing criteria. We conclude with brief remarks in Sect. 5.

2 Literature

Model selection is an important and well-investigated topic [8, 26]. Model selection methods can be generally grouped into two subsets: statistical measures [4] and information criteria [10]. Statistical metrics work well when the underlying distributional requirements are satisfied, while information metrics do not depend on any prior assumptions. Given their flexibility, information criteria are often used to capture nonlinear relationships in the model. The importance of correct model selection has led to a significant amount of research in this field [32, 35]. Despite a plethora of model selection approaches, there remains room for improvement.

AIC is one of the most commonly used information criteria. In [30], the authors use AIC and BIC to evaluate three potential models for bioconvective nanofluid flow. In medical science, risk prediction models for post-operative mortality were evaluated based on AIC and BIC [21]. In epidemiology, AIC has been used to evaluate epidemic models that describe the spread of Covid-19 [37]. There exist several attempts to extend AIC for various applications. In [2], the authors propose an Akaike-type generalization for model selection in logistic regression, multilevel regression, and structural equation model. Another AIC-based model selection criterion was proposed in [16] for structural equation models. An extension of AIC based on regression convolutional neural network was proposed in [18] for acoustic emission signals onset-time selection. In [19], the authors propose an AIC-based approach to identify the most relevant variables in eco-efficiency evaluation. An adjusted AIC similar to adjusted R^2 was proposed in [24]. Information-based criteria are frequently used in phylogenetic model selection, where computational complexity is one of the primary concerns [12, 17]. The authors in [33] used AIC-based approach to identify the important variables in modeling hotel profitability.

Bayesian criterion has also been widely used in the literature. It has been shown that BIC is an effective substitute for out-of-sample criteria [31]. A comparison of various BIC-based approaches can be found in [23]. A number of studies have used and extended BIC in different applications. BIC has been widely used in model averaging in ecology as an alternative to null hypothesis [6, 9]. Bayesian information gain was applied to interaction tasks [20]. BIC was used in time series analysis in [15]. In [25], the authors used BIC as a fitness function to evaluate classification models. An extended BIC was proposed in [7] to deal with small-n-large-p sparse GLM. The issue of evaluating models with order constraints on the covariates is tackled in [22], and a truncated unit information prior is utilized under the order-constrained model.

Other information criteria such mutual information have also been used in model selection [27]. The authors employ mutual information to determine the optimal model in case of limited labeled data [28]. Model selection in the form of feature selection has also been an active research area. The authors in [13] combine several criteria into a single vector metric to identify the optimal subset of features. The L2 norm between covariates and the target variable was used the metric for feature selection in [34]. Variance decomposition according to individual covariates was used in [14] to identify the optimal features for model selection.

The main drawback of the existing information-based model selection approaches is the requirement for large quantity of data. Since information criteria are inherently non-parametric, they require a significant amount of data to produce correct estimations of model effectiveness. The amount of required data increases exponentially with the number of variables in the model.

3 Regularized Information Criterion

The proposed method is based on AIC and BIC. It aims to measure the generalization capacity of a statistical model using regularized information criterion. First, we give a brief background on AIC and BIC. Then, we provide the details of the newly proposed approach.

3.1 AIC

The Akaike information criterion (AIC) was originally proposed by Hirotugu Akaike [1]. It uses information theory to estimate the predictive error of a statistical model trained on a given dataset. In particular, the AIC estimates the amount of information loss by a model. The information loss can be used to evaluate the quality of the model, where a model with lower information loss is considered better. In estimating the information loss, the AIC balances between the model bias and variance by penalizing models with greater number of attributes. The AIC is calculated according to the following formula

$$\mathrm{AIC} = -2\log L(\theta|D) + 2k, \tag{1}$$

where θ is the vector of model parameters, $L(\theta|D)$ is the maximum likelihood of the model for the given dataset D, and k is the number of estimated parameters in the model. To calculate the AIC, first the maximum likelihood of the model with parameters θ conditional on a given dataset D is computed. Then, twice the negative log-likelihood is added to twice the number of variables in the model to obtain the value of AIC. As can be seen from Eq. 1, AIC decreases as the likelihood of the model parameters increases, i.e., as the model fits better to the given data. On the other hand, AIC increases as the number of model parameters increases. Thus, AIC penalizes models for additional explanatory variables to avoid overfitting. However, it may be argued that the penalty for the number of attributes is not steep enough to disqualify the models that overfit. To address this issue, we propose a modified version of AIC which will be discussed in the next section.

3.2 BIC

The Bayesian information criterion (BIC) was originally proposed by Gideon E. Schwarz [29]. It also uses information theory to estimate the predictive error of a statistical model. In particular, it estimates the amount of information loss by a model for a given dataset, thereby quantifying the quality of a model. The BIC is developed using a Bayesian argument. It is given by the following formula

$$\text{BIC} = -2\log L(\theta|D) + k\ln(n), \tag{2}$$

where $L(\theta|D)$ and k are the same as above, and n is the number of samples in the data, i.e., the size of D. Similar to AIC, BIC penalizes models for additional attributes. However, unlike in AIC, the amount of penalty in BIC depends on the size of the data. In particular, the penalty increases with respect to the number of samples. The $k\ln(n)$ term acts as a regularizer. On the other hand, a large number of samples in data also automatically acts as a regularizer preventing the model from overfitting. Thus, BIC imposes additional regularization in terms of n.

3.3 RIL

The existing information criteria do not adequately measure the performance of statistical models on unseen data. To address this issue, we propose a new information criterion for evaluating statistical models. The proposed criterion is based on modifying AIC by increasing the penalty for additional model parameters.

The AIC assigns a linear penalty for the number of parameters which may lead to model overfitting and poor out-of-sample performance. We suggest to increase the penalty for additional parameters while maintaining the same likelihood reward as in AIC. In particular, we propose to replace the linear penalty with a semi-polynomial penalty. A semi-polynomial penalty has a higher rate of increase with respect to the model size which would favor sparse models.

The proposed criterion is called regularized information loss (RIL) and is given by the following formula

$$\text{RIL} = -2\log L(\theta|D) + k\sqrt{k}. \tag{3}$$

When applying RIL in model evaluation, the model with the lower RIL value is preferred to the model with the higher RIL value. As can be seen from Eq. 3, the penalty for additional attributes is increased from $2k$ to $k\sqrt{k}$. For instance, for $k = 9$, the AIC and RIL penalties are 18 and 27, respectively. The difference between AIC and RIL increases as k increases. The increased penalty for the number of parameters helps avoid model overfitting leading to improved out-of-sample performance.

4 Numerical Experiments

To measure the effectiveness of the RIL, we perform a range of numerical experiments using a fund raising dataset. The dataset presents a regression problem of predicting the amount of donors' contributions from a direct mail campaign based on their demographics and history of past contributions (max contribution, average contribution, last contribution, gender, etc.). It is a modified subset of the original dataset collected by the United States National Veterans' Organization to develop a predictive model to capture donors expected contributions.

We consider models with varying number of parameters and analyze the corresponding RIL values in relation to the model test errors. We use AIC and BIC as the benchmarks for the effectiveness of RIL in predicting the out-of-sample model performance. The results show that the RIL is more correlated with the test errors than AIC and BIC.

4.1 Methodology

The initial fundraising dataset consists of 11 continuous attributes $\{x_i\}$ and 1 continuous target variable y. We expand the number of covariates by including the pairwise products of the original attributes. Thus, we add the predictors

$$x_1x_1, x_1x_2, x_1x_3, \ldots, x_{10}x_{11}, x_{11}x_{11}$$

to the initial dataset which results in a total of 77 attributes—11 original features plus 66 pairwise products.

The numerical experiments consist of fitting the least squares linear regression model to data with different number of features starting with 1 predictor to 77 predictors. For each subset of predictors, we conduct 100 bootstrap simulations to evaluate model performance. Concretely, for each bootstrap simulation, we select a random sample of observations which is used as the train data. The unselected samples are used as the test data. The bootstrap sample is used to fit the linear regression model, and the corresponding train mean squared error (MSE), AIC, BIC, and RIL are calculated. The unselected samples are used to calculate the test MSE. Finally, the average train MSE, test MSE, AIC, BIC, and RIL values are calculated over the 100 runs.

Our goal is to determine whether RIL values provide more accurate evaluation of the model quality than AIC and BIC. This is done primarily by comparing RIL values to the bootstrap test errors. Since the test error represents the performance on unseen data, it indicates the quality of the model. The information criterion that matches closely to the test errors is deemed to be the optimal.

4.2 Results and Analysis

The graphs of the bootstrap train and test errors over the range of model complexity are given in Fig. 1. It shows that the train and test errors diverge as the number of predictors increases which indicates model overfitting. As the number of parameters increases, the model learns to fit the noise inside the data and achieve low train error. On the other hand, it leads to the model producing high test errors because it overfits on the train set. As shown in Fig. 1, there is an initial sharp drop in the test error.

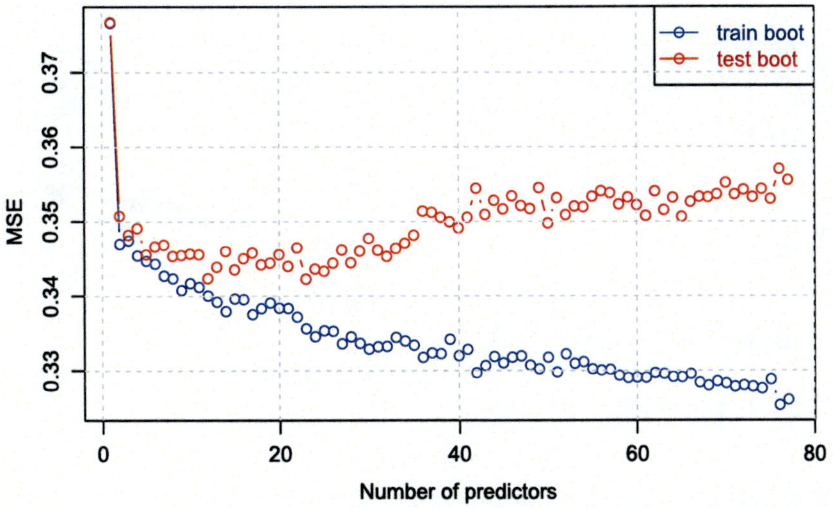

Fig. 1 Train and test errors over the range on model complexity. Each value is calculated as the mean over 100 bootstrap sample sets

Then, the train and test errors produce similar results until $k = 10$, after which the two graphs begin to diverge. Beyond approximately $k = 23$, the divergence between the test and train MSE becomes very sharp.

The values of RIL, AIC, and BIC for models with different number of predictors are shown in Fig. 2. Recall that for each subset of features, the values of the index are calculated as the average over 100 bootstrap samples sets. It can be seen that both the BIC and the RIL graphs approximately match the graph of the test error in Fig. 1. As shown in Fig. 2, there is an initial sharp drop in the value of RIL. Then, RIL remains relatively stable until $k = 20$, after which it begins to increase. It is a similar pattern as the test error in Fig. 1 albeit the test error begins to increase earlier around $k = 10$. Nevertheless, the overall pattern of the graphs of RIL and test error is similar. We also note the RIL graph reflects the test error graph significantly better than the AIC graph. Given the low regularization in the AIC, it fails to adequately penalize the models with excessive number of parameters. The difference between the graphs of the AIC and RIL in Fig. 2 with respect to the graph of the test error in Fig. 1 highlights the advantage of the RIL.

To quantify and summarize the relationship between the test error and the information criteria, we calculate their correlations. In Table 1, we present the Pearson correlations between the AIC, BIC, and RIL values and the bootstrap train and test errors. As can be seen from the table, RIL achieves the highest correlation with the test error among the three information criteria. In particular, the RIL correlation is 0.88, while the AIC and BIC correlations are 0.14 and 0.87, respectively. It shows that the RIL is more consistent than other information criteria in predicting out-of-sample error.

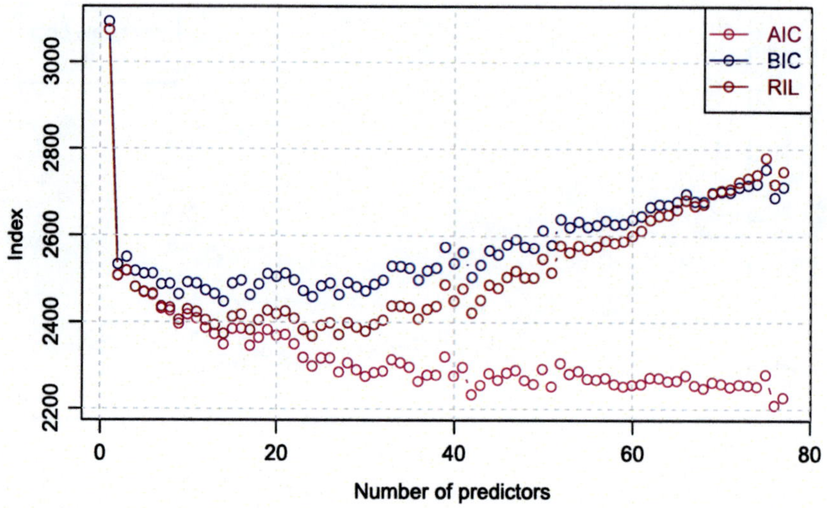

Fig. 2 RIL, AIC, and BIC values over the range on model complexity. Each value is calculated as the mean over 100 bootstrap sample sets

Table 1 Pearson correlations between the information criteria and the test MSE

	Train	Test	AIC	BIC	RIL
Train	1.00	−0.04	0.98	−0.05	−0.09
Test	−0.04	1.00	0.14	0.87	0.88
AIC	0.98	0.14	1.00	0.16	0.11
BIC	−0.05	0.87	0.16	1.00	0.98
RIL	−0.09	0.88	0.11	0.98	1.00

Table 2 Spearman correlations between the information criteria and the test MSE

	Train	Test	AIC	BIC	RIL
Train	1.00	−0.78	0.95	−0.73	−0.67
Test	−0.78	1.00	−0.73	0.82	0.83
AIC	0.95	−0.73	1.00	−0.58	−0.51
BIC	−0.73	0.82	−0.58	1.00	0.97
RIL	−0.67	0.83	−0.51	0.97	1.00

Similarly, in Table 2, we present the Spearman rank correlation between the errors and information criteria. The results are in line with the Pearson's correlation. The RIL achieves the highest correlation with respect to the test error among the three information criteria. In particular, the RIL correlation is 0.83, while the AIC and BIC correlations are −0.73 and 0.82, respectively.

The results demonstrate that RIL provides a better reflection of the out-of-sample performance of a model than AIC or BIC. The increased penalty for the number of parameters produces an improved measure of model performance. As shown in Figs. 1 and 2, the graph of the RIL is similar to the graph of the test MSE. In addition, Tables 1 and 2 show that RIL achieves higher correlation with the test MSE than AIC and BIC. Thus, the RIL values are more in line with the test MSE values than other information criteria.

5 Conclusion

In this paper, we proposed a new method, called the regularized information loss (RIL), for evaluating model quality using information theory. The proposed method is based on the traditional information loss criteria but with an increased penalty for additional model parameters. Numerical experiments demonstrate the effectiveness of RIL over the traditional measures AIC and BIC.

The initial results show that RIL can be a useful metric for statistical model evaluation. However, as with other information-based criteria, there exist practical limitations. In particular, if the model parameters take values on the boundary of the parameter space, the asymptotic distribution of the likelihood ratio statistic will not be that of χ^2 distribution with 1 degrees of freedom. This boundary problem can affect the effectiveness of RIL.

In future research, further studies with greater number of datasets are warranted to fully analyze the effectiveness of the proposed criterion. In addition, the use of RIL in other applications such as feature selection and classification models can be explored in future.

References

1. Akaike H (1998) Information theory and an extension of the maximum likelihood principle. In: Selected papers of Hirotugu Akaike. Springer, New York, NY, pp 199–213
2. Altinisik Y, Van Lissa CJ, Hoijtink H, Oldehinkel AJ, Kuiper RM (2021) Evaluation of inequality constrained hypotheses using a generalization of the AIC. Psychol Methods 26(5):599. Chicago
3. Bai Z, Choi KP, Fujikoshi Y (2018) Consistency of AIC and BIC in estimating the number of significant components in high-dimensional principal component analysis. Ann Stat 46(3):1050–1076
4. Barron AR (2020) Predicted squared error: a criterion for automatic model selection. In: Self-organizing methods in modeling. CRC Press, pp 87–103
5. Bozdogan H (1987) Model selection and Akaike's information criterion (AIC): the general theory and its analytical extensions. Psychometrika 52(3):345–370
6. Burnham KP, Anderson DR, Huyvaert KP (2011) AIC model selection and multimodel inference in behavioral ecology: some background, observations, and comparisons. Behavioral Ecol Sociobiol 65(1):23–35

7. Chen J, Chen Z (2012) Extended BIC for small-n-large-p sparse GLM. Stat Sin 22(2):555–574. http://www.jstor.org/stable/24310025
8. Ding J, Tarokh V, Yang Y (2018) Model selection techniques: an overview. IEEE Sign Process Mag 35(6):16–34
9. Dormann CF, Calabrese JM, Guillera-Arroita G, Matechou E, Bahn V, Bartoń K, Hartig F (2018) Model averaging in ecology: a review of Bayesian, information-theoretic, and tactical approaches for predictive inference. Ecol Monogr 88(4):485–504
10. Dziak JJ, Coffman DL, Lanza ST, Li R, Jermiin LS (2020) Sensitivity and specificity of information criteria. Briefings Bioinform 21(2):553–565
11. Heinze G, Wallisch C, Dunkler D (2018) Variable selection-a review and recommendations for the practicing statistician. Biometrical J 60(3):431–449
12. Kalyaanamoorthy S, Minh BQ, Wong TK, Von Haeseler A, Jermiin LS (2017) ModelFinder: fast model selection for accurate phylogenetic estimates. Nat methods 14(6):587–589
13. Kamalov F, Thabtah F (2017) A feature selection method based on ranked vector scores of features for classification. Ann Data Sci 4(4):483–502
14. Kamalov F (2021) Orthogonal variance decomposition based feature selection. Expert Syst Appl 182:115191
15. Khan FM, Gupta R (2020) ARIMA and NAR based prediction model for time series analysis of COVID-19 cases in India. J Saf Sci Resilience 1(1):12–18
16. Kuiper R (2022) AIC-type theory-based model selection for structural equation models. Struct Eq Model Multidisc J 29(1):151–158
17. Lefort V, Longueville JE, Gascuel O (2017) SMS: smart model selection in PhyML. Mol Biol Evol 34(9):2422–2424
18. Li H, Yang Z, Yan W (2022) An improved AIC onset-time picking method based on regression convolutional neural network. Mech Syst Sign Process 171:108867
19. Li Y, Zhang Q, Wang L, Liang L (2021) An AIC-based approach to identify the most influential variables in eco-efficiency evaluation. Expert Syst Appl 167:113883
20. Liu W, Rioul O, Beaudouin-Lafon M (2023) Bayesian information gain to design interaction
21. Mahmud N, Fricker Z, Hubbard RA, Ioannou GN, Lewis JD, Taddei TH, Kaplan DE (2021) Risk prediction models for post-operative mortality in patients with cirrhosis. Hepatology 73(1):204–218
22. Mulder J, Raftery AE (2022) BIC extensions for order-constrained model selection. Sociol Methods Res 51(2):471–498
23. Piironen J, Vehtari A (2017) Comparison of Bayesian predictive methods for model selection. Stat Comput 27(3):711–735
24. Pham H (2019) A new criterion for model selection. Mathematics 7(12):1215
25. Qasim OS, Algamal ZY (2018) Feature selection using particle swarm optimization-based logistic regression model. Chemometr. Intell Lab Syst 182:41–46
26. Raschka S (2018) Model evaluation, model selection, and algorithm selection in machine learning. arXiv preprint arXiv:1811.12808
27. Rajab K, Kamalov F (2021) Finite sample based mutual information. IEEE Access 9:118871–118879
28. Schnapp S Sabato S (2021) Active feature selection for the mutual information criterion. In: Proceedings of the AAAI conference on artificial intelligence, vol 35, no 11, pp 9497–9504
29. Schwarz G (1978) Estimating the dimension of a model. Ann Stat 461–464
30. Shafiq A, Lone SA, Sindhu TN, Al-Mdallal QM, Rasool G (2021) Statistical modeling for bioconvective tangent hyperbolic nanofluid towards stretching surface with zero mass flux condition. Sci Rep 11(1):1–11
31. Sharma PN, Shmueli G, Sarstedt M, Danks N, Ray S (2021) Prediction-oriented model selection in partial least squares path modeling. Decis Sci 52(3):567–607
32. Solorio-Fernández S, Carrasco-Ochoa JA, Martínez-Trinidad JF (2020) A review of unsupervised feature selection methods. Artif Intell Rev 53(2):907–948
33. Taylor DC, Snipes M, Barber NA (2018) Indicators of hotel profitability: model selection using Akaike information criteria. Tour Hosp Res 18(1):61–71

34. Thabtah F, Kamalov F, Hammoud S, Shahamiri SR (2020) Least loss: a simplified filter method for feature selection. Inf Sci 534:1–15
35. Tredennick AT, Hooker G, Ellner SP, Adler PB (2021) A practical guide to selecting models for exploration, inference, and prediction in ecology. Ecology 102(6):e03336
36. Wagenmakers EJ, Farrell S (2004) AIC model selection using Akaike weights. Psychon Bull Rev 11(1):192–196
37. Yang W, Zhang D, Peng L, Zhuge C, Hong L (2021) Rational evaluation of various epidemic models based on the COVID-19 data of China. Epidemics 37:100501

Review on Use of Agile Techniques in Software Development over Traditional Customs

Himanshu Srivastava and Nitish Ojha

Abstract Software development using agile techniques is gaining popularity nowadays just because of its benefits over traditional development. Many software practitioners and researchers are taking interest in agile development, but they need to be aware about all agile techniques currently available and about all the challenges they will face while migrating to the agile technologies. It has been seen that only big companies are getting benefits of agile, but not small or startup companies, while agile can be beneficial more for such type of organization. More awareness for agile development should be spread. This paper presents agile goals, principles, team ethics, and moral values with different agile methodologies which will help to create a truly agile team for developing software.

Keywords Agile software development · Scrum · Feature driven development · Kanban · Dynamic system software · Extreme programming · Adaptive software · Lean · Crystal

1 Introduction

Background and progress to software development: Initially software development gets up as the professional concerned about how to create software and how to increase its quality. Quality can refer to speed, usability, size, minimum cost, less development time, on time delivery, customer satisfaction, security, and bug free. It requires rules and regulation for writing quality code. It can be done by hiring good working practices, professionals, and a good working environment. During the 1960's software development was affected by the so called software crisis [1]

H. Srivastava
Department of CSE, Bennett University, Greater Noida, India

N. Ojha (✉)
Department of CSE, University of Stirling, Ras Al Khaimah, UAE
e-mail: nitishkumarojha@gmail.com

G. Rajakumar et al. (eds.), *Intelligent Communication Technologies and Virtual Mobile Networks*, Lecture Notes on Data Engineering and Communications Technologies 171,
https://doi.org/10.1007/978-981-99-1767-9_59

and it will continue till next two more decades. At the time of the software crisis many problems were identified in developing software. Result was that user requirements were not supported by the software, unreliable, and poor quality product was produced, it was exceeding the budget and delivery time. Software was hardly delivered. If it is delivered then it would be difficult to maintain it. Demand was increasing to develop complex software as hardware capability was increasing than software developing ability. Software became too dangerous as it was harming human beings, even killing in case of radiotherapy machines [2] and much more [3]. After that, several researchers and practitioners find out the nature of software development and make a well-defined, disciplined systematic approach to develop software known as software engineering whose aim was to produce quality software [4].

Traditional development methods: Traditional approaches, also called engineering approaches, are well defined and systematic. Indeed, the software development process needs to be controlled somehow. So software engineers came into existence along with engineering methods through which software development processes can be controlled. Common traditional methodologies include waterfall, iterative and incremental development, prototyping rapid application development, and spiral model development. All these are bounded by some discipline where stages of development are the requirement analysis, design, build, test, deployment, and maintainability are well predictable. These methods, also known as heavyweight processes [5], are rich in documentation and complex to apply. In fact, this is the main reason to discuss the disadvantages of traditional software development methods. It can be said that traditional methodologies are bureaucratic. They use systematic approaches where the design phase is always followed by the perfect build phase [6]. Because of its main features about detailed planning and designing phases, it is useful if the project is very large and has a high level of risk factor [7]. But if a project is very dynamic, requirements change day by day, it makes it hard to predict, in this case traditional methodologies will fail and agile methodologies come into existence to handle such types of projects.

Emergence to Agile methodology: In 2001, 17 software development practitioners gathered in Snowbird, Utah [8], to talk about their shared ideas and various methods to develop softwares. Their main aim was to emphasize close collaboration between the stakeholder and development team with frequent delivery of partially finished business valued products. For this, they combined old and new ideas of software development and created some principles and values which are manifested in Agile Manifesto [9]. The agile methods are different from traditional methods by their level of adaptability, low level of bureaucracy, high level of face to face communication, customer interaction at all the phases of development, quick feedback, high responsiveness, quality assurance, lower defects, timely delivery, and welcome changes [10]. Basically, an agile approach is an incremental and iterative approach. Many big companies like Microsoft, Intel, Yahoo, Shopzilla, and Siemens [11] are using agile methodologies.

This paper is divided into six sections. First section having the introduction related to the starting of software development, software crisis, systematic software development, and origin of agile approach. Second section shows the literature survey done before writing this paper. Third section contains concepts regarding agile manifesto, agile values, agile principles, agile team ethics and values, agile life cycle, and its comparison with traditional software development approaches. Fourth section having information regarding various famous agile techniques and burn charts used for monitoring software development. Fifth section contains challenges to agile migration and the last section contains the conclusion of this paper.

2 Literature Survey

In 2015 Georgios [12] suggested the benefits of moving to agile software development from traditional software development. It provides evidence that working with agile is much better than working with a traditional approach. This paper claims that an agile approach is best in practice if working in a distributed environment. Agile team uses better collaboration and communication that enhances relation among team members and improved employee and customer satisfaction. Sletholt et al. [13] conducted a review for knowing the effects of agile practices over traditional software development. They find that software companies migrating to agile have an improved testing process as compared to traditional software development.

Begel and Nagappan [14] conducted a web-based survey on Microsoft employees for disclosing advantages and disadvantages of agile practices. Around one-third employees were using an agile approach and most of the users have a positive mind about agile. Some researchers say that agile is best suited for small projects but not suited for very large projects [15]. Traditional processes are suitable for larger projects but not an agile approach. Parreiras and Campanelli [16] presented research on agile methods tailoring. The term tailoring means selecting an agile technique for an organization. In 2015, Software Advice, conducted a survey on project managers regarding facing challenges in agile practices and around 49% say that agile training is a very difficult and challenging task, they face while adopting agile culture [17]. One more literature [18] suggests current challenges in agile and focused on agility in a distributed environment, requirement prioritization, organization culture, frequent change in requirement, and lack of skilled persons in agile techniques.

3 Agile Approaches

Agile has an iterative view of software development. Instead of keeping all the planning upfront, agile focuses on being lean and generating minimum workable product over a set period of time while renovating each passage. Agile approach is also known as lightweight approach [19].

Manifesto/Goals: Researchers and practitioners derived new values for agile development [9]. They decided that an agile environment needs only motivated individuals with fresh minds and maximize communication among them and their locations does not matter. Agile assumes that stakeholders can wish for changes in the requirement many times and it is the responsibility of the developer.

Principles: All the agile methodologies follow agile principles [9] that are written by practitioners and researchers. Agile software developers must follow agile practices. According to agile practices, sustained delivery of precious products is required and customers must be highly satisfied at any cost. Sometimes it may allow delay in development, but always welcome changes done by the stakeholder without taking any headache. It allows daily communication between stakeholder and developer to satisfy the customer. According to agile, the best form of communication is only face to face communication, even technology can be used to fulfill this principle. It delivers partially finished products in some weeks or any other bounded time period, but does not deliver complete products at first time. Only trusted and motivated individuals are allowed to work in an agile team. They all must have high moral values. If not so then the team is not truly agile and cannot be succeeded. Project progress can be measured only if the team is delivering workable products to stakeholders and this constant pace should be maintained by the team with continuous development. Team must maintain technical excellence at all times. Steadily, the team is revealed to be more effective and problem solving quickly. Last principle says that just think simple to solve a problem rather than thinking complex.

Agile Team Ethics and Values: Human beings are social elements and get accidently affected by emotions and sensibility which affect their work [20]. A team's success highly depends on the team member behavior, working environment, and organization culture. As developers, our first crucial moral obligation is to confer value to our customers for which the team were hired. It is the only value that holds the team truly agile, after the trainer leaves the team. Some researchers define values as [21]: *Commitment*–It means 'do or do not do', there is no try if you have committed then you must fulfill it else no need to do a commitment. *Focus*–You do not need to be sidetracked just remember whatever you have committed and focus your energies on saturating your promise. *Openness*–It means always keeping project status visible to everyone by displaying it. *Communication*–Face to face real time communication is required. Geographically separated teams can use technology like 'Google Meet', and 'Microsoft Teams'. *Simplicity*–It is better to develop something very simple that may be extended later, rather than spending a long time on finding a solution for a small problem that is more complicated and may not be necessary. *Feedback*–The team must obtain feedback from stakeholders to verify the product built by the team and must match with their willingness. *Courage*–All team participants should have courage to 'say no'. This is the basic value that will give rise to all others. *Respect*–Each team member must start with agreeing that each member deserves to be treated with respect. These all above values help to work in an ethical manner. Finally, an agile team must have a better technical and social cohesion to get success.

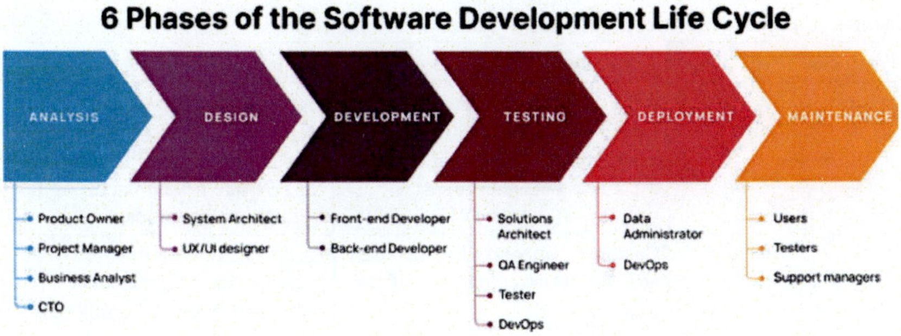

Fig. 1 Agile life cycle

Agile Life Cycle: Software development using agile approach is based upon incremental and iterative process and mainly focuses on adaptability of changing product requirement and higher customer satisfaction by frequent delivery of partially finished shippable product [22], continuous feedback and client participation in all the phases. There exist several agile methods varieties; it depends upon the developer to choose one. Agile team members are experienced and have a great level of expertise in activities like designing, building, testing, and quality assurance. These activities go in parallel [23] in the case of agile. At the end of each sprint, a working product is shipped to the customer for verification purposes. If the customer is satisfied then the team plans for the next sprint, if not then the customer can change it. It means customers can add or remove features from the product and it may waste the work of hardly some hours or some days which is acceptable by the agile team without any headache thus it reduces the risk [24]. This cycle continues until the final product is prepared with higher quality. Agile Life Cycle is shown in Fig. 1.

Difference Between Traditional Development and Agile Development: The software development strategy can be subdivided into two categories [25], the traditional process and the agile process. Traditional process made by a sequential series of steps like requirement gathering, planning, implementing, testing, and deployment. Firstly stakeholder needs are carefully noted down in a document and then designing, coding, and various types of testing are performed. The main idea here is that extensive visualization of the product at each stage. Agile progressives admit that software is not a big block structure, but a beautifully organic entity with sophisticated moving parts communicating with each other [25]. Hence, they produce more importance to constant, compatibility, testing, and adaptability. Whatever is the difference between these two, it can be shown in the Table 1.

Table 1 Comparison between Agile and Traditional methodologies

Parameter	Traditional methodology	Agile methodology
Process	It follows a sequential approach. It uses very poor traditional mechanisms. Traditional methods use a heavyweight process	It follows an iterative and incremental approach. It uses a modern, truly dynamic approach for software development. Agile is a lightweight process
Feedback mechanism	Customers only provide feedback after the 100% completion of the product	At the end of each iteration of a partially finished product is handed over to the customer for review
Customer involvement	Very less customer involvement only at the starting and ending of the product	During software development a customer is involved at all the time as a stakeholder representative
Adaptability	It supports very less adaptability. Team members are afraid to do so	It welcomes changes, even delay is accepted in development
Product release	Product is released only after the completion of the entire project	At the end of each iteration, the product is released to the customer for verification
Focus	It focuses only on following the plan	It focuses on satisfying customer wishes
Value	It gives value to the completion of the project	It is a customer-oriented approach
Cycles	Limited cycles	Unlimited cycles
Customer satisfaction	Customer satisfaction is lean	Highly satisfied customers
Product quality	It only gives value to delivery of the product	It provides high quality products
Project size	Suitable for large projects	Suitable for medium and small projects
Phases	All development phases go sequential	All development phases go parallel
Expenses	It involves customers only at the starting of the development process. Therefore, the number of mistakes can be higher so extra money is required to rework on them	Agile involves customers at each phase so corrections are easy, as soon as possible. This will help to save money
Team size	It is large	It is smaller like 6 to 9 persons
Developing problem encountered	When a problem occurs, the manager of the project is contacted immediately	Team itself tries to resolve issues and get solutions itself
Serious about	More serious about process then product	Agile methods are less serious about processes but more serious about product quality

(continued)

Table 1 (continued)

Parameter	Traditional methodology	Agile methodology
Nature	Predictive and static	Visionary, uncertain, and adaptive
Initially requirements	Maximum requirements are cleared at the starting of the project	All requirements are not clear at the time of starting of project. It can be added or removed at any time
Communication	Limited communication and more stress on documentation. It uses hierarchical and diplomatic types of communication	Face to face communication over a short period of time among all team members and less importance on documentation. It uses an integrated and transparent type of communication
Requirement change	Not flexible for requirement change. A lot of rework effort is involved in that case. Cost of change is higher	More flexible to change requirements at any stage. Cost of change is very low
Testing	It involves only after the completion of development	Testing is done parallel with the development phase, and it helps to find defects soon
Leadership style	More controlling and commanding leadership style	No one works as a leader. Team itself plans and finds solutions. It uses a collaborative process
Culture	Individual focuses, blame shifting, competition, fear of failure, risk aversion	Team performance, trust, transparency, collaboration, risk management, failure as a new learning opportunity, and welcome change
Team members	Employees are inflexible and ignorant to changes	Multi-skilled, self-motivated, multi-functional, great learner, and self-committed
Return on investment	Only at the end of the project, when completed, product is delivered	Early in the project, sprint is delivered
Developmental model	It follows a capability maturity model	It follows the story card maturity model
Risk factor	Risks in traditional models are higher as it delivers the entire project at the end of development	It reduces the risk by frequent delivery of the workable product

4 Agile Techniques and Monitoring Tool

Agile development is an incremental and iterative development approach [26]. Popular agile methodologies include Scrum, Feature Driven Development, Kanban, Dynamic System Development Method, Extreme Programming, Adaptive Software Development, Lean, and Crystal Development [27]. All these methodologies are

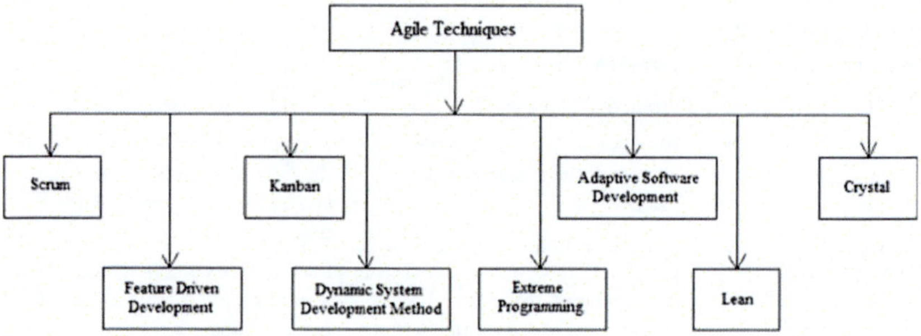

Fig. 2 Agile techniques

unique to its specific approach, but they all share basic values and generic vision that comes from agile manifesto. They all include continuous planning, continuous integration and continuous testing, and other forms of sustained growth. Agile Techniques are shown in Fig. 2.

Scrum: Scrum framework is an agile methodology that divides the project development into iterations known as sprints [28]. In designing any software, companies employ 'Scrum Lead' who are responsible for managing all the sprints, making all the team members work together and achieving the goal together. Scrum is a popular framework for designing complex software. Sprint is always time boxed, no more than a month (2–4 weeks). Sprint duration is decided by the team based on capability and requirements. Every iteration, should attempt to build shippable product increments that are tested and work well. Since sprint is prepared in a short duration, only important features would be delivered first and then customers can test and provide feedback.

As shown in Fig. 3, we start the process with the product owner, it is nothing but stakeholder who provides requirements. Product backlog is a set of all requirements broken down into small features and prioritized into a list. Scrum methodology advocates for a sprint planning meeting at the start of the sprint, where team members find out how many features or items they can commit within a sprint and then create a sprint backlog from product backlog which contains a list of tasks to perform during the sprint. On each day, during the sprint, all team members must attend a daily stand-up meeting. This meeting is also time-boxed around 15 min in which work barriers and its solutions are discussed. At the end of the sprint, the team points out new functionality to the product owner that influences the next sprint. Each sprint ends with a meeting, named retrospective meeting, attended by team members, scrum master, and product owner where they discuss what we have done in the current sprint and how the next sprint can be improved. Scrum Team is shown in Fig. 4.

Feature Driven Development: This methodology was developed by Jeff De Luca and Peter Coad with M. A. Rajashima, Stephen Palmer, and Jon Kern [29, 30]. As the name suggests, feature driven development works on a set of rules by organizing the software development process into several steps and checking its progress. It started

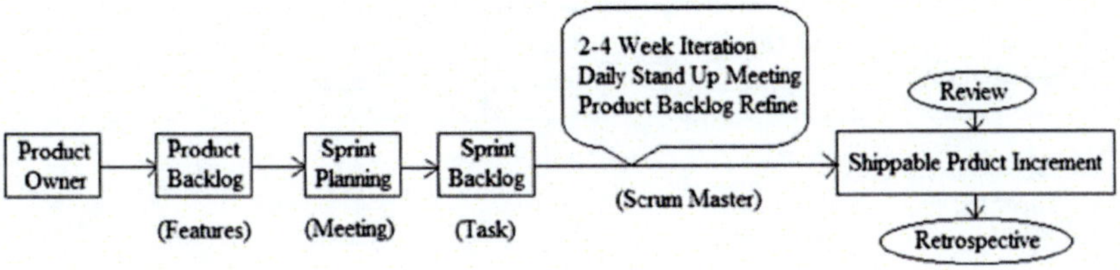

Fig. 3 Scrum working

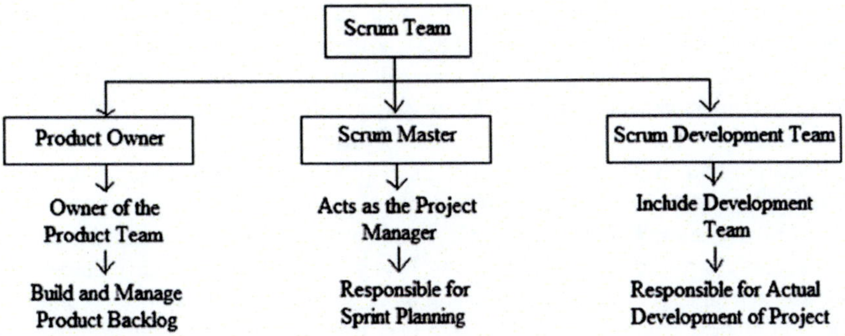

Fig. 4 Scrum team

primarily in 1999, while designing a java-based software with unified modeling language. In a feature driven model the steps are gradual, i.e., developing a model, building a feature list, planning by feature, designing by feature, and finally delivering the end product, which has been designed. This concept was very thriving in the software industry, but many times it was not found suitable for small projects as requirement of feature makes it adverse. Feature driven development begins with a walk through of the introduced system and then detailed domain is elected, later on, it can be merged with an overall model. Features are miniature pieces of client valued functions given in the form as <action> <result> for example–calculating the total of a sale. After finishing the creation of the feature list, the next step is to produce a product development plan. It should not take more than two weeks. A design must be produced for each feature (class). After successful design creation, each design is converted into code. After unit testing and successful code inspection, a completed feature is added to the main build. This process continues [30]. Feature Driven Development Process is shown in Fig. 5.

Kanban: This method is used for making team workflow, awesome improvements in performance, continual improvement without the overhead and risk. Here you can do a piece of work based on trust, not based on pressure. Basically, this concept comes from Walmart [31] improvement in the 1950's. At that time, they want to improve their performance by reducing waste in Walmart. Workers noted down the future requirements on some cards and submitted them to stockholders. Stockholder fulfills

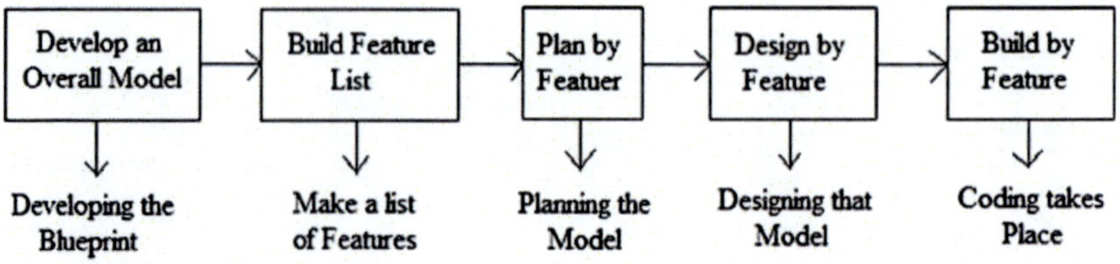

Fig. 5 Feature driven development process

the requirements to the mart and submitted the same requirements to the supplier; later on the supplier fulfills the needs of the stockholder. This means Walmart having enough goods to fulfill the present requirements but not too many goods that become complex to manage and resultant to the wastage. This concept originated from lean technology [32] but with different functionality as it is using cards concept to pass information and this card passing system is known as Kanban concept in agile. Each card contains information regarding to do list, work in progress and work done tasks [33]. Basically Kanban is a visual system, based on three basic principles, first it visualizes the workflow, second it helps the flow-based approach so team members do not start and finish too much unmanageable work, third when something is committed the next highest rating task is pulled into play from the backlog. Kanban Board is shown in Fig. 6.

Dynamic System Development Method: This methodology came into light in 1994 as a more disciplined and structured approach than its predecessor namely rapid application development [34]. Thus, this consortium serves as a possible replacement for unstructured rapid application development which is an incremental and iterative approach. This technology fixes cost, time, and quality. It uses MoSCoW prioritization [35] concept and is a popular prioritization technique for managing requirements. This acronym represents four categories like must have, should have, could have, and won't have this time, to fulfill project delivery for meeting the stated time, cost, and quality constraints. This method has five phases as feasibility study, business study, functional model iteration, design and build iteration, and implementation. Within each phase this technology relies on different techniques and activities based on eight rules as it must focus on business need, project must be delivered on time, team members must be collaborative, never compromise with quality, always prepare incremental product, develop iteratively and improve it, must communicate clearly and continuously, and transparent visibility of the project.

Extreme Programming: Extreme programming took effective practices and principles to the extreme level. It was developed by Kent Beck and Ward Cunnigham in 1999 [36]. Here planning is effective because short iterations are used. Design is always effective as everyone needs to perform refactoring daily [37]. Refactoring is the process of simplifying and clarifying the design of existing code, but without changing its behavior. Code reviews are effective because the code is reviewed at all

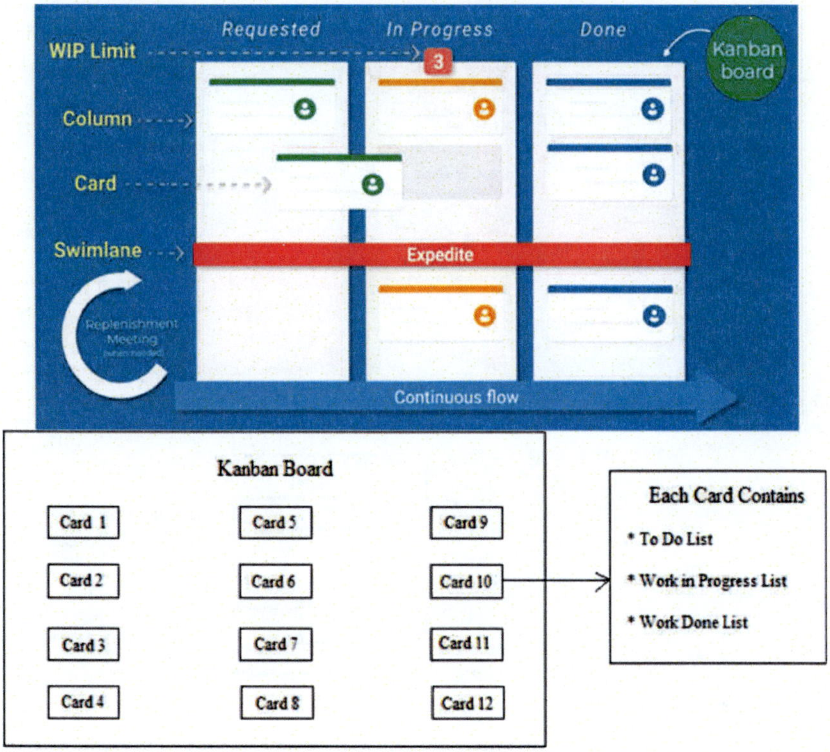

Fig. 6 Kanban board

the time and it uses pair programming concepts. Pair programming [38] is a technique in which two programmers work together at one machine. One (driver) writes the code while the other (observer or navigator) observes each line of code. The two programmers can switch roles frequently. Testing is very effective like unit testing, continuous integration testing [39], and acceptance testing for each small release. Extreme Programming Process is shown in Fig. 7.

Adaptive Software Development: This technology is best suited for the projects where higher levels of uncertainty lies. An adaptive approach encourages changes throughout the project to optimize the overall solution. The adaptive software development grew out of the rapid software development method with slight changes by Jim Highsmith and Sam Bayer [40]. This contains a repeating series of speculate, collaborate, and learn cycles that replaces traditional waterfall cycles [41]. The lack of pre-planning steps allows us to make software quickly. The developmental cycle is so short that a new version with new features can come out quickly. The learning cycle is based on short iterations with design, build, and testing. During these iterations one can attain knowledge by making small mistakes thus leading to maximizing experience [41]. Adaptive Software Development Process is shown in Fig. 8.

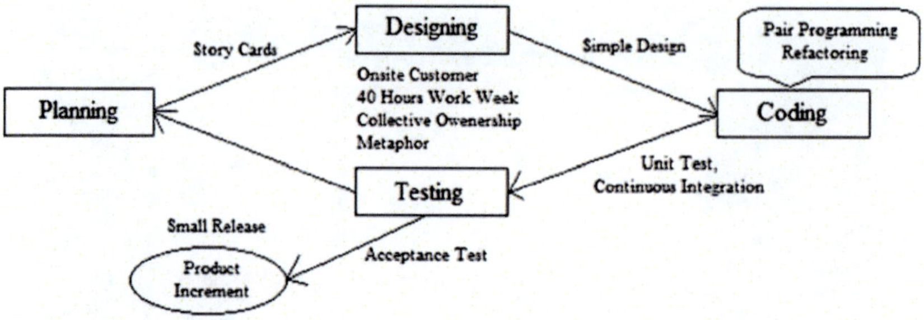

Fig. 7 Extreme programming process

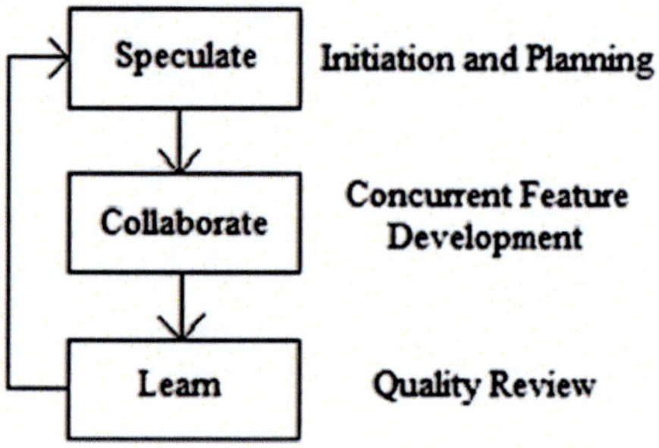

Fig. 8 Adaptive software development process

Lean Development: This methodology was originally developed for Toyota automobile manufacturing company. Engineer Ohno [42] developed lean development principles which mainly focus on optimizing efficiency, empowering workers, eliminating waste, reducing inventory, and improving productivity. By maximizing the use of multi-skilled workers, companies can respond to change quickly to the market demand than their competitors. Software industry also adopted the principles of Toyota lean production [43]. The main principle of lean methodology includes, do it right at first time, eliminate waste, minimizing inventory, meet customer demand, pull production from customer requirements, maximum flow, empower workers, adaptability, and culture of sustained improvements.

Crystal: Crystal is made up of a family of agile methodologies such as crystal clear, crystal orange, crystal yellow, and others [44]. Their unique characteristics are derived by many factors such as size of developmental team, system criticality, project priorities, and project risk. Mainly crystal methodology focuses on community, people interaction, talents, skills, and communication.

Agile Monitoring using Burn Charts: Burn charts are graphical representations of a work progress. There exist two types of burn charts: the first one is a burnup chart and second one is a burndown chart as shown below in Fig. 9 [45]. Burnup chart depicts how much work of a committed day (or week, month, or other time period) has been completed while burndown chart shows how much remaining work of commitment

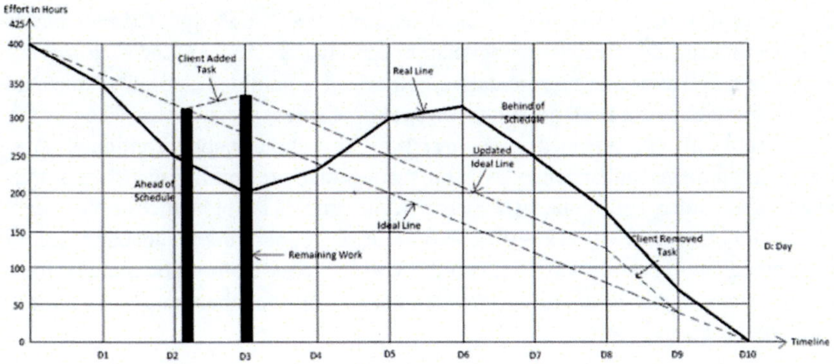

Fig. 9 Burn down chart

is left. In burn charts x-axis represents timeline and y-axis represents amount of work or efforts in the form of days or hours or story points. Burn chart having two lines such as ideal line that shows expected progress and real line that shows current progress. In the burnup chart, the real line moves toward upward while in the burndown chart, the real line moves toward downward.

Burn charts can be used as a forecasting tool. Let us take the example of the burndown chart in Fig. 9, suppose one sprint is of 10 days, team consists of 8 members working 5 h daily. Above diagram shows the ideal line in dotted format and real line in bold format. Within the first four days the real line is below the ideal line; it means that the team is working ahead of schedule. During the third day more features are added by the stakeholder so the ideal line is updated to the new ideal line. On the fifth day the real line starts to go above the ideal line that means current work is behind the schedule. During the ninth day the client removed some features and again the ideal line was updated. At the end of tenth day the project team met their goal. One sprint is completed on time with the help of a burndown chart because the team can visualize its performance at all the time during the sprint.

5 Migrating to Agile

Benefits of agile methodologies, all the time, motivate the developers for migrating to agile from traditional but they should be aware of the challenges they will face while migrating. Migration affects all dimensions of an organization [46]. Main challenges are in organization culture, process, people, and management. Enough training, time, effort, and patience can help them to grasp it. These challenges can be explain as change in: *Management style*–Traditional system uses control and command style, while agile uses collaborative environment style, *New tool*–Agile technology is in favor of object oriented so development requires new knowledge, *Current literature*–It is not sufficiently exist for agile approach, *Availability of CRACK customer*–Agile needs collaborative, representative, authorized, committed, and knowledgeable customer which is not at all the time, *Power shift*–Critical decisions are taken by the team, not by only one person, *Technology issues*–Tool should supply continuous integration, incremental enhancement, version management, reworking, distributed environment, and other agile concept, *Process*–Agile process is not predictive but uncertain and requires creativity at any time, *People*–Agile requires maximum communication and collaboration among team members which required trust and moral values, *Elitist culture*–Agile needs no difference among team members and it is possible with good team members.

6 Conclusion

This research paper is prepared for the learners, practitioners, and researchers so that they have some knowledge regarding agile software development. There exist a number of traditional methods for developing software even then agile techniques come into light and become popular. These agile techniques are used by a number of big companies, but not by the small companies. This paper illuminates the advantages of agile software development over traditional development so that small companies as well as startup companies can take advantage of agility while developing software.

References

1. NATO software engineering conference 1968. http://homepages.cs.ncl.ac.uk/brian.randell/NATO/NATOReports/index.html
2. Leveson TCS (1993) An investigation of the Therac-25 accidents. Comp 26(7):18–41. ISSN 0018-9162
3. http://catless.ncl.ac.uk/Risks/ accessed on 01/12/2022 at 1400 Hrs
4. IEEE STD 610.12-1990 (1990) IEEE standard glossary of software engineering terminology
5. Nikiforova O, Nikulsins V, Sukovskis U (2009) Integration of MDA framework into the model of traditional software development. Frontiers in artificial intelligence and applications, databases and information systems V, vol 187. IOS Press, Amsterdam, pp 229–239
6. Fowler M (2005) The new methodology
7. Wysocki RK, McGary R (2003) Effective project management, 3rd edn. Wiley
8. https://www.theatlantic.com/technology/archive/2017/12/agile-manifesto-a-history/547715/, accessed on 01/12/2022 at 1400 Hrs
9. Beck K, Beedle M, van Bennekum A, Cockburn A, Cunningham W, Fowler M, Grenning J, Highsmith J, Humt A, Jeries R, Kern J, Marick B, Martin RC, Mellor S, Schwaber K, Sutherland J, Thom D (2001) Manifesto for agile software development. http://agilemanifesto.org/, accessed on 01/12/2022 at 1400 Hrs
10. Highsmith J, Cockburn A (2001) Agile software development: the business of innovation. Computer 34:120–127
11. Schwaber K (2004) Agile project management with scrum. Microsoft Press

12. Papadopoulosa G (2015) Moving from traditional to agile software development methodologies also on large, distributed projects. Procedia Soc Behav Sci 175:455–463
13. Sletholt MT, Hannay J, Pfahl D, Benestad HC, Langtangen HP (2011) A literature review of agile practices and their effects in scientific software development. In: Proceedings of the 4th International Workshop on Software Engineering for Computational Science and Engineering, Wuhan, 9–11 December, 1–9
14. Begel A, Nagappan N (2007) Usage and perceptions of agile software development in an industrial context: an exploratory study. In: 1st International Symposium on Empirical Software Engineering and Measurement, Madrid, 20–21 Sep, M. Al-Zewairi et al. 96 tember 2007, 255–264. https://doi.org/10.1109/ESEM.2007.12
15. Nerur S, Balijepally V (2007) Theoretical reflections on agile development methodologies. Comm ACM 50(3):79–83
16. Campanelli AS, Parreiras FS (2015) Agile methods tailoring—a systematic literature review. J Syst Soft 110:85–100. https://doi.org/10.1016/j.jss.2015.08.035, accessed on 01/12/2022 at 1400 Hrs
17. SoftwareAdvice (2015) Agile project management software user report. http://www.softwareadvice.com/resources/agile-project-management-user-trends-2015/, accessed on 01/12/2022 at 1400 Hrs
18. Resti W, Rahayu P, Indra D (2016) Challenges in agile software development: a systematic literature review. In: Computer Science and Information Systems (ICACSIS), 2016 International Conference. IEEE Xplore. Malang, Indonesia. pp 155–164. https://doi.org/10.1109/ICACSIS.2016.7872736
19. https://www.arrkgroup.com/thought-leadership/lightweight-software-development-explained/, accessed on 01/12/2022 at 1400 Hrs
20. https://www.agileconnection.com/article/agile-ethics-and-values, accessed on 01/12/2022 at 1400 Hrs
21. Triandis HC (1994) Culture and social behavior. McGraw-Hill Book Company
22. Beck K (1999) Embracing change with extreme programming. Computer 32(10):70–77. https://doi.org/10.1109/2.796139
23. Martin RC (2002) Agile software development, principles, patterns, and practices. Prentice Hall
24. Moran A (2014) Agile risk management. Springer Verlag, ISBN, p 3319050079
25. http://www.itsandpm.com/traditional-vs-agile-software-development/, accessed on 01/12/2022 at 1400 Hrs
26. Larman C (2003) Iterative and incremental development: a brief history (PDF). Computer 36(6):47–56. https://doi.org/10.1109/MC.2003.1204375.ISSN0018-9162
27. Dyba T (2008) Empirical studies of agile software development: a systematic review
28. Raising L, Janoff NS (2000) The Scrum software development process for small teams. IEEE Softw 17(4):26–32
29. Palmer SR, Felsing JM (2002) A practical guide to feature-driven development. Prentice Hall
30. Coad P, Lefebvre E, De Luca J (1999) Java modelling in color with UML: enterprise components and process. Prentice Hall International (ISBN 0-13-011510-X)
31. https://leankit.com/learn/kanban, accessed on 01/12/2022 at 1400 Hrs
32. Womack JP (2007) The machine that changed the world. ISBN 978-1847370556
33. Boeg J (2012) Priming Kanban. Info. Retrieved 2014-02-17
34. Abrahamsson P, et al (2003) New directions on agile methods: a comparative analysis. Software Engineering. In: Proceedings. 25th International Conference on IEEE
35. Clegg D, Barker R (2004) Case method fast-track: a RAD approach. Addison-Wesley. ISBN 978-0-201-62432-8
36. Interview with Kent Beck and Martin Fowler. http://www.informit.com/articles/article.aspx?p=20972, accessed on 01/12/2022 at 1400 Hrs
37. Fowler M, Beck K, Brant J, Opdyke W, Roberts D (1999) Refactoring: improving the design of existing code. Addison Wesley, Reading, MA

38. Williams L, Ramsey D, Bourque P, Dupuis R (2001) Integrating pair programming into a software development process. In: 14th Conference on Software Engineering Education and Training. Charlotte. pp 27–36. https://doi.org/10.1109/CSEE.2001.913816. ISBN 0-7695-1059-0
39. https://www.thoughtworks.com/continuous-integration, accessed on 01/12/2022 at 1400 Hrs
40. Highsmith JM, Messy E, Anxiety-Ridden (1997) Adaptive software development. American Program 10:23–29
41. Messy E, Anxiety-Ridden, Adaptive software development. https://en.wikipedia.org/wiki/Adaptive_software_development accessed on 01/12/2022 at 1400 Hrs
42. Ohno T (1988) Toyota production system. Productivity Press. p 8. ISBN 0-915299-14-3
43. Monden Y (1998) Toyota production system, an integrated approach to just-in-time, 3rd edn. Engineering & Management Press, Norcross, GA. ISBN 0-412-83930-X
44. Cockburn A (2004) Crystal clear: a human-powered methodology for small teams. Addison-Wesley, ISBN 0-201-69947-8
45. Cohn M (2006) Agile estimating and planning. Upper Saddle River, NJ, Prentice Hall
46. Yang C, Liu J, Chen H, Luo X (2009) Adaptive optimization of agile organization of command-and-control resource. J Syst Eng Elect 20:558–564

Smart Traffic Signal Control System Using Artificial Intelligence

G. R. P. Kumari, M. Jahnavi, M. Harika, A. Pavani, and C. Venkata Lakshmi

Abstract One of the biggest issues in metropolitan areas is traffic congestion, despite having well-planned road systems and adequate infrastructure. The main cause of this issue is the 40% annual increase in the number of cars on the road. Most current traffic control systems are fixed cycle types, which always cycle through red, yellow, and green. The deployment of these pilots is accompanied by the deployment of traffic police officers to maintain order in the streets. Unlike human traffic cops, these inflexible systems cannot adjust to changing circumstances on the fly. Intelligent traffic management systems are needed immediately. In order to measure traffic volume, our proposed system will use AI and image processing to analyse live feeds from cameras placed at intersections. The amount of vehicles passing through the intersection is predicted to increase by around 32% based on simulation results, which is a substantial gain over the status quo. More training and calibration of the model with actual CCTV data can bring about significant improvements in the system's performance.

Keywords Traffic monitoring · Traffic congestion detection · Image processing · Intelligent transport systems · Automated sensing · Object detection

1 Introduction

There are now so many cars on the road that traffic congestion is one of the world's biggest issues. Congestion on the roads is caused by a number of factors [1], for example, a static signalling system and a dearth of necessary infrastructure. These predetermined intervals cannot meet the needs of today's dynamic web traffic [2]. They perform optimally only in steady-state traffic conditions. Today's traffic management systems make use of time-sharing and assign an overall cycle period to

G. R. P. Kumari · M. Jahnavi (✉) · M. Harika · A. Pavani · C. V. Lakshmi
Vignan's Nirula Institute of Technology and Science for Women, Guntur, Andhra Pradesh, India
e-mail: jahnavichowdary9876@gmail.com

G. Rajakumar et al. (eds.), *Intelligent Communication Technologies and Virtual Mobile Networks*, Lecture Notes on Data Engineering and Communications Technologies 171,
https://doi.org/10.1007/978-981-99-1767-9_60

each phase in order to adapt to actual traffic conditions in real time. Today's traffic lights typically use one of three methods to regulate vehicular flow [3].

In this modern era of rapid population and traffic expansion fuelled by industrialization, traffic management has emerged as one of the most pressing issues [4]. Congestion, infractions of traffic laws, and other issues are just some of the negative outcomes of a traffic boom. Congestion and poor planning also cause unnecessary delays, which waste time and money. Therefore, a quick, heap, and efficient controlling traffic system is necessary for national development [5]. Applying mechanisation and intellectual control methods to highway infrastructure and vehicles is one way to enhance flow of traffic and security in current transportation systems.

With a growing population, it is getting harder for current transportation infrastructure to keep up [6]. Road and highway expansion is highly unlikely due to limited resources and available land, making intelligent systems like advanced traffic control crucial for maximising the use of existing infrastructure [7]. Wasted time, energy, and resources can be the result of mistimed signals. About 40% of the fuel consumed by vehicles on road systems with badly timed traffic lights occurs when vehicles are stopped and idling.

There are currently three widely used methods of traffic control:

For example, in Human Management, people are actively involved in directing traffic. The traffic police is assigned to control the areas that need it the most.

The second type is the traditional traffic signal with a static clock, which has a timer whose value is always the same. The lights will cycle among red and green at predetermined intervals, as set by the timer. Thirdly, electronic sensors are another state-of-the-art approach. These can take the form of various sensors or motion detectors placed on the road. A traffic sensor like this one can provide useful insights into vehicular movements. As a result of sensor information, traffic lights can be managed more effectively.

Use live footage from surveillance cameras at intersections to count the number of cars waiting at the lights and adjust the timing of the green lights in real time based on the detected traffic density [8]. For precise estimates of available green time, vehicles are sorted into car, bike, bus/truck, and rickshaw categories [9]. With the help of YOLO, you can monitor traffic volumes at any given moment and adjust the lights accordingly.

2 Literature Review

Kumar et al. [1] proposed a model to circumvent the constraints of the ideal facility, and intelligent transit has emerged as a pressing concern for the smart city. In this work, we propose a flexible and intelligent traffic light system that uses real-time traffic data as input to optimise the stoplight's performance and address the issues it currently faces.

Yang et al. [2] proposed a model effective algorithms for detecting objects in images which have been transferred directly to video. Motion blur, lack of focus, and unusual poses are all degraded by these frame-by-frame processing techniques, making them less than ideal.

Verma et al. [3] proposed a model which plays an important role in attempting to address demanding data managerial functions in intelligent transportation systems (ITS). However, issues with sensor node battery life, uptime, and "security trade-offs" are becoming increasingly problematic.

Lee et al. [4] proposed a model graphic analytics system that can investigate, monitor, and forecast traffic congestion using data from vehicle detectors in a dynamic and intuitive way. In order to better understand what causes traffic congestion and where it is headed, an information visualisation system was built.

Ke et al. [5] proposed a model coordination of traffic lights and the identification of traffic collisions. This article presents a new method for detecting congestion that does not rely on the cumbersome, expensive, and potentially damaging to road surfaces traditional methods.

Liu et al. [6] proposed a model to detect and keep tabs on traffic accidents, and modern traffic management systems collect and process vast amounts of video data in real time. Sending collected data to a road network (TMC) for analysis is a common practise, but doing so can add complexity to the network path to the TMC.

3 Proposed Model

The proposed system gathers data from CCTV cameras placed at intersections and uses image analysis and object detection to determine the volume of traffic in real time [10]. As shown in Fig. 1, this picture is sent to a YOLO-based vehicle detection algorithm. The traffic density is determined by tallying the total number of vehicles as well as the total number of different types of vehicles. This density is just one of many inputs used by the algorithm that controls the timing of the green lights at intersections. Time spent in red has been adjusted accordingly [11]. In Fig. 1, to further prove the system's worth and allow for comparison with pre-existing static systems, simulations will be developed as well.

Module for Detection of Motor Vehicles—the proposed system employs the principle of You only look once (YOLO) to identify passing vehicles. This will provide the precision and speed of processing that you require. Vehicles of various sizes and weights (cars, bikes, buses, trucks, and rickshaws) can all be identified thanks to a specially made YOLO model trained for the purpose. YOLO is a method that provides real-time object detection using neural networks [12]. The popularity of this algorithm is due to its accuracy and quickness. It has been applied in a variety of ways to identify animals, humans, parking metres, and traffic lights.

YOLO is an advanced CNN that can recognise objects in real time. Using a single neural network, this algorithm processes the entire image before segmenting it into smaller regions to make predictions about bounding boxes and probabilities [13]. These outlines are given a value based on the predicted probabilities. One of YOLO's selling points is its ability to function in real time while still maintaining a

Fig. 1 Composition of a smart traffic light

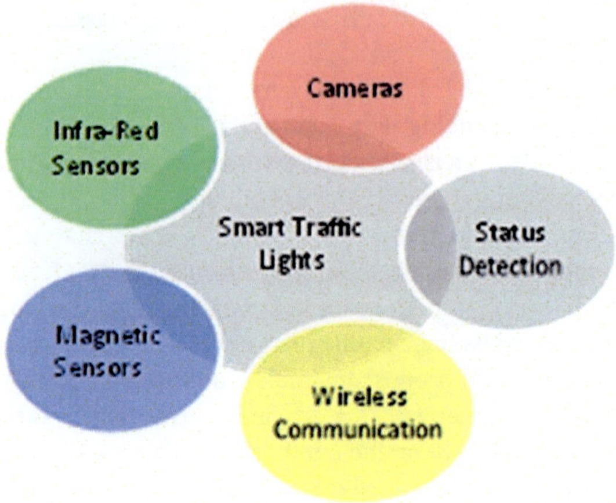

high degree of precision. As shown in Fig. 2, the algorithm only performs a forward transmission through to the neural network in order to make a prediction, meaning that it does not perform any backpropagation [14]. The detected objects and their bounding boxes should be output after non-maximal suppression has been applied (this ensures that the attribute detection algorithm only finds each object once). For multiple boundary frames, YOLO uses a single CNN to predict their probabilities of belonging to different classes.

In order to reduce processing time, the YOLO backbone CNN can be simplified even further. Open-source software runs the DarkNet. It is quick, it is simple to set up, and it works with both CPUs and GPUs [15]. DarkNet improves YOLO's accuracy on this ImageNet from 72.9% in the top 1% to 91.2% in the top 5%. In order to extract features, DarkNet typically employs 33 filters, while 11 filters are used to minimise the number of output channels. As a second method of prediction, it employs pooling data from all over the world.

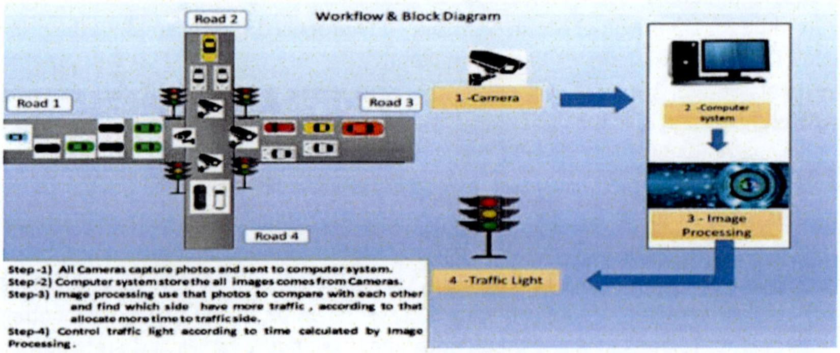

Fig. 2 Smart control of traffic light system

Fig. 3 Vehicle detection

It is necessary to adjust the amount of filters by the Eq. 5*(5 + number of courses), which in our case is 45. been schooled up until. In the example image shown in Fig. 3, shows the object detection of vehicles. In first picture, it is taking the cctv record, and in second picture, it is detecting the images in a box shape and identifying the vehicles. Our traffic surveillance model has been applied. The message switching algorithm updates the red-light times of other signals in accordance with the traffic levels restored either by vehicle detection module. In addition, it uses a timer to alternate signals at regular intervals.

We described how the detection module feeds data about detected vehicles into the algorithm. The file is in JSON format, with the object label serving as the identifier and the self-belief and coordinates serving as the values. With this data, we can determine how many automobiles fall under each category. Assigning the signal's green time and adjusting the red times of all the other signals follows. This algorithm is scalable, meaning that it can be applied to an arbitrary number of traffic signals. When designing the algorithm, we took into account the following details:

The initial run of the algorithm establishes the default time for the first signal in the first cycle, and the algorithm then establishes the time for the remaining transmissions in the first cycle and all signals in subsequent cycles. Each direction's vehicle detection is handled by a different thread, while the main thread keeps track of the signal's timer.

When 5 s remain until the next green light, this is the image that is captured. In this way, the system can analyse the image, count how many vehicles fall into each category, determine how long the green light will be on, and then adjust both the green and red lights to the incoming signal. It is a quick process that only needs 10 s of your time. By calculating the length of time required for each vehicle type to cross the intersection, we were able to determine the optimised green-light time given the number of vehicles in each subject at the traffic light. The discovered intersection points have been calculated. Time till the green light is determined.

In some places, you can adjust how long it typically takes for a given vehicle type to cross a given intersection. H. Manage traffic at each intersection differently depending on its unique characteristics, taking into account the surrounding area and major roads in the vicinity. This can be done by analysing information provided by the relevant shipping company.

Fig. 4 Traffic signals

The direction of the signal is not taken into account at the outset, but rather it alternates at regular intervals. That is in line with the current system, where lights change to green sequentially and in a predictable sequence, so drivers do not have to change their routes or get confused. The current order of traffic signals is maintained, and the yellow lights are considered as well.

Order of signals: Red → Green → Yellow → Red.

We built an entirely new traffic simulation using Pygame simulation module. The time remaining until the light adjustments from green to yellow, yellow to red, or red to green is displayed on a timer at each traffic signal. Every traffic light also shows the number of automobiles that have passed through the intersection. Automobiles, bicycles, buses, trucks, and rickshaws (among other vehicles) arrive from all directions. So that the simulation is more accurate, a few vehicles in the far-right lane are making a U-turn to go across the intersection. When the motor is generated, a random number also determines whether or not the vehicle will turn. This also includes a time limit that suggests how so much time has elapsed since the modelling started. Figure 4 it displaying the signals, provides a quick look at the simulation's final results.

In the event of an emergency, it is essential to get to your destination as quickly as possible to minimise losses of life and property and ensure that urgent situations receive the prompt, appropriate care they need. Despite the fact that numerous studies propose various methods to allow a green light or a clear pathway to EVs, as alluded to earlier, they all assume that only one oncoming vehicle is arriving from a single direction. The following are required for the management of such a dynamic system: an attempt to speculate on the nature of the incident. It involves gathering incident details and storing them in a real-time traffic management system, collecting data on the current state of traffic, identifying the urgency of EVs in response to an incident based on the nature and available data. By using priority levels and current traffic circumstances, the controller determines the best possible route and approves appropriate set of traffic lights' digital signal along and other appropriate roads. Once the event has been cleared, the controller restores normal signalling.

4 Result

Evaluating the Vehicle Detection Subsystem using a variety of test images with varying numbers of vehicles, we determined that the vehicle sensor had an accuracy of 75–80%. This works well enough, but it is not ideal. The scarcity of adequate training data is primarily to blame for the subpar performance. As a solution, we can use data from actual traffic video evidence to fine-tune the model and boost the reliability of the system.

Analysis of the Adaptive System Under Consideration Using a various traffic distribution for each of the 4 quadrants, 15 simulation results of both the proposed approach and the current static system were run over the course of 5 min. The number of cars that could make it through the intersection in a given amount of time was used as a performance metric. Control time at a light, or the period of time when no vehicles are moving through a crossing point because the light is green. Assuming that the existing static system is simulated, Fig. 5.

Intelligent business lights, also known as smart business lights, are a companies' outsource control system that intelligently routes vehicles and ramblers by combining conventional business lamps with a network of sensors and computer vision technology. By providing reliable public transportation, severe penalties for breaking company rules, automated marking systems, etc., a clever business operation system offers a distinct advantage.

In an automated setting, our system should be able to resolve the congestion problem at the intersection, while a manual setting would require human intervention. In Fig. 6, in our framework, both automated and human traffic controls are available (for illustration on the off chance that any rally comes at that point set manual mode

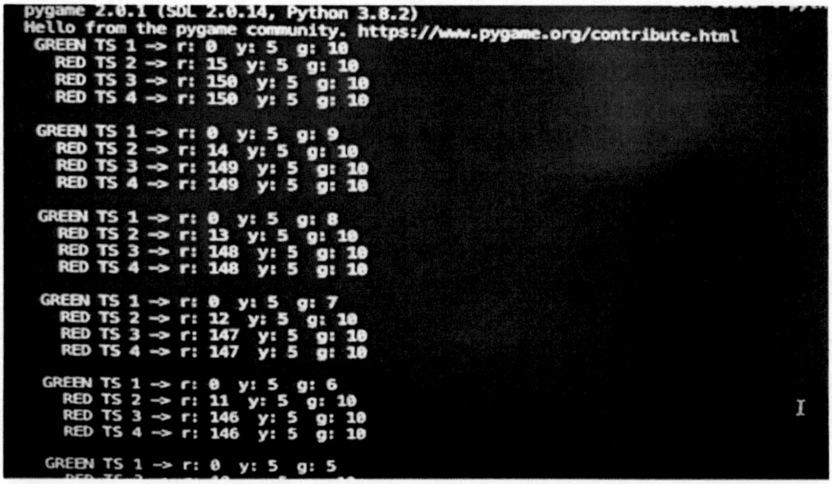

Fig. 5 Output data of simulation

Fig. 6 Simulation output

by setting time). Therefore, give way to emergency vehicles if detected by a sensor, and leave the activity system as is. This means that the Smart Action Light Control System is a simple way to shed light on the activity problem.

For any given distribution, the proposed approach system performs better than the current static system. The degree towards which performance is enhanced is proportional to the degree of lane-to-lane variation in traffic. Figure 7 represents distance and object detection for existing model and proposed model. This holds true for all four of the aforementioned simulations. Here, we see a performance boost of around 9%.

Whenever the traffic dispersion is slightly skewed, the current proposal performs much better than the existing system. In fact, this holds true for simulations 5, 6, 7, 8, 14, and 15. The increased efficiency amounts to around 22%. This distribution of vehicles is representative of what you had seen in the real world. The proposed system significantly outperforms the existing system in the case of highly skewed traffic distribution. Both simulations 9 and 13 exhibit this behaviour; this same red line drops precipitously and a sizable chasm opens up between the two colours. About 36% more efficiency is gained here.

Fig. 7 Accuracy level

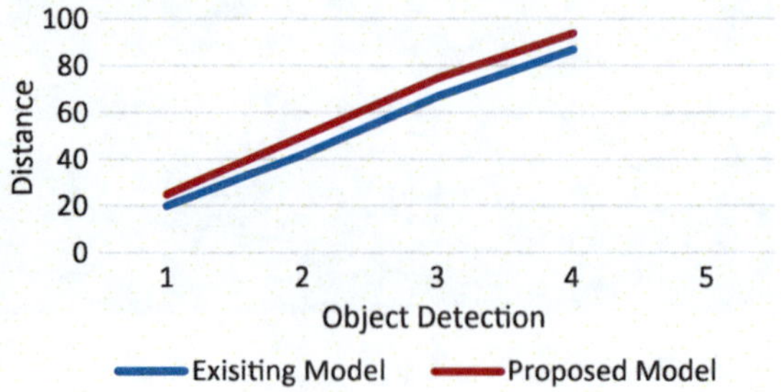

Each simulation ran for 1 h, 15 min, and 300 s, assuming the same traffic allocation, automotive components, transform possibility, length between vehicles, etc. In other words, we discovered that the proposed solution performance improvements by about 23% on median after around 5 min of shipping, in contrast to the current transition stage. There will be less time for cars to sit at a green light and wait. When compared to other adaptive systems, our proposed system performs better in a variety of categories. Our accuracy is 70%, whereas the proposed system achieves 80%. The proposed system improves performance by 23% over the static system on average, while the reference only improves by 12%.

5 Conclusion

In conclusion, the proposed system dynamically adjusts the length of time that traffic lights are green based on the volume of traffic, giving higher priority to directions that see more frequent use. increase. Delays are cut, congestion and wait times are shortened, and pollution and fuel use are both reduced as a result. According to simulation results, the system significantly increases the amount of vehicles trying to cross the intersection by about 23%. This system can be made more effective by readjusting and teaching the model with his actual CCTV data. This means that the proposed framework can be combined with his CCTV in major cities to improve traffic flow management.

References

1. Kumar N, Rahman SS, Dhakad N (2021) Fuzzy inference enabled deep reinforcement learning-based traffic light control for intelligent transportation system. IEEE Trans Intell Transp Syst 22(8):4919–4928. https://doi.org/10.1109/TITS.2020.2984033
2. Wu Y, Zhang H, Li Y, Yang Y, Yuan D (2020) Video object detection guided by object blur evaluation. IEEE Access 8:208554–208565. https://doi.org/10.1109/ACCESS.2020.3038913
3. Verma S, Zeadally S, Kaur S, Sharma AK (2022) Intelligent and secure clustering in wireless sensor network (WSN)-based intelligent transportation systems. IEEE Trans Intell Transp Syst 23(8):13473–13481. https://doi.org/10.1109/TITS.2021.3124730
4. Lee C, et al (2020) A visual analytics system for exploring, monitoring, and forecasting road traffic congestion. IEEE Trans Visualization Comp Graph 26(11):3133–3146. https://doi.org/10.1109/TVCG.2019.2922597
5. Ke X, Shi L, Guo W, Chen D (2019) Multi-dimensional traffic congestion detection based on fusion of visual features and convolutional neural network. IEEE Trans Intell Transp Syst 20(6):2157–2170. https://doi.org/10.1109/TITS.2018.2864612
6. Liu G et al (2022) Smart traffic monitoring system using computer vision and edge computing. IEEE Trans Intell Transp Syst 23(8):12027–12038. https://doi.org/10.1109/TITS.2021.3109481
7. Derrouz H, Elbouziady A, Abdelali HA, Thami ROH, El Fkihi S, Bourzeix F (2019) Moroccan video intelligent transport system: vehicle type classification based on three-dimensional and two-dimensional features. IEEE Access 7:72528–72537. https://doi.org/10.1109/ACCESS.2019.2920740

8. Yilmaz Ö, Görgü L, O'grady MJ, O'hare GMP (2021) Cloud-assisted mobile crowd sensing for route and congestion monitoring. IEEE Access 9:157984–157996. https://doi.org/10.1109/ACCESS.2021.3129932
9. Ghasempour T, Nicholson GL, Kirkwood D, Fujiyama T, Heydecker B (2020) Distributed approximate dynamic control for traffic management of busy railway networks. IEEE Trans Intell Transp Syst 21(9):3788–3798. https://doi.org/10.1109/TITS.2019.2934083
10. Lei X, Liu C, Li L, Wang G (2020) Automated pavement distress detection and deterioration analysis using street view map. IEEE Access 8:76163–76172. https://doi.org/10.1109/ACCESS.2020.2989028
11. Narayana VL, Sirisha S, Divya G, Pooja NLS, Nouf SA (2022) Mall customer segmentation using machine learning. Inter Conference Elect Renew Syst (ICEARS) 2022:1280–1288. https://doi.org/10.1109/ICEARS53579.2022.9752447
12. Pavani V, Sri S, Krishna KSP, Narayana VL (2021) Multi-level authentication scheme for improving privacy and security of data in decentralized cloud server. In: 2021 2nd International Conference on Smart Electronics and Communication (ICOSEC), pp 391–394. https://doi.org/10.1109/ICOSEC51865.2021.9591698
13. Sravanthi GL, Devi MV, Sandeep KS, Naresh A, Gopi AP (2020) An efficient classifier using machine learning technique for individual action identification 11(6):513–520
14. Shaik S, Sudhakar P, Mohiddin SK, A novel framework for image inpainting. Inter J Comp Trends Techn (IJCTT) 14:141–147
15. Gadde S, Sri K, Dumala A, Krishna P, Professor A (2020) A secured cloud environment with user validation method for data communication. Inter J Adv Sci Tech 29:9448–9456

Strategy for Charging of Battery and Supercapacitor Combined Storage System

R. Shalini, P. S. Manoharan, N. KumaraSabapathy, and G. Kannayeram

Abstract This paper presents a strategy for charging the combined energy storage (CES) system that contains supercapacitor and battery. When battery suddenly charges or discharges very quickly within a few seconds, reduction in the life of the battery occurs. To enhance the battery life, supercapacitor is used along with the battery. Whenever the hybrid storage system is connected to the load, it should give power to the load only if it has enough energy. When the battery-state of charge is below a defined level it has to be charged. Likewise, whenever the voltage across the supercapacitor is below the certain level, it has to be charged in order to supply the load. So, here charging strategy is based on battery-state of charge (SOC) and voltage across the supercapacitor.

Keywords Battery · Supercapacitor · Hybrid storage · Charging · Combined storage · Discharge

1 Introduction

As the depletion of fossil fuels occurs in a faster rate, it is important to shift to renewable side quickly [1]. Renewable energy includes the solar, wind, hydro, geothermal, tidal, etc. these sources cannot be used directly instead they are converted from its form to the electrical form and it is stored. For example, in case of hydro, when the

R. Shalini (✉) · P. S. Manoharan
Thiagarajar College of Engineering, Madurai, India
e-mail: shalinijesus99@gmail.com

P. S. Manoharan
e-mail: psmeee@tce.edu

N. KumaraSabapathy
Anna University, Tirunelveli, India

G. Kannayeram
Sri Ramakrishna Institute of Technology, Coimbatore, India

G. Rajakumar et al. (eds.), *Intelligent Communication Technologies and Virtual Mobile Networks*, Lecture Notes on Data Engineering and Communications Technologies 171,
https://doi.org/10.1007/978-981-99-1767-9_61

water is stored in the dam, it is stored in the vein of potential energy. While runoff, the potential energy which was set aside is converted to kinetic energy (KE) [2]. Then, the KE is used to run the turbine and the generator generates energy. The generated energy is either sent to grid by maintaining the power quality of the system [3] or it is stored in batteries for further use. In batteries the energy is stored in the chemical guise. Not only from hydro, several energies obtained is finally stored in a chemical form that is in a battery. Indicated energy density is high but power density is very small. Suppose if the battery supplies the load, sudden change in the load would increase the stress on the battery. It would lead to reduction in the life of the battery [4, 5]. If we need to prolong the life of the battery, then another component should be used along with the battery. The other component is the supercapacitor. Supercapacitor has characteristics opposite to that of the battery. It has high power density and its energy density is very low [6, 7]. The supercapacitor and battery combination is known as the CES system. This type of storage system is used in several applications such as Electric Vehicles, distributed generation systems discussed in [8, 9], respectively. The design of HE storage system and its usage is a complicated process. In this paper only a small part of it is discussed. Charging strategy for battery and supercapacitor hybrid system is discussed. Several papers [10–12] has discussed about the control algorithm for the HE storage system that consists of battery and supercapacitor. Many papers discussed about charging strategy for battery. In [13], they have implemented charging method (Model predictive control) only for battery but the procedure that they have used has lot of calculations, which is complicated. In [14], they have discussed about charging using constant voltage method and also using PI controller and used optimization method which always does not yield best global solution rather than sometimes it gives local solution. So, charging strategy for hybrid system has been proposed in this paper, which is simple. The chosen algorithm is simple compared to other algorithms discussed in various literature papers. The computational procedure compared to other papers is uncomplicated and very straightforward. The control effect of proposed algorithm is better compared to other algorithms. It does not any human intelligence, which is very much needed in soft computing methods.

2 Methodology

2.1 Problem Formulation

Figure 1 shows the strategy for charging. Here, there is a battery which is connected to DC source and also a supercapacitor connected to the charging source through a block which has the code or algorithm for charging.

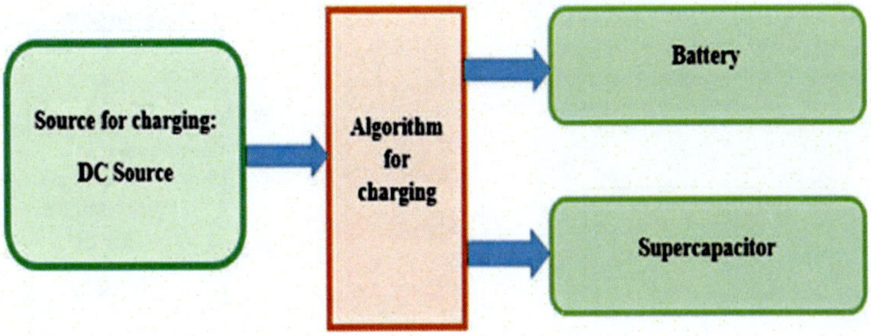

Fig. 1 Charging strategy—block diagram

2.2 *Simulink Model*

Figure 2 shows the simulation diagram of the proposed method. As we can see in the figure that there is DC source. DC source can be a battery or a constant voltage source. Here in this simulation, let us consider the DC source as battery. In Electric Vehicles, the battery used is charged by another source, it can be DC or AC. If we use AC source, then proper design of the inverter should be done according to the battery rating. The battery which is to be charged, is connected to the DC source through the switches. The switch used here has the control signal where the command to close or open can be given. One switch is connected to the Battery's positive side while other switch is connected to negative side of battery. So, for a single storage source, there are two switches connected to the two terminals.

Batteries are of different types. Lithium ion batteries are widely used. Cells are combined together to form batteries. Figure 3 shows battery used in the Simulink diagram.

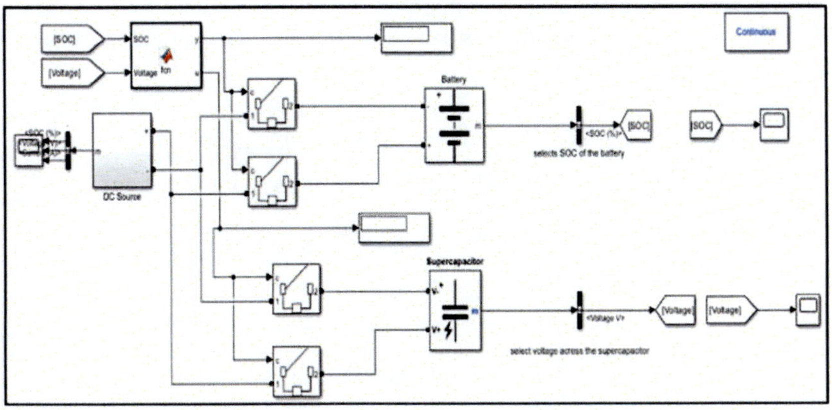

Fig. 2 Simulation diagram

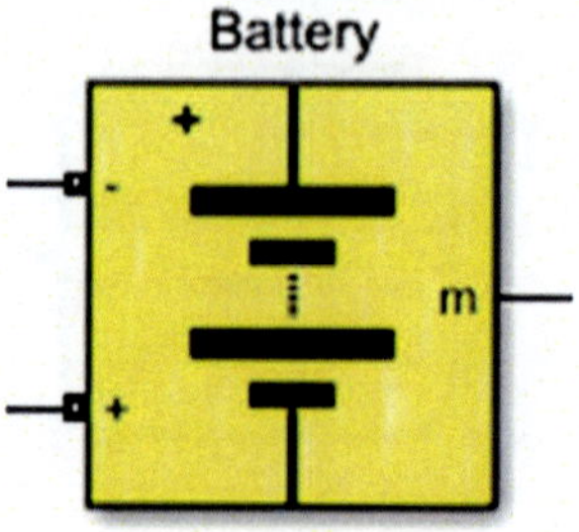

Fig. 3 Battery

Table 1 Cell details

S. No.	Cell formula	Cell ratings
1	$LiFePO_4$	3.3 V, 2.3 Ah
2	$LiCoO_2$	3.6 V, 2.05 Ah
3	$LiNiO_2$	3.6 V, 3.6 Ah
4	$LiNiO_2$	3.6 V, 48 Ah
5	$LiCoO_2$	7.4 V, 5.4 Ah
6	$LiCoO_2$	11.1 V, 6.6 Ah
7	$LiFeMgPO_4$	12.8 V, 40 Ah

Various types of cells were used in the simulation. Table 1 shows different types of cells used in the simulation.

It is important to understand discharge characteristics of the battery for understanding the charging process. Figures 4 and 5 shows the battery's discharge characteristics.

Next, storage element is supercapacitor. The supercapacitor which is to be charged, is connected to the DC source through the switches. The switch used here has the control signal where the command to close or open can be given. One switch is connected to the supercapacitor positive side while the other switch is connected

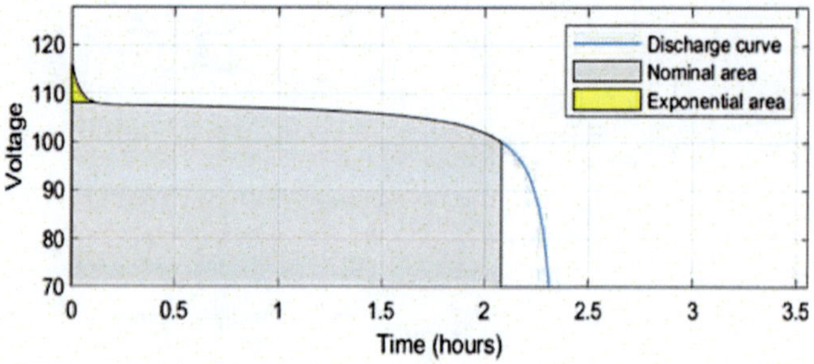

Fig. 4 Nominal current discharge feature

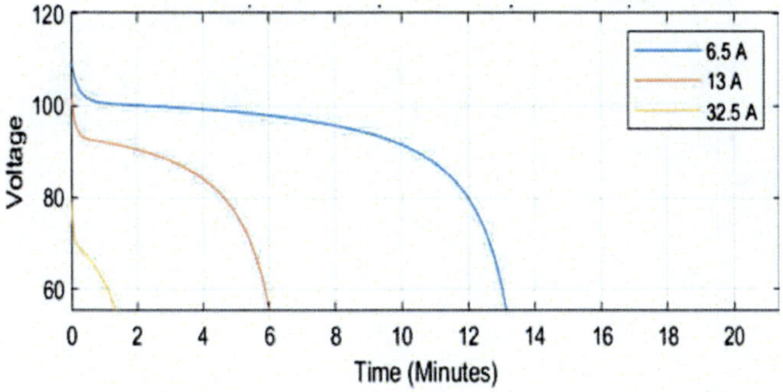

Fig. 5 Battery discharge features

Fig. 6 Supercapacitor

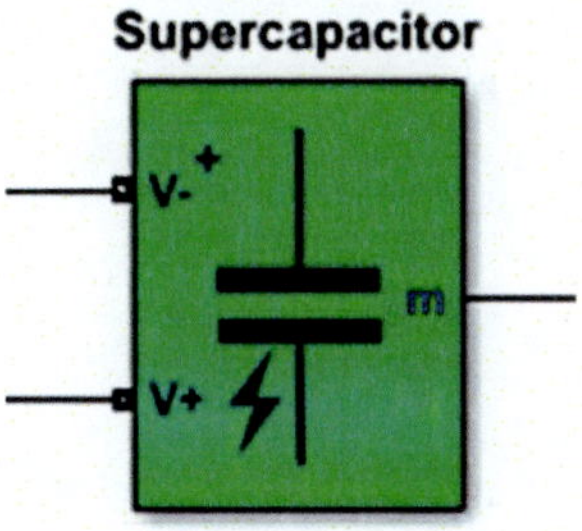

to the supercapacitor negative side. So, for a single storage source, there are two switches connected to the two terminals.

Supercapacitors are costly. But for particular applications, the necessity of supercapacitor is proved in various literatures. So let us charge both battery and supercapacitor with help of algorithm presented in this paper. Figure 6 shows supercapacitor used in the Simulink diagram while Fig. 7 shows charge characteristics of supercapacitor.

From the Fig. 7 we can infer that supercapacitor can charge faster (within seconds).

2.3 *Charging Strategy*

Figure 8 shows the MATLAB function where the code for charging strategy is entered. Here *y* is the control signal given to the circuit breakers which are linked to the batteries terminal while *u* is the control signal given to the switches which are connected to the terminals of the circuit breaker.

The code is given below:

```
function [y, u] = fcn(SOC, Voltage)
persistent y1
```

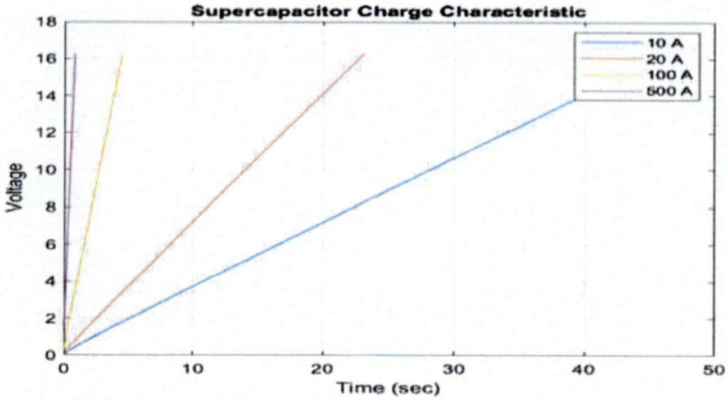

Fig. 7 Supercapacitor charge characteristics

Fig. 8 MATLAB function

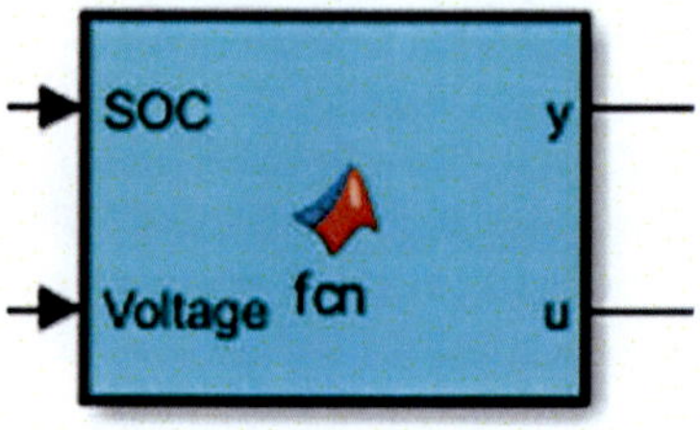

persistent $d1$
$d1 = 0;$
$y1 = 0;$
if (SOC < 40)

$y1 = 1;$

elseif (SOC == 100)

$y1 = 0;$

end

if Voltage ≤ 12

$d1 = 1;$

elseif Voltage == 16

$d1 = 0;$

end
$y = y1;$
$u = d1;$

The charging strategy is based on the battery-state of charge and voltage across the supercapacitor. The lower limit for battery-state of charge is 40%. Whenever, the state of charge of the battery is lesser than the lower limit, then charging of battery occurs. When the state of charge is full which is equal to 100%, then the charging of battery stops. Voltage is the function of state of charge. So for supercapacitor, voltage is considered as governing factor to implement the charging strategy. The lower limit of the voltage is 12 V. When the voltage across the supercapacitor is lesser than the mentioned lower threshold limit, charging of supercapacitor occurs. When the voltage across supercapacitor is equal to the full voltage across supercapacitor, the charging stops. Charging of both battery and supercapacitor takes place when their respective parameter reaches their respective lower limit. Upon reaching the lower limit, based on the code battery or supercapacitor or both may get charged.

3 Results and Discussion

By using the charging strategy, Simulink model is run for 1800s and the supercapacitor result is shown in Fig. 9 and the state of charge of the battery is shown in Fig. 10. From Fig. 9 we can infer that the initial voltage across the capacitor is 16 V. As here no load is attached, and supercapacitor voltage is 16 V, it starts to discharge due to its self-discharge capabilities. As in Fig. 9, when the voltage across the supercapacitor reaches threshold at 800th s, it begins to charge. Self-discharge occurs very slowly. This is because supercapacitor has self-discharge characteristics. Current prior open circuit is 10 A. Voltage starts to decrease for every seconds. At 10th s voltage decreases by 0.2 V and so on. The state of charge is a function of open circuit voltage. As a result state of charge decreases, and thus the self-discharge occurs. Due to its self-discharge characteristics,

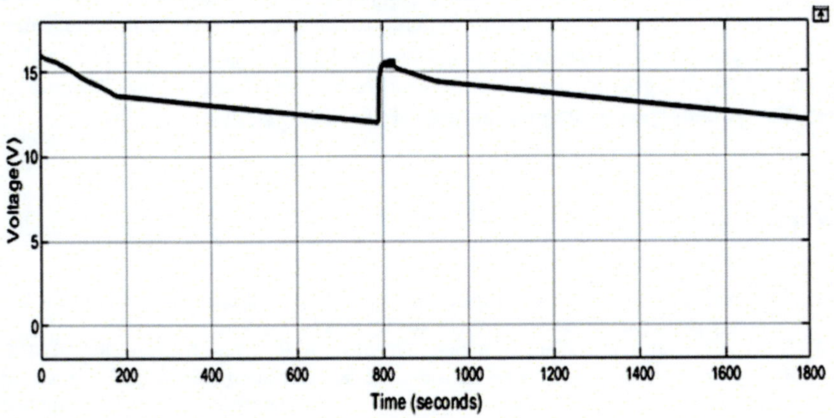

Fig. 9 Supercapacitor voltage

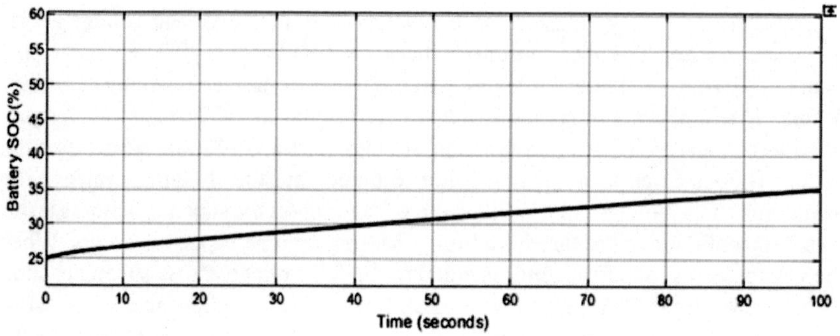

Fig. 10 SOC of the battery

The initial SOC given for battery is 25%. In case of battery, the simulation result is shown for 100 s, as it takes loads of time for a battery to charge and the result is shown in Fig. 10.

4 Conclusion

Thus the charging strategy for the supercapacitor and battery CES system is dealt and the components required for the implementation of this strategy is arranged and the connections are given. Then, from the results we can infer that the whenever the storage device (battery or the supercapacitor) energy content is very less, then it should be immediately charged in order to meet the demand. So, a charging strategy considering only the two parameters of the two devices is run in the Simulink. The future scope of this work is that battery is an essential component in upcoming days. Battery plays an important role in storing charge that we get from the renewable side. So as to increase life span of battery, supercapacitor is used along with battery that is known as the combined energy storage system. This storage system can be used in Electric Vehicles, distribution side storage systems, in uninterrupted power supplies, etc. This paper explains charging strategy of the CES system in a simpler way.

References

1. Xu Q, Niu Y, Li J, Yang Z, Gao J, Ding L, Ni H et al (2022) Recent progress of quantum dots for energy storage applications. Carbon Neutrality 1(1):1–18
2. Worku AK, Ayele DW, Habtu NG, Admasu BT, Alemayehu G, Taye BZ, Yemata TA (2022) Energy storage technologies; recent advances, challenges, and prospectives. In: Planning of hybrid renewable energy systems, electric vehicles and microgrid: modeling, control and optimization, 22 May 2022. Springer Nature, Singapore, pp 125–150

3. Kumarasabapathy N, Manoharan PS (2015) MATLAB simulation of UPQC for power quality mitigation using an ant colony based fuzzy control technique. Sci World J 2015:1–9
4. Pay S, Baghzouz Y (2003) Effectiveness of battery-supercapacitor combination in electric vehicles. In: IEEE power tech conference proceedings
5. Kouchachvili L, Yaici W, Entchev E (2018) Hybrid battery/supercpacitor energy storage system for the electric vehicles. J Power Sour 374:237–248
6. Guoju Z, Xisheng T, Zhiping Q (2010) Research on battery supercapacitor hybrid storage and its application in microgrid. In: Power and energy engineering conference. IEEE, pp 1–4
7. Bai Y, He H, Li J, Li S, Wang Y-X, Yang Q (2019) Battery anti-aging control for a plug-in hybrid electric vehicle with a hierarchical optimization energy management strategy. J Clean Prod 237:117841
8. Sun L, Feng K, Chapman C, Zhang N (2017) An adaptive power-split strategy for battery–supercapacitor powertrain—design, simulation, and experiment. IEEE Trans Power Electron 32(12):9364–9375
9. Garcia-Torres F, Valverde L, Bordons C (2016) Optimal load sharing of hydrogen-based microgrids with hybrid storage using model-predictive control. IEEE Trans Ind Electron 63(8):4919–4928
10. Morato M, Normey-Rico J, Sename O (2020) Model predictive control design for linear parameter varying systems: a survey. Annu Rev Control 49:64–80
11. Hu Y, Chen C, He T, He J, Guan X, Yang B (2020) Proactive power management scheme for hybrid electric storage system in EVs: an MPC method. IEEE Trans Intell Transp Syst 21(12):5246–5257
12. Keerthana M, Manoharan PS, Ravi A (2022) Design of control strategy for battery-supercapacitor hybrid storage system. In: 2022 7th international conference on communication and electronics systems (ICCES), Coimbatore, India, pp 798–803. https://doi.org/10.1109/ICCES54183.2022.9835989
13. Wang Y, Zhou C, Chen Z (2022) Optimization of battery charging strategy based on nonlinear model predictive control. Energy 241:122877
14. Pavković D, Premec A, Krznar M, Cipek M (2022) Current and voltage control system designs with EKF-based state-of-charge estimator for the purpose of $LiFePO_4$ battery cell charging. Optim Eng:1–29

Swarm Flight of UAV in Virtual Rigid Formation Using Olfati-Saber Algorithm

Y. Zhu and V. P. Shkodyrev

Abstract In this work, using the flocking algorithm based on reaction directly controls agents in the swarm on a virtual structure as a single control target, achieve obstacle avoidance and flying in a swarm. The human operator controls the self-organized swarm with a higher level of autonomy, flying in fixed formations or switching between predefined formations, thus allowing each quadcopter follows the desired flight trajectory throughout the maneuver processing and completes actions to maintain, move, or switch formations.

Keywords Flocking algorithm · Virtual structure · Artificial potential field

1 Introduction

In the field of unmanned research, UAV technology entered the stage of rapid development. The quadcopter is easy to operate and hand-to-fly characteristics, which significantly reduced the operating threshold of small drones, the traditional small drones, such as small unmanned helicopters and fixed-wing drones, had been quickly squeezed out of the public view. However, facing the increasingly complex and diverse requirements, UAVs still have certain limitations due to their software and hardware conditions.

For a single UAV, fuel, weight, and size play a crucial limiting role; due to the limitations of on-board sensors and communication equipment, a single UAV is also unable to achieve multidimensional and large-scale coverage of the mission area; when performing high-risk missions, a single UAV may fail due to accidents or its faults, resulting in low fault tolerance of the mission system, etc., especially so for small UAVs.

According to different demands, researchers have conducted systematic research and proposed various research directions for technological upgrades to improve the

Y. Zhu (✉) · V. P. Shkodyrev
Peter the Great Saint Petersburg Polytechnic University, 29 Politekhnicheskaya St., Saint Petersburg 195251, Russia

G. Rajakumar et al. (eds.), *Intelligent Communication Technologies and Virtual Mobile Networks*, Lecture Notes on Data Engineering and Communications Technologies 171,
https://doi.org/10.1007/978-981-99-1767-9_62

performance of quadcopters and expand application scenarios, such as enhancing quadcopters' autonomy. Due to the rapid development of artificial intelligence technology and stunning results in this research direction, which is easier to achieve and favored by many researchers.

In the Unmanned Systems Integrated Roadmap FY2017-2042 released by the U.S., the highest level of autonomous control capability for UAVs is called autonomous cluster control, the swarm UAV is a concrete case of the UAV possessing a high level of autonomy.

The swarm robotic system concluded the following characteristics [1]:

- robots are autonomous;
- robots are located in the environment and can act to modify it;
- robots' sensing and communication capabilities are local;
- robots do not have access to centralized control and/or to global knowledge;
- robots cooperate to tackle a given task.

Therefore, UAVs with high autonomy can form swarms and work collectively by collaborating and communicating to accomplish tasks that a single UAV cannot perform. The UAVs that can form swarms must be built based on possessing a certain level of autonomy.

In possible application scenarios, UAVs will work in multi-aircraft coordination. There will probably be homogeneous or heterogeneous UAV swarms, mixed swarms of homogeneous or heterogeneous UAVs with human-crewed aircraft, and mixed heterogeneous UAVs with other manned or unmanned vehicles to operate in coordination to accomplish the task together. It can maximize the advantages of UAVs, improve the efficiency of carrying out tasks, and avoid losses caused by poor mission execution or accidents. In order to control multiple UAVs that perform tasks cooperatively and improve the effectiveness of the functions, a reasonable and efficient cooperative control method is indispensable. Collaborative control of swarm robots during actual mission execution is a highly complex and challenging process.

The survey by Coppola et al. [2] presents an overview of the challenges that need to be solved before UAV swarms can be successfully deployed for real-world operations. From the lowest layer (local for single UAVs) to the highest layer (swarm drones), the tight and complex relationship between the different layers and research areas needs to be appreciated.

The research on autonomous cooperative control of swarm robots can include many aspects, such as control system architecture, distributed information fusion, multi-agent optimization and decision-making, and even image processing [3–5]. Distributed autonomous cooperative control in complex environments is an essential problem for swarm robots.

2 Previous Work

Swarm robotics is still in an early stage, and its development for different branches is not homogeneous. And also, some topics, such as design, analysis, and real-world applications, have already received attention. Other topics, such as maintenance and performance measurement, have received less attention.

Numerous swarm algorithms have been proposed for autonomous cooperative control of swarm robotics. Brambilla M., Dorigo M., and others proposed two taxonomies for swarm algorithms and behaviors: methods and collective behaviors [1]. The mainstream research ideas for controlling UAV swarms mainly include "Top-down" and "Bottom-up" approaches by taxonomies, which means cooperative control of multiple UAVs based on the hierarchical recursive approach and cooperative control of multiple UAVs based on the self-organization approach, also called bird-inspired and insect-inspired swarm control algorithm.

The insect-inspired swarm control algorithm. The swarm algorithm based on reaction is mainly a self-organization-based approach of autonomous cooperative control strategy, focusing on individual perception of the environment, judgment, decision-making, dynamic response, and rule-based behavior coordination among multiple individuals [6–10].

The classical swarm algorithm based on reaction often describes global coordinated motion as the combination of local interactions that happen at the individual layer. Despite their explanatory success, these models might fail to guarantee rapid and safe collective motion when applied to aerial robotic swarms flying in cluttered environments of the real world, such as forests and urban areas.

One of the most valuable works has been shown in the paper [11], Gábor Vásárhelyi and others proposed a flocking algorithm and use for real drones, the result confirmed the adequacy of the approach and flocking algorithm, but also showed problems might exist in practical use, such as optimization and compulsory long distance among drones in the swarm.

The bird-inspired swarm control algorithm. As shown in the paper [12], this approach means the algorithm is mainly based on hierarchical recursive solutions to reduce the difficulty of problem-solving. The autonomous cooperative control problem can be divided into several layers, such as task assignment, trajectory planning, and formation control, and then for each sub-problem of each layer, establish the mathematical model of the sub-problem according to the characteristics of the UAV system and the task requirements, and research on suitable solution methods and coordination methods between sub-problems.

The hierarchical architecture for cooperative control of swarm UAVs decomposes the problem into four layers or even more, inter-formation task allocation, intra-formation task coordination, multi-aircraft trajectory planning, and trajectory optimization.

This method of problem decomposition enables the UAV to achieve hierarchical coordination for complex tasks and related research problems including wide-area target search, cooperative task allocation, cooperative operation, and trajectory planning for multi-UAVs.

As shown in the paper [13], Zhou Xin and others have developed a new method based on trajectory planning that allows several vehicles to fly efficiently as a swarm. The algorithm receives a target or sequence of targets and plans a trajectory for UAVs to move toward the current target. This method shows higher optimality and flexibility.

This approach is relatively mature but complex. Even without considering the influence of environmental disturbances, it is more difficult to calculate the dynamic model of swarm, because of the complexity of dynamics in a UAV swarm. This approach not only means high costs for individual drones, but also means that organizing large-scale is very difficult.

The advantage of the swarm algorithm based on reaction mainly stems from its self-organization and resource efficiency, especially in terms of power, mass, and cost [14]. The swarm algorithm based on trajectory planning shows higher optimality and flexibility, but also maturity.

It can be envisaged that in a multi-UAV system, where some of them have the capability of intelligent decision-making including trajectory planning while others in the swarm follow the leader by using the swarm algorithm based on reaction, or even following their higher level, only by tracking algorithms. This approach helps with cutting costs by reducing the power consumption and processing requirements of each individual in a swarm. The swarm, as a whole agent, cannot only maintain the desired formation and navigation but also avoid collisions with obstacles and other drones.

In this paper, we study and apply both kinds of swarm algorithms to the multi-UAV autonomous cooperative control and mainly use the insect-inspired swarm control algorithm (flocking algorithm) for UAV formation control and obstacle avoidance. The formation points are used to track the virtual leader of the swarm—navigational points; and they are also points on the virtual rigid structure [15]. The trajectory points are under the control of the artificial potential field of the swarm algorithm and form various formations by following the formation points. The UAV tracks its corresponding trajectory point and forms the same shape around the virtual leader. The group of UAVs thereby forms the formation through this three-layer structure.

3 Virtual Rigid Formation

3.1 Navigational Points

The position vector and velocity of the navigation point represent the desired direction and velocity of the swarm movement, denoted by $q \in R^3$, $p \in R^3$, respectively, in the inertial coordinate system, and the kinematic equation is expressed as follows:

$$\begin{cases} \dot{q} = p \\ \dot{p} = u \end{cases}, \quad (1)$$

where u—denotes the control input and the control law associated with the planning of the formation trajectory.

3.2 Formation Points

Formation points on the virtual rigid structure form the rigid formation. In the coordinate system established with the virtual structure, the kinematic equations are denoted as follows:

$$\begin{cases} \dot{q}_i^v = p_i^v \\ \dot{p}_i^v = u_i^v \end{cases}. \quad (2)$$

In the inertial coordinate system, the i-th point in the virtual structure, i.e., the i-th formation point, its position and velocity vector denote $q_i \in R^3$, $p_i \in R^3$. $q \in R^3$, $p \in R^3$ denote the position vector and velocity of the navigation point.

The coordinate transformation matrix can be written as

$$q_i = q + R(\psi)Lq_i^v, \quad (3)$$

where q_i^v is a set of position vectors of the formation points, represents the shape of the virtual solid body structure,

$R(\psi)$—matrix of coordinate transformation from the virtual structural coordinate system to the inertial coordinate system,

L—scaling coefficient matrix of the formation,

$$R(\psi) = \begin{bmatrix} \cos\psi & \sin\psi \\ -\sin\psi & \cos\psi \end{bmatrix}, L = \begin{bmatrix} l & 0 \\ 0 & l \end{bmatrix}. \quad (4)$$

Changing UAV formation can be classified into the following types: changing the shape of formation, rotating formation, and scaling of formation.

q_i^v is a set of position vectors for the formation points, represents the structural form of the virtual solid.

Different values q_i^v mean that the structural shapes of the virtual solid are different, i.e., the shapes of the UAV formation are different.

Consequently, $q_i^v(t)$ describes the process of changing the formation structure. The value change $R(\psi)$ represents the rotation of the virtual solid centered at the navigation point, which is a function of the course angle of the virtual structure ψ.

L represents the size of the virtual structure. If the value of L changes, the size changes.

Position of formation point q_i, velocities of formation point $p_i \in R^3$

$$p_i = p + \dot{R}(\psi)Lq_i^v + R(\psi)\dot{L}q_i^v + R(\psi)L\dot{q}_i^v. \tag{5}$$

4 Using the Flocking Algorithm Based on Reaction in Multi-UAV Autonomous Cooperative Control System

4.1 The Trajectory Point

The previous section introduced the multi-UAV formation hierarchical cooperative control approach. The research trend of multi-UAV autonomous cooperative control technology mainly focuses on centralized hierarchical control. In this paper, we study the swarming model and apply the multi-agent swarming algorithm to the multi-UAV autonomous cooperative control. As the shape of the formation changes, the virtual rigid body shape in this dynamic process, from to with step change, the flight trajectories of corresponding formation points also swift to new trajectories with step change.

Introduce the flocking model of swarm and trajectory points, enable the trajectory points tracking the corresponding formation points, use the trajectory points as a medium for the introduction of the flocking model, and study the trajectory for UAV swarm tracking the formation points, meanwhile avoid the collision among agents and collision with obstacles.

Through the mechanism of trajectory points, the position and velocity of the trajectory points on trajectories denoted as $q_i^p \in R^3$ and $p_i^p \in R^3$. UAVs tracks trajectory points on their trajectories. As a result that the trajectory points track the formation points of virtual rigid body, thus, while agents coincide with the trajectory points, the agents finish the change of formation and form the formation represented by the virtual rigid body.

The motion model of the trajectory point i can be written as follows:

$$\begin{cases} \dot{q}_i^p = p_i^p \\ \dot{p}_i^p = u_i^p \end{cases}, \tag{6}$$

where $q_i^p \in R^3$ and $p_i^p \in R^3$ are the position vector and velocity vector of the i-th trajectory point in the inertial coordinate system.

$u_i^p \in R^3$ is the control input of the i-th trajectory point, can be written as follows::

$$u_i^p = c_q(q_i^p - q_i) + c_p(p_i^p - p_i) + \sum_{j \in N_i} f_{i,j} + \sum_{k \in K_i} f'_{i,k}, \tag{7}$$

where $q_i \in R^3$, $p_i \in R^3$—position vectors and velocity vector of the i-th formation point corresponding to the i-th point of virtual solid structure in the inertial coordinate system, and c_q is the feedback coefficient for position error, c_p is the feedback coefficient for velocity error.

4.2 The Function of Artificial Potential Field in Swarm Algorithm

$F_{i,j}$—forces of the artificial potential field for trajectory points to avoid collisions among trajectory points.

Where N_i represents the set of neighboring path points for the i-th path point, and the index j represents the j-th path point in the set of neighboring points N_i.

It is supposed that the i-th path point is in the artificial potential field generated by the j-th path point, and the potential function can be written as follows:

$$U_{i,j}(q_i, q_j) = \varphi_{hf}(\|q_j - q_i\|), \tag{8}$$

where $\varphi_{hf}(*)$ is the hyperbolic function of artificial potential field and $*$ is the vector modulus value. The structure $\varphi_{hf}(*)$ can be seen as follows:

$$\varphi_{hf}(r) = \begin{cases} c_{hf}\left(\frac{1}{r} - \frac{1}{r_0}\right) 0 < r \le r_0 \\ 0 r > r_0 \end{cases}, \tag{9}$$

where r_0 is radius of the artificial potential field. If the relative distance between path points is greater than r_0, the interaction of artificial potential field forces would not occur to each other.

c_{hf} is the coefficient of potential field. For selection of c_{hf}, the inertial size of agent and the size of control output should be taken into account.

For the function of artificial potential field $U_{i,j}(q_i, q_j)$ at position q_i, the i-th path point obtained by negative gradient. The force of artificial potential field generated by i-th path point at the j-th path point can be written as follows:

$$f_{i,j}(q_i, q_j) = -\nabla_{q_i} U_{i,j}(q_i, q_j) = \dot{\varphi}_{hf}(\|q_j - q_i\|) n_{i,j}, \tag{10}$$

where $n_{i,j} = (\|q_j - q_i\|)/q_j - q_i$ denotes positive direction of the potential field force, which is from the i-th path point to the j-th path point.

The forces for the i-th path point are supposed to be in artificial potential fields

$$\sum_{j \in N_i} f_{i,j}(q_i, q_j) = -\nabla_{q_i} \sum_{j \in N_i} U_{i,j}(q_i, q_j). \tag{11}$$

Substituting the formula into the control input of path point

$$u_i^p + c_p p_i^p + c_q q_i^p = c_p p_i + c_q q_i + \sum_{j \in N_i} f_{i,j}(q_i, q_j). \tag{12}$$

The distance between the points in virtual structure of formation should be larger than range of the repulsion function.

$$\|q_j - q_i\| \geq R \forall i, j \in N_i.$$

$f'_{i,k}$—the forces of artificial potential field for obstacle avoidance, which are used to prevent collisions from other trajectory points or obstacles. A K_i represents the set of obstacles, it is supposed to obtain different function, and it can be discussed in detail depending on various types of the obstacles. The potential field function is defined in the objective function.

5 Result of Simulation

5.1 Olfati-Saber Algorithm

Parameter Setting and the objective function. The Olfati-Saber algorithm [6] reflects the joint effect of three sub-control term from α, β, γ, which denote three agents according to Olfati-Saber's flocking theory.

$$u_i = u_i^\alpha + u_i^\beta + u_i^\gamma, \tag{13}$$

where α—agent denotes any individual agent in swarm, β—agent denotes the physical obstacle to be avoided, and γ—agent has been used to construct navigation feedback, and it denotes the target to be tracked. u_i^α denotes interaction term (α, α), u_i^β denotes interaction term (α, β), and u_i^γ is a distributed navigation feedback. The objective functions are defined and shown as follows:

$$J = \sum_i J_i = \sum_i J_i^\alpha + J_i^\beta + J_i^\gamma. \tag{14}$$

Without considering the presence of obstacles, the objective function can be written in following forms:

$$J = \sum_i J_i = \sum_i \sum_j f_i^\alpha(q_i, p_i, q_j, p_j) + \sum_i f_i^\gamma(q_i, p_i, q_\gamma, p_\gamma), \tag{15}$$

$$J_i = c_q^\alpha \sum\nolimits_{j \in N_i^\alpha} \rho_H(q_i)\varphi_\alpha(q_i) + c_p^\alpha \sum\nolimits_{j \in N_i^\alpha} a_{ij}(q_i)(p_i - p_j)$$
$$+ c_q^\gamma \sigma_1(q_i - q_\gamma) + c_p^\gamma(p_i - p_\gamma). \tag{16}$$

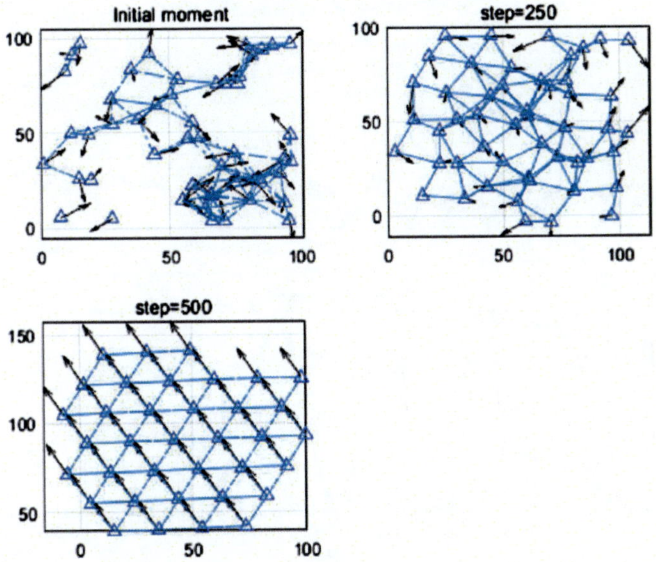

Fig. 1 Agents at initial moment and at step 250, 500

As shown in paper [6], $\varphi_\alpha(q_i)$ denotes function of potential field, $a_{ij}(q_i)$ is a scalar function that smoothly varies between 0 and 1. Fragmentation is the trap of the Olfati-Saber flocking algorithm, and introduce $\rho_H(q_i)$ can effectively avoid fragmentation. The value of $\rho_H(q_i)$ also increases rapidly as the distance between the agents increases.

In this section, first we present several simulation results (Figs. 1 and 2), reproduced the results of Olfati-Saber algorithm. And the objective function, the objective functions influenced by velocity, and influenced by position vector, are shown, respectively. Figures 1 and 2 demonstrate that flocking algorithm is responsible for creation of spatial-order in flocks. This algorithm embodies all three Reynolds' rules.

These parameters remain fixed throughout all simulations:

50 agents in free-space, range of interaction $r = 16$, time size of step $t = 1$ and the positions, velocity direction, and velocity at initial moment are chosen randomly. The objective function is shown in Fig. 3.

5.2 *Swarms of UAVs Using Flocking Model and Method of Virtual Structure*

Formation change in flocking algorithm. The multi-UAVs flocking algorithm enable the UAVs tracking the trajectory points which are introduced for tracking

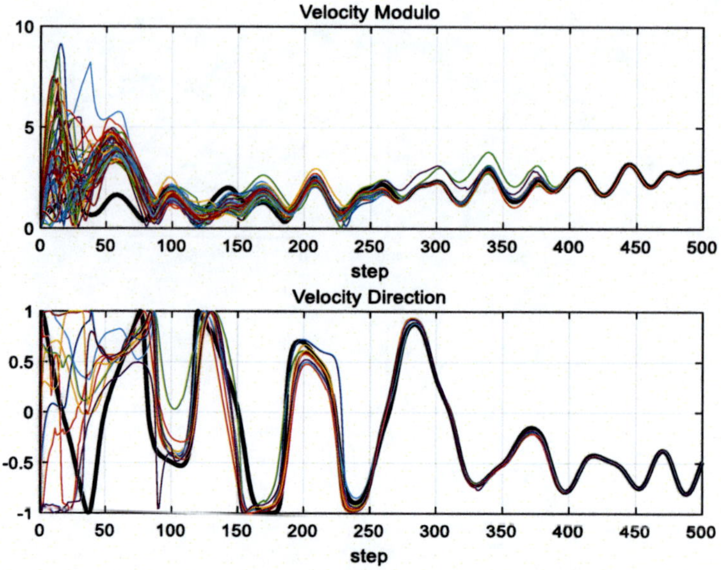

Fig. 2 Velocity direction and velocity of agents and the virtual leader

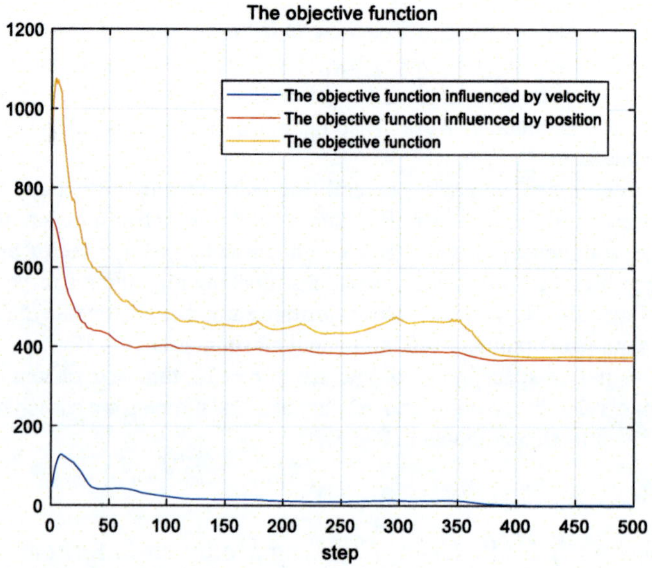

Fig. 3 Objective function

the formation points, the formation points form a fixed formation and converging the velocity vectors to navigation point.

Formation points, also represent u_i^γ and velocities of formation point $p_i \in R^3$—tracking targets for trajectory points which in flocking algorithm, indirectly control the trajectory points follow the virtual navigation point in the swarm.

The coordinate transformation matrix for navigation term in flocking algorithm can be written as:

$$q_i = q_\gamma = q + R(\psi)Lq_i^v, \tag{17}$$

$$p_i = p_\gamma = p + \dot{R}(\psi)Lq_i^v + R(\psi)\dot{L}q_i^v + R(\psi)L\dot{q}_i^v. \tag{18}$$

The objective function. Without considering the presence of obstacles, the objective function can be written in the following formula:

$$J = \sum_{i \in N}\left(c_q q_\gamma - q_i^p + c_p p_\gamma - p_i^p + \sum_{j \in N_i} f_{i,j}\right), \tag{19}$$

$$\begin{aligned} J = \sum_{i \in N}\Big(& c_q\left\|q_\gamma - q_i^p\right\| + c_p\left\|p_\gamma - p_i^p\right\| + c_q^\alpha \sum\nolimits_{j \in N_i^\alpha} \rho_H\left(q_i^p\right)\varphi_\alpha\left(q_i^p\right) \\ & + c_p^\alpha \sum\nolimits_{j \in N_i^\alpha} a_{ij}\left(q_i^p\right)\left(p_i^p - p_j^p\right)\Big) \end{aligned} \tag{20}$$

$$J = \sum_{i \in N}\begin{pmatrix} c_q\left\|\left(q + R(\psi)Lq_i^v\right) - q_i^p\right\| + c_p\left\|\left(p + \dot{R}(\psi)Lq_i^v + R(\psi)\dot{L}q_i^v\right.\right. \\ \left.\left. + R(\psi)L\dot{q}_i^v\right) - p_i^p\right\| + c_q^\alpha \sum\nolimits_{j \in N_i^\alpha} \rho_H\left(q_i^p\right)\varphi_\alpha\left(q_i^p\right) \\ + c_p^\alpha \sum\nolimits_{j \in N_i^\alpha} a_{ij}\left(q_i^p\right)\left(p_i^p - p_j^p\right) \end{pmatrix}. \tag{21}$$

In 1500 steps, the trajectory points switch from random initial positions to three different formations in succession

$$q_i^v = q + \Delta q_i^v = q + R(\psi)Lq_i^v. \tag{22}$$

Formations set as follows:

In the virtual structure coordinate system, formation 1 shape—trapezoid, formation 2 shape—line, and formation 3 shape—rectangle.

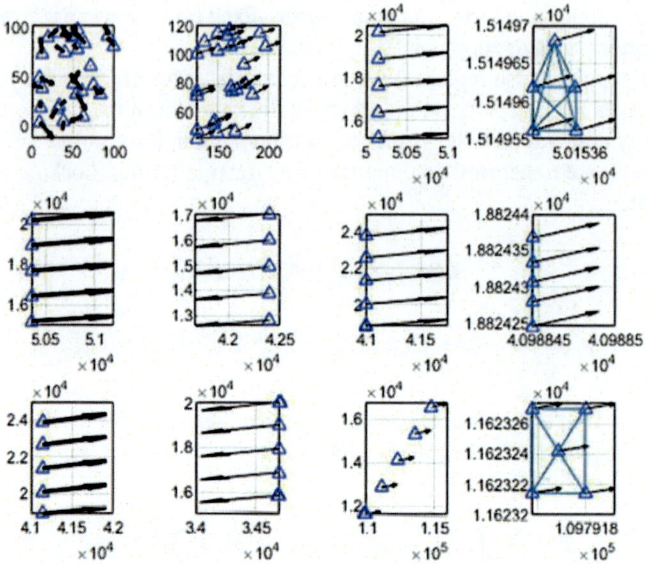

Fig. 4 Trajectory points in 1500 steps

5.3 *Results*

Using the flocking algorithm with virtual structure, 25 agents formed 5 swarms in a fixed shape. The simulation results and objective functions are shown below.

The objective functions for switching the formation are shown as follows.

Figure 4 has shown the swarm motion and changes the shape of the formation, Figs. 5, 6 and 7 have shown the objective functions while individuals in the swarm move, Fig. 5 has shown the individuals quickly form swarms from random locations and speeds, Figs. 6 and 7 have shown that after the shape of the formation formed, the objective function have almost only influenced by position, it means the UAV swarm implement hierarchical coordination of complex tasks, quickly form, and change the formation while significantly reducing the needs of online computation and obstacle avoidance among the UAVs in the swarm.

6 Conclusion

Usually, algorithm for multi-UAV control system, the problem is decomposed into several sub-problems by hierarchical distributed architecture and carry out the planning trajectories. These methods enable the UAV swarm to implement hierarchical coordination of complex tasks and reduce the difficulty of online computation, but

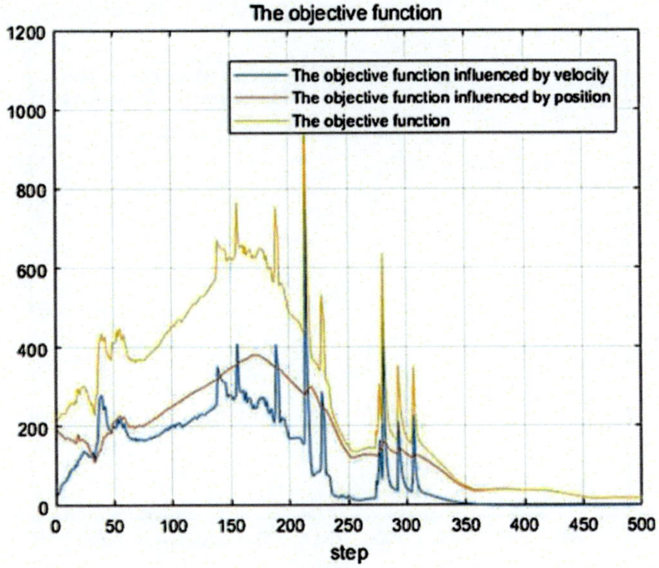

Fig. 5 Objective function on switching to formation 1

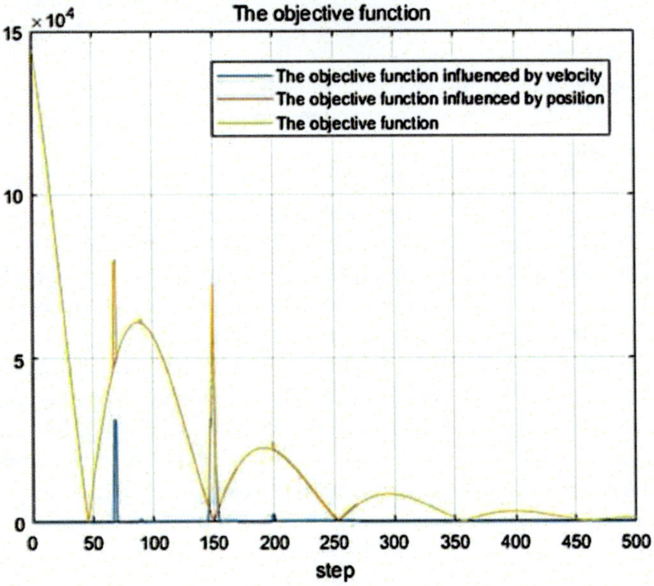

Fig. 6 Objective function on switching to formation 2

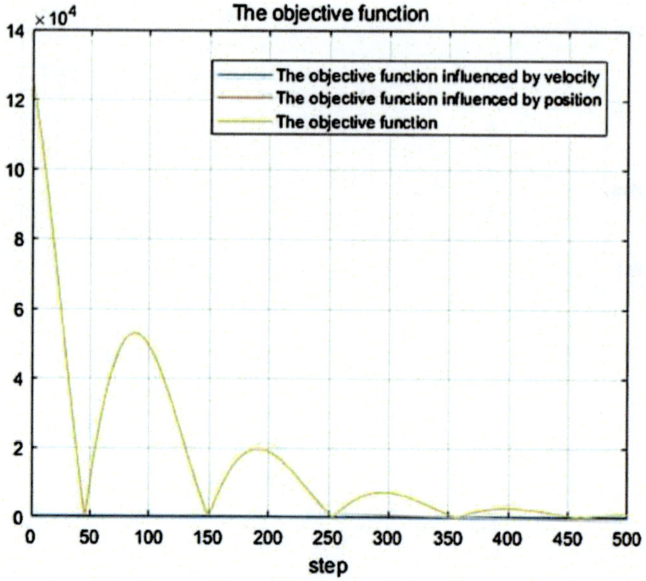

Fig. 7 Objective function on switching to formation 3

still remains problems such as fail planning of trajectories. The classical swarm algorithm based on reaction might fail to guarantee fast and safe collective motion when applied to aerial robot swarms flying in cluttered real-world environments, such as forests and urban areas.

In this paper, the algorithm based on trajectory planning is introduced into the flocking algorithms based on reaction, which enables the UAV swarm to implement hierarchical coordination of complex tasks while significantly reducing the difficulty of online computation and obstacle avoidance among the UAVs in the swarm, the results of the simulation are shown in Sect. 5. This paper studies using both of the flocking algorithms based on reaction and the method virtual structures for swarm control, which provides a new idea for the design and analysis of distributed cooperative control algorithms for self-organized UAV swarm and flocking motion. The swarms can easily change the shape of the formation and fly in a fixed formation while retaining the ability to avoid obstacles, at low cost, and have the potential to organize super-large swarms in a hierarchical structure.

References

1. Brambilla M, Ferrante E, Birattari M, Dorigo M (2013) Swarm robotics: a review from the swarm engineering perspective. Swarm Intell 7(1):1–41
2. Coppola M, McGuire KN, De Wagter C, de Croon GCHE (2020) A survey on swarming with

micro air vehicles: fundamental challenges and constraints. Front Robot AI 7:18. https://doi.org/10.3389/frobt.2020.00018

3. Guo X, Lu J, Alsaedi A, Alsaadi FE (2018) Bipartite consensus for multi-agent systems with antagonistic interactions and communication delays. Physica A Stat Mech Appl 495(1):488–497
4. Wong EM, Bourgault F, Furukawa T (2005) Multi-vehicle Bayesian search for multiple lost targets. In: Proceedings of the 2005 IEEE international conference on robotics and automation. IEEE, pp 3169–3174
5. Dai F, Chen M, Wei X et al (2019) Swarm intelligence-inspired autonomous flocking control in UAV networks. IEEE Access 7:61786–61796
6. Olfati-Saber R (2004) Flocking for multi-agent dynamic systems: algorithms and theory. IEEE Trans Autom Control 51(3):401–420
7. Qiu H, Duan H (2017) Pigeon interaction mode switch-based UAV distributed flocking control under obstacle environments. ISA Trans 71(1):93–109
8. Zhao W, Chu H, Zhang M et al (2019) Flocking control of fixed-wing UAVs with cooperative obstacle avoidance capability. IEEE Access 7:17798–17808
9. Zhu P, Dai W, Yao W et al (2020) Multi-robot flocking control based on deep reinforcement learning. IEEE Access 8:150397–150406
10. Bezcioglu MB, Lennox B, Arvin F (2021) Self-organised swarm flocking with deep reinforcement learning. In: 2021 7th international conference on automation, robotics and applications (ICARA). IEEE, pp 226–230
11. Vásárhelyi G, Virágh C, Somorjai G et al (2018) Optimized flocking of autonomous drones in confined environments. Sci Robot 3(20):eaat3536
12. Zhou Y, Rao B, Wang W (2020) UAV swarm intelligence: recent advances and future trends. IEEE Access 8:183856–183878
13. Zhou X, Wen X, Wang Z et al (2022) Swarm of micro flying robots in the wild. Sci Robot 7(66):eabm5954
14. de Croon GCHE, Dupeyroux JJG, Fuller SB, Marshall JAR (2022) Insect-inspired AI for autonomous robots. Sci Robot 7(67):eabl6334
15. Zhou D, Wang Z, Schwager M (2018) Agile coordination and assistive collision avoidance for quadrotor swarms using virtual structures. IEEE Trans Rob 34(4):916–923

Military Aircraft Detection Using YOLOv5

P. Ajay Kumar Goud, G. Mohit Raj, K. Rahul, and A. Vijaya Lakshmi

Abstract In making strategic decisions in the military, military aircraft detection has become increasingly crucial. The identification of military aircraft continues to be a problematic issue. In operations and wars, military aircraft detection is crucial for detecting unknown aircraft. The difficulty is always in accurately recognizing the unfamiliar aircraft, regardless of class and orientation. Military aircraft, such as stealth aircraft, are still difficult to detect because stealth aircraft are more challenging to detect or track using conventional radar. Still, these aircraft can be detected using object detection. In this article, we proposed identifying five types of airplanes independent of class or direction using object detection. The You Only Look Once version 5 (YOLOv5) method and the PyTorch military aircraft dataset were used to identify various aircraft. The identification of different aircraft was discovered using the YOLOv5 algorithm and the PyTorch military aircraft dataset. Bounding boxes for the dataset, data pre-processing, and data augmentation are made using Roboflow. The goal is to employ computer vision and object identification to identify whether a particular aircraft is a military aircraft. This military aircraft detection may be used in the border area, air force, and marine force.

Keywords Object detection · YOLOv5 · Data pre-processing · Data augmentation class · Orientation

1 Introduction

Aircraft are used for transportation for [1] general purposes, goods, and services. In contrast, military aircraft are employed to patrol borders. Military aircraft play a crucial role in war and are used in conflicts for offensive and defensive purposes [2]. These aircraft damaged crucial infrastructure, airstrips, ammunition storage facilities, and supplies during the war. Military aircraft support the land-based army through

P. Ajay Kumar Goud · G. Mohit Raj · K. Rahul · A. Vijaya Lakshmi (✉)
Vardhaman College of Engineering, Hyderabad, India
e-mail: vijayalakshmivardhan@gmail.com

G. Rajakumar et al. (eds.), *Intelligent Communication Technologies and Virtual Mobile Networks*, Lecture Notes on Data Engineering and Communications Technologies 171,
https://doi.org/10.1007/978-981-99-1767-9_63

its defensive role. It also discourages potential enemy airstrike threats. In naval warfare, military aircraft are used to find and destroy warships and submarines to protect the seacoast from enemy attack. Military aircraft often support military bases with logistics and participate in national disaster relief efforts. It is well recognized that sustaining sovereignty and winning a war depends on having a strong air force and military dominance. We are all aware of how crucial military aircraft are to arm troops. [3] Since World War II, military aircraft detection has been a difficult challenge in developing defense technologies [4]. Radars are often used to capture aircraft through electromagnetic wave reflection. Studies that have been done in this sector so far have mostly centered on airport or airbase aircraft detection. Notably, the aircraft carrier serves as a vital conduit for military aircraft, which are harder to detect, especially in chaotic conditions. Finding military aircraft is difficult every time, and utilizing radar signals to identify them is challenging. For instance, radar equipment has difficulty detecting airplanes. Because the frequency of various planes varies, military aircraft may be detected utilizing [5] the most recent developments in object detection technology, regardless of class or orientation.

Modern systems in a variety of fields, [6, 7] advances in computer vision, and object identification have considerably benefited [8] the development of technologies like intelligent robotics, automated control systems, and human–computer interaction. Automatic target detection, on the other side, is a military application and the primary technology for military activities and surveillance tasks [9, 10]. Object detection in computer vision has made impressive strides recently. In commercial and consumer networks, object detection is one use of computer vision that is growing in popularity. Object detection involves measuring its area and size to identify each object in an image from a collection of distinct objects.

This paper used PyTorch libraries, TensorFlow, and Roboflow to perform object detection and recognition [11]. The raw images are converted using Roboflow into annotated images, which are then augmented and pre-processed. The PyTorch libraries offer the most significant and simple approach to training the YOLOv5-based custom object detection model.

2 Related Work

Object detection relies on convolution neural networks in computer vision and object detection [12]. The convolution neural network is split into two types, the primary type could be a two-stage method Fast R-CNN, Faster R-CNN, and R-CNN. These techniques have high detection and low speed [13]. The two-stage method involves manual feature extraction and classification [14]. Whereas the second type could be a one-stage method like SSD and YOLO, these methods traditionally do not require feature extraction and classification. These methods have a faster detection speed [15]. With different aspect ratios, the SSD algorithm compares items in default boxes. To detect objects with better precision at varying sizes, it uses multi-scale feature maps. The SSD object detection algorithm is divided into feature map extraction and object detection using small convolutional filters. The filters are applied to feature maps to predict class scores and boxes for a set of fixed default boxes, but [16] YOLO uses a compound scaling mechanism that simultaneously modifies the resolution, depth, and breadth for all backbone, feature network, and box/class prediction networks to ensure the greatest accuracy and efficiency with the least amount of computational power. With various algorithms, deep learning techniques for aircraft detection have been examined. Wei and Zhang [17] such as YOLOv3, SSD, and Faster R-CNN discovered that the algorithms accurately identify aircraft from the images. The researchers and scholars found detecting using the YOLO [18] algorithm is the speed in detection and has given significant results in detecting the target [13, 19] based on the YOLOv3 algorithm. In 2020, the YOLOv4 algorithm was released. Over YOLOv3, the AP and FPS have risen by 10% and 12%, respectively. Sozzi et al. [20] YOLOv4 has a 244 MB size, 50 frames per second, and Darknet architecture. YOLOV5 is only 27 MB in size, and [21] is equally accurate as YOLOV4. The earlier versions of YOLO have been developed using the Darknet framework [22]. The new YOLOv5 version is developed PyTorch framework [23, 24]. In YOLOv5, PyTorch is more easily configurable compared to Darknet. YOLOv5 is faster than the previous version of YOLO, and YOLOv5 inference time is 140 frames per second. All PyTorch libraries can be used in YOLOv5.

3 Proposed Methodology

As shown in Fig. 1, the proposed system used custom object detection to detect the different types of military aircraft. For custom object detection, the YOLOv5 model is used, and for annotating the dataset, pre-processing, and augmentation, the Roboflow is used. The initial step is to gather the dataset of various military aircraft, and then the dataset is uploaded to Roboflow. In Roboflow, the dataset is labeled using bounding boxes of the different classes of military aircraft. A training, valid, and test dataset are created after the labeling process. We then pre-process and augment the dataset before deploying it in the YOLOv5 model. After pre-processing and data augmentation, the dataset is deployed in the YOLOv5 model and trained. After training the data, test images were shown along with an evaluation of how well the trained model performed in identifying various military aircraft.

3.1 Dataset

Raw images from Kaggle datasets are gathered to create a dataset. There are 41 distinct types of aircraft in the Military Aircraft Dataset, and the images in the collection include the Mirage 2000, V22, A10, and A400M. This dataset is deployed to train the model.

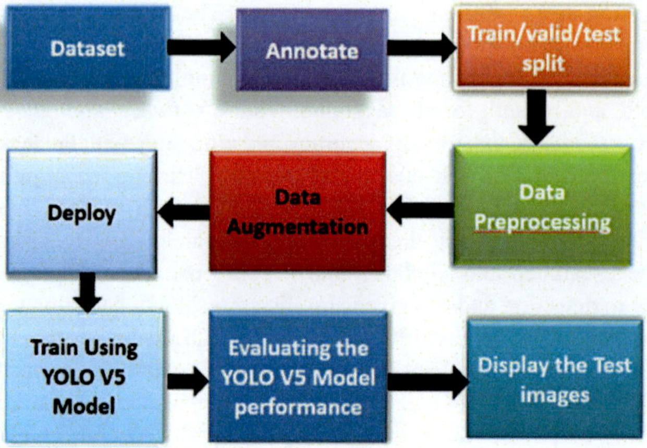

Fig. 1 Proposed method of the military aircraft detection

3.2 *Annotate*

The collected dataset must be annotated and labeled. In annotation, all the images are given annotated. A bounding box annotation could be a rectangle or polygon surrounding an object that specifies its position. Because of the high aircraft density, each image may contain numerous aircraft to be identified, resulting in many bounding boxes. Each column of data relates to the specific image ID, image width, height, and bounding box data [class, $x_{\min}$, $y_{\min}$, width, height]—the object detection dataset in typical bounding box format. The bounding box data in YOLO is represented as [class, x_{center}, y_{center}, width, height] as shown in Fig. 2.

Using the smallest point ($x_{\min}$, $y_{\min}$), the center point (x_{center}, y_{center}) is calculated as follows:

Fig. 2 YOLO format bounding box

$$X_{\text{center}} = X_{\min} + \frac{\text{width}}{2}. \tag{1}$$

$$Y_{\text{center}} = Y_{\min} + \frac{\text{height}}{2}. \tag{2}$$

However, because YOLO requires a label argument, the bounding boxes must be labeled with the appropriate classes. PyTorch reads the bounding box data in a text file (.txt) in YOLOv5, and the data of all bounding boxes in a single image should be clustered and written in the exact file text as that image.

3.3 Splitting Data into Training, Validation, and Test

The dataset has been divided into the train, valid, and test images. The majority of the images 70% are utilized for training, followed by 20% for validation, and 10% for testing.

3.3.1 Training Set

The model is taught and trained using the training set of data to find hidden traits and patterns. The neural network architecture receives training data at each epoch, which enables the model to continue acquiring input properties. Inputs from various sources should be included in the training set so that the model may be trained on every scenario conceivable and be capable of forecasting any future random data sample.

3.3.2 Validation Set

A different set of data than the training set is utilized for the validation set, which is used to validate the model's performance during training. Information gathered throughout the validation process is employed to change the model's hyperparameters and settings. Similar to getting criticism on how well your training is going. Each epoch ends with a test on the validation set once the model has been trained on the training set. To prevent overfitting, which occurs when our model performs very well at identifying samples in the training set but struggles to generalize to new data and offer accurate classifications.

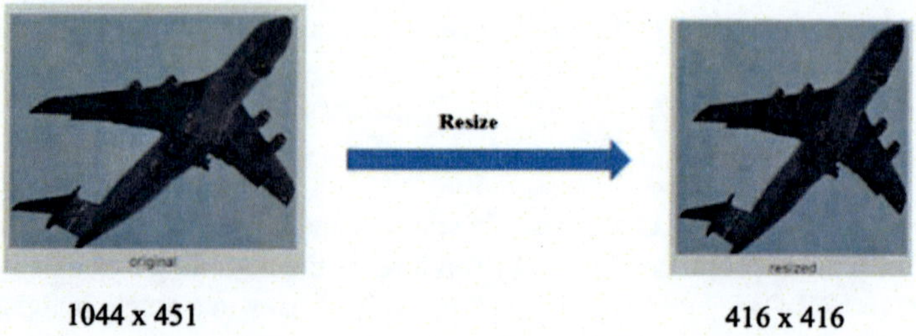

Fig. 3 Resizing the original image

3.3.3 Test Set

The model is tested after training using a distinct dataset known as the test set. Regarding accuracy, precision, and recall, it provides an unbiased final model performance metric.

3.4 Data Pre-processing

Data pre-processing is a series of procedures on data to prepare it to be utilized as input during the training phase. Image data pre-processing comes in a variety of forms. That is feasible. Pre-processing can increase inference and decrease training time. Applying pre-processing to the training, validation, and testing sets will ensure that learning and inference occur on the same image properties. In this paper, the pre-processing is performed as follows.

3.4.1 Resize

The dimensions of the images are resized. Annotations are proportionally changed, as shown in Fig. 3

3.4.2 Grayscale

Using the grayscale technique, an image with RGB channels is converted to an image with a single grayscale channel and is shown in Fig. 4.

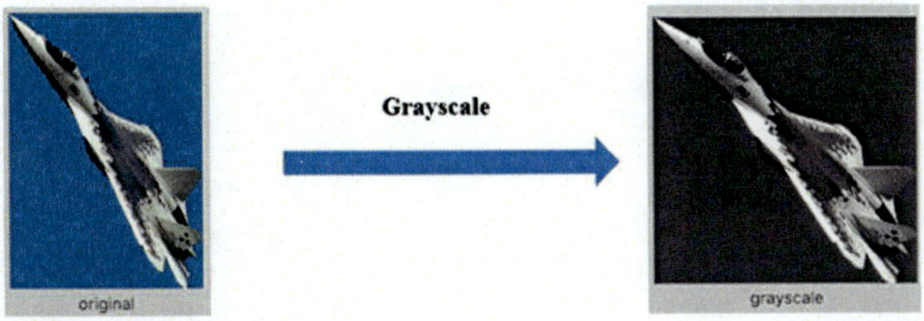

Fig. 4 RGB to grayscale conversion

3.4.3 Auto-Adjust Contrast

Using this auto-adjust contrast enhances a low-contrast image by using either contrast stretching, histogram equalization, or adaptive equalization.

i. **Contrast Stretching**: All intensities between the images' 2nd and 98th percentiles are included after rescaling, and it is shown in Fig. 5.
ii. **Histogram Equalization**: In an image, it "spreads out the most frequent intensity values." All pixel colors are evenly represented in the equalized picture, which has a relatively uniform distribution. Histogram equalized picture is shown in Fig. 6.
iii. **Adaptive Equalization**: A local contrast enhancement approach that uses histograms calculated over several tile sections of the image. Thus, local details are improved even in darker or brighter areas than in most images, as shown in Fig. 7.

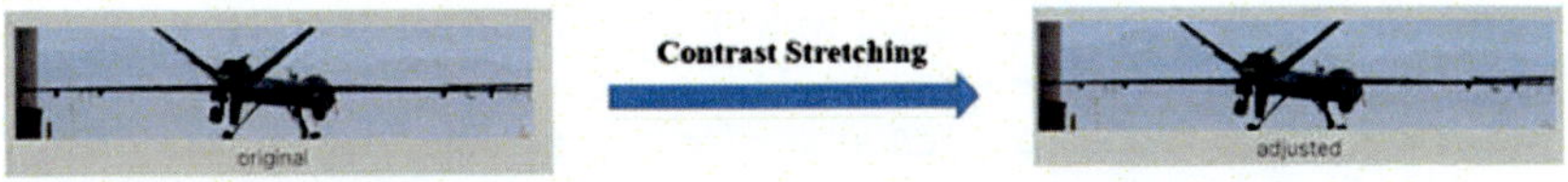

Fig. 5 Contrast stretching

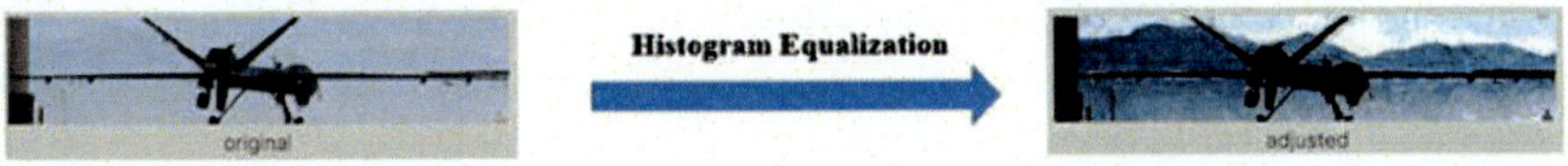

Fig. 6 Histogram equalization

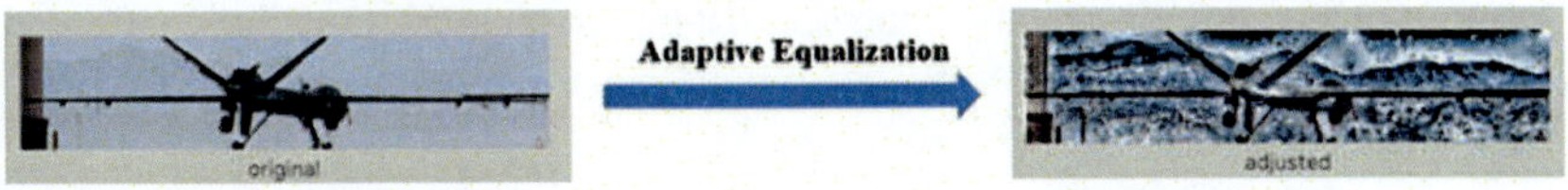

Fig. 7 Adaptive equalization

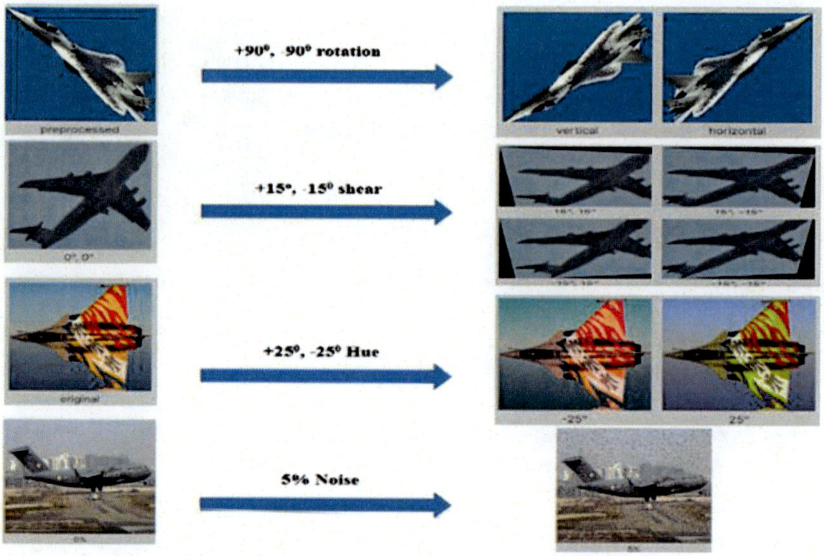

Fig. 8 Data augmentation technique

3.5 Data Augmentation

The military aircraft dataset collection is minimal. The data should be sufficiently large for object detection to provide high accuracy. It would be quite challenging to manually gather such a vast amount of data, so data augmentation is necessary. We use a different data augmentation technique to increase the dataset. While data augmentation images can be changed without changing what they depict because image objects have specific invariances. In Fig. 8, + 90°, − 90° rotation, + 15°, − 15° shear, hue, and noise are performed. Despite the object being in a different orientation in each of the four images, we can still recognize it. A neural net will consider each object unique instead of having each unique image in the original dataset. In this case, four different data augmentation methods have been applied to the main image and then translated to different areas of the image.

3.6 Deploy

The pre-processed and data augmentation dataset framework must be configured with TXT annotation and YAML configuration, then the ".yaml" dataset is exported to the YOLOv5 model.

3.7 Training Model

Using PyTorch and PyTorch libraries, the YAML file is configured and then set up in the YOLOv5 algorithm. Once setup is completed using torch 1.11.0 + cu113 (Telsa T4). The dataset gets trained using GPU with 100 epochs and batch 16. According to the given batch size and epochs, the model is trained.

3.8 Evaluating the Performance of the Trained Model

Following model training, tensor board graphs are used to evaluate the model's performance. In addition to the confusion matrix, precision, recall, mAP@0.5, and mAP@0.5:95, the model also produces train loss/valid box, train loss/obj loss, and train loss/cls loss. Metrics are analyzed and evaluated for graphs.

Evaluation Metrics

Various metrics have been used to measure the efficiency of the YOLOv5 detection model. The precision, recall, and mAP are used for the evaluation. Precision is the ratio of correctly identified positive samples to all positively classified samples. The calculation formula is written as follows:

$$\text{Precision} = \frac{\text{True Positive}}{\text{True Positive} + \text{False Positive}} \times 100. \tag{3}$$

The ratio of detected positives among all ground true positives is known as recall. The calculation formula is written as follows:

$$\text{Recall} = \frac{\text{True Positive}}{\text{True Positive} + \text{False Negative}} \times 100\%\,, \tag{4}$$

where TP is a true positive result occurs when the model accurately anticipates the positive class. FP is a false positive as a result when the model anticipates the positive class wrongly. FN is a false negative result when the model anticipates the negative class wrongly.

The performance of the detection network's algorithm is evaluated using the mAP. It is suitable for image detection and calculation of single-label and multiple labels. The formula is as follows:

$$\text{mAP} = \frac{\sum_{a=1}^{N} \text{Precision}(a)\Delta\text{Recall}(a)}{M}. \tag{5}$$

N is the total number of samples in the test set, and Precision(a) evaluates the magnitude of the precision rate when samples are recognized simultaneously. Δ Recall(a) quantifies the change in recall rate when the number of identified samples goes from

Fig. 9 Detected military aircraft test images

$a - 1$ to a. The total number of classes involved in the multi-class detection process is M.

3.9 Displaying the Test Images

The 10% of test images in the dataset are predicted by the trained YOLOv5 model with precision, which is located on the upper left of the bounding box. The detected military aircraft test images are shown in Fig. 9.

The YOLOv5 model gives out awe-inspiring predictive results and almost detects all the military aircraft in the image.

4 Experimental Results and Analysis

The YOLOv5 model is trained by setting the network epochs to 100, batch size to 16, and image size to 416 × 416 and performing different techniques to obtain the best training results. After completing 100 iterations with YOLOv5, the values of different parameters are recorded while undergoing training, as shown in Fig. 10.

The three columns of the first and second rows are YOLOv5 model loss components, box loss, objectness loss, and classification loss of the training and validation. These loss components explain how effectively an algorithm predicts an object. The

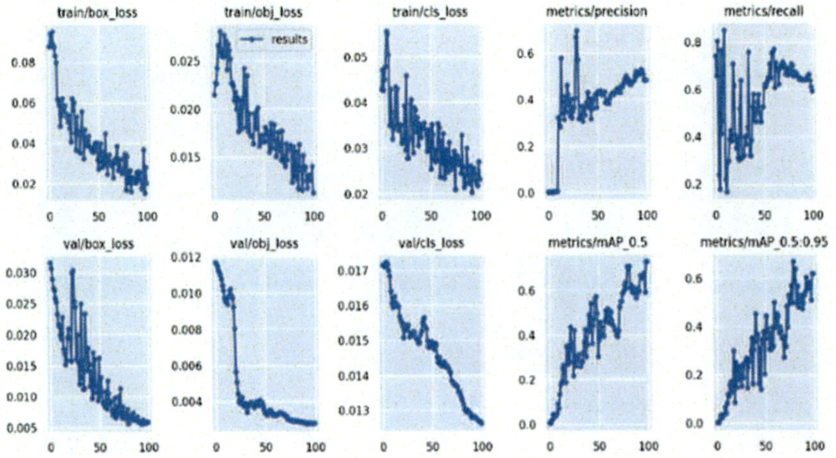

Fig. 10 Results graph

box loss function, objectness loss function, and classification loss function are illustrated in Fig. 10. It clearly states whether the model can distinguish between all classes. In this training and validation, all loss components initially have a high loss value; as the number of iterations increases, the learning rate increases the model will learn and makes the loss close to zero.

The model has achieved 50.1% precision, 70.4% recall, and 70.4% mAp@0.5 of all the classes. The Rafale aircraft has achieved 100% precision and the recall is 100% for A10 and Mirage2000 aircraft. Individual aircraft such as Rafale and mirage2000 aircraft have achieved mean average precision of 99.5%. The graphs of precision, recall, mAp@0.5, and mAp@0.5:0.95 values in Fig. 10 are increasing as the number of epochs rises. It shows that the model is getting trained and differentiating the different aircraft. Since mAp@0.5 is 70.4% the model can successfully detect unique aircraft. The metric analysis for all the classes is shown in Table.1.

One of the evaluation metrics, represented in Fig. 11, is the confusion matrix. A confusion matrix is a table that shows the model's performance in giving some truth values. A cell color intensity represents each class probability.

Table 1 Metric analysis table

Classes	Precision	Recall	mAp@0.5	mAp@0.5:0.95
All	0.501	0.704	0.704	0.67
A10	0.424	1	0.685	0.646
A400M	0.297	0.75	0.325	0.325
Mirage2000	0.351	1	0.995	0.995
Rafale	1	0	0.995	0.895
V22	0.434	0.77	0.521	0.486

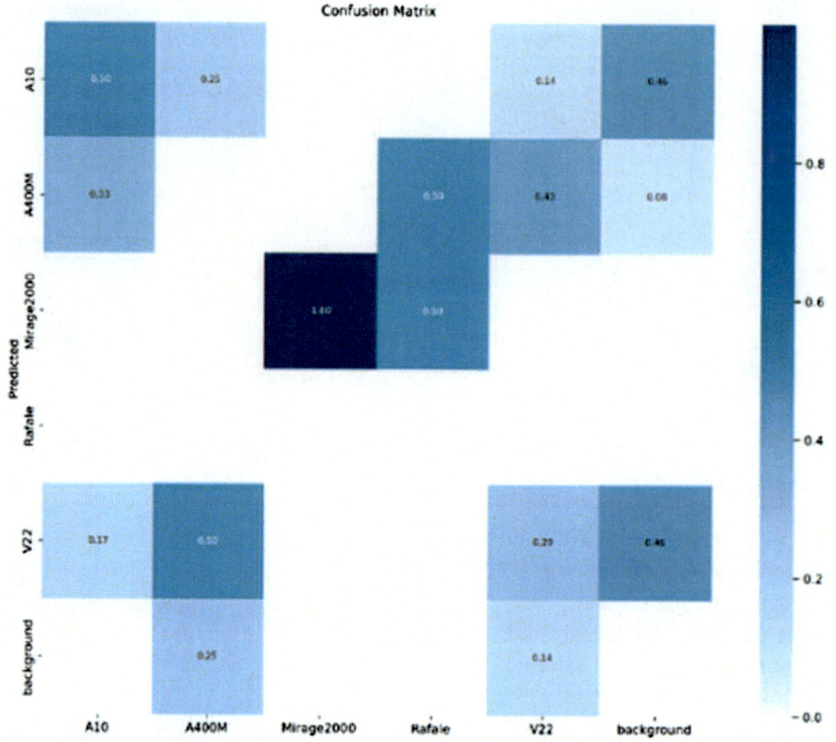

Fig. 11 Confusion matrix

5 Conclusion

This paper proposed a YOLOv5-based military aircraft detection algorithm with various military aircraft images, which was evaluated. The performance of the YOLOv5 military aircraft was presented, and it displayed improvement in terms of precision (50.1%), recall (70.4%), and mAP@0.5 (70.4%) in extreme conditions, including partial visibility, lighting fluctuations, and shadow. Because of this, the YOLOv5 military aircraft provided a reliable aircraft detection algorithm in remote sensing images. However, various test trials revealed that elements like light and blur have a special impact on detection results. During the subsequent training phase, more image data in a highly complicated environment must be gathered to improve the classification performance of the YOLOv5 military aircraft model.

References

1. Ji et al F (2021) Aircraft detection in high spatial resolution remote sensing images combining multi-angle features driven and majority voting CNN. Remote Sens 13(11):2207. https://doi.org/10.3390/RS13112207
2. Biswas K, Biswas K (2019) Military aviation principles. Military Eng. https://doi.org/10.5772/INTECHOPEN.87087
3. Zhu H, Lung H, Lin N (2020) Carrier-based aircraft detection on flight deck of aircraft carrier with simulated 3-D model by deep neural network. In: ACM international conference proceeding series, pp 96–101, May 2020. https://doi.org/10.1145/3403746.3403911
4. Pattanaik B, Chauhan A (2021) A study of stealth technology. Mater Today Proc. https://doi.org/10.1016/J.MATPR.2021.03.705
5. Azam B et al (2022) Aircraft detection in satellite imagery using deep learning-based object detectors. Microprocess Microsyst 94:104630. https://doi.org/10.1016/J.MICPRO.2022.104630
6. Surrisyad H, Gadjah Mada U, Wahyono I (2020) A fast military object recognition using extreme learning approach on CNN. Int J Adv Comput Sci Appl 11(12):210–220. https://doi.org/10.14569/IJACSA.2020.0111227
7. Chen Z, Li X, Wang L, Shi Y, Sun Z, Sun W (2022) An object detection and localization method based on improved YOLOv5 for the teleoperated robot. Appl Sci 12(22):11441. https://doi.org/10.3390/APP122211441
8. Zhou X, Gong W, Fu W, Du F (2017) Application of deep learning in object detection. In: Proceedings—16th IEEE/ACIS international conference on computer and information science, ICIS 2017, pp 631–634, Jun 2017. https://doi.org/10.1109/ICIS.2017.7960069
9. Janakiramaiah B, Kalyani G, Karuna A, Prasad LVN, Krishna M (2021) Military object detection in defense using multi-level capsule networks. Soft Comput 2021:1–15. https://doi.org/10.1007/S00500-021-05912-0
10. Gali M, Dhavale S, Kumar S (2022) Real-time image based weapon detection using YOLO algorithms. Commun Comput Inf Sci 1614:173–185. https://doi.org/10.1007/978-3-031-12641-3_15/COVER
11. Malta A, Mendes M, Farinha T (2021) Augmented reality maintenance assistant using YOLOv5. Appl Sci 11(11):4758. https://doi.org/10.3390/APP11114758
12. Luo S, Yu J, Xi Y, Liao X (2022) Aircraft target detection in remote sensing images based on improved YOLOv5. IEEE Access 10:5184–5192. https://doi.org/10.1109/ACCESS.2022.3140876/REFERENCES
13. Yang Y, Xie G, Qu Y (2021) Real-time detection of aircraft objects in remote sensing images based on improved YOLOv4. In: IEEE advanced information technology, electronic and automation control conference (IAEAC), pp 1156–1164. https://doi.org/10.1109/IAEAC50856.2021.9390673
14. Wang Y et al (2022) TransEffiDet: aircraft detection and classification in aerial images based on EfficientDet and transformer. Comput Intell Neurosci 2022. https://doi.org/10.1155/2022/2262549
15. Tarimo W, Sabra M, Hendre S (2020) Real-time deep learning-based object detection framework. In: 2020 IEEE symposium series on computational intelligence, SSCI 2020, pp 1829–1836, Dec 2020.https://doi.org/10.1109/SSCI47803.2020.9308493
16. Carrasco DP, Rashwan HA, Garcia MA, Puig D (2021) T-YOLO: tiny vehicle detection based on YOLO and multi-scale convolutional neural networks. IEEE Access. https://doi.org/10.1109/ACCESS.2021.3137638
17. Wei W, Zhang J (2019) Remote sensing image aircraft detection technology based on deep learning. In: Proceedings—2019 11th international conference on intelligent human-machine systems and cybernetics, IHMSC 2019, vol 1, pp 173–177, Aug 2019. https://doi.org/10.1109/IHMSC.2019.00048

18. Tan S, Lu G, Jiang Z, Huang L (2021) Improved YOLOv5 network model and application in safety helmet detection. ISR 2021—2021 IEEE international conference on intelligence and safety for robotics, pp 330–333, Mar 2021. https://doi.org/10.1109/ISR50024.2021.9419561
19. Sozzi M, Cantalamessa S, Cogato A, Kayad A, Marinello F (2022) Automatic bunch detection in white grape varieties using YOLOv3, YOLOv4, and YOLOv5 deep learning algorithms. Agronomy 12(2):319. https://doi.org/10.3390/AGRONOMY12020319
20. Yang G et al (2020) Face mask recognition system with YOLOV5 based on image recognition. In: 2020 IEEE 6th international conference on computer and communications, ICCC 2020, pp 1398–1404, Dec 2020. https://doi.org/10.1109/ICCC51575.2020.9345042
21. Zhou F, Zhao H, Nie Z (2021) Safety helmet detection based on YOLOv5. In: Proceedings of 2021 IEEE international conference on power electronics, computer applications, ICPECA 2021, pp 6–11, Jan 2021. https://doi.org/10.1109/ICPECA51329.2021.9362711
22. Abhinu CG, Aswin P, Krishnan K, Baby B, Viji KSA (2021) Multiple object tracking using deep learning with YOLO V5. Int J Eng Res Technol 9(13).https://doi.org/10.17577/IJERTCONV9IS13010
23. Maity M, Banerjee S, Sinha Chaudhuri S (2021) Faster R-CNN and YOLO based vehicle detection: a survey. Proceedings—5th international conference on computing methodologies and communication, ICCMC 2021, pp 1442–1447, Apr 2021. https://doi.org/10.1109/ICCMC51019.2021.9418274
24. Zhang X, Qian B, Si H, Zeng L, Wang H (2021) Research on the computer intelligent recognition of electric appliance by Yolo V5 algorithm. J Phys Conf Ser 2083(3):032078. https://doi.org/10.1088/1742-6596/2083/3/032078

Developing a Curriculum for Ethical and Responsible AI: A University Course on Safety, Fairness, Privacy, and Ethics to Prepare Next Generation of AI Professionals

Ashraf Alam

Abstract In this scientific paper, the researcher develops a course on "Safety, Fairness, Privacy, and Ethics of Artificial Intelligence" (SFPE-AI) designed for university students. The course aims to provide students with a comprehensive understanding of the technical and ethical issues associated with the development and deployment of AI systems. The course is designed to be interdisciplinary, drawing on concepts and techniques from computer science, philosophy, and law. The curriculum is divided into four modules: safety, fairness, privacy, and ethics. To facilitate student learning, the course employs a variety of pedagogical tools, such as interactive lectures, case studies, group discussions, and hands-on projects. The case studies used in the course include real-world examples of AI applications and their associated ethical and societal implication, thus providing students with a diverse perspective on the challenges and opportunities associated with AI. After the completion of this course, students are expected to understand the technical and ethical issues associated with AI, design and develop AI systems that are safe, fair, private, and ethical, and critically evaluate the societal implications of AI. The SFPE-AI course is expected to prepare the next generation of AI professionals to build responsible and trustworthy AI systems. The course will also serve as a model for other universities and educational institutions looking to integrate the discussion of AI safety, fairness, privacy, and ethics into their curriculum.

Keywords Classroom · Curriculum · Teacher · University · Pedagogy · Artificial intelligence · Safety · Fairness · Privacy · Ethics · Educational technology · ICT in education · Learner

A. Alam (✉)
Rekhi Centre of Excellence for the Science of Happiness, Indian Institute of Technology Kharagpur, Kharagpur, West Bengal, India
e-mail: ashraf_alam@kgpian.iitkgp.ac.in

G. Rajakumar et al. (eds.), *Intelligent Communication Technologies and Virtual Mobile Networks*, Lecture Notes on Data Engineering and Communications Technologies 171,
https://doi.org/10.1007/978-981-99-1767-9_64

1 Introduction

Artificial intelligence (AI) is a rapidly growing field that has the potential to revolutionize many aspects of our lives [1]. However, as AI becomes more advanced and more widely adopted, it is important to consider the safety, fairness, privacy, interpretability, human-AI interaction, and ethical implications of these technologies. Safety is a critical concern for AI systems, as they are increasingly being deployed in high-stakes situations such as transportation, health care, and finance. Ensuring the safety of AI systems requires a combination of technical measures, such as testing and validation, as well as governance and regulatory frameworks. For example, techniques such as "explainable AI" (XAI) can help make AI systems more transparent and interpretable, which can improve their safety by making it easier to understand and anticipate their behavior [2]. Fairness is another important consideration for AI systems, as they can perpetuate and even amplify existing biases in society [3]. For example, biased training data or algorithms can lead to unfair outcomes for certain groups of people. To address this, techniques such as "fairness-aware AI" have been developed, which aim to ensure that AI systems are fair and do not discriminate against certain groups of people [4]. Privacy is also a key concern for AI systems, as they often involve the collection and processing of sensitive personal data. Ensuring the privacy of individuals in the context of AI requires a combination of technical measures such as data anonymization and differential privacy, as well as robust governance and regulatory frameworks. Interpretability is a crucial aspect of AI, as it allows us to understand how an AI system is making its decisions. This is important for ensuring the safety and fairness of AI systems as well as for building trust in these technologies [5]. Techniques such as XAI can help make AI systems more interpretable and transparent, which can improve their accountability and reduce the risk of unintended consequences.

Human-AI interaction is another important area of research, as it deals with how people interact with and understand AI systems [6]. This includes issues such as user interface design, natural language processing, and human-AI collaboration. Ensuring that AI systems are easy to use and understand is crucial for building trust in these technologies and making them accessible to a wide range of people. Finally, ethics is a crucial area of consideration for AI, as these technologies have the potential to have a profound impact on society [7]. Ethical issues that need to be considered include issues such as autonomy, accountability, and social responsibility. Developing a robust ethical framework for AI is important for ensuring that these technologies are developed and used in a responsible and beneficial way [8].

In conclusion, AI is a rapidly growing field with enormous potential to improve our lives, but it is important to consider the safety, fairness, privacy, interpretability, human-AI interaction, and ethical implications of these technologies. Addressing these concerns will require a combination of technical measures, governance, and regulatory frameworks, as well as ongoing research and collaboration across multiple disciplines.

2 Robust Ethical Framework

A robust ethical framework for AI should be based on a set of guiding principles that can be used to evaluate the design, development, and deployment of AI systems [9]. Such a framework should take into account the unique characteristics of AI, such as its ability to learn and adapt, and should be flexible enough to be applied to a wide range of applications [10]. One possible set of principles for an ethical framework for AI include.

Respect for autonomy: AI systems should respect the autonomy of individuals and not be used to control or manipulate them without their consent.

Fairness and non-discrimination: AI systems should be designed to be fair and not discriminate against any individuals or groups based on their race, gender, age, class, caste, or other characteristics.

Responsibility and accountability: AI systems should be designed to be transparent and interpretable, and the creators and users of these systems should be held responsible for their actions.

Privacy: AI systems should be designed to protect the privacy of individuals and not be used to collect or process sensitive personal data without their consent.

Safety and security: AI systems should be designed to be safe and secure and should not be used to cause harm or put individuals at risk.

Transparency: AI systems should be designed to be transparent and explainable, and the creators and users of these systems should be open and honest about how they work.

Human-AI interaction: AI systems should be designed to be easy to use and understand and should be designed to complement and enhance human capabilities, not replace them.

Social benefit: AI systems should be designed to be beneficial to society and should not be used to perpetuate or amplify existing societal problems.

Ethical decision-making: AI systems should be designed to make ethical decisions or to support human decision-making in an ethical way.

Continual evaluation: The ethical implications of AI should be continually evaluated as the technology and its applications evolve, and the framework should be updated as necessary.

It is important to note that these principles are not exhaustive and may not be applicable to all context or scenarios, but they provide a good starting point for creating an ethical framework for AI. Additionally, the implementation of these principles should involve active participation from multiple stakeholders such as technologists, policymakers, legal experts, ethicists, and representatives from affected communities to ensure that the framework is inclusive and comprehensive.

3 Explainable Artificial Intelligence (XAI)

Explainable artificial intelligence (XAI) is a rapidly growing field that aims to make AI systems more transparent and interpretable, so that their behavior can be better understood and anticipated. This is important for ensuring the safety, fairness, and accountability of AI systems, as well as for building trust in these technologies [11]. One of the main challenges in XAI is that many modern AI systems, such as deep neural networks, are highly complex and their behavior can be difficult to understand. One approach to addressing this is to develop techniques that can generate human-interpretable explanations of the decisions made by AI systems [12]. This can be done in a variety of ways, such as by identifying the most important features used by the AI system in making its decisions or by generating natural language explanations [13].

Another approach to XAI is to develop techniques that can directly modify the behavior of AI systems to make them more interpretable. This can be done by designing the AI system to be more transparent in how it makes its decisions or by developing techniques that can "debug" the AI system to identify and correct any errors or biases [14]. Additionally, XAI can be used in the context of interpretability of models, where it is essential to understand the model's decision-making process. It can be achieved by using techniques such as feature importance, model-agnostic interpretability, and also post-hoc interpretability [15].

There are also various methods for evaluating the interpretability of AI models, including user studies, human-in-the-loop evaluations, and metrics based on information theory [16]. These methods can be used to measure the effectiveness of XAI techniques and to identify areas where further research is needed. Another area of XAI is the generation of synthetic data or counterfactual examples, which can be used to understand the decision-making process of models and also to identify potential biases in the model [17].

In conclusion, XAI is an important field that aims to make AI systems more transparent and interpretable and has the potential to improve the safety, fairness, and accountability of AI systems. The field is rapidly evolving, and ongoing research is needed to develop new techniques and methods for generating human-interpretable explanations, modifying AI systems to make them more interpretable, evaluating interpretability, and identifying potential biases [18]. Additionally, collaboration between experts in AI, cognitive psychology, human–computer interaction, and other relevant fields is important to advance the field of XAI [19].

4 Fairness-Aware Artificial Intelligence (FAAI)

Fairness-aware artificial intelligence (FAAI) is a rapidly growing field that aims to ensure that AI systems are fair and do not discriminate against certain individuals or groups based on their race, gender, caste, region, religion, age, or other characteristics.

This is a particularly important issue in sensitive applications such as criminal justice, health care, and finance, where AI systems have the potential to perpetuate existing societal biases [20]. One of the main challenges in FAAI is that many modern AI systems are highly complex and their behavior can be difficult to understand. This makes it difficult to identify and correct any biases in these systems. One approach to addressing this is to develop techniques that can identify and quantify biases in AI systems. This can be done in a variety of ways, such as by analyzing the training data used to train the AI system or by measuring the performance of the AI system on different groups of individuals [21]. Another approach to FAAI is to develop techniques that can directly modify the behavior of AI systems to make them more fair. This can be done by designing the AI system to be more transparent in how it makes its decisions or by developing techniques that can "debias" the AI system to correct any errors or biases [22].

One widely used technique for FAAI is the "pre-processing" technique, where the data is transformed or modified before training a model to remove or reduce any biases present in the data. Another technique is "in-processing" where the model is designed to be fair during the training process, for example, by adding a fairness constraint to the optimization problem [23].

Additionally, there are various methods for evaluating the fairness of AI models, including demographic parity, equal opportunity, and equalized odds [24]. These metrics can be used to measure the effectiveness of FAAI techniques and to identify areas where further research is needed. Another area of FAAI is the use of interpretable models, which can be used to understand the decision-making process of models and also to identify potential biases in the model [25]. Collaboration between experts in AI, statistics, sociology, and other relevant fields is important to advance the field of FAAI.

5 Developing Robust "Governance and Regulatory Frameworks" to Address Privacy Concerns for AI Systems

As artificial intelligence (AI) becomes more prevalent in society, it is increasingly important to develop robust governance and regulatory frameworks to address privacy concerns [26]. Privacy is a complex and multifaceted issue that is closely tied to the development and deployment of AI systems [27]. It is critical to ensure that these systems are designed, built, and deployed in a way that respects individuals' privacy and protects their personal information. One key aspect of privacy in AI is the collection, storage, and use of personal data. This includes not just data that individuals provide directly, but also data that is inferred or generated by AI systems [28]. It is crucial to ensure that personal data is collected and used in a way that is transparent, fair, and lawful. This includes providing individuals with clear and concise information about what data is being collected, how it will be used, and who it will be shared with. It also includes ensuring that data is stored and processed securely and that individuals have the right to access, correct, and delete their personal data [29].

Another important aspect of privacy in AI is the use of personal data to make decisions that affect individuals. This includes decisions made by AI systems in areas such as credit, employment, housing, and health care. It is important to ensure that these decisions are fair, unbiased, and transparent, and that individuals have the right to contest and correct any decisions that are made about them. To address these concerns, it is necessary to develop governance and regulatory frameworks that are tailored to the unique challenges posed by AI [30]. This may include laws and regulations that govern the collection, use, and storage of personal data, as well as guidelines and best practices for the design and deployment of AI systems. Additionally, it may involve creating independent oversight bodies to ensure that these laws and regulations are being followed, and that any violations are quickly identified and addressed. In addition to the traditional legal frameworks, technical solutions like differential privacy, federated learning, and homomorphic encryption can also be used to protect personal data while still allowing AI models to be trained and used [31].

6 Interpretability Aspect of AI

Interpretability is a crucial aspect of artificial intelligence (AI) that refers to the ability to understand and explain the behavior and decisions of AI systems [32]. This is particularly important in sensitive applications such as health care, finance, and criminal justice, where the decisions made by AI systems can have a significant impact on individuals and society. One of the main challenges in interpretability is that many modern AI systems, such as deep neural networks, are highly complex and their behavior can be difficult to understand [33]. This makes it difficult to identify and correct any errors or biases in these systems. Additionally, the black box nature of these models makes it hard to understand how the model arrived at a particular decision and what features of the data were important in making that decision [34].

To address these challenges, researchers have developed a variety of techniques for making AI systems more interpretable. One approach is to use techniques that can identify and quantify the features of the data that are most important in making a decision [35]. For example, feature importance techniques like permutation feature importance, SHAP, and LIME can be used to understand which features of the data were most important in making a decision. Another approach is to use techniques that can directly modify the behavior of AI systems to make them more interpretable. This can be done by designing the AI system to be more transparent in how it makes its decisions, or by developing techniques that can "debias" the AI system to correct any errors or biases [36]. For example, decision trees and rule-based models are more interpretable than deep neural networks because they make decisions based on a set of simple if–then rules, which are easy to understand and explain [37]. Additionally, there are various methods for evaluating the interpretability of AI models, such as the model's global and local interpretability. Global interpretability refers to the extent to which the overall behavior and decision-making process of the model can be understood, while local interpretability refers to the extent to which the model's behavior for a specific input can be understood [38].

In the medical field, interpretability is crucial as it allows medical experts to understand the decisions made by the AI system and ensure that they align with the medical knowledge. For example, in the case of radiology, interpretable models can help radiologists understand the model's decision-making process and identify the features that are important in making the diagnosis. This can help radiologists identify any errors or biases in the model and make adjustments accordingly.

7 Why Teach a Course on Safety, Fairness, Privacy, and Ethics" to Students?

Teaching a course on Safety, Fairness, Privacy, and Ethics of Artificial Intelligence (AI) to students is important for several reasons:

Preparing for the future: AI is rapidly advancing and is expected to play an increasingly important role in society. By teaching students about the safety, fairness, privacy, and ethics of AI, they will be better prepared for the future and the potential impact of AI on their lives.

Developing critical thinking skills: Understanding the safety, fairness, privacy, and ethics of AI requires students to think critically about the implications of AI on society. This can help students develop the critical thinking skills they need to evaluate information, make informed decisions, and solve problems.

Promoting responsible use of AI: By teaching students about the safety, fairness, privacy, and ethics of AI, they will be more likely to use AI responsibly and to recognize when AI is being used in a way that is harmful or unethical.

Promoting digital literacy: Understanding the safety, fairness, privacy, and ethics of AI is an important aspect of digital literacy. This will enable students to navigate the digital world and to understand the impact of technology on society.

Creating a more inclusive society: AI systems can perpetuate societal biases and discrimination if not designed and implemented carefully. Teaching students about the fairness and ethical considerations of AI can help create a more inclusive society by encouraging them to design and use AI in a way that is fair and unbiased.

Fostering a culture of responsibility: By teaching students about the safety, fairness, privacy, and ethics of AI, we can foster a culture of responsibility where individuals take responsibility for the impact of AI on society and work to ensure that AI is used in a way that is safe, fair, and respects privacy.

8 Curriculum Outline for a Course on "Safety, Fairness, Privacy, and Ethics of Artificial Intelligence"

I. Introduction

- Overview of the course
- Importance of understanding the safety, fairness, privacy, and ethics of AI
- Key concepts and definitions related to AI safety, fairness, privacy, and ethics.

II. Safety of Artificial Intelligence

- Overview of safety concerns in AI
- Techniques for evaluating and ensuring the safety of AI systems
- Case studies of AI safety incidents and lessons learned

III. Fairness in Artificial Intelligence

- Overview of fairness concerns in AI
- Techniques for evaluating and ensuring fairness in AI systems
- Case studies of AI fairness incidents and lessons learned.

IV. Privacy in Artificial Intelligence

- Overview of privacy concerns in AI
- Techniques for evaluating and ensuring privacy in AI systems
- Case studies of AI privacy incidents and lessons learned.

V. Ethics of Artificial Intelligence

- Overview of ethical concerns in AI
- Techniques for evaluating and ensuring the ethical use of AI systems
- Case studies of AI ethical incidents and lessons learned.

VI. Human-AI Interaction

- Overview of the interactions between humans and AI
- Techniques for designing human-AI interactions that are safe, fair, private, and ethical
- Case studies of human-AI interactions and lessons learned.

VII. Governance and Regulation of Artificial Intelligence

- Overview of governance and regulatory frameworks for AI
- Techniques for designing governance and regulatory frameworks that promote safety, fairness, privacy, and ethics in AI
- Case studies of governance and regulatory frameworks for AI and lessons learned.

VIII. Interpretability of Artificial Intelligence

- Overview of the interpretability of AI

- Techniques for making AI systems more interpretable
- Case studies of interpretable AI systems and lessons learned.

IX. Conclusion

- Summary of key concepts and takeaways
- Discussion of future directions in AI safety, fairness, privacy, and ethics.

Throughout the course, students will engage in hands-on exercises and projects, as well as discussions and debates to help them apply the concepts and techniques covered in the course to real-world scenarios.

9 Learning Outcomes

The learning outcomes for a course on "Safety, Fairness, Privacy, and Ethics of Artificial Intelligence" for university students include the following:

1. *Understanding of the technical and societal implications of AI*: Students will be able to understand the technical capabilities and limitations of AI systems, as well as the potential impact of AI on society, including issues related to safety, fairness, privacy, and ethics.
2. *Knowledge of AI safety, fairness, privacy, and ethics*: Students will develop a comprehensive understanding of the key concepts and principles related to AI safety, fairness, privacy, and ethics.
3. *Ability to analyze and evaluate AI systems*: Students will be able to analyze and evaluate AI systems based on safety, fairness, privacy, and ethics criteria and to identify potential risks and ethical issues.
4. *Skills in designing and developing safe, fair, and ethical AI systems*: Students will be able to design and develop AI systems that are safe, fair, and ethical and to implement appropriate measures to mitigate risks and ensure compliance with relevant regulations and standards.
5. *Understanding of governance and regulatory frameworks*: Students will be able to understand the governance and regulatory frameworks related to AI and to analyze the effectiveness of different approaches.
6. *Ability to communicate effectively about AI safety, fairness, privacy, and ethics*: Students will be able to communicate effectively about AI safety, fairness, privacy, and ethics both with technical and non-technical audiences and to participate in interdisciplinary discussions and collaborations.
7. *Ability to use storytelling method to understand and communicate about AI safety, fairness, privacy, and ethics*: Students will be able to use storytelling method to understand and communicate about AI safety, fairness, privacy, and ethics and to evaluate the effectiveness of different storytelling methods.
8. *Critical thinking and problem-solving skills*: Students will develop critical thinking and problem-solving skills and will be able to apply them in real-world scenarios related to AI safety, fairness, privacy, and ethics.

In conclusion, the course will provide students with a solid foundation in the field of AI safety, fairness, privacy, and ethics and will equip them with the knowledge and skills needed to work with AI systems in a safe, fair, and ethical manner.

10 Pedagogical Tools and Teaching–Learning Materials

Pedagogical tools and teaching–learning materials (TLMs) that shall be used to teach a course on "Safety, Fairness, Privacy, and Ethics of Artificial Intelligence" for university students include.

Lectures and presentations: Use of lectures and presentations to introduce key concepts and provide an overview of the course material. This can include slides, videos, and interactive elements such as quizzes and polls.

Case studies: Use of real-world case studies to illustrate the importance of safety, fairness, privacy, and ethics in AI and to provide students with an opportunity to apply the concepts and techniques covered in the course to real-world scenarios.

Hands-on projects: Use of hands-on projects, such as programming assignments or design projects, to give students the opportunity to apply the concepts and techniques covered in the course in a practical context.

Guest lectures: Inviting experts in the field of AI safety, fairness, privacy, and ethics to give guest lectures on specific topics, to provide students with a broader perspective on the field.

Simulation and role-playing exercises: Use of simulation and role-playing exercises to help students understand the implications of AI in different scenarios.

Online resources: Use of online resources, such as videos, articles, and tutorials, to supplement the course material and provide students with additional information and resources on the topics covered in the course.

Discussion forums and debates: Encourage students to discuss the topics covered in the course and to express their opinions and ideas through discussion forums and debates.

Reading materials: Provide students with a recommended reading list of relevant books, articles, and research papers on the topic of AI safety, fairness, privacy, and ethics.

Ethical dilemma: Provide students with ethical dilemmas related to AI and have them work through resolving the situations.

Collaborative activities: Encourage students to work in teams or groups on projects and assignments, to foster collaboration and teamwork.

These pedagogical tools and teaching–learning materials can be used to create an interactive and engaging learning experience for students and help them understand the importance of safety, fairness, privacy, and ethics in AI, and how to apply the concepts and techniques covered in the course to real-world scenarios.

11 Games and Activities

Games and activities that shall be used to teach a course on "Safety, Fairness, Privacy, and Ethics of Artificial Intelligence" for university students include

1. *Ethical dilemma game*: Create a game that presents students with ethical dilemmas related to AI and have them work through resolving the situations.
2. *Scenario-based role-playing*: Create scenarios that depict the potential consequences of AI and have students role-play different characters in the scenario, to help them understand the implications of AI in different contexts.
3. *Debate*: Organize debates on current controversies related to AI safety, fairness, privacy, and ethics and have students participate in the debates and express their opinions on the topic.
4. *Escape room*: Create an escape room experience that simulates a complex AI scenario where students need to use their understanding of safety, fairness, privacy, and ethics to solve the puzzles and escape.
5. *Board game*: Develop a board game that simulates the decision-making process in AI development and have students play the game to understand the importance of safety, fairness, privacy, and ethics in AI.
6. *AI simulation*: Create a simulation of an AI system and have students experiment with different settings and configurations to understand the impact of different choices on safety, fairness, privacy, and ethics.
7. *Ethical hackathon*: Organize an ethical hackathon where students work in teams to identify and fix ethical issues in existing AI systems.
8. *Group discussions*: Divide students into small groups and give them a case-study or scenario related to AI safety, fairness, privacy, and ethics and have them discuss the situation and come up with solutions or recommendations.
9. *Case-study analysis*: Have students analyze real-world case studies related to AI safety, fairness, privacy, and ethics and have them present their findings and recommendations to the class.
10. *Jeopardy game*: Create a Jeopardy-style game where students answer questions on AI safety, fairness, privacy, and ethics and get points based on their answers.

These games and activities can be used to create an interactive and engaging learning experience for students and help them understand the importance of safety, fairness, privacy, and ethics in AI, and how to apply the concepts and techniques covered in the course to real-world scenarios.

12 Storytelling Method

Storytelling can serve as a powerful tool for teaching a course on "Safety, Fairness, Privacy, and Ethics of Artificial Intelligence" to university students. Here are a few ways storytelling can be used in such a course:

1. *Case studies*: Share real-world case studies of AI systems that have faced ethical challenges. Use storytelling to present the background, the situation, the challenges faced and the resolution.
2. *Scenarios*: Create fictional scenarios that depict the potential consequences of AI, both positive and negative, and use storytelling to present the scenarios in an engaging and relatable way.
3. *Stories from experts*: Invite experts in the field of AI safety, fairness, privacy, and ethics to share their experiences and stories of challenges they have faced and how they overcame them.
4. *Role-playing*: Have students act out scenarios in small groups, to help them understand the implications of AI in different contexts.
5. *Interactive storytelling*: Use interactive storytelling tools such as virtual reality, augmented reality, and gamification to create an immersive learning experience for students.
6. *Storytelling through data*: Use data storytelling techniques like data visualization, infographics, and videos to help students understand the impact of AI on society and how to measure safety, fairness, privacy, and ethics.
7. *Storytelling through analogies*: Use analogies and metaphor to help students understand the complex concepts of AI safety, fairness, privacy, and ethics.
8. *Ethical storytelling*: Use storytelling to explore ethical dilemmas related to AI and guide students through the decision-making process.

By using storytelling, students can relate to the material in a more personal way and better understand the implications of AI on society. Storytelling can also be used to create an engaging and interactive learning experience, making the course more memorable and effective.

13 Case-Study Method

The case-study method can be used to teach a course on "Safety, Fairness, Privacy, and Ethics of Artificial Intelligence" for university students in a number of ways, including the following:

1. *Real-world examples*: Case studies can provide students with real-world examples of AI systems and applications, including both successes and failures, and can help to illustrate the key concepts and principles related to AI safety, fairness, privacy, and ethics.

2. *Analysis and evaluation*: Students can analyze and evaluate the case studies using the concepts and principles learned in class and can identify the potential risks and ethical issues associated with each case.
3. *Problem-solving*: Case studies can serve as a starting point for problem-solving exercises, where students can work in groups to develop solutions to the issues identified in the case studies and to propose recommendations for addressing the risks and ethical issues.
4. *Debate and discussion*: Case studies can be used to spark debate and discussion among students, as they can have different perspectives and opinions on the issues and risks identified in the case studies.
5. *Guest speakers*: Inviting experts or practitioners who have been involved with the specific case studies being discussed can provide students with valuable insights and perspectives on the real-world challenges of AI safety, fairness, privacy, and ethics.
6. *Current events*: Case studies can be selected to reflect current events and the most recent developments in the field of AI, which can make the course more engaging and relevant for the students.

In conclusion, the case-study method can be an effective way to foster problem-solving skills, promote critical thinking and debate, and stay current with the field's developments.

14 Example of a Case-Study: The Case of Amazon's AI-Powered Recruitment Tool

In this case-study, the students can learn about:

The application of AI in recruitment: Amazon developed an AI-powered recruitment tool that uses machine learning algorithms to analyze resumes and job applications, in order to identify the most qualified candidates for open positions.

Bias in AI systems: Amazon's recruitment tool was found to have a gender bias, as it was trained on resumes submitted to the company over a 10-year period, which were mostly from men. As a result, the tool was less likely to recommend female candidates for open positions.

Fairness in AI: The case raises important questions about fairness in AI systems, and the potential risks and challenges associated with bias in AI-powered decision-making.

Ethical considerations: The case highlights the ethical considerations that must be taken into account when developing and deploying AI systems, particularly in relation to issues of fairness, privacy, and transparency.

Mitigating bias: The case-study can be used to discuss ways to mitigate bias in AI systems, such as using diverse training data, and developing techniques for identifying and addressing bias in machine learning models.

Current events: This case-study is based on a real event that occurred in 2018, it is still relevant today as it highlights the importance of preventing bias in AI systems and the need for AI governance and regulations.

Guest speaker: A guest speaker from Amazon, who was involved in the development and deployment of the recruitment tool, could be invited to speak about their experience and the lessons learned, providing valuable insights into the challenges of implementing AI in practice. Overall, this case-study can be used to teach students about the practical challenges and ethical considerations associated with the application of AI in recruitment and to encourage them to think critically about the implications of bias in AI systems and the need for fairness-aware AI.

References

1. Alam A (2022) A digital game based learning approach for effective curriculum transaction for teaching-learning of artificial intelligence and machine learning. In: 2022 international conference on sustainable computing and data communication systems (ICSCDS). IEEE, pp 69–74
2. Currie G, Hawk KE (2021, March) Ethical and legal challenges of artificial intelligence in nuclear medicine. Sem Nucl Med 51(2):120–125
3. Alam A (2022) Investigating sustainable education and positive psychology interventions in schools towards achievement of sustainable happiness and wellbeing for 21st century pedagogy and curriculum. ECS Trans 107(1):19481
4. Saghiri AM, Vahidipour SM, Jabbarpour MR, Sookhak M, Forestiero A (2022) A survey of artificial intelligence challenges: analyzing the definitions, relationships, and evolutions. Appl Sci 12(8):4054
5. Alam A (2022) Social robots in education for long-term human-robot interaction: socially supportive behaviour of robotic tutor for creating robo-tangible learning environment in a guided discovery learning interaction. ECS Trans 107(1):12389
6. Vellido A (2019) Societal issues concerning the application of artificial intelligence in medicine. Kidney Dis 5(1):11–17
7. Alam A, Mohanty A (2022) Metaverse and Posthuman animated avatars for teaching-learning process: interperception in virtual universe for educational transformation. In: international conference on innovations in intelligent computing and communications. Springer, Cham, pp 47–61
8. Al-Hwsali A, Al-Saadi B, Abdi N, Khatab S, Solaiman B, Alzubaidi M, Abd-alrazaq A, Househ M (2022) Legal and ethical principles of artificial intelligence in public health: scoping review
9. Alam A (2020) Pedagogy of calculus in India: an empirical investigation. Periódico Tchê Química 17(34):164–180
10. Siala H, Wang Y (2022) SHIFTing artificial intelligence to be responsible in healthcare: a systematic review. Soc Sci Med:114782
11. Alam A (2022) Positive psychology goes to school: conceptualizing students' happiness in 21st century schools while 'minding the mind!' Are we there yet? Evidence-backed. School-Based Positive Psychol Intervent ECS Trans 107(1):11199
12. Nasim SF, Ali MR, Kulsoom U (2022) Artificial intelligence incidents & ethics a narrative review. Int J Technol Innov Manage (IJTIM) 2(2)

13. Alam A (2022) Mapping a sustainable future through conceptualization of transformative learning framework, education for sustainable development, critical reflection, and responsible citizenship: an exploration of pedagogies for twenty-first century learning. ECS Trans 107(1):9827
14. Huang C, Zhang Z, Mao B, Yao X (2022) An overview of artificial intelligence ethics. IEEE Trans Artif Intell
15. Alam A (2022) Employing adaptive learning and intelligent tutoring robots for virtual classrooms and smart campuses: reforming education in the age of artificial intelligence. In: Shaw RN, Das S, Piuri V, Bianchini M (eds) Advanced computing and intelligent technologies. Lecture notes in electrical engineering, vol 914. Springer, Singapore
16. O'Sullivan S, Nevejans N, Allen C, Blyth A, Leonard S, Pagallo U, Holzinger K, Holzinger A, Sajid MI, Ashrafian H (2019) Legal, regulatory, and ethical frameworks for development of standards in artificial intelligence (AI) and autonomous robotic surgery. Int J Med Robot Comput Assist Surg 15(1):e1968
17. Alam A (2022) Platform utilising blockchain technology for eLearning and online education for open sharing of academic proficiency and progress records. In: Asokan R, Ruiz DP, Baig ZA, Piramuthu S (eds) Smart data intelligence. Algorithms for intelligent systems. Springer, Singapore
18. del Pero AS, Wyckoff P, Vourc'h A (2022) Using artificial intelligence in the workplace: what are the main ethical risks?
19. Alam A (2023) Cloud-based e-learning: scaffolding the environment for adaptive e-learning ecosystem based on cloud computing infrastructure. In: Satapathy SC, Lin JCW, Wee LK, Bhateja V, Rajesh TM (eds) Computer communication, networking and IoT. Lecture notes in networks and systems, vol 459. Springer, Singapore
20. Köbis L, Mehner C (2021) Ethical questions raised by AI-supported mentoring in higher education. Front Artif Intell 4:624050
21. Alam A (2022) Cloud-based E-learning: development of conceptual model for adaptive e-learning ecosystem based on cloud computing infrastructure. In: Kumar A, Fister Jr I, Gupta PK, Debayle J, Zhang ZJ, Usman M (eds) Artificial intelligence and data science. ICAIDS 2021. Communications in computer and information science, vol 1673. Springer, Cham
22. Juric M, Sandic A, Brcic M (2020, Sept) AI safety: state of the field through quantitative lens. In: 2020 43rd international convention on information, communication and electronic technology (MIPRO). IEEE, pp 1254–1259
23. Alam A (2021) Possibilities and apprehensions in the landscape of artificial intelligence in education. In: 2021 international conference on computational intelligence and computing applications (ICCICA). IEEE, pp 1–8
24. Jobin A, Ienca M, Vayena E (2019) The global landscape of AI ethics guidelines. Nat Mach Intell 1(9):389–399
25. Alam A (2022) Impact of university's human resources practices on professors' occupational performance: empirical evidence from India's higher education sector. In: Inclusive businesses in developing economies. Palgrave Macmillan, Cham, pp 107–131
26. Ashok M, Madan R, Joha A, Sivarajah U (2022) Ethical framework for artificial intelligence and digital technologies. Int J Inf Manage 62:102433
27. Alam A (2022) Educational robotics and computer programming in early childhood education: a conceptual framework for assessing elementary school students' computational thinking for designing powerful educational scenarios. In: 2022 international conference on smart technologies and systems for next generation computing (ICSTSN). IEEE, pp 1–7
28. Dhirani LL, Mukhtiar N, Chowdhry BS, Newe T (2023) Ethical dilemmas and privacy issues in emerging technologies: a review. Sensors 23(3):1151
29. Alam A (2020) Challenges and possibilities in teaching and learning of calculus: a case study of India. J Educ Gifted Young Sci 8(1):407–433
30. Arslan J, Ounissi M, Jiménez G, Kar A, Racoceanu D (2022) Responsible artificial intelligence: a review of current trends. Winter School AI4Health

31. Alam A (2020) Possibilities and challenges of compounding artificial intelligence in India's educational landscape. Int J Adv Sci Technol 29(5):5077–5094
32. Alam A, Mohanty A (2022) Business models, business strategies, and innovations in EdTech companies: integration of learning analytics and artificial intelligence in higher education. In: 2022 IEEE 6th conference on information and communication technology (CICT). IEEE, pp 1–6
33. Lukkien DR, Nap HH, Buimer HP, Peine A, Boon WP, Ket JC, Minkman MM, Moors EH (2023) Toward responsible artificial intelligence in long-term care: a scoping review on practical approaches. Gerontologist 63(1):155–168
34. Alam A, Mohanty A (2022) Foundation for the future of higher education or 'misplaced optimism'? Being human in the age of artificial intelligence. In: International conference on innovations in intelligent computing and communications. Springer, Cham, pp 17–29
35. Kaur D, Uslu S, Rittichier KJ, Durresi A (2022) Trustworthy artificial intelligence: a review. ACM Comput Surv (CSUR) 55(2):1–38
36. Alam A (2020) Test of knowledge of elementary vectors concepts (TKEVC) among First-semester bachelor of engineering and technology students. Periódico Tchê Química 17(35):477–494
37. Bakiner O (2022) What do academics say about artificial intelligence ethics? An overview of the scholarship. AI Ethics:1–13
38. Alam A (2021) Should robots replace teachers? Mobilisation of AI and learning analytics in education. In: 2021 international conference on advances in computing, communication, and control (ICAC3). IEEE, pp 1–12

The Video Conferencing Technology Implications as an Educational Method on Undergraduate Students Learning Outcomes

Jason Fonseca Ng, Andrew, Stanley Nathanael Irawan, Ford Lumban Gaol, and Tokuro Matsuo

Abstract The COVID-19 outbreak is the greatest global health catastrophe and the largest threat facing humanity ever since the Second World War. It started in the Asia region in late 2019 and since then it has spread to almost every country on the planet, causing millions of people to be sick. The deadly disease has now reached the tragic milestone of one million deaths. It has surely impacted the world in most factors, especially in the education factor. The government has introduced social-distancing protocols that affect the teaching and learning process to decrease the infection rate. To cope with the situation, schools and universities implement long-distance teaching with video conference methods. This long-distance learning and teaching with video conference methods has several impacts on undergraduate students because, they are not accustomed to learning using this method, and that some courses require direct practices in the field. Therefore, this study is to gain a conclusion if long-distance teaching with video conference methods is effective, if it brings positive effects to the students in their learning process and if it motivates to study. This study's sampling method is submitted through Google form wherein the questionnaires are aimed at undergraduate students to obtain their outcomes about the video conference method used in their learning process.

J. F. Ng · Andrew · S. N. Irawan
Department of Information Systems, School of Information Systems, Bina Nusantara University, Jakarta, Indonesia 11480
e-mail: jason.fonseca1@binus.ac.id

Andrew
e-mail: andrew028@binus.ac.id

S. N. Irawan
e-mail: stanley.irawan1@binus.ac.id

F. L. Gaol (✉)
Department of Computer Science, BINUS Graduate Program—Doctor of Computer Science, Bina Nusantara University, Jakarta, Indonesia 11480
e-mail: fgaol@binus.edu

T. Matsuo
Advanced Institute of Industrial Technology, Tokyo, Japan
e-mail: matsuo@aiit.ac.jp

G. Rajakumar et al. (eds.), *Intelligent Communication Technologies and Virtual Mobile Networks*, Lecture Notes on Data Engineering and Communications Technologies 171,
https://doi.org/10.1007/978-981-99-1767-9_65

Keywords Video conferencing · Educational method · Undergraduate students learning outcomes

1 Introduction

Technology has only recently achieved a degree of stability, accessibility, and affordability that enables it to be used more in actual teaching scenarios than in research ventures. Using video is being hailed as the next electronic communication breakthrough. In order to promote concepts such as virtual teams, telecommuting and online conferencing, several enterprises are developing systems [12]. Recently, due to the current global pandemic, which requires everyone to carry out their daily activities from their home, including students studying, school has shifted the traditional learning methods into video conferencing methods.

The World Health Organization (WHO) has announced that the current coronavirus can be categorized as a global pandemic. In connection with the announcement of the pandemic in anticipation of the spread of the COVID-19 virus in Indonesia, President Joko Widodo stated that, to minimize the transmission of the coronavirus or COVID-19 disease, strategies for productive practices at home need to be implemented. To reduce the spreading of COVID-19, regulations for studying and working from home need to be intensified. In this home learning condition, schools, through the ranks of educators and educational staff, still must do learning even though they are not in school. The teacher must be proactive and creative to be able to hold teaching and learning activities as effectively as face-to-face [1].

It needs to be comprehended that studying at home is not a holiday, and it does not mean there are no literacy activities. Keep learning with targets that are already in the curriculum. Home learning is not just a teacher giving questions or assignments that must be done by students that are sent by the teacher to students via social media. The teacher must continue to provide material to explain the learning they have been doing in classroom face-to-face activities. The philosophy according to Alkaher [2], determines how the preferred teaching methods are perceived and handled, including how (or if) e-learning technologies are chosen and used.

Teachers must use the new era of technology and information development, which is very powerful, to make it easier to learn in any situation, anytime, and anywhere. One of them is distance immersive learning through video conferencing. Even if the places are far away and not in one city, it helps teachers and students to communicate through video conferencing, which is linked to the Internet. However, the teaching quality and strategy impacted learning more than the type of technology utilized [29].

1.1 Background of Study

Communication is one of the ways to interact with each other, and when long distances do not become a barrier for each person to communicate, the need for it increases in time. The rise in technology development is becoming very helpful in changing communication patterns for personal or industrial interests, or even in terms of education and learning. It can be used and supported by the teaching and learning process, given the agile developments in the field of web-based Internet technology. With the Internet as an online learning tool between resource people or lecturers and participants or students who can share data without having to meet face-to-face, the use of this platform for distance lectures is still not frequently used. The basic method of teaching and learning practices is face-to-face or conventional learning. The approach has been questioned because passive learning is promoted, individual differences and needs of learners are overlooked and problem-solving, logical thinking, or other higher order thinking skills are not taken into account [4, 10]. On the other hand, online learning also runs into opposition. Many educators and trainers do not endorse online schooling because they do not think it addresses challenging teaching and learning issues [7].

There are often instances of face-to-face learning where the lecturer is unable to attend, creating disruptions in the learning process. This encourages long-distance learning between lecturers and students. Therefore, with video conferencing technology, the long-distance learning process is necessary, where each student and lecturer can meet face-to-face even in different places through real-time online audio and video rather than postponing and moving the study session to another day. This makes learning more productive and engaging. This remote lecture system service can use application services from video conferencing service providers such as Zoom, Microsoft Teams, and Google Classroom. On the other hand, online learning reduces a lecture's environment and reduces the degree of interaction during the lecture [18, 26, 30].

1.2 Problem Statement

From the above background, the problem formulation is as follows:

- Does the implementation of teaching and learning activities on undergraduate students in Indonesia using the video conference applications (Zoom, Microsoft Teams, Google Meet, etc.) have a positive effect on student learning outcomes when assessed by final scores?

1.3 Research Questions

Is there a statistically significant difference between the mean scores of students using video conferencing as an educational tool relative to the mean scores of those students using conventional teaching only?

1.4 Hypotheses

There is a statistically significant difference between the mean scores of students using video conferencing as an educational tool relative to the mean scores of those students using conventional teaching only.

1.5 Relevance and the Importance of Research

This research was conducted in line with the widespread use of the video conferencing method as a learning method during the COVID-19 pandemic. This research is expected to be a benchmark reference regarding the effectiveness of this learning method when compared with conventional learning methods such as face-to-face learning.

2 Literature Review

Online learning has been made feasible by the Internet, and many educators are fascinated in online learning classes to improve and enhance the quality of student learning while combating the lack of resources, facilities, and equipment, particularly in higher education institutions. Due to its ability to provide more convenient access at any time and from any place to content and instruction, online learning has become prominent [22].

In higher education, some of the first applications of video conferencing methods in education are identified, where the technical ability to transmit lectures directly to a wide audience, whether locally (e.g., to accommodate a significant number of students) or more, allows students to participate concurrently at a distance [23, 28, 32]. Although this initial pilot was an educational transmission and learning model [6] in which technology allowed distant audiences to 'receive' conventional face-to-face lectures, it also proved successful in promoting less didactic higher education teaching features, such as tutorials or small group seminars [3, 24].

Heinich [11] characterized distance learning as the learner's physical seclusion from data, but with two-way communication given by a few innovations, so that it was conceivable to take after a guidelines program. The benefits of distance learning are perceived to be versatility in addressing the needs of learners and therefore a more equal situation for remote learners and cost savings [15]. The advent of communication technology has modernized indirect learning methods and the provision of data to further areas. It is done by a range of electronic devices, including gadget devices, Internet, email, and the video conferencing system. Subsequently, video conferencing is one conclusion of a range of conveyance modes, but is unique in its capacity to facilitate immediate contact between learners and instructors [31], compared to the offbeat learning generally related with mail or talk sheets, for example, a benefit in providing a 'social presence and relaxing learning environment' over other distance learning technologies [27], so that 'the synchronous nature of video conferencing makes a preferred distance learning technology for business, school, and other social sciences, where interpersonal skills are an essential element of student education' [33].

The technology of video conferencing is a broadcasting medium that allows associated clients to share visual and sound offices in real time. It also permits registered clients to exchange records, slides, inactive pictures, and content (such as desktop and web) via utilized platforms [17]. While the accessibility of bandwidth, interconnected systems and electronic device speed has expanded overtime, it has gotten to be more conceivable and reasonable for proficient networks, schools, and colleges to utilize video conferencing. Even with a high-speed network, in accordance with the requirements of use and environmental conditions, the use of certain video conferencing systems may appear to suggest different experiences. Interaction between fellow learners or learners-instructors in videoconference situations has created freshly developed possibilities in progressing the conveyance of conventional teaching methods, according to Lawson [19]. Numerous instructors utilize video conferencing platforms to facilitate the development and expertise of students and themselves to solve problems [20]. Simultaneous video conferencing setup, however, does not inherently provide users with the requisite collection of education results and an improved teaching method, where new higher education challenges increase [21].

The key benefit of the format 'lecture-at-a-distance' is often quoted as the opportunity to cover a greater number of learners in significant content publicizing activities to reduce wasted work, saving time and expense. Therefore, it is built as a versatile method for meeting growing university courses, satisfying wider audiences, modularizing, and raising part-time and mature students [16]. This emphasis on the financial and geographical benefits was echoed by students, describing reduction in travel expense and time [8]. Bates [5] showed how video conferencing provided a more effective cost to link remotely located US students, introducing learning opportunities that would otherwise be difficult to provide to them, such as clinical procedures. However, the organizational challenges faced by video conferencing, such as schedule preparation, booking specifications, and support staff resources, are increasing [24] and not all researchers agreed that lectures were a suitable environment for the use of two-way video conferencing.

Even so, in using this technology, there were many benefits for lecturers, in that they were seen as being able to use the saved time to organize more easily and arrange more focused content [9]. The possibility for document sharing and exchanges that video conferencing offered was also appreciated by lecturers, although these were not always taken up by the lecturers in Freeman's case study. To analyze and strengthen it, they have found it practical to provide a video log of their production. In addition, Jacobs and Rodgers [13] found that the more video conferencing was used by lecturers, the more likely they were to implement the improvements in the technique of lecturing that Kristiansen argues are required to optimize learning. Schiller and Mitchell [28] argued that lecturers needed help to establish a variety of teaching methods using video conferencing from more seasoned users.

2.1 Key Concepts

Educating is a social collaborative learning process that often entails community and peer engagement for the dynamic creation of new or even modern ideas, information, and comprehension through individual learning. It means that the main learning ability is that of communication. Key criteria for viable collective learning are clear communication, effective communication tools, productive communication methods, and networks. Dialogue is one of the many methods of communication. Dialogue alludes not as it were the interaction between both the students and lecturers but moreover also the interaction between students [14]. There are mainly three categories of learning. Traditional learning is the conventional learning method in which students turn up to their lectures, take notes, attend classes, seminars, laboratories, complete given assignments, etc. Every student's learning schedule and materials are arranged by the school institution. Distance learning is the process of learning in which the course and degree are independent of the position. Students from either area may undertake the course but they have limited opportunities to communicate with others or their peers. Lastly, open learning does not rely on time or place for students completing an open learning curriculum. Same as distance learning, students in this learning program have limited opportunity in interacting with others.

Video conferencing is one of the types of digital communication methods that involves the electronic transmission of sound, vision, and information between two or more places that are delivered electronically to facilitate concurrent collaborative communication. In this way, people using video conferencing methods can deliver a much more unique and productive experience than the audio-conferencing method, since everyone participating is able to see facial expressions and body language which is particularly essential toward the way people communicate. Using a few distinctive tools make video conferencing able to function. Hardware as well as applications are the innovations used. A video conference can be held between two locations, i.e., regions that are related to each other through the video conference, or the conference can interface different areas. The communication may occur in an unconventional video conferencing office or on a standard home computer prepared with a web camera or a video call on a handheld like the third era portable smartphone, is categorized in this scope.

Distance learning is a moderately new educational learning method that can deliver classroom materials to students that does not require them to be physically on-site. Instead, both students and teachers/lecturers communicate on their chosen schedule via email or messaging application or utilizing technologies that assist them, such as video conferencing applications, to connect at the same moment. The most general approach of offering distance learning classes today is web-based classes. These classrooms use different online applications, such as email services, online messaging, upload or download files throughout the school's website or application platform, and message boards which functions for sharing materials in the classroom and notifications for upcoming news to help students and teachers communicate with each other throughout the web-based learning method. For instance, there will be a live classroom video feed that enables students to stream over it on the Internet. Utilizing the different video conferencing technology available, educators may provide their students with a more immersive distance learning experience by providing their students with real time, bi-directional (functioning in two directions), audio, and data communications, instead of using the traditional electronic media.

The distinction between conventional and remote teaching is blurred by modern communication technologies. In all cases, it has possible applications. One of them is video conferencing that can be used to encourage distance learning and traditional teaching at its best. More class materials and improved planning of teaching materials are usually correlated with distance learning. Video conferencing is a way to bring students and tutors, although virtually, to a single venue [12].

Using video conferencing for educational purposes allows teachers more versatility so that they can share thoughts and subjects without distance restriction. The frequent shuttle that some educators normally make to meet their place of employment is reduced as their opportunities to reach students are improved. For example, in three separate school buildings around campus, one professor might concurrently instruct all fourth-year university students. With class not having to be individually taught on the same subject courses, cost and time will be saved.

2.2 *Key Debates and Controversies*

Online learning using video conferencing has become quite common these days. The prohibition of activities that increase the transmission of COVID-19 makes every educational institution implement video conferencing in teaching and learning activities. Activities that should be done at school now are only done through a student gadget screen. After a while, it cannot be denied that learning through video conferencing can affect student learning intentions and outcomes both positively and negatively.

Learning, which is usually done formally and rigidly, can now be done in a way that invites students who are less active during face-to-face learning because one or two things, tend to be more active. Students are embarrassed to ask questions or express their opinions in face-to-face learning, because they think the questions asked are trivial or embarrassing. With video conferencing, students can ask the teacher through the private message feature, so that questions are known only to the teacher and questioner. In addition, learning activities that were previously rigid can now be carried out anywhere, anytime, without having to come to school, wear uniforms, and other regulations applied by schools. Students are also required to be independent so that they can be motivated to search for material on the Internet, find many sources and references, and be more curious.

Not all schools are ready to implement this video conferencing system, but it is already due to regulations that require schools to reduce or eliminate face-to-face learning activities. Students who live in places far from cities tend to find it difficult to get a good signal, or even no signal at all, to take classes held online. In addition to signals, the absence of compatible gadgets for attending classes is also felt by many students, especially students with low economic conditions. On the other hand, the attendance system is ineffective and invalid, making students only fill in attendance but do not pay attention to the material provided. Many cases where students will only attend class when attendance is done, for example, at the end of class or at the beginning of class, and do not follow the rest. In other cases, attendance is done automatically, where the system will fill the attendance based on how many minutes the student's account or gadget has taken the class. If the class lasts 100 min but the student's account is only detected having attended the class for 50 min, the system will not automatically be filled in. In this way, students must attend class from start to finish to fill in the attendance list. However, with this method, students can leave the classroom or what is commonly known as 'AFK' or Away from Keyboard.

Every effort is made to fight the transmission rate of COVID-19, one of which is by applying this video conference learning method. However, another important thing to consider is student learning outcomes before and after the implementation of online learning.

2.3 Gaps in Existing Knowledge

The knowledge gap from this study is whether there is a correlation between the application of online learning using video conferencing and student learning outcomes. To close this gap, data is collected from the Internet related to the topic being discussed, for example, a graph of the increase or decrease in student learning outcomes before and after the implementation of online learning using video conferencing.

3 Research Design and Methods

This research uses quantitative research methods. Agreeing to Sugiyono, quantitative research strategies can be translated as inquiring about positivism-based strategies that utilizes investigating certain populaces or tests. The inspecting technique for the most part is carried out arbitrarily, collecting information and utilizing investigate rebellious, quantitative/statistical information examination with the objectives of testing the foreordained speculation. According to Kerlinger, survey research is a study of large and small populations (universe) by selecting and examining the selected sample from that population, to find the relative incidence, distribution, and interrelation of the variables. The survey strategy was chosen since the survey's result displayed information in a strong and effective way so that it might clearly depict the subject.

3.1 Aims

- To determine whether the implementation of teaching and learning activities using the video conference method brings positive effect on the student in the learning process.
- To determine the learning outcome of the student whether it increases if the video conference method is implemented in the learning process.
- To describe the effect of video conference methods in teaching and learning activities.
- To determine whether using video conferencing is the most effective in long-distance learning.

3.2 Objectives

1. Theoretical Benefits
    - Generally, research can provide knowledge whether using video conferencing methods in online teaching and learning activities has a positive effect and provide knowledge of its effectiveness for its students.
2. Practical Benefits
    - The benefit for schools:
      Increases the awareness and concern for the development of students during online learning activities.
    - The benefit for students:
      Increases interest and motivation in this long-distance learning process.

3.3 Methods

This research is basically conducted by analyzing, questionnaire, and observation methods. The first stage is done by analyzing the needs giving out questionnaires to students. The results obtained are the student's average score whether the long-distance learning process using video conference methods sustains a positive effect. Afterward, researchers will observe the learning activities and processes of students in the environment.

3.4 Resources

The analytical data will be obtained using questionnaire, documentations, and observations technique that will be submitted by respondents, namely students. The research will be openly conducted by researchers who are directly present to observe student's learning activities, distribute questionnaires to students, and lastly the evidence will be documents that are considered important within the research scope. For the questionnaire, each variable and question is identified by name, and the results are obtained with a valid range of values, as in general the variables will be distributed, categorized, and given values.

The research designed provides answers to the following questions.

- Are students motivated if the long-distance learning process is carried out using the video conference method?
- Does the video conference method bring a positive effect and be considered effective in carrying out the learning subject?
- Are student's grades improving or decreasing?
- What are the advantages of learning using the video conferencing method when compared to face-to-face learning?

3.5 Potential Obstacles

The following are challenges and limitations of methodological problems that can potentially influence the research conclusion.

- Difficulty in observing students due to the social-distancing protocols during this pandemic.
- Problems with sampling and sample selection which can result in errors in sampling that might occur when the selected sample does not reflect the general population or the appropriate population which can result in limitation for study usually known as sample bias or selection bias.

3.6 Implications and Contributions to Knowledge

The implication of this research is the increasing knowledge of researchers and readers about the effect of using video conferencing as a learning tool on student learning outcomes. The knowledge that has been obtained is anticipated to be a reference material for instructors regarding video conference learning methods that can be used as a learning method for students. The contributions to knowledge are about how various factors in the delivery of lessons through video conferencing applications can affect the convenience and motivation of students which ultimately affect their learning outcomes.

3.7 Practical Implications

With the knowledge obtained from this research, teachers and teaching staff can consider the learning method that will be chosen according to the results they want by considering various factors according to their respective circumstances. That way students can use learning methods that are deemed appropriate if the teachers use either face-to-face learning methods or learning methods using video conferencing.

3.8 Theoretical Implications

This research provides theoretical knowledge for teachers and teaching staff regarding the implications of the application of video conferencing methods on learning outcomes of undergraduate students in Indonesia. The theoretical implications are how various factors in the delivery of lessons through video conferencing applications can affect the convenience and motivation of students which ultimately affect their learning outcomes. Online learning may have the convenience and the flexibility of study both for educator and student. For offline or face-to-face learning, due to the classroom atmosphere, students can concentrate better since there is less distraction than studying at places besides formal education areas.

4 Results

The survey results obtained show that as many as 32 respondents have filled out the questionnaire. Figure 1 shows that there were 20 respondents using Zoom (62.5%), 7 respondents using Microsoft Teams (21.9%), 3 respondents using Google Meet (9.4%), 1 person (3.1%) who use Adobe Connect, and 1 person (3.1%) who

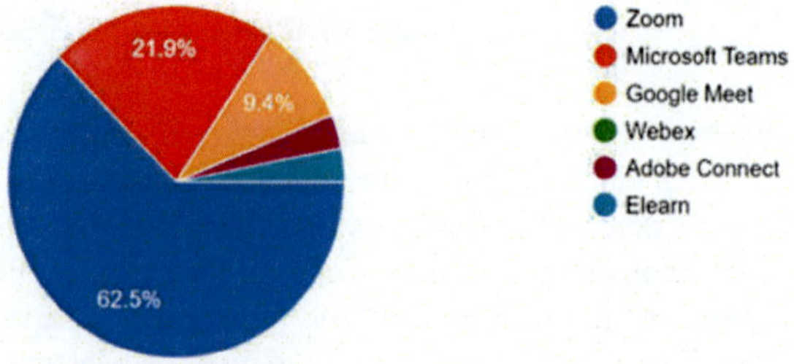

Fig. 1 Pie chart showing video conferencing applications used by respondents

use eLearn. This shows that most respondents use Zoom as a video conferencing application for learning activities.

Table 1 shows that there are 20 respondents (62.5%) who felt uncomfortable when learning activities are carried out through the video conferencing method compared to face-to-face learning, while there are 12 respondents (37.5%) who felt more comfortable when learning activities are conducted via video conferencing method. This shows that most respondents feel uncomfortable when learning activities are carried out through the video conferencing method rather than face-to-face learning.

Table 2 shows that there are 25 respondents (78.1%) who felt unmotivated when learning activities are carried out through the video conferencing method compared to face-to-face learning, while there are 7 respondents (21.9%) who felt more motivated when learning activities are carried out through the video conferencing method. This indicates that most respondents feel unmotivated when learning activities are carried out through the video conferencing method rather than face-to-face learning.

Table 3 shows that there are 21 respondents (65.6%) who feel that learning activities carried out through the video conference method are less effective when compared to face-to-face learning, while there are 11 respondents (34.4%) who feel that learning activities are more effective if done via the video conferencing method.

Table 1 Respondents' answers to the question about whether they agree that the video conferencing method increases learning comfort and convenience

Variable A (comfort and convenience)	Answer to the questionnaire	Percentage (%)
	Disagree	62.5
	Agree	37.5

Variable A (comfort and convenience) Defintiion: The situation of comfort and convenience

Table 2 Respondents' answers to the question about whether they agree that the video conferencing method increases learning motivation

Variable B (motivation)	Answer to the questionnaire	Percentage (%)
	Disagree	78.1
	Agree	21.9

Variable B (motivation) Defintiion: The situation of motivation

Table 3 Respondents' answers to the question about whether they agree that the video conferencing method increases learning efficiency

Variable C (efficiency)	Answer to the questionnaire	Percentage (%)
	Disagree	65.6
	Agree	34.4

Variable C (efficiency) Defintiion: The situation of efficiency

Table 4 Respondents' answers to the question about whether they agree that the video conferencing method increases learning outcomes

Variable D (learning outcomes)	Answer to the questionnaire	Percentage (%)
	Agree	62.5
	Disagree	37.5

Variable D (learning outcomes) Defintiion: The situation of learning outcomes)

This shows that most respondents feel that learning activities carried out through the video conference method are more effective than face-to-face learning.

Table 4 shows that there are 20 respondents (62.5%) who felt that their learning outcomes has increased when learning activities are carried out through the video conference method compared to face-to-face learning, while there are 12 respondents (37.5%) who felt their learning outcomes has decreased when learning activities conducted through video conferencing method compared to face-to-face learning. This indicates that most respondents believe that their learning results have increased when learning activities are carried out through the video conference method when compared to applying face-to-face learning.

4.1 Discussion

The study found out that using video conferencing as an educational method resulted in positive learning outcomes for students in Indonesia. The students' samples were accurate in their agreement about the efficiency of video conferencing. The effect of the video conferencing method on the student learning process refers to the rise or fall of student learning outcomes. Indeed, with limited time and space for direct teaching and learning activities, the video conferencing learning method using Zoom may be one of the most successful and effective ways to help students learn from any place.

However, this research is not so relevant when applied to groups of students who have never been exposed or are accustomed to using technology such as computers, smartphones, and the use of the Internet. Additional research needs to be done to determine the effect of using the video conferencing method with the Zoom application as a learning tool on the learning outcomes of students who have never been exposed and are not accustomed to using technology. Especially, in areas where there are no good infrastructure facilities such as an adequate Internet network or where most of the residents do not have devices such as smartphones and computers.

5 Conclusion

Video conferencing technology can be applied in educational institutions as a method by which teachers and students can carry out teaching and learning activities in separate places. This study documents that using video conferencing as an educational method results in positive learning outcomes for students in Indonesia. Although many students felt that they were more enthusiastic and motivated when they could learn face-to-face, most respondents felt that their scores increased when they were learning using the video conferencing method. Hence, it can be understood that although the motivation and enthusiasm of the students did not increase when studying with the use of video conferencing applications, their scores show positive and improving outcomes.

References

1. Abdillah L (2019) Empowering digital social media in organizing computer Science Students Higher Education Assignments. In: The 4th international conference of islamic education (ICIED2019) pp 426–433
2. Alkaher I, Dolan E (2011) Instructors' decisions that integrate inquiry teaching into undergraduate courses: How do I make this fit? Inter J Teach Learn 5:1–24
3. Andrews T, Klease G (2002) Extending learning opportunities through a virtual faculty—the video conference option. Inter J Educ Technol 3:1–11
4. Banathy B (1994) Designing educational systems: Creating our future in a changing world. In: Reigeluth CM, Garfinkle RJ (eds) Systemic change in education (pp 27–34). Englewood Cliffs, NJ: Educ Technol Publ
5. Bates AW (2005) Technology, e-learning and distance education. London: Routledge
6. Boldt S, Devine B, MacDevitt D, Morgan M (1998) Educational disadvantage and early school-leaving. Dublin: Combat Poverty Agency
7. Conlon T (1997) The internet is not a panacea. Scott Educ Rev 29:30–38
8. Dallat C, Salmon PM, Goode N (2015) All about the teacher, the rain and the backpack: The lack of a systems approach to risk assessment in school outdoor education programs. Procedia Manufact 3: 1157–1164
9. Freeman R, Lewis R (1998) Planning and implementing assessment. London: Kogan Page
10. Hannum W, Briggs L (1982) How does instructional system design differ from traditional instruction? Educ Technol 22(1):9–14
11. Heinich R (1993) Instructional media and technologies for learning. Englewood Cliffs, N.J: Merril
12. Hunsaker PL, Hunsaker JS (2008) Virtual teams: a leader's guide. Team Perform Manage 14(1/2):86–101
13. Jacobs G, Rodgers C (1997) Remote teaching with digital video: a trans-national experience, Br J Educ Technol 28(4):293–304

14. Jochems W, van Merriënboer J, Koper R (2004) An introduction to integrated e-Learning. In: Jochems W, vanMerriënboer J, Koper R (eds) Integrated e-Learning: Implication for pedagogy, technology and organization. London: Routledge Falmer
15. Kerka S (1996) Distance learning, the internet and the worldwide web. Columbus, OH: ERIC Clearinghouse
16. Knipe D, Lee M (2002) The quality of teaching and learning via video. Br J Educ Technol 33(3):301–311
17. Krutka DG, Carano KT (2016) As long as I see you on Facebook I know you are safe: Social Media Experiences as Humanizing Pedagogy. In: Crowe A, Cuenca A (eds) Rethinking social studies education in the twenty-first century. (pp 207–222). Switzerland: Springer International Publishing
18. Larkin H (2010) "But they won't come to lectures ...": The impact of audio recorded lectures on student experience and attendance. Australas J Educ Technol 26(2):238–249
19. Lawson T, Comber C, Gage J, Collum-Hanshaw A (2010) Images of the future for education? Videoconferencing: A literatur review. Technology, Pedagogy, and Education 19(3):295–314
20. Lawson T, Comber C, Gage J, Cullum-Hanshaw A (2010) Images of the future for education? Videoconferencing: a literature review. Technology. Pedagogy and Education, 295–314. https://doi.org/10.1080/1475939X.2010.513761. Retrieved from https://www.researchgate.net/publication/232832362_Images_of_the_future_for_education_Videoconferencing_A_literature_review. Accessed 3 Dec 2020
21. Mason C, Berson M (2000) Computer mediated communication in elementary social studies methods: An examination of student's perceptions and perspectives. Theory Res Soc Edu 28(4:527–545
22. Morrison FJ, Ponitz CC, McClelland MM (2010) Self-regulation and academic achievement in the transition to school. In: Calkins SD, Bell M (eds), Child development at the intersection of emotion and cognition (pp 203–224). Washington, DC: American Psychological Association
23. Pimmer C (2016) Mobile learning as boundary crossing: an alternative route to technology-enhanced learning? Interact Learn Environ 24:979–990. https://doi.org/10.1080/10494820.2015.1128211
24. Pitcher N, Davidson K, Goldfinch J (2000a) Videoconferencing in higher education. Innovations Educ Training Inter 37:199–209
25. Pitcher CR, Poiner IR, Hill BJ, Burridge CY (2000b) The implications of the effects of trawling on sessile megazoo-benthos on a tropical shelf in northeastern Australia. ICES J Mar Sci 57:1359–1368
26. Preston G, Phillips R, Gosper M, McNeill M, Woo K, Green D (2010) Web-based lecture technologies: Highlighting the changing nature of teaching and learning. Australas J Educ Technol 26(6):717–728
27. Sabry K, Baldwin L (2003) Web-based learning interaction and learning styles. Br J Educ Technol 34(4):443–454
28. Schiller J, Mitchell J (1993) Interacting at a distance: Staff and student perceptions of teaching and learning via video conferencing. Australas J Educ Technol 9(1):41–58
29. Schramm W (1977) Big media, little media. Beverly Hills, CA: Sage
30. Skene J, Cluett L, Hogan J (2007) Engaging Gen Y students at university: What web tools do they have, how do they use them and what do they want? In: Proceedings first year in higher education conference
31. Sorensen C, Baylen DM (1999) Interaction in interactive television instruction: Perception or reality. In: Paper presented at the conference of the American educational research association (AERA), Montreal, Canada
32. Storck J, Sproull L (1995) Through a glass darkly: What do people learn in videoconferences? Hum Commun Res 22(2):197–219. https://doi.org/10.1111/j.1468-2958.1995.tb00366.x
33. Sun X, Zhan W (2006) An exploration for influencing factors of interaction learning in web-based instruction. In: Paper presented at the 10th international conference on computer supported cooperative work in design, Nanjing, China

Transformer-Based Medical Abbreviation Disambiguation—A Comparative Study

Krzysztof Pałczyński, Magda Czyżewska, Marta Gackowska, Damian Ledziński, and Tomasz Andrysiak

Abstract One abbreviation could have multiple meanings, depending on the area in which it is used. Of course, within a single domain, its meaning can be ambiguous. This is also the case with medical data, where the challenge for artificial intelligence techniques is to identify them correctly and efficiently, as any mistake can cost human lives. This paper disambiguates medical abbreviations from Medical Dataset for Abbreviation Disambiguation (MeDAL). It compares the DistilBERT Transformer network with bag-of-words encoding-based classification and typical machine learning algorithms such as XGBoost, Random Forest, and Decision Tree. The study shows that DistilBERT Transformer achieved worse accuracy and F1 than the best baseline model when each tested shortcut was examined. The study shows that DistilBERT Transformer achieved worse accuracy and F1 than the best baseline model when each tested shortcut was examined. The basic model was better in the categories of shortcuts that have less than 500 examples and 500–5000 examples. The results in the group with 5000–20,000 examples are relatively equal for both approaches.

Keywords Abbreviation disambiguation · Transformer neural networks · DistilBERT · TF-IDF · Bag-of-words

1 Introduction

Medical informatics has progressed rapidly in the last twenty years. One example is the use of electronic medical documentation, which is faster than paper form. However, despite this, medical abbreviations are often used in these documents to

K. Pałczyński (✉) · M. Gackowska · D. Ledziński · T. Andrysiak
Faculty of Telecommunications, Computer Science and Electrical Engineering, Bydgoszcz University of Science and Technology, Bydgoszcz, Poland
e-mail: krzysztof@palczynski.dev

M. Czyżewska
Faculty of Mechanical Engineering, Bydgoszcz University of Science and Technology, 85-796 Bydgoszcz, Poland

G. Rajakumar et al. (eds.), *Intelligent Communication Technologies and Virtual Mobile Networks*, Lecture Notes on Data Engineering and Communications Technologies 171,
https://doi.org/10.1007/978-981-99-1767-9_66

minimize the time to fill them out. Deciphering abbreviations used in medicine is difficult, as evidenced by their translation into different languages. Knowledge of the medical field is fundamental, but it is equally important to know the professional language. Medical abbreviations are often written in a specific way, are not always grammatically correct, and can even have several meanings. This involves the context and understanding of the abbreviation used. The task of the abbreviation disambiguation is treated as a classification problem, where the classes are all possible expansions.

Text classification is an application domain for natural language processing systems. This is a fundamental problem behind many research topics, such as categorizing topics and relations extraction. An important issue in text classification is learning how to represent it. Deep Neural Networks are more effective for models that use bag-of-words, and they also use word order information and find more semantic features. Therefore, natural language Processing (NLP) can decode medical abbreviations.

Transformers are used in various NLP tasks, but the Convolutional Neural Network (CNN) attracts the least attention. CNN does not capture sequential and long-distance dependencies and other transformers but is characterized by extracting local characteristics. In article [1], a convolutional transformation (ACT) was introduced, which is a combination of the features of the transformation and CNN, in order to use it for the satisfactory classification of the text. ACT is much smaller and faster than Transformer, with 56% fewer parameters and a 2.7-time faster inference speed. Detailed analyzes show that ACT is a lightweight and efficient universal text classifier that achieves consistently good results over different text classification tasks, outperforming CNN-based and attentive models, including Transformer. The novel attentive convolution mechanism was proposed that utilizes the semantic meaning of convolutional filters attentively to transform text from complex word space to a more informative convolutional filter space where important grams are captured [1].

There are papers [2] that use extreme multi-label text classification. The method based on transformers, i.e., LightXML, performs better than other methods. XR-Transformer was developed to have procedures by which it is possible to recursively tune models to be performed on a set of multi-target targets related to the XMC target function. The described method showed that XR-Transformers are characterized by a shorter learning time than XMC models, which are based on transformers. In addition, they show better and more modern results, which can be seen on the basis of the XR-Transformer, which is 20x faster, and the Precision @1 increases by 3%. The novel recursive approach, XR-Transformer, was proposed to accelerate the procedure through recursively fine-tuning transformer models on a series of multi-resolution objectives related to the original XMC objective function.

The article presents [3] mortality forecasting based on the LSTM, LSTM + Self Attention, and Transformers models. Pre-training was used in the study, but the absolute highest accuracy was achieved by Transformers (84.43%), then LSTM + Self Attention (82.98%), and LSTM (82.80%). The results show that the initial training procedure increases convergence and improves performance in medical tasks. It can be seen that LSTM and LSTM + show greater accuracy if trained beforehand, and the two models' performance increased by more than 70% relatively.

There are seven state-of-the-art transformer-based text classification models: BERT, DistilBERT, RoBERTa, DistilRoBERTa, XLM, XLM-RoBERTa, and XLNet. The Chatbot Interaction with artificial intelligence can be presented in this article as a tool to classify human–machine communication tasks through an architecture learning approach similar to a transformer-based chatbot. Using the Transformer model, the training set is paraphrased and performs text-to-text transfer (T5). In this article [4], models that the T5 model improves achieve 4% better accuracy in text classification. On the other hand, the RoBERT model, which was trained using the extended data of the T5 model, turned out to be better than the described model. The described RoBERT model achieved an accuracy of 98.96% in text classification. However, the set of transformer models showed the most accurate classification of the text at the level of 99.95%. This result was achieved thanks to the logistic regression of the output label predictions.

In language processing, the Transformer model is used to interpret sentences. In the paper [5], two approaches are introduced to improve the performance of Transformers. In order to separate a word that contains a practical meaning, the attention score was used. To calculate it, the correlation coefficient was multiplied by a weight vector, which is estimated according to the importance of the part of speech. Each layer has been combined in such a way as to obtain more accurate results of the task representation. Based on the tests performed, it can be seen that the presented Transformer-F model is effective and improves the results by 5.28% in relation to the vanilla transformer. It can be concluded that the classification of the text has been significantly improved. Acronyms and abbreviations are used in all types of writing. They represent shortened phases, and therefore, they are more accessible and are used in everyday life. As a result, systems are built that allow abbreviations and acronyms recognition to find their true meaning. No work would introduce a uniform solution dealing with their processing at this stage. In the paper [6], an online acronym identification (MadDog) and disambiguation system were created to process acronyms from various scientific, biomedical, and general fields. The presented system is written in Python 3, which allows it to be used in other applications.

The study aims to determine the usability of Transformer Deep Neural Networks compared to classification based on bag-of-words encoding and typical machine learning algorithms like XGBoost or Random Forest for abbreviation disambiguation. The Transformer Neural Network is considered the state-of-the-art solution for many NLP-related tasks due to its ability to focus on particular phrases in sentences and extract the context of the utterances. In comparison, the bag-of-words algorithm for sentence encoding is one of the most basic feature extractions from text, as it relies on the statistical properties of words in sentences. Due to its simplicity, the algorithm is much faster in inference and does not require hardware with high computational capacity like GPU servers, as opposed to a Transformer network. This research aims to establish a clear comparison between these highly computationally different methods and determine whether the additional cost of required hardware and computational time required for using a Transformer network is justified for the task of abbreviation disambiguation. The study was conducted on Medical Dataset for

Abbreviation Disambiguation (MeDAL). The Transformer neural network employed for this task is DistilBERT model [7] pretrained on the sentiment analysis task using dataset [8].

2 Materials and Methods

Figure 1 shows a general overview diagram of the method which will be described in detail in the following sections.

2.1 Database

The Medical Dataset for Abbreviation Disambiguation for Natural Language Understanding (MeDAL) was used in this study. The presented collection contains data from medical texts that have been selected to disambiguate the skins used in medicine. Therefore, the set used relates to natural language understanding. In addition, it consists of 14,393,619 articles with an average of 3 skins per article.

2.2 Neural Network

In this research, the examined network was DistilBERT Transformer. DistilBERT is a distilled version of the original BERT Transformer [9] achieving 60% faster inference compared to the original while losing only 3% of its language understanding capabilities. The network is an encoder-type transformer capable of extracting fea-

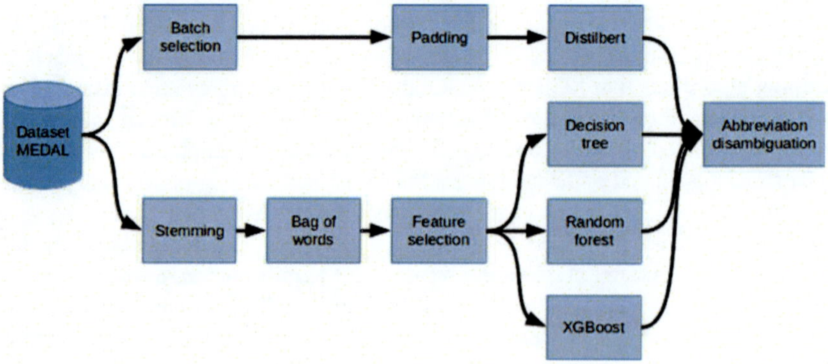

Fig. 1 General overview diagram of the method

tures from raw text due to its ability to attenuate to specific phrases across long spans of unrelated words. The Transformer was pretrained on the BookCorpus dataset. Thus, transfer learning has been applied.

In order to obtain maximum effectiveness of abbreviations for each abbreviation, the new pretrained DistilBERT network was trained. On top of the Transformer network, the linear layer was added with a number of neurons equal to the possible explanations of the abbreviation. This approach interpreted the abbreviation disambiguation task as a text classification task. In order to perform the classification, the output of the linear layer was processed with one of the two computational paths depending on the number of possible abbreviation explanations.

If the examined abbreviation contained only two possible explanations, then the result was processed by the Binary Cross-Entropy loss function given by the equation:

$$\mathrm{BCE}(Y, Y') = \frac{1}{|Y'|}\sum_{i=1}^{|Y'|}(Y_i \log Y'_i - (1 + Y_i)\log(1 + Y'_i)) \tag{1}$$

where Y' is the vector of obtained values from the linear layer and Y is the vector of expected results.

However, if the abbreviation has more than two possible explanations, then the output of the linear layer is processed by the softmax function, and its result is used to compute the cross-entropy loss function. The softmax function is given by the equation:

$$\sigma(Y')_i = \frac{e^{Y'_i}}{\sum_{j=1}^{|Y'|} e^{Y'_j}} \tag{2}$$

The cross-entropy loss function is given by the equation:

$$L(p, y) = -\sum_{c=1}^{M} y_{o,c} \ln p_{o,c} \tag{3}$$

where p—the probability that sample o is associated with the class c computed by application of softmax function on the output of the linear layer. y—binary value equal to 1 if sample o is associated with the class c and 0 in if not.

2.3 Bag-of-Words-Based Approach

In contrast to the previous approach, bag-of-words encoding is the most basic and well-known method of extracting features from the text. It uses the frequency of word occurrence to determine keywords and encodes the sentences as a vector of keyword existence in the analyzed text.

This method neither uses the context of the text nor understands temporal text structures like the addition of the word "not" changing sentence meaning into its complete opposite. Compared to the Transformer Neural Network approach, this method has a simplistic nature and trades off the ability to find complex text interconnections for low computational requirements. Because of that, this method was used for comparison with the Transformer Neural Network. Evaluating the simplest possible solution provides insight into the justification of the most computationally expensive method of neural network application usage.

In the first step, the text is stemmerized. Stemming is the process of converting a word into its linguistic "root", by stripping it of its prefixes and suffixes. As a result, similar words most likely have the same stem. For example, the words "care", "cared", and "caring" has the same stem "care". The application of stemmerization allows to reduce the number of unique words for the keyword detection algorithm. Without stemmerization, the aforementioned words "care", "cared" and "caring" would be interpreted as distinct words without any affiliation to the common meaning. However, after conducting them to stemmerization process, they all are now represented by the same stem "care". Hence, the keyword detection algorithm considers them the exact words, and as a result, the likelihood of the word "care" being deemed as a keyword is increased. The stemmer used in this task is the SnowballStemmer [10] from NLTK library [11].

In the next step, stemmerized words were used to compute the result matrix of the TF-IDF algorithm. TF-IDF algorithm (Term Frequency-Inverse Document Frequency) is a statistic depicting how often a certain term appears in the examined part of the text and how rare it is in the whole corpus. This methodology aims to find frequent terms for the examined part of the corpus rather than the whole one. For example, the word "I" frequently appears in the works of belles-lettres, so in the case of classifying belles-lettres texts, the word "I" would not have any significant impact due to its lack of distinctiveness for any specific group. However, the word "research", although rare in human-created texts, is widespread in scientific literature, thus having a high TF-IDF score regarding corpus containing various types of literature.

The TF-IDF algorithm was used to establish which words are keywords regarding abbreviation disambiguation. Paragraphs of texts representing the particular abbreviation were extracted from the MeDAL dataset, stemmed, and conducted to the TF-IDF algorithm. The goal was to determine how frequent the term is to the input paragraphs and how rare in the corpus made out of every paragraph in the MeDAL dataset using this particular abbreviation. As a result, the algorithm returned the score of how likely these stemmed words are keywords.

The computation of the TF-IDF metric was conducted using the formula:

$$\mathrm{tfidf}_{i,d} = \frac{f_{i,d}}{\sum_{k \in d} f_{k,d}} \cdot \frac{N}{|\{d \in D | i \in d\}|} \tag{4}$$

where $f_{i,d}$ stands for frequency how many times term i occurred in document d. The first term describes term frequency, and it is a measurement of how often the term appears in the document in comparison to other terms in that document. The second term depicts inverse document frequency, and it measures how many documents in the corpus has at least one occurrence of term i.

At this point, the words have been converted into their TF-IDF metric. However, due to the enormous number of words in each abbreviation-related corpus, the selection of terms is required to reduce computations and prevent overfiting. Each term for each abbreviation-related corpus has its maximum TF-IDF score examined, and the terms with the highest maximum amounts of TF-IDF metric are selected as keywords for this particular abbreviation disambiguation. Three numbers of terms have been evaluated: 20, 100, and 1000 and results based on the number of words selected have been presented in the Results section.

After the keyword selection, the paragraphs of text from the MeDAL dataset have been encoded in n-dimensional vectors containing the TF-IDF score of each established keyword, where n is the number of selected terms. In the last step, these vectors are used to form training, validation, and test datasets for training several machine learning models: XGBoost, Random Forest, and Decision Tree.

2.4 Training

The DistilBERT Transformer instance was created for every abbreviation in the MeDAL dataset. Each of these transformers was pretrained on a sentiment analysis task using the BookCorpus dataset and adjusted to perform the disambiguation of only one abbreviation. The raw, unprocessed text was encoded using a tokenizer. Due to the restriction of the PyTorch library requiring all GPU-computable tensors to be of a homogenous size, all paragraphs computed by the network must be of the same length. Due to that limitation, the size of 512 tokens per paragraph was chosen. As a result, the texts with less than 512 tokens were padded to reach the desired length, and paragraphs with more tokens were trimmed. The reduction of size was performed every epoch. It involved randomly choosing the subset of tokens from the original paragraph that contained 512 tokens. Each token was a direct successor of the previous token in the subset (except for the first one), thus keeping a coherent text structure.

The batch size of 32 was used for training DistilBERT, and Adam [12] was selected as an optimizer of the network's parameters. The data was split into training, validation, and test datasets containing 70%, 15%, and 15% of available batches. The training dataset was used to adjust the weights of the Transformer network. A validation set was used to select the best network variant and determine whether it was time to perform an early stopping [13]. The training was stopped if one hundred epochs elapsed or there was no improvement on the validation dataset after 10 epochs, or the accuracy on the validation dataset outreached 95%, which was considered the limit to which measurements still carried comparable and interpretable results instead of network artificial adjusting to the dataset. Also, if the accuracy on the training dataset reached 100% then training was stopped due to the network's inability to generalize training further. These restrictions were introduced due to the necessity of training several thousands of networks for each abbreviation.

2.5 *Metrics*

Neural networks were evaluated using the metrics described below [14]. For the purpose of simplicity of equations, certain acronyms have been created, as follows: TP—True Positive, TN—True Negative, FP—False Positive, and FN—False Negative. Metrics used for network evaluation are:

- Accuracy: $\mathrm{Acc} = (\mathrm{TP} + \mathrm{TN})/(\mathrm{TP} + \mathrm{FP} + \mathrm{TN} + \mathrm{FN})$,
- $\mathrm{Precision} = \mathrm{TP}/(\mathrm{TP} + \mathrm{FP})$,
- $\mathrm{Recall} = \mathrm{TP}/(\mathrm{TP} + \mathrm{FN})$,
- $\mathrm{F1} = 2 * \mathrm{Precision} * \mathrm{Recall}/(\mathrm{Precision} + \mathrm{Recall})$,
- AUC—Area under ROC—Area Under Receiver operating characteristic.
  ROC is a curve determined by calculating True-Positive Rate = $\mathrm{TFP} = \mathrm{TP}/(\mathrm{TP} + \mathrm{FN})$ and False-Positive Rate = $\mathrm{FPR} = \mathrm{FP}/(\mathrm{TN} + \mathrm{FP})$. False-Positive Rate describes the x-axis and True-Positive Rate the y-axis of a coordinate system. By changing the threshold value responsible for classification of an example as belonging to either the positive or negative class, pairs of TFP-FPR are generated, resulting in the creation of the ROC curve. AUC is a measurement of the area below the ROC curve.

3 Results

This section presents the evaluation results of the DistilBERT Transformer network and bag-of-words-based models presented. For the remainder of the article, these models using TF-IDF scores of keywords are called "baseline models". Tables present four metrics: accuracy, precision, recall, and F1 score. Accuracy is a metric depicting model performance in all the classes. However, precision, recall, and F1 score describe the model's performance in only one class. Because of that, these metrics have been averaged among all the classes. As a result, every abbreviation evaluation contains the accuracy of the model and its average precision, recall, and F1 taken from all the possible disambiguations.

Table 1 contains averaged results of the DistilBERT network and the best baseline model from all examined abbreviations. The "best baseline model" means that for every abbreviation, there were three algorithms tested: XGBoost, Random Forest, and Decision Tree, on the matrices of TF-IDF scores of three different amounts of keywords: 20, 100, and 1000. As a result, nine baseline models were trained and examined for every abbreviation. The model deemed the best was the model

Table 1 Comparison between DistilBERT and best baseline model average performance on the whole dataset

Name	Accuracy (%)	Precision (%)	Recall (%)	F1
DistilBERT	85.3	63.2	65.6	0.633
Best baseline model	86.8	69.7	75.8	0.708

Table 2 Results of the baseline models' performances on the whole dataset

Algorithm	Terms	Accuracy (%)	Precision (%)	Recall (%)	F1
Random Forest	1000	86.9	67.6	73.6	0.686
XGBoost	1000	86.6	70.3	76.4	0.717
Decision Tree	1000	83.1	69.2	71.9	0.697
Random Forest	100	76.4	52.8	59.8	0.530
XGBoost	100	75.7	51.6	58.7	0.516
Decision Tree	100	75.0	52.4	58.7	0.525
Random Forest	20	69.6	39.7	43.8	0.365
Decision Tree	20	69.6	39.7	43.8	0.366
XGBoost	20	69.6	39.1	40.8	35.4

achieving the highest F1 metric in this particular abbreviation disambiguation task. The motivation for choosing the F1 metric for model selection is to use a score that reflects the model's ability to classify the examples that do not discriminate between a small number of examples.

Table 2 shows results of the baseline models' performances on the whole dataset.

Abbreviations in the MeDAL dataset differ in both the amount of the examples and number of possible disambiguations. Figure 2 depicts the distribution of the examples among abbreviations. Due to such uneven spread of examples, there are results of DistilBERT and baseline models presented for gropus of abbreviations that:

- have less than 500 examples
- have between 500 and 5000 examples
- have between 5000 and 20,000 examples
- have above 20,000 examples.

Tables 3, 5, 7, and 9 show the comparison between DistilBERT and best baseline model average performance on abbreviations disambiguation for different examples ranges. Tables 4, 6, 8, and 10 show results of the baseline models' performances on abbreviations disambiguation for different examples ranges.

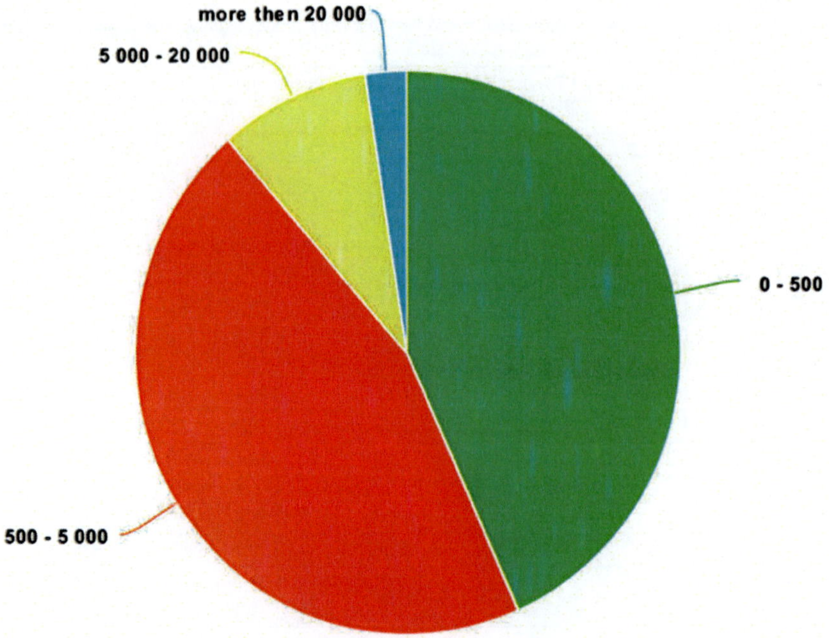

Fig. 2 Distribution of the examples among abbreviations

Table 3 Comparison between DistilBERT and best baseline model average performance on abbreviations disambiguation, which contains less than 500 examples

Name	Accuracy	Precision (%)	Recall (%)	F1
DistilBERT	84.9	67.4	69.4	0.673
Best baseline model	85.9	74.0	77.8	0.742

4 Discussion

The best baseline model achieved better accuracy and F1 score than DistilBERT Transformer when examined on every tested abbreviation. The baseline model was better in categories of abbreviations that have below 500 examples and between 500 and 5000 examples. The results in the group with examples between 5000 and 20,000 examples are relatively even for both of the approaches. The Transformer has better results only in the group of abbreviations containing above 20,000 examples. It turns out that the additional computational complexity of the Transformer Neural Network able to extract utterance's context and multi-word terms does not provide significant benefit above the simplest possible method of extracting keywords and performing classification using machine learning algorithms.

Table 4 Results of the baseline models' performances on abbreviations disambiguation, which contains less than 500 examples

Algorithm	Terms	Accuracy (%)	Precision (%)	Recall (%)	F1
Random Forest	1000	86.3	71.4	75.1	0.712
XGBoost	1000	85.7	74.8	78.9	0.754
Decision Tree	1000	82.3	73.6	75.9	0.738
Random Forest	100	80.8	67.3	71.8	0.674
XGBoost	100	79.0	65.3	69.5	0.655
Decision Tree	100	78.2	66.6	70.0	0.665
Random Forest	20	70.1	49.0	52.4	0.458
Decision Tree	20	70.0	49.0	52.4	0.458
XGBoost	20	69.9	47.9	48.3	0.439

Table 5 Comparison between DistilBERT and best baseline model average performance on abbreviations disambiguation, which contains between 500 and 5000 examples

Name	Accuracy (%)	Precision (%)	Recall (%)	**F1**
DistilBERT	86.1	61.4	64.1	0.619
Best baseline model	89.0	70.5	77.2	0.720

Table 6 Results of the baseline models' performances on abbreviations disambiguation, which contains between 500 and 5000 examples

Algorithm	Terms	Accuracy (%)	Precision (%)	Recall (%)	F1
Random Forest	1000	89.1	68.9	76.6	0.706
XGBoost	1000	88.5	71.4	77.8	0.732
Decision Tree	1000	85.0	70.2	73.1	0.708
XGBoost	100	74.2	44.9	54.2	0.451
Random Forest	100	74.1	45.7	54.8	0.462
Decision Tree	100	73.5	45.6	54.0	0.460
Random Forest	20	69.5	34.8	39.5	0.314
XGBoost	20	69.4	34.5	37.0	0.308
Decision Tree	20	69.5	34.8	39.3	0.314

Table 7 Comparison between DistilBERT and best baseline model average performance on abbreviations disambiguation, which contains between 5000 and 20,000 examples

Name	Accuracy (%)	Precision (%)	Recall (%)	F1
DistilBERT	85.0	56.6	59.3	0.571
Best baseline model	84.1	54.5	65.5	0.572

Table 8 Results of the baseline models' performances on abbreviations disambiguation, which contains between 5000 and 20,000 examples

Algorithm	Terms (%)	Accuracy (%)	Precision (%)	Recall (%)	F1
XGBoost	1000	84.2	53.5	65.0	0.565
Random Forest	1000	83.1	53.4	61.5	0.556
Decision Tree	1000	80.4	53.0	56.7	0.541
XGBoost	100	71.2	28.6	38.6	0.274
Random Forest	100	70.4	28.7	37.5	0.278
Decision Tree	100	70.4	28.7	37.4	0.278
XGBoost	20	70.0	26.5	30.4	0.239
Random Forest	20	69.6	26.5	30.7	0.240
Decision Tree	20	69.6	26.5	30.7	0.240

Table 9 Comparison between DistilBERT and best baseline model average performance on abbreviations disambiguation, which contains above 20,000 examples

Name	Accuracy (%)	Precision (%)	Recall (%)	F1
DistilBERT	78.5	44.3	48.3	0.453
Best baseline model	74.2	36.1	53.4	0.396

The Transformer Neural Network is significantly more computationally expensive for training and inference than any baseline models. It requires specially dedicated hardware like GPUs to perform inference in the production environment. In contrast, the baseline methods examined in this work require only stemming the text, computing TF-IDF statistics, and performing one of the lightweight machine learning algorithms. Such a system can be processed on any general-purpose computational unit like CPU and does not require special libraries like NLTK or sklearn.

Every baseline model using 1000 keywords achieved significantly better results than models using 100 keywords, which in fact, scored better than models using 20 keywords in nearly every example. It suggests a strong positive trend regarding the

Table 10 Results of the baseline models' performances on abbreviations disambiguation, which contains above 20,000 examples

Algorithm	Terms (%)	Accuracy (%)	Precision (%)	Recall (%)	F1
XGBoost	1000	74.4	33.7	50.5	0.366
Random Forest	1000	71.5	33.5	40.4	0.355
Decision Tree	1000	70.0	32.7	38.5	0.345
XGBoost	100	64.7	17.6	25.7	0.163
XGBoost	20	64.2	16.6	19.0	0.146
Random Forest	100	63.8	17.6	23.9	0.164
Decision Tree	100	63.7	17.6	23.8	0.164
Random Forest	20	63.5	16.7	19.6	0.147
Decision Tree	20	63.5	16.6	19.6	0.147

number of keywords and quality of inference, and a larger amount of keywords may improve the classification even more.

XGBoost algorithm scored the best F1 value in every group. However, Random Forest and Decision Tree model results are not significantly worse. Nonetheless, the XGBoost algorithm is the best model for this particular task. Metrics get progressively worse for both Transformer network and baseline models as the number of examples for abbreviations increases. This phenomenon is surprising due to the common sense in artificial intelligence and statistics suggesting that more data provides more accurate results. The authors plan on further investigating this issue. Baseline models achieved comparable, if not better, results to the Transformer Neural Network. It suggests that keyword extraction may be sufficient, and because of that, the Transformer also learned terms extraction instead of sentence context analysis. If such is the case, then Transformer computation may generate quasi-TF-IDF-matrices and find patterns in them. As a result, the task of abbreviation disambiguation may not require analyzing text context and can be reduced to simple keyword extraction and application of statistical models. The authors plan on further investigating this issue.

5 Conclusions

The artificial intelligence tools may successfully perform disambiguation of the medical abbreviations. However, the benefit of the more complicated, context-sensitive Transformer Neural Network was absent compared to simpler keyword extraction-based models. Due to that and the significantly increased computational complexity of the Transformer Neural Networks above models relying on computing words' frequency of occurrence, it seems that basic bag-of-words-based models are still preferable. The tendency between fewer examples for abbreviation and better classification quality is interesting. It motivates further examination due to its contradiction of common sense that more data translated to better classification.

References

1. Li P, Zhong P, Mao K, Wang D, Yang X, Liu Y, Yin J, See S (2021) Act: an attentive convolutional transformer for efficient text classification. In: Proceedings of the AAAI conference on artificial intelligence, vol 35, no 15, pp 13261–13269
2. Zhang J, Chang WC, Yu HF, Dhillon I (2021) Fast multi-resolution transformer fine-tuning for extreme multi-label text classification. Adv Neural Inf Process Syst 34:7267–7280
3. Wen Z, Lu X, Reddy S (2020) MeDAL: Medical abbreviation disambiguation dataset for natural language understanding pretraining. arXiv preprint 2020, arXiv:2012.13978
4. Bird J, Ekárt A, Faria, D (2021) Chatbot Interaction with Artificial Intelligence: human data augmentation with T5 and language transformer ensemble for text classification. J Amb Intell Hum Comput, pp 1–16
5. Shi Y (2021) Transformer-F: a transformer network with effective methods for learning universal sentence representation. arXiv preprint 2021, arXiv:2107.00653
6. Veyseh A, Dernoncourt F, Chang W, Nguyen T (2021) MadDog: a web-based system for acronym identification and disambiguation. arXiv preprint 2021, arXiv:2101.09893
7. Sanh V, Debut L, Chaumond J, Wolf T (2021) DistilBERT, a distilled version of BERT: smaller, faster, cheaper and lighter. arXiv preprint 2021, arXiv:1910.01108
8. Maas A, Daly R, Pham P, Huang D, Ng A, Potts C (2011) Learning word vectors for sentiment analysis. In: Proceedings of the 49th annual meeting of the association for computational linguistics: Human language technologies, pp 142–150
9. Devlin J, Chang M, Lee K, Toutanova K (2018) Bert: pre-training of deep bidirectional transformers for language understanding. arXiv preprint 2018, arXiv:1810.04805
10. Porter M (2001) Snowball: a language for stemming algorithms
11. Hearst M (2005) Teaching applied natural language processing: Triumphs and tribulations. In: Proceedings of the Second ACL Workshop on Effective Tools and Methodologies for Teaching NLP and CL, pp 1–8
12. Kingma D, Ba J (2012) Adam: a method for stochastic optimization. arXiv preprint 2012, arXiv:1412.6980
13. Caruana R, Lawrence S, Giles C(2000) Overfitting in neural nets: Backpropagation, conjugate gradient, and early stopping. Adv Neural Inf Process Syst 13
14. Śmigiel S, Pałczyński K, Ledziński D (2021) ECG signal classification using deep learning techniques based on the PTB-XL dataset. Entropy 23(9):1121

Vehicle Number Recognition Using General Surveillance Camera

Nivarthi Tushara, Voraganti Sahithi, Bushgari Haripriya, and U. Chandrasekhar

Abstract With the increase of automobiles on roadways, a vehicle system for recognizing numbers is necessary which can be implemented in the most simple way possible for effective traffic control, surveillance, smart parking management systems, and tollbooth record management. The system should be able to recognize number plates from live video feeds so that it can be used for automated survelliance and other practical purposes. The system takes a live video/image and localizes the license plate. License plate localization is performed using a trained YOLOv4 or YOLOv7 object detection models. Real-ESRGAN, a super-resolution technique is applied on the obtained license plate to enhance the image quality. Next, an OCR such as Tessaract and Easy OCR are used to perform image-to-text conversion extracting the required characters present on the number plate. The resultant output is validated as per the limitations of the Indian vehicle registration number format to improve accuracy.

Keywords Object detection · Vehicle license plates · OCR—Optical character recognition · Real-ESRGAN—Real enhanced super-resolution generative adversarial network · YOLO—you only look once · Tesseract

---

N. Tushara · V. Sahithi · B. Haripriya · U. Chandrasekhar (✉)
CSE Department, BVRIT Hyderabad College of Engineering for Women, Hyderabad, India
e-mail: chandrasekhar.u@bvrithyderabad.edu.in

N. Tushara
e-mail: 19wh1a0519@bvrithyderabad.edu.in

V. Sahithi
e-mail: 19wh1a0555@bvrithyderabad.edu.in

B. Haripriya
e-mail: 20wh5a0502@bvrithyderabad.edu.in

G. Rajakumar et al. (eds.), *Intelligent Communication Technologies and Virtual Mobile Networks*, Lecture Notes on Data Engineering and Communications Technologies 171,
https://doi.org/10.1007/978-981-99-1767-9_67

1 Introduction

1.1 Background

Vehicle Number Plate Recognition (VNPR) is a technology that uses computer vision and algorithms and machine learning to automatically recognize and extract information from vehicle number plates. This information can include the license plate number, as well as other details such as the country or state of origin, the make and model of the vehicle, and so on [1].

VNPR systems are frequently employed for a number of purposes, such as traffic control, toll collecting, and vehicle tracking. They can be applied in many different ways, such as by using specialized hardware like cameras and image sensors or by using software algorithms that are performed on a computer or mobile device. Overall, VNPR is a strong and efficient technique for automating the identification of and information extraction from vehicle number plates, and it has many uses in a range of industries. The identification of Indian car serial numbers is the topic of this research project.

In India, vehicle number plates are issued by the respective state transport departments and are required to be displayed on all vehicles. Indian number plates come in five different color combinations, which are used to distinguish between different categories of vehicles.

These colors are:

- White: This color is used for private vehicles, including cars, motorcycles, and other non-commercial vehicles.
- Yellow: This color is used for commercial vehicles, such as taxis, autorickshaws, and goods carriers.
- Blue: This color is used for government-owned vehicles, including those owned by the central government, state government, and local authorities.
- Green: This color is used for vehicles that are owned by defense forces and paramilitary organizations.
- Black: This color is used for diplomatic vehicles, such as those owned by foreign embassies and consulates.

In addition to these colors, Indian number plates also feature a series of alphanumeric characters that identify the vehicle and its owner. The specific format and layout of these characters may vary depending on the state or region in which the vehicle is registered:

- The state code: This is a two-letter code that indicates the state or union territory in which the vehicle is registered.
- The district code: This is a one- or two-digit code that indicates the district or region in which the vehicle is registered.
- The serial number: This is a combination of one or more digits that uniquely identifies the vehicle within the state or district.

- The registration mark: This is a combination of letters and numbers that identifies the issuing authority and the date of registration.

In addition to these features, Indian number plates may also include other identifying information, such as the make and model of the vehicle, the type of fuel it uses, and so on. Some number plates may also feature additional security features, such as holograms or UV-sensitive ink, to prevent counterfeiting. There are also different formats followed for the Vehicles of Foreign Missions, vehicles of Indian Armed Forces and BH (Bharat) series registration.

This research work intends to create a system that recognizes the vehicle license plate on live surveillance camera feed. The videos are input to YOLO, which is trained to detect the number plates directly instead of vehicle detection followed by number plate detection. The system includes three major modules namely license plate recognition, super-resolution, and recognition of text from images.

- License plate detection: YOLO v4 and v7 library is used to obtain the license plate from a live video/recorded video/image [2–6].
- Super-resolution: Real-ESRGAN, a GAN architecture variant, is used to improve image quality by obtaining With a low-resolution photograph, create a high-resolution image [1, 7].
- Text recognition from images: Paddle OCR, Easy OCR, and tesseract OCR are used for extracting the characters from the vehicle license plate.

Easy OCR is a software tool that allows users to extract text from images. Easy OCR uses advanced OCR technology to recognize and extract text from images and PDF files, even if the text is distorted, blurry, or otherwise difficult to read. It is able to handle English, Spanish, French, and German are among the many languages spoken there, and can recognize both printed and handwritten text.

Paddle OCR is a suite of optical character recognition (OCR) tools developed by Baidu PaddlePaddle, an open-source deep learning platform. It includes a range of OCR models that can be used to extract text from images and documents, as well as tools for training and deploying custom OCR models. One of the key features of Paddle OCR is its ability to handle a wide range of languages and scripts, including Chinese, English, Korean, Japanese, and many others. It also provides a number of pre-trained models that can be used for tasks including document layout analysis, scene text recognition, and text recognition in natural images.

Tesseract optical character recognition (OCR) is an open-source project software tool that is widely utilized in converting scanned images of text into editable text documents. It was developed by Hewlett-Packard (HP) in the 1980s and has been maintained and updated by Google since 2006. Tesseract OCR uses machine learning algorithms to recognize and extract text from images, and is able to handle a wide range of languages and scripts. It is highly accurate and can handle text that is distorted, blurry, or otherwise difficult to read. One of the key features of Tesseract OCR is its ability to be trained to recognize new languages and scripts, making it highly flexible and adaptable to a wide range of OCR tasks. It also provides a number of advanced features, such as layout analysis and text recognition in natural images, and has a number of customization options that allow users to fine-tune its performance.

The final step involves validation of the misclassified images using domain knowledge.

1.2 Objectives

In order to automatically detect and retrieve license plate data from photos or videos of automobiles, the following objectives need to be achieved.

- Acquiring an image or video of a vehicle from general surveillance camera. Images may be obtained from videos using some convenient frame rate.
- Identification of the registration plate region: The algorithm locates the license plate on the given input at this stage.
- Extraction of a license plate from the rest of the image: Next, the system should extract the license plate from it. Techniques including cropping, thresholding, and morphological processes should be used for this.
- Character recognition: This is accomplished through the use of optical character recognition (OCR).
- Compare the efficiency of different number plate recognition techniques.

The aim of this research is to effectively and efficiently recognize the license plate information from pictures or videos of automobiles.

2 Related Work

In the literature, numerous systems for reading license plates have been suggested and put into practice.The fundamental steps in the recognition systems include digital image enhancement, number plate area detection from the acquired image, segmentation of each character, and character recognition [1]. This work describes the development of a system for recognizing Bangladeshi vehicle license plates. The template matching algorithm is used to extract the number plate region from the vehicle's input image frame. The spatial super-resolution technique is used to obtain a clear and high-resolution number plate, and the bounding box method is used to segment each character of the number plate. The method extracts the city, type, and number of the vehicle from the plate region. The system used CNN to train on 700 number plate images, and it provided 4096 features for each character to correctly recognize.

In work [2] YOLO and its variant YOLO-9000 were customized to handle license plate detection effectively. The primary customizations are changes to estimate of the bounding box parameter, and grid size, as well as the creation of a more difficult

Application-Oriented License Plate Extended (AOLPE) performance database evaluation. Because the original YOLO and YOLO-2 were not designed for LPD, they were unable to handle LPD on the AOLPE to be handled without customizations.

YOLO is used in [3] to detect vehicles and license plates on image frames obtained from videos. The license plate image is segmented into characters, and text is recognized using tesseract OCR. The ImageAI library is used to facilitate the instruction procedure. Images from Tamil Nadu license plates are utilized to evaluate the model's performance.

Image processing techniques are used in [8] for license plate detection. An innovative image processing technique for the detection and identification of Indian number plates is developed that can handle lowing and filtered based on character dimensions and spatial localization. Character recognition is done using the K-nearest neighbor technique once the region of interest has been filtered and de-skewered.

YOLO is used as the object detector in [4]. Convolutional Neural Networks (CNNs) are prepared and refined at each ALPR stage by training to ensure their robustness under varying conditions (e.g., variations in camera, lighting, and background). A two-stage approach is designed specifically to recognize and split characters, employing simple data augmentation techniques like reversed letters and inverted license plates (LPs).

Convolutional Neural Network (CNN) is trained and tuned for detecting the license plate of Bangladeshi automobiles and recognizing characters using tesseract from the detected license plates in the YOLOv4 object identification model, which is shown in [5]. Tkinter was used to create a graphical user interface (GUI).

Enhanced Super-Resolution Generative Adversarial Network (ESRGAN) is a type of machine learning model developed for the task of super-resolution of images, It includes enhancing an image's resolution while trying to preserve as much detail as possible. ESRGAN uses a generative adversarial network (GAN) architecture, which comprises of a generator network and a discriminator network, two neural networks. The discriminator network is taught to tell the difference between generated high-resolution photos and actual high-resolution photographs, while the generator network is trained to create high-resolution images from low-resolution input images.

During training, the generator network produces a discriminator and a high-resolution image from a low-resolution image network, which tries to determine whether the generated image is real or fake. If the offenders network determines the picture is a fake, it provides feedback to the network of generators, which adjusts its parameters to produce a more realistic image. This process continues until the generator network is able to produce high-resolution images that cannot be distinguished from real images [7].

YOLOv4 is known for its fast and accurate object detection capabilities. One of the key features of YOLOv4 is its ability to achieve high accuracy while still maintaining a fast processing speed. This is achieved through the use of a number of techniques, including multi-scale training, feature pyramid networks, and a novel cross-stage partial connection (CSP) architecture [9].

Compared to YOLOv4, YOLOv7 uses 36% less processing, reduces the number of parameters by 75%, and produces 1.5% greater average precision (AP). YOLOv7-tiny decreases the amount of parameters by 39% and computation by 49% while maintaining the same AP as the edge-optimized version YOLOv4-tiny.

To more effectively support the context of Korean automobiles in surroundings with many lanes of traffic and metropolitan areas, Park et al. [6] includes license plate recognition, vehicle type recognition, and license plate character detection. Two YOLOv4 custom detectors are being built. A VT LP detector, the first one, can identify seven classes, in the input image (i.e., six different Korean vehicle types and LPs). The Korean LPs' LPC detector, which recognizes 68 distinct digits and characters is the second detector. On a h-resolution image, the character size of LPs is small in comparison to the entire image, making character identification more difficult. To address this issue, the LPC detector receives segmented LP regions from the LP cropping procedure applied to the KVT-LPR [6].

Real-ESRGAN is an extension of ESRGAN that is specifically designed for the task of training on pure synthetic data and then applying the model to real-world images. One of the challenges in using machine learning models for image super-resolution is the requirement for substantial high-resolution training data. Collecting and labeling real-world images at high resolutions can be time-consuming and expensive, so using synthetic data can be a useful alternative. However, there can be a discrepancy between the appearance of synthetic and real-world images, which can make it difficult to transfer the model trained on synthetic data to real-world images. We retained the data and then focused on a specific portion of real-world data to adjust for the differences between the two types of images. This allows the model to learn from a larger and more diverse dataset, while still being able to generalize to real-world images.

Overall, Real-ESRGAN is a promising approach for training image super-resolution models using synthetic data, and has the potential to significantly reduce the cost and effort required to create high-quality training data for this task [10].

2.1 License Plate Recognition Using YOLO

Compared to other object detection methods, YOLO is seen as being more effective object detection methods [11]. Following are the characteristics which demarcate YOLO from others.

- A single convolutional network creates area suggestions in the YOLO algorithm, whereas the R-CNN algorithm creates region proposals (or Region Of Interest) using the selective search algorithm.
- When compared to other object recognition methods used by YOLO, the backbone network is a compact and flexible speedier attribute extractor, but in R-CNN it is time-consuming and difficult.

- A single-stage network in YOLO forecasts the class conditional probabilities, confidence score, and bounding box offsets. In contrast, the first stage of the R-CNN algorithm collects region suggestions, while the second stage extracts feature vectors, followed by detections.
- In R-CNN, it has higher accuracy and slower speed, whereas YOLO has faster detection and accuracy that is closer to two-stage object detectors.
- computing in YOLO is less expensive and uses limited resources. Coming to R-CNN, they demand strong computing resources and are costly to compute.
- YOLO is an end-to-end object identification approach, it can analyze an image and identify objects without the requirement for post-processing or refinement [12].

3 Proposed Methodology

Figure 1 shows the over all process followed to vehicle number recognition from Survelliance videos.

The training process involves feeding the **YOLOv4** and **YOLOv7** models with 2000 images, along with the corresponding labels indicating the location and class of the objects in the images (in this case, the vehicle number plates). Through this process, the model learns to recognize and sort the things in the images.

Once the model has been trained, it can be used to find and recognize vehicle number plates in new images or video streams. To do this, we would feed the model an input image or video frame and it would output the location and class of any detected objects in the image (in this case, the vehicle number plate). Figure 2 shows

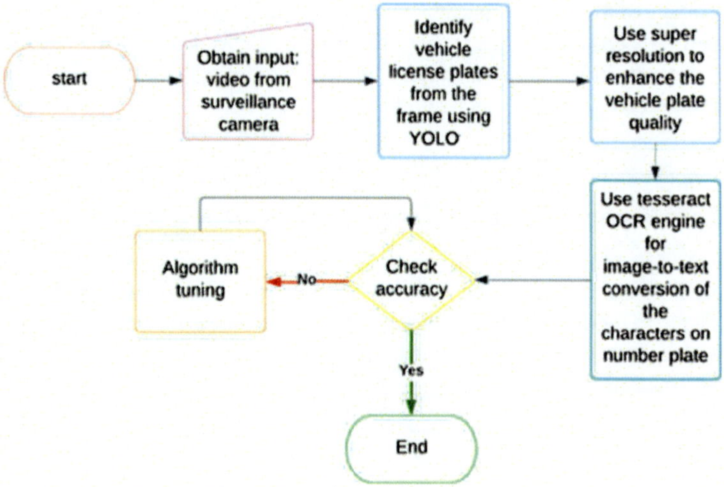

Fig. 1 Flow chart of the proposed system for vehicle number recognition

Fig. 2 Detection of number plates from video frame: **a** Rear view, **b** Partial number plate detection, **c** Front view

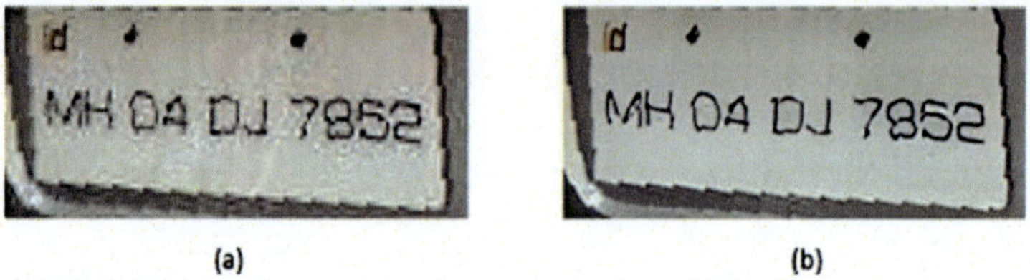

Fig. 3 Improving image quality by super-resolution using ESRGAN **a** Input image, **b** Real-ESRGAN output

the number plate recognized using YOLO tool after the image frames have been retrieved from the videos using an acceptable frame rate.

3.1 Super-Resolution Using Real-ESRGAN

To crop image of the license plate, the coordinates of the bounding boxes that enclose it are used. The cropped image is then subjected to the Real-ESRGAN technique in order to improve its quality and obtain a high-quality image [7, 10]. Figure 3, shows better resolution of numbers on a number plate.

3.2 Text Recognition from Images Using OCR

Easy OCR, Tesseract OCR, and Paddle OCR are used to perform text recognition on the obtained super-resolution images. To compare the performance of these OCR engines, they are run on the test super-resolution images and the accuracy of the results are measured.

3.3 *Handling Missing Values or Inaccuracies*

The first two parts of an Indian registration plate can be used to locate vehicles to their districts. In general, the first part is the state code (AP—Andhra Pradesh, AS—Assam, MH—Maharashtra, etc.). The first part can only contain the alphabet, so if any character in this part is recognized as a number, it can be immediately corrected to an alphabet, which it most likely is, by using the district code from the second part or the other letter from the first part. The second part is the district code, which is given to each district in a state or territory of the union. It can only contain numeric characters, so if a character is recognized as "O" or "D" it is thereafter corrected to "0".

4 Experimental Setup

The YOLOv4 and YOLOv7 models are trained on the Google Open Images V7 dataset—Vehicle Number Registration Dataset with 2000 train size and 386 validation size. The dataset includes foreign vehicles with license plates in english. The OIDv4 toolkit was used to generate labels in YOLO format.

The model was trained for maximumBatches = 6000 (classes * 2000). But not less than the training data size and not less than 6000. The following are the parameters used [9, 12, 13].

- classes = 1
- batch size = 64
- subdivisions = 16
- network size width = 416, height = 416
- Filters = 18 in the last convolutional layers before YOLO layers ((classes + 5) * 3) [9].

5 Results

As shown in Table 1, mean average precision (mAP) value was calculated for every 1000 iterations, with the model performing best at 3000 iterations. Overfitting has resulted in a slight decrease in mAP values after 3000 iterations. The use of super-resolution on images has significantly improved text recognition accuracy. The mAP was shown for YOLOv4 and YOLOv7. Figure 4 shows the confusion matrix for the recognition accuracy of the system.

Figure 5, shows the various goodness parameters obtained with YOLOv7. Figure 6 shows various clean outputs on hazy number plates after applying the super-resolution technique.

Table 1 mAP values for every 1000 iterations on Google Open Images V7 dataset

S. No.	Iterations	mAP(v4)	epochs	mAP(v7)
1	1000	81.35	25	0.909
2	2000	88.35	50	0.905
3	3000	90.09	75	0.917
4	4000	90.03	100	0.918
5	5000	89.97		
6	6000	89.43		

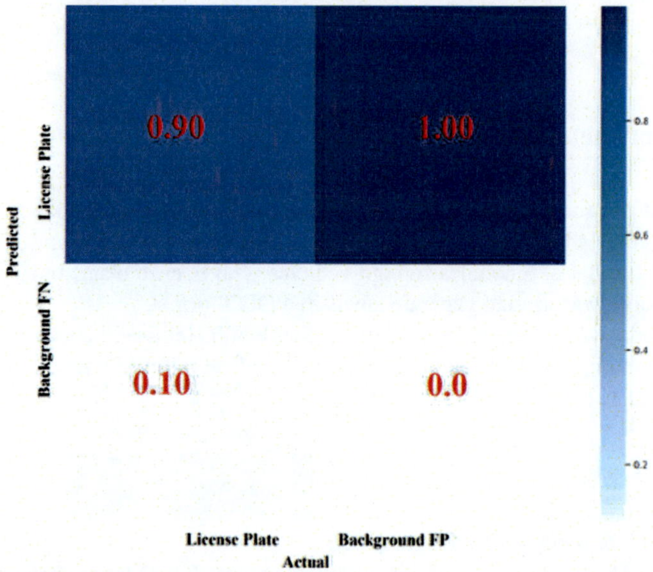

Fig. 4 Confusion matrix for recognition accuracy

Figure 7, is the input given to the text conversion OCR.

Output format—lists of text box coordinates, text, model confidence level.
([[31,73], [385,73], [385,189], [31,189]],
'PB 37F',
0.6883575759789258),
([[473,69], [731,69], [731,179], [473,179]],
'8999',
0.9999973773956299).

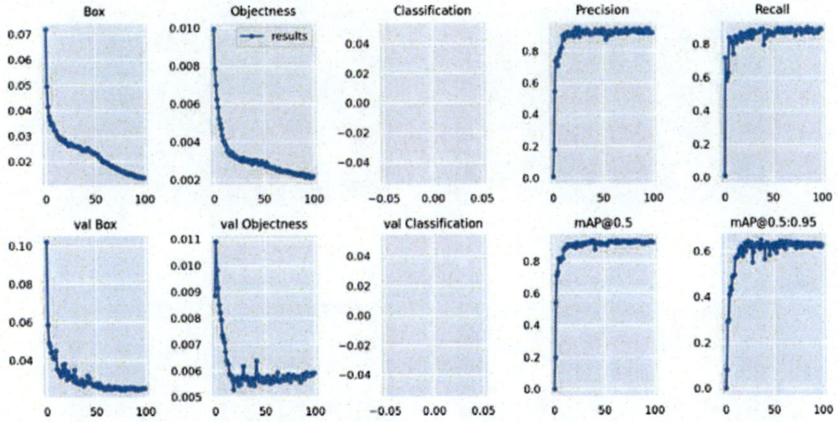

Fig. 5 YOLOv7 results

6 Conclusion and Future Scope

This research work suggests a system for recognizing license plates for a vehicle number in India that are written in English. The live video/image is fed into a YOLO-trained model, which creates a bounding box around the license plate. The picture is cropped to remove the license plate or video frame using the bounding box coordinates. To improve the image quality, Real-ESRGAN, a super-resolution technique, is then used on the license plate images. The enhanced image is then fed into an OCR engine, which extracts text from the image. In the event of inconsistencies or missing values, the Indian number plate format rules are used to determine the most likely values, and improve accuracy. The overall accuracy on a few real-world videos is above 90 percent, which is very subjective to conclude given the presence of various challenges. It was found that YOLVOv7 gave better precision compared with YOLLOv4 and was also efficient.

This system is not foolproof, and work continues to include use cases such as mist, rain, fog, smoke, and improper capture due to heavy traffic. Domain knowledge and analytical reasoning will be used to address these challenges.

Fig. 6 High quality images by Super-resolution **a** Input image, **b** Real-ESRGAN output

Fig. 7 High resolution number plate for image-to-text conversion

References

1. Alam NA, Ahsan M, Based MA, Haider J (2021) Intelligent system for vehicles number plate detection and recognition using convolutional neural networks. Technologies 9(1):9. https://doi.org/10.3390/technologies9010009
2. Hsu G-S, Ambikapathi A, Chung S-L, Su C-P (2017) Robust license plate detection in the wild.In: 2017 14th IEEE international conference on advanced video and signal based surveil-

lance (AVSS). https://doi.org/10.1109/AVSS.2017.8078493
3. Gnanaprakash V, Kanthimathi N, Saranya N (2021) Automatic number plate recognition using deep learning. In: IOP conference series: materials science and engineering. https://doi.org/10.1088/1757-899X/1084/1/012027
4. Laroca R et al (2018) A robust real-time automatic license plate recognition based on the YOLO detector. In: 2018 International joint conference on neural networks (IJCNN), pp 1–10. https://doi.org/10.1109/IJCNN.2018.8489629
5. Onim MS, Akash MI, Haque M, Hafiz RI (2020) Traffic surveillance using vehicle license plate detection and recognition in Bangladesh. In: 2020 11th International conference on electrical and computer engineering (ICECE), pp 121–124. https://doi.org/10.1109/ICECE51571.2020.9393109
6. Park S-H, Yu S-B, Kim J-A, Yoon H (2022) An all-in-one vehicle type and license plate recognition system using YOLOv4. Sensors 22(3):921. https://doi.org/10.3390/s22030921
7. Wang X, Xie L, Dong C, Shan Y (2021) Real-esrgan: Training real-world blind super-resolution with pure synthetic data. In: Proceedings of the IEEE/CVF international conference on computer vision, pp 905–1914
8. Ravi Kiran Varma P, Srikanth Ganta, Hari Krishna B, Svsrk P (2020) A novel method for indian vehicle registration number plate detection and recognition using image processing techniques. Procedia Comput Sci 167:2623–2633. ISSN 1877-0509. https://doi.org/10.1016/j.procs.2020.03.324
9. Bochkovskiy A, Wang CY, Liao HY (2020) Yolov4: optimal speed and accuracy of object detection. arXiv preprint arXiv:2004.10934
10. Wang X, Yu K, Wu S, Gu J, Liu Y, Dong C, Qiao Y, Change Loy C (2018) Esrgan: enhanced super-resolution generative adversarial networks. In: Proceedings of the European conference on computer vision (ECCV) workshops, pp 0–0
11. Diwan T, Anirudh G, Tembhurne JV (2022) Object detection using YOLO: challenges, architectural successors, datasets and applications. Multimedia Tools Appl, pp 1–33
12. https://viso.ai/deep-learning/yolov7-guide/
13. Wang CY, Bochkovskiy A, Liao HYM (2022) YOLOv7: Trainable bag-of-freebies sets new state-of-the-art for real-time object detectors. arXiv preprint arXiv:2207.02696

Author Index

A

Aarthi, V. P. M. B., 259
Abdillah, Abdillah, 75
Abdullayev, Khagani, 153
Abhishek Goyal, 315
Aditya Vijay Kavatage, 749
Ahmed, Md Apu, 219
Ajay Kumar Goud, P., 203, 865
Akhmedova, Nozima, 687
Akula Triyan Subramanyan, 761
Amit Ganatra, 663
Ande Vijaya Krishna, 139
Andrew, 895
Andrysiak, Tomasz, 911
Angeline Esther, A., 489
Animesh Giri, 315, 357
Anjum, Shaik Shabana, 119
Ankur Raut, 373
Apoorv Kumar Sinha, 651
Apurba Sinha, 539
Ardi Satyo Ramadhan, 715
Arpit Kogta, 315
Arun Kumar, T., 259
Arya, L. D., 509
Ashok Kumar, P. M., 593
Ashraf Alam, 571, 879
Asmanurhidayani, 619
Atayev, Nadir, 153
Atharva Purohit, 373
Atharv Natu, 299
Avinash, N. J., 233
Avi, Shuva Dasgupta, 219

B

Biranav Kumar, P., 459
Balaji, B. V., 357
Balaji, K., 43
Berg, Rozen, 401
Bhaskar, N., 347
Bhavin Patil, 299
Bhoomika, P., 233
Bhoomika P. Bhavimath, 357
Buela Likhitha, K., 593
Buhari, Rd Ahmad, 75
Bushgari Haripriya, 925
Butta Singh, 163

C

Castillo, Antonio Cortés, 27
Chandraprabha, K. S., 281
Chandrasekhar, U., 925
Chithra, S. M., 13
Chowanda, Andry, 729
Czyżewska, Magda, 911

D

Danilov, Igor Val, 443
Deepamangai, P., 489
Dennis Joseph, 739
Dhivya, M., 233, 387
Dipak Ramoliya, 663
Durai Murugan, A., 327
Durgalakshmi, V., 357
Durga Mahato, 539

G. Rajakumar et al. (eds.), *Intelligent Communication Technologies and Virtual Mobile Networks*, Lecture Notes on Data Engineering and Communications Technologies 171,
https://doi.org/10.1007/978-981-99-1767-9

G
Gackowska, Marta, 911
Gaol, Ford Lumban, 501, 607, 715, 895
Gaurav B. Patil, 509
Ghosh, Aditi, 219
Glad, Jerald, 715
Govind Shivaji, G., 259
Gunawan, William, 501
Gurbanova, Gulnar, 153
Gurulakshmi, A. B., 387

H
Habib, Md. Rawshan, 219
Harika, M., 829
Hasanov, Mehman, 153
Himali Sarangal, 163
Himanshu Srivastava, 813
Huang, Kendrew, 501

I
Ikrar Gama Raditya, 715
Irawan, Stanley Nathanael, 895

J
Jabez, J., 419
Jahnavi, M., 829
Jaisaran, N., 459
Jalajakshi, V., 189
Jangam Deepthi, 269

K
Kalluri Shanmukha Sai, 761
Kamalov, Firuz, 677, 801
Kannayeram, G., 839
Kartheeswari, K. R., 631
Karthick Ram, V., 459
Kashtanov, V., 93
Kavitha, C., 761
Khamarjaha, S. K., 593
Kranthi, S., 139
Krishna Reddy, D., 105
Kuldeep Vayadande, 373
KumaraSabapathy, N., 839
Kumari, G. R. P., 829

L
Lakshmi Priya, K., 593
Lasmy, 729
Ledziński, Damian, 911
Lingamaiah, D., 105
Lingamaneni Sri Nidesh, 761

M
Mahalakshmi, P., 327
Mambu, Joe Y., 619
Manjit Singh, 163
Manoharan, P. S., 489, 839
Matsuo, Tokuro, 501, 607, 715, 895
Maulana, Fairuz Iqbal, 247
Md. Sohaib, 259
Meiryani, 247
Mohit Raj, G., 203, 865
Morozov, V., 93
Munish Gupta, 559
Myna, N., 189

N
Nabilah, Fannisa Hanin, 607
Nabilah Putri Intaka, 607
Naga Laxmi, M., 259
Naveenkumar, G., 57
Nemtinov, K., 93
Nemtinov, V., 93
Ng, Jason Fonseca, 895
Nitish Ojha, 813
Nivarthi Tushara, 925
Nivedita, J., 489
Nur Asitah, 247

P
Pagadala Ganesh Krishna, 139
Pałczyński, Krzysztof, 911
Pandiaraja, P., 459
Pangestu, Gusti, 247
Pavani, A., 829
Pavithra, K. N., 177
Perumalla Naveen Kumar, 105
Pihoo Verma, 315
Pineng, Martina, 619
Pooja Chopra, 559
Poonkuzhali, S., 699
Prabanand, S. C., 327
Prabu, U., 433
Pravanya, P., 593
Preethisree, V., 233
Prem Anand, T. P., 57, 699
PriyaDharshini, A., 475
Priyansh Jain, 315
Priya Vinod, 523
Protasova, Yu, 93

Purnomo, Agung, 247

R
Rachman, Suci Rahma Dani, 619
Radhakrishna Bhat, 177
Raghavender Sharma, M., 347
Rahaman, K. M. Monzur, 219
Rahul, B., 357
Rahul, K., 203, 865
Rahunathan, L., 475
Rajesh, G., 387
Rakesh, K. M., 43
Ramana Murthy, M. V., 347
Ramesh, M., 259
Reyes, Jorge Avante, 677, 801
Rismayani, 619
Rivaldi, 501
Roopesh, R. S., 43
Roshita Bhonsle, 373
Rosyidah, Elsa, 247
Roy, Sibaji, 219
Rusliadi, Rusliadi, 75

S
Sameerunnisa, S. K., 419
Sandeep, B. S., 401
Sangeetha Archana, J., 699
Sanjeev Sharma, 387
Sankari Subbiah, 43
Sanskriti Sanjay Kumar Singh, 651
Santosh S. Raghuwanshi, 509
Saranya, S., 43
Saravanan, D., 739
Satveer Kour, 163
Sekhar, M., 1
Selvi Rajendran, P., 631
Shalini, R., 839
Shanika, K. S., 233
Shankar, R., 593
Shawmee, Tahsina Tashrif, 219
Sheeja, S., 523
Sherif Moussa, 677, 801
Shibghatullah, Adbul Samad Bin, 119
Shilpa, S., 749
Shishir Menon, 749
Shkodyrev, V. P., 849
Shreesh Ravindra Devi, 749
Shreyas Sai, 651
Siddhi Patil, 749
Sindhu, C., 761
Sivabalaselvamani, D., 475
Sivagami, M., 651
Sneha Palwe, 299
Solleti Ramana, 347
Sridevi, S., 13
Sriharish, T., 327
Srinivasan Kathiravan, 781
Sudha, B., 781
Sudha, G., 43
Sumalatha, M. R., 401
Suman Nelaturi, 1
Sunjaya, Daniel Ryan, 119
Suriyaprakash, M. V., 57
Swati Jadhav, 299

T
Tagiyev, Ali, 153
Tanvir, Md Shahnewaz, 219
Tarunbalaji, S., 327
Thanabal, M. S., 327
Tharunraj, M., 401
Trisha Chatterjee, 233

U
Uttkarsh Patel, 299

V
Vadher, Abhishek, 219
Varad Ingale, 373
Vasuki, A., 129
Veeramani, V., 13
Venkata Lakshmi, C., 829
Venu Gopal, T., 269
Vignesh, M., 475
Vijayakumar Ponnusamy, 129
Vijaya Lakshmi, A., 203, 865
Vikas Kumar Aharwal, 539
VinithKumar, G., 475
Vithika Pungliya, 373
Voraganti Sahithi, 925

W
Wahyuni, Sri, 619
Widianingsih, Ida, 75

Z
Zhu, Y., 849